DEVELOPMENT OF THE OCULAR LENS

Development of the Ocular Lens provides a current view of research in lens developmental biology, emphasizing recent technical and molecular breakthroughs. Elucidation of the mechanisms that govern lens development has enabled us to understand how the normal lens forms and how developmental processes, namely, cell proliferation and differentiation, are involved in the maintenance of its normal structure, function, and growth throughout life. This knowledge is fundamental to our understanding of many lens disorders. The ocular lens has also become a model for understanding the developmental biology of more complex organ systems. In this book, leading experts in lens cell biology and development discuss lens evolution, induction, and morphology; the regulation of the lens cell cycle and fiber cell differentiation; and lens regeneration. This book is a broad and authoritative treatment of the subject that will serve as a reference for graduate students and research scientists in developmental biology and the visual sciences as well as for ophthalmologists.

Frank J. Lovicu is Senior Lecturer at the Save Sight Institute and in the Department of Anatomy and Histology, Institute for Biomedical Research, at the University of Sydney, Australia. He currently heads the Lens Research Laboratory in the Department of Anatomy and Histology, where he studies the role of growth factors in regulating normal and aberrant lens cell behavior.

Michael L. Robinson is Assistant Professor in the Department of Pediatrics at The Ohio State University, USA. He also heads the Transgenic and Embryonic Stem Cell Core Facility at Columbus Children's Research Institute. Dr. Robinson's research is focused on the role of FGF receptor signaling during lens fiber cell differentiation.

DEVELOPMENT OF THE OCULAR LENS

Edited by

FRANK J. LOVICU
University of Sydney

MICHAEL L. ROBINSON
Ohio State University and Columbus Children's Research Institute

CAMBRIDGE UNIVERSITY PRESS
Cambridge, New York, Melbourne, Madrid, Cape Town, Singapore,
São Paulo, Delhi, Dubai, Tokyo, Mexico City

Cambridge University Press
The Edinburgh Building, Cambridge CB2 8RU, UK

Published in the United States of America by Cambridge University Press, New York

www.cambridge.org
Information on this title: www.cambridge.org/9780521184236

First published 2004
First paperback edition 2010

A catalogue record for this publication is available from the British Library

Library of Congress Cataloguing in Publication data

Development of the ocular lens / edited by Frank J. Lovicu, Michael L. Robinson.
p.; cm.
Includes bibliographical references and index.
ISBN 0-521-83819-3 (HB)
1. Crystalline lens – Molecular aspects. 2. Crystalline lens – Cytology. I. Lovicu, Frank J. (Frank James), 1966– II. Robinson, Michael L., (Michael Lee), 1965–
[DNLM: 1. Lens, Crystalline – cytology. 2. Developmental Biology. WW 260 D489 2004]

QP478.D485 2004
612.8′44–dc22 2004040411

ISBN 978-0-521-83819-1 Hardback
ISBN 978-0-521-18423-6 Paperback

Additional resources for this publication at www.cambridge.org/9780521184236

Contents

List of Contributors — *page* ix
Preface — xiii
Acknowledgments — xv

Part 1. Early Lens Development

1 The Lens: Historical and Comparative Perspectives — 03
MICHAEL L. ROBINSON AND FRANK J. LOVICU
1.1. Lens Anatomy and Development (Pre-1900) — 03
1.2. Comparative Ocular Anatomy — 15
1.3. Development of the Vertebrate Lens — 23

2 Lens Induction and Determination — 27
MARILYN FISHER AND ROBERT M. GRAINGER
2.1. Introduction — 27
2.2. Historical Overview — 29
2.3. Current Model of Lens Determination — 36
2.4. Inducing Signals — 45
2.5. Conclusions and Future Directions — 47

3 Transcription Factors in Early Lens Development — 48
GUY GOUDREAU, NICOLE BÄUMER, AND PETER GRUSS
3.1. Introduction — 48
3.2. The Key Transcriptional Regulators Involved in Eye Development Are Conserved in Different Species — 49
3.3. Transcription Factors from Different Classes Are Involved in Lens Development — 51
3.4. Concluding Remarks — 68

Part 2. The Lens

4 The Structure of the Vertebrate Lens — 71
JER R. KUSZAK AND M. JOSEPH COSTELLO
4.1. Introduction — 71
4.2. Lens Development — 71
4.3. Different Types of Lenses as a Function of Suture Patterns — 75
4.4. Lens Gross Anatomy — 86

4.5. Lens Ultrastructure 94
4.6. Summary 115

5 Lens Crystallins 119
MELINDA K. DUNCAN, ALES CVEKL, MARC KANTOROW, AND JORAM PIATIGORSKY
5.1. Introduction 119
5.2. Structure and Function of Crystallins 120
5.3. Control of Crystallin Gene Expression 128
5.4. Lessons from Transcriptional Control of Diverse Crystallin Genes: A Common Regulatory Mechanism? 146
5.5. Current Questions 148
5.6. Conclusion 150

6 Lens Cell Membranes 151
JOERG KISTLER, REINER ECKERT, AND PAUL DONALDSON
6.1. Introduction 151
6.2. An Internal Circulation Is Generated by Spatial Differences in Membrane Proteins 152
6.3. Membrane Conductances Vary between Lens Regions 154
6.4. Lens Cells Are Connected by Gap Junction Channels 157
6.5. Na^+ Pump Activity Is Greatest at the Lens Equator 160
6.6. Water Flow across Lens Cell Membranes Is Enhanced by Aquaporins 161
6.7. Specialized Transporters Serve Nutrient Uptake 163
6.8. Changes in Membrane Channel- or Transporter-Activity May Result in Cataract 165
6.9. Some Membrane Receptors Have the Potential to Regulate Lens Homeostasis 168
6.10. Multiple Membrane Receptors May Control the Highly Organized Lens Tissue Architecture 169
6.11. Proteins with Adhesive Properties Further Support the Crystalline Lens Architecture 170
6.12. Conclusion 172

7 Lens Cell Cytoskeleton 173
ROY QUINLAN AND ALAN PRESCOTT
7.1. Introduction 173
7.2. Major Components of the Lenticular Cytoskeleton 173
7.3. Microtubule Networks in the Lens 179
7.4. Actin in the Lens 183
7.5. Conclusion 187

Part 3. Lens Development and Growth

8 Lens Cell Proliferation: The Cell Cycle 191
ANNE E. GRIEP AND PUMIN ZHANG
8.1. Introduction 191
8.2. Regulation of the Cell Cycle 191
8.3. Cellular Proliferation in the Lens 197

8.4. Expression Patterns of Cell Cycle Regulatory Genes in the Developing Lens 200
8.5. Cell Cycle Regulation during Fiber Cell Differentiation 202
8.6. Regulation of Proliferation in the Lens Epithelium 210
8.7. Significance of Understanding Cell Cycle Control for Clinical Issues 211
8.8. Key Questions for Future Investigation 212

9 Lens Fiber Differentiation 214
STEVEN BASSNETT AND DAVID BEEBE
9.1. Introduction 214
9.2. The Stages of Fiber Cell Differentiation 214
9.3. Organization of Cells at the Lens Equator 216
9.4. The Initial Events in Lens Fiber Cell Differentiation 219
9.5. The Elongating Fiber Cell 225
9.6. The Maturing Fiber Cell 228
9.7. The Mature Fiber Cell 231
9.8. How Is the Process of Fiber Cell Differentiation Related to the Overall Shape of the Lens? 241
9.9. Lens Pathology: Cataracts Caused by Abnormal Fiber Cell Differentiation 242
9.10. Concluding Remarks 244

10 Role of Matrix and Cell Adhesion Molecules in Lens Differentiation 245
A. SUE MENKO AND JANICE L. WALKER
10.1. Extracellular Matrix 245
10.2. Integrin Receptors 250
10.3. Cadherins 257
10.4. Other Lens Cell Adhesion Molecules 259
10.5. Summary 260

11 Growth Factors in Lens Development 261
RICHARD A. LANG AND JOHN W. MCAVOY
11.1. Lens Induction and Morphogenesis 261
11.2. Lens Differentiation and Growth 271
11.3. Overview 289

12 Lens Regeneration 290
KATIA DEL RIO-TSONIS AND GORO EGUCHI
12.1. Introduction 290
12.2. General Background on the Process of Lens Regeneration 290
12.3. Problems Involved in the Study of Lens Regeneration 293
12.4. Classic Approaches to the Problems 294
12.5. Modern Approaches to Lens Regeneration 297
12.6. Lens Regenerative Capacity of Vertebrates 305
12.7. Transdifferentiation of PECs as the Basis of Lens Regeneration 306
12.8. Future Prospects 311

Bibliography 313
Index 387

Contributors

Bassnett, Steven, Department of Ophthalmology and Visual Science, Washington University School of Medicine, 660 S. Euclid Ave. CB 8096, St. Louis, MO 63110-1093, USA, phone: (1-314) 362-1604, fax: (1-314) 362-3638, e-mail: bassnett@vision.wustl.edu

Bäumer, Nicole, Department of Medicine, Hematology and Oncology, University of Münster, Domagstr. 3, 48129 Münster, Germany, phone: (49-251) 8357147, fax: (49-251) 8352673, e-mail: nbaeumer@uni-muenster.de

Beebe, David, Department of Ophthalmology and Visual Science, Washington University School of Medicine, 660 S. Euclid Ave. CB 8096, St. Louis, MO 63110-1093, USA, phone: (1-314) 362-1621, fax: (1-314) 747-1405, e-mail: beebe@wustl.edu

Costello, M. Joseph, Department of Cell and Developmental Biology, University of North Carolina at Chapel Hill, CB7090, Chapel Hill, NC 27599-7090, USA, phone: (1-919) 966-6981, fax: (1-919) 966-1856, e-mail: mjc@med.unc.edu

Cvekl, Ales, Department of Ophthalmology and Visual Science and Department of Molecular Genetics, Albert Einstein College of Medicine, 713 Ullmann, 1300 Morris Park, Bronx, NY 10467-2490, USA, phone: (1-718) 430-3217, fax: (1-718) 430-8778, e-mail: cvekl@aecom.yu.edu

Del Rio-Tsonis, Katia, Department of Zoology, Miami University, Oxford, OH 45056-1400, USA, phone: (1-513) 529-3128, fax: (1-513) 529-6900, e-mail: delriok@muohio.edu

Donaldson, Paul, Department of Physiology, School of Medical Sciences, University of Auckland, Private Bag 92019, Auckland, New Zealand, phone: (64-9) 3737599 ext. 84625, fax: (64-9) 3737499, e-mail: p.donaldson@auckland.ac.nz

Duncan, Melinda K., Department of Biological Sciences, University of Delaware, 327 Wolf Hall, Newark, DE, 19716, USA, phone: (1-302) 831-0533, fax: (1-302) 831-2281, e-mail: duncanm@udel.edu

Eckert, Reiner, Department of Biophysics, Institute of Biology, University of Stuttgart, Pfaffenwaldring 57, D-70550 Stuttgart, Germany, phone: (49-711) 685 5028, fax: (49-711) 685 5096, e-mail: reiner.eckert@po.uni-stuttgart.de

Eguchi, Goro, Chairman and President, Shokei Educational Institution, 2-6-78, Kuhonji, Kumamoto, 862-8678, Japan, phone: (81-96) 364 0116, fax: (81-96) 363-6520, e-mail: shokei@shokei-gakuen.ac.jp

Fisher, Marilyn, Department of Biology, University of Virginia, Gilmer Hall, P.O. Box 400328, Charlottesville, VA 22904-4328, USA, phone: (1-434)-982-5606, fax: (1-434)-982-5626, e-mail: mf4b@virginia.edu

Goudreau, Guy, Department of Molecular Cell Biology, Max-Planck-Institute of Biophysical Chemistry, Am Fassberg 11, 37070 Göttingen, Germany, phone: (49-551) 201 1361, fax: (49-551) 2011712, e-mail: guy.goudreau@sympatico.ca

Grainger, Robert M., Department of Biology, University of Virginia, Gilmer Hall, P.O. Box 400328, Charlottesville, VA 22904-4328, USA, phone: (1-434) 982-5495, fax: (1-434) 982-5495, e-mail: rmg9p@virginia.edu

Griep, Anne E., Department of Anatomy, University of Wisconsin Medical School, 1300 University Avenue, Madison, WI 53706, USA, phone: (1-608) 262-8988, fax: (1-608) 262-7306, e-mail: aegriep@facstaff.wisc.edu

Gruss, Peter, Department of Molecular Cell Biology, Max-Planck-Institute for Biophysical Chemistry, Am Fassberg 11, 37070 Göttingen, Germany, phone: (49-551) 201 1361, fax: (49-551) 201 1504, e-mail: peter.gruss@mpg-gv.mpg.de

Kantorow, Marc, Biomedical Sciences, Florida Atlantic University, Biomedical Sciences, 777 Glades Rd., P.O. Box 3091, Boca Raton, FL 33431-0991, USA, phone: (1-561) 297-2910, e-mail: mkantoro@fau.edu

Kistler, Joerg, School of Biological Sciences, University of Auckland, Private Bag 92019, Auckland, New Zealand, phone: (64-9) 3737599 ext. 88250, fax: (64-9) 3737415, e-mail: j.kistler@auckland.ac.nz

Kuszak, Jer R., Departments of Pathology and Ophthalmology, Rush Presbyterian–St. Luke's Medical Center, 1653 W. Congress Parkway, Chicago, IL 60612, USA, phone: (1-312) 942-5630, fax: (1-312) 942-2371, e-mail: jkuszak@rush.edu

Lang, Richard A., Divisions of Developmental Biology and Ophthalmology, Cincinnati Children's Hospital Research Foundation, 3333 Burnet Avenue, Cincinnati, OH 45229, USA, phone: (1-513) 636-7030 (assistant), (1-513) 636-2700 (office), fax: (1-513) 636-4317, e-mail: Richard.Lang@cchmc.org

Lovicu, Frank J., Save Sight Institute and Department of Anatomy and Histology (F13), Institute for Biomedical Research, University of Sydney, NSW, 2006, Australia, phone: (61-2) 9351 5170, fax: (61-2) 9351 2813, e-mail: lovicu@anatomy.usyd.edu.au

McAvoy, John W., Save Sight Institute and Department of Anatomy and Histology, University of Sydney, Sydney Eye Hospital, Macquarie Street, GPO Box 4337, Sydney 2001, Australia, phone: (61-2) 9382 7369, fax: (61-2) 9382 7318, e-mail: johnm@eye.usyd.edu.au

Menko, A. Sue, Department of Pathology, Anatomy and Cell Biology, Thomas Jefferson University, 571 Jefferson Alumni Hall, 1020 Locust St., Philadelphia, PA 19107-6799, USA, phone: (1-215) 503-2166, fax: (1-215) 923-3808, e-mail: Sue.Menko@jefferson.edu

Piatigorsky, Joram, Laboratory of Molecular and Developmental Biology, National Eye Institute, National Institutes of Health, Building 7, Room 100A, 7 Memorial Drive MSC 0704, Bethesda, MD 20892-2730, USA, phone: (1-301) 496-9467, fax: (1-301) 402-0781, e-mail: joramp@nei.nih.gov

Prescott, Alan, School of Life Sciences, MSIWTB, University of Dundee, Millers Wynd, Dundee DD1 5EH, UK, phone: (441) 382-344884, fax: (441) 382-345893, e-mail: a.r.prescott@dundee.ac.uk

Quinlan, Roy A., Department of Biological Sciences, The Science Park, South Road, Durham DH1 3LE, UK, phone: (+44) 0191-334-1331, fax: (+44) 0191-334-1207, e-mail: r.a.quinlan@durham.ac.uk

Robinson, Michael L., Division of Molecular and Human Genetics, Children's Research Institute, Wexner 492, 700 Children's Drive, Columbus, OH 43205, USA, phone: (1-614) 722-2764, fax: (1-614) 722-2716, e-mail: RobinsoM@pediatrics.ohio-state.edu

Walker, Janice L., Department of Pharmacology, University of Pennsylvania, 167 Johnson Pavillion, 3620 Hamilton Way, Philadelphia, PA 19104, USA, e-mail: janicelwalker 2000@yahoo.com

Zhang, Pumin, Department of Molecular Physiology and Biophysics, Baylor College of Medicine, One Baylor Plaza, Houston, TX 77030, USA, phone: (1-713) 798-1866, fax: (1-713) 798-3475, e-mail: pzhang@bcm.tmc.edu

Preface

Like the beginning of the 20th century, the dawn of the 21st century is witnessing tremendous advances in developmental biology. Research on ocular lens development remains at the forefront of these advances, just as it was in 1900. The difference between then and now lies in the current synergism between developmental biology, genetics, and molecular biology that has led to the identification of the very molecules responsible for many of the inductive processes only descriptively defined earlier.

The period from the first descriptions of mice with transgenes targeted specifically to the lens until the present day, when lens-specific gene deletion has become almost routine, spans only about the past 20 years. It is in the context of this breathtaking influx of information that we thought it time to devote a text specifically and exclusively to lens development. Because of its simplicity and its predictable pattern of development and differentiation, the lens, arguably the sparkling jewel of anatomy, has attracted the attention of many developmental biologists. All stages of lens differentiation, from the proliferation of lens epithelial cells to the differentiation of mature, organelle-deficient fiber cells, are represented in the lens of any individual. Analogous to the rings of a tree, the components of the lens, from the newest cells born moments before tissue collection to the oldest cells originating during embryogenesis, offer a key to its life history. This fact, combined with the ability of researchers to observe and manipulate the lens in a living animal, ensures that this remarkable organ will remain a paradigm for fundamental biological investigations for decades to come.

It has been our good fortune to bring together for the first time an international panel of well-known developmental biologists currently working on research problems specifically related to aspects of lens induction, development, differentiation, growth, and maintenance. Each of these scientists has contributed to this up-to-date text covering virtually all aspects of lens development. The 12 chapters of *Development of the Ocular Lens* provide a current view of research in lens developmental biology, with an emphasis on recent technical and molecular breakthroughs. The introductory chapter covers historical and evolutionary aspects of lens development and structure. It is followed by chapters describing lens induction and formation as well as the transcription factors that provide the molecular control of these processes. Further chapters discuss the development of the lens structure, including its ultrastructure and fiber organization; crystallin synthesis and function; and lens cell membranes and cytoskeleton. The final chapters provide insights into the processes of lens development and growth at the molecular level, including the cell cycle, fiber cell differentiation, cell adhesion, and extracellular matrix and growth factor signaling. The last chapter describes the fascinating field of lens regeneration, bringing lens development full circle, from induction to regeneration. Although all the chapters are written from a developmental perspective, where appropriate, discussion of lens pathology related to developmental abnormalities complements the description of basic lens research.

Past texts have dealt with the lens in the context of cataract and molecular biology, and a few recent books have concentrated on eye development as a whole. *Development of the Ocular Lens* is unique in its unsurpassed depth of information specifically focused on lens development. Written by many of the leading researchers in the field, this text is an invaluable resource for anyone interested in lens biology, providing, among other things, a springboard into the primary research literature. We have been humbled and honored to have taken part in this endeavor, and it is our sincere hope that graduate students, residents, postdoctoral fellows, principal investigators, and clinicians alike will enjoy reading and using this book as much as we have enjoyed assembling it.

We owe a tremendous debt of gratitude to all of the contributing authors. We would sincerely like to thank them for their hard work as well as their patience during the long process of getting this work into its present form. Most importantly, we would like to thank our families, our wives (Maria Lovicu and Julia Robinson) and children (Christopher, Alexander, and Matthew Lovicu and Emily, Eric, Jessica, and Ian Robinson), for giving us the time to assemble and edit a book that we are very proud of.

Frank J. Lovicu
Michael L. Robinson

Acknowledgments

Acknowledgments for specific chapters are as follows: *Chapter 1*: FJL acknowledges support from the Sydney Foundation for Medical Research and the National Health and Medical Research Council (NHMRC) of Australia. MLR also acknowledges support from the National Eye Institute, USA (EY12995). The authors thank Barbara Van Brimmer of the Medical Heritage Center at the Ohio State University, USA, for her assistance in gathering information on the history of ophthalmology, as well as Professor John W. McAvoy (Save Sight Institute, Australia) for critically reviewing this chapter. *Chapter 2*: Work cited from the Grainger Lab was funded by NIH Grants EY06675, EY10283, and RR13221. We thank Dr. Nicolas Hirsch for critically reviewing the manuscript. *Chapter 4*: The authors gratefully acknowledge the contributions of K. J. Al-Ghoul, Ph.D.; H. Brown, M.D., Ph.D.; R. K. Zoltoski, Ph.D.; J. Chrisman; C. F. Freel; K. O. Gilliland; A. J. Kuszak; C. W. Lane; H. Mekeel; R. Nordgren; L. Novak; and K. L. Peterson. JRK has been supported by NIH NEI Grant EY06642, the Regenstein Foundation, the Louise C. Norton Trust, and the Alcon Research Institute. MJC has been supported by NIH NEI Grants EY08148 and EY05722. Human lenses were obtained from the NDRI and the North Carolina Eye Bank. Primate lenses were obtained from the Regional Primate Center in Seattle, Washington, and the Yerkes Primate Center in Atlanta, Georgia. *Chapter 6*: Research work in the authors' laboratories was supported by the Health Research Council of New Zealand, the Marsden Fund, the Lotteries Grant Board, the University of Auckland Research Committee, and the Auckland Medical Research Foundation. *Chapter 11*: We thank Tina Coburn for assistance in the preparation of this chapter. This work was supported by Grants EY03177 (JMcA.) and EY11234, EY10559, EY12370, and EY12370 (RAL) from the National Institutes of Health and by a grant from the National Health and Medical Research Council (NHMRC) of Australia. JMcA acknowledges support from the Sydney Foundation for Medical Research and an Infrastructure Grant from the NSW Department of Health. The Lang Laboratory is also supported by funds from the Abrahamson Pediatric Eye Institute Endowment at Children's Hospital Medical Center of Cincinnati.

Acknowledgments

Part One

Early Lens Development

1

The Lens: Historical and Comparative Perspectives

Michael L. Robinson and Frank J. Lovicu

1.1. Lens Anatomy and Development (Pre-1900)

The past decade has witnessed a tremendous increase in the basic understanding of the molecules and signal transduction pathways required to initiate embryonic lens development. Other advances in this time period have elucidated structural and physiological properties of lens cells, often in an evolutionary context, making it possible to frame many pathological conditions of the lens as errors of specific developmental events. All of these recent advances rest on the fundamental observations of talented investigators in previous decades and centuries. While several texts describe the history of ophthalmology as a clinical discipline, the conceptual history of basic eye research as a science, and in particular the history of lens development research, is a much less traversed subject. Though it is inevitable that we cannot include all of the many important experiments and personalities that have played fundamental roles in shaping the field of lens development, we hope to stimulate appreciation for those pioneers, both past and present, to whom we owe a debt of gratitude for their contributions to the field.

Throughout human history, the sense of sight has been both treasured and revered. Without doubt, visual loss resulting from lens dysfunction has always plagued the human family. In the early years of lens development research, investigations of the eye were intertwined with the genesis of the field of ophthalmology. Two valuable texts, extensively cited in this chapter, provide much more detail on the origins of this medical discipline than we are able to offer here. For those particularly interested in the history of ophthalmology, we recommend *The History of Ophthalmology*, edited by Daniel Albert and Diane Edwards (1996), as well as Julius Hirschberg's eleven-volume series *The History of Ophthalmology*, translated by Frederick C. Blodi. We also highly recommend Howard B. Adelmann's *Marcello Malpighi and the Evolution of Embryology* (1966). Adelmann's text presents a good history of ocular embryology in volume 3 under Excursus XII, "The Eyes."

For many, the history of research in eye lens development largely dates back to the famous experiments of Hans Spemann and his work on lens induction at the turn of the twentieth century. However, descriptive knowledge of all the basic ocular structures was well established by the time Spemann began his experiments. Spemann's fundamental experiments on lens induction, along with those of Mencl and Lewis, are reviewed in subsequent chapters (see, e.g., chap. 2). One of the aims of the present chapter is to review the major recorded advances in the understanding of the anatomy, pathology, and development of the ocular lens from antiquity up to 1900.

The ancient Egyptians may have been the first to document cases of cataract, as this was likely the disease state referred to under the descriptions 'darkening of the pupil' and 'white disease of the eye' (Edwards, 1996). The Greek philosopher Alcmaeon conducted the first recorded human dissections about 535 BC (Weisstein, 2003). Although these dissections included examinations of the human eye, no specific mention was made of the lens (Magnus, 1998). There is some debate as to when the lens was first recognised as a distinct anatomical entity. In the Hippocratic book entitled *Fleshes*, written about 340 BC, roughly 35 years after the death of Hippocrates, there is a description of the internal contents of the human eye that reads thus: 'The fluid of the eye is like jelly. We frequently saw from a burst eyeball a jelly-like fluid extrude. As long as it is warm it remains fluid, but when it cools, it becomes hard and resembles transparent incense; the situation is similar in man and animals' (Hirschberg, 1982, p. 72). The jelly-like fluid is often interpreted to be the vitreous, but the hard, transparent remnant described in the quotation is thought to be the lens. Prior to and during this period, it was believed that the lens was liquid and that it only became solid as the result of disease. The observation that the internal contents of human and animal eyes were similar suggests that the fundamentals of comparative anatomy were already familiar to the ancient Greeks. Some followers of Hippocrates performed detailed studies of chick development and discovered that the eyes were visible early in embryogenesis. This finding contradicted the belief, later expressed by Pliny the Elder (23–79 AD), that human eyes were the last of all human organs to develop in the uterus (Magnus, 1998). Aristotle (384–322 BC), often cosidered the founder of biology, performed numerous dissections of mature and embryonic animals. In describing the anatomy of a ten-day-old chicken embryo, Aristotle wrote, 'The eyes about this time, if taken out, are larger than beans and black; if their skin is removed the fluid inside is white and cold, shining brightly in the light, but nothing solid' (Magnus, 1998). Again, the failure to appreciate the solid nature of the lens suggests a general lack of knowledge of its precise structure (Fig. 1.1). While Aristotle recognised that the eyes begin forming early in embryogenesis, he also believed that the eyes were the last structures 'to be formed completely' and mistakenly thought that they shrink during later embryonic development (Adelmann, 1966). Neither the ancient Egyptians nor the pre-Alexandrian Greeks had an anatomical term to describe the lens of the eye. As stated by Magnus (1998), 'Of a certain knowledge of the lens, nothing is to be found in the writings of the pre-Alexandrian era' (p. 54). Alexander the Great founded the Alexandrian School in Egypt in approximately 331 BC. This school became a great centre of Greek learning, and it was here that the dissection of corpses became a regular practice.

Roman medicine was profoundly influenced by Greek medical philosophy, and the most complete surviving Roman medical text was written by Aulus Cornelius Celsus, who lived from approximately 25 BC to AD 50. Celsus's book *De Medicina* was written about AD 30, and his anatomical descriptions of the eye were likely based on descriptions provided by earlier Greek authors (Albert, 1996a). Celsus did specifically mention the lens as resembling egg white, and he expressed what would become a long-held belief that the lens was the organ from which visual function originated (Albert, 1996a). In his diagrams of the eye, Celsus also mistakenly placed the lens in the centre of the globe (Fig. 1.2). Although couching, a surgical procedure to treat cataracts, likely originated in Asia or Africa prior to the birth of Hippocrates (Hirschberg, 1982), Celsus's writings provide clear documentation of this procedure, which was either unknown to or rarely practiced by the pre-Alexandrian Greeks. Couching, the only form of cataract surgery prior to the eighteenth century, involves displacing lens opacities by inserting a needle into the eye and depressing the lens against the vitreous until the opacity no longer obscures the pupil (Albert, 1996b). Cataract surgery in the days of Celsus, prior to the advent of anaesthesia, was obviously not for the faint of heart. In the seventh book of *De Medicina*, Celsus described the characteristics of a

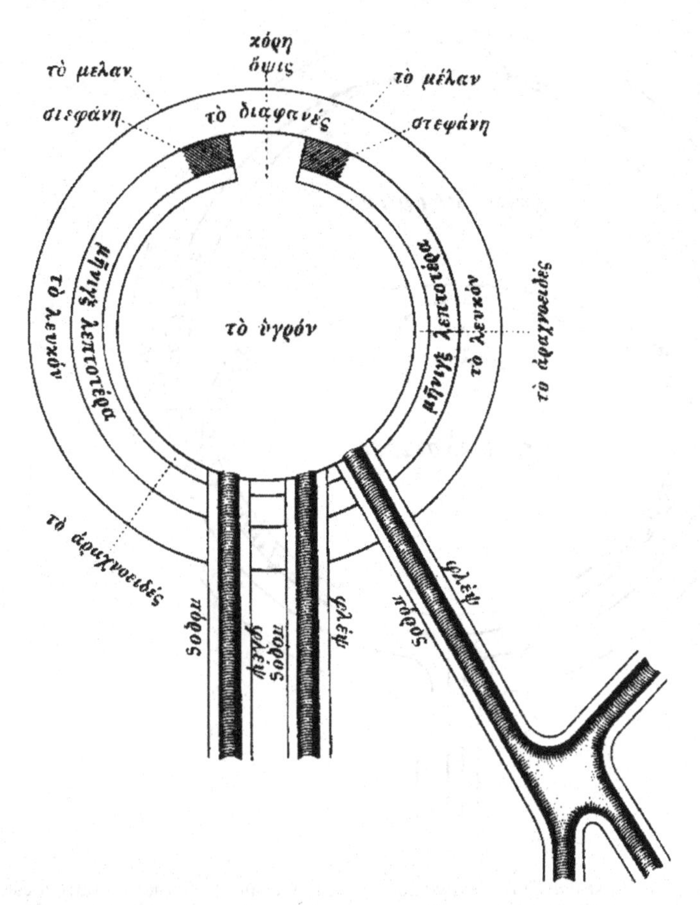

Figure 1.1. Reconstruction of the eye according to Aristotle. Note the absence of a structure representing the lens and the inclusion of three different vessels thought to transport fluids to and from the eye. (Reprinted from Wade, 1998b, after Magnus, 1901.)

desirable surgeon, as 'a young man with a steady hand who could remain unmoved by the crying and whining of his patients' (Albert, 1996a, p. 24).

Another important figure in early ophthalmology, and indeed all medical disciplines, was Claudius Galen (AD 130–200). Galen was born in Pergamon (currently in Turkey), was educated in Alexandria, and practised medicine in Rome. According to Galen, '1. Within the eye the principal organ of sensation is the crystalline lens; 2. The sensation potential comes from the brain and is conducted via the optic nerves; 3. All other parts of the eyeball are supporting structures' (Hirschberg, 1982, p. 280). It was Galen's view that the lens formed from the vitreous and that the function of the retina was to nourish the vitreous and lens as well as to transmit the visual information gathered from the lens to the brain (Albert, 1996a). Galen's anatomical description of the eye, in contrast to that of Celsus, placed the lens in the proper location, in the ocular anterior near the pupil, which was identified by Rufus of Ephesus (Fig. 1.3) several years before Galen's birth (Albert, 1996a). While Rufus had described the lens as 'lentil- or disc-shaped with the same curvature on the front and back surfaces', Galen recognised that the lens was more flattened on the anterior than on the posterior surface (Fig. 1.4; Magnus, 1998). According to Galen, cataracts, which

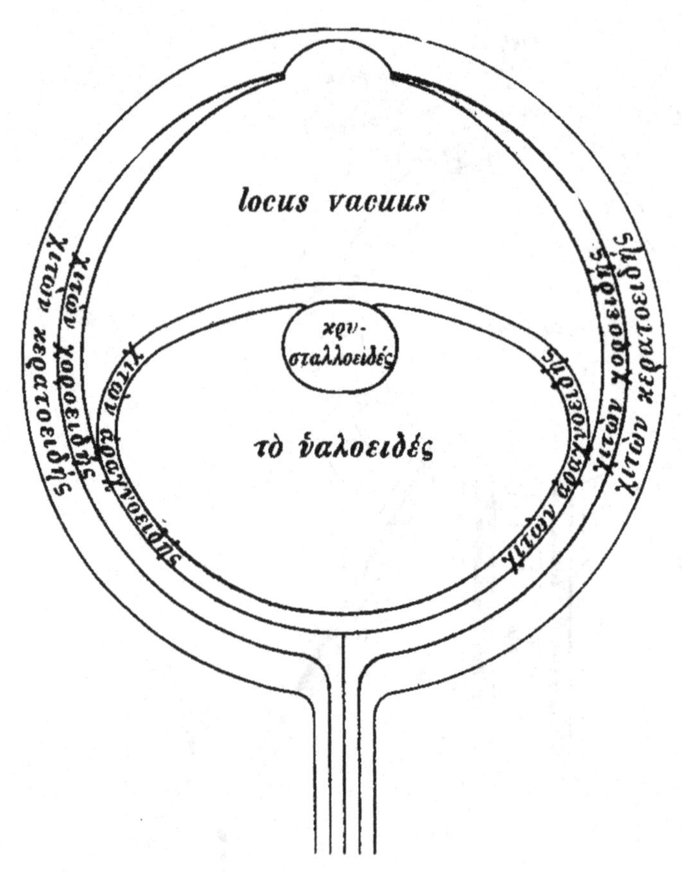

Figure 1.2. Reconstruction of the eye according to Celsus. Note the centrally located lens 'chrystalloides' within the vitreous 'hyaloides'. (Reprinted from Wade, 1998b, after Magnus, 1901.)

he called 'hypochyma', were the result of a thickening or condensation of aqueous fluid that clots and lies between the iris and the lens (Hirschberg, 1982). Galen wrote over a hundred surviving books, and his influence on European medicine was so profound that he was referred to as the 'final authority' for nearly fourteen centuries after his death (Albert, 1996a). After Galen, only minor changes in the understanding of lens anatomy, development, function, and physiology occurred for the next 1,250 years. Certainly some advances were made in clinical ophthalmology during this period. Notably, around AD 1000, an Arabian ophthalmologist named Ammar produced a manuscript entitled *Choice of Eye Diseases* in which he described the removal of soft cataracts by a modification of the couching procedure: the opacity was removed by suction through a hollow needle rather than by depressing the lens against the vitreous (Albert, 1996a). This procedure was the first step toward the treatment of cataracts by lens extraction pioneered by Jacques Daviel (1696–1762), a method that would ultimately replace couching (Albert, 1996a).

Leonardo da Vinci (1452–1519) has often been called the first great modern anatomist, and he did not ignore the eyes in his work. He made drawings from cadavers and is credited with devising the 'earliest technique for embedding the eye for sectioning by placing it within an egg white and then heating the embedded specimen until it became hardened

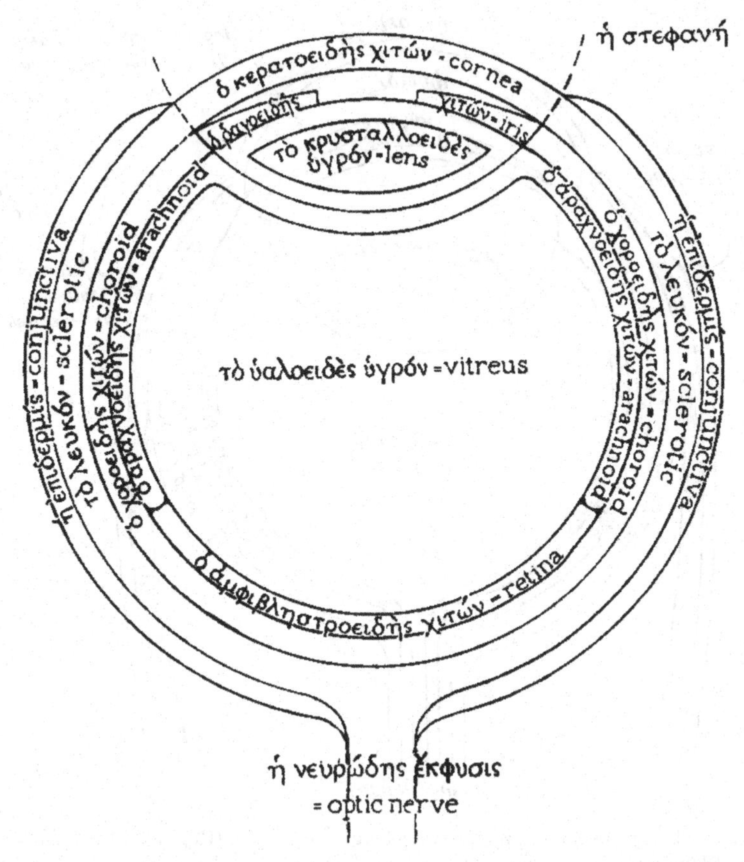

Figure 1.3. Reconstruction of the eye according to Rufus of Ephesus. Note that the lens is now represented directly behind the iris. (Reprinted from Wade, 1998b, after Magnus, 1901, with English labels added by Singer, 1921.)

and could be cut transversely' (Albert, 1996c, p. 47). While not always correct in his interpretation of the results of his ocular studies, da Vinci did reject the view, then current, that the lens was the primary sensory organ of the eye. In his drawings, da Vinci depicted the lens as focusing incident light directly onto the optic nerve (Albert, 1996c). His drawings also showed the lens as overly large and placed in the center of the eyeball. The work of da Vinci was largely unappreciated by ocular anatomists for more than 250 years after his death.

Andreas Vesalius (1514–64) published an anatomical work, *De Humani Corporis Fabrica*, in 1543 while teaching anatomy in Padua. This work was widely copied throughout Europe and succeeded in becoming the anatomical standard, replacing the writings of Galen, which had dominated European medicine for more than 1,300 years. Vesalius continued the misconception that the lens was in the centre of the eyeball (Fig. 1.5), but he did demonstrate that the isolated lens acted 'like a convex lens made of glass' (Albert, 1996c, p. 48). Georg Bartisch (1535–1606) correctly positioned the lens behind the iris in his 1583 publication *Ophthalmodouleia: das ist Augendienst*, considered by Albert (1996c) to be 'the first modern work on ophthalmology' (p. 49). Bartisch's work was also notable for being published in the vernacular German rather than Latin, as had been the tradition for medical texts until

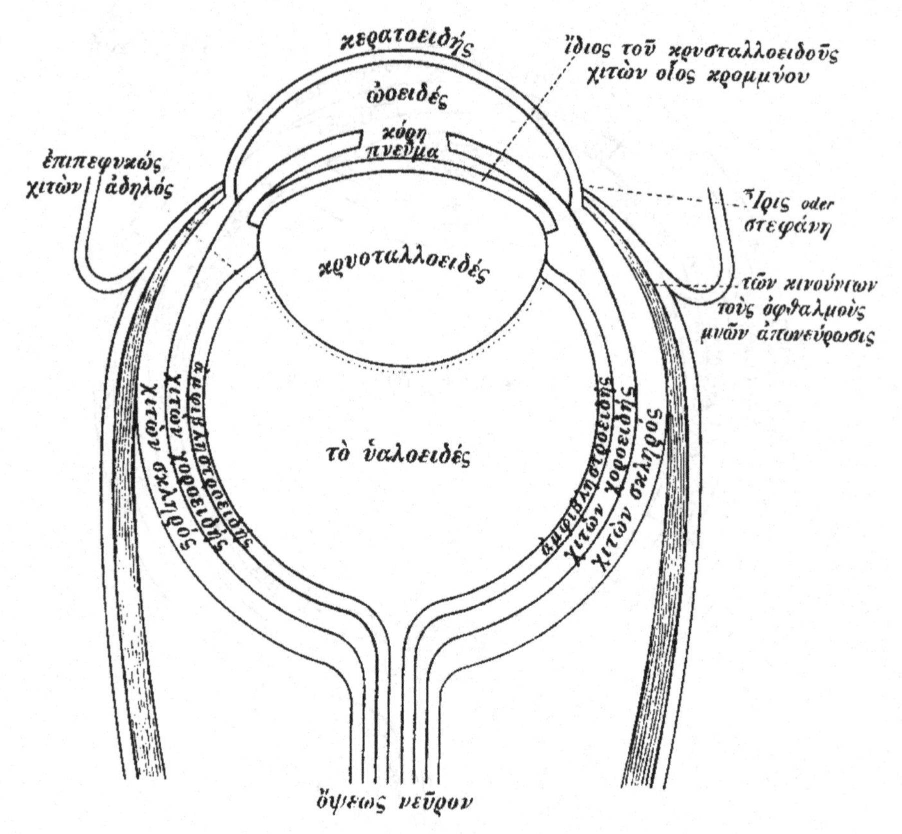

Figure 1.4. Reconstruction of the eye according to Galen. (Reprinted from Wade, 1998b, after Magnus, 1901.)

that time. Falloppio Hieronymus Fabricius ab Aquapendente (1537–1619), like Vesalius, became a professor of anatomy at Padua. Fabricius studied anatomy, embryology, muscular mechanics, and surgery. He is also known for being a mentor of William Harvey, who later discovered the process of blood circulation. Fabricius's work in embryology mostly concentrated on chick development. Fabricus incorrectly believed that the chicken embryo was derived from neither the egg white nor the yolk but from the chalazae, the rope-like strands of egg white that anchor the yolk. He constructed several arguments to support his belief. He asserted, for example, that the three visible nodes in the chalazae are the precursors of the brain, heart, and liver and that 'the eyes are transparent, so are the chalazae, therefore the latter must give rise to the former' (Needham, 1959, p. 108). To his credit, Fabricius, in the *Tractatus de Oculo Visuque Organo*, published in 1601, depicted the lens directly behind the iris and not separated from the pupillary margin by the 'cataract space' (Albert and Edwards, 1996c, p. 49). Harvey himself investigated the developing embryos of the chick and other animals, such as deer and sheep. He made the observation that 'the eye in embryos of oviparous animals is much larger and more conspicuous than that of viviparous animals' (Adelmann, 1966, p. 1238).

Felix Platter (1536–1614) was an anatomist in Basel who carried out the first public dissections of the human body in a Germanic country (Albert, 1996c). In 1583, Platter published *De Corporis Humani Structura et Usu*, relying heavily on previous illustrations

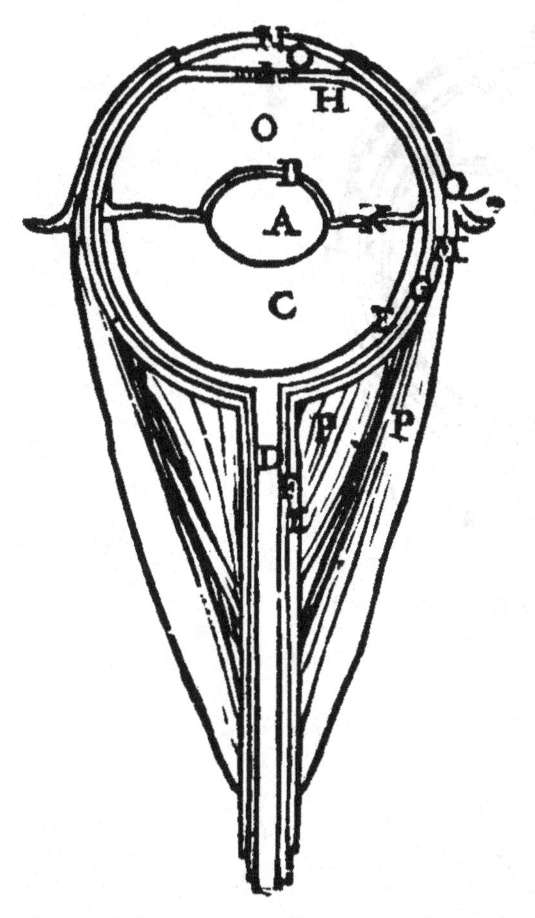

Figure 1.5. The anatomy of the eye according to Vesalius. A, crystalline lens; B, portion of the capsule; C, vitreous body; D, optic nerve; E, retina; F, pia-arachnoid coat of optic nerve; G, choroid; H, iris; I, pupil; K, ciliary processes; L, dural coat of the optic nerve; M, sclera; N, cornea; O, aqueous humour; P, ocular muscles; Q, conjunctiva. (Reprinted from Wade, 1998b, after Saunders and O'Malley, 1950.)

from Vesalius (Fig. 1.6). Notably, in *De Corporis*, Platter concluded that it was the retina, rather than the lens, that was the primary visual sensory organ in the eye, and he emphasised this view in a subsequent publication, *Praxeos Medicae*. Platter's nephew, Felix Platter II (1605–71), disseminated his uncle's view that the retina was the primary visual structure, and the publication of his dissertation, *Theoria Cataracta*, was responsible for 'finally displacing the crystalline lens as the true seat of vision' (Albert, 1996c, p. 50). The primacy of the retina in visual perception was confirmed by the work of Johannes Kepler (1571–1630). Kepler was an astronomer whose work with glass optical lenses shaped his views on the functional anatomy of the eye. Kepler published *Ad Vitellionem Paralipomena* in 1604 and *Dioptrice* in 1611, and both of these works had a substantial impact on the field of ophthalmology. In *Dioptrice*, 'Kepler convincingly demonstrated for the first time how the retina is essential to sight and explained the part that the cornea and lens play in refraction' (Albert, 1996c, p. 51). In 1619, Christoph Scheiner (1575–1650) supported Kepler's beliefs about the retina with his work *Oculus hoc est*, in which he presented diagrams of the eye

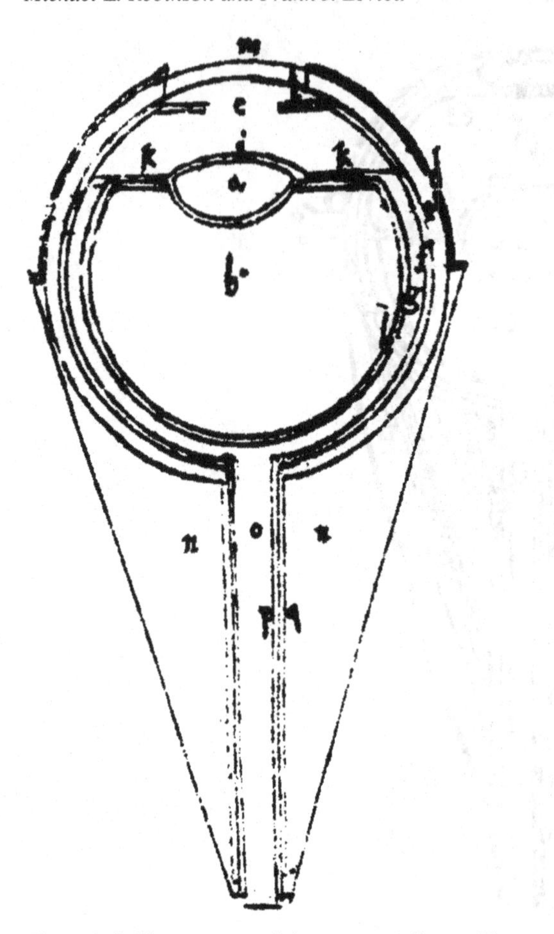

Figure 1.6. The anatomy of the eye according to Platter. a, crystalline humour; b, vitreous humour; c, aqueous humour; d, related coat; e, opaque part of the sclerotic; f, choroid; g, retina; h, hyaloid; i, crystalline capsule; k, ciliary processes; l, boundary of the choroids on the sclerotic; m, cornea; n, ocular muscles; o, optic nerve; p, thin nerve membranes; q, thick nerve membranes. (Reprinted from Wade, 1998b.)

showing how images are projected onto the retina. Scheiner is also given credit for drawing the first anatomically correct diagrams of the eye (Fig. 1.7) and for understanding that the optic nerve head is not in the optical axis but enters the right eye on the left side and the left eye on the right side (Wade, 1998a, pp. 78–80). He also described how the curvature of the lens could change during accommodation and devised a pinhole test for illustrating accommodation and refraction (Albert, 1996c).

Until the seventeenth century, all the investigations of the adult and developing eye were the result of indirect or direct observations of living or postmortem specimens with the naked eye. The course of eye research changed dramatically with the invention of the microscope. The invention of this instrument also set the stage for the emerging fields of embryology and developmental biology to diverge from more medically related disciplines, such as anatomy and pathology. According to Albert (1996c), the first microscopic investigation of an eye was by Giovanni Battista Odierna (1597–1660), who extensively described the fly eye in his treatise *L'Occhio della Mossca*, published in 1644. Marcello Malpighi (1628–94),

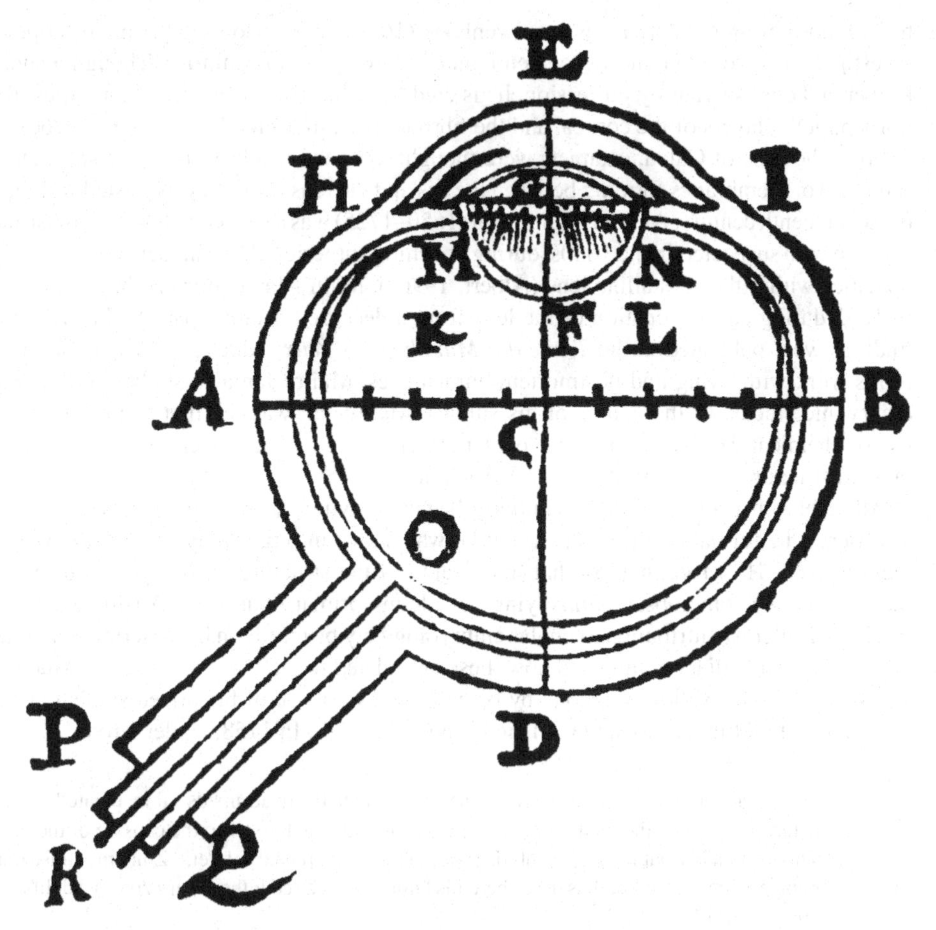

Figure 1.7. Anatomy of the eye according to Scheiner. Note that, in contrast to the previous illustrations of ocular anatomy, the optic nerve is not drawn in a direct line with the lens and cornea. (Reprinted from Wade, 1998b, after Scheiner, 1619.)

known as 'the founder of histology', was a professor of anatomy at Bologna, Pisa, and Messina and became a physician to Pope Innocent XII in Rome shortly before his death (Albert, 1996c). In 1672, Malpighi submitted two dissertations describing the embryonic development of the chick – *De Formatione Pulli in Ovo* and *Appendix Repetitas Auctasque De Ovo Incubato Observationes Continens* – to the Great Royal Society of England (Adelmann, 1966). The former described the development of the chick with the blastoderm lying on the yolk. In the latter, Malpighi described an isolated embryonic blastoderm mounted on a piece of glass and viewed under a microscope (Adelmann, 1966). Malpighi made several fundamental observations regarding the development of the eyes. He did not recognise their connection to the forebrain, but he did discover the optic vesicles, and he made many detailed illustrations of the developing chick eye that would be unsurpassed for decades after his death (Adelmann, 1966). While Nicolaus Steno (1638–86) was probably the first person to identify the choroid fissure, during his observation of the developing chick in 1665 (Steno, 1910), he did not publish his manuscript *In Ovo et Pullo Observationes* until 1675 (Adelmann, 1966). Therefore, Malpighi's illustrations of the choroid fissure were published before those of Steno. Anton van Leeuwenhoek (1632–1723) also used the microscope to investigate the eye in the late seventeenth and early eighteenth centuries. Though a shopkeeper and civil servant by profession, he is credited with discovering the retinal rods, the fibroepithelial layers of the cornea, and the fibrous structure of the lens (Albert, 1996c).

From the time of Galen, cataracts were thought to be the result of insoluble substances (humors) or membranes forming between the iris and the lens. This view persisted well into the seventeenth century. Antoine Maître-Jan (1650–1725) was a French ophthalmologist and surgeon who suspected in the 1680s, during couching surgeries, that cataracts were actually opacities within the crystalline lens (Albert, 1996c). Maître-Jan confirmed his suspicions in 1692 during an examination of the lens from a deceased cataract patient. In 1707, his findings were published in the *Traité des Maladies des Yeux* (Albert, 1996c). In addition to his work with living and postmortem human eyes, Maître-Jan also studied the eyes of embryonic chicks. In the course of his studies, Maître-Jan was the first to introduce the use of chemical fixatives to preserve ocular structures and for discovering the 'onionlike' layered structure of the crystalline lens (Adelmann, 1966).

Albrecht von Haller (1708–77) initially began studying the development of the chick to observe the formation of the heart, but he was drawn into the study of ocular development as well. He wrote in 1754 that 'the beauty of the structure of the eye has beguiled me into making some observations lying outside my primary purpose' (Adelmann, 1966, p. 1245). Haller's contribution to ocular embryology lay primarily in his descriptions of the ciliary body and ciliary zonule and how these are related to the lens and vitreous. Much of the necessary observation was done by or with the assistance of his prized student at the University of Göttingen, Johann Gottfried Zinn (1727–59). In 1758, Haller wrote,

> Some very careful anatomists have seen in man and in the quadrupeds a thin, pleated lamina detach itself from the membrane of the vitreous and attach itself to the capsule of the lens. This lamina is what forms the anterior wall of the circle [canal] of Petit. Zinn, my illustrious pupil and my successor, has described this lamina and called it the ciliary zone. (Adelmann, 1966, p. 1247)

Zinn would ultimately die before Haller, but he did publish *Descripto*, which was the first complete publication on ocular anatomy and which remained a standard atlas of the eye well into the nineteenth century (Albert, 1996c).

While descriptions of the anatomy of the vertebrate embryonic and adult eye had progressed a great deal in the 2,122 years between Aristotle's death and 1800, a true understanding of the embryonic origins of the lens or other ocular tissues was still lacking at the dawn of the nineteenth century. The descriptive embryology of the eye was to blossom in the 1800s, providing the framework necessary for Spemann and others to carry on the work of experimental rather than descriptive developmental biology in the twentieth century.

Karl Ernst von Baer (1792–1876) was a Prussian embryologist who laid many of the foundations of comparative embryology. Von Baer is often recognized for discovering that the optic vesicles were indeed an outgrowth of the embryonic forebrain. This discovery, however, was first published in the 1817 Latin dissertation of one of von Baer's friends, Christian Pander (1794–1865), who is best known for discovering the three embryonic germ layers (Adelmann, 1966). Von Baer extended the observations of Pander, suggesting that fluid pressure from inside the central nervous system was the motive force for the outgrowth of the optic vesicles. He also believed that the optic vesicle opened at the end to form the future pupil and that the vitreous body and lens were formed by the coagulation of fluid within the optic vesicle (Adelmann, 1966). Von Baer's views on eye development were presented in *Entwickelungsgeschichte der Thiere*, published in two parts in 1828 and 1837.

Emil Huschke (1798–1858) in many ways demonstrated a better fundamental understanding of vertebrate eye development than von Baer. Among his most important contributions was the recognition in 1830 that the lens forms not from the fluid of the optic vesicle but as an invagination of the integument (surface ectoderm). In his 1832 manuscript *Ueber die erste Entwinkenlung des Auges und die damit zusammenhängende Cyklopie*, he stated 'The lens capsule is a piece of the outer integuments which separates off and retreats inward, to be covered again later by several membranes, for example, by the cornea' (Adelmann, 1966, p. 1272). However, he mistakenly believed that the substance of the lens was a fluid secreted by the walls of the lens vesicle. Huschke is also given credit for discovering that the optic vesicle forms the double-layered optic cup, though he misinterpreted the ultimate fate of the individual optic cup layers (Adelmann, 1966). His description of the formation of the optic cup and choroid fissure also corrected von Baer's erroneous views of the formation of the pupil.

Wilhelm Werneck (d. 1843) announced in his 1837 publication *Beiträge zur Gewebelehre des Kristallkörpers* that the internal substance of the lens was not fluid: 'The contents of this capsule [the lens] are not fluid, as I earlier believed, but are of a more pulpy character' (Adelmann, 1966, p. 1277). Werneck also realised that lens fibers grow during embryogenesis – 'the fibers, continuing to grow from the periphery toward the centre, become increasingly visible' – but he apparently did not recognise that elongation of the primary lens fibers occurred in a posterior to anterior direction (Adelmann, 1966, p. 1277).

In 1838, Matthias Jakob Schleiden (1804–81) and Theodor Schwann (1810–82), both students of the German physiologist Johannes Müller (1801–58), formulated what would become known as the 'cell theory'. According to this theory, all living things are formed from cells, the cell is the smallest unit of life, and cells arise from preexisting cells. The cell theory had a profound impact on all aspects of biological study, and Schwann himself made several important contributions to the understanding of lens development in his book *Mikroskopische Untersuchungen über die Uebereinstimmung in der Struktur und dem Wachsthum der Thiere and Pflanzen*, published in 1839.

> The lens is known to be composed of concentric layers, made up of characteristic fibres, which, not to go into details, may be said to pursue a general course from anterior to posterior surface. In order to become acquainted with the relation which these fibres bear to the elementary cells of organic tissues, we must trace their development in the foetus. . . . In a foetal pig, three and a half inches in length, the greater part of the fibres of the lens is already formed; a portion, however, is still incomplete; and there are many round cells awaiting their transformation. . . . The fibers may readily be separated from each other, and proceed in an arched form from the anterior towards the posterior side of the lens. . . . Nuclei are also frequently found upon the fibres of the foetal pig. Some of the fibres are flat. I have, also, several times observed an arrangement of nuclei in rows; but I do not know what signification to attach to the fact. (Henry Smith's 1847 English translation, Adelmann, 1966, pp. 1278–79).

Robert Remak (1815–65) was also an embryologist who studied under Johannes Müller at the University of Berlin. Remak was the first to apply the cell theory to the three primary embryonic germ layers first described by Christian Pander. It was Remak who gave these germ layers their current names: ectoderm, mesoderm, and endoderm. His descriptions of embryonic eye development were the finest of the mid-nineteenth century. With respect to the lens, Remak recognised that the ectoderm (*Hornblatt*) thickened to form what we now refer to as the lens placode as it is contacted by the optic vesicle. Remak also described the fate of the lens vesicle in previously unparalleled detail in his 1855 book *Untersuchungen über die Entwickelung der Wirbelthiere*:

> That the lens invaginates into the optic vesicle from the outside, Huschke, as we know, discovered. But what this investigator did not and in the state of the science at his time could not know, is the fact which I have discovered, namely that this invagination proceeds from the upper germ layer, which later furnishes the cellular (epidermal) coverings of the body. . . . The bulk of the wall of the lens, after being constricted off, consists of cylindrical, radially arranged cells, which also resemble strongly the cells of a columnar epithelium in that they appear very sharply delimited on their free surface facing the cavity. Each cell contains one, in rare cases even two, nuclei. These nuclei do not, however, lie at the same level. . . . This layer of cells is surrounded by a . . . very thin, apparently 'structureless' membrane. . . . This is the anlage of the lens capsule. . . . All fibers pass without visible interruption from the posterior wall of the lens capsule to the anterior almost parallel to the visual axis; the fibers are, consequently, shorter the farther away they lie from the visual axis. Their anterior and posterior ends are cut off sharply. At some distance from its anterior end each fiber contains a nucleus, but no trace of nuclei can be detected at the posterior end. . . . The posterior end of each fiber is directly in contact with the lens capsule; the anterior end, on the other hand, is separated from it by an epithelium consisting of nucleated cells which adheres to the capsule. Hence it follows that the cells of the posterior wall of the lens vesicle form the lens fibers, those of the anterior wall, on the contrary, form the epithelium. (Adelmann, 1966, pp. 1293–4)

With these observations by Remak, the basic descriptive embryology of lens formation was virtually complete, though many fine points, such as the origin of the lens capsule, would remain subject to debate and investigation for several more decades. Remak was a rather tragic figure in the history of nineteenth-century embryology. He received his medical degree in 1838 but was initially barred from teaching by Prussian law because of his Jewish faith. After graduation, he remained as an unpaid assistant in Johannes Müller's laboratory, where he conducted basic research on the nervous system and supported himself with his medical practice. Remak was eventually granted a lectureship in 1847, becoming the first Jew to teach at the University of Berlin (Enersen, 2003). Remak's descriptive work on the development of the eye was a very small part of his substantial body of research, but despite this he only succeeded in attaining the rank of assistant professor in 1859, six years before his death. Remark was the father of neurologist Ernst Julius Remak (1849–1911) and the grandfather of mathematician Robert Remak (1888–1942), who was killed in the Nazi concentration camp at Auschwitz (Enersen, 2003).

After Remak, who studied eye development in the chick, frog, and rabbit, others added details concerning lens formation in other species. For example, in 1877, Paul Leonhard Kessler described the development of the mouse lens in his work *Zur Entwickelung des Auges der Wirbelthiere*. Carl Rabl also published a marvelous book in 1900, *Uber den Bau und die Entwicklung der Linse*, which describes and illustrates lens development in fish, mammals, reptiles, amphibians, and birds. While these publications added details, they still built on the common theme elegantly described by Remak. Thus, at the close of the nineteenth century, the stage had been set for the descriptive embryology of the lens to give way to the experimental embryology of the lens that continues through the twentieth and into the twenty-first centuries. In particular, the groundwork had been laid for Hans Spemann (1869–1941) to perform his classic experiments revealing the induction of the lens by the optic vesicle and establishing the theory of embryonic induction, which would become part of the foundation of developmental biology and lead to the discovery of the Spemann–Mangold organiser. Embryonic induction was an idea whose time had come.

1.2. Comparative Ocular Anatomy

Over time, developmental biologists have focused much of their attention on understanding the mechanisms underlying cell and tissue differentiation and the orderly manner in which they grow, developing to the size and shape appropriate for the body's requirements as well as forming in the correct anatomical relationship with each other. These events are thought to depend on inductive cell and tissue interactions. As will be appreciated in the following chapters, for over a century the lens of the vertebrate eye has provided numerous researchers with a means of examining inductive tissue interactions involved in tissue differentiation throughout development and growth. The lens has readily been adopted as a model for such studies because, as will be described later, it is a relatively simple tissue made up of cells from a single cell lineage. The lens retains all the cells that are produced throughout its life, and it is isolated from a nerve and blood supply. The positioning of the lens in the eye, on the surface of the body, has made it easily accessible for experimentation, but most importantly the lens has proved to be suitable for the study of cell differentiation *in vitro* and *in vivo*, as not only do differentiating lens cells synthesise uniquely defined proteins such as the crystallins, they also undergo very distinct morphological changes, including cell elongation, cell membrane specialisation, and the loss of cytoplasmic organelles and nuclei.

The remainder of this chapter will therefore be devoted to introducing the ocular lens by reviewing in brief its structural diversity in a range of organisms and highlighting its adoption as an ideal tissue model for cell and developmental biologists alike. Sections of this part of the chapter summarise the wealth of information documented by Sir Stewart Duke-Elder (1958) in "The Eye in Evolution" a book volume on the ontogeny and phylogeny of the invertebrate and vertebrate eye, which the reader is encouraged to read in its entirety.

For many animals, the sense of vision is the most important link to the environment. Animals have adapted for survival in a variety of climatic conditions and terrains and thus have evolved a diversity of eye designs. Despite this diversity, each of the visual organs has a common functional role, the perception of light. The sensation of light is the most fundamental of the visual senses. The acquisition and development of vision has stemmed from the dependence of living organisms on light, with light influencing many aspects of survival, such as general metabolism and the control of movement (characteristic of the most primitive of animals), as well as influencing the behaviour and consciousness (through the visual senses) of higher animals.

Unicellular organisms such as ciliate and flagellate protozoa (e.g., *Euglena*) provide us with examples of the earliest stage in the evolution of an eye. These organisms contain a small region of protoplasm that has differentiated into a photosensitive 'eyespot'. This specialised light-sensitive area, partially covered by a pigmented shield close to the root of its motile flagella, not only can receive a visual stimulus but is also utilised to orient the organism and direct it to more favourable regions in its environment. With the evolution of multicellular organisms came the differentiation of light-sensitive cells that allowed these higher organisms to distinguish between light and dark and even determine the direction of light. From this, eyes went on to become more specialised, evolving further to detect motion, form, space, and color.

The visual organs of invertebrates show a much greater diversity in structure than those found in vertebrates, varying in complexity from the simple eyespot to the vesicular eyes of cephalopods and the compound eyes of insects. Despite this diversity, many of the eyes of these unrelated invertebrate species comprise analogous photoreceptive cells. The entire bodies of primitive invertebrates, such as jellyfish, coral, sea anemones, worms, and echinoderms, are sensitive to light. Their eyes are no more than a collection of eyespots or photosensitive cells, frequently associated with pigment, which serves as a light-absorbing agent. These light-sensitive cells are ectodermally derived and can be found alone or in association with other cells to form an eye. Depending on the structural organisation of these cells – whether they form an organ singly or as part of a community – an invertebrate eye can be classified as either a simple eye (or ocellus) or a compound eye. Intermediate forms are referred to as aggregate eyes and are usually composed of a cluster of ocelli packed so closely that they resemble a compound eye. The major distinction is that each ocellus in an aggregate eye is anatomically and functionally separate.

1.2.1. The Simple Eye

The ocellus, or simple eye, can be defined as a light-sensitive cell or a group of such cells that are not functionally associated but each act independently. The simple eye has many different forms, from its primitive beginning as a single cell to a more complex structure represented by the vesicular eye (Figure 1.8). The most primitive association of light-sensitive cells is seen in the 'flat eye', which comprises a number of specialised contiguous surface cells that form a plaque (found in some unsegmented planarian worms and leeches).

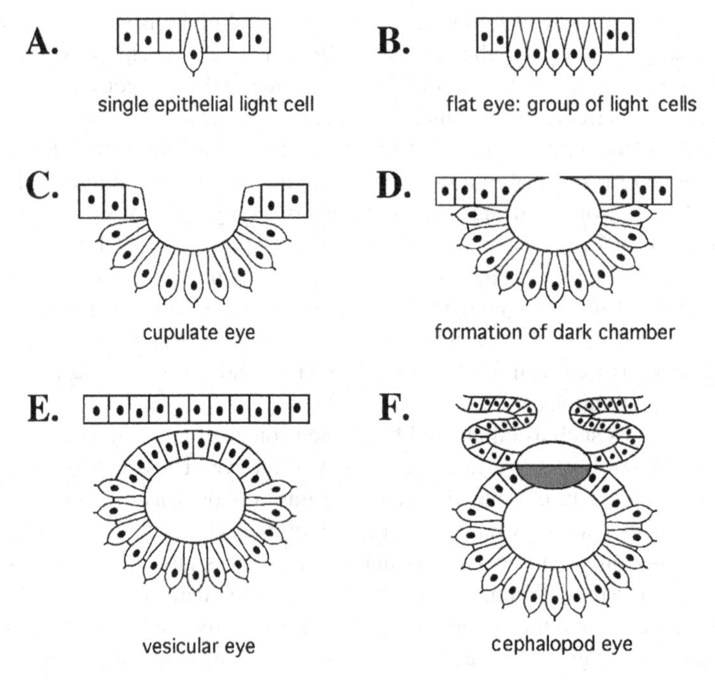

Figure 1.8. Schematic diagram depicting the comparative anatomy of the simple eye of invertebrates. (Adapted from Duke-Elder, 1958.)

In more advanced organisms, these patches of photosensitive epithelium indent to form a depression, giving rise to the 'cupulate eye' or 'cup eye'. This structural change had the functional advantage of allowing for the development of a primitive sense of direction. Cupulate eyes have several forms, depending on the degree of invagination of the light-sensitive epithelium. The most primitive form can be found in the larva of the common housefly (*Musca*), present as a shallow pit in the epithelium. A deeper invagination, together with an increase in the number of photosensitive cells, converts this depression into a cavity with a small opening. As the opening of the depression continues to narrow, a dark chamber with a pinhole opening is formed. Such an eye is found in the chambered nautilus (a primitive cephalopod). Although the photoreceptors of the nautilus are indented to form an optic cup, the cup does not contain a lens, and its operation is based on the same principles as a pinhole camera. The pinhole is used to focus images, and although this provides excellent depth of field, a lot of light is required to provide an image of any quality. The optic cup can be filled with sea water, as in the case of the nautilus, or with secretions, as found in the ear shell (*Haliotis*). The final form of the cupulate eye is characterised by the closure of the cavity by the growth of an overlying transparent acellular cuticle which will one day go on to form the lens. The enclosed secretory mass forms the vitreous body, as seen in *Nereis*, the marine polychaete worm. Improvements on this design are found in some insects. For example, hypodermal cells might form a thickened cuticular layer which acts as a refringent apparatus. The optical arrangements of such an eye may further be improved, as seen in *Peripatus* (a caterpillar-like arthropod), in which the hypodermal cells form a large lens in place of the vitreous. These hypodermal cells, usually continuous with the surface ectoderm or with the sensory cells of the cupula, may also edge themselves underneath the cuticle and go on to form a transparent refractile mass below the cuticular lens, thereby constituting a primitive lens or vitreous. Overall, the lens of a simple eye may be either acellular and cuticular or cellular.

The vesicular eye may be considered the final stage in the development of the simple eye. This type of eye is marked by the closure of the invaginated light-sensitive epithelium, which gives rise to an enclosed vesicle separated entirely from the surface ectoderm by mesenchyme. In its simplest form, the vesicular eye is spherical and lined with ectodermal cells, as found in the edible snail *Helix pomatia*. The vesicle has a specific polarity, with the more posterior cells being partly light sensitive and partly secretory while the more anterior cells remain relatively undifferentiated. The cavity of the vesicle is filled with a refractile mass of secreted material, homologous with the vitreous of higher organisms.

In a further stage of complexity, the vesicular eye takes the form of a camera-like eye through the addition of a lens and now resembles the eye of vertebrates. The best example of this can be found in cephalopods (e.g., octopus), which have the most elaborate eyes in the invertebrate kingdom. The eye vesicle of cephalopods is filled with a vitreous secretion. The posterior cells lining it form the retina while the anterior cells fuse with an invagination of the surface epithelium to form a composite spheroidal lens (see Fig. 1.8F). The posterior half of the lens is thus made up of vesicular epithelium while the anterior half is derived from the surface epithelium. Encircling the lens, the fusion of the vesicular and surface epithelium gives rise to a 'ciliary body', with an 'iris' derived from the surface epithelium. This type of cephalopod eye is highly complex, is capable of image formation, and has the ability to accommodate. In contrast to other invertebrates with fixed-focus lenses, cephalopods can focus for near and far vision by changing the position of the lens relative to the retina. Although at the morphological level these eyes rival those of vertebrates, they are simple

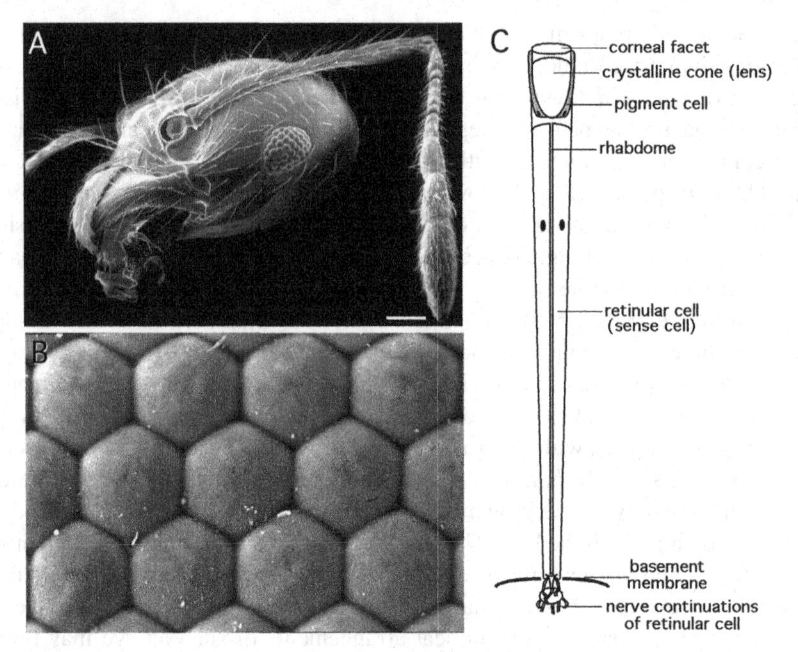

Figure 1.9. (A) Scanning electron micrograph of an ant, showing its compound eye comprising relatively few ommatidia. (B) Scanning electron micrograph of a representative region of ommatidia from the compound eye of a moth. (C) Schematic diagram of a longitudinal section of a single ommatidium from a compound eye. Scale bar: (A), 100 μm; (B), 10 μm.

in type. They are derived solely from the epithelium and combine a vertebrate-like optical system with invertebrate photoreceptors.

1.2.2. The Compound Eye

The compound or faceted eye has evolved along a different path from that of the simple eye. Unlike the simple eye, characterised by functionally independent light-sensitive cells, the sensory elements of the compound eye are structurally and functionally associated in groups (Fig. 1.9). The sensory elements that make up the compound eye, each referred to as an ommatidium, are separated from their neighbours by a mantle of pigment cells. The number of ommatidia can vary greatly, from as few as 9, as found in some ants, to 30,000, as found in dragonflies. The visual field of the compound eye is composed of images from individual ommatidia, with each acting as a single retinal cell of the simple eye. Each image does not represent the same overall picture but a small portion of the visual field within the ommatidium's angular range. These separate parts of the field are transmitted to the receptive cells of the ommatidium, where the overall image is fused; the large number of images increases the acuity of vision by a mosaic effect.

Each ommatidium has a relatively simple structure (Fig. 1.9C). Superficially, the cuticle is composed of corneal facets which fit into each other to form a mosaic (hence the name *faceted eye*). Underneath this lies a refractive device, the crystalline cone, which transmits light through the action of its two convex surfaces. The crystalline cone is made up of concentric lamellae, with its refractive index increasing from the periphery to its central axis.

This allows it to act as a lens cylinder in which incident light is progressively refracted until brought to its central axis. Unlike the lens of cephalopods and vertebrates, the crystalline cone has a fixed focus and is incapable of adjustment. The remainder of the ommatidium is made up of the sensory cells arranged in a tubular form, referred to as the *retinule*. These nerve cells (rhabdomeres) are supported by a fenestrated basement membrane and are radially arranged so that their differentiated inner borders join to form a centralised refractile rod, the rhabdome.

One of the earliest examples of the compound eye can be found in the long-extinct trilobite, an ancient marine creature. Trilobites had a faceted eye that was limited in its view: it could see sideways but not upward. Although extinct, the trilobite has many living arthropod relatives. While most arthropods have lenses made up of relatively soft, unmineralized cuticle (similar to the rest of their exoskeleton), the eye of the trilobite, which is continuous with its solid armour, is made up of the abundant mineral calcite. Trilobites used the transparency of clear calcite as a means of transmitting light. The trilobite eye varied from species to species, with the types differing in the number of lenses they contained (from one to over a thousand). One of the most common trilobites, *Phacops*, had eyes that contained lenses lined up in conspicuous rows, each lens distinguished as a tiny, perfectly formed sphere (or a slightly drop-shaped sphere). These lenses were often slightly sunken, with a wall between adjacent lenses, stopping light from overlapping onto neighbouring lenses. Other trilobite eyes contained larger biconvex lenses, designed to focus light.

To date, nature has provided animals with two very different types of image-forming eyes, which have developed independently: the compound eye and the camera-like eye of vertebrates. As described above, the compound eye is a fixed-focus eye composed of ommatidia, each responsible for different portions of the visual field. The camera-like eye of vertebrates contains a pigment-shielded cup, elaborated by the presence of a single lens as well as muscles for focusing. The lens (and the cornea, in land dwellers) has developed to focus an image on a continuous receptor surface, the retina. Common to these two types of image-forming eyes is the presence of a lens, whose role is to transmit and refract light. As the cornea is primarily responsible for this in land creatures, the lens also serves to adjust the focus for near and distant vision. There are exceptions to this. In the simple eye of invertebrates, for instance, the lens may be employed, not to form an image, but to concentrate light in order that the eye may function in low light. Furthermore, a lens is not essential for an image to be formed on the retina, for a pinhole can play this role, as it does in the chambered nautilus, described earlier.

1.2.3. The Vertebrate Lens

The vertebrate ocular lens is suspended in the eyeball, situated behind the cornea and iris, and is supported posteriorly by the vitreous body (Fig. 1.10). The slightly convex anterior surface of the lens is in contact with the pupillary margins of the iris, and its more convex posterior surface occupies the hyaloid fossa of the vitreous. In most vertebrates, the primary support of the lens is provided by numerous suspensory ligaments, commonly known as zonular fibers, that extend from the equatorial rim of the lens to the surrounding muscle of the ciliary body. With relaxation of the ciliary muscle, the lens flattens, losing some of its convexity. This is the basic principle underlying accommodation: focusing the lens to give a clear view of a specific object.

Although the structure of the lens at first glance appears relatively simple, it is remarkably complex (Fig. 1.11). The lens substance consists entirely of densely packed lens fibers and

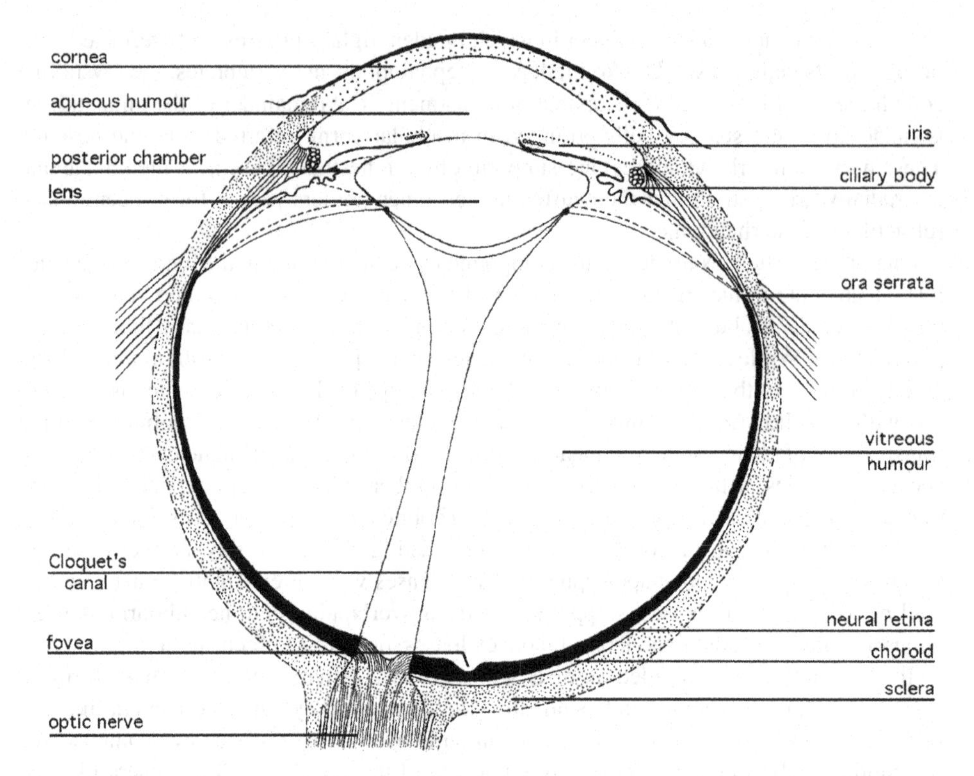

Figure 1.10. Schematic diagram showing the topographic anatomy of the eye.

epithelium, enclosed by a thick membranous lens capsule. The orderly aligned elongate fiber cells, which extend circumferentially from an anterior to a posterior point, make up the mass of the lens. These fiber cells are arranged in many layers, and fibers of any one layer, running in diametrically opposite meridians, meet end to end along radial planes known as *lens sutures.* The anterior half of the fiber cell mass is covered by a monolayer of cuboidal cells, the lens epithelium, which extends to the lens equator. The epithelial cells lie directly beneath the lens capsule. This basic structure is found in all vertebrates, and any variations are incidental in nature and have evolved essentially as adaptations to differences in function or habitat.

The simplest vertebrate eyes belong to cyclostomes (e.g., lampreys); these contain a large primitive lens not attached to the walls of the eyeball but fixed in place by the cornea and the vitreous humour. As mentioned earlier, in terrestrial vertebrates, the cornea does most of the focusing, strongly refracting most of the incident light rays. In marine vertebrates, such as fish, this is not the case, and so all of the focusing is performed by the lens. As a result, the lenses of fish are very large and almost spherical and have a higher refractive index than that of any other vertebrate lens. In fish, lens sutures are present, although simplified as a single line. As the flattened cornea of fish is optically ineffective, the lens must serve to gather as well as focus light. Because of this, the lens needs to be situated far forward in the ocular globe, bulging through the pupil and residing close to the cornea. In contrast, in land vertebrates, the lens is less rounded (relatively flat in higher vertebrates) and is situated farther back in the eyeball. The spherical shape of the fish lens and its close association with the cornea persist in amphibians, but only in their larval stage (e.g., tadpole). With

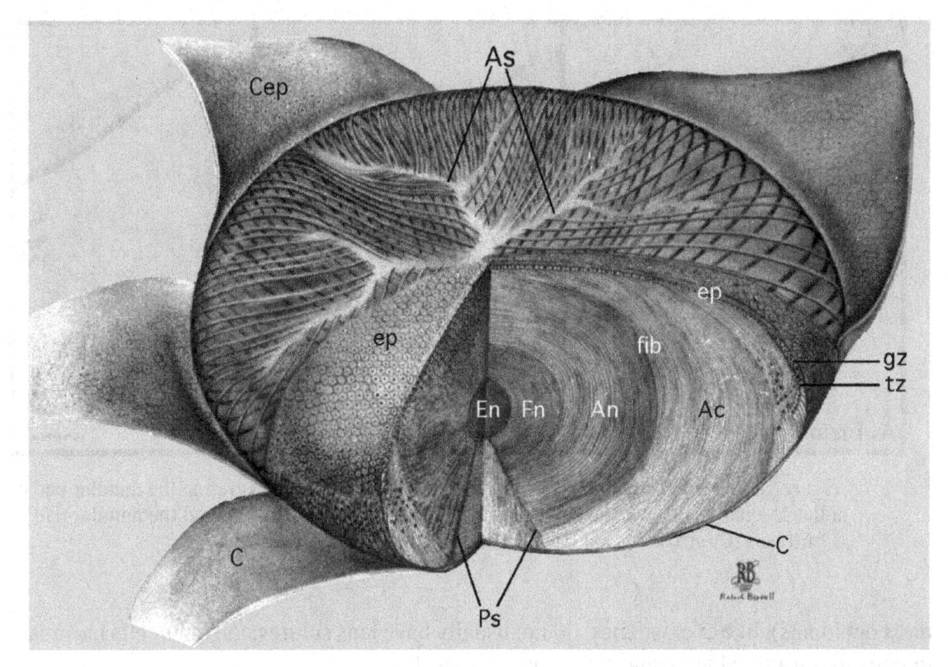

Figure 1.11. Anatomy of the human lens. The lens capsule (C) or lens capsule with attached epithelium (Cep) has been peeled back to expose the anterior surface. The internal anatomy of the lens is evident following removal of a lens wedge. ep, lens epithelium; fib, lens fibers; En, embryonic nucleus; Fn, fetal nucleus; An, adult nucleus; Ac, adult cortex; gz, geminative zone; tz, transitional zone; As, anterior suture; Ps, posterior suture. (Adapted from Worgul, 1982.)

adaptation to land, the primary burden of focusing light fell to the cornea, and the lens became situated posteriorly, behind the iris, and also became somewhat flattened in an anteroposterior direction. Accommodation in amphibians is achieved by bringing the lens forward for close-up focus. This occurs through the action of a small ventral muscle which pulls the lens forward or backward, although anurans (frogs and toads) utilise a dorsal muscle as well. The lens of reptiles is also flattened anteroposteriorly and has a greater convexity posteriorly. The ability to change the shape of the lens, or accommodate, was first seen in reptiles. Reptiles accommodate for near and far vision by making the lens rounder (for near sight) or flatter (for far sight) through the utilisation of muscles that do not exist in amphibians or fish. It is this modification of the capacity to accommodate that is responsible for the incredible range of focus shared by birds and mammals.

Variation in the structure and positioning of the subcapsular lens epithelium occurs among vertebrates. In contrast to the lens epithelium of fish, which extends beyond the lens equator, the epithelium in amphibians conforms to the higher vertebrate plan, extending only as far as the equator. With reptiles, the lens epithelium is further modified: the equatorial epithelial cells lose their cuboidal morphology, elongating in a radial direction to form an annular pad (Fig. 1.12). As a result, the lens now abuts the ciliary body. Variations on this theme are observed in other reptiles, such as the ophidians, which include snakes. Instead of an equatorial annular pad, the lens of a snake contains an anterior pad, and the anterior epithelial cells are elongated instead of cuboidal (Fig. 1.12B). A further distinction between the lens of a snake and that of other reptiles is the presence of sutures. Reptiles (with the exception of

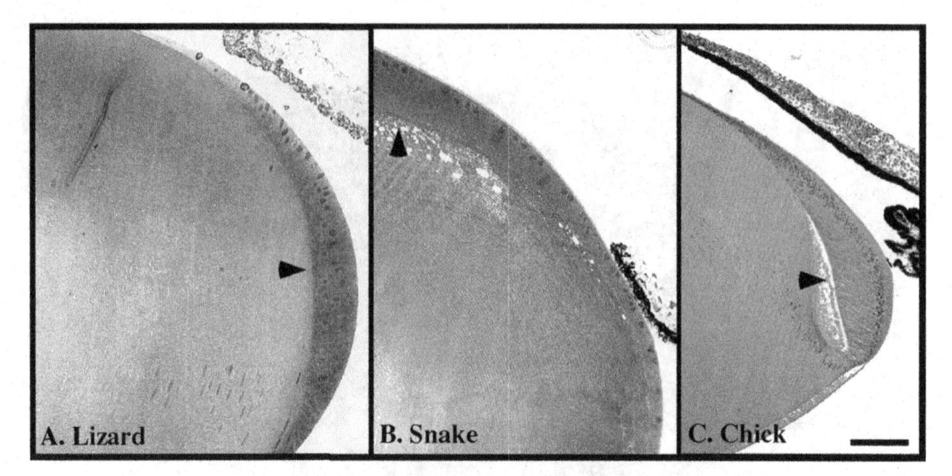

Figure 1.12. Representative histological sections of eyes demonstrating the annular pad of a lizard lens (A, arrow), the anterior pad of a snake lens (B, arrow), and the annular pad of a bird lens (C, arrow). Scale bar: (A) and (B), 40 μm; (C), 160 μm.

most ophidians), like cyclostomes, do not usually have lens sutures; the fiber cells terminate in one circumscribed area anteriorly and posteriorly.

Birds are amongst the descendants of primitive reptiles and have evolved on diverging lines from mammals. Throughout their evolution, they have attained the highest degree of vertebrate eye specialisation, a distinction shared only with mammals. The eyes of birds are designed on the same general plan of reptiles. They contain a lens with a well-defined annular pad (Figure 1.12C), the size of the pad dependent on the species. The annular pad is retained, albeit in a much smaller version, in the lens of some mammals, including monotremes (e.g., platypus) and marsupials. In placentals, which make up the majority of mammals, no annular pad is found in the lens. Instead, the lens contains an anterior monolayer of cuboidal epithelial cells which end at the equator.

One of the most significant improvements in the eyes of birds (and of mammals) is the positioning of the lens, which is brought forward toward the cornea, allowing for an increased image size on the retina. Furthermore, because of its flattened shape, the avian eye is capable of maintaining an entire visual field in focus at any one time – a major advantage over the rounder eye of mammals, where the point of focus is limited to a small area of the visual field. The mechanism of accommodation in reptiles and birds is essentially similar to that of mammals. Reptiles and birds have a well-developed ciliary body containing a ring of pad-like processes which make contact with the lens periphery. With contraction of the ciliary body musculature, these processes push on the lens, forcing it to acquire a more rounded shape, suited for near vision. In more advanced land mammals, the lens is suspended by zonular fibers that attach to the ciliary muscles, so the ciliary body is not in direct contact with the lens. Contraction and relaxation of the ciliary muscles modify the tension on the zonular fibers, altering the shape of the lens for near and far vision, respectively.

This brief and by far incomplete review of comparative lens structure and function indicates the extent of adaptive variation seen in vertebrate eyes. One thing that has remained constant amongst vertebrates is the manner in which the different parts of the eye – in particular, the lens – develop, differentiate, and continue to grow throughout life.

1.3. Development of the Vertebrate Lens

The establishment of the unique architecture of the lens in the embryo is thought to depend on a series of inductive interactions that initiate its differentiation from head ectoderm, as discussed in Chapter 2. These inductive influences are thought to emanate initially from the surrounding neural and non-neural tissues with which the ectoderm interacts throughout the course of development. Once established in the embryo, the lens is thought to maintain its distinct architecture and polarity throughout life by means of its interaction with the surrounding ocular media (Coulombre and Coulombre, 1963).

In this part of the chapter, we briefly describe the morphological changes that result in the transformation of a sheet of ectodermal cells into a highly specialised light-transmitting organ, the lens. This topic has been covered in great detail in many excellent books, including those by Mann (1928, 1950, 1964), Duke-Elder (1958), and Jakobiec (1982), which continue to serve as good resources. These texts primarily describe lens morphogenesis during human development. The goal of this text is to provide a more generalised and comprehensive view of lens development across a variety of the more commonly used animal models (see Fig. 1.13). The lens and the skin share a common embryonic origin; both are derived from the embryonic surface ectoderm. Consistent with this, the lens, like skin, continues to grow throughout life. Although skin is continually being renewed, with its superficial cells being replaced with the progeny of the more basally located stem cells, the cells of the lens are all retained, resulting in the lens becoming larger.

Unlike the eye in most invertebrates, which is solely derived from ectoderm, the vertebrate eye requires the contribution of three primordial tissues: the ectoderm of the neural tube, surface ectoderm, and mesoderm. The vertebrate lens is derived from the surface ectoderm, which also gives rise to many other tissues, including the corneal epithelium. In vertebrates, the eyes first appear as flattened areas (optic areas) at the anterior end of the embryonic plate on either side of the neural groove. At the stage when the neural groove deepens into the underlying mesoderm, the optic areas 'dimple' to form the optic pits. With the closure of the neural tube (presumptive central nervous system), the optic pits deepen, hollowing out to form the optic vesicles (presumptive neural retina and retinal pigmented epithelium). The optic vesicles are simply bilateral evaginations of the neural ectoderm of the forebrain (anterior end of the neural tube) embedded in mesoderm. This evagination process ultimately brings the outer surface of the optic vesicles into contact with the surface ectoderm (presumptive lens) at either side of the head. Thus the presumptive lens and retina originate from separate groups of cells situated within an ectodermal sheet. In the course of development, neural tube formation results in the folding of this ectodermal sheet so that the presumptive retina is internalised as part of the neural tube, with the basal aspects of the presumptive lens and neural cells now facing each other.

In most vertebrates, lens formation is initiated by the proliferation and palisading of the layer of ectodermal cells overlying the optic vesicle to form a thickened lens disc or plate (placode) at the area of contact. The lens placode remains as a single layer of columnar cells, although it appears stratified due to the staggered positioning of its cell nuclei. With further growth, the central part of the lens placode indents, invaginating from this small depression, together with the optic vesicle, to form the lens pit and double-layered optic cup, respectively. With increased cell crowding, the lens pit continues to deepen and remains open at the surface only by a small pore. This pore is closed off rapidly, creating a spherical lens vesicle of columnar epithelial cells attached to the overlying ectoderm via the lens stalk. The lens stalk gradually degenerates, completely separating the lens tissue from the

Stages of Lens Morphogenesis	Embryonic Ages			
	Frog* (hours)	Chick (hours)	Mouse (days)	Human (days)
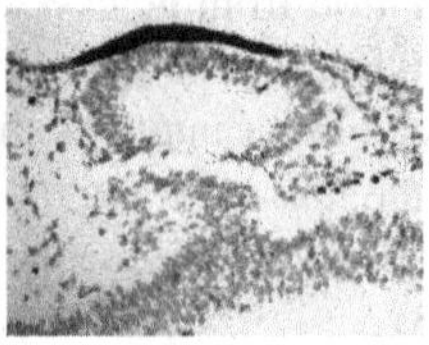 Thickened ectodermal placode overlying optic vesicle	35	48-52	9.5	28
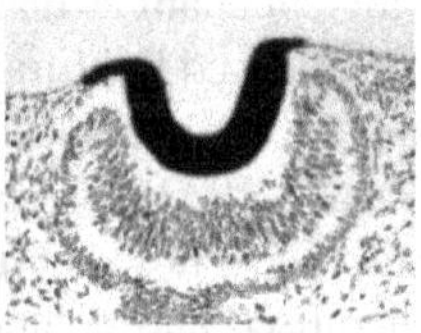 Invagination of the lens placode and optic vesicle	40	50-53	10	29-30
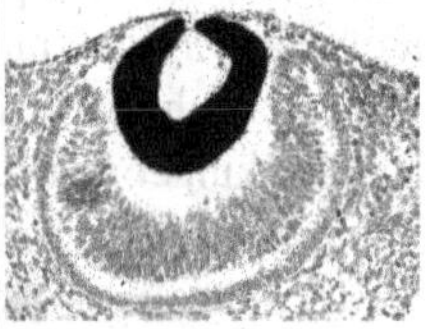 Deepening of lens pit and optic cup	45	50-56	10.5-11	31-32
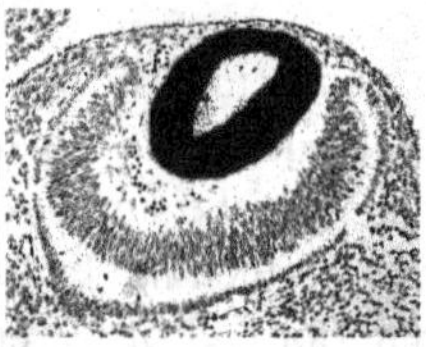 Closure of lens pit Lens vesicle separates from surface ectoderm	50	52-64	11.5	33-34
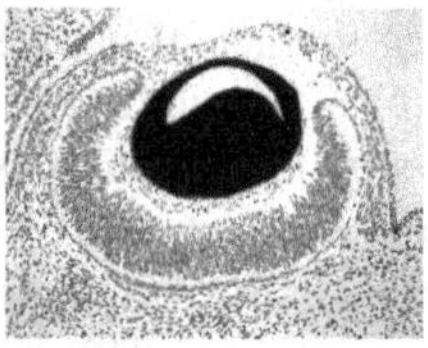 Developed lens vesicle and optic cup	55	65-69	12-12.5	35

Figure 1.13. Comparative timetable of lens morphogenesis in frog, chick, mouse, and human. *In frog, developmental stages following fertilisation are dependent on temperature and species. The times stated are only approximate and are based on the developing *Xenopus laevis* at 22–24°C.

overlying ectoderm (presumptive corneal epithelium). As the lens vesicle represents an invagination of the surface ectoderm, the basal portions of the cells face outward while the apical portions face the lumen. As a result, components of the basal lamina, secreted by the lens cells throughout life, are deposited externally, encasing the lens in a membranous envelope which through appositional growth acquires more layers and eventually forms the lens capsule. During lens morphogenesis, the anterior capsule thickens more slowly than the posterior capsule, but subsequently anterior capsule synthesis overtakes the remainder, and the anterior capsule becomes several-fold thicker in the mature lens.

Concomitant with the initiation of lens capsule formation, we see the appearance of the first hyaloid vessels of the tunica vasculosa lentis, a vascular net that encompasses the lens and is connected proximally to the hyaloid artery of the retina (Fig. 1.14). The main trunk

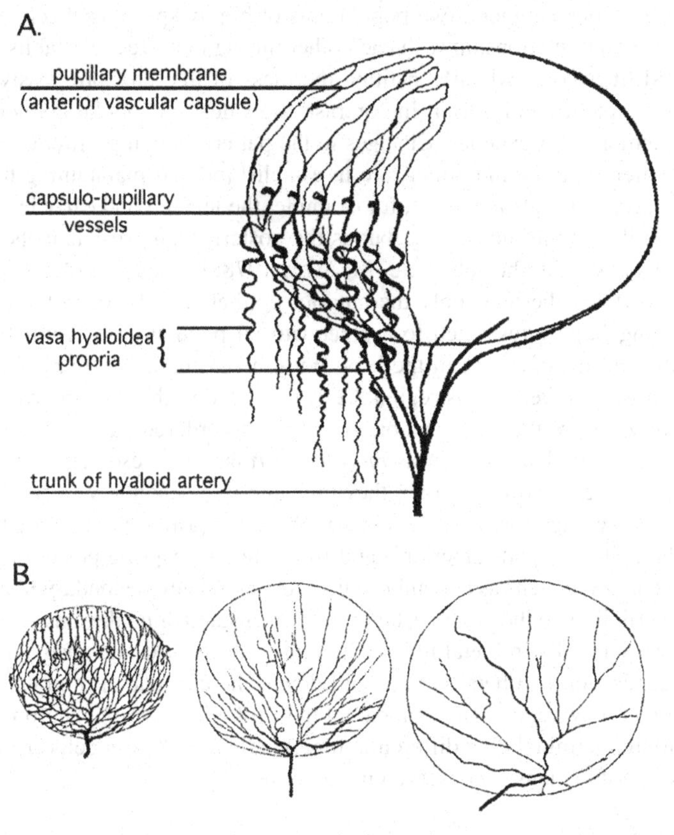

Figure 1.14. (A) Blood vessels on the lens prior to atrophy of the vasa hyaloidea propria. (B) Retrogression of the blood vessels on the lens. (Adapted from Mann, 1928.)

of the hyaloid artery leads into the vasa hyaloidea propria, which branches on to the back of the lens, anastomosing to form the posterior vascular capsule. These vessels lead on to the straight capsulopupillary vessels of the lateral part of the vascular capsule, which once again anastomose with loops of the anterior vascular capsule (pupillary membrane) as well as the choroidal vessels (see Fig. 1.14). The tunica vasculosa lentis is thought to provide nutritional exchange for the lens until the anterior chamber and aqueous humour are operational. With embryonic development, the tunica vasculosa lentis eventually retrogress and disappears. It is usually absent at birth in humans, but in lower vertebrates, such as the rat and the mouse, it persists for a short period postnatally.

The positioning of the cells of the lens vesicle ultimately determines their fate. By the time that the lens vesicle has separated from the overlying surface ectoderm, the posterior cells of the lens vesicle are slightly elongated. The more posteriorly situated vesicle cells facing the optic cup elongate first, advancing ahead of the equatorially placed cells and encroaching on the circular cavity of the lumen, rendering it crescentic. These elongating cells, which constitute the primary fibers, eventually make contact with the apical surface of the anterior lens epithelium, obliterating the lumen of the lens vesicle. The monolayer of anterior cells that face the developing cornea increase in number with the growth of the lens and differentiate into the lens epithelium, assuming a low columnar or cuboidal shape.

At this stage of lens development, two populations of highly specialised cells that will maintain a defined spatial relationship with each other throughout life are established. The posterior elongated fiber cells, which constitute the mass of the lens, progressively lose their intracellular organelles and nuclei. In contrast, the anterior cuboidal epithelial cells retain a full complement of organelles. Changes in the pattern of cell proliferation during early lens development play an important role in establishing and maintaining this well-defined lens architecture. In rats and mice, for example, the lens pit and early lens vesicle have an evenly distributed mitotic activity, but as the posterior lens vesicle cells begin to elongate, these cells exit from the cell cycle (McAvoy, 1978; Lovicu and McAvoy, 1999). With continued growth of the lens, only the anterior epithelial cells retain the ability to proliferate, becoming largely restricted to a preequatorial population of epithelial cells, which make up the germinative zone of the lens. With increasing age, a marked decrease in proliferation is also observed in this region. Cells produced in the germinative zone are displaced posteriorly toward the lens equator, giving rise to ordered parallel rows of cells lying in the meridia of the lens, known as *meridional rows*. As these cells move below the lens equator, into the transitional zone, they elongate into secondary fibers. With the continual addition and elongation of the secondary fibers, the primary fiber mass becomes separated from the epithelial cells anteriorly and from the lens capsule posteriorly, and it now occupies the core of the lens as the embryonic nucleus. As new secondary fibers form, they are added superficially to the previous layer of fibers, establishing the lens cortex. This process proceeds very rapidly in foetal life and at a much decreased rate postnatally. With each successive generation of fibers that are laid down, the cell nuclei migrate forward, establishing a defined 'bow region' (most apparent in histological sections) before they are eventually lost through terminal fiber differentiation. As all the lens fiber cells are retained, the entire life history of the lens is conserved in the tissue.

2

Lens Induction and Determination

Marilyn Fisher and Robert M. Grainger

2.1. Introduction

Just as the ocular lens gathers and focuses light, so too has it captured and focused the attention of developmental biologists. Since Spemann's first experiments introduced the concept a century ago, the vertebrate lens has served as a model for the phenomenon of embryonic induction. Figure 2.1 provides a diagrammatic representation of the major steps in vertebrate lens determination to illustrate the physical relationships among developing tissues during stages pertinent to this review. The figure is based on the chick embryo, as its relatively flat topology during the earliest stages of development is particularly convenient for illustrative purposes. The lens differentiates from a region of head ectoderm that early in development lies adjacent to the region of the neural plate from which the retina will form (Fig. 2.1A). As development proceeds, the region of presumptive lens ectoderm (PLE) is not in contact with the retinal rudiment, as the neural plate folds up into a closed tube (Fig. 2.1C), but it is brought into close proximity to the retinal anlage by virtue of the outgrowth of the optic vesicle (OV) from the forebrain (Fig. 2.1D). The first overt signs of lens formation appear only after the OV establishes close contact with the PLE. After contact is made, the PLE thickens to form a placode (Fig. 2.1E) that subsequently invaginates simultaneously as the inward collapse of the OV forms the double-layered optic cup (Fig. 2.1F). Shortly thereafter, the lens vesicle detaches from the surface ectoderm and begins to acquire characteristic lens morphology, with elongated fiber cells differentiating from the retinal side of the vesicle and a thin layer of epithelial cells along the opposite side of the vesicle (Fig. 2.1G).

Observations of the normal sequence of eye development and of malformed embryos, including cyclopic monsters whose malpositioned eyes nevertheless contained perfectly situated lenses, led to speculation that the developing retinal rudiment played an important role in lens determination. Spemann (1901) was the first to experimentally test the role of the eye rudiment by destroying the presumptive retina of neural plate stage frog embryos without damaging the nonneural head ectoderm, and he found that these embryos subsequently lacked both a retina and a lens. While Spemann's initial observations supported a simple 1-step model in which interaction with the forming eye rudiment induces the formation of a lens in the closely apposed ectoderm, the ensuing century of experimentation has revealed that lens determination involves more than a simple 1-step inductive interaction with the optic vesicle and begins well before the retinal rudiment contacts the presumptive lens tissue. In this chapter, we provide a historical overview of lens induction studies, especially early chick and mouse studies, which are generally supportive of and, when taken together with the large body of amphibian studies, provide the basis for a revised 5-step model of

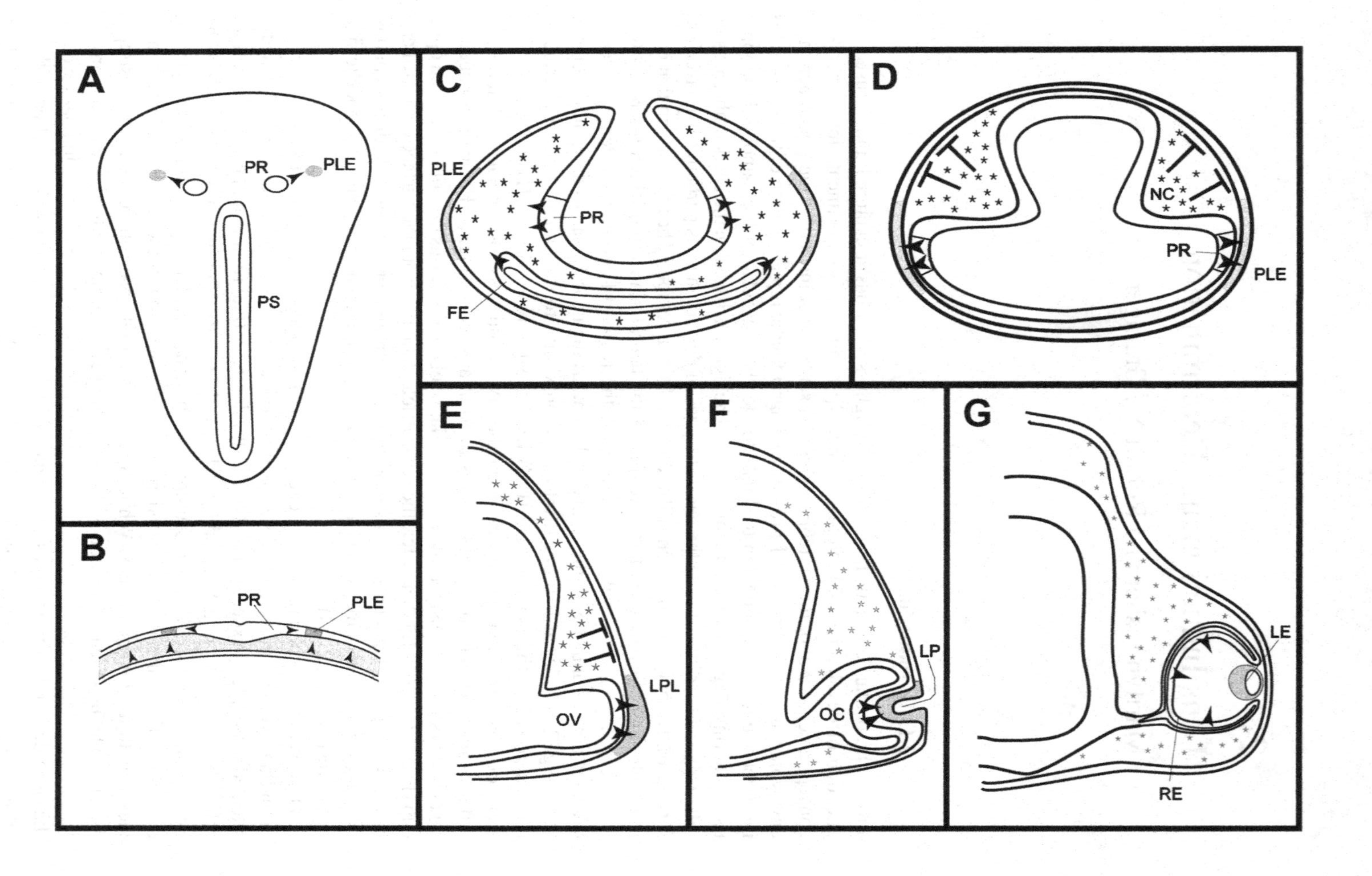
A
PR
PLE
PS
B
PR
PLE
C
PLE
PR
FE
D
NC
PR
PLE
E
OV
LPL
F
LP
OC
G
LE
RE

lens determination. We discuss the elements of the current model, the general applicability of the model, and directions for future studies.

2.2. Historical Overview

2.2.1. Methodological Considerations

In order to test Spemann's model of lens induction and demonstrate the role of the optic vesicle in lens determination, investigators have sought to test both the necessity and the sufficiency of the optic vesicle as a lens inductor. Their experiments have generally taken two forms:

1. separation of the PLE from the influence of the OV either by explantation *in vitro*, transplantation to a remote site, or ablation of the optic rudiment to determine whether lens differentiation could occur in its absence;
2. transplantation of the OV to remote sites, transplantation of foreign ectoderm to the PLE site, or recombination of tissues with the optic rudiment *in vitro* to test its ability to induce lens formation in tissues not normally fated to undergo lens formation.

These latter experiments address the sufficiency as well as the specificity of the OV's inductive properties.

As other investigators began to follow up and expand on Spemann's intriguing observations, conflicting results soon appeared, and conclusive affirmation of Spemann's model of lens induction was not immediately forthcoming, although the majority of observations were generally supportive. Reviewing the literature from our current perspective, a number of technical limitations and methodologic issues, common especially among the earlier studies, can be identified as contributing to the often conflicting results.

The bulk of lens induction studies have been done with amphibians, but a number of investigators have used chick and, more recently, mouse embryos to test the generality of the emerging principles. Each organism has its own strengths and limitations as an experimental model. Saha et al. (1989), in their critical review of amphibian lens induction studies, raised the two most serious problems, namely, end point definition and host-donor marking.

Of paramount importance to studies of lens determination are the criteria used to ascertain a positive lens response. The early investigators, whether studying amphibian or chick embryos, relied exclusively on the morphology of the responding tissue. If lens differentiation had proceeded to the point where characteristic elongated lens fibers were present,

Figure 2.1. (facing page) Lens development. The steps of lens development are illustrated diagrammatically using the chick as a representative vertebrate embryo. At the primitive streak stage (A), the PR and PLE are located anterior to the PS. At this stage, early positive inductive signals (arrowheads) travel within the plane of the ectoderm from the neural plate to the outerlying nonneural ectoderm. The transverse section in (B) shows early inductive signals from neural tissue as well as underlying mesendoderm that continue through neurulation (C). After neural tube closure, the optic vesicles grow out from the diencephalon and establish close contact with the PLE (D). At about this same time, NC cells migrating throughout the head produce signals that inhibit lens-forming bias (⊥) in head ectoderm outside of the PLE (D and E). Inductive signals from the OV elicit LPL thickening and the subsequent steps in lens morphogenesis and differentiation (E, F, and G). FE, foregut endoderm; LE, lens; LP, lens pit; LPL, lens placode; NC, neural crest; OV, optic vesicle; OC, optic cup; PLE, presumptive lens ectoderm; PR, presumptive retina; PS, primitive streak; RE, retina.

interpretation was straightforward. Often, however, investigators were confronted with less clear-cut ectodermal thickenings or small vesicles, end points that, while superficially resembling early placode or vesicle stages of lens differentiation, could equally well represent artifacts resulting from the wounding of the epithelium during the experiment. In the case of chick experiments, some authors acknowledged this problem, while others interpreted even ectodermal thickenings as a positive lens response.

McKeehan (1951) was the first to try to circumvent this problem by applying rigorous cytological criteria to define the end points of his experiments. He described a number of cytological changes that normally occur in the PLE of chick embryos from the time the OV first contacts it until there is a well-defined lens placode. These changes include decreased vacuolization of the cytoplasm, reorientation, reshaping, and alignment of the nuclei, features that distinguished the PLE from surrounding ectoderm. A significant advance in this area came when Mizuno and Katoh (1972a, 1972b) introduced the use of antisera to lens crystallins to assay for lens response in their studies of chick lens induction. Since they found crystallins were synthesized already at the placode stage, the presence of these lens-specific proteins served as a reliable indicator that the initial stages of lens differentiation had begun. This same criterion was adopted in amphibian studies beginning with Henry and Grainger (1990). Recently, Sullivan et al. (1998) cautioned against relying exclusively on the immunohistochemical detection of δ-crystallin as the sole indicator of chick lens induction, as δ-crystallin is expressed in some nonlens tissues. They recommended instead an assay combining crystallin expression and cell elongation.

The second technical limitation common to both amphibian and chick lens induction studies was a lack of proper host and donor labeling to enable unambiguous recognition of the source of induced lenses in transplantation experiments. In all vertebrates studied to date, there is extremely tight adherence between the OV and overlying PLE during the stage when most investigators believed the critical inductive interaction was occurring. This tight adherence made clean separation of the OV from the PLE, either for ablation or transplantation purposes, quite difficult, especially when investigators were using purely mechanical separation techniques. Thus, in many cases there was the possibility that a lens found associated with a transplanted OV might actually have arisen from PLE cells carried with the OV rather than from induction of the overlying nonlens ectoderm. This very real possibility was documented by Grainger et al. (1988) in experiments with the frog *Xenopus laevis*. The problem was not addressed practically in chick studies until Karkinen-Jääskeläinen (1978a) employed the distinctive quail nucleolar organization popularized by Le Douarin (1969) to label grafted tissues in her experiments. In amphibians, apart from a small number of studies that relied on pigmentation differences in interspecies grafts or Nile blue sulfate labeling, the host and donor–marking problem remained until Henry and Grainger (1987) applied the horseradish peroxidase lineage–labeling technique introduced by Jacobson and Hirose (1978) and Weisblat et al. (1978).

A general concern that particularly applies to chick experiments is the great diversity among the experimental approaches of investigators who were trying to answer basically the same questions. Chick embryos present some unique experimental challenges that made certain types of manipulation more difficult than in amphibian embryos. For example, Danchakoff (1926) found that it was very difficult to ablate an OV in a chick embryo employing the same method that Lewis (1904) had used in frog embryos, that is, by lifting a flap of ectoderm, cutting out the underlying OV, and replacing the ectoderm without damaging the PLE. She therefore used a ventral approach to ablate OVs, but she also recognized that in the process she was damaging other tissues that might be (and indeed have subsequently been shown to be) important for lens induction. Other investigators have neither acknowledged this problem nor have given clear descriptions of how they performed OV ablations.

Similarly, in making transplants of OVs to regions away from the normal lens-forming ectoderm, investigators used everything from half the forebrain of neural plate–stage embryos to fragments of the well-developed optic cup. To circumvent the problem of the strong adherence between OV and PLE, Alexander (1937) cut off the distal part of the OV with the PLE attached and used the remainder, primarily the presumptive pigmented retina and the optic stalk with attached forebrain fragment. The underlying assumption that these diverse starting materials would have the same inductive potential as an intact OV is likely not valid.

Finally, there is much variation in the culture media used for *in vitro* studies of chick lens induction. Some investigators used solid culture media, while others used a variety of different formulations supplemented with varying amounts of horse, fetal calf, or chick serum in addition to chick embryo extract. All of these variations in methodology can make it difficult to compare or relate the results of different investigators.

2.2.2. Necessity of the Optic Vesicle for Lens Induction

Although Spemann's (1901) first experiments suggested that the OV was required for lens formation in frog embryos, there soon followed reports of lens formation in the absence of the OV in several amphibian species (King, 1905; Spemann, 1907, 1912) as well as in fishes (Mencl, 1903; Lewis, 1909; see Jacobson and Sater, 1988 for an extensive list). These "free" lenses presumably arose without any contact with the late neural stage OV. While not generally subject to rigorous authentication either as legitimate lenses or as having arisen in the complete absence of eye tissue, they nevertheless cannot all be discounted (discussed by Saha et al., 1989). Spemann concluded from the existence of these free lenses, which in no case resembled normal, fully differentiated lenses, that each species had some degree of innate tendency toward lens formation but that only under the influence of the OV could this lens-forming potential be fully realized (Spemann, 1938).

Experiments involving OV ablations in chick embryos similarly gave mixed results. Danchakoff (1926) and van Deth (1940) reported complete absence of lens formation following OV ablations in embryos from 7–10 and 11–20 somite stages, respectively. In contrast, Waddington and Cohen (1936) reported that slight ectodermal thickenings were present 27 hours after ablation of the OV in 4–5 somite stage embryos. These thickenings may be similar to some of the weaker cases of free lenses in amphibians, and Waddington and Cohen interpreted them as reflecting a "tendency to self differentiation" of lenses. Their interpretation might have been strengthened by waiting longer, as they apparently terminated the experiment at a stage when unoperated controls would be expected to have only a well-defined placode. Somewhat later experiments with chick embryos addressed both the necessity of the OV and the mechanism of its influence on the PLE. McKeehan (1951) blocked contact between the OV and the PLE by interposing a strip of cellophane. He found that lens response was blocked in that part of the PLE that was separated from the OV by cellophane but not in nearby areas where the two tissues were still in contact. In contrast, interposing agar strips did not block induction (McKeehan, 1958; van der Starre, 1977, 1978). In fact, van der Starre (1978) reported that agar strips, after being held in contact with optic cup tissue for 4 hours, could act in place of the OV to induce lens response in head ectoderm of stage-18 embryos (staging according to Hamburger and Hamilton [H&H] 1951).

Following up on Muthukkaruppan's (1965) observations on lens development in mice, Karkinen-Jääskeläinen (1978a) compared the lens-forming ability of PLE taken from chick and mouse embryos at a stage just prior to contact by the OV. Muthukkaruppan had explanted PLE around the time of initial contact with the OV and found that PLE alone did not form lenses but did so readily when combined with the optic cup, either in direct contact or separated by a millipore filter. In similar experiments, Karkinen-Jääskeläinen found a very low percentage of mouse PLE (23%) and no chick PLE formed lenses when cultured alone, but both formed lenses readily (80–90%) when cocultured with the OV, either in direct contact or separated by a millipore filter. Interestingly, Karkinen-Jääskeläinen reported that if fetal calf serum was replaced in the culture medium with chick serum, lenses formed in over 50% of isolated, precontact chick PLE samples, and precontact mouse PLE developed recognizable lenses more than 50% of the time if cultured in human amniotic fluid. The combination of morphology and positive crystallin immunohistochemistry suggests that the lens response of isolated PLE in culture is legitimate and argues that, in both chick and mouse, lens determination is well advanced by the time the OV contacts the PLE. Barabanov and Fedtsova (1982) explanted pieces of ectoderm from all over the head to assess lens-forming potential and found that between H&H stages 5 and 11 all of the head ectoderm had an equal ability to manifest lens differentiation despite not being in intimate contact with the OV. In contrast, they found that H&H stage 10 trunk ectoderm had no ability to form lenses when isolated in culture. These results have recently been confirmed by C. Sullivan and R. Grainger (unpublished data), who found lens-forming ability in ectoderm explants from H&H stage 8 and 10 embryos even in serum-free medium. Furthermore, Barabanov and Fedtsova (1982) found that non-PLE head ectoderm had lost lens-forming ability by H&H stage 12. The conclusion that can be drawn from the variety of evidence is that the OV is not necessary for the earliest stages of lens development (i.e., placode and vesicle formation) but is necessary for the later stages of development (i.e., the differentiation of the lens vesicle into a mature lens).

2.2.3. Sufficiency of the Optic Vesicle as a Lens Inductor

The other side of the coin with regard to the role of the OV as the primary lens inductor is the question of whether the OV can by itself direct lens development. When Lewis (1904) transplanted OVs from late neurula stage embryos underneath the ectoderm posterior to the PLE and observed lens formation, he concluded that the OV is not only necessary but also sufficient for directing lens differentiation in tissues that would normally never form lenses. There have been many variations on Lewis's experiment, including experiments in which foreign tissues as diverse as gastrula stage ectoderm and late neurula stage trunk ectoderm were transplanted over various staged optic rudiments as well as experiments in which the OV was transplanted to a variety of foreign locations. The amphibian studies, which were reviewed by Saha et al. (1989), show that there are some circumstances in which the OV apparently induces lens development from non-lens ectoderm, lending support to Lewis's conclusion. However, when all the results are taken together, it becomes clear that the OV is capable of inducing a lens response in some embryonic ectoderm but not all and that the closer the ectoderm source is to the PLE, the better the lens response.

The results of similar experiments with chick embryos likewise support the conclusion that there are limited circumstances in which the OV is sufficient to induce lenses in nonlens ectoderm. In the chick experiments, there was even more diversity among the sources of ectoderm tested; these sources ranged from primitive streak blastoderm to chorioallantoic ectoderm of 8- to 9-day-old embryos. Van Deth (1940) reported that the OV from the 20 somite stage embyro had the ability to induce a lens in ectoderm from all regions of a similarly staged embryo when recombined in culture. However, in light of the reliance on a morphologic assay, the absence of host-donor marking, and his own assertion of the difficulty of separating the PLE from the OV during this stage (when they are tightly adherent), one cannot be confident that the observed lentoids and vesicles did not come from PLE cells that contaminated the OV. Alexander (1937) transplanted whole or partial optic cups, free of lens tissue, from 25–40 somite stage embryos to the body wall of hosts ranging from 1 to 13 somites. He concluded that the younger the host and the closer to the normal eye region the transplant is located, the greater the probability of lens formation, although even at best the frequency of lens response remained quite low (15 of 118 cases overall). There are two points to consider here: (1) The young hosts used were at an age when the head ectoderm is already likely to be strongly predisposed toward lens formation (Grainger et al., 1997; Barabanov and Fedtsova, 1982), so it is not surprising that lenses formed close to the eye region (12 of 15 positive cases); (2) most of Alexander's transplants were partial optic cups rather than whole cups, and most showed some neural and pigmented retinal differentiation by the end of the experiment. Since he used no host-donor marking in these experiments, one cannot rule out the possibility that some or all of the lenses that formed, especially in the trunk region (3 of 15 positive cases), were the result of contamination. Thus, Alexander's extensive experiments do not provide strong evidence in support of the sufficiency of the optic rudiment as a lens inducer.

McKeehan (1951) also made transplants of the optic rudiment from embryos, primarily at the 6 somite stage, placing them in such locations as the coelom or beneath the ectoderm of the head, neck, or flank. The ages of the hosts are not specified in all cases, but some were as old as 14 somite stage. Using his rigorous cytological criteria for assessing positive lens response, he found only 6 of 29 cases that were positive, and he discounted 4 of these 6 as having arisen from contamination of the graft with donor PLE, since those grafts were taken from donors where the OV and PLE had already formed a strong adherence, making complete separation impossible. In the remaining 2 positive cases, the grafts consisted not of the OV alone but of one-half of the forebrain from embryos at the 4 somite stage. Thus, McKeehan's results likewise argue against the sufficiency of the OV as a lens inducer.

The strongest case for the sufficiency of the OV as an inducer comes from Karkinen-Jääskeläinen's (1978a, 1978b) reports of the ability of the OV to induce lens formation in trunk ectoderm from embryos at 5–14 somite stages (the illustrated case is an embryo at 8 somites). Despite the fact that she used quail and chick chimeric grafts for host-donor marking and crystallin immunohistochemistry as a marker for lens differentiation, she does not use both together in any of the illustrated cases. There is one case where the induced structure looks obviously lenslike and shows positive crystallin immunoreactivity, but it does not have host-donor marking. In another case, the induced structure obviously shows trunk ectoderm origin but does not look lenslike and was not immunostained to demonstrate crystallin expression. Following up on these observations, Karkinen-Jääskeläinen (1978b) studied the inductive influence of the OV using trunk ectoderm as a responding tissue. In those studies, she reported that lens response could be elicited in up to 60% of cases, even when the OV and the trunk ectoderm were separated by a millipore filter. However, as in her previous study, here too the illustrated cases do not simultaneously show host-donor marking and crystallin expression, leaving the interpretation open to question. Furthermore, since the antibody used by Karkinen-Jääskeläinen was apparently raised against the whole lens and was not specific to crystallins, it is possible that the positive immunoreactivity she observed did represent some response of the trunk ectoderm to factors provided by the OV but not necessarily lens crystallin production. Unfortunately, this apparently striking demonstration of the inductive power of the OV has proven difficult to confirm (C. Sullivan and R. Grainger, unpublished data). Thus, in light of the results described above, which raise questions about the necessity and sufficiency of the OV as an inductor, it is doubtful that during the course of normal development the OV of any vertebrate species is the primary inductor of lens development.

2.2.4. Role of Tissues Other Than the Optic Vesicle in Lens Induction

The early investigations focused on the properties of the OV as an inductor, but the results forced recognition of the fact that in any induction system the properties of the responding component must also be considered. It is clear from the studies already discussed that the ability of embryonic ectoderm to respond to induction by the OV varies in a temporal and spatial manner. In general, during the neurula stages, the head ectoderm is most responsive, and within the head ectoderm, that closest to the actual PLE retains its responsiveness the longest. That the PLE prior to contact with the OV already has some potential for lens differentiation was apparent from the development of free lenses in the absence of contact with the OV and inspired some to look for other potential lens-inducing factors. A number of early investigators addressed this problem (see discussion in Saha et al., 1989), but the most extensive contribution consisted of the studies of Jacobson (1958, 1966), who tested the lens-inducing capabilities of foregut endoderm and presumptive heart mesoderm, both of which he noted came to lie close beneath the PLE as a result of gastrulation movements (see Figs. 2.1B and C). He showed that these tissues had additive lens-inducing effects when recombined *in vitro* with PLE from early neurula stage amphibian embryos. However, since he tested PLE from neural plate stage embryos, tissue we now recognize as already biased (Grainger et al., 1997), he did not show that endoderm and mesoderm could act as primary inducers to elicit a lens response in young tissue that had not already been exposed to early lens-inducing signals. Mizuno (1972) tested the lens-inducing ability of early endomesoderm in chick embryos. After unexpectedly finding lens differentiation during his study of feather germ induction, he demonstrated that it was possible to induce readily recognizable lenses having elongated lens fibers in cephalic epiblast from primitive streak stage embryos by recombining it with cephalic hypoblast plus dermis from the back of 6.5-day-old embryos. While a variety of embryonic mesenchymes worked in this assay, dermis was most effective, and neither dermis nor hypoblast alone with epiblast was sufficient. In these cultures, there was no apparent neural differentiation, certainly not a recognizable OV. Using this same system, Mizuno (1973) further showed that epiblast taken from the trunk region or extraembryonically from the area opaca of H&H stage 4 embryos could form crystallin-positive lens vesicles when recombined with cephalic hypoblast plus dermis. Again, although a number of mesenchymes could substitute for back dermis, trunk hypoblast could not substitute for cephalic hypoblast. These results support the conclusion that, in chicks as well as amphibians, embryonic foregut endoderm is capable of contributing to the early stages of lens development.

The possibility that other anterior structures such as placodal ectoderm and early anterior neural plate might also have some lens-inducing potential has been raised. Several investigators noted that free lenses formed after optic rudiment ablation generally were more advanced when they were in contact with the nasal pit (e.g., LeCron, 1907; Spemann, 1912; Jacobson, 1958; Reyer, 1962). The idea that early neural tissue, possibly the presumptive eye rudiment itself, might influence the adjacent ectoderm containing the PLE, arose from Nieuwkoop's (1952, 1963) hypothesis that placodal ectoderm (including lens) is specified as an extension of neural induction by signals passing laterally through the plane of the ectoderm (see Figs. 2.1A and B). There was only questionable experimental evidence to support the idea that planar signals passing from neural to nonneural ectoderm might play a role in lens determination until Henry and Grainger (1990) tested it directly. They found that PLE from Nieuwkoop and Faber (N&F) stage 14 (Nieuwkoop and Faber, 1956) *Xenopus laevis* embryos cultured by itself could not form even rudimentary lens structures, but when cultured in combination with anterior neural plate of the same stage, crystallin-positive ectodermal thickenings formed. This result was enhanced when the PLE and anterior neural tissue were isolated and maintained as a continuous sheet, even when isolated as early as N&F stage 11.5–12. These results confirm that the neural plate is an early lens inducer. The importance of the anterior neural plate was further suggested by the demonstration that the ability to form free lenses was greatly reduced if anterior neural tissue was removed at N&F stage 14. Henry and Grainger (1990) found that free lenses formed in only 15% of cases following early ablation of the optic rudiment, compared with 94% of cases when the optic rudiment was ablated later (N&F stage 17–18).

While the PLE appears to increase steadily its lens-forming ability from the early neural plate stage onward, non-PLE head ectoderm has been shown by numerous investigators to lose its lens-forming responsiveness shortly after the stage when contact has been established between the PLE and the OV. While one could imagine a number of mechanisms by which this could happen, there is some evidence from experiments carried out by von Woellwarth (1961) that the neural crest might play an important role in restricting the region of lens-forming responsiveness. He found that if he extirpated the anterior neural plate in newt embryos, no free lenses formed, but if he extirpated the posterior neural plate as well, eliminating the source of the head neural crest, free lenses did form. The neural crest is normally blocked from interacting with the PLE by virtue of its tight adherence to the underlying OV but would have access to the rest of the head ectoderm.

2.2.5. A Revised Model of Lens Determination

The persistence of interest in and experimentation on the problem of lens induction resulted in reevaluation and revision of the model of lens determination, which came to be viewed, not as a simple process elicited by interaction between head ectoderm and the optic vesicle, as envisioned by Spemann and Lewis, but a more protracted multistep process, as first suggested by Jacobson (1966). In Jacobson's model, lens determination involves a series of inductive interactions beginning late in gastrulation as the PLE sequentially comes under the influence of endoderm, mesoderm, and finally the optic vesicle. Jacobson suggested that the influence of the separate inductors were qualitatively similar but that their relative strengths varied among species, explaining why free lenses form much more readily in some species than in others. Combining some of the elements of Jacobson's model and some of Nieuwkoop's ideas on the importance of signaling within the ectoderm during the induction of placodal structures (including the lens), Saha et al. (1989) proposed a new model of lens determination as a process with four distinguishable stages (see also Grainger, 1992, 1996). As the result of an additional decade of rigorous testing, the model has been further revised to recognize five important stages in lens determination: competence, bias, specification, inhibition, and differentiation (Hirsch and Grainger, 2000). In the next sections, we review the evidence for the current understanding of the molecular bases of these stages and discuss likely sources of and candidates for lens-inducing signals.

2.3. Current Model of Lens Determination

The current model of lens determination is based on the vast body of amphibian lens induction studies done over the past century, including more recent studies from the last decade specifically designed to test and refine the elements of the model. We have reviewed lens induction studies of chick and mouse embryos to assess the general applicability of the principles of lens determination across vertebrate classes. There is a much smaller body of literature dealing with chick and mouse embryos than with amphibian embryos, owing to their being somewhat more challenging as experimental models. Because they need complex growth media, specialized atmospheric conditions, and elevated temperature, mouse embryos are particularly difficult to maintain *in vitro*, especially for periods longer than 1 or 2 days at a time, as would be required to observe the entire period of lens development from the earliest inductive interactions through the differentiation of lens fibers. Adding to the challenge of working with mammalian embryos are the dramatic size and shape changes these embryos undergo during gastrulation and early neurulation. Amphibian embryos are relatively large from the beginning and change very little in size between fertilization and lens placode formation. Mouse embryos, on the other hand, grow considerably in size during that period and undergo a dramatic change in configuration: during the gastrula stages, they are U-shaped, with the endoderm on the outer surface and the ectoderm on the inner surface, then they rotate along the body axis during the neurula stages to take on a more conventional configuration, with the ectoderm fully enclosing the embryo's internal structures. For these reasons, we do not have the same depth of background information, especially for the earliest period of development, when, as we know, the first steps of lens determination are occurring in amphibian embryos. However, fate-mapping studies of various tissues from amphibian, chick, and mouse embryos show striking similarities in the relative positions of early rudiments in these organisms (see comparative fate maps of gastrula stage embryos in Lawson et al., 1991 and Quinlan et al., 1995 and of the prosencephalic neural plate in Inoue et al., 2000). So, despite the rather different sizes and configurations of these organisms during the blastula, gastrula, and early neural stages of development, it seems reasonable to postulate that, at least from the mid to late gastrula stages onward, the origins of presumptive retina and lens rudiments in the various vertebrate classes have a similar relationship to one another, as exemplified by the well-documented amphibian model. From the time the OV begins to grow out toward the overlying PLE through the maturation of the lens, the morphologic changes appear virtually identical across the vertebrate classes considered here. Furthermore, the data that are available from lens induction studies in chick and mouse embryos are generally consistent with the elements of the amphibian-based model, as noted earlier. To emphasize the common features that we believe are shared by these organisms, we have chosen to illustrate diagrammatically the steps in lens determination using the chick embryo (Figure 2.1).

Briefly, the current view is that lens induction begins during mid to late gastrulation, when ectoderm acquires the ability (competence) to respond to the earliest lens-inducing signals that arise from the anterior neural plate (Figs. 2.1A and B). These early signals elicit within the entire head ectoderm a predisposition (bias) to form lens. Under the continued influence of the anterior neural plate as well as the underlying endomesoderm (Figs. 2.1B and C), the head ectoderm becomes increasingly biased toward lens formation until reaching a threshold level that allows the PLE to begin differentiation even if isolated from further lens inductors. At this point the ectoderm is said to be "specified." While a rather large area of head ectoderm initially acquires lens-forming bias, around the time of specification of the PLE there is a loss of lens-forming bias in head ectoderm outside of the PLE, possibly due to inhibition by the neural crest (Figs. 2.1D and E). Finally, under the influence of the OV, the PLE becomes fully determined and differentiates into a mature lens. In the following sections, we primarily discuss recent studies of lens induction that provide data relevant to individual components of the model while making reference to some earlier studies that were instrumental for developing the key concepts of the model.

2.3.1. Competence

Waddington (1932) was the first to use the term "competence" to describe the ability of a tissue to respond to a particular inducing stimulus, and he was the first to try to define experimentally the conditions necessary for establishing lens competence in amphibian embryonic ectoderm. Specifically, he hypothesized that competence for lens induction might depend on the prior induction of neural tissue or on some process controlled by mesoderm. He tested this idea by isolating early gastrula stage ectoderm (prior to the possibility of its being influenced by mesoderm or neural tissue) and culturing the ectoderm until control embryos reached the early neural plate stage. He then recombined the ectoderm isolates with presumptive retinal tissue from neural plate stage embryos and observed that in the cases where lenses developed there was no correlation between lens formation and neural or mesodermal induction. He concluded that apparently lens competence arose independently of any other specific inductive interactions, requiring only that the ectoderm take on the form of a thin-walled vesicle (not unlike the animal cap ectoderm of a mid to late gastrula).

More recently, Henry and Grainger (1987) and Servetnick and Grainger (1991) reexamined the issue of lens-forming competence. For this purpose, they defined competence operationally as the ability of a tissue to form a lens when exposed to the full range of normal inductive influences by being placed in the presumptive lens region of a young open neural plate stage embryo (for *Xenopus*, N&F stage 14). Using this assay system, they showed that there is a period from the mid to late gastrula stage (N&F stage 10.5–12) when animal cap ectoderm is competent to respond to lens-inducing signals and that the peak of lens responsiveness is at N&F stage 11–11.5. At these early stages, nearly all ectoderm shows some level of lens competence.

When competent ectoderm from mid gastrula stage embryos was transplanted over the outgrowing OV of an older neurula stage embryo (N&F stage 19), the lens-forming response was greatly diminished both qualitatively and quantitatively. On the other hand, neurula stage ectoderm (N&F stage 18) responded well to inductive influences of the OV. These results indicate that the OV is not a sufficient inductor of lens differentiation in tissues that have not been exposed to the full range of inductive signals occurring during gastrulation and early neurulation, prior to OV outgrowth. Henry and Grainger's experiments also showed, just as Waddington had noted, that the sequential acquisition and loss of neural and then lens-forming competence by the ectoderm was tissue autonomous, occurring in isolated cultured ectodermal pieces on the same time scale as in the intact embryo. Although early investigators used the term "competent," virtually all of the studies that addressed the regional ability of ectoderm to form a lens used the OV or the optic cup as the inductor and

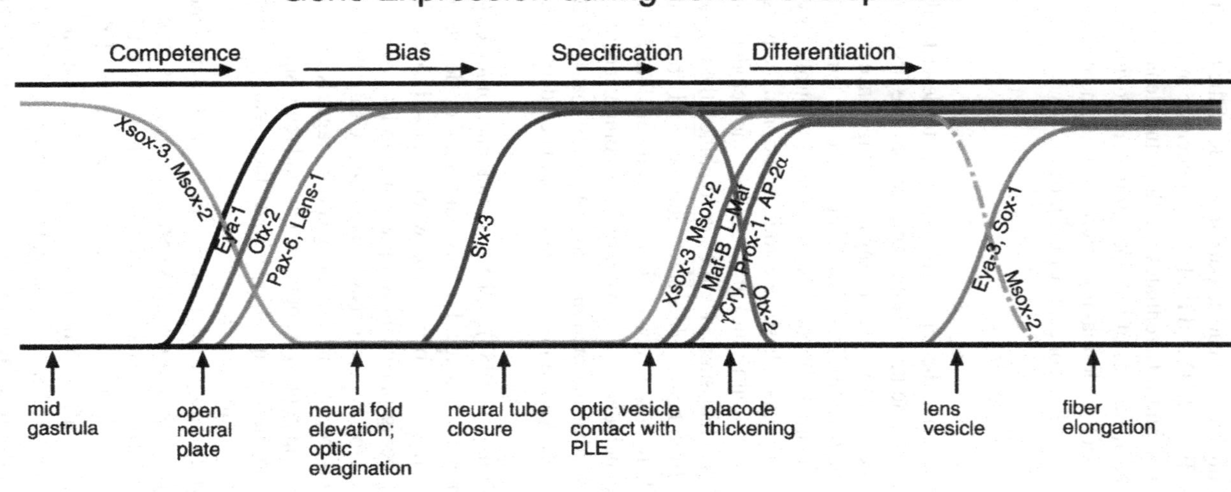

Figure 2.2. (See color plate I.) Gene expression during lens determination. The dynamic expression in the PLE of several genes involved in lens determination is graphically represented in relation to the timing of the major phases of lens determination (shown across the top) and to key developmental stages (shown across the bottom). The various amphibian, chick, and mouse genes included are those for which there is the most complete expression information available for the developmental stages of interest. The genes are discussed individually (with references) throughout the text.

therefore were measuring not competence but a later stage in the lens-forming process, as discussed in the next section. There have been no studies of chick or mouse embryos to define a period of lens competence comparable to that established for *Xenopus*, although Servetnick et al. (1996) demonstrated a window of lens competence during gastrulation in axolotls.

2.3.1.1. Molecular Correlates of Competence

In *Xenopus*, animal cap ectoderm passes sequentially through periods of competence for mesoderm, neural, and lens induction. This sequence occurs even in animal cap tissue isolated and maintained *in vitro* (Grainger and Gurdon, 1989; Servetnick and Grainger, 1991), suggesting that these periods are somehow controlled by an autonomous timer (or timers) whose elements are at present unknown. While the molecular basis of lens competence has not yet been systematically investigated, some candidate regulatory genes are known to be expressed in animal cap ectoderm during the period of lens competence. One such candidate is *Sox3*, an HMG-box gene whose close relative *Sox2* has been shown to increase the responsiveness of ectoderm to fibroblast growth factor (FGF) signaling during the neural competence period (Mizuseki et al., 1998). As illustrated in Figure 2.2, which shows gene expression in PLE during lens determination, *Sox3* is expressed in animal cap ectoderm during the lens competence period (Zygar et al., 1998) and could similarly affect tissue responsiveness to lens-inducing signals. FGF has been suggested as a competence factor because of its role in enabling cells to respond to mesoderm induction (Cornell et al., 1995; Slack et al., 1996; Isaacs, 1997). Curran and Grainger (2000) showed that a shift from cytoplasmic to nuclear localization of activated MAP-kinase, an integral component of the FGF-signaling pathway, coincides with the onset of mesoderm competence. So, although one might hypothesize that competence is determined by the presence or functionality of receptors for inducing signals, it must be recognized that many components of the relevant signaling pathway could serve as target sites for modulating specific competencies.

2.3.2. *Bias*

A consistent observation in early studies of lens induction in amphibian and chick embryos was that the lens response elicited from the PLE by the OV improved as the age of the PLE used approached the stage at which close contact with the OV normally occurred. Grainger et al. (1997) documented this in studies with *X. laevis*. They showed that PLEs taken from progressively older embryos give increasingly better lens response when transplanted over the OV of N&F stage 18 hosts. Similarly, Enwright and Grainger (unpublished manuscript) found that in mice, just as in amphibians, there is a period during early neurulation when head ectoderm exhibits lens-forming bias. Head ectoderm taken from mouse embryos at several stages – from the open neural plate stage through the closed neural tube stage – shows increasing lens responsiveness when recombined *in vitro* with the OV from an embryo at embryonic day 9.5 (E9.5), the stage when tight apposition between these two tissues normally occurs. Experiments with chicks have also demonstrated a strong lens-forming bias in head ectoderm from neurula stage embryos. In fact, according to Barabanov and Fedtsova (1982) and C. Sullivan and R. Grainger (unpublished data), chick embryonic head ectoderm is so strongly biased by the early neural plate stage that it is already capable of lens differentiation in the absence of interaction with the OV (i.e., it is already specified).

Interestingly, the existence of a period during which a rather large region of ectoderm is predisposed toward a particular differentiation is not unique to the process of lens determination. In his studies on the determination of placode-derived tissues (lens, ear, and nasal tissues) in newts, Jacobson (1963a, 1963b, 1963c) described a broad strip of ectoderm adjacent to the neural plate that, during the early neural stages, was able to give rise to each of those structures when provided with region-specific inductive signals. That is, presumptive nasal ectoderm could give rise to the otic vesicle (and vice versa) when the ectoderm strip was transplanted with reversed anteroposterior (A-P) orientation. Indeed, sometimes when this "placodal strip" was transplanted with reversed A-P orientation, structures appropriate to both the origin and the new position differentiated adjacent to one another, such as the ear and the nose. Similarly, lenses could differentiate adjacent to the ear and the nose. Jacobson interpreted these results to indicate that determination of all three structures had begun already at the neural plate stage throughout the placodal zone and that the regional positioning of these structures normally depends on a combination of signals from neural, mesodermal, and endodermal tissues, which differ slightly in character along the A-P axis. These observations fit well with the broad area of lens-forming bias observed in frog, chick, and mouse embryos and are consistent with the observation of several investigators, mentioned earlier, that free lenses forming in contact with the nasal pit were more advanced than those forming in other locations.

2.3.2.1. Molecular Correlates of Bias

To understand the basis for the progressive change in lens-forming ability of head ectoderm, Zygar et al. (1998) looked for changes in the expression of some genes that seemed likely candidates to be involved in regulating lens formation. They found that changes in the expression patterns of three transcription factors, *Otx2, Pax6*, and *Sox3*, during the early stages of *X. laevis* development are consistent with the involvement of these genes during one or more of the early steps of lens determination. The dynamic expression patterns of these and additional genes from amphibian, chick, and mouse embryos are represented in Figure 2.2, illustrating that, to the extent that homologous genes have been examined across the vertebrate classes, there is remarkable similarity in both temporal and spatial expression patterns. Zygar et al. (1998) found that *Sox3* is expressed during the competence period but down-regulated at the onset of the bias period and that *Otx2* and *Pax6* expression is up-regulated during the bias phase. Furthermore, they showed that these genes were activated in lens-competent (N&F stage 11–11.5) but not postcompetent (N&F stage 12) ectoderm when transplanted to the PLE region but not when transplanted to other sites of N&F stage 14 hosts.

Another transcription factor that is first expressed in PLE during the bias period is *Xlens1*, a member of the forkhead family. Overexpression of *Xlens1* in *Xenopus* embryos represses lens differentiation, suggesting that it might play some role in regulating the shift between proliferation and differentiation, as is further suggested by its later restriction to anterior lens epithelium and its loss from differentiating fiber cells (Kenyon et al., 1999).

A survey of the literature and our own unpublished observations suggest that during the bias period in mice (E8.5–9.5), just as in *Xenopus*, a number of potential regulatory genes show dynamic expression patterns. Figure 2.2 illustrates the temporal pattern of expression of several genes within the PLE during this period. In addition, several genes exhibit a dynamic spatial expression pattern within the larger context of the head ectoderm. *Pax6, Otx2*, and *Sox2* (the mouse gene whose expression most closely resembles that of the

Xenopus Sox3), genes that are expressed broadly in head ectoderm at the open neural plate stage (E8.0), become restricted to PLE and/or presumptive nasal epithelium (PNE) by early E9.5 (Walther and Gruss, 1991; Simeone et al., 1993; Ang et al., 1994; Grindley et al., 1995; Collignon et al., 1996). *Six3* is already expressed in the anterior neural plate at E8 but is expressed in the PLE, and PNE by E9.5 (Oliver et al., 1995). While there is no evidence to establish specific roles for these genes as "bias factors," their expression patterns nonetheless serve as an indication that the head ectoderm is undergoing continuous change during this period.

One of the most intensely studied genes with regard to lens determination is *Pax6*, the paired-type homeobox gene homologous to the *Drosophila* eyeless gene (Quiring et al., 1994). Mutations in the *Pax6* locus cause serious eye defects, including human aniridia (Ton et al., 1991) and the "small eye" phenotype in mice and rats (Hill et al., 1991; Fujiwara et al., 1994). *Pax6* is expressed in PLE during the bias period before tight contact between the OV and PLE in amphibian, chick, and mammalian embryos (Li et al., 1994; Grindley et al., 1995; Zygar et al., 1998). Homozygous small eye mice and rats fail to form either lens or nasal placodes (Hogan et al., 1986; Fujiwara et al., 1994) and show defects in the expression of *Sox2, Six3, and Otx2* in PLE and/or PNE. Furthermore, following up on the study of small eye rats by Fujiwara et al. (1994), J. Enwright and R. Grainger (unpublished data) found that head ectoderm from either homozygous or heterozygous small eye mouse embryos at E9.5 does not exhibit normal lens-forming bias, giving no or very poor lens response when recombined in culture with wild-type OVs (0–10% crystallin positive vs. 82% crystallin positive for wild-type recombinants).

While a number of genes have been identified whose expression within the PLE or surrounding tissues change during bias, most of the changes occur after the onset of bias and thus are likely not themselves responsible for, but rather contribute to, the increasing predisposition of the tissue toward lens formation. One might expect that a bias factor would be turned on and broadly expressed in the ectoderm during the mid to late gastrula stage (i.e., during or toward the end of the lens competence period). On the other hand, genes that come on in more restricted expression domains during the bias phase might be likely candidates for triggering a later stage of lens determination, such as specification or early differentiation. As mentioned above, *XSox3* and *MSox2* are both broadly expressed during the competence period. Similarly, *Otx2* in both frog and mouse embryos and *Pax6* in mouse embryos are broadly expressed in head ectoderm around the onset of bias, but it is not clear that their expression domains are coextensive with the area of head ectoderm that has demonstrated lens-forming bias. Other potential bias factors are members of the *Eya* and *Dlx* families, homologs of genes involved in *Drosophila* eye development and body patterning, respectively. *Dlx3/dll2* (mouse, chick, zebrafish, *Xenopus*) is first expressed in ectoderm during gastrulation (zebrafish and *Xenopus*), before or near the time of onset of bias, and later expression excludes the PLE but includes both the otic and nasal placodes (mouse, chick, zebrafish, and *Xenopus*), suggesting a role in differentiation of those tissues but not the lens (Akimenko et al., 1994; Dirksen et al. 1994; Robinson and Mahon, 1994; Pera and Kessel, 1999). *Eya1* in the mouse and frog (Xu et al., 1997b; Offield and Grainger, unpublished data) is expressed broadly around the onset of bias, at least coextensively with the placodal zone, and is later restricted to structures that include the lens and nasal placodes.

It is worth noting that PNE and PLE regions have overlapping but not identical molecular phenotypes, suggesting that they share certain features of determination. This is consistent with Jacobson's (1963a, 1963b, 1963c) observations on the placodal zone of head ectoderm during the early lens bias period and with the effects of *Pax6* mutation on both lens and nasal placode development (Grindley et al., 1995).

2.3.3. *Inhibition*

Since a relatively large area of ectoderm initially acquires lens-forming bias, one might hypothesize that there is a mechanism to ensure that lenses form only in that ectoderm that becomes properly positioned with respect to the developing eye rudiment. One could imagine that positive signals coming from the OV stabilize the lens-forming commitment in closely apposed ectoderm, while the lack of such positive signals in more distant ectoderm could contribute to destabilization and loss of lens-forming bias. Alternatively, or in addition to positive regulatory signals from the OV, there could be regulatory signals that come from some other source(s) and preferentially influence ectoderm not in contact with the OV, so as to alter its lens-forming bias and cause (or allow) it to differentiate along other pathways.

The progressive restriction of the ectodermal expression domains of a number of transcription factors (e.g., Otx2, MSox2, XSox3, and Pax6) likely to be important for lens formation bears testimony to the fact that the ectoderm is undergoing changes at the molecular level during the bias period, but it does not suggest any particular mechanism. As mentioned earlier, however, there is some experimental evidence that points to a potential inhibitory role for head neural crest. Von Woellwarth's (1961) experiments demonstrated that free lenses were more likely to form if posterior neural tissue, including the source of head neural crest, was removed along with the retinal rudiment. Recently this hypothesis has been tested more directly in both mouse and chick embryos. J. Enwright and R. Grainger (unpublished data) found a strong inhibitory influence of head mesenchyme on the ability of specified head ectoderm from mouse embryos to form lenses *in vitro* (5% of early and 37% of late E9.5 head ectoderm plus head mesenchyme recombinants formed crystallin-positive lenses, compared with 37% of early head ectoderm and 71% of late E9.5 head ectoderm cultured alone). Similarly, C. Sullivan and R. Grainger (unpublished data) found that specified head ectoderm from outside the PLE region of H&H stage 10 chick embryos, when cultured alone, formed crystallin-positive lenses more than 90% of the time. In contrast, crystallin-positive lenses never formed when stage 10 specified head ectoderm was left attached to underlying mesenchyme, and they formed 30% of the time when the ectoderm was partially released from the head mesenchyme by brief trypsinization. Using the neural crest cell marker HNK-1, Sullivan and Grainger showed that, at H&H stage 10, neural crest is found throughout the head mesenchyme but not between the PLE and the OV.

Two additional observations from mouse mutants are consistent with an inhibitory role for head neural crest. First, in both small eye mutants and *Lhx2* knockout mice, the OV fails to achieve the tight contact with PLE that would normally preclude PLE interaction with neural crest, and no lens placodes form (Grindley et al., 1995; Porter et al., 1997). Second, J. Enwright and R. Gainger (unpublished data) found that head ectoderm from these mutants shows reduced level of lens-forming bias, perhaps partly because these tissues have more extensive interaction than normal with neural crest and/or because of deficiencies in the positive signals from the OV.

Further evidence that lens-forming bias in non-PLE is actively repressed comes from a zebrafish mutant (*yot*) with a defect in Gli2-mediated hedgehog signaling (Kondoh et al., 2000). In this mutant, Rathke's pouch, which has been shown in chicks to have a strong bias toward lens differentiation, even transiently expressing high levels of δ-crystallin

(Barabanov, 1977; Fedtsova and Barabanov, 1978; Barabanov and Fedtsova, 1982), differentiates into normal-looking lenses while the adenohypophysis fails to form. These ectopic lenses have elongated fibers and express β-crystallin, but they degenerate by 120 hours of development, presumably due to the absence of necessary sustaining factors that would be supplied to a lens by the optic cup. A similar phenomenon is seen in the chick mutant *talpid*^{3}. Ede and Kelly (1964) reported abnormalities, including the absence of Rathke's pouch and the presence of ectopic lenses in its place. It has recently been shown that *talpid*^{3} mutants are unable to respond to sonic hedgehog signaling (Lewis et al., 1999). These results suggest that, during normal development, hedgehog signaling from the ventral midline of the CNS in the region of the diencephalon shifts the lens-forming bias of Rathke's pouch toward adenohypophysis formation.

2.3.4. Specification

The PLE becomes increasingly biased toward lens formation throughout the neurula stage until around the time when the outgrowing OV establishes tight adherence with it. Around this time, the PLE has reached a level of determination that allows it to form a small, simple crystallin-positive lens even when removed from further inductive influences, as when explanted *in vitro*. At this point, the PLE is said to be "specified" to form a lens in the absence of other external cues. Specification is operationally distinct from determination, in that "specified" tissue might be induced to undergo alternative differentiation if placed in an appropriate environment whereas "determined" tissue is committed to follow a particular developmental course regardless of external cues. This point is clearly illustrated in chick embryos, where a large portion of epiblast is specified for lens formation even as the neural plate is forming (H&H stage 5, according to Barabanov and Fedtsova, 1982), at a stage when, in frog and mouse embryos, head ectoderm is only beginning to acquire lens-forming bias. Despite being able to form lens tissue *in vitro* at a very early stage (compared with other vertebrates), most of the epiblast ultimately will not differentiate as lens tissue if left *in situ*, where it will receive cues that will direct differentiation along several alternative pathways, including nasal, otic, and epidermis pathways. In amphibians and mice, PLE specification occurs at a very similar developmental stage, just before or right at the time when close contact with the OV is established (Karkinen-Jääskeläinen, 1978a; Henry and Grainger, 1990; J. Enwright and R. Grainger, unpublished manuscript).

So far, the relatively early specification of head ectoderm in chick embryos is the most striking difference we have found in the process of lens determination among the vertebrates studied, and it raises the point that although we can distinguish operationally between tissue bias and specification, we still do not know what is required at the molecular level for either condition. In amphibian and mouse systems, where the bias period lasts many hours, it is possible to demonstrate its progressive nature by testing ectoderm at several time points. Specification could represent attainment of a threshold level of some factor(s) activated during the bias phase (i.e., it could be a quantitative distinction rather than a qualitative one), or there may be distinct "specification factors" acting near the end of the bias phase. Early specification of chick head ectoderm does not necessarily signify a fundamental difference in the mechanism underlying lens determination. As far as can be judged from reviewing the literature, lens determination in chicks appears in all other regards to proceed similarly to that in mice and frogs. Because specification is defined operationally by the tissue's ability to undergo at least rudimentary lens differentiation, we discuss the molecular correlates in the next section.

2.3.5. *Differentiation*

The end point of determination is the irreversible commitment to overt lens differentiation. It now seems clear that this is the step where interaction with the closely apposed optic rudiment is crucial. The first steps in overt lens differentiation (early crystallin expression and placode and vesicle formation) can occur in the absence of the OV, as demonstrated both by the existence of free lenses *in vivo* and the small lenses that form *in vitro* from isolated specified PLE. However, such lenses do not attain the same degree of maturation as normal lenses, and in some cases free lenses degenerate in the absence of a sustained interaction with the optic cup. One might then hypothesize that there is an early phase of differentiation inextricably linked, by definition, to the event of tissue specification and a later phase encompassing the final maturation of the lens, including the formation of elongated lens fibers and the expression of the full range of lens-specific proteins. Recently, Offield et al. (2000) provided evidence that the onset of differentiation in specified ectoderm *in vivo* is not controlled by events intrinsic to lens ectoderm. Using a line of *Xenopus tropicalis* that carries a transgene construct linking γ-crystallin promoter elements with the coding sequence for green fluorescent protein (GFP), they were able to directly assay the onset of lens differentiation by montoring GFP expression in living animals and tissue explants. They found that specified lens ectoderm transplanted into young, neural plate stage hosts delayed GFP expression by about 4.5 hours (the time required for the hosts to reach the developmental stage of the donor tissue), and they also found that when specified lens ectoderm was explanted *in vitro*, GFP expression was delayed by around 16 hours relative to the control embryos. It is not yet known whether the signals in the embyro controlling the timing of onset of differentiation come from the optic vesicle or other adjacent tissues.

2.3.5.1. Molecular Correlates of Lens Differentiation

A number of transcriptional regulators and growth factors have already been implicated in the initiation and/or maintenance of lens differentiation. The timing of expression of many of these factors fits well with the idea that there are early and late phases to differentiation. For more detailed discussion of the regulation of crystallin expression and factors regulating lens fiber differentiation and maintenance, see chapters 5 and 11 in this volume.

As mentioned in a previous section, *Pax6* is expressed in both the PLE and OV during the bias phase, and in homozogyous small eye mice the early phase of differentiation is blocked so that placodes do not form and other potential regulators or indicators of early differentiation (e.g., γ*FCry, Sox2, Eya1*, and *Six3*) fail to be expressed in the PLE around the time of placode formation (Oliver et al., 1995; Xu et al., 1997; Furuta and Hogan, 1998; M. Fisher and R. Grainger, unpublished manuscript). Clearly, *Pax6* is important to the early phase of differentiation, and there is evidence from two sources that the block in small eye mutants involves a failure of retinoic acid (RA) signaling (discussed in the next section).

Another group of regulatory factors that appear to be involved at several levels of lens determination consist of Sox proteins, members of the *Sry*-related HMG domain family. At least three Sox proteins are believed to have roles in activating crystallin gene expression in chick and mouse embryos (Kamachi et al., 1998). As noted previously, in *Xenopus*, *Sox3* is expressed during the lens competence period but is turned off at the onset of bias and reexpressed in PLE at the time of specification (see Fig. 2.2). PLE expression of its mammalian counterpart, M*Sox2* (also expressed during gastrulation) is dependent on bone morphogenetic proteins (BMPs) 4 and 7 and Pax6 (Furuta and Hogan, 1998; Wawersik et al., 1999). *Sox1* expression is required for lens fiber differentiation, and in fact *Sox2* in both chick and mouse embryos is down-regulated in lenses when *Sox1* is turned on (see Fig. 2.2; Nishiguchi et al., 1998; Kamachi et al., 1998).

Like the Sox family of proteins, members of the Maf family of basic leucine zipper transcription factors, including *c-Maf, Mafβ*, and *L-Maf*, also play roles in the early and late phases of lens differentiation by regulating expression of crystallins and other lens-specific proteins (Sakai et al. 1997; Yoshida et al., 1997; Ogino and Yasuda, 1998). *Maf* gene expression is up-regulated in specified ectoderm just prior to placode formation and early crystallin expression (Fig. 2.2; Ogino and Yasuda, 2000). Ectopic expression of *L-Maf* was shown to convert cultured neural retina cells into lens cells and to produce ectopic patches of δ-crystallin expression in the head ectoderm of chick embryos (Ogino and Yasuda, 1998). In homozygous *c-Maf* knockout mice, lens differentiation is blocked at the stage of fiber formation (Kawauchi et al., 1999; Kim et al., 1999).

A number of genes first expressed during the lens placode stage are implicated in the later differentiation and/or maintenance of either lens fiber cells or the anterior lens epithelium. *Prox1*, a homeobox gene homologous to *Drosophila prospero*, is first expressed in the placode stage in chick and mouse embryos and also appears to be important for the differentiation of lens fibers (Oliver et al., 1993; Tomarev et al., 1996; Wigle et al., 1999). *Sw3–3*, a putative basic-leucine zipper transcription factor expressed in chick lens placode and anterior lens epithelium, is suggested to have a role in lens differentiation (Wang and Adler, 1994). Similarly, *FoxE3*, a member of the forkhead family of transcription factors, is expressed in mouse lens placode and later is restricted to anterior lens epithelium. This factor maps to the murine *dysgenetic lens* locus, mutations in which result in the failure of lens vesicle separation from the corneal epithelium (Blixt et al., 2000; Brownell et al., 2000).

Thus, a number of factors are required for the onset of overt differentiation (i.e., placode and vesicle formation), and there is a second crucial point – that lens fiber differentiation requires a different set of factors. Interestingly, some of the factors required for later differentiation, such as Sox1 and c-Maf, are close relatives of factors required for early differentiation (i.e., Sox2/3 and L-Maf).

2.4. Inducing Signals

Presently we can only speculate as to the nature of the early lens-inducing signals that we now know must be arising first from the anterior neural plate and somewhat later from the foregut endoderm and head mesoderm. However, there is experimental evidence suggesting that RA, BMPs, and FGFs play important roles as signals during the stages of overt lens differentiation. Bavik et al. (1996) showed that RA deficiency beginning at 10–12 somite stages in mouse embryos leads to the failure of lens placode formation and optic vesicle invagination as well as to the interposition of mesenchyme between the OV and PLE – all features common to the small eye phenotype (Grindley et al., 1995). In addition, Enwright and Grainger (2000), using a transgenic mouse line carrying a RA response element (RARE)–β-galactosidase fusion construct, detected altered retinoid signaling in the heads of small eye mouse embryos. The RARE transgene expression pattern reveals that, during the development of wild-type embryos, RA-mediated gene activation is first apparent at late E8.5 in the OV and forebrain and by E8.75 in the PLE. By E9 the transgene is expressed in the OV, PLE, forebrain, PNE, and head mesenchyme (all structures that show defects in small eye mice and that express *Pax6* at this time). Enwright and Grainger also showed that RARE transgene expression was dramatically reduced in the PLE and PNE of small eye mice and that this was due to defects in the abilities of these tissues to respond to exogenously applied RA. Furthermore, the PNE and PLE did not produce RA, unlike their wild-type counterparts. These results suggest that, among other things, *Pax6* acts within the head ectoderm to influence both response to and production of RA. In contrast to the PLE, the OV of small eye mutants appeared to have normal responsiveness to exogenous RA and only a mild reduction in RA production.

Other factors that are likely to be involved in the early differentiation phase are BMP4 and BMP7, members of the transforming growth factor-β superfamily. Furuta and Hogan (1998) showed that, in mice, BMP4 is expressed in the OV but not the PLE and becomes further restricted to the distal tip of the OV underlying the PLE just before placode formation. They further showed that, in homozygous BMP4 knockout embryos that survive to E10.5, no lens placode forms, and *Sox2* expression is not turned on in the PLE as it normally is, just prior to placode formation. In contrast, *Pax6* and *Six3* expression appears normal in the PLE of BMP4 mutant embryos. Lens formation in cultured eye rudiments from these mutants could be rescued by loading the OV with BMP4-coated beads, but the beads alone could not substitute for the OV. Furuta and Hogan also demonstrated that BMP4 was acting through the OV by showing that recombinants of wild-type OV and BMP4 mutant head ectoderm formed lenses at a high frequency whereas the reciprocal of mutant OV with wild-type head ectoderm formed lentoids at a low frequency. These results demonstrate that BMP4 acts within the OV to elicit the OV's lens-inducing capabilities. Interestingly, they found that small eye mutants showed normal expression of BMP4 and its receptors, Alk3 and Alk6, suggesting that BMP4 and *Pax6* are regulated and function independently with regard to lens determination.

BMP7 was first implicated in the late stages of lens differentiation (Dudley et al., 1995), but a more recent study by Wawersik et al. (1999) suggests that it is also involved in the early stages of differentiation. They showed that, on the C3H genetic background, a BMP7 knockout mutation blocked lens placode formation in homozygous embryos. They also demonstrated that both the BMP7 antagonist follistatin and a BMP7-neutralizing antibody could block lens-forming ability in cultured wild-type eye rudiments, reducing it to the same level as found in cultured eye rudiments from BMP7 mutants. Like Furuta and Hogan, they found that *Sox2* was not expressed in the PLE of the mutants at E9.5, and in contrast to the BMP4 mutants, the BMP7 mutants also lost *Pax6* expression from E9.5 PLE. Like BMP4, BMP7 was expressed normally in the PLE of small eye mutants.

Already mentioned earlier as potential candidates for competence factors, members of the FGF family are also implicated in lens fiber differentiation and maintenance. In chicks, FGF-8 is expressed in the distal optic vesicle when it contacts the PLE, and implantation of FGF-8–coated beads near the eye, soon after the OV contacts the PLE, leads to expansion of the lens field, as evidenced by the expanded domain of *L-Maf* expression (Vogel-Höpker et al., 2000). The importance of FGF activity in fiber cell differentiation and maintenance is further underscored by a number of recent transgenic studies (see the review in chap. 11 of this volume). Targeted overexpression of FGFs to the developing lens of transgenic mice resulted in inappropriate differentiation of the lens epithelium (Robinson et al., 1995a, 1998; Lovicu and Overbeek, 1998). Furthermore, transgenic mice expressing dominant-negative FGF receptors specifically in the lens fibers, beginning at the earliest stages of overt differentiation, developed small lenses with disorganized fibers (Robinson et al., 1995b; Stolen and Griep, 2000).

2.5. Conclusions and Future Directions

After a century's worth of investigative efforts directed toward understanding the seemingly simple phenomenon of induction of lens development by interaction with the OV, we have come to appreciate the complexity of this process. We have developed a model of lens determination that draws heavily on amphibian studies but appears to be generally applicable to other vertebrates based on the more limited information available from chick and mouse studies.

Our current model envisions a process that begins during gastrulation, when, through some intrinsic timing mechanism, ectoderm becomes competent to respond to signals from the anterior neural plate. These early signals impart a lens-forming bias to a large portion of the ectoderm. Additional signals, likely coming from mesoderm and/or endoderm, during the early neurula stages enhance the lens-forming bias in the ectoderm of the head until it reaches the point of specification, at which time the earliest stages of overt lens differentiation can occur, even if the ectoderm is removed from further inductive influences. At this point, the OV establishes an intimate relationship with a region of specified head ectoderm and provides the inductive signals necessary for final commitment and maturation of the lens. At this same point, head ectoderm not in close contact with the OV loses its lens-forming bias, likely due at least in part to signals from the underlying migrating neural crest and ventral neural midline. Interference at any of these steps can disrupt lens determination. We do not yet know the nature of the several inductive signals that ultimately act on the PLE, although RA, BMPs, and FGFs have all been implicated in one or more phases of the process. Our present understanding of the timing and likely participants in interactions with the PLE should help to focus the search for the key inductive signals. Likewise, elucidating the relationships among the emerging array of transcription regulatory factors, many of which are also known to function in *Drosophila* eye development, will help to further clarify the complex process of lens determination.

Although mammalian embryos are less well suited to the classical type of embryo manipulation that has proven so useful in defining critical tissue interactions during lens induction, they are proving quite valuable for their genetic manipulability. Naturally occurring and engineered mutations that affect lens determination provide a very useful resource for elucidating the molecular mechanisms that underlie each phase of lens determination. In this regard, the promise of *X. tropicalis*, the diploid, rapidly maturing relative of *X. laevis*, as a genetic system (Amaya et al., 1998; Offield et al., 2000) could give amphibians the clear edge, as far as experimental organisms go, by combining the best of both worlds – classical embryological manipulability combined with the potential for genetic analysis. In addition, new technologies for genetic analysis, such as DNA microarray analysis, are opening up the possibility of rapid identification of previously unrecognized genes that are involved in specific phases of lens determination.

3

Transcription Factors in Early Lens Development

Guy Goudreau, Nicole Bäumer, and Peter Gruss

3.1. Introduction

The study of lens development provides a useful experimental system for investigating fundamental processes in developmental biology. The vertebrate lens develops from a series of interactions between the surface ectoderm, the optic vesicle, and the surrounding mesoderm, and these interactions involve successive steps of bias, competence, specification, and differentiation (see chap. 2 of this volume; see also McAvoy et al., 1999; Hirsch and Grainger, 2000). In recent years, these cellular and morphogenetic processes have been subject to investigation focusing on the molecular events underlying them (Weaver and Hogan, 2001). In particular, important insights were gained through genetic studies performed on the development of the eye in *Drosophila* (Treisman, 1999) and by comparisons of gene expression and function between the eyes of invertebrate and vertebrate species (Hill et al., 1991; Quiring et al., 1994; reviewed in Wawersik and Maas, 2000; Wawersik et al., 2000). These studies have led to the identification of conserved regulatory pathways mediating eye formation in both the fly and vertebrates.

Additional insight into these molecular events has been provided by the evaluation of mouse or human syndromes in which morphogenesis is defective (Freund et al., 1996; Graw, 2000). The eye is frequently affected by inherited eye disorders: roughly one-quarter of the phenotypes listed in *Mendelian Inheritance in Man* involve the eye (Boyadijiev and Jabs, 2000), and several candidate genes implicated in these phenotypes have so far been identified. Both in *Drosophila* and in mammals, transcriptional regulators have been the most common genes uncovered by these genetic approaches, in particular those belonging to the homeobox group.

It is the aim of this review to summarize these recent findings in the context of the development of the lens. The different classes of transcription factors involved in lens formation will be evaluated and their roles examined, particularly when targeted or natural mutations of the genes are available or when misexpression experiments provide insight into their role in lens formation. Despite the importance of transcription factors in eye and lens development, it should be emphasized that several other factors (e.g., secreted factors and their receptors; Furuta and Hogan, 1998; McAvoy et al., 2000; Rasmussen et al., 2001) also play critical roles in lens formation (see chap. 11). Although the description of these factors falls outside the compass of this chapter, they will be mentioned in relation to their influence on transcription factor regulation. Lastly, transcription factors involved in lens formation often have additional developmental roles, such as regulating other aspects of eye formation, and a detailed analysis of these roles is also beyond the scope of this chapter.

Note, however, that such roles have been the subject of two recent reviews (Jean et al., 1998; Morrow et al., 1998).

3.2. The Key Transcriptional Regulators Involved in Eye Development Are Conserved in Different Species

An important development in past years has been the realization that the genetic chain of events leading to eye formation has, to a great extent, been conserved in species ranging from the fruit fly *Drosophila* to vertebrates (reviewed in Desplan, 1997; Treisman, 1999; Wawersik and Maas, 2000). This finding is all the more striking considering the vast diversity of structures and light-gathering strategies found in the visual sense organs of these different species (Fernald, 2000). In *Drosophila*, interest has mainly focused on a small group of transcription factors and nuclear proteins. These factors are *eyeless (ey)* (Quiring et al., 1994), *twin of eyeless (toy)* (Czerny et al., 1999), *sine oculis (so)* (Cheyette et al., 1994), *optix* (Toy et al., 1998), *eyes absent (eya)* (Bonini et al., 1993), and *dachshund (dac)* (Mardon et al., 1994). Among these factors, *ey* and the closely related gene *toy* are thought to function at the top of a gene hierarchy, since their expression does not depend on the presence of the remaining genes, and their ectopic expression induces the expression of other factors as well as the formation of ectopic eyes (Czerny et al., 1999; Halder et al., 1998). For these reasons, *ey* and *toy* have been proposed as key regulators of eye development in *Drosophila*, with current evidence indicating that *toy* may act as an upstream regulator of *ey* (Czerny et al., 1999; Hauck et al., 1999). The remaining factors regulate each other's expression and participate in gene regulatory loops with *ey* (Fig. 3.1) (Bonini et al., 1997; Chen et al., 1997a; Pignoni et al., 1997; Shen and Mardon, 1997), an important exception being the *so*-related *optix* gene, which is regulated by an *ey*-independent mechanism (Seimiya and Gehring, 2000). These factors can physically interact with one another, and protein complexes have been demonstrated between *so* and *eya* and between *da* and *eya* (Chen et al., 1997a; Pignoni et al., 1997). Misexpression of these factors also leads to ectopic eye formation in *Drosophila*, although at a frequency that is lower than in the case of *ey*; in

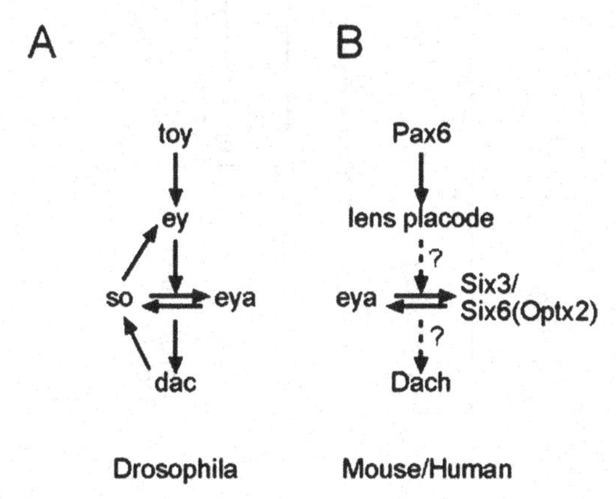

Figure 3.1. Conserved cascade of transcription factors involved in compound eye development (A) and in eye development of vertebrates (B). For details, see text and accompanying references. (After Wawersik and Maas, 2000.)

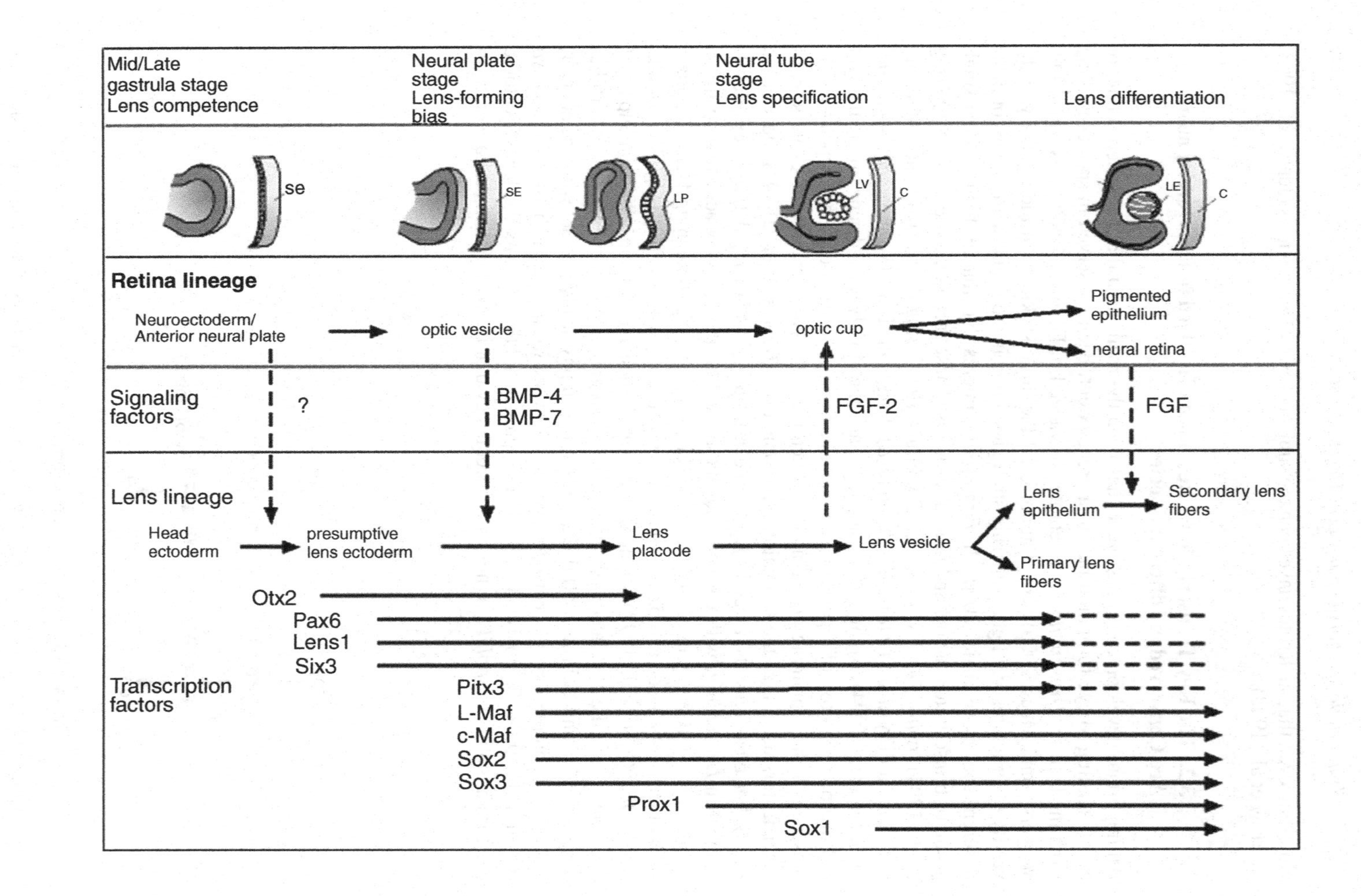

Mid/Late gastrula stage Lens competence
Neural plate stage Lens-forming bias
Neural tube stage Lens specification
Lens differentiation
se
SE
LP
LV
C
LE
C
Retina lineage
Neuroectoderm/ Anterior neural plate
optic vesicle
optic cup
Pigmented epithelium
neural retina
Signaling factors
?
BMP-4
BMP-7
FGF-2
FGF
Lens lineage
Head ectoderm
presumptive lens ectoderm
Lens placode
Lens vesicle
Lens epithelium
Primary lens fibers
Secondary lens fibers
Transcription factors
Otx2
Pax6
Lens1
Six3
Pitx3
L-Maf
c-Maf
Sox2
Sox3
Prox1
Sox1

addition, a combined misexpression of factors is often required (Bonini et al., 1997; Chen et al., 1997a; Pignoni et al., 1997; Shen and Mardon, 1997).

Remarkably, the vertebrate orthologs of these different factors have also been shown to be involved in eye development. *Pax6*, the vertebrate ortholog of *ey*, is required for eye development in most species so far studied (reviewed in Gehring and Ikeo, 1999; Wawersik et al., 2000), and experiments involving misexpression of *Pax6* in *Xenopus laevis* have led to ectopic lens and eye formation (Altmann et al., 1997; Chow et al., 1999). Like in *Drosophila*, vertebrate *Pax6* is therefore sufficient to initiate the chain of events leading to eye formation, indicating a widespread role for *Pax6* as a key regulator of eye development. In vertebrates, the function of the *ey*-related *toy* gene is assumed to have been taken by *Pax6*, since an additional *Pax6*-like gene has only been identified in zebrafish (Nornes et al., 1998). Vertebrate orthologs for genes belonging to the remaining gene families have been isolated and shown, in some instances, to be expressed during eye formation (Borsani et al., 1999; Hammond et al., 1998; Heanue et al., 1999; Oliver et al., 1995a; Oliver et al., 1995b; Toy et al., 1998; Xu et al., 1997b) and to adopt gene regulatory strategies similar to those seen in *Drosophila* (Heanue et al., 1999; Loosli et al., 1999; Ohto et al., 1999; Zuber et al., 1999; reviewed in Relaix and Buckingham, 1999). Details concerning targeted mutations for the latter genes are, for the most part, either unavailable or inconclusive regarding their role in vertebrate eye development (Davis et al., 2001; Ozaki et al., 2001; Xu et al., 1999b). However, gain-of-function studies (Loosli et al., 1999; Oliver et al., 1996) and information gathered from cases of human mutations (Azuma et al., 2000; Gallardo et al., 1999; Wallis et al., 1999) indicate that these genes do participate in vertebrate eye formation.

3.3. Transcription Factors from Different Classes Are Involved in Lens Development

Recent studies have revealed additional transcription factors involved in eye and lens formation (see Fig. 3.2 and Table 3.1). These additional factors have been uncovered, for the most part, through reverse genetic approaches from existing *Drosophila*, mouse, or human mutations or on the basis of their capacity to bind to the regulatory sequences of lens-specific genes. These genes will be described according to the transcription factor classes to which they belong. Although members of most classes of transcription factors are expressed in the lens, transcription factors belonging to the homeodomain group are by far the must abundant and best studied regulatory factors involved in eye and lens formation and will therefore be described first.

3.3.1. Homeobox Genes

The homeodomain (Gehring et al., 1994) consists of a 60–amino acid core motif encoded by a 180–base pair sequence called the homeobox, which functions as a dimerization as

Figure 3.2. (facing page) Inductive phases and expression of transcription factors during lens development. *Top*: Schematic views on the different stages of eye development with respect to the lens inductive events. *Middle*: Links between retina lineage, secreted factors from the optic vesicle/retina, and lens lineage. *Bottom*: Expression timetable of transcription factors during lens development. C, cornea; LE, lens epithelial cells; LF, lens fiber cells; LP, lens placode; LV, lens vesicle; NR, neural retina; RPE, retinal pigmented epithelium; SE, surface ectoderm. For references, see corresponding chapters in the text. (After Ogino and Yasuda, 2000.)

Table 3.1. *Transcription Factors Involved in Lens Development*

Gene Family	Subfamily	Transcription Factor	Characteristic Domains	Species	Loss-of-Function Ocular Phenotype	Gain-of-Function Ocular Phenotype	Putative Function/ Target Genes	Reference
Homeobox (DNA-binding)	Pax	Pax6	Paired-box (DNA-binding)	D (*eyeless*), M, Hu. Me, Xe	Heterzygous: small eye homozygous: arrest at optic vesicle stage, no lens	Ectopic eye structures, ectopic lenses (medaka)	Lens placode formation/ + crystallins, +Pax6, Sox2, Six3, Prox1, Eya1, Eya2	Walther and Gruss, 1991
	Six	Six3	Six-domain	D (*sine oculis*), M, Me, Hu	nd	Ectopic eyes, including ectopic lenses	/+ Pax6 dac	Oliver et al., 1995a
	Prox	Prox1	Novel homeobox	D (*prospero*), Xe, Ze, M, Hu	Defects in lens fiber formation	nd	Cell cycle arrest, cell elongation/ $+\gamma$D-crystallin	Oliver et al., 1993
	Pitx	Pitx3		Xe, M, Hu	Aphakia	nd	Lens differentiation	Semina et al., 1997
	Otx	Otx1	*Bicoid*-like homeodomain	D (*orthodenticle*), Xe, Ch, M, Hu	Eye defects, no lens defect	nd		Simeone et al., 1992; 1993
		Otx2			Lethal at E 9.5	nd	Specification of eye field	
		Otx1/Otx2			Otx1+/−/ Otx2+/−: no lens or impaired lens differentiation	nd	Control of morphogenetic movements	

	LIM	Lhx2	LIM-box (protein-protein-interaction)	M, R, Hu	No eye differentiation beyond optic vesicle stage; no lens placode formation	nd	(/+ in surface ectoderm Pax6 via secreted factor)	Xu et al., 1993
	Rx	Rx	paired-like homeodomain	D (*Drax*), Xe, Ch, M, R, Hu	No optic vesicle	Hyperproliferation of NR and RPE repressing lens tissue		Eggert et al., 1998
	Sox	Sox1	*Sry*-box (DNA-binding)	Ze, Ch, M, Hu	No lens fiber elongation	+ on all δ-/γ-crystallins	Cell cycle arrest/ + Pax6 + on all δ-/γ-crystallins	Kamachi et al., 1995
		Sox2		Ze, Ch. M, Hu	Preimplantation lethality			
		Sox3		Ze, Xe, Ch, M, Hu	nd			
bZIP	Maf	L-Maf		Xe, Ch	nd	+ δ1-crystallin	+ αA-crystallin + δ1-crystallin	Ogino and Yasuda, 1998
		c-Maf		Ze, Ch, M, R, Hu	No lens fiber elongation, due to less γF-/β-crystallins	nd	+ γF -/β-crystallins	Kim et al., 1999
HSH (protein-protein-interaction)	AP	AP2α	Basic region (protein-protein-interaction)	M, Hu	Anophthalmia or lens defects	nd	Cell adhesion	West-Mays et al., 1999

Key: nd = not determined; D = *Drosophila*; Ch = chicken; Hu = human; M = mouse; R = rat; Xe = *Xenopus*; Ze = zebrafish well as a sequence-specific DNA-binding element. Additional motifs are often found in homeodomain proteins, and these have also been shown to participate in protein-protein interactions and in DNA binding. Of the more than 25 classes of homeodomain identified to this date, the members from 7 classes of homeodomain proteins have been shown to participate in lens development.

3.3.1.1. Pax6

Pax genes were isolated on the basis of the presence of a 384–base pair sequence called the *paired box*, which shares sequence similarities to the *Drosophila paired* gene (Noll, 1993). The resulting 128-amino-acid-long protein structure, the paired domain, acts as a sequence-specific DNA-binding domain (Wilson et al., 1995). Members of the *Pax* gene family have been subdivided primarily according to the presence or absence of a paired-type homeobox, which may be partial or complete, or according to whether they encode a conserved octapeptide (Noll, 1993; Walther et al., 1991). The latter motif corresponds to the consensus sequence (H/Y)S(I/V)(N/S)G(I/L)LG (Noll, 1993) and is related to the core (eh1) region of the engrailed repressor domain (Eberhard et al., 2000). *Pax* genes have been isolated from several vertebrate and invertebrate species, are widely expressed during embryogenesis, and are involved in developmental processes and in tumorigenesis (Barr, 1999; Dahl et al., 1997; Mansouri et al., 1999; Nutt et al., 2001). In vertebrates, nine members of the *Pax* gene family have been identified. Of these nine *Pax* genes, only *Pax6* is expressed in the lens (Walther and Gruss, 1991). In addition to its well-described function during eye development (Gehring and Ikeo, 1999; Mathers and Jamrich, 2000; Wawersik et al., 2000) in vertebrate species, *Pax6* is also involved in the development of the central nervous system (Bishop et al., 2000), nose (Quinn et al., 1996), pituitary gland (Kioussi et al., 1999), and pancreas (Dohrmann et al., 2000). These different requirements for *Pax6* function are indicative of the multiple functions that *Pax6* is likely to carry out during development.

Structurally, human and murine Pax6 proteins are 422 amino acids in length and contain a paired domain and a paired-type homeodomain but do not contain an octapeptide (Callaerts et al., 1997). Pax6 encodes a C-terminal proline-serine-threonine (PST)–rich domain, which is thought to function as a transactivating domain. Both the homeodomain (Sheng et al., 1997a; Sheng et al., 1997b) and the paired domain (Czerny and Busslinger, 1995; Epstein et al., 1994a) of Pax6 appear to function independently as well as in cooperation to bind specific DNA sequences. Pax6 can also bind DNA either through the N-terminal or C-terminal portion of the paired domain, increasing the number of possible DNA-binding sites (Epstein et al., 1994b; Xu et al., 1999a). Additionally, a splice variant (Pax6(5a)) has been identified in several vertebrate *Pax6* genes, and it results in a 14–amino acid insertion in the N-terminal portion of the paired domain (Epstein et al., 1994b; Kozmik et al., 1997). *In vitro* experiments have indicated that this insertion disrupts the ability of the paired domain to bind DNA, since the extended paired domain only recognizes DNA through its C-terminal portion, therefore limiting its DNA-binding capacity (Duncan et al., 2000; Epstein et al., 1994b; Kozmik et al., 1997).

Pax6 Expression during Lens Formation. All *Pax6* genes so far studied have been found to be expressed in the developing eye (Gehring and Ikeo, 1999), and detailed analysis of Pax6 expression has been done in the developing ocular structures of several species (Callaerts et al., 1997). In the developing mouse eye (Walther and Gruss, 1991), Pax6 mRNA is first detected at embryonic day 8.0–8.5 (E8.0–8.5) on the head surface ectoderm and in the neural ectoderm, including the optic pit. Pax6 expression in the surface ectoderm initially covers a broad area, but following contact between the optic vesicle and the surface ectoderm (E9.0), Pax6 transcripts in the surface ectoderm become restricted to the lens placode (E9.5). At these same stages, Pax6 is also expressed in the optic vesicle and stalk. At later stages (E10.0), Pax6 mRNA expression is detected in the lens vesicle and the optic cup. At stages E13.5 and later, Pax6 expression in the lens is located in the proliferating anterior epithelial cells and the transition zone. Although it is initially present in the differentiating lens fiber cells, Pax6 mRNA is not detected in lens fiber cells beyond stage E13.5. At these same stages, Pax6 transcripts are expressed on surface ectodermal structures that give rise to the cornea, conjunctiva, and eyelids and in the ciliary marginal zone and ganglion and the amacrine layers of the retina. In the adult eye, Pax6 is expressed in the anterior epithelial cells and bow region of the lens, the cornea and conjunctiva, the iris and ciliary body, and the retina (Koroma et al., 1997).

Transgenic studies have recently identified some of the regulatory elements responsible for the tissue-specific expression of the *Pax6* gene (Kammandel et al., 1999; Williams et al., 1998; Xu et al., 1999c). The genomic structure of the *Pax6* gene has been extensively evaluated in mice (Fig. 3.3A). The gene encodes 11 exons, and its transcripts are initiated from 2 distinct sites (P0 and P1) that are thought to have originated from intragenic duplication (Xu et al., 1999c). Within the *Pax6* gene, regulatory elements for tissue-specific expression have been identified for the eye (lens and retina), pancreas, and brain. Remarkably, it has been shown that *Pax6* eye regulatory regions can function as enhancers in the *Drosophila* visual system and that eye enhancer regions from the *Drosophila* eyeless gene can direct eye- and CNS-specific expression in transgenic mice, indicating an evolutionary conservation of the mechanisms regulating the expression of both genes (Xu et al., 1999c). *Pax6* expression in the lens appears to be regulated from a 341-bp element situated approximately 3.5 kb upstream of the 5′ transcription initiation site (P0; Kammandel et al., 1999; Williams et al., 1998). Extensive homologies have been found within this fragment between mouse, human, and medaka fish *Pax6* genomic sequences. Within this DNA fragment, a 107-bp region has been shown in transient transgenic assays to be sufficient to direct the expression of a reporter gene in the lens (Kammandel et al., 1999). Additional studies should allow the identification of transcription factors that regulate *Pax6* expression through this enhancer locus.

Regulation of Target Gene Expression. Consensus DNA-binding sequences for the Pax6 paired domain have been well established (Epstein et al., 1994a; Xu et al., 1999a), and *in vitro* assays have demonstrated that Pax6 can activate transcription by interacting with paired domain consensus DNA-binding sequences (Czerny and Busslinger, 1995). Moreover, most homeodomains recognize a TAAT core motif, and paired-type homeodomains have been shown to cooperatively dimerize on palindromic DNA sequences composed of two TAAT half-sites (Sheng et al., 1997b; Wilson et al., 1993). However, the accurate prediction of Pax6-binding sites in genomic DNA has proven difficult, and several mismatches have been reported between natural Pax6-responsive DNA sites and artificial consensus binding sequences. A satisfactory explanation for this discrepancy remains to be found, although it has been proposed that cellular cofactors (Kamachi et al., 2001) or combinatorial binding from different protein subdomains (Duncan et al., 2000) may be required for efficient DNA-binding of Pax proteins to occur. Despite these limitations, more than a dozen genes have been proposed to act as downstream effectors of *Pax6* function. Genes acting downstream of *Pax6* have been described in the eye (Duncan et al., 2000; Marquardt

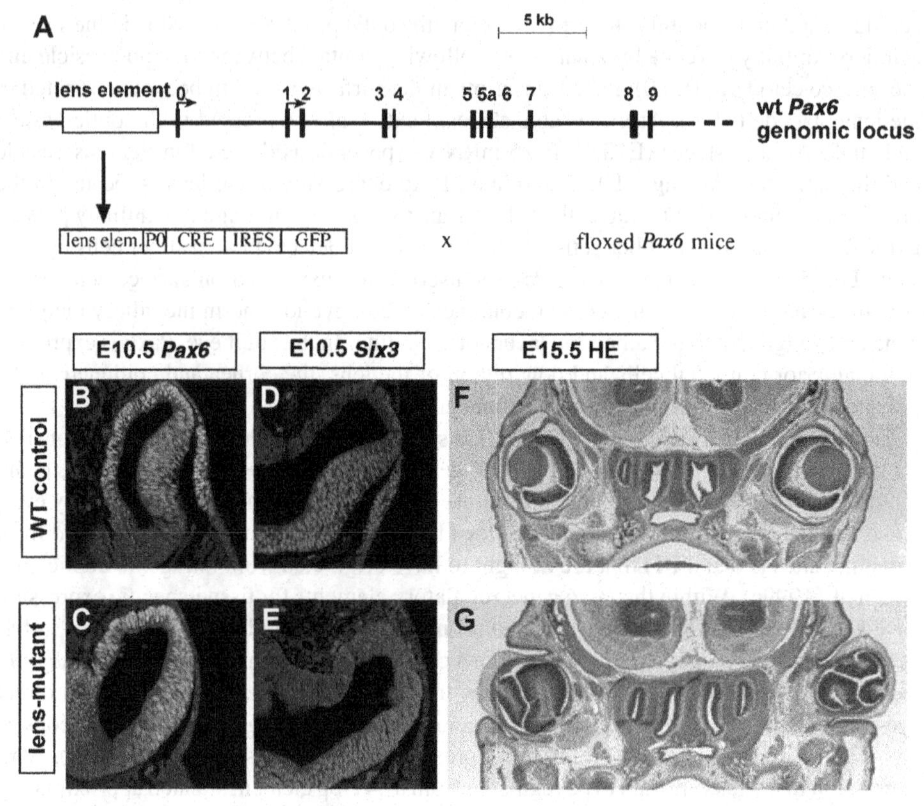

Figure 3.3. (See color plate II.) Pax6 functions in eye and lens development. (A) The regulatory element reponsible for Pax6 expression in the surface ectoderm and subsequently in the lens is located upstream of Exon 0 of the Pax6 genomic locus. This so-called lens element was used to drive Cre-recombinase and GFP expression in transgenic mice. When intercrossed with flox-Pax6 mice, mosaic animals ("lens-mutants") carrying a tissue-specific deletion of Pax6 function in the prospective lens placode were generated (after Ashery-Padan et al., 2000). Lens-mutants express neither Pax6 (C) nor Six3 (E) in the surface ectoderm at E10.5, in contrast to eyes from wild-type mice (B and D). At E15.5, coronal sections of lens-mutants reveal an absence of lens formation. Multiple optic cups appear, due to defective signaling between the optic vesicle and surface ectoderm (F, G) (hematoxylin and eosin).

et al., 2001; reviewed in Cvekl and Piatigorsky, 1996), brain (Götz et al., 1998; Kim et al., 2001; Meech et al., 1999), and pancreas (Dohrmann et al., 2000), and they include genes for transcription factors, structural proteins, and adhesion molecules. In addition, it has been suggested that the human (Okladnova et al., 1998) and quail (Plaza et al., 1993) *Pax6* genes may positively regulate their own expression. In the lens, the target gene regulation of Pax6 has mainly focused on crystallin genes (Cvekl and Piatigorsky, 1996). Pax6 has been shown to regulate the expression of the chicken δ1-crystallin gene (Cvekl et al., 1995b), mouse αB-crystallin gene (Gopal-Srivastava et al., 1996; Haynes et al., 1996), guinea pig ζ-crystallin gene (Richardson et al., 1995), and chicken and mouse αA-crystallin genes (Cvekl et al., 1995a; Cvekl et al., 1994). The evidence that Pax6 may regulate the expression of these crystallin genes has been derived, for the most part, from *in vitro* reporter gene assays. However, in the case of the guinea pig ζ-crystallin gene (Richardson et al., 1995), a transgenic study done in mice has clearly shown that transcriptional activity is reduced following the mutation of a Pax6-binding site located within the gene promoter sequences. Interestingly, it has been suggested that Pax6 may have a dual role during lens development, acting either as a transcriptional activator or repressor, since *in vitro* assays have shown that Pax6 could down-regulate chicken βB1-crystallin gene expression (Duncan et al., 1998). βB1-crystallin is a late differentiation marker in lens fiber cells, and its expression follows an inverse spatial relationship with Pax6, which supports the hypothesis that Pax6 may act as a transcriptional repressor for this crystallin gene.

Few studies have focused on the identification of Pax6 target genes at earlier stages of lens formation. Up-regulation of the transcription factor *Sox2* is an early response of the presumptive lens ectoderm to the inductive signal of the optic vesicle (Furuta and Hogan, 1998; Wawersik et al., 1999). Sox2 expression is not up-regulated on the surface ectoderm of Pax6 null mutants (Furuta and Hogan, 1998), indicating a possible dependence on Pax6 activity. The conditional disruption of Pax6 function in the preplacodal surface ectoderm has, however, shown that, although Pax6 seems initially required to up-regulate Sox2 expression, it is not required to maintain its activity in the specified surface ectoderm (Ashery-Padan et al., 2000). The same study has also indicated that the expression of *Six3* (Figs. 3.3D and 3.3E) and *Prox1* in the specified surface ectoderm is dependent on Pax6 function. Expression of the novel *forkhead* gene *FoxE3* in the head ectoderm and midbrain appears regulated by Pax6 (Brownell et al., 2000). Lastly, defects in retinoic acid signaling (Enwright and Grainger, 2000) and lack of expression of *secreted Frizzled-related protein-2*, a component of the Wnt signaling pathway (Wawersik et al., 1999), have been reported in the head ectoderm of Pax6 mutant mouse embryos, indicating that these signaling pathways may be involved in the pathogenesis of the Pax6 eye phenotype.

Upstream regulators of the vertebrate *Pax6* gene remain to be properly characterized. The secreted factor Bmp7, which is expressed in the ectoderm of the developing eye and in the optic vesicle, appears to be required for the maintenance of *Pax6* expression in the lens placode, although the initial induction of *Pax6* expression is independent of Bmp7 (Wawersik et al., 1999). Conversely, it has been shown that *Pax6* function is not required for either *Bmp4* or *Bmp7* to be expressed in the optic vesicle or the surface ectoderm (Furuta and Hogan, 1998; Wawersik et al., 1999). The transcription factor *Six3*, which shares sequence homologies with the *Drosophila so* gene, may also regulate *Pax6* gene expression (Ashery-Padan et al., 2000; Loosli et al., 1999).

Effects of *Pax6* Mutations and Overexpression. The effects of *Pax6* mutations on eye development have been extensively studied. Spontaneous *Pax6* mutations have been described in mice (*Small eye [Sey]*; Hill et al., 1991; Hogan et al., 1986) and rats (*rSey*; Fujiwara et al., 1994), and both conventional (St-Onge et al., 1997) and tissue-specific (Ashery-Padan et al., 2000; Marquardt et al., 2001) targeted mutations of the *Pax6* gene have been done in the mouse. *Pax6* mutant mice do not develop ocular structures, although optic vesicle remnants are present, and these mice die in the neonatal period (Hogan et al., 1988; Hogan et al., 1986). In heterozygotes, a semidominant eye phenotype is observed (Collinson et al., 2001; van Raamsdonk and Tilghman, 2000). The lens is smaller than in wild-type controls (Hogan et al., 1986). This defect is observed from early embryogenesis and is associated with a delay in lens placode formation (van Raamsdonk and Tilghman, 2000). Other eye anomalies in *Pax6* heterozygous mice include microphthalmia, vacuolated fiber cells, small or absent irises, persistent ectodermal tissue linking the cornea to the anterior lens, a thickened cornea, and cataracts in adults (Collinson et al., 2001). *Pax6* mutations leading to eye abnormalities have also been described in humans (Prosser and van Heyningen, 1998).

To examine the requirements for the *Pax6* gene during lens formation, reconstitution studies have been done in the mouse and rat. An initial study (Fujiwara et al., 1994) used the surface ectoderm and optic vesicle of wild-type and mutant rat embryos from early stages of lens development and showed that lens formation is dependent on the presence of Pax6 in the surface epithelium, since the defect was not rescued using the optic vesicle of wild-type embryos. A similar conclusion was later reached in a study using *Sey*/wild-type mouse chimeras (Quinn et al., 1996). In this study, lens development only occurred when the surface ectoderm was of a wild-type origin. The results of both studies are consistent with a lens-autonomous defect, since the lens formation defect in *Sey/Sey* embryos was not rescued by the presence of a wild-type optic vesicle. The exact nature of this defect remains to be determined. However, a more recent evaluation of chimeras between *Sey/Sey* and wild-type embryos showed that *Pax6*$^{-/-}$ cells are nearly completely removed from the lens placode at stage E9.5 (Collinson et al., 2000). This process is cell autonomous, and the behavior of *Pax6*$^{-/-}$ cells in chimeras suggests a change in the adhesive properties of mutant cells. Interestingly, the same defect was also demonstrated using *Pax6*$^{+/-}$ cells in chimeras (Collinson et al., 2001). *Pax6*$^{+/-}$ cells were less readily incorporated in the lens placode than wild-type cells, while heterozygous cells contributed evenly to the formation of the remainder of the embryo. The latter findings would suggest that the abnormalities in the heterozygous eye arise as a consequence of a primary lens defect.

The analysis of the behavior of Pax6 mutant cells in chimeras has been complemented by experiments involving the tissue-specific deletion of Pax6 function in the lens and in the retina (Ashery-Padan et al., 2000; Marquardt et al., 2001). The results of these studies indicate that Pax6 has multiple roles in lens and eye development. For example, it is implicated in (1) the maintenance of lens competence, (2) the interaction between the lens placode and the optic vesicle, and (3) the subsequent differentiation of the lens and retina. These multiple roles are, in part, illustrated by the consequences of the tissue-specific deletion of Pax6 function in the lens placode (Figs. 3.3B, 3.3C, 3.3F, and 3.3G) (Ashery-Padan et al., 2000). Lens development in the resulting mouse embryos does not progress beyond the lens placode stage, and multiple fully differentiated retinas form, likely as a failure of the prospective lens to instruct the correct placement of the retina in the eye.

Experiments involving the ectopic expression of *Pax6* have been done in *Xenopus laevis*, and these have resulted, under certain experimental conditions, in the appearance of ectopic lenses and eyes (Altmann et al., 1997; Chow et al., 1999). Under other experimental conditions, *Pax6* overexpression does not lead to ectopic formation of eye structures but increases the effect of other eye-regulatory genes (Zuber et al., 1999). The effects of *Pax6* were found to be cell autonomous (Altmann et al., 1997; Chow et al., 1999), and the activation of *Six3* and other eye-specific developmental genes was demonstrated (Chow et al., 1999), suggesting a conservation of mechanisms regulating eye formation in vertebrates and in *Drosophila* (Wawersik et al., 2000). Results from both gain- and loss-of-function experiments therefore confirm that Pax6 acts as a key transcriptional regulator of eye and lens formation in vertebrate species.

3.3.1.2. Six Genes

The transcription factor *Six3* (Oliver et al., 1995a) was initially isolated in a screen for mouse homologs of the *Drosophila so* gene, which is essential for eye formation in the fly

(Cheyette et al., 1994). In addition to the homeodomain, *so* encodes a 60–amino acid motif called the *Six* domain, and both domains are present in all members of the *Six* gene family. Coincidentally, 6 members of this gene family are found in most vertebrate species (reviewed in Kawakami et al., 2000). According to the degree of similarity of their homeodomain and *Six* domain, *Six* genes have been separated into three groups (Seo et al., 1999). Members of the first two groups, which include *Six1/2* and *Six4/5*, have a high nucleic acid homology to *so* and are expressed preferentially in muscles and limb tendons (Ohto et al., 1998; Oliver et al., 1995b), although *Six2, 4,* and *5* are also expressed in the eye (Kawakami et al., 1996; Niiya et al., 1998; Winchester et al., 1999). A second group of genes includes *Six3* and *Six6* (also called *Optx2*) (Toy et al., 1998). These have a lesser homology to *so* and are preferentially expressed in the eye (Lagutin et al., 2001; Oliver et al., 1995a; Toy and Sundin, 1999; Toy et al., 1998). This apparent contradiction was further complicated when it was realized that *Six3* and *6* both have a high nucleic acid homology to the *Drosophila optix* gene (Toy et al., 1998). Recent studies have indicated that *Six6* appears to represent a vertebrate ortholog of the *optix* gene (Seimiya and Gehring, 2000). However, the precise evolutionary relationship between *Six3*, *so*, and *optix* remains unclear (Halder et al., 1998; Lagutin et al., 2001; Seimiya and Gehring, 2000).

Expression of *Six3* in the developing eye and lens has been extensively studied in several species (Kawakami et al., 2000). In the mouse embryo (Lagutin et al., 2001; Oliver et al., 1995a), *Six3* is first expressed at E7.0–7.5 in the anterior neural ectoderm. At E8.0–8.5, *Six3* transcripts are detected in the developing retinal field, and at approximately E9.0, *Six3* is strongly expressed in the head ectoderm region that will give rise to the lens placode. From this stage onward, *Six3* expression in the developing eye closely overlaps that of *Pax6* (Oliver et al., 1995a). *Six3* is expressed in the lens pit and vesicle and the developing neuroretina. From E13.5 onward, *Six3* mRNA remains confined to the anterior epithelial cells and the transition zone, and the differentiated lens fiber cells do not contain *Six3* transcripts. At these later stages, *Six3* expression in the retina is confined to the inner layer. Although the ocular expression of the *Six6* gene has been well studied, it has not been found to be expressed in the lens in most species so far evaluated (Toy and Sundin, 1999; Toy et al., 1998).

A role for *Six3* in eye development has been supported by studies involving gain- or loss-of-function approaches. Overexpression of *Six3* in medaka fish results in the formation of an ectopic lens and retina (Loosli et al., 1999; Oliver et al., 1996), indicating a decisive role for *Six3* in vertebrate eye development. Furthermore, misexpression of *Six3* in fish results in the ectopic expression of *Pax6* (Loosli et al., 1999), which is analogous to the pattern of gene activation observed in *Drosophila* following the ectopic expression of *so*. Reciprocally, *Pax6* may regulate *Six3* expression, since Six3 protein is not detected in the lens placode following the tissue-specific deletion of Pax6 function in this structure (Ashery-Padan et al., 2000) (Figs. 3.3D and 3.3E). The precise developmental role of *Six3* remains unclear, pending the results of the targeted deletion of this gene. However, the consequences of *Six3* overexpression in zebrafish embryos (Kobayashi et al., 1998) and the phenotypes identified in patients with *SIX*3 gene mutations (Wallis et al., 1999) would indicate an essential role for this gene in the formation of the eye and of the central nervous system.

The *SIX*5 gene has been implicated in the pathogenesis of the human disease myotonic dystrophy (Klesert et al., 2000; Sarkar et al., 2000; Winchester et al., 1999). This autosomal dominant disease is associated with a CTG trinucleotide repeat expansion in the 3′ UTR of the *DMPK* gene. Altered expression of neighboring genes, such as *SIX5*, has been proposed to explain the features of the disease, since targeted deletion of the murine ortholog of

DMPK does not reproduce all the clinical features of myotonic dystrophy (Jansen et al., 1996; Reddy et al., 1996). Targeted deletion of the murine *Six5* gene has shown that mutant mice are viable and that both $Six5^{-/-}$ and $Six5^{+/-}$ animals develop cataracts (Klesert et al., 2000; Sarkar et al., 2000). Other features of the human disease, however, were not observed in mutant mice. These findings establish that *SIX5* deficiency does contribute to the cataract phenotype in myotonic dystrophy, and they imply that additional genes participate in the pathogenesis of the disease.

3.3.1.3. Prox1

The *Prox1* gene, a vertebrate homolog of the *Drosophila* panneural patterning gene *prospero*, encodes for a novel class of homeodomain proteins (Oliver et al., 1993).

Prox1 is highly conserved in all species from which it has been isolated so far, among others zebrafish, *Xenopus*, chicken, mouse, and human. Its expression is also similar in these species in several developing organs, including the eye, CNS, liver, and lymphatic system. In newts, Prox1 protein is apparently the first transcription factor to show a specific regulation in the intact dorsal epithelial cells of the iris and during the process of lens regeneration (see chap. 12; see also Del Rio-Tsonis et al., 1999).

In the mouse eye, *Prox1* is first detected specifically in the lens placode at E9.5. After E10.5, it is expressed in the lens vesicle, and by E12.5 onward in the anterior epithelium, the lens fiber cells, the surface ectoderm, and the developing retina (Glasgow and Tomarev 1998; Oliver et al., 1993; Tomarev et al., 1996; Wigle et al., 1999). *Prox1* expression is maintained in the lens and retina at postnatal stages (Tomarev et al., 1998).

Prox1 mutant mice die around E14.5 due to a defect in lymphatic vessel development. These mice show a lens differentiation phenotype consisting of a defect in fiber cell elongation that is associated with the specific absence of γB- and γD-crystallin, while other crystallins are expressed (Wigle et al., 1999). Consistent with the absence of γD-crystallin in *Prox1* mutant mice, a Prox1-responsive element was recently identified on the murine γD-crystallin gene *crygd* that is activated in vitro by Prox1 (Lengler et al., 2001). Additionally, lens fiber cells of mutant mice fail to express the cell-cycle inhibitors $p27^{KIP1}$ and $p57^{KIP2}$ and produce E-cadherin, a marker for epithelial cells that is down-regulated in wild-type lens fiber cells (Wigle et al., 1999).

In summary, Prox1 seems to be required for differentiation from the lens vesicle stage to lens fiber cells and their elongation but not for the initial induction of lens fiber formation.

3.3.1.4. Pitx3

The *Pitx* genes encode bicoid-related homeobox transcription factors thus far found in frogs, mice, and humans. These genes were first isolated through binding assays of pituitary-specific genes (*Pit*uitary homeobo*x* genes) and as a candidate gene for Rieger's syndrome (Semina et al., 1997). Members of this family are expressed in pituitary, midbrain, eye, and mesenchyme. Although Semina et al. (2000) mentioned as unpublished data that *Pitx1* and *Pitx3* overlap in the developing lens, so far *Pitx3* is the only member of this gene family shown to be expressed in the lens. *Pitx3* is present from E10.0 in the thickening lens placode and later on in the lens vesicle. After E14, *Pitx3* is abundant in the lens epithelial cells and in fiber cells, with highest levels in the bow regions of the lens (Semina et al., 1998; Semina et al., 2000; Semina et al., 1997).

Although the exact function of *Pitx3* still has to be determined, several lines of evidence suggest an important functional role for this gene in lens development. First, the natural mouse mutant *aphakia* (*ak*), which lacks lens differentiation after E11, seems to be a *Pitx3* gene double-deletion mutant, since it was demonstrated that these mice have a *Pitx3* enhancer deletion and abnormal *Pitx3* expression (Semina et al., 2000), and Rieger and collaborators (2001) found a deletion of *Pitx3* exon 1 in the same mutant. Second, human *PITX3* is mutated in two human sydromes that manifest cataracts (anterior segment mesenchymal dysgenesis and autosomal dominant congenital cataracts; Semina et al., 1998). Last, an analysis of *aphakia* mutant mice revealed the cell-autonomous function of *Pitx3* (Liegeois et al., 1996).

3.3.1.5. Otx1 and Otx2

Otx genes are vertebrate homologs of the *Drosophila* anterior segmentation gene *orthodenticle* (*otd*). The three members of this gene family, *Otx1*, *Otx2*, and *Crx*, contain a bicoid-like homeodomain (Finkelstein and Boncinelli, 1994) and have been isolated in *Xenopus*, chicken, mouse, and human. *Otx1* and *Otx2* are relevant for lens development whereas *Crx* is only involved in photoreceptor cell differentiation (Chen et al., 1997b; Furukawa et al., 1997).

In the mouse, *Otx2* expression starts at the onset of gastrulation and becomes later restricted to midbrain and forebrain neuroepithelium, including the eye domain. From these stages onward, it widely overlaps with *Otx1* expression (Simeone et al., 1992; Simeone et al., 1993). In the eye, both genes are expressed in the dorsal optic vesicle and the surface ectoderm (Martinez-Morales et al., 2001). Later, Otx1 and Otx2 transcripts are restricted to the retinal pigmented epithelium. *Otx2* is additionally found in postmitotic retinal neuroblasts in chick embryos (Bovolenta et al., 1997) and transiently in differentiating neural retina cells in mice (Baas et al., 2000).

Otx1 mutant mice are viable but have abnormalities of the dorsal telencephalic cortex accompanied by spontaneous epileptic seizures (Acampora et al., 1996). *Otx2* knockout mice die at the gastrulation stage and exhibit a deletion of the rostral brain (Acampora et al., 1995). Interestingly, *Otx2* heterozygous mice reveal eye defects with varying penetrance in different genetic backgrounds. These defects include lack of lens, cornea, and iris; microphthalmia; and hyperplastic neural retina and retinal pigmented epithelium (Acampora et al., 1995; Ang et al., 1996; Matsuo et al., 1995). According to their high structural similarity, *Otx1* and *Otx2* share functional equivalence and functional conservation with *Drosophila otd* and human *OTX1* and *OTX2*, as demonstrated in different null phenotype – rescuing knock-in approaches (Acampora et al., 1999b). These knock-in studies done in the mouse indicate that the divergent *Otx1* and *Otx2* null phenotypes originate mostly from their different expression domains and not from their different functions (Acampora et al., 1999a).

This apparent functional complementarity of *Otx* genes was confirmed in the context of mouse eye development. Compound $Otx1^{-/-}$ and $Otx2^{+/-}$ mutants and 30% of double heterozygous embryos had severe eye defects, including aphakia or microphthalmia and lens misplacement (Martinez-Morales et al., 2001). Low levels of αA-crystallin were detected in the lens remnant, implying a delay in or failure of lens development progression.

Experiments on *Xenopus* in which unspecified tissue was transplanted into the presumptive lens-forming region revealed that *Otx2* is activated by lens-inductive signals followed by Pax6 and Sox3, suggesting the early function of Otx2 in the induction of lens-forming competence (Zygar et al., 1998).

More general, *Otx2* is required for inducing the competence of the anterior neuroepithelium for ocular specification, as overexpression of *Pax6* or *Six3* leads to ectopic eyes only in the *Otx2*-positive anterior neuroectoderm (Chow et al., 1999; Loosli et al., 1999).

In summary, *Otx1* and *Otx2* could be directly involved in the specification of lens competence and in the control of morphogenetic movements. But it still has to be investigated whether the lens defects in the compound mutants are an autonomous *Otx* function or a secondary effect due to the misspecified neuroepithelium and the nonrelease of lens-inductive signals.

3.3.1.6. Lhx2

Lhx2, formerly also known as *LH2*, is a member of the *L*IM *h*omeobo*x* gene family characterized by the LIM domain, which mediates protein-protein interactions (Hobert and Westphal, 2000). It belongs to the LIM family subgroup of homologs of the *Drosophila apterous* gene, and it was identified in human, mouse, and rat.

Besides the eye, *Lhx2* is expressed in developing B-cells and the forebrain. In the murine eye, *Lhx2* expression starts in the early optic vesicle and remains expressed in the neural retina up to birth, where it is restricted to the inner nuclear layer (Xu et al., 1993).

Lhx2 null mutant mice die in utero due to severe anemia. These mutants are anophthalmic because the development of the eye arrests prior to optic cup formation (Porter et al., 1997). The surface ectoderm of those eyes are devoid of *Pax6* expression, which leads to the assumption that Lhx2 controls the induction or maintenance of *Pax6* in the surface ectoderm by signaling molecules released from the optic vesicle.

3.3.1.7. Rx1

Rx1, also called *Rax* for *r*etinal and *a*nterior neural fold homeobo*x* gene (Mathers and Jamrich, 2000), contains a *paired*-like homeodomain and is highly conserved between different species (Eggert et al., 1998; Mathers et al., 1997). In mice, *Rx* defines the eye field by E8.5 and is later expressed in the neural retina. *Rx* knockout mice show the earliest eye phenotype found so far, as they do not evaginate an optic vesicle from the neuroectoderm, which is even formed in *Pax6* and *Lhx2* null mutations (Mathers et al., 1997). Overexpression of *Rx1* in *Xenopus* embryos induces hyperproliferating neural retina and retinal pigmented epithelium, while the lens fails to differentiate properly (Mathers et al., 1997; Mathers and Jamrich, 2000). Whether this demonstrates a direct effect of Rx1 on lens differentiation as a result of a loss of lens inductive signals or whether the lens ectoderm loses its bias toward retinal differentation remains to be determined.

3.3.2. Sox Genes

Sox genes were initially isolated on the basis of their homology to a DNA-binding motif, the high mobility group (HMG) box, present in the sex-determining factor Sry (reviewed in Kamachi et al., 2000; Pevny and Lovell-Badge, 1997; Wegner, 1999). *Sox* (*S*ry b*ox*) genes have been isolated from several species, and they appear in general to be involved in the developmental processes leading to cell fate decisions (Kishi et al., 2000; Pevny et al., 1998). *Sox* genes are classified according to a number of criteria, most importantly according to the similarity of their HMG box (Wegner, 1999); the *Sox* gene family has been subdivided into seven groups, A to G, group A belonging to Sry. HMG domains interact with the minor DNA groove and therefore differ from most other DNA-binding domains, which instead bind to the major DNA groove. The interaction of the HMG domain with the minor groove causes a sharp bend of the bound DNA, and it has been hypothesized that HMG proteins may have a general role as architectural components in the assembly of multiprotein transcription factor complexes (Werner and Burley, 1997).

The closely related Sox1, 2, and 3 proteins, which make up the group B1 Sox proteins (Uchikawa et al., 1999), are expressed in an overlapping fashion during lens development (Kamachi et al., 1998; Nishiguchi et al., 1998). In chick embryos, *Sox2* is detected in a wide domain of the head ectoderm prior to lens induction. After the optic vesicle is apposed to the head ectoderm, *Sox2* expression is strongly up-regulated in the vesicle-facing head ectoderm, and *Sox3* is first activated. *Sox2* and *3* are then strongly expressed in the lens placode. In contrast, in chick embryos, *Sox1* is initially expressed at the lens pit stage. Chicken *Sox1*, *2*, and *3* genes are expressed in the lens vesicle, and at late embryonic stages their transcripts are mainly detected in the fiber cells. Group B1 *Sox* genes are also expressed in other areas of the developing embryo, in particular, the nervous system (Kishi et al., 2000; Pevny et al., 1998). Species-related variations in the expression patterns of group B1 *Sox* genes have been reported, indicating a functional redundancy between *Sox* genes in different species (Koster et al., 2000; Nishiguchi et al., 1998; Zygar et al., 1998).

In addition to their DNA-bending ability, Sox proteins participate in sequence-specific gene activation. Studies involving Sox proteins belonging to different subfamilies indicate that these proteins require tissue-specific partners for their transactivating effect. This interaction permits the high-affinity DNA-binding of the Sox protein on the regulatory DNA sequences, resulting in transcriptional activation (Kamachi et al., 1999; Kamachi et al., 2000). A well-studied example involves the activation of the mouse γF-crystallin and chicken δ1-crystallin genes by the Sox1, 2, and 3 proteins (Kamachi et al., 1995; Kamachi et al., 1998; Nishiguchi et al., 1998). δ1-Crystallin is the earliest marker of lens differentiation in the chick, and the evaluation of factors that transactivate this gene has proven critical in understanding this process. In the case of the group B1 Sox proteins, the *in vitro* activation of the δ1-crystallin minimal (DC5) enhancer requires the participation of a cofactor called δEF3 (Kamachi et al., 1995; Kamachi et al., 1998). Activation of this minimal enhancer, which consists of a 30-bp DNA fragment, is dependent on the protein-protein interaction between the Sox protein and δEF3. δEF3 has recently been isolated, and this partner protein appears to be Pax6 (Kamachi et al., 2001); *Pax6* is coexpressed with *Sox2* and *3* in areas of δ1-crystallin expression (Kamachi et al., 1998). The Pax6 protein also binds to 2 additional functional sites, both located approximately 50 bp upstream of the DC5 enhancer (Cvekl et al., 1995b). These additional Pax6 sites appear dispensable for the core activity of the enhancer but may be important for the full activity of the δ1-crystallin enhancer (Cvekl et al., 1995b; Kamachi and Kondoh, 1993). Additional factors such as *Maf* genes, which have specific expression in the lens and have been shown to activate lens-specific genes, including the δ1-crystallin gene, could participate with group B1 Sox proteins and Pax6 to provide tissue-specific expression of crystallin genes (Kondoh, 1999; Ogino and Yasuda, 2000).

Several lines of evidence indicate that *Sox* genes are activated from an inductive signal from the optic cup and that these genes are required for the subsequent differentiation and formation of the lens (Kondoh, 1999). Chicken *Sox2* and *3* are activated in the region of ectoderm directly apposed to the optic vesicle, and after ablation of the optic vesicle in chick embryos, *Sox2* and *3* are not expressed and the lens does not form (Kamachi et al., 1998). Similar results were also obtained following the ablation of the optic vesicle in *Xenopus* embryos (Zygar et al., 1998). Clues to the possible identity of this inductive signal have come from the study of *Bmp4* homozygous mutant mouse embryos (Furuta and Hogan, 1998). *Bmp4* is strongly expressed in the optic vesicle and to a lesser extent in the head ectoderm of embryos prior to lens induction. In *Bmp4* null embryos, *Sox2* expression is not up-regulated, and the lens does not form, although the presumptive lens ectoderm contacts the optic vesicle in a normal fashion. Interestingly, inserting BMP4 beads into a mutant optic vesicle in an explant culture restores *Sox2* expression in the presumptive lens ectoderm and allows the lens to form. Substituting BMP4 for the optic vesicle, however, does not restore *Sox2* expression and does not rescue the lens defect, implying that the inductive signal corresponds to a Bmp4-responsive factor originating from the lens vesicle (Furuta and Hogan, 1998).

Evidence for the role of *Sox* genes in lens formation and differentiation has come, in part, from misexpression studies. When electroporated in the ectoderm of chick embryos, Sox2 activates δ-crystallin gene expression and leads to morphological changes suggestive of lens placode formation (Kamachi et al., 2001). These effects, however, are only observed when Sox2 was expressed in combination with Pax6, suggesting that these factors form protein complexes on lens-specific DNA sequences to trigger the process of lens formation. Furthermore, when overexpressed in the head ectoderm of medaka fish embryos, *Sox3* leads to the formation of ectopic lenses in a cell-autonomous fashion (Koster et al., 2000). These morphological changes were accompanied by the ectopic activation of *Pax6* gene expression.

Despite the apparent importance of the *Sox2* gene in lens formation, the targeted disruption of this gene in mice has not helped to reveal its role in lens development, since it causes a preimplantation lethal phenotype (Pevny et al., 1998). However, the targeted deletion of the *Sox1* gene has shown that *Sox1* mutant mice are viable and have microphthalmia and cataracts (Nishiguchi et al., 1998). Analysis of the lens phenotype revealed a failure of lens fiber cell elongation associated with a down-regulation of γ-crystallin genes. These defects are consistent with the late onset of *Sox1* gene expression in mice (Kamachi et al., 1998) and with its proposed effect in the regulation of γ-crystallin gene expression (Nishiguchi et al., 1998). Up-regulation of *Sox2* expression was not observed in the lens of mutant mice, and therefore a possible functional redundancy related to these genes does not account for the late onset of the phenotype.

3.3.3. Winged Helix/Forkhead Genes

Transcription factors belonging to the winged helix/forkhead class are characterized by a 100–amino acid DNA-binding domain named the *winged helix domain*, and over 100 members of this gene family have been identified since the isolation of the first member, the *Drosophila forkhead* gene (Kaestner et al., 2000; Kaufmann and Knochel, 1996). Recently, novel members of the winged helix/forkhead gene family showing a highly restricted expression in the developing lens have been described (Blixt et al., 2000; Kenyon et al., 1999). In *Xenopus*, the *Xlens1* gene (Kenyon et al., 1999) represents an early marker of lens formation and would function to maintain the specified ectoderm in an undifferentiated state. The related *FOXE3* and *FoxE3* genes were subsequently reported in humans and mice (Blixt et al., 2000; Brownell et al., 2000; Semina et al., 2001). Expression of *FoxE3* in the mouse lens is first detected at E9.5 (Blixt et al., 2000; Brownell et al., 2000). Following lens vesicle formation, *FoxE3* expression ceases in lens fiber cells but persists in the anterior epithelial cells. Transient *FoxE3* expression is also detected in the developing nervous system (Blixt et al., 2000; Brownell et al., 2000).

FoxE3 has been shown to map to the *dysgenetic lens* locus (Blixt et al., 2000; Brownell et al., 2000). This autosomal recessive phenotype is characterized by a fusion between lens and cornea and an arrest of development of anterior epithelial and lens fiber cells (Sanyal and Hawkins, 1979). Analysis of these mutants suggests that *FoxE3* is essential for the proliferation and survival of the anterior epithelium (Blixt et al., 2000; Brownell et al., 2000). Anterior segment defects and cataracts have also been reported with FOXE3 mutations affecting the C-terminal portion of the protein, providing added proof for the role of this gene in lens formation (Semina et al., 2001).

3.3.4. Basic Region/Leucine Zipper (bZIP) Genes

3.3.4.1. Maf *Genes*

Proteins encoded by the *Maf* gene family are basic region-leucine zipper factors that contain a bZIP domain responsible for DNA-binding and protein-protein interactions (reviewed in Blank and Andrews, 1997). The founding member of this gene family, the avian oncogene *v-maf* (Nishizawa et al., 1989), can efficiently form homo- or heterodimers with other bZIP factors through their leucine zipper structures, the dimers differing in their DNA-binding specificity (Kataoka et al., 1994b). Similar dimerization studies have been carried out with different members of the *Maf* gene family (Kataoka et al., 1994a; Kerppola and Curran, 1994a, 1994b; Matsushima-Hibiya et al., 1998), and a potentially wide repertoire of transcriptional regulation appears possible through *Maf* family genes. The consensus DNA-binding sequence of *Maf* genes has been established (MARE; Kataoka et al., 1994b; Kerppola and Curran, 1994b), and these genes may prove to be important regulators of cell differentiation (Blank and Andrews, 1997).

Maf genes are subdivided into large and small *Maf* subfamilies depending on the presence of a putative activation domain at the protein N-terminus, which is absent in small Maf proteins (Blank and Andrews, 1997). To date, four members of the large *Maf* gene subfamily have been reported to be expressed in the developing lens. Although *Maf* genes are generally expressed in several tissues, the large *Mafs*, *L-maf* (Ishibashi and Yasuda, 2001; Ogino and Yasuda, 1998) and *c-maf* (Kawauchi et al., 1999; Ring et al., 2000), are almost exclusively expressed in the lens and are therefore likely to contribute to the tissue-specific expression of lens genes (Kondoh, 1999; Ogino and Yasuda, 2000). *L-maf* (lens-specific maf; Ogino and Yasuda, 1998) is a *Maf* gene that was isolated on the basis of its binding to a motif (αCE2) contained in the chicken αA-crystallin enhancer. *L-maf* appears to be a potent activator of the αA-crystallin promoter, since its activity is abolished in αA-crystallin reporter gene constructs following mutations of αCE2. In the chick embryo, *L-maf* transcripts are first detected at the lens placode stage and precede the expression of δ1-crystallin (Ogino and Yasuda, 1998). Expression is later localized to the developing lens vesicle. Although it has been isolated in the chicken (Ogino and Yasuda, 1998) and *Xenopus* (Ishibashi and Yasuda, 2001), a *L-maf* homolog remains to be identified in the mouse (Kawauchi et al., 1999). When electroporated in chick embryos, *L-maf* was found to ectopically induce the expression of δ1-crystallin in embryonic head ectoderm, suggesting a role for *L-maf* in lens induction and differentiation (Ogino and Yasuda, 1998). Similar results were recently obtained following injection of XL-maf RNA into *Xenopus* (Ishibashi and Yasuda, 2001).

Another member of the large *Maf* subfamily, the mouse *c-maf* gene (Kataoka et al., 1993), is first expressed in the head ectoderm and later in the lens placode and vesicle (Kawauchi et al., 1999; Ring et al., 2000). At later stages, it is expressed strongly in the fiber cells. It has also been shown to transactivate the γF-crystallin promoter in transient reporter gene assays (Kim et al., 1999). The mouse *c-maf* gene has been disrupted (Kawauchi et al., 1999; Kim et al., 1999; Ring et al., 2000), and mutant embryos have small and hollow lenses due to a failure of lens fiber cell elongation. This defect is apparent at E12, indicating redundant functions of the *c-maf* gene in the lens before this stage. In mutant embryos, expression of members of the γ-crystallin gene cluster was severely reduced (Kawauchi et al., 1999; Kim et al., 1999; Ring et al., 2000). The failure of lens differentiation in the absence of *c-maf* suggests that this gene directly activates several classes of crystallin genes (Kim et al., 1999; Ring et al., 2000). However, additional roles for *c-maf* in lens formation cannot be ruled out.

Additional members of the large *Maf* subfamily that are expressed in the lens include *NRL* (neural retina leucine-zipper) and *mafB/Kreisler*. *NRL* is expressed in the anterior epithelial cells of the lens and in the developing retina (Liu et al., 1996). The closely related rat *mafl* (Yoshida et al., 1997), mouse *mafB/Kreisler* (Cordes and Barsh, 1994), and *Xenopus XmafB* (Ishibashi and Yasuda, 2001) genes are principally expressed in anterior epithelial lens cells. Further evaluation remains to be done regarding the role of these genes in lens development. No lens defect has been reported in the mouse mutant *Kreisler* (Cordes and Barsh, 1994), although *Kreisler* could represent a hypomorphic allele (Eichmann et al., 1997).

3.3.4.2. CREB-2

CREB-2 is a member of the CREB/ATF family of mammalian transcription factors (Hai et al., 1989; Ivashkiv et al., 1990; Karpinski et al., 1992; Tsujimoto et al., 1991; Vallejo et al., 1993). Members of this gene family show a conserved bZIP domain and can form homo- or heterodimers with other CREB/ATF proteins or with related bZIP proteins (Hai et al., 1989; Ivashkiv et al., 1990). Although CREB-2 is ubiquitously expressed, CREB-2 deficient mice are viable and have a phenotype that is restricted to the lens (Hettmann et al., 2000; Tanaka et al., 1998). In mutant mice, anterior epithelial cells undergo apoptosis at approximately stage E14.5, followed by resorption of the developing lens. This defect is p53-dependent and is rescued by breeding CREB-2 mutant mice with p53-deficient animals (Hettmann et al., 2000). These findings implicate CREB-2 as an essential survival factor for anterior epithelial cells. Further studies should determine the cell survival pathways regulated by CREB-2 in these cells.

3.3.4.3. Sw3-3

Sw3-3 is a basic leucine zipper-like gene that differs from other members of the bZIP family of transcription factors because of the presence of two leucine zipper motifs flanking the basic domain (Wang and Adler, 1994). In chick embryos, transcripts are first detected in the lens placode and later in the anterior epithelial cells but not in the lens fiber cells. *Sw3-3* is also strongly expressed in the developing retina and in the chick neural tube. Further work should determine the function of this gene.

3.3.4.4. c-fos and c-jun

Both c-fos and c-jun have been detected in the developing lens (Spiewak Rinaudo and Zelenka, 1992). The precise role that these transcription factors play in the developing lens remains to be determined.

3.3.5. Basic Region/Helix-Span-Helix Genes

The *AP2* gene family consists of three genes (*AP2α*, *β*, and *γ*) that contain a basic DNA-binding domain that overlaps with a helix-span-helix dimerization domain (Bosher et al., 1996; Chazaud et al., 1996; Moser et al., 1995; Williams et al., 1988). AP2 proteins can form homo- or heterodimers between themselves or with fos (Williams and Tjian, 1991). *AP2α* and *β* are expressed in the lens placode and in anterior epithelial lens cells and in other organs, while *AP2γ* is not expressed in the developing eye (Ohtaka-Maruyama et al., 1998a; West-Mays et al., 1999). Recent studies have demonstrated that 4 of the 5 isoforms of the *AP2α* gene are expressed in the lens (Ohtaka-Maruyama et al., 1998a). These isoforms encode activating or repressor forms of the AP2α protein. AP2α has been shown to bind to the MIP gene promoter (Ohtaka-Maruyama et al., 1998b), and putative binding sites have been identified in upstream regions of the *Pitx3* gene (Semina et al., 2000).

Varying degrees of eye defects are seen in *AP2α* mutant embryos (Nottoli et al., 1998; Schorle et al., 1996; West-Mays et al., 1999; Zhang et al., 1996), ranging from anophthalmia to lens defects (e.g., adhesion of the lens to the overlying surface ectoderm). The precise role of AP2 in lens development remains to be determined, although a possible role in the regulation of cell adhesion has been proposed (West-Mays et al., 1999).

3.3.6. Others

3.3.6.1. Eya1 and Eya2

Eya genes are mammalian orthologs of the *Drosophila eya* gene, which encodes a novel nuclear protein essential for eye formation in the fly (Bonini et al., 1993). The *eya* gene forms transcriptional complexes with *dac* or with *so* (Chen et al., 1997a; Pignoni et al., 1997) and can direct ectopic eye formation in *Drosophila* (Bonini et al., 1997). Four *eya* homologs have been isolated in the mouse (Borsani et al., 1999; Xu et al., 1997a; Xu et al., 1997b), and transcripts from both *Eya1* and *Eya2* have been detected in the developing lens (Xu et al., 1997b). Whether Eya proteins form complexes with the Six3 protein in the lens remains unclear. However, both *in vitro* and *in vivo* assays suggest that Eya proteins may interact with other Six proteins (Heanue et al., 1999; Ohto et al., 1999).

The role of *Eya* genes in eye formation remains to be determined. The *EYA1* gene is mutated in the human branchio-oto-renal syndrome, in which the eye is unaffected (Abdelhak et al., 1997), and *Eya1*-targeted mice do not display an eye phenotype (Xu et al., 1999b). However, anterior segment defects and cataract formation have been reported in patients with *EYA1* gene mutations (Azuma et al., 2000), suggesting of a role for this gene in human eye formation.

3.3.6.2. Retinoic Acid Receptors

A role for retinoic acid (RA) signaling in lens formation has been suggested by experiments in which antisense oligonucleotides to retinol-binding protein mRNA were injected in yolk sacs of cultured 10- to 12-somite stage mouse embryos (Bavik et al., 1996). RA-depleted embryos did not form lens placodes, and adding RA to the medium restored normal eye development, thereby demonstrating that the defect was due to RA-deficiency. Lens fiber abnormalities and aphakia have been reported in RA receptor double-null mutant embryos, although at a low frequency (Lohnes et al., 1994), and defects in RA signaling have been reported in the head ectoderm of *Pax6* mutant embryos (Enwright and Grainger, 2000). In addition, RA receptors may participate in the transcriptional activation of crystallin genes (Gopal-Srivastava et al., 1998; Li et al., 1997; Tini et al., 1995). In the developing eye, RA receptors appear to be principally distributed in the periocular mesenchyme (Grondona et al., 1996). Further studies should provide better understanding of the interactions between RA signaling and lens formation. See Table 3.1.

3.4. Concluding Remarks

The field of developmental eye research has witnessed substantial advances since the critical role of transcription factors in eye formation was first determined nearly a decade ago (Halder et al., 1995; Hill et al., 1991; Quiring et al., 1994; Walther and Gruss, 1991). Additional progress should occur in the next few years as additional transcription factors are found to be involved in eye and lens formation. Their discovery should be driven, in part, by information gathered from the sequencing of the human genome (IHGS Consortium, 2001; Venter et al., 2001) and the soon-to-be-completed sequencing of the mouse genome. In addition, a better understanding should soon be reached of the interactions between transcriptional regulators and other factors involved in eye development, such as secreted factors and their receptors (Kumar and Moses, 2001; Pichaud et al., 2001; Rasmussen et al., 2001) and cell cycle regulators (Zuber et al., 1999). The latter information may provide us with a basic framework for a better understanding of the development of the vertebrate eye and lens.

Part Two

The Lens

4

The Structure of the Vertebrate Lens

Jer R. Kuszak and M. Joseph Costello

4.1. Introduction

The cornea and the lens are the principal refractive elements of the eye responsible for, respectively, stationary and variable refraction. However, while both the cornea and the lens must be transparent to function properly, the basis of their transparency is quite different. In general, the cornea relies on the continuous pumping of interstitial fluid across its semipermeable surface membranes and a supramolecular organization of collagen fibrils for clarity. In contrast, lens transparency is presumed to be the result of a highly ordered arrangement of its unique fiberlike cells, or fibers, and a gradient of refractive index produced by a variable crystallin protein concentration within the fibers. However, while lens gross anatomy has a major role in determining lens optical quality (variable focusing power), lens ultrastructure is the principal factor in determining lens transparency. Furthermore, while all vertebrate lenses have a similar form, or structure, their anatomy is not identical, and thus their optical quality varies from species to species and as a function of age. In fact, on the basis of structure, four types of lenses can be distinguished. Key differences in lens morphology, caused during specific periods of development and growth, result in quantifiable variations in optical quality. Furthermore, lens structural anomalies, caused during the same periods of development and growth, result in quantifiable degradation in optical quality. Thus, vertebrate lenses are a prime example of form following function and malformation leading to malfunction.

In this chapter we describe the gross and fine structure of vertebrate lenses throughout development, growth, and aging and define the anatomical differences that result in four distinct types of lenses. In addition, we discuss how variations in the structure of lenses influence their optics, physiology, and pathology.

4.2. Lens Development

4.2.1. Primary Fiber Cell Formation

Lens formation begins early in embryogenesis. As described in chapter 1, a group of surface ectodermal cells immediately overlying the developing optic vesicle are induced to thicken and form the lens placode (Fig. 4.1). As the optic vesicle begins to fold inward and form the optic cup, the lens placode invaginates toward the developing optic cup until it eventually pinches off as the inverted (inside-out) lens vesicle. Note that the apical surfaces of the lens vesicle cells are directed toward the lumen while their basal surfaces are directed toward the outer surface. Throughout life, the basal surfaces of lens cells will produce a basement membrane that envelops the lens. This basement membrane is known as the lens capsule.

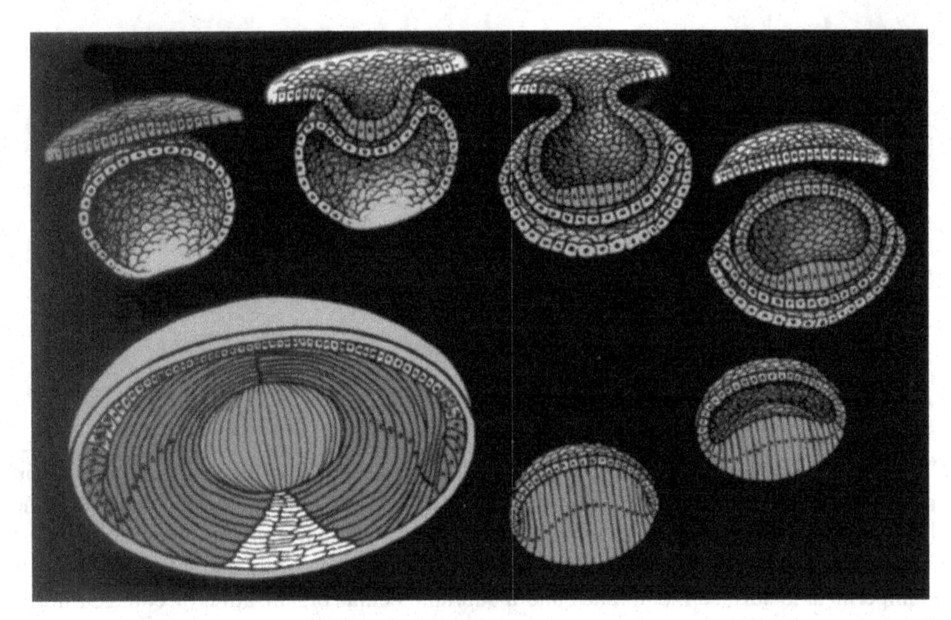

Figure 4.1. (See color plate III.) Schematic diagram of the key structural events in lens development (clockwise from top left): The thickening of the lens placode (green). The invagination of the lens placode (green) toward the developing optic cup (orange). The inverted, or "inside out," lens vesicle (green). The elongation of posterior lens vesicle cells as they terminally differentiate into primary lens fibers (green). The obliteration of the lumen of the lens vesicle by fully elongated primary fibers (green). The continuous formation of secondary lens fibers (dark blue), resulting in successive growth shells being continuously added onto the primary fiber mass (the embryonic lens nucleus; green). From Kuszak and Brown, 1994, courtesy of W. B. Saunders Co., Philadelphia, PA.

Lens development proceeds as cells of the lens vesicle approximating its retinal half are induced to terminally differentiate. A significant structural consequence of lens terminal differentiation is that the originally cuboidal lens vesicle cells are transformed into long fiber cells. The elongation of these cells proceeds until they fill in the lumen of the lens vesicle. Since these fibers are considered to be the first lens cells to be transformed into fibers, they are referred to as "primary fibers."

4.2.2. Secondary Fiber Formation

The cells of the lens vesicle that were not induced to form primary fibers remain as a monolayer of epithelial cells covering the anterior surface of the primary fiber mass. This monolayer is referred to as the lens epithelium. Throughout life, lens development and growth continues in a similar manner to other stratified epithelia, with the lens epithelium constituting the basal layer. However, whereas typically stratified epithelia have their stem cells distributed throughout the basal layer, the lens is unique in that its stem cells are sequestered as a distinct subpopulation within the lens epithelium known as the germinative zone (gz). The remainder of the lens epithelial cells are also sequestered into three additional regions known as the central zone (cz), pregerminative zone (pgz), and transitional zone (tz). The cz epithelial cells compose a broad polar cap of the lens epithelium covering most (approximately 80%) of the anterior surface of the lens. These cells are arrested in the G_0

stage of the cell cycle and are not recruited to terminally differentiate into fibers (Harding et al., 1971). The pgz cells form a narrow, latitudinal band that comprises approximately 5% of the lens epithelium and is peripheral to the cz. While a small number of these cells undergo mitotic division, only rarely are any of their daughter cells induced to terminally differentiate and become additional fibers. Rather, it has been proposed that these daughter cells add to the population of the lens epithelium as the anterior surface of the lens increases in size as a function of growth and age (Rafferty and Rafferty, 1981; Kuszak, 1997). The gz cells form a narrow, latitudinal band that comprises approximately 10% of the lens epithelium and is peripheral to the pgz. These cells undergo mitotic division, and a number of the daughter cells are induced to terminally differentiate and become additional fibers. Since these are the second fibers to develop, they are referred to as "secondary fibers." Finally, the tz cells form a narrow, latitudinal band that comprises approximately 5% of the lens epithelium and is peripheral to the gz. These cells are nascent fibers, having already begun the process of elongation.

4.2.3. Fiber Elongation

The secondary fibers compose the layers, or strata, of the lens. An understanding of the manner in which the fibers are added onto the existing lens throughout life is necessary to comprehend the complex structure of the lens paramount for the establishment and maintenance of its transparency.

As the tz cells migrate posteriorly, they rotate 90° about their polar axis and begin to elongate bidirectionally. As fiber elongation continues, the anterior ends are insinuated between the lens epithelium and the primary fiber mass while at the same time the posterior ends are insinuated between the primary fiber mass and the lens capsule (Fig. 4.2). Elongation is complete when the anterior and posterior ends of newly formed fibers break contact with, respectively, the epithelium anteriorly and the capsule posteriorly. The fact that the basal ends of all fibers detach from their basement membrane, the lens capsule, is direct evidence that the lens is a "stratified" rather than a "simple" epithelium. In simple epithelia, all cells retain contact with the basement membrane.

The anterior ends of fully elongated fibers abut and overlap with one another, as do the posterior ends, to form a growth shell rather than a simple layer or simple strata. Since secondary fiber formation, or stratification, occurs throughout life, the net result is the establishment of successive growth shells surrounding the original primary fiber mass. Thus, a cross section through a lens equator reveals concentric growth shells and/or radial cell columns formed throughout life (Fig. 4.3). The inside-out development scheme of the lens dictates that the growth shells become progressively more internalized and therefore cannot be sloughed off. In fact, all primary and secondary fibers formed are retained and must be supported for a lifetime. Failure to preserve the viability of any fiber is presumed to lead to pathology.

After fiber elongation is complete, lens fiber terminal differentiation continues as newly formed fibers routinely eliminate their nuclei, Golgi bodies, rough endoplasmic reticulum, as well as most of their smooth endoplasmic reticulum and mitochondria (Modak et al., 1968; Jurand and Yamada, 1967; Papaconstantinou 1967; Modak and Bollum 1972; Kuwabara and Imaizumi, 1974). The removal of these organelles is often described as necessary because their retention would cause a significant diffraction of light and thereby compromise lens function. However, other stratified epithelia (e.g., skin) also routinely eliminate these same organelles from cells in upper or older layers as a function of terminal differentiation.

Figure 4.2. Low-magnification scanning electron micrograph montages demonstrating the transformation of cuboidal pregerminative zone (PGZ) and germinative zone (GZ) lens epithelial cells (a, upper left) into columnar transitional zone (TZ) cells (a, lower left; b, upper left), which rotate a total of 90° about their long axis (asterisk) as they bidirectionally elongate into fibers (b, lower left). The variation in the size, shape, and extent of anterior end flare of the forming fibers is apparent as they elongate toward their sutural terminations beneath the different zones (TZ, GZ, PGZ, and central zone) of the lens epithelium. From Kuszak and Rae, 1982, courtesy of Academic Press, London, UK.

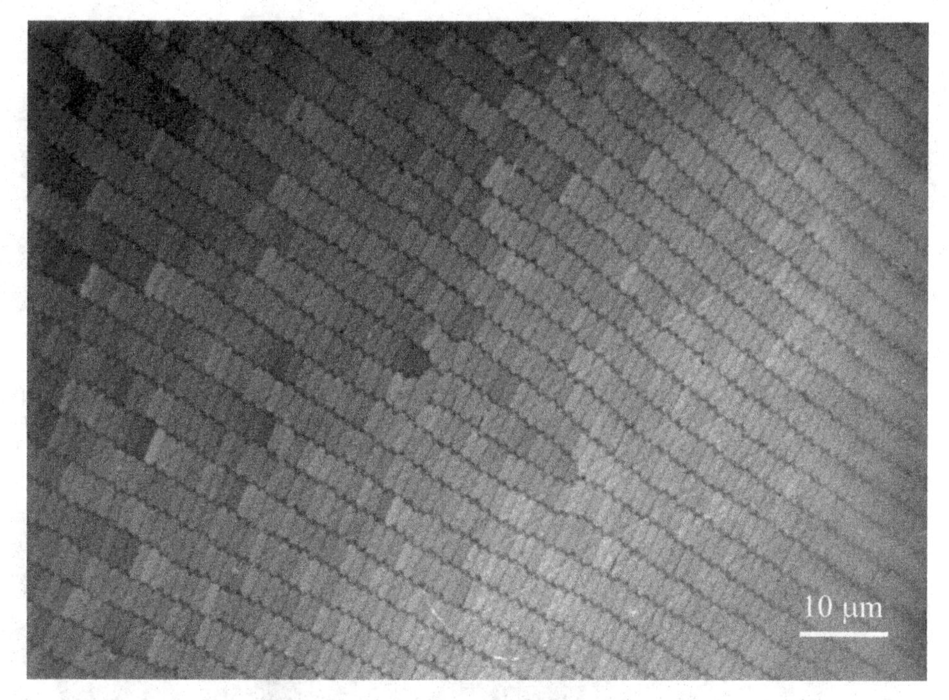

Figure 4.3. Low-magnification light micrograph of a thick-sectioned adult frog lens showing the uniformly hexagonal fibers arranged into radial cell columns and/or growth shells. The direction of the radial cell columns extends from the lower right to upper left.

4.3. Different Types of Lenses as a Function of Suture Patterns

A common misconception is that all fully elongated fibers are meridians, that is to say, crescent-shaped, fusiform cells that are widest at their midportion and taper to a point at their ends. A logical extension of this misconception is that the ends of all secondary fibers extend to confluence at the anterior and posterior lens poles. While such a description of fiber shape, length, and end-to-end organization is fairly accurate for some vertebrate lenses (most notably avian lenses), it grossly oversimplifies the intricate shape and variable length that together dictate the highly ordered arrangement of fibers necessary for a lens to function.

Fibers are not meridians. They neither extend from pole to pole, nor do they have tapered ends. In fact, in primate lenses, fiber ends are more than 3 times wider than their midportions (i.e., they are flared). Furthermore, the entire length of most fibers does not lay within a plane coincident with the visual axis defined by its equatorial location. Instead, most fibers have 2 types of curvature: the convex-concave curvature that gives to a fiber its crescent appearance and the opposite-end curvature that gives it a three-dimensional (3D) S-shape (Fig. 4.4). "Opposite-end curvature" refers to the curving of the anterior and posterior fiber segments in opposite directions, away from the polar axis, within a growth shell.

Fiber length is normally less than one-half the polar circumference of a lens. Thus, the ends of most fibers do not extend to confluence at the poles. Instead, they abut and overlap within and between growth shells as distinct latitudinal arc segments. Individually these arc segments are referred to as "suture branches," and collectively they make up defined "suture patterns" within each growth shell. Fiber shape, end taper or flare, and the extent

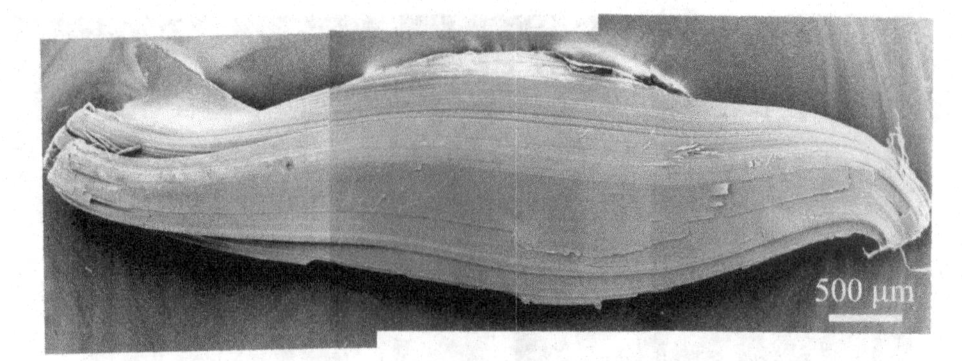

Figure 4.4. Low-magnification scanning electron micrograph montage demonstrating opposite end curvature of fibers. As a result of 3D S-shape of the fibers, their ends abut and overlap within and between growth shells to form suture branches.

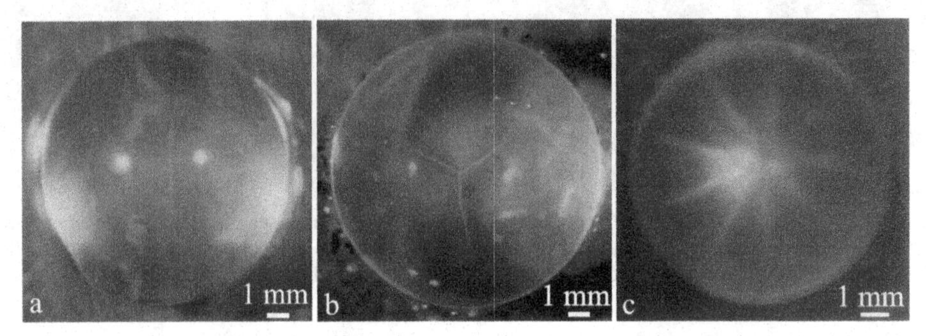

Figure 4.5. Low-magnification surgical dissecting microscope photographs of (a) a normal rabbit lens anterior line suture, (b) a normal cat lens anterior Y suture, and (c) a normal human lens star suture. From Al-Ghoul et al., 2003, courtesy of Wiley-Liss, Inc.

of opposite-end curvature combine to determine the number of suture branches and the resulting suture patterns (Fig. 4.5). These structural parameters are the basis for defining four distinct types of lenses.

4.3.1. Umbilical Suture Lenses

The "umbilical suture" lens is the simplest type of lens. This lens is typical of birds and some reptiles (e.g., snakes and turtles). The fully elongated secondary fibers of these lenses are essentially meridians. Their shape is crescent and fusiform, being widest at their middle (equatorial segment) and tapered to almost a point at their ends (Fig. 4.6a). Thus, the entire length of any of these fibers lies in a plane passed through the visual axis defined by their equatorial location. In umbilical suture lenses, each growth shell consists of essentially equally long secondary fibers whose anterior and posterior ends extend to confluence at the anterior and posterior poles, respectively (Fig. 4.6b). The sites where the fiber ends abut and overlap within and between the growth shells are referred to as the anterior and posterior "umbilical sutures." Each successive growth shell consists of essentially radially longer and uniformly shaped fibers. Thus, highly ordered radial cell columns (Fig. 4.6c) and/or concentric growth shells (Fig. 4.6d) extending from the primary fiber mass to the lens

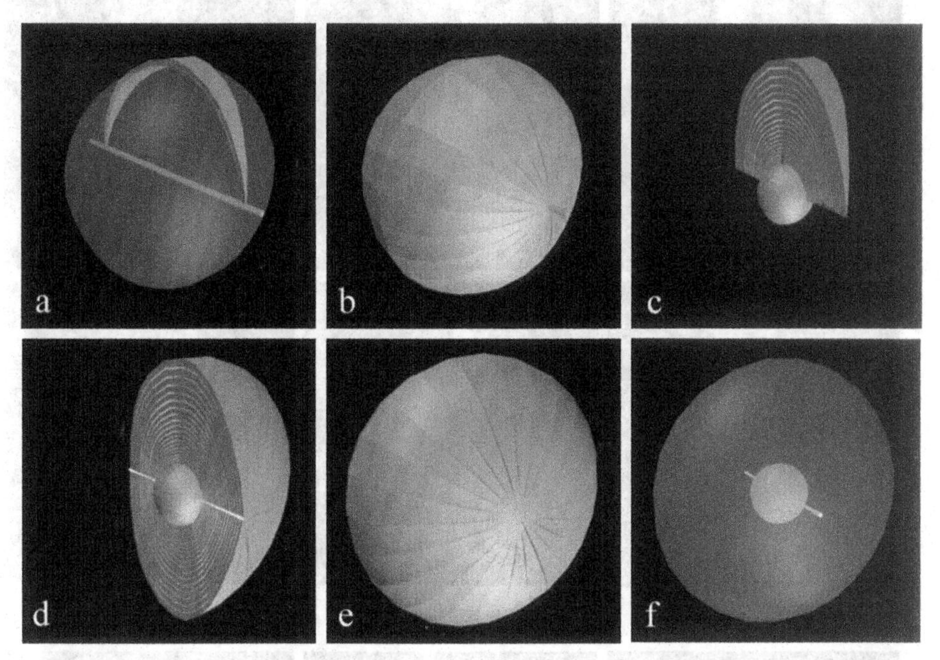

Figure 4.6. Scale 3D-CADs showing key structural elements in the formation of an umbilical suture lens. (a) A meridian secondary fiber added onto the primary fiber mass extends from pole to pole (rod positioned along the visual axis to facilitate orientation). (b) All other secondary fibers are also meridians, resulting in a growth shell with all fiber ends extending to confluence at the poles (umbilicus). As additional secondary fibers are added throughout life, radial cell columns (c) and concentric growth shells (d and e) are formed; these consist of identically shaped but longer fibers as a function of radial location. Since the ends of all secondary fibers become aligned at the poles, continuous rodlike umbilical suture planes extending from the embryonic nucleus to the lens periphery are formed.

periphery are produced. Since the umbilical sutures of each growth shell become overlain in successive growth shells (Fig. 4.6e), narrow, cylindrical umbilical suture poles that extend from the embryonic nucleus to the anterior and posterior lens periphery are also formed (Fig. 4.6f).

4.3.2. Line and Y Suture Lenses

In all other vertebrate lenses, growth shells are composed of two types of secondary fibers: straight and S-shaped fibers. A straight fiber is crescent shaped, with its entire length lying within a plane passing through the visual axis defined by its equatorial location (Figs. 4.7a_1 and a_2). However, only one end of a straight fiber extends to a pole. In frog and rabbit lenses, each growth shell normally contains 4 straight fibers (Fig. 4.7a_1). In the anatomical position, these straight fibers are normally positioned equidistantly around the lens equator beginning with the most superior equatorial location. Thus, each growth shell is effectively subdivided into four quadrants. By comparison, in mouse, rat, hamster, guinea pig, dog, cat, bovine, and sheep lenses, each growth shell normally contains six straight fibers (Fig. 4.7a_2). These straight fibers are also normally positioned equidistantly around the lens equator beginning with the most superior location. Thus, each growth shell in these lenses is effectively subdivided into sextants.

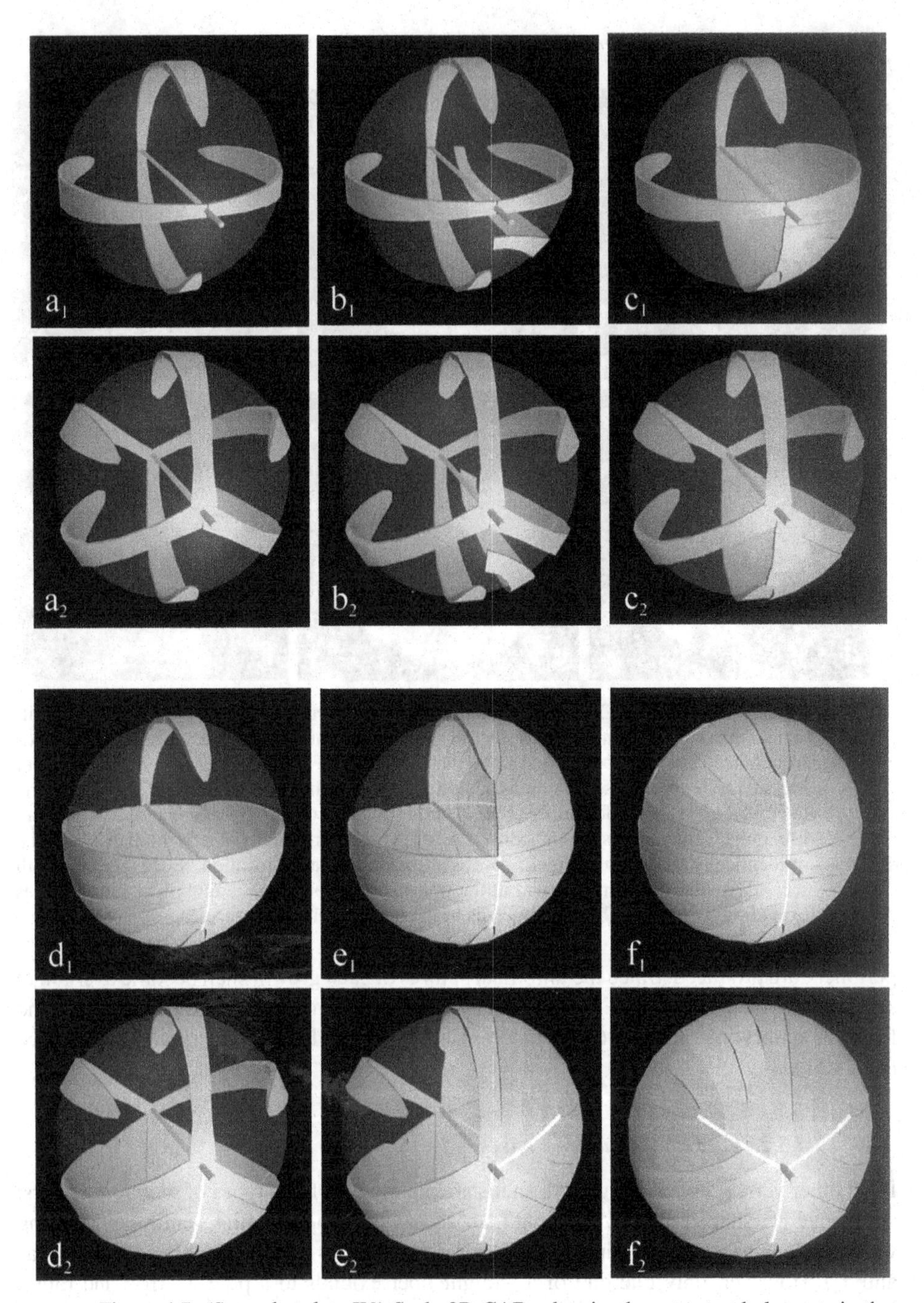

Figure 4.7. (See color plate IV.) Scale 3D-CADs showing key structural elements in the formation of line (1st and 3rd rows, a_1–f_1), and Y suture lenses (2nd and 4th rows, a_2–f_2). After forming the primary fiber mass (embryonic nucleus), four and six straight fibers (a_1, a_2), normally positioned equidistantly around the equator, separate growth shells into equal quadrants and sextants composed of S-shaped fibers (b_1, b_2, c_1, c_2). The ends of the S-shaped fibers abut and overlap to form suture branches (white lines) that extend to confluence at the poles (d_1, d_2, e_1, e_2, f_1, f_2). Note that the length and location of suture branches are defined by the ends of straight fibers. The end result is a vertical anterior line suture (f_1) or, because of opposite end curvature, a horizontal posterior line suture (partially shown in e_1), an upright Y anterior suture (f_2), or an inverted posterior Y suture (partially shown in e_2).

S-shaped fibers have a crescent shape but in addition have anterior and posterior end segments that exhibit precise curvature away from the poles in opposite directions (Figs. 4.7b_1 and b_2). Neither of the ends of an S fiber extends to a pole, nor does its entire length lie within a plane passing through the visual axis as defined by its equatorial location. S fibers are arranged as distinct groups positioned between straight fibers. Because of opposite-end curvature, within any growth shell the anterior and posterior ends of S fibers become aligned as offset anterior and posterior latitudinal arc lengths (Figs. 4.7c_1 and c_2). The anterior ends of proximal groups of S fibers abut and overlap within and between successive growth shells to produce anterior suture branches (Figs. 4.7d_1 and d_2). However, while the anterior ends of fibers in proximal groups form anterior suture branches, opposite-end curvature dictates that their posterior ends form precisely "offset" posterior branches with different groups (Figs. 4.7e_1 and e_2). The origin of anterior suture branches is defined by the ends of straight fibers that extend to confluence at the posterior pole but not to the anterior pole. Similarly, the origin of posterior suture branches is defined by the ends of straight fibers that extend to confluence at the anterior pole but not to the posterior pole. All suture branches extend to confluence at the poles and combine to form discrete anterior and posterior suture patterns (Figs. 4.7f_1 and f_2). In frog and rabbit lenses, the two anterior suture branches are normally oriented longitudinally 180° to one another to form a vertical line suture pattern (Figs. 4.5a and 4.8). It follows then that the two posterior suture branches are also normally oriented 180° to one another but because of opposite-end curvature are offset 90° to the anterior suture branches to form a horizontal line suture pattern. In mouse, rat, hamster, guinea pig, dog, cat, bovine, and sheep lenses, the three anterior suture branches are normally oriented 120° longitudinally to one another to form an upright Y suture pattern (Figs. 4.5b and 4.8). It follows then that the three posterior suture branches are also normally oriented 120° to one another but because of less opposite-end curvature are only offset 60° to the anterior suture branches, forming an inverted Y suture pattern.

As additional growth shells are added throughout life in line and Y suture lenses, each shell normally consists of radially longer, identically shaped secondary fibers (Figs. 4.8a_1 and a_2). Thus, as the line and Y suture branches of each growth shell become overlain in successive growth shells, the anterior and posterior ends of all fibers become arranged as right triangle–shaped suture planes extending from the primary fiber mass to the lens periphery (Figs. 4.8b_1–f_1 and 4.8b_2–f_2). In a line suture lens, the anterior suture planes are normally oriented 180° to one another. Thus, they combine to form a single vertically oriented, isosceles triangle–shaped suture plane. The two posterior suture planes are also normally oriented 180° to one another but because of opposite-end curvature combine to form a single horizontally oriented, isosceles triangle–shaped posterior suture plane that is offset 90° to the anterior suture plane. In a Y suture lens, the three anterior suture planes are normally oriented 120° to one another, as are the three posterior suture planes. As a result of opposite-end curvature, all six of the Y suture planes are normally offset 60° to one another.

4.3.3. Star Suture Lenses

Primate lenses also feature straight and S-shaped fibers. However, after birth and continuing throughout life, the way in which the straight and S-shaped secondary fibers are normally arranged in successive growth shells results in the more structurally intricate star suture lens.

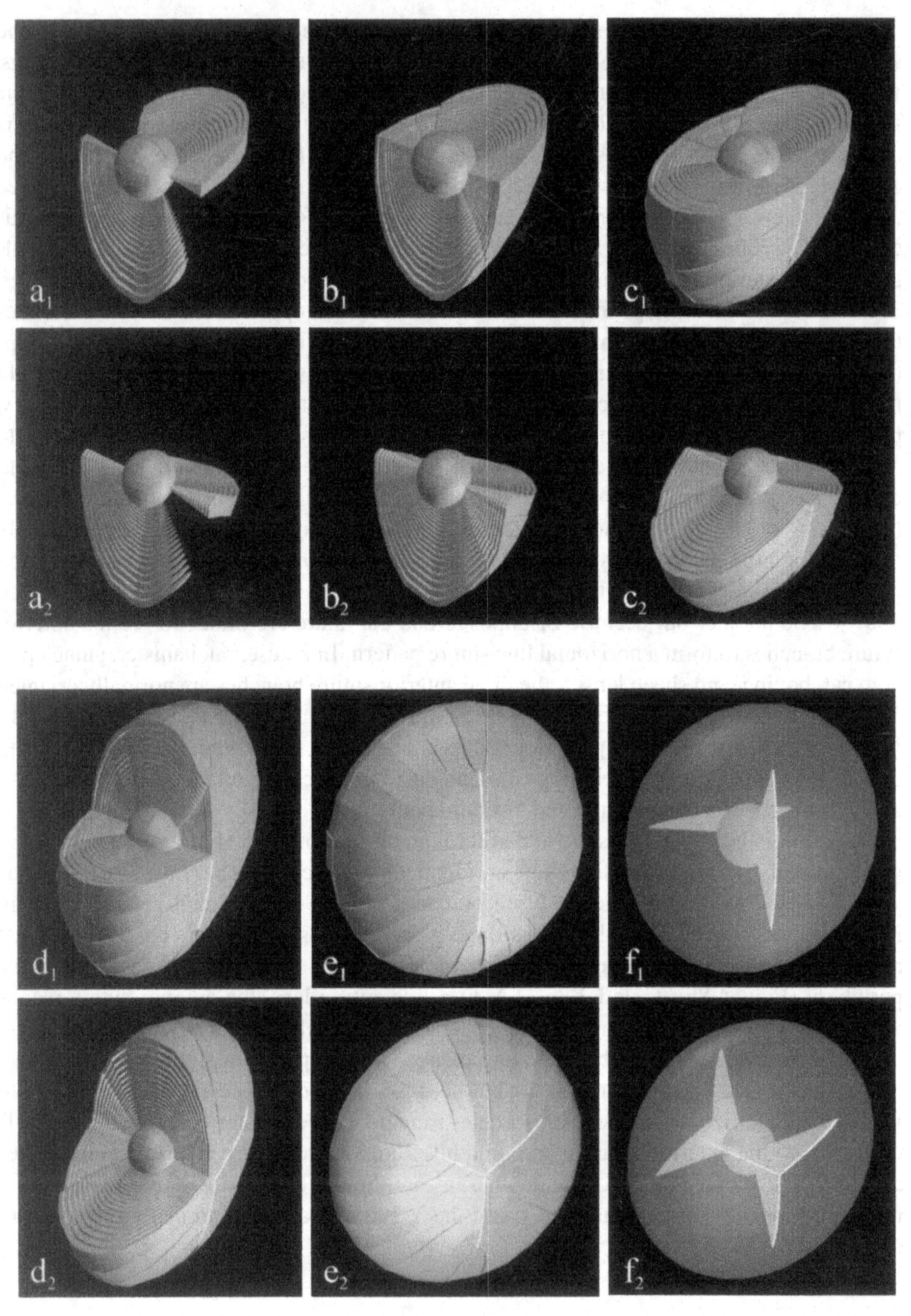

Figure 4.8. Scale 3D-CADs showing key structural elements in the production of continuous suture planes in line (1st and 3rd rows, a_1–f_1), and Y suture lenses (2nd and 4th rows, a_2–f_2). As additional secondary fibers are added throughout life, radial cell columns (a_1, a_2) and concentric growth shells (b_1, b_2, c_1, c_2, d_1, d_2, e_1, e_2) are formed; these consist of identically shaped but radially longer fibers. Since the ends of all secondary fibers become aligned at suture branches that are overlain in successive growth shells, continuous triangular suture planes extending from the embryonic nucleus to the lens periphery are formed (f_1, f_2).

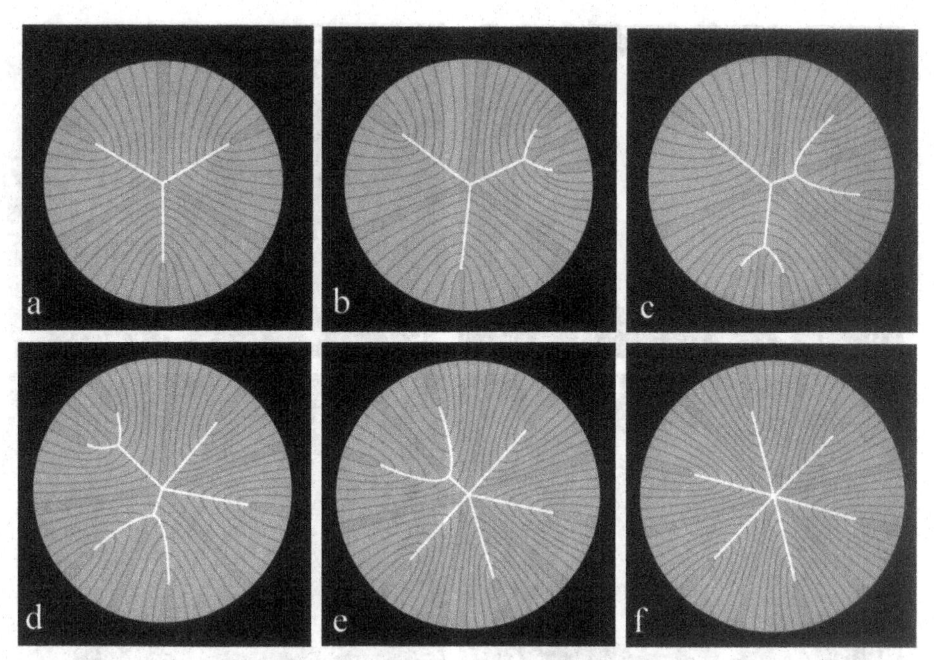

Figure 4.9. Scale 3D-CADs showing key structural elements in the production of a primate lens six-branch star suture from birth through infancy. (a) At birth, 6 straight fibers (darker blue), normally positioned equidistantly around the equator, separate growth shells into equal groups of S-shaped fibers with ends that abut and overlap to form suture branches (white lines) that extend to confluence at the poles. Note that the length and location of suture branches are defined by the ends of straight fibers. (b–e) In successive growth shells, additional straight fibers positioned nonequidistantly around the equator separate growth shells into unequal groups of S-shaped fibers with ends that abut and overlap to form additional suture branches that extend to confluence at the original suture branches. (f) By the end of the infantile period, 12 straight fibers, positioned equidistantly around the equator, separate growth shells into equal groups of S-shaped fibers with ends that abut and overlap to form a simple star suture. From Kuszak et al., 1994, courtesy of Academic Press, London, UK.

Throughout the fetal period of primate lens development, a Y suture lens is formed. However, primate lenses normally form three progressively more complex star suture patterns during the periods of infancy, adolescence, and adulthood. The evolution of the simple star suture produced during the infantile period is depicted in Figure 4.9. As additional growth shells are added after birth, specific S-shaped fibers are overlain by a new pair of straight fibers. Since the ends of straight fibers delimit suture branches, new groups of S-shaped fibers with less opposite-end curvature are defined on either side of the new straight fibers. The ends of these fibers abut and overlap to form two new evolving suture branches that extend to confluence at one of the original three Y suture branches (Fig. 4.9b). At this point, S-shaped fibers are arranged in eight unequal groups positioned between eight straight fibers.

As more growth shells are added (Fig. 4.9c), the new pair of straight fibers are overlain by longer straight fibers, and a second new pair of straight fibers are also added. The S-shaped fiber groups on either side of the original new pair of straight fibers are longer, forming two longer evolving suture branches that extend to confluence farther along one of the original Y suture branch. At the same time, the ends of the second new pair of straight fibers

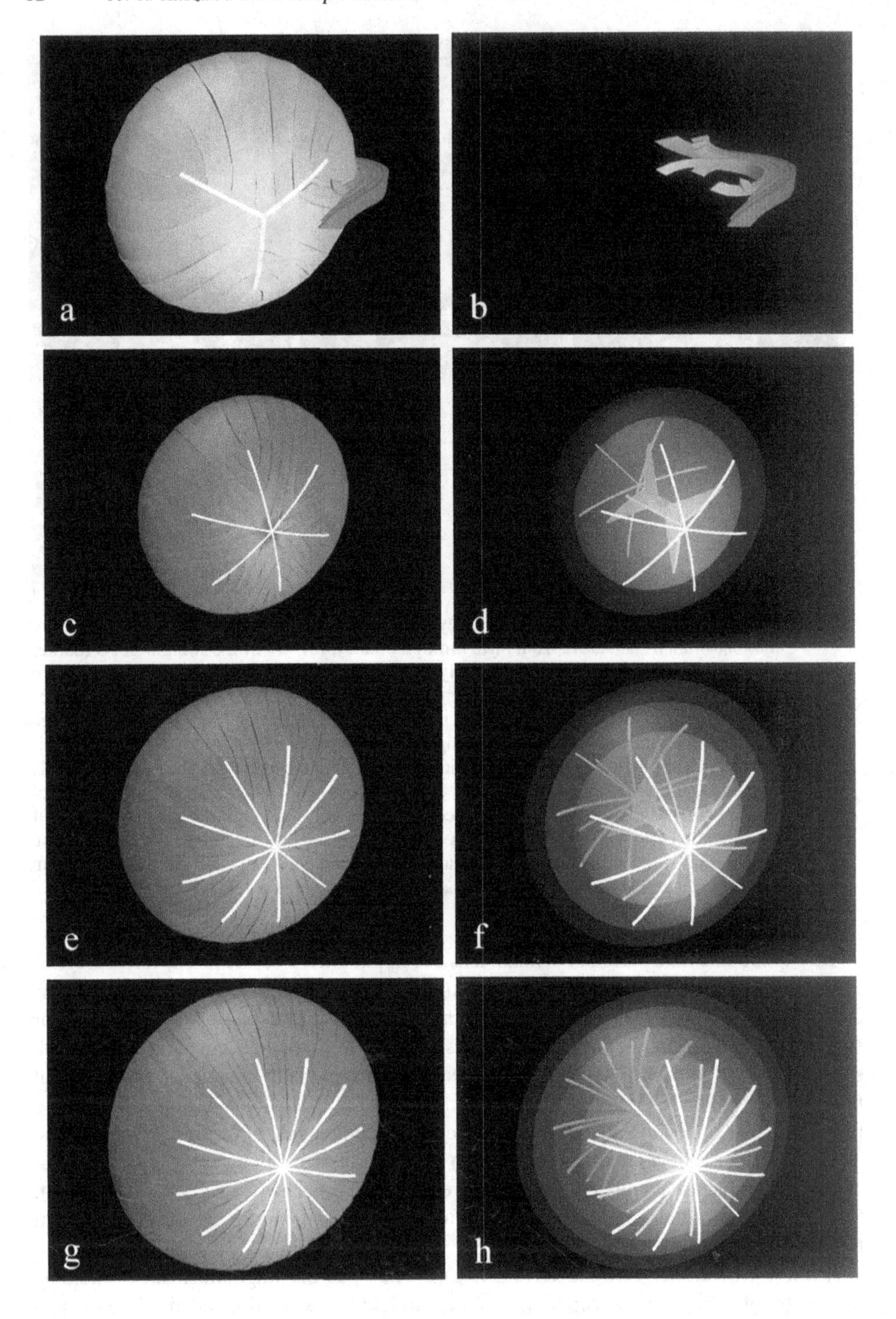
a
b
c
d
e
f
g
h

delimit additional new evolving suture branches that extend to confluence at a second of the original three Y suture branches. At this point, the S-shaped fibers are arranged into 10 unequal groups positioned between 10 straight fibers.

As more growth shells are added (Fig. 4.9d), the first new pair of straight fibers are overlain by even longer straight fibers, while the second new pair of straight fibers are overlain by longer straight fibers, and a third new pair of straight fibers are added. The ends of the S-shaped fibers positioned on either side of the first pair of straight fibers now form two complete new suture branches that extend to confluence at the poles. At the same time, the ends of the S-shaped fiber positioned on either side of the second pair of new straight fibers continue to form evolving suture branches that extend to confluence farther along the second of the original three Y suture branches. The ends of the third new pair of straight fibers delimit additional new evolving suture branches that extend to confluence at the third of the original three Y suture branches. At this point, S-shaped fibers are arranged into 12 unequal groups positioned between 12 straight fibers.

As more growth shells are added (Fig. 4.9e), the second new pair of straight fibers are overlain by longer straight fibers, and the third new pair of straight fibers are overlain by longer straight fibers. The groups of S-shaped fibers positioned on either side of the second new pair of straight fibers now form two additional new suture branches that extend to confluence at the poles. At the same time, the ends of the third pair of new straight fibers form longer evolving suture branches that extend to confluence farther along the third of the original three Y suture branches. As the final infantile growth shells are added (Fig. 4.9f), all three new pairs of straight fibers continue to be overlain by straight fibers. All the groups of S-shaped fibers positioned on either side of all the new pairs of straight fibers form a symmetrical, six-branch, simple star anterior and posterior sutures that are offset 30°.

Because a simple star suture has formed over the period of infantile development, radial cell columns now contain fibers that have neither an identical shape nor lengths that are a simple radial function (Figs. 4.10a and b). The ends of fibers and by extrapolation the suture branches are out of register in successive growth shells. Thus, "discontinuous" suture planes are formed from the fetal lens nucleus to the lens periphery (Figs. 4.10c and d).

Throughout the periods of adolescence and adulthood, star sutures (Figs. 4.10e and f) and complex star sutures (Figs. 4.10g and h), respectively, the third and fourth generations of a primate lens, are formed in a similar manner. The fully formed anterior and posterior star sutures, offset 20° because of opposite-end curvature, are normally characterized by 18 groups of S-shaped fibers that have even less opposite-end curvature and are positioned on either side of 18 straight fibers. Finally, the fully formed anterior and posterior complex star sutures, now only offset 15° because of the small amount of opposite-end curvature, are normally characterized by 24 groups of S-shaped fibers positioned on either side of 24 straight fibers.

Figure 4.10. (facing page) (See color plate V.) Scale 3D-CADs showing key structural elements in the production of discontinuous suture planes in primate lenses. (a and b) As shown in Figure 4.9, from birth through infancy, fibers in radial cell columns are neither of uniform shape nor simply longer as a function of radial location. (c and d) Consequently, the 12 suture branches of the simple star suture form "discontinuous" suture planes extending from the fetal nucleus to the lens periphery. Similarly, the 18 suture branches of the star suture form "discontinuous" suture planes extending from the juvenile nucleus to the lens periphery (e and f), and the 24 suture branches of the complex star suture form "discontinuous" suture planes extending from the adult nucleus to the lens periphery (g and h). Modified from Kuszak and Costello, 2003, courtesy of Lippincott Williams and Wilkins.

The factor or factors that determine whether a fiber will be straight or exhibit opposite-end curvature and to what degree fiber ends will curve are not known. However, it is known that the evolution of star suture branches normally commences in the inferonasal quadrant and then proceeds to occur in turn in the superotemporal, inferotemporal, and superonasal quadrants. The quadrant-related specificity is not without developmental or pathologic precedent. Colobomas of the eye (e.g., retina, ciliary body, iris, choroid, lens, and zonules), the failure or arrest of normal embryonic fissure closure during embryogenesis, typically occur in the inferonasal quadrant (Schwartz et al., 1989). In fact, this embryologic defect is considered atypical if it occurs in another quadrant.

4.3.4. Zones of Discontinuity

It is also important to note that the anatomic location and measure of the normal zones of discontinuity, seen in human lenses with transillumination slit-lamp biomicroscopy, are coincident with the four progressively more complex generations of sutures. This leads to the compelling argument that the zones of discontinuity are these four generations of sutures. Indeed, the sharply demarcated zones of discontinuity are coincident neither with alterations in human fiber membrane surface structure (Kuwabara, 1975; Kuszak et al., 1988) nor with variations in the concentration and density of fiber crystallin and cytoskeletal proteins (Alcala and Maisel, 1985; Hoenders and Bloemendal, 1981) as a function of development, growth, and age. Furthermore, because the aforementioned age-related changes in lens morphology are common to both primate and nonprimate lenses, all lenses should develop zones of discontinuity throughout life. In fact, only primate lenses have zones of discontinuity. Finally, in all lenses, transillumination slit-lamp biomicroscopy reveals a thin equatorial band, or region of minimal light scatter, commonly referred to as the "central sulcus." The thickness of this region is equivalent to the thickness of the embryonic nucleus. Thus, it only includes embryonic nuclear fibers and the midportions of fetal, infantile, juvenile, and adult nuclear and cortical fibers that are aligned parallel to the anteroposterior axis. Note that the central sulcus is characterized by a complete lack of sutures within its boundaries.

Thus, the abnormal slit-lamp profiles of cataractous lenses (e.g., diabetic and cortical) are likely to contain a record of abnormal suture development. Correlative SEM and 3D computer-aided design (3D-CAD) analyses of human cortical cataract lenses surgically removed by the extracapsular technique confirm this premise (Kuszak et al., 1998).

4.3.5. Optical Significance of Umbilical, Line, Y, and Star Suture Lenses

The functional significance of umbilical, line, Y, and star suture lens architecture is as follows: As light rays pass through the highly ordered radial cell columns of any lens, they repeatedly encounter fiber membrane, cytoplasm filled with crystallin proteins, fiber membrane, extracellular space, etc., in relatively constant and precisely defined increments. Cumulatively, the lens is transformed into a series of coaxial refractive surfaces – that is, into a structure that will cause a minimum of light diffraction (Trokel, 1962). This theory of lens function has been proven to be accurate by correlating the focusing power of lenses ascertained by low-power helium neon laser scan analysis with lens structure as assessed by light microscopy (LM) and electron microscopy (EM) analysis. In a series of experiments, laser beams were passed through precisely mapped regions of umbilical (Priolo et al., 1999a; Priolo et al., 1999b), line (Kuszak et al., 1991), and Y suture lenses (Sivak et al., 1994).

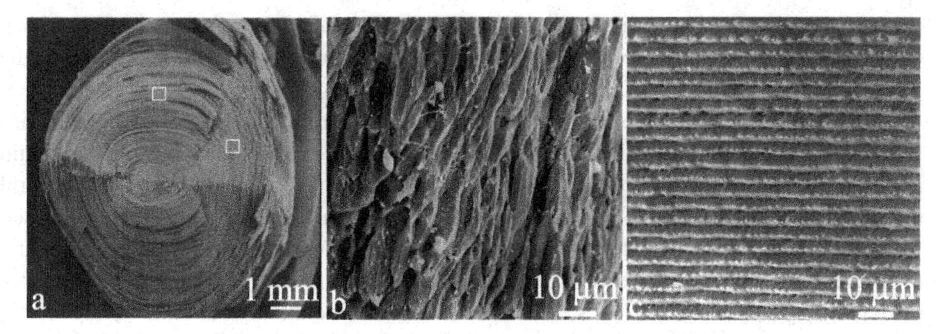

Figure 4.11. (a) A baboon lens fetal nucleus split along the visual axis reveals two of the three triangular anterior suture planes. (b) Higher magnification of right white boxed area in A showing the nonuniform fiber ends that constitute a suture plane. (c) Higher magnification of left white boxed area in A showing the uniform fiber lengths that constitute a radial cell column. (a) From Kuszak et al., 1988, courtesy of Academic Press, London, UK. (b), (c) From Kuszak, 1995, courtesy of Elsevier Science, Oxford, UK.

Then, the focal length of each beam was accurately and reproducibly quantified (Sivak et al., 1986). Correlative LM and EM analysis of the mapped regions confirmed that focal length variability (FLV) was least when lasers were transmitted through the highly ordered radial cell columns and greatest when the beams were transmitted through sutures. The reason for this naturally occurring variation in lens optical quality as a function of structure is as follows: The ends of fibers are neither uniform in shape or size (Fig. 4.11). Thus, as light rays pass through suture planes, they repeatedly encounter fiber membrane, cytoplasm with crystallins, fiber membrane, extracellular space, etc., in inconstant and ill-defined increments (the ultrastructure of fiber ends, as shown in greater detail in a later section). In this manner, the arrangement of uniformly shaped fibers as ordered radial cell columns aligned directly along the visual axis minimizes FLV. The arrangement of nonuniformly shaped fiber ends into continuous, disordered suture planes also aligned directly along the visual axis produces increased FLV. Furthermore, as lenses age, fiber structure at the fiber ends as well as along the fiber length becomes less ordered, and as a result the structure-function relationship becomes even more pronounced (Sivak et al., 1994; Priolo et al., 1999a).

In contrast, the primate fetal lens has only small triangular suture planes within the nucleus, overlain by discontinuous suture planes extending from the nucleus to the lens periphery. Thus, as light rays pass through the primate lens, they encounter fewer of the inconstant or ill-defined increments of fiber membrane, cytoplasm with crystallins, fiber membrane, extracellular space, etc., that are characteristic of nonprimate lens suture planes. This explains, at least in part, why primate lenses are optically superior to nonprimate lenses.

4.3.6. Physiological Significance of Umbilical, Line, Y, and Star Suture Lenses

Lenses lack a blood supply. Their nourishment, contained in the aqueous and vitreous humors, has been presumed to enter the lens mass through simple diffusion. However, Fischbarg et al. (1999) showed that the effective depth for nutrient diffusion into lenses is restricted to the outer 10% of the lens radius. This means that only the lens epithelial and the peripheral superficial cortical fibers, both of which feature full complements of organelles, have effective access to critical nutrients. In contrast, all other cortical and nuclear fibers, which lack organelles, have very limited or no access to nutrients. Of course, as lenses grow in size, this nutrient delivery problem is further amplified. Lens structure, in the form of sutures, may play an important role in supporting an internal circulatory system in the lens.

Mathias et al. (1997) theorized that current circulates around and through the lens, inward at the poles and outward at the equator (see chap. 6). The inward movement of fluid is described as occurring along the polar intercellular clefts of fibers (i.e., by structural definition, the sutures) and as convecting glucose to the innermost fibers, where it is used for anaerobic metabolism. The outward movement of fluid is described as intracellular and as convecting waste products of metabolism out of the lens at the equator, presumably through gap junctions. Fiber ends, or those portions of fibers that abut and overlap within and between growth shells to form suture branches, make up the apicolateral and basolateral membrane of fibers. If these fiber end segments are characterized by pumps, transporters, and channels, as is typical of other epithelia, then diffusional transport throughout a lens may be enhanced at suture branches. This leads to speculation that the level of sutural complexity is a major factor in determining how efficiently fluids can move into a lens. Lenses with more extensive suture patterns (e.g., star sutures) would be the more efficient at fluid transport than lenses with less extensive suture patterns (e.g., line sutures). This may explain, at least in part, why primates develop progressively more complex suture patterns as a function of age.

It has been proposed that the numerous gap junctions connecting fiber midportions (i.e., those segments of fibers not involved in sutures) are responsible for the outward movement of fluid (the fiber ultrastructure is described in greater detail in a later section). However, the density of fiber gap junctions varies considerably among species (Kuszak et al., 1978; Kuszak et al., 1985; Lo and Harding, 1986). The midportions of chick, rat, frog, and human fibers have, respectively, 65%, 33%, 12%, and 5% of their membrane specialized as gap junctions. This implies that fiber gap junction density will also be a factor in determining how efficiently fluids can move out of lenses in different species and/or as a function of age.

Note that the lenses with the most complex suture patterns have the lowest density of fiber gap junctions while the lenses with the highest density of fiber gap junctions have the simplest suture patterns. It remains to be determined how and/or if these two structural parameters are balanced to ensure the optimal efficiency of internal fluid circulation networks in different lens types.

4.4. Lens Gross Anatomy

4.4.1. Lens and Fiber Shape

A lens is an asymmetric, oblate spheroid. (A spheroid is a 3D figure resembling a sphere but with the polar surface curvature flatter than that of a sphere; the degree of flatness is defined by standard mensuration formulas [Eves, 1978].) The differences in the amount of anterior and posterior lens surface curvature are species and age dependent (Kuszak et al., 1996). As an example, while the anterior surface of most fish lenses is the equivalent of one-half of an 80° spheroid, the posterior surface is the equivalent of one-half of a 90° spheroid (i.e., a sphere). In contrast, the anterior surface of a normal adult human lens is the equivalent of one-half of a 15° spheroid, and its posterior surface is the equivalent of one-half of a 30° spheroid. Lens thickness is the measure of a lens from anterior to posterior poles along the visual axis, and lens width is the span of its equator perpendicular to the visual axis. Individual fibers are generally hexagonal in cross section and have two broad and four narrow faces (Fig. 4.3). Thus, in growth shells and/or radial cell columns, fibers are arranged so that their broad faces are oriented parallel to the lens surface. The measure between broad faces constitutes fiber thickness, and the measure spanning the distance defined by the angles formed by opposite paired narrow faces constitutes fiber width. In spite of considerable differences in lens size, there are only minor differences in fiber thickness and width among species. At the equator, fiber thickness is approximately 1.0–1.5 μm and fiber width measures on average 10.0–12.0 μm.

4.4.2. Lens Cortex and Nucleus

A cortex can be defined as the outer part or external layers of an internal organ, and a nucleus can be defined as anything serving as the center of development or growth. Given the inside-out development and growth scheme of vertebrate lenses, it is entirely appropriate to consider any lens to consist of a cortex and a nucleus, each composed of, respectively, cortical and nuclear fibers. However, a review of the literature reveals a lack of general agreement on what constitutes the boundaries of a lens cortex and a lens nucleus in different types of lenses or in different species. Indeed, while some consider age to be the determining factor in designating a fiber as cortical or nuclear, others argue that time of development and/or state of maturation are the determinants of cortical and nuclear fibers. For example, there is generally no dispute that in a human lens at birth the most peripheral growth shells of mature fibers consist of cortical fibers. However, at age 50, these same growth shells, now overlain by thousands of additional growth shells formed over the course of a half century, are considered to consist of nuclear fibers because of their age. At what age did the cortical fibers become nuclear fibers? Indeed, aged fibers are structurally and biochemically different from young fibers (Garland et al., 1996). Also, common laboratory animals (mice, rats, rabbits, etc.) have considerably shorter life spans than primates. Irrespective of different rates of metabolism, does this imply that the ratio of cortical to nuclear fibers varies as a function of species?

In this section we address the various structural parameters that determine when a cortical fiber becomes a nuclear fiber and define the boundaries of the cortex and the nucleus in all species. By definition, primary fibers, the center of lens development and growth, are the original nuclear fibers in any lens. Similarly, as secondary fibers are formed, they compose the external layers of any lens. But because the lens grows throughout life, all secondary fibers become nuclear fibers upon completion of the maturation process. Fibers are not mature until they have completely elongated, have eliminated most of their internal organelles (Bassnett and Beebe, 1992; Bassnett, 1992, 1995, 1997), and have completed the production of specialized cytoplasmic proteins, the crystallins, as well as specialized cytoskeletal and plasma membrane components, including lens gap junctions and square array membrane. Thus, which type of nuclear fiber a secondary fiber becomes is a function of when it originated during development, growth, and aging.

The embryonic and fetal nuclei are composed of all of the fibers formed during the periods of embryonic and fetal development, respectively. These regions are recognizable in any lens because of the unique structural features of representative fibers. For example, the primary fibers of the embryonic nucleus do not form sutures. Their long axis is oriented essentially parallel to the anteroposterior axis. Their average cross-sectional area is large and has a high standard deviation. Thus, within any one equatorial cross section, embryonic nuclear fibers with a large cross-sectional area are seen adjacent to embryonic nuclear fibers with a very small cross-sectional area (Fig. 4.12; Taylor et al., 1996; Shestopalov and Bassnett, 2000a). As a result, the embryonic nucleus is also the only lens region where fibers are not organized into ordered radial cell columns.

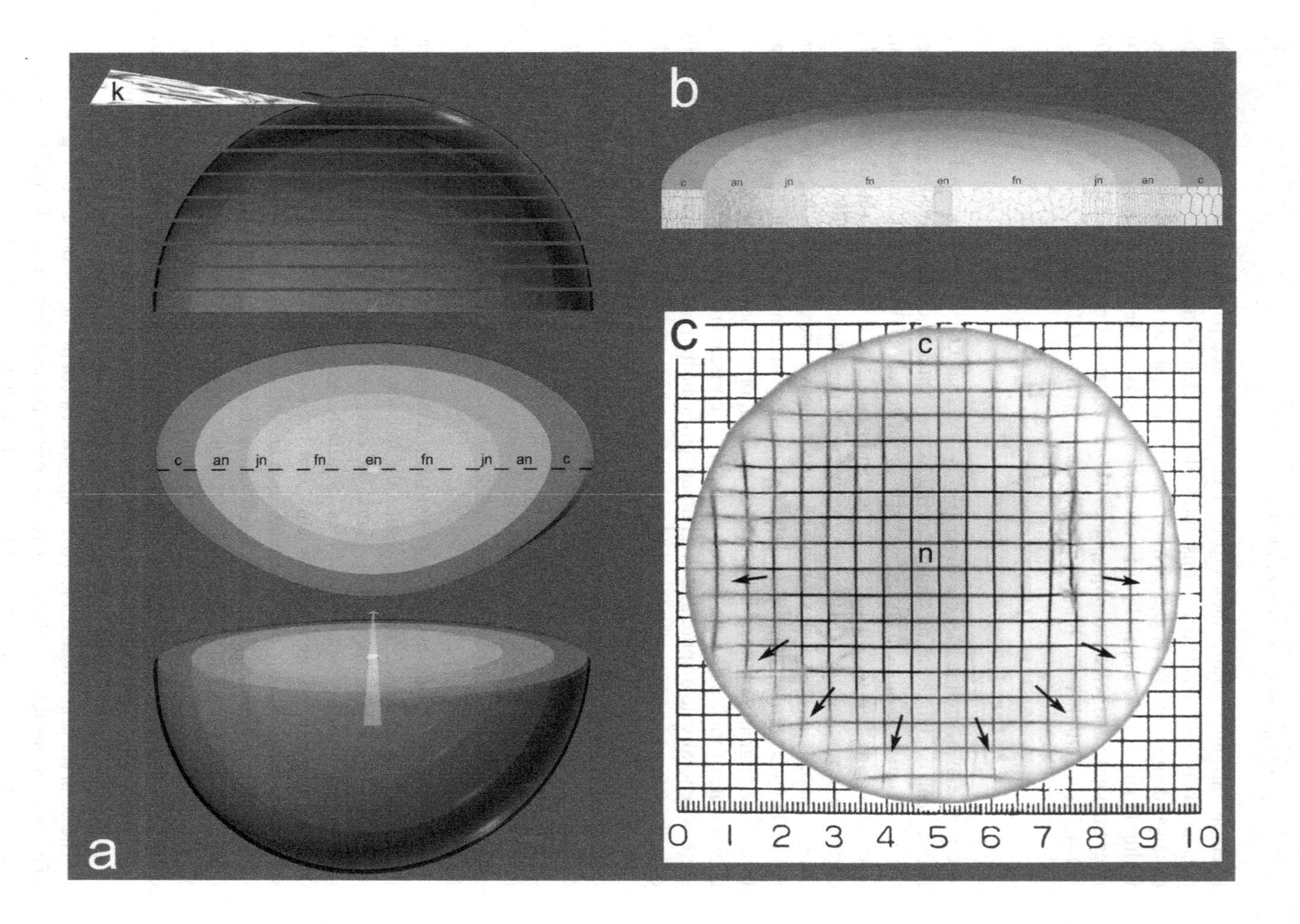
k
a
c an jn fn en fn jn an c
b
c an jn fn en fn jn an c
C
c
n
0 1 2 3 4 5 6 7 8 9 10

Fetal nuclear fibers, in contrast, are smaller on average and more uniform in cross-sectional size and shape (Fig. 4.12; Taylor et al., 1996). The organization of fetal nuclear fiber evolves over the period of fetal development. As the initial fetal nuclear fibers are added onto the embryonic nuclear fiber mass, their shape is irregularly polygonal (e.g., rhombic, pentagonal, hexagonal, etc.), and thus near the peripheral border of the embryonic nucleus they are arranged in short radial rows or cell columns. By the end of fetal nuclear fiber formation, their shape and size are relatively uniform (hexagonal). Thus, by birth they are arranged in ordered radial cell columns. Because fetal nuclear fibers are either straight or S-shaped, their ends abut and overlap to form sutures. Therefore, the beginning of suture formation signals the beginning of the fetal nucleus.

The fibers formed after birth and during childhood have been described using various terms (Duke-Elder and Cook, 1963). It is helpful to designate all of the secondary fibers formed after birth and through sexual maturation as juvenile nuclear fibers (Fig. 4.12; Taylor et al., 1996). Because different species have different maturation rates, the juvenile nucleus is variable in thickness. Furthermore, within one species, such as humans, the age of maturation is variable, as is the length of time to complete the maturation process. Thus, the boundary of the juvenile nucleus is broader and less distinct than the boundaries of the fetal and embryonic nuclei. In humans, the average cross-sectional area of juvenile nuclear fibers is smaller and more uniform than that of fetal nuclear fibers. A defining characteristic of juvenile nuclear fibers is the flattened hexagonal cross-sectional profile possessed by the vast majority of these cells. A minority of juvenile nuclear fibers are larger pentagonal cells that represent the fusion of two or more fibers from adjacent radial cell columns (Kuszak et al., 1985; Shestapalov and Bassnett, 2000b). The presence of fused fibers in part explains why the number of radial cell columns increases significantly as a radial function from the center of the lens (see Fig. 4.3) whereas the width of the radial cell columns remains nearly constant. In humans, the juvenile nuclear fibers terminate in suture patterns that transition from three to six main branches through infancy and from six to nine branches through adolescence. Primates are the only species shown to have evolved more complex sutural patterns during defined periods of development, growth, and aging.

The adult lens nucleus comprises all the fibers formed after sexual maturation minus the fibers within the cortex. This definition is only precise if the boundary between the cortex and nucleus is clearly defined. Structural examination of fibers reveals many parameters

Figure 4.12. (facing page) (See color plate VI.) Dissection of fresh lenses for ultrastructural examination reveals internal organization in all developmental regions. (a) Vibratome sectioning of fresh lenses. Central section is rotated 90° to display developmental regions approximately to scale. c, cortex; an, adult nucleus; jn, juvenile nucleus; fn, fetal nucleus; en, embryonic nucleus. For orientation, a beam of light is drawn along the optic axis. After immersion fixation and embedding, the central section is cut along the dashed line to expose cells in the equatorial plane. k, knife. (b) Central section split along the equator to reveal fibers cut in cross section. Fibers in the en region and the regions to the left are taken from digitized light micrograph images, and those to the right are idealized diagrams of fibers. Although each developmental region are drawn roughly to scale, the interfaces between regions is artificially abrupt. Many layers of cells have been omitted in each region for clarity. Note that the fibers in the light micrograph image of the adult nucleus are too compacted to be recognized as individual cells. (c) Image of whole fresh human transparent lens (67 years old). The nucleus (n) is characterized by yellow pigmentation absent from the cortex (c). The cortex-nucleus interface is marked by arrows in the lower hemisphere. The surface blemish at 7.5 on the mm scale was caused by a scalpel during extraction. (a), (b) From Gilliland et al., 2001, courtesy of *Molecular Vision.*

that serve to define a fiber as either a cortical or a nuclear fiber. First, all elongating fibers are superficial cortical fibers. Second, fully elongated fibers in the process of eliminating their organelles are intermediate cortical fibers. Third, after elimination of the organelles, fiber terminal differentiation continues as deep cortical fibers alter their cytoplasm and plasma membrane. Cytoskeletal changes include the loss of selected cytoskeletal elements (Alcala and Maisel, 1985; Blankenship et al., 2001) and increased condensation through the modification and cross-linking of crystallins and through the loss of water (Tardieu et al., 1992; Kenworthy et al., 1994). These processes result in fiber compaction, especially dramatic in primate lenses, in which fiber cross-sectional area is decreased by a factor of three (Fig. 4.12; Taylor et al., 1996). Fourth, since the loss of water occurs without a concomitant elimination of membrane surface area, the increased complexity of fiber surface membrane signals the transition of a fiber from cortical to nuclear (Al-Ghoul et al., 2001). A redistribution of the integral membrane proteins of nuclear fibers creates regions rich in MIP (aquaporin0) in the form of large areas of square array membrane (Kuwabara, 1975; Zampighi et al, 1989; Costello et al., 1989). Square array membrane in nuclear fibers is characterized by furrowed membrane domains, or undulating membrane (Taylor et al., 1996). As a consequence of all of the above factors, the nucleus becomes a hardened region as compared with the cortex. Dissection of lenses from various species at any age reveals a peripheral region of softer tissue, which represents the cortex. The boundary between the cortex and nucleus can be visualized in slit-lamp images or darkfield photography. Even when using white transmitted light, the nucleus of aged humans has a nearly uniform yellow coloration that is clearly distinct from the colorless cortex (Fig. 4.12; Taylor et al., 1996). Based on the diameter of the chromatic region, the average thickness of the cortical region in adult human lenses, age range 49–73, is 1.13 ± 0.15 mm ($n = 10$; unpublished data). The thickness of the cortical region does not appear to increase with age.

The above convention for developmental regions applies particularly well to aged human lenses. Slight modifications are necessary when considering younger human lenses or lenses of animals. A three-year-old bovine lens is a young adult lens that does not display significant compaction (Al-Ghoul and Costello, 1997). Thus, the boundaries between the cortical and nuclear layers are more difficult to recognize. In addition, progressively more complex suture patterns are not formed throughout the periods of fetal, juvenile, and adult nuclear development, unlike in a primate lens. However, histologically the juvenile nucleus contains many large pentagonal cross-sectional profiles. These fibers are interspersed among radial cell columns composed of typically flattened hexagonal fibers. This pattern distinguishes the juvenile from the fetal and adult nuclear regions. The boundary between the adult nucleus and cortex is not as pronounced histologically but can be consistently identified by the differences in mechanical properties.

Another example is a two-month-old mouse lens. In this case, there are no adult nuclear fibers, as the juvenile nuclear fibers are still forming. However, the cortex will still have the same layered features as in an adult lens. As the animal ages, the cortical fibers formed within three to six months will age and become part of the nuclear mass, contributing to the formation of the juvenile nucleus and beginning the formation of the adult nucleus.

Therefore, as described earlier, cortical and nuclear regions can be reproducibly defined in human and animal lenses of all ages. It is worth emphasizing that within the fetus, or at any age after birth, any lens has cortical fibers that will be incorporated into the nucleus in due time. Also, because the rate of adult lens growth is very slow, it is important to recognize that the nucleus of any lens makes up a much larger proportion of the lens mass than the cortex (Fig. 4.13).

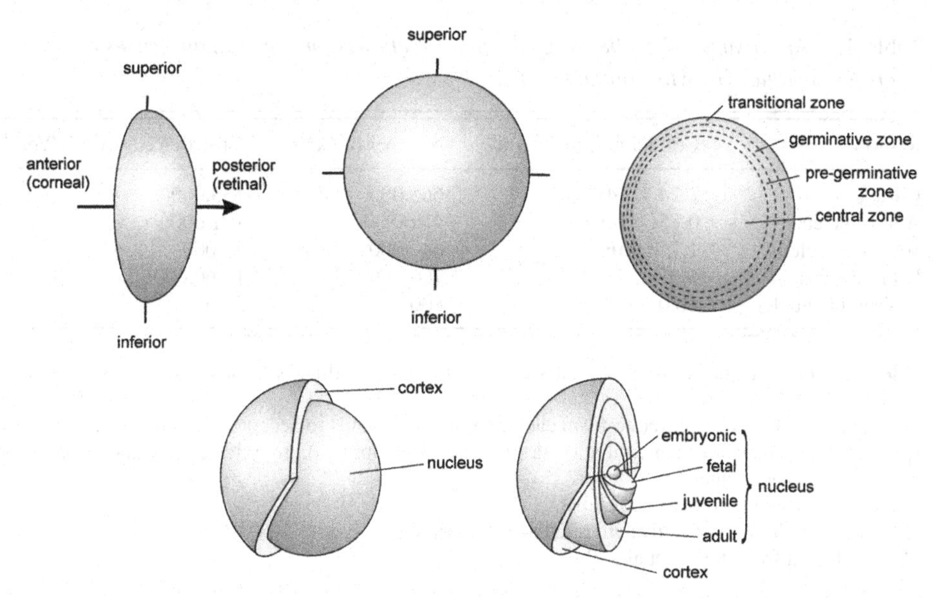

Figure 4.13. Scaled diagrammatic representation showing the gross shape, anatomic orientation, and developmentally defined regions of the epithelium and fiber mass in an adult human lens. Shown clockwise from upper left: the asymmetric, oblate, spheroid shape of the lens viewed along the equatorial axis; the anterior surface viewed along the polar axis; the locations of the lens epithelial zones; the locations of the lens cortex and nucleus; the locations of the lens cortex and nuclear regions. From Kuszak et al., 1988, courtesy of Academic Press, London, UK.

By relating lens and fiber dimensions, reasonable estimates of the number of fibers as a function of age can be calculated. The data presented in Table 4.1 clearly show that lens growth continues throughout life.

4.4.3. Lens Epithelium and Fiber Homeostasis

The size of any tissue cell population is determined by both the rate at which new cells are added (the cell birth rate, K_B) and the rate at which cells are lost (the cell loss rate, K_L; Budtz, 1994). While K_B can be readily assessed in most tissues by such methods as metaphase blocking (Wright and Appleton, 1980), K_L is more difficult to measure because of exfoliation (cell death and migration). In simple terms, if $K_B > K_L$, then the tissue is growing; if $K_B < K_L$, then the tissue is regressing; and if $K_B = K_L$, then the tissue is in kinetic equilibrium.

4.4.3.1. Fiber Homeostasis

As described earlier, any lens sectioned along its equator reveals all fibers formed throughout life arranged in age-defined growth shells (Wanko and Gavin, 1959; Kuszak and Rae, 1982; Rae et al., 1982; Taylor et al., 1996). By sectioning lenses in this manner, it can be easily demonstrated that in any lens the cumulative fiber total increases without a concomitant loss of the previously formed fibers. Thus, with regards to the fiber mass, the lens grows throughout life ($K_B > K_L$).

Table 4.1. *Approximate Number of Cells in Different Regions of Human Lenses and the Average Regional Growth Rate per Year*

Region	Cell Dimensions (μm)	Number of Cells	Growth Rate[†] (Cells/Year)
Cortex	2.24 × 14[‡]	665,000	133,000
Adult nucleus	0.75 × 7.5[‡]	4,460,000	101,000
Juvenile nucleus	1.7 × 9.0[‡]	640,000	53,000
Fetal nucleus	5.0[§]	700,000	1,360,000
Embryonic nucleus	10.0[§]	800	3,500

[*]Measurements of fiber cross-sectional size were taken from thick sections through the equatorial plane.
[†]The number of years required to form each region corresponds to accepted periods of development (primary and secondary fiber formation during fetal development), growth (infancy and childhood), and aging (beyond puberty).
[‡]Thickness × width of an average fiber.
[§]Diameter of an average equivalent circular cell perimeter.
Note. Adapted from Taylor et al., 1996.

Table 4.2. *Average Central Zone (cz) and Germinative Zone (gz) Lens Epithelial Cell Size and Number as a Function of Development, Growth, and Age*

	At Birth	Adult	Old Age
Total anterior surface area	$31.65 \times 10^6\ \mu m^2$	$57.15 \times 10^6\ \mu m^2$	$81.76 \times 10^6\ \mu m^2$
Cz surface area	$25.32 \times 10^6\ \mu m^2$	$45.72 \times 10^6\ \mu m^2$	$65.41 \times 10^6\ \mu m^2$
Gz surface area	$3.165 \times 10^6\ \mu m^2$	$5.715 \times 10^6\ \mu m^2$	$8.176 \times 10^6\ \mu m^2$
Avg. cz epithelial cell size	$77.43\ \mu m^2$	$92.21\ \mu m^2$	$193.97\ \mu m^2$
Avg. gz epithelial cell size	$77.43\ \mu m^2$	$83.76\ \mu m^2$	$99.58\ \mu m^2$
Cz epithelial cell #	287,701	435,390	290,696
Gz epithelial cell #	121,718	203,316	254,744
Total epithelial cell #	408,799	638,069	545,44

Note. Estimates of average lens epithelial zone surface area and cell size are derived from examination of two newborn, eight 7-year-old, and seven 25-year-old baboon (*Papio anubis*) lenses. Individual lens axial dimensions were assessed under a Zeiss surgical dissecting microscope equipped with a millimeter reticule. Measurements were made prior to any chemical preservation of lenses for microscopy.

4.4.3.2. Lens Epithelial Cell Homeostasis

At birth, gz and cz cells are essentially identical in size. But as the lens grows and ages, epithelial cell size varies as a function of zonal location (Fagerholm and Philipson, 1981; Karim et al., 1987; Kuszak, 1997). The variations in monkey lens anterior surface area, size of the gz and cz, and epithelial cell size as a function of age are shown in Table 4.2. From birth to adulthood, gz cell size increases on average by approximately 10%. From adulthood to old age, it increases by approximately 20%. In contrast, the gz anterior surface area increases in size by 80% from birth to adulthood and by 50% from adulthood to old age. Consequently, gz K_B is greater than gz K_L, and therefore the gz is a growing cell population throughout life. Gz growth is likely due to the addition of daughter cells at the periphery that did not enter into terminal differentiation to become fibers. This premise is supported indirectly by results from a study by Rafferty and Rafferty (1981) that estimated that on average the number of daughter cells produced exceed the number of secondary fibers added to a lens.

The average age-related increase in cz cell size is more pronounced than the increase in gz cell size. By adulthood, the average cz cell has increased in size by approximately 20%. However, by old age, the average percentage size increase from birth is 110%. In contrast, while the cz anterior surface area increases in size by approximately 80% from birth to adulthood, it only increases on average by 50% from adulthood to old age. Consequently, while cz K_B is greater than cz K_L from birth through adulthood, it is less from adulthood through old age. In other words, the cz is a growing cell population throughout adulthood and a regressing cell population for the rest of life. Failure to consider zonal variation in lens epithelial cell density leads to a gross overestimation of the total cell number (Kuszak and Brown, 1994).

4.4.4. Lens Epithelial Cell Apoptosis

Given that the lens epithelium eliminates some cells throughout life, the question remains, how is lens epithelial cell loss accomplished and regulated? There is no reason a priori to discount the notion that normal elimination of some lens epithelial cells is the result of apoptosis. Indeed, Ishizaki et al. (1993) showed that in mature rat lenses some cz cells are normally eliminated via apoptosis. In addition, LM and EM studies have revealed that occasional dark cells exist among normal cz and gz cells. These cells, often characterized by fragmented nuclei and condensed chromatin, have usually been treated as artifacts of preparation. It is more likely that they are apoptotic cells. In fact, Gorthy and Anderson (1980) showed that as cz cells increase in size there is a concomitant increase in the number of their intracellular lysosomal bodies as a function of age. Perhaps these increasingly large and prominent lysosomal bodies represent the breakdown of apoptotic cells and fragments. Indeed, macrophages could not play such a role in the lens because it is completely enclosed by the capsule.

An increasing body of cell biological research is elucidating the importance of apoptosis in normal development, tissue homeostasis, cell aging, and phenotypic fidelity (Tomei et al., 1994) and strongly suggests placing this form of cell death in the context of the normal cell cycle. For example, cz cells do not normally divide. Are these cells held in a state of reversible growth arrest or are they senescent cells, as suggested by von Sallman (1957). If senescent, can the decrease in cz cell density be accounted for by apoptosis? With regards to gz cells, during lens development and growth, a number of gz cells enter the mitotic cycle and produce two daughter cells. Some but not all of the daughter cells terminally differentiate into fibers. How many of the remaining gz cells do not become fibers? Since the cz and gz areas increase throughout life, are both of these enlarged areas populated by gz daughter cells not selected to become fibers? And are these cells kept in a state of reversible growth arrest regulated by extrinsic controls (soluble growth inhibitory substances) and intrinsic controls (signal transducers) (Barrett and Preston, 1994; Boyd and Barrett, 1990)? If so, the increasing number of gz cells produced over a lifetime can be accounted for, and as the progenitor cells of the gz reach their doubling potential limit, they might be eliminated by apoptosis and replaced by releasing other reversibly growth arrested gz daughter cells. In fact, if the above speculations are correct, then both the lens epithelium and fiber mass are populations of variably aged cells.

Although at this time these are merely speculations, they are not unreasonable, because lens epithelial cells are known to have the gene program necessary for apoptosis. In addition, occasionally some lens epithelial cells display characteristic programmed cell death morphology in various stages of apoptosis. Furthermore, it has even been shown that external factors known to induce apoptosis can result in experimental cataract (Li et al., 1995). However, questions have been raised as to how valid this type of animal cataract may be in relation to human cataracts (Harocopos et al., 1996). Nevertheless, the fact that a loss of normal apoptotic control in tissues, either negative or positive, leads to a range of pathologies suggests that studies are warranted to determine if apoptosis has any role in cataractogenesis.

4.5. Lens Ultrastructure

Detailed descriptions of lens epithelial and fiber cell membrane, junctional specialization, and cytoskeleton are presented in chapters 6 and 7, respectively. For the purpose of this chapter, we present a basic overview of lens epithelial and fiber structure as a function of development and aging.

4.5.1. Lens Epithelium

The lens epithelium is composed of low cuboidal cells (Fig. 4.14). These cells have large indented nuclei with two nucleoli and numerous pores. As the lens ages, these cells become more flattened, although they cannot be accurately described as squamous. There are nominal numbers of ribosomes and polysomes, smooth and rough endoplasmic reticulum, Golgi bodies, lysosomes, dense bodies, and glycogen particles. The mitochondria of these cells are small and have irregular cristae.

Cytoskeletal elements found in lens epithelial cells include actin; intermediate filaments (vimentin); microtubules; and the proteins spectrin, actinin, and myosin (Alcala and Maisel, 1985). It is presumed that, as in all eukaryotic cells, these elements are interconnected to produce a well-defined cytoskeleton or microtrabecular latticework that compartmentalizes the components of the cell's interior. Lens epithelial cells are remarkable in that they have a prominent, well-characterized cytoskeletal network consisting of actin filaments and myosin in the form of polygonal arrays, or "geodomes," located subjacent and attached to their apical membrane (Figs. 4.14a and d; Rafferty and Scholz, 1984, 1985, 1989). All the cytoskeletal components become more dense as the cells age.

The polarized lens epithelial cells have apical, lateral, and basal membranes. The smooth basal membranes underlie and produce the lens capsule, a replicated basal lamina. The lateral membranes of lens epithelial cells are markedly enfolded. Localization of NaK-ATPases and acid phosphatase varies within the cz (Garner and Kong, 1999). Acid phosphatase, involved in cellular breakdown and removal, increases with age. The lateral membranes also feature a small number of gap junctions composed of connexin 43 (Fig. 4.15). Desmosomes also exist between lateral membranes of lens epithelial cells. An uncommon feature of lens epithelial cell apicolateral membrane is that it lacks effective tight or occluding junctions. Simple interlaced, linear arrays of intramembrane particles are present between some lens epithelial cells at the apicolateral border. However, physiologically they present no significant barrier to extracellular flow and are thus considered to be "very leaky" tight junctions (Rae and Stacey, 1979; Zampighi et al., 2000).

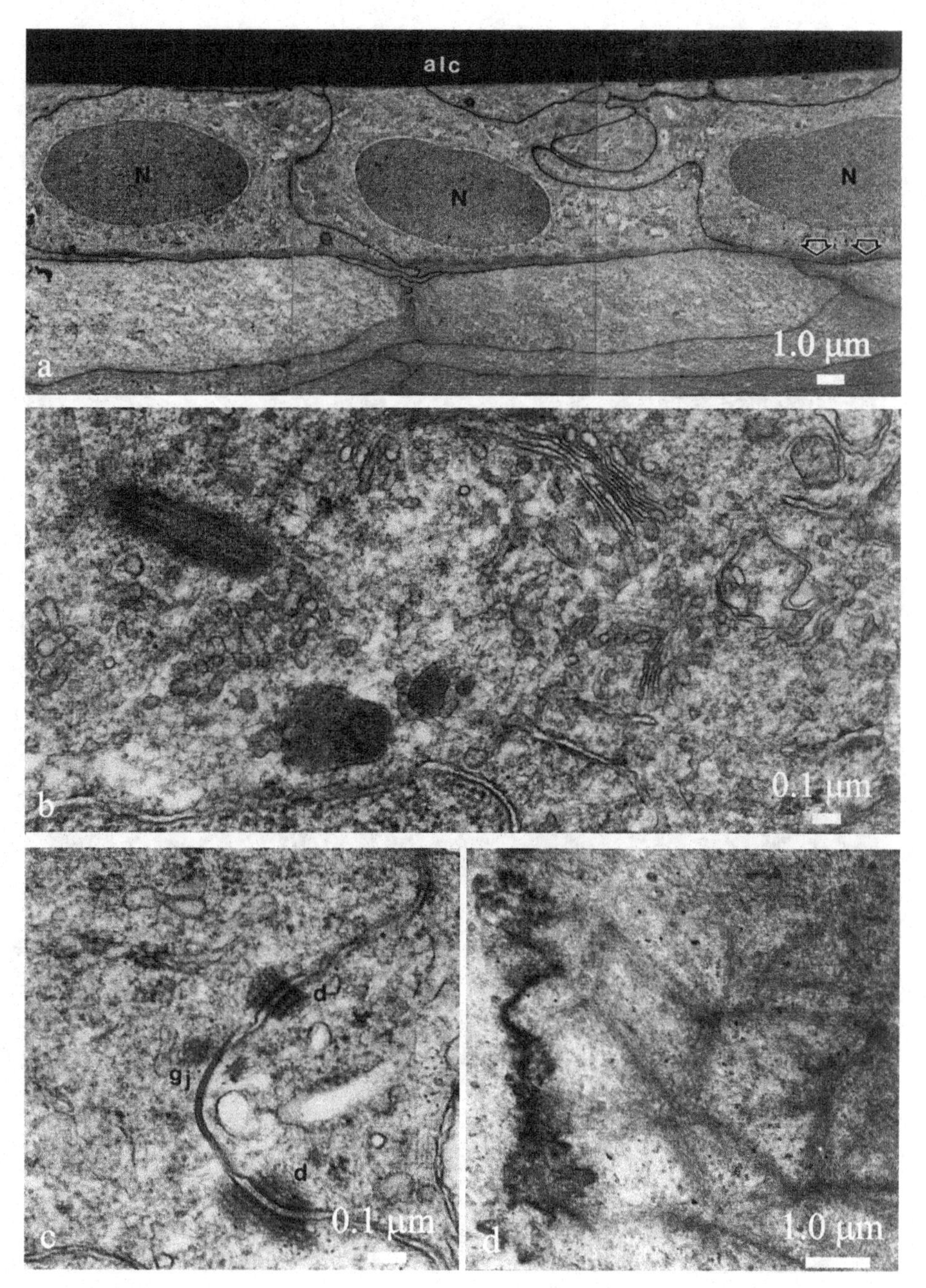

Figure 4.14. Transmission electron micrographs of central zone epithelial cells from an adult monkey lens. (a) The low cuboidal cells lying beneath the anterior lens capsule (alc) show markedly enfolded lateral plasma membrane and perpendicularly sectioned polygonal domains of actin bundles, or "geodomes," immediately subjacent to the apical membrane (open arrowheads). (b) Detail of a Golgi body, centriole, and rough endoplasmic reticulum. (c) Detail of lateral membrane desmosomes (d) and a gap junction (gj). (d) Detail of geodomes sectioned *en face*. Note their attachment to the apical and apicolateral plasma membrane. (Micrograph d courtesy of N. S. Rafferty, Ph.D.) From Kuszak and Brown, 1993, courtesy of W. B. Saunders Co., Philadelphia, PA.

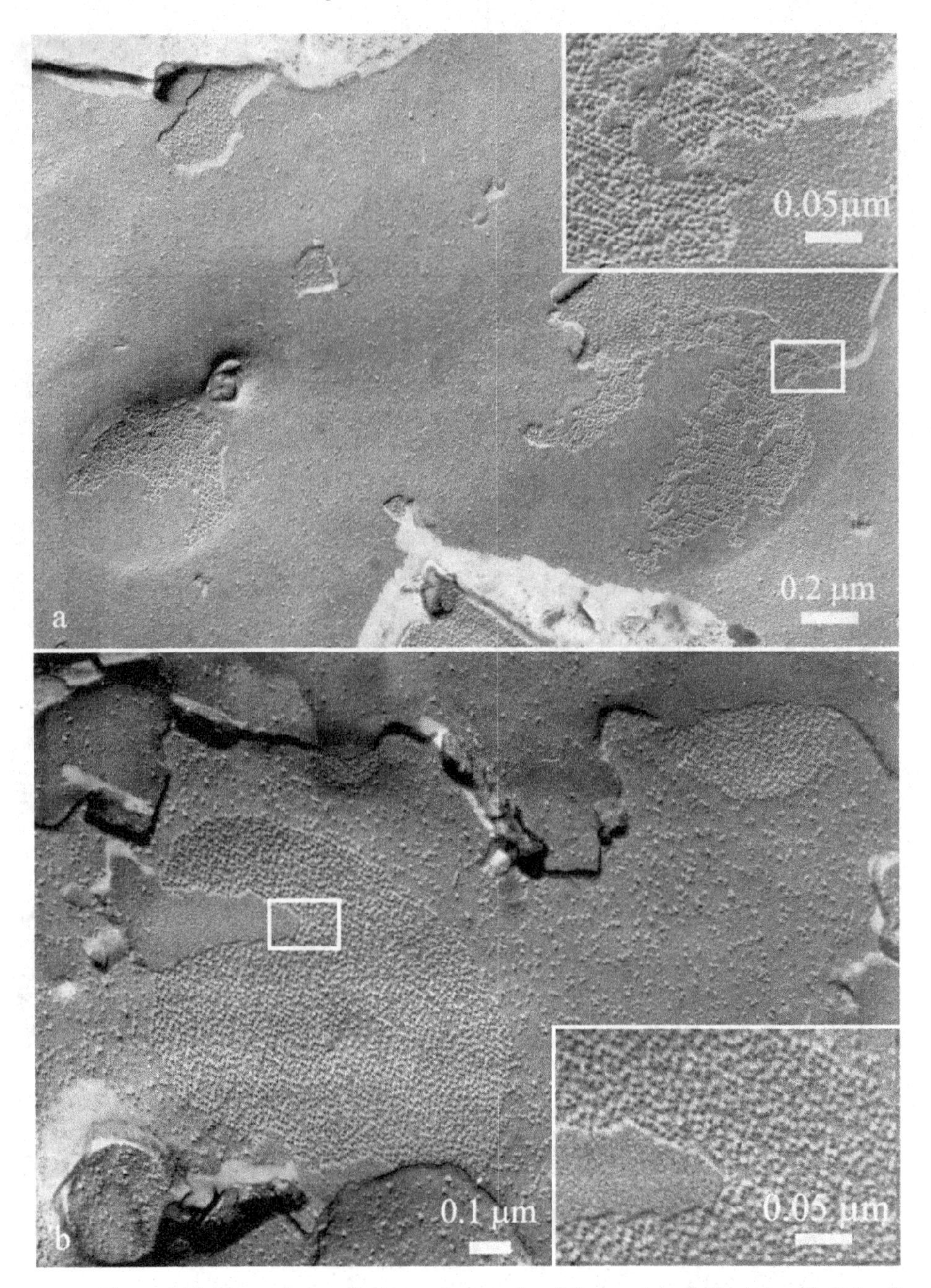

Figure 4.15. Transmission electron micrographs of freeze-etch–replicated epithelial cell (a) and cortical fiber (b) lateral membrane. Note the difference in particle packing between epithelial and fiber gap junctions (white boxed areas shown at higher magnification in insets). (a) From Kuszak, Novak and Brown, 1995, courtesy of Academic Press, London, UK. (b) From Kuszak and Brown, 1993, courtesy of W. B. Saunders Co., Philadelphia, PA.

The apical membrane of lens epithelial cells is planar and interfaces with the apical membranes of elongating fiber cells as they migrate to their sutural locations. This epithelial-fiber interface (efi) is characterized by transcytotic events (Brown et al., 1990; Kuszak et al., 1995). Numerous examples of micropinocytosis and clathrin-coated vesicles can be found at and immediately subjacent to the efi. Thus, nutrients, ions, essential metabolites, and presumably other receptor-mediated substances can be exchanged across the efi via transcytosis (Figs. 4.16 and 4.17). Both gap junctions and "square array membrane," areas of membrane characterized by groups of orthogonally arranged aquaporins, are extremely rare at the efi. The function of square array membrane is not known. One hypothesis is that it regulates fluid movement in and out of the extracellular compartment (Zampighi et al., 1989; Costello et al., 1989).

4.5.1.1. Physiologic Significance of the EFI as a Function of Lens Size and Age

The lens has been described as a "functional 3-dimensional syncytium" (Mathias et al., 1981). Though it is clearly not "a multinucleate mass of protoplasm produced by the merging of all cells" (the literal definition of a syncytium), such a characterization is nevertheless not inappropriate when one considers the following: The elongating fibers, or superficial cortex, of any lens form incomplete shells on the lens periphery. As mentioned, when the lens fibers elongate, their apical ends are in close apposition with the apical surfaces of lens epithelial cells, giving rise to the efi. SEM analysis of complete radial cell column exposure along the polar axis reveals that a fiber that breaks contact with either the cz epithelium at the anterior pole or the posterior capsule at the posterior pole is approximately 200 μm deep at the equator (Kuszak and Rae, 1982). Thus, Kuszak and Rae (1982) proposed that this arrangement provides a transport "short circuit" to the lens interior. Substances that enter the above-described fiber can traverse the 200 μm into the lens interior by diffusing in the cytoplasm of this cell rather than by crossing myriad cell boundaries through gap junctions. Such an intracellular diffusion pathway is likely to be at least an order of magnitude less resistive than the gap junctional pathway. Since elongating fibers are electrically coupled to epithelial cells at the efi (Rae and Kuszak, 1983), it has been presumed that small molecules transported by the epithelial cells can diffuse into these fibers through a single set of gap junctions and are thereafter free to diffuse deeper into the lens through the fiber cytoplasm. Indeed, lens fibers are coupled by an unusually high density of fiber-fiber gap junctions as compared with other epithelia (Benedetti et al., 1976; Goodenough, 1979; Kuszak et al., 1978; Kuszak et al., 1982; Costello et al., 1985; Lo and Harding, 1986). Thus, the efi has been considered to be an important component in the transport of essential components into the lens.

4.5.1.2. Correlative Structure Function Analysis of the EFI

Both structural studies (TEM and freeze-etch; Brown et al., 1990; Bassnett et al., 1994; Kuszak et al., 1995) and physiological studies (electrotonic and dye-coupling; Schuetze and Goodenough, 1982; Rae and Kuszak, 1983; Miller and Goodenough, 1986; Prescott et al., 1991; Prescott et al., 1994; Bassnett et al., 1994), which have quantified intercellular communication at the efi, have consistently found that the extent of the coupling at the efi is markedly less than either fiber–fiber or epithelial cell–epithelial cell coupling.

Brown et al. (1990) were the first to define the fiducial markers necessary for unequivocally identifying the efi in freeze-etch replicas, perhaps the most appropriate methodology

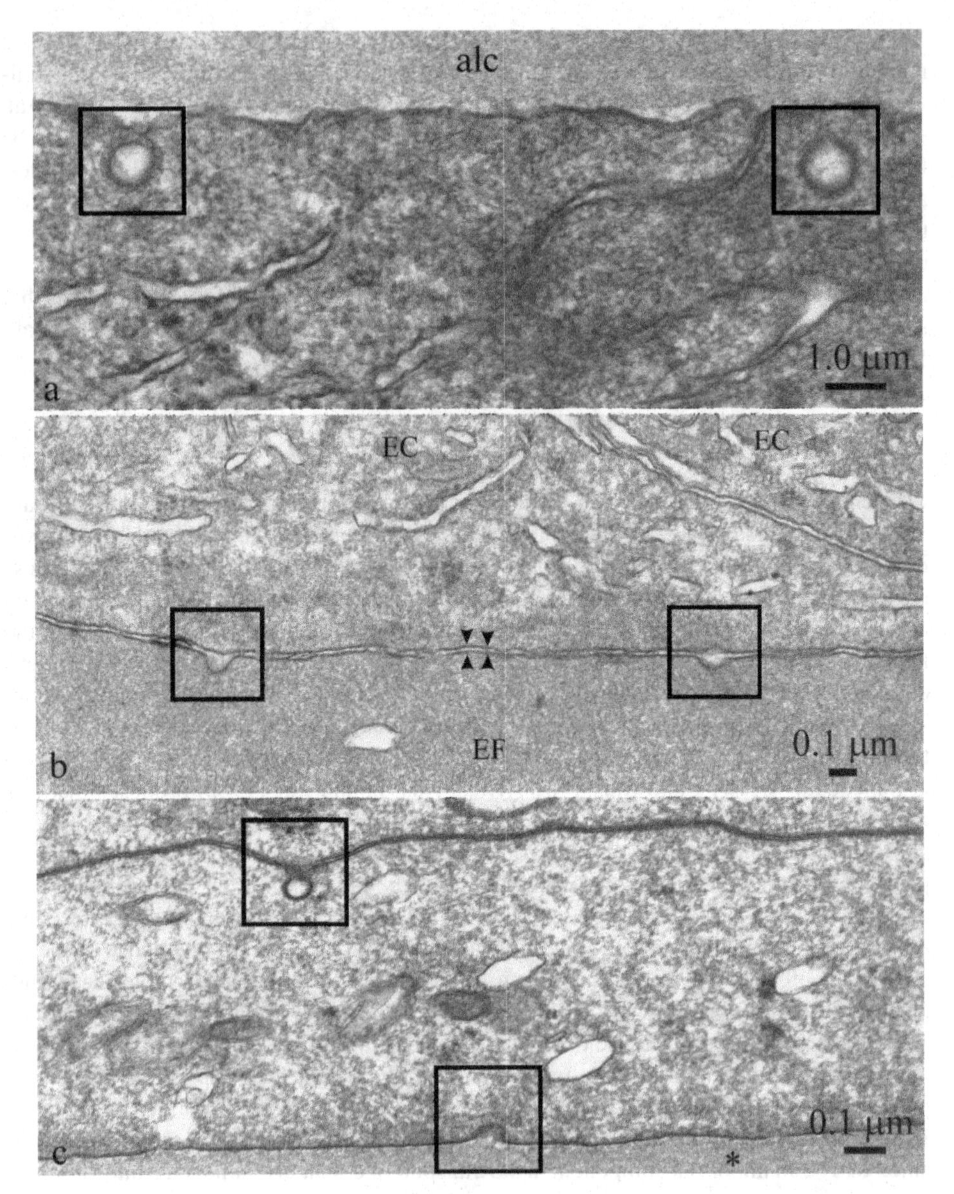

Figure 4.16. Transmission electron micrographs demonstrating endocytotic events (black squares) at different locations in a lens. (a) Clathrin-coated vesicles at the basal membranes of central zone epithelial cells beneath the anterior lens capsule (alc). (b) Forming endocytotic vesicles in the apical membrane of an elongating fiber at the central zone (cz) epithelial-fiber interface (efi; opposed arrowheads). Note the lack of intercellular junctions conjoining the apical surfaces of epithelial cells (EC) and an elongating fiber (EF) opposed in this region of the cz efi. (c) A clathrin-coated vesicle at the basolateral membrane of an elongating fiber and a forming endocytotic vesicle at the basal membrane of the same elongating fiber immediately subjacent to the posterior lens capsule (black asterisk). From Kuszak and Brown, 1993, courtesy of W. B. Saunders Co., Philadelphia, PA.

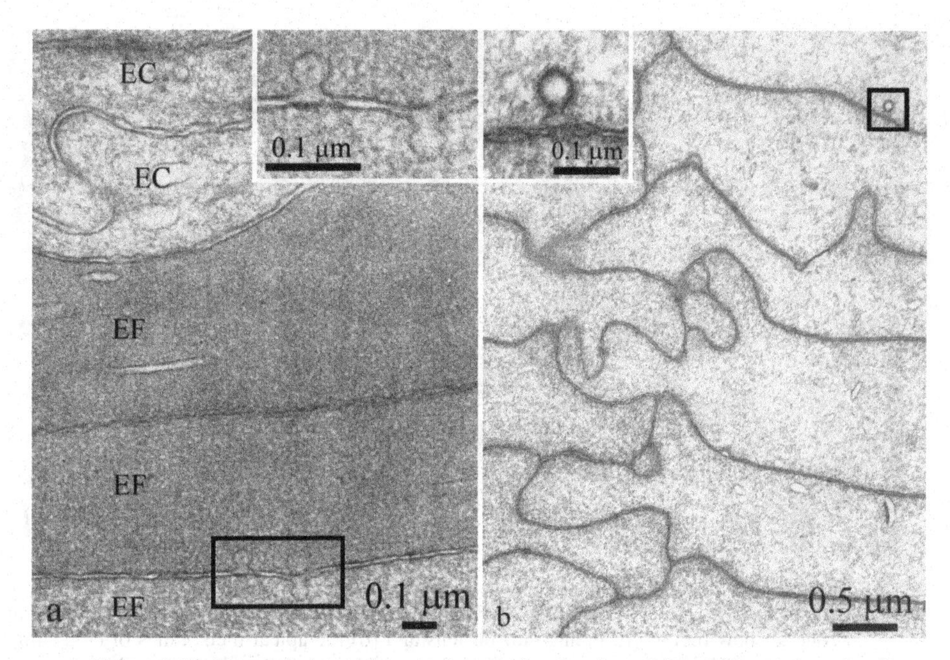

Figure 4.17. Transmission electron micrographs showing additional examples of endocytotic events (black squares) at different locations in a lens. (a) Forming endocytotic vesicles (shown enlarged in upper right inset) at the apicolateral membranes of elongating fibers (EF) two and three layers beneath the central zone epithelium (EC). (b) A clathrin-coated vesicle (shown enlarged in upper left inset) at the apicolateral membrane of a cortical fiber terminated at an anterior suture plane. From Kuszak and Brown, 1993, courtesy of W. B. Saunders Co., Philadephia, PA.

for studying membrane ultrastructure (Kreutziger, 1968; Chalcroft and Bullivant, 1970; McNutt and Weinstein, 1970). In this study, freeze-etch replicas containing >20,000 μm^2 of adult chick lens efi revealed only two gap junctions. This analysis included the ultrastructural characterization of numerous complete apical surfaces of elongating fibers that had interfaced with cz epithelial cells *in situ*. Thus, the possibility that gap junctions went undetected due to limited membrane exposure, as can be the case with thin-section analysis (Revel et al., 1971; Gabella, 1979; Ryerse and Nagel, 1991), was minimized. Additional correlative structural and physiological studies of embryonic chick (Bassnett et al., 1994) and adult frog lens efi (Prescott et al., 1991; Prescott et al., 1994) also confirmed that intercellular communication via gap junctions at the efi is very limited. In a study of primate lenses (Kuszak et al., 1995), quantitative analysis of >10,000 μm^2 of cz efi revealed no gap junctions. Correlative TEM thin-sections of >1,500 linear microns of cz efi from this region confirmed the presence of epithelial-epithelial gap junctions and elongating fiber–elongating fiber gap junctions but an extreme paucity of epithelial–elongating fiber gap junctions. In contrast, TEM thin-sections of >1,000 linear microns of pgz, gz, and tz efi revealed a number of epithelial–elongating fiber gap junctions. This finding is confirmed by freeze-etch analysis of mouse lens tz efi (Figs. 4.18 to 4.20).

The results of studies using electrotonic and dye-coupling techniques to assess intercellular coupling at the efi also reveal very limited communication at the efi. Bassnett et al. (1994) demonstrated that fluorescent dye (carboxyfluorescein diacetate) is retained by lens epithelial cells over an extended period of time without any significant evidence of dye

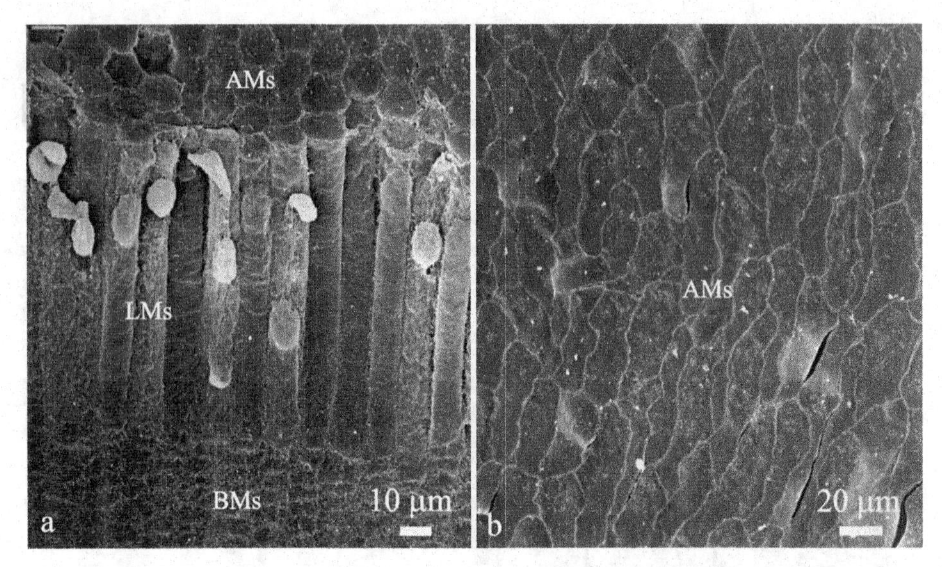

Figure 4.18. Low-magnification scanning electron micrographs of (a) the apical (AMs), lateral (LMs), and basal membranes (BMs) of nascent elongating fibers at the transitional zone, or bow region efi, in an adult rat lens and (b) the apical membranes of elongating fibers in an adult monkey lens at the central zone efi. Note that while the apical membranes of nascent elongating fibers are of relatively uniform size and shape (hexagonal), the apical membranes of elongating fibers at the cz efi are of neither uniform shape nor size. (a) From Kuszak, Brown and Peterson, 1996, courtesy of John Wiley & Sons, Inc. New York, NY. (b) From Kuszak, Novak and Brown, 1995, courtesy of Academic Press, London, UK.

uptake by the underlying elongating fibers. However, there was a small amount of dye transfer recorded from elongating fibers to epithelium. These results are consistent with comparable physiologic analyses that have shown either a lack of or restricted dye coupling at the efi (Goodenough et al., 1980; Rae and Kuszak, 1983; Miller and Goodenough, 1986). Finally, a study that specifically targeted dye coupling at the lens efi (Rae et al., 1996) presented evidence suggesting that less than 1 in 10 epithelial cells were coupled to underlying elongating fibers. This study also confirmed previous findings of Baldo and Mathias (1992) that showed gz epithelial cells are better coupled to underlying nascent elongating fibers than cz epithelial cells are to elongating fibers in the latter stages of terminal differentiation or to almost fully elongated immature fibers just prior to their detaching from the epithelium anteriorly and the capsule posteriorly.

In spite of the results of numerous studies that have shown only limited communication across the efi, it can still be reasonably argued that cell-cell coupling across this interface is important to lens physiology (Rae et al., 1996). Epithelial-epithelial and fiber-fiber coupling has been shown by correlative structural and functional studies to be essentially 1:1. Therefore, every epithelial cell does not need to be conjoined directly to an underlying elongating fiber.

4.5.1.3. Percentage of Fibers in Direct Contact with the EFI

Although the equatorial diameters of adult chicken (*Gallus domesticus*), frog (*Rana pipiens*) and rat (Wistar and Sprague-Dawley) lenses are, respectively, 7, 4.5, and 3 mm, SEM

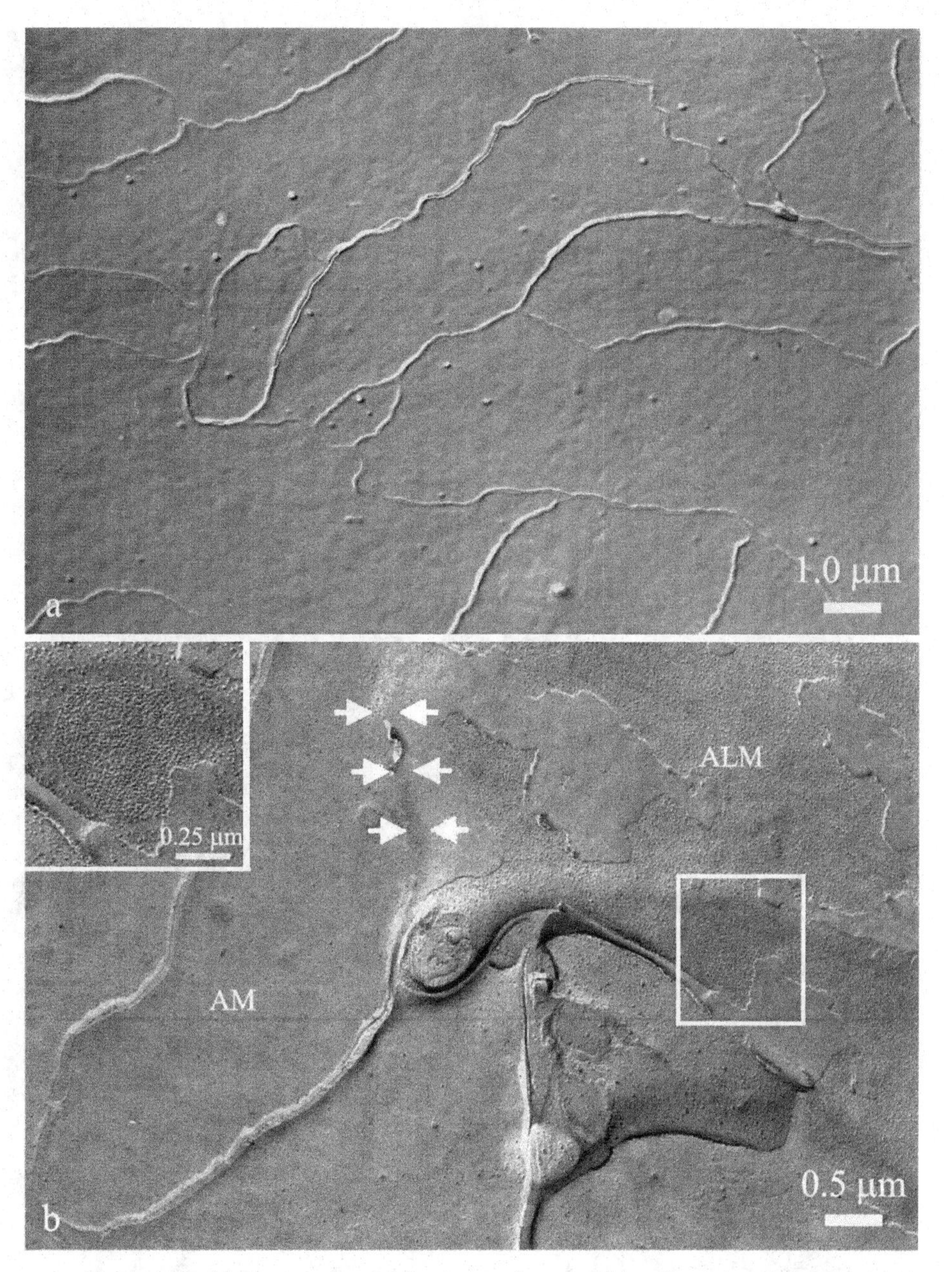

Figure 4.19. (a) A comparison of the shapes, sizes, and three-dimensionality of membrane surfaces shown in this low-magnification transmission electron micrograph of a freeze-etch replica with the appearance of adult monkey cz elongating fiber apical membranes (AM) shown by SEM in Figure 4.18b unequivocally confirms that the freeze-fracture plane has exposed the ams of elongating fibers at the cz efi. (b) In contrast to ams the tz, or bow region efi, the opposed ams of cz epithelial cells and elongating fibers are rarely conjoined by gap junctions. The only gap junctions found in this region were between elongating fiber apicolateral membranes (alms) near their apicolateral borders (opposed arrows; white boxed area shown at higher magnification in inset). From Kuszak, Novak and Brown, 1995, courtesy of Academic Press, London, UK.

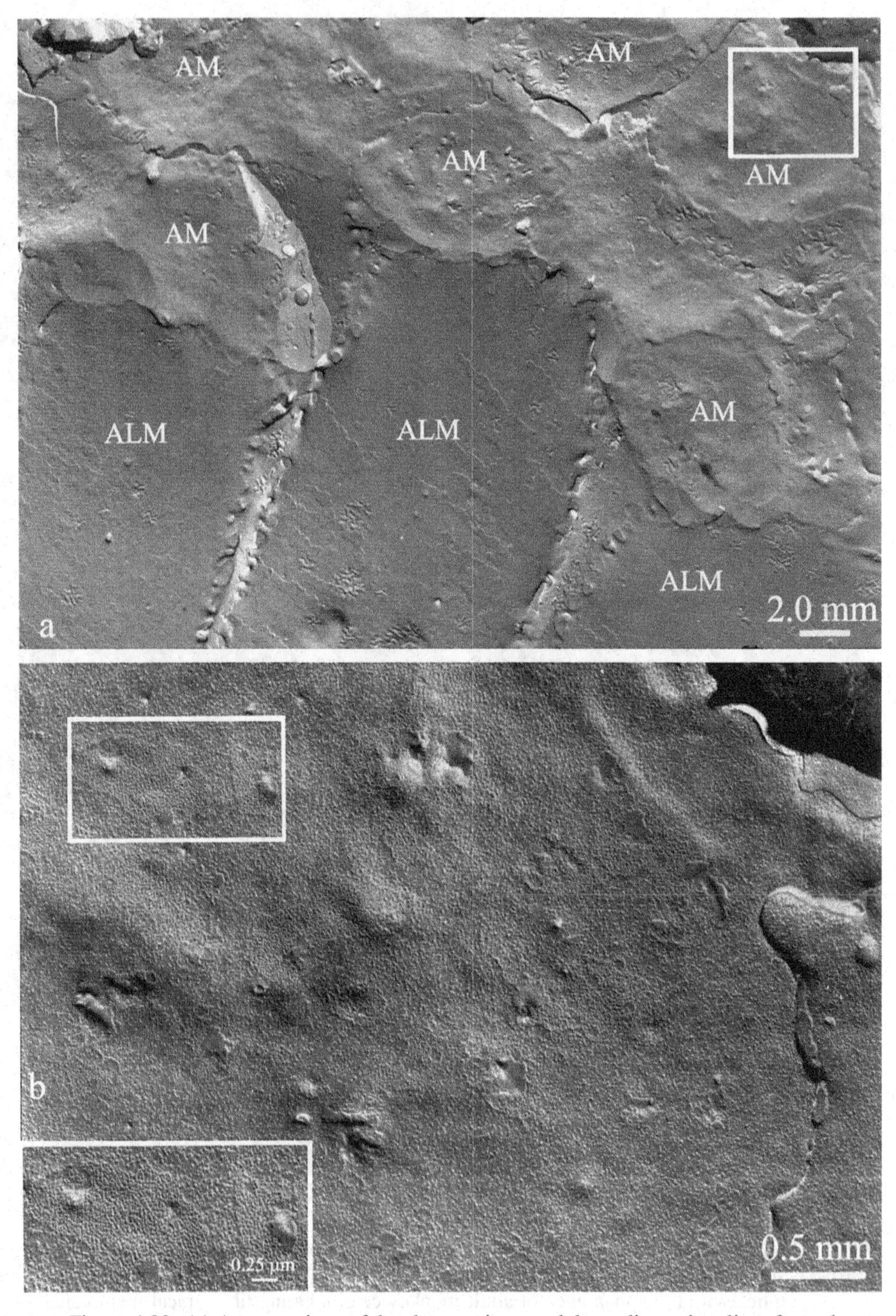

Figure 4.20. (a) A comparison of the shapes, sizes, and three-dimensionality of membrane surfaces shown in this low-magnification transmission electron micrograph of a freeze-etch replica with the appearance of adult mouse (tz) nascent elongating fibers shown by SEM in Figure 18a unequivocally confirms that the freeze-fracture plane has exposed the apical membranes (AM) and apicolateral membranes (ALM) of elongating fibers at the tz efi. (b) Higher magnification of white boxed area from (a). In contrast to the cz efi, the opposed AM of tz epithelial cells and nascent elongating fibers are conjoined by a number of gap junctions (gap junctions within white boxed area shown at higher magnification in inset).

analysis of all of these lenses shows that a fiber that breaks contact with either the cz epithelium at the anterior pole or the posterior capsule at the posterior pole is approximately 200 μm deep at the equator (Kuszak and Rae, 1982; Kuszak et al., 1986). The consistent depth of the elongating fiber zone, or superficial cortex, is a function of the fact that elongating fibers are thinnest in their midportions, widest and most flared at their anterior ends, and intermediate in these parameters in their posterior segments (Kuszak et al., 1986). It has also been demonstrated in the primate lens that elongating fibers extend to a depth of approximately 200 μm at the equator (Bassnett and Beebe, 1992; Bassnett, 1992). The equatorial diameter of the adult primate lens is approximately 9 mm. Thus, the percentage of fibers in direct contact with the epithelium at the efi will vary as a function of size and/or age. Estimates of the percentage of fibers in direct contact with the epithelium at the efi can be derived from standard mensuration formulas for spheroidal geometry (Eves, 1978). When calculating such estimates, it is important to take into account that lenses are asymmetrical, oblate spheroids and not spheres. Failure to accurately factor in lens shape results in a gross overestimation of lens volume.

The shapes and relevant axial dimensions of mouse and human lenses used to estimate the percentage of fibers in direct contact with the epithelium at the efi are shown in Figure 4.21.

Since all vertebrate lenses are asymmetrical, oblate spheroids, the total volume estimate for any lens can be calculated as the sum of the anterior lens volume, defined as $(4/3\ \pi a_1{}^2 b_1)/2$, and the posterior lens volume, defined as $(4/3\ \pi a_1 b_2{}^2)/2$, where a_1 is the major equatorial radius, b_1 is the minor anterior radius, and b_2 is the minor posterior radius. The same formulae can be used to estimate the percentage of fibers in direct contact with the epithelium at the efi. However, in this case, a_1 is replaced by a_2, the major equatorial radius minus the 200 μm contribution of the elongating fibers. Of course, in the case of a lens like the mouse lens at birth, subtracting out the 200 μm contribution of the elongating fibers effectively changes the major and minor axes of the lens. Therefore, in this instance, the volume estimate for the percentage of fibers not in direct contact with the epithelium at the efi must be calculated as the sum of the anterior and posterior lens volume. Calculations of the volumes and percentages for human, and mouse lenses are summarized in Table 4.3.

If the size of the elongating fiber zone is constant throughout life and between species, as suggested by morphological studies to date (Kuszak and Rae, 1982; Kuszak et al., 1986; Bassnett and Beebe, 1992; Bassnett, 1992), then more fibers are in direct contact with epithelial cells at the efi in smaller and rounder lenses (e.g., frog, mouse, and rat) than in larger and flatter lenses (e.g., cat, dog, monkey, baboon, and human). It is important to note that presuming a lens such as a mouse lens to be essentially spherical results in a gross overestimation of the percentage of fibers that are in direct contact with epithelial cells at the efi (the spheroidal estimates are 45% at birth and 29% in an adult lens; the spherical estimates are 69% at birth and 59% in an adult lens).

4.5.2. Fibers

Because every fiber is maintained for a lifetime, the lens presents an ideal model for studies of cellular senescence. In this section, the morphology of the oldest fibers, the primary fibers of the embryonic nucleus, is compared to the morphology of the younger secondary fibers.

In normal transparent lenses, the cytoplasm of fibers in all the developmental regions is smooth and homogeneous. This fact can be confirmed by Fourier transform analysis. A selected region of an adult human embryonic nuclear fiber in Fig. 4.22a is displayed as a transform and a radially averaged plot in Figure 4.22b. The smoothly varying curve and minimal

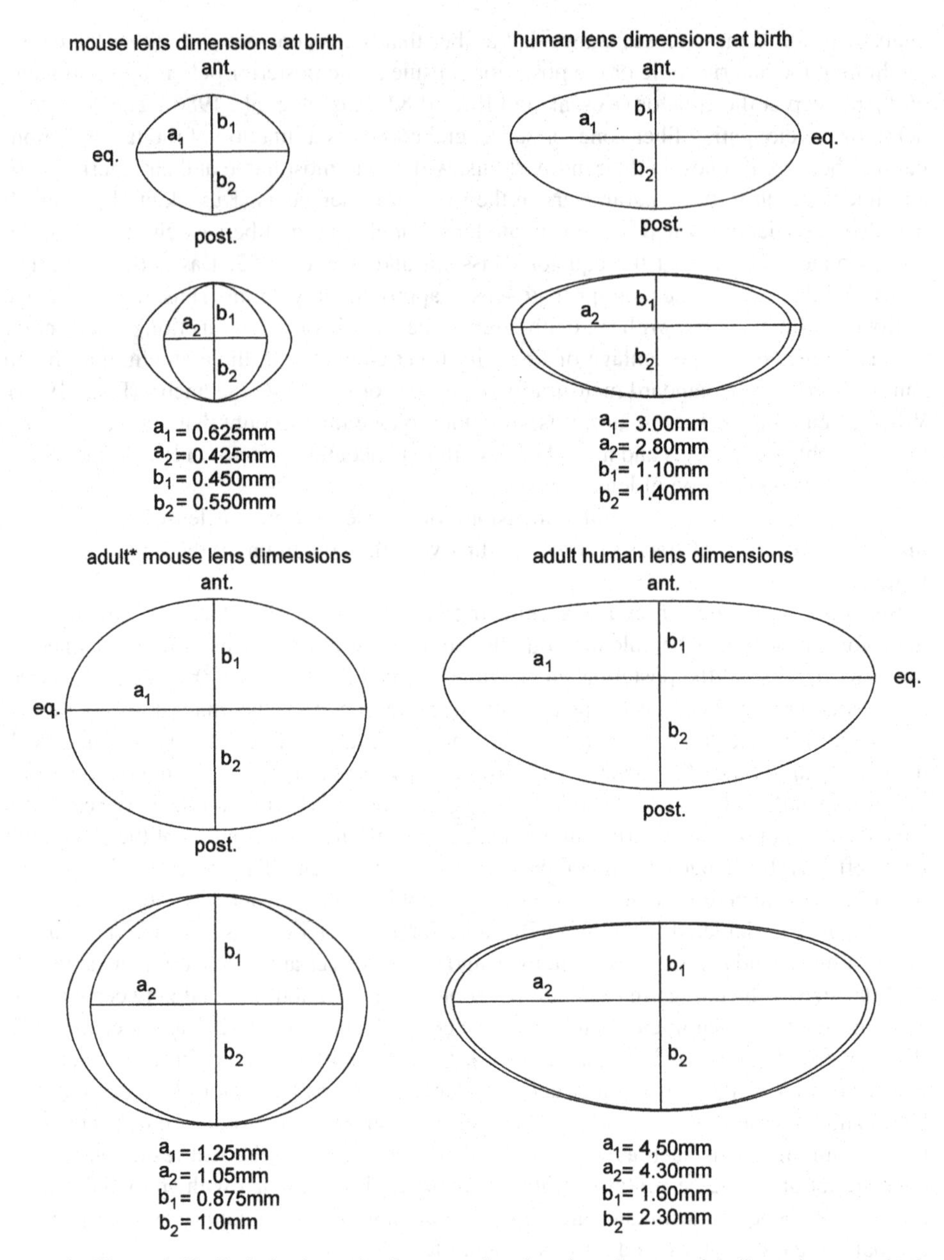

Figure 4.21. Scaled schematic diagrams showing the relevant axial dimensions required to estimate the percentage of the lens superficial cortex in direct contact with the lens epithelium in juvenile and adult mouse and human lenses.

intensity near the center suggest that the cytoplasmic proteins and water are packed into an amorphous medium with minimal fluctuations in refractive index (Freel et al., 2002; Taylor and Costello, 1999). The principal components of fiber cytoplasm are the specialized lens crystallin proteins and the lens cytoskeleton. The crystallins provide a medium of high refractive index. With increased age, a variable concentration of the different types of crystallins is related to the higher water content in the lens cortex compared with the

Table 4.3. *Percentage of Lens Volume Containing Superficial Cortical Fibers in Direct Contact with the Lens Epithelium as a Function of Development and Growth*

	Mouse Lenses		Human Lenses	
	At Birth	Adult	At Birth	Adult
Total lens volume	0.82 mm^3	6.14 mm^3	47.12 mm^3	165.38 mm^3
Percentage of lens volume containing fibers "not in direct contact" with the epithelium at the EFI	55% (0.45 mm^3)	71% (4.33 mm^3)	87% (41.04 mm^3)	91% (151.01 mm^3)
Percentage of lens volume containing fibers "in direct contact" with the epithelium at the EFI	45% (0.37 mm^3)	29% (1.81 mm^3)	13% (6.07 mm^3)	9% (14.37 mm^3)

nucleus. The breakdown of crystallins and their subsequent aggregation or cross-linking are believed to be responsible for some opacities.

The fiber cytoskeleton consists of actin; intermediate filaments; beaded filaments; and the proteins spectrin, alpha-actinin, myosin, and tropomyosin. It is presumed that these components are collectively arranged into a supportive network that pervades the cell to organize the lens crystallins and maintain fiber shape as a function of age and accommodation. With increased age, most formed cytoskeletal elements are modified and are not found in the lens nucleus (Alcala and Maisel, 1985).

The interfaces between fibers are consistently composed of plasma membrane pairs from adjacent fibers. During differentiation, the fibers are separated by a well-defined extracellular space except at gap junctions. Due to terminal differentiation, the extracellular space is minimized as the paired membranes become transformed into a complex topology involving a variety of intercellular junctional contacts. In a low-magnification overview (Fig. 4.22a), it is possible to locate gap junctions (white arrows), high-amplitude undulating membranes (arrowheads), and cytoplasmic profiles (black arrows). At high magnification, the thickness of a fiber gap junction typically measures 16 nm; the gap, which measures 1–2 nm, is not visible in thin-section images (Fig. 4.22c). The fiber gap (communicating) junctions are thinner than gap junctions from other tissues, such as liver and stomach, which have visible gaps of 2–4 nm (Peracchia et al., 1985). The high-amplitude undulating membranes at intermediate magnification often display the overlap regions as pentalamellar profiles (Fig. 4.22d, arrows), similar in appearance to gap junctions. At high magnification one such overlap region (Fig. 4.22e, black arrows) measures 14 nm, noticeably thinner than a 16-nm-thick gap junction occurring at the end of a domain of high-amplitude undulating membranes (Fig. 4.22e, white arrow). The thinner junctions have been termed "thin symmetric junctions" (Costello et al., 1989). In addition, the curved crystalline arrays of MIP/aquaporin0 (arrowheads) can be followed as they merge with the short segment of a thin symmetric junction (Fig. 4.22e, black arrows). Circular profiles (Fig. 4.22f) and edge processes are composed of paired membranes, often with curvatures similar to the undulating membranes. These objects are associated with and most likely derived from projections of the plasma membrane pairs of adjacent fibers.

The large and irregular embryonic fibers are overlain with progressively smaller and more regular fibers, as noted earlier (Fig. 4.12b) and evident in the images of bovine lenses (Fig. 4.23).

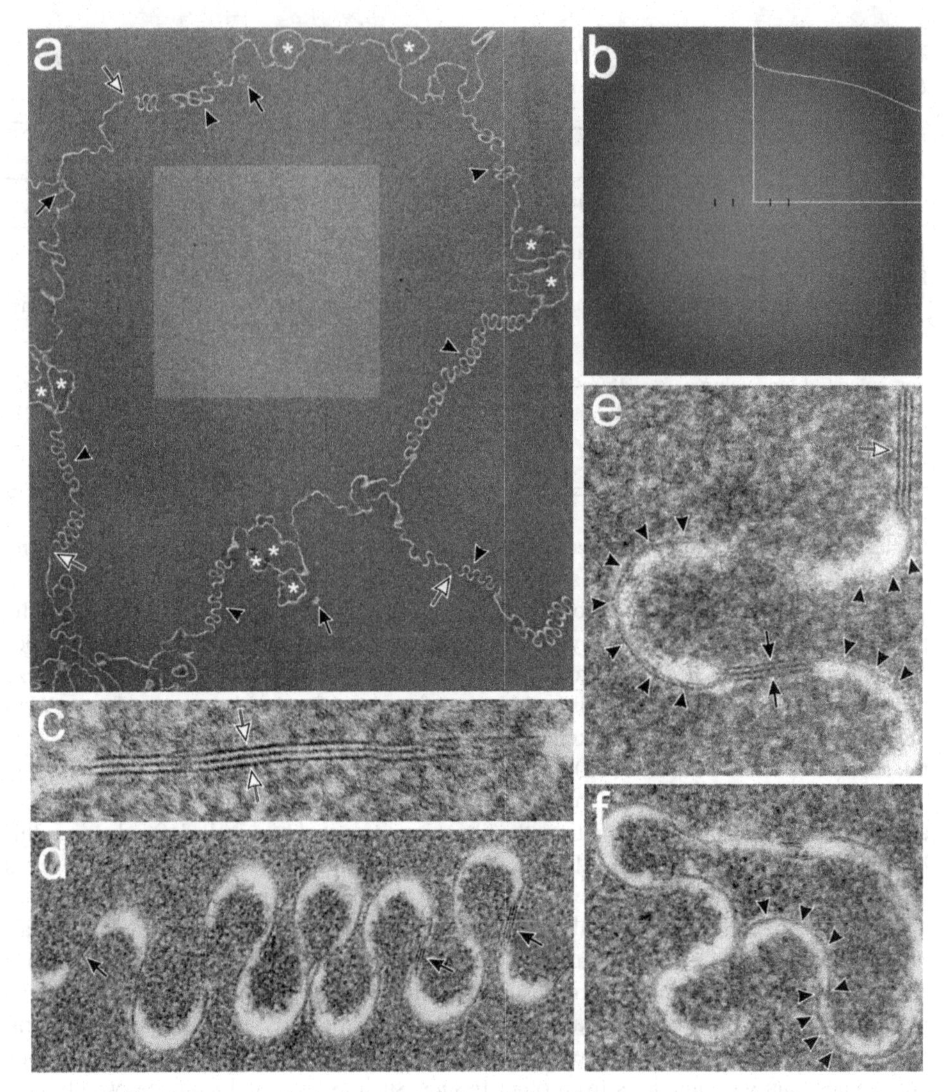

Figure 4.22. Transmission electron microscopy of human embryonic fiber. (a) Low-magnification overview of a typical fiber from the embryonic nucleus (age 67). The numerous undulating membranes (arrowheads), edge processes (asterisks), and cytoplasmic profiles (black arrows) are obvious. The gap junctions (white arrows) appear as breaks in the intercellular interface (white lines), and their structure can only be confirmed at higher magnification. (b) Fourier transform of the region marked in panel a. Note the smoothly varying transform and the radially averaged plot characteristic of homogeneous cytoplasm. (c) Gap junction that consistently measures 16 nm thick (arrows). (d) High-amplitude undulating membranes responsible for the furrowed surfaces seen by SEM. Overlap regions often display pentalamellar profiles (arrows). (e) High magnification of a gap junction (16 nm; white arrow) leading into undulating membranes and a thinner pentalamellar profile (14 nm; black arrows) where curved square arrays (arrowheads) of MIP/aquaporin0 overlap. (f) Cytoplasmic profile at high magnification reveals that cytoplasm is composed in part of undulating membranes (arrowheads). Pairs of nonjunctional membranes are also observed. All of the profiles that display membranes are surrounded by paired membranes, indicating that they are sectioned interdigitations between adjacent fibers.

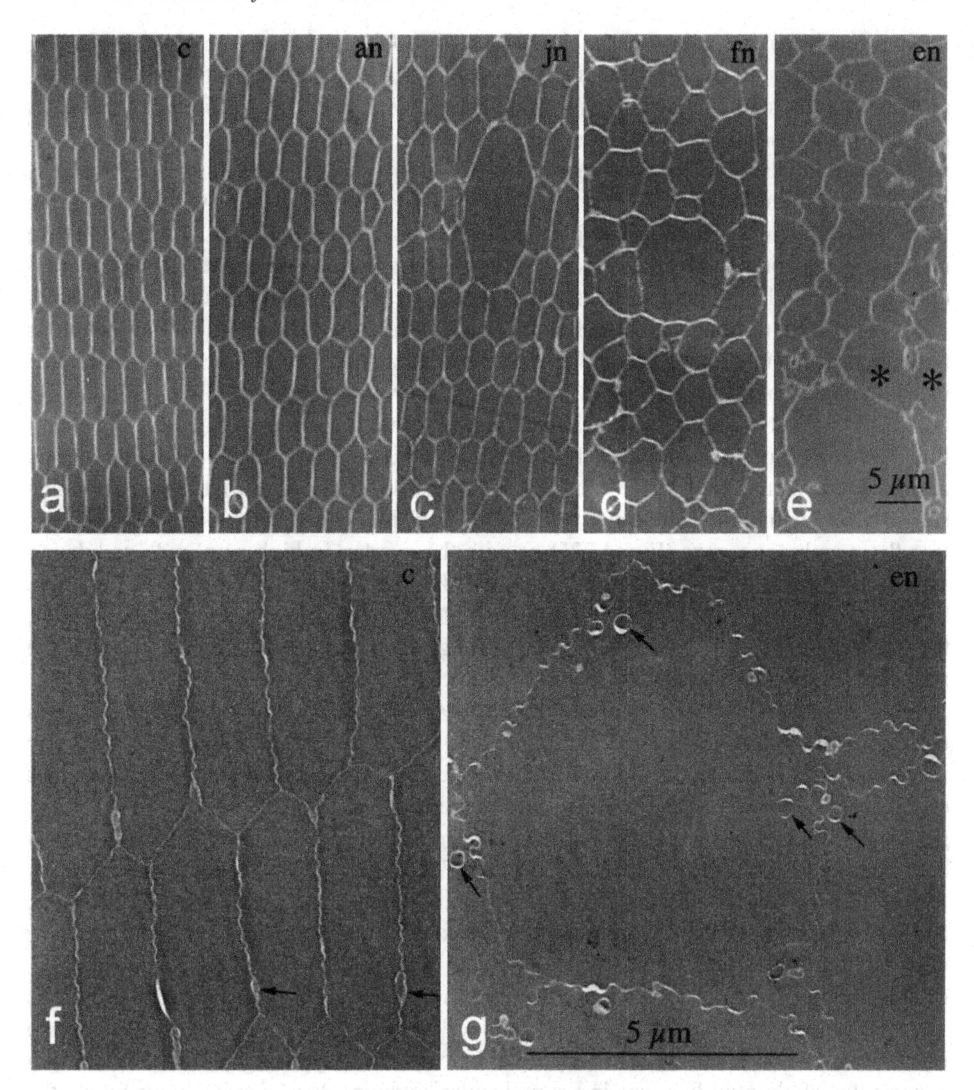

Figure 4.23. Light and electron microscopy of bovine fibers. (a–e) Light micrographs of toluidine blue–stained, 1 μm–thick sections showing the cortex (c), adult nucleus (an), juvenile nucleus (jn), fetal nucleus (fn), and embryonic nucleus (en), respectively. Note the ordered radial cell columns in the cortex and adult nucleus and the progressively more disordered patterns in the fetal and embryonic nuclei. The fibers in panel e marked with asterisks may represent the fusion of two fibers. (f) Transmission electron micrograph of cortical fibers. Note the undulating membranes and a few edge processes (arrows). (g) Transmission electron micrograph of embryonic fibers showing many edge processes and cytoplasmic profiles (arrows), probably derived from edge processes that project into neighboring fibers. (c)–(e) From Al-Ghoul and Costello, 1997, courtesy of *Molecular Vision.*

Note that the bovine embryonic nuclear fibers (Figs. 4.23e and g) are similar in shape, size, and arrangement to those seen in human lenses (Fig. 4.22a). Bovine intercellular junctions are shown in Figure 4.24. High magnification of the bovine fiber gap junctions from intact or isolated membranes (Figs. 4.24a–c and e) are 16 nm thick whereas thinner membrane pairs with a similar pentalamellar staining profile, the "thin symmetric junctions" (Figs. 4.24b, d, and e), are 14 nm thick (Costello et al., 1989). Still thinner junctions within undulating membranes have an asymmetric appearance and are 11 nm thick (Figs. 4.24b–e). These undulating membrane pairs have been shown to contain square arrays of MIP/aquaporin0 (Zampighi et al., 1989) and thus are termed "square array junctions" (Costello et al., 1989).

It is important to understand the key structural differences between fiber gap junctions and square array junctions because these differences lead to different functions. The molecular interpretation of thin-section TEM images is shown diagrammatically in Figure 4.24f (Costello et al., 1993). Previous evidence suggests that the offset of square array membrane proteins correlates with the thin (11-nm) asymmetric junctions seen conjoining low-amplitude undulating membrane pairs (Costello et al, 1989). In contrast, the overlap of two square array membranes correlates with the 14-nm thin symmetric junctions conjoining two flat membrane arrays (Fig. 4.22f) or with the overlap in high-amplitude undulating membrane pairs (Fig. 4.24f, arrows). These results imply that, in addition to the possible role of MIP/aquaporin0 in water transport (Mulders et al., 1995; Chandy et al., 1997), the crystalline arrays of this abundant fiber integral membrane protein may promote the adhesion of adjacent fibers (Fotiadis et al., 2000).

Fiber gap junctions, composed of connexins 46 and 50, conjoin the lateral membranes of neighboring fibers (Figs. 4.15, 4.22, and 4.24). Freeze-etch analysis reveals fiber gap junctions to be composed of complementary aggregates of transmembrane protein particles (connexins) conjoined across a narrowed intercellular space or "gap" (Fig. 4.15). Fiber gap junctions occur primarily between the midsegments of fibers, and their density varies considerably between species (Kuszak et al., 1978; Kuszak et al., 1985; Lo and Harding, 1986). The midsegments of chick, rat, frog, and human fibers have, respectively, 65%, 33%, 12%, and 5% of their membrane specialized as fiber gap junctions (Fig. 4.25). Fiber gap junctions differ from "typical" gap junctions (e.g., gap junctions between hepatocytes or lens epithelial cells; Peracchia et al., 1985) in that neither their amino acid sequences nor their ultrastructures are identical. The minor structural differences between fiber gap junctions (as well as cardiac and otocyst gap junctions) and typical gap junctions strongly suggest that slight modifications in the molecular components of gap junctions reflect the terminal differentiation process rather than a major physiological distinction.

The membrane junctions and edge processes have a major influence on the fiber surface topology and are most clearly visualized in SEM images. The lateral membranes of fibers feature interdigitations arrayed along their length. These outpocketings and enfoldings of lateral membrane, known as "balls and sockets" and "flaps and imprints," occur at the angles made by the six faces of fibers (Fig. 4.26).

In all lenses, after the elongation and terminal differentiation processes are completed, the smooth lateral membranes of fibers are radically altered. SEM reveals that mature and aged fiber membranes become characterized by numerous polygonal domains of furrowed membrane (Fig. 4.27). These domains are the low- and high-amplitude undulating membranes seen in thin-section EM. Freeze-etch analysis reveals that the domains are characterized by patches of square array membrane (Fig. 4.24; Costello et al., 1985; Lo and Harding, 1984; Zampighi et al., 1989) alternating with protein-free patches aligned along the furrows and crests of the ridged membrane (Fig. 4.28).

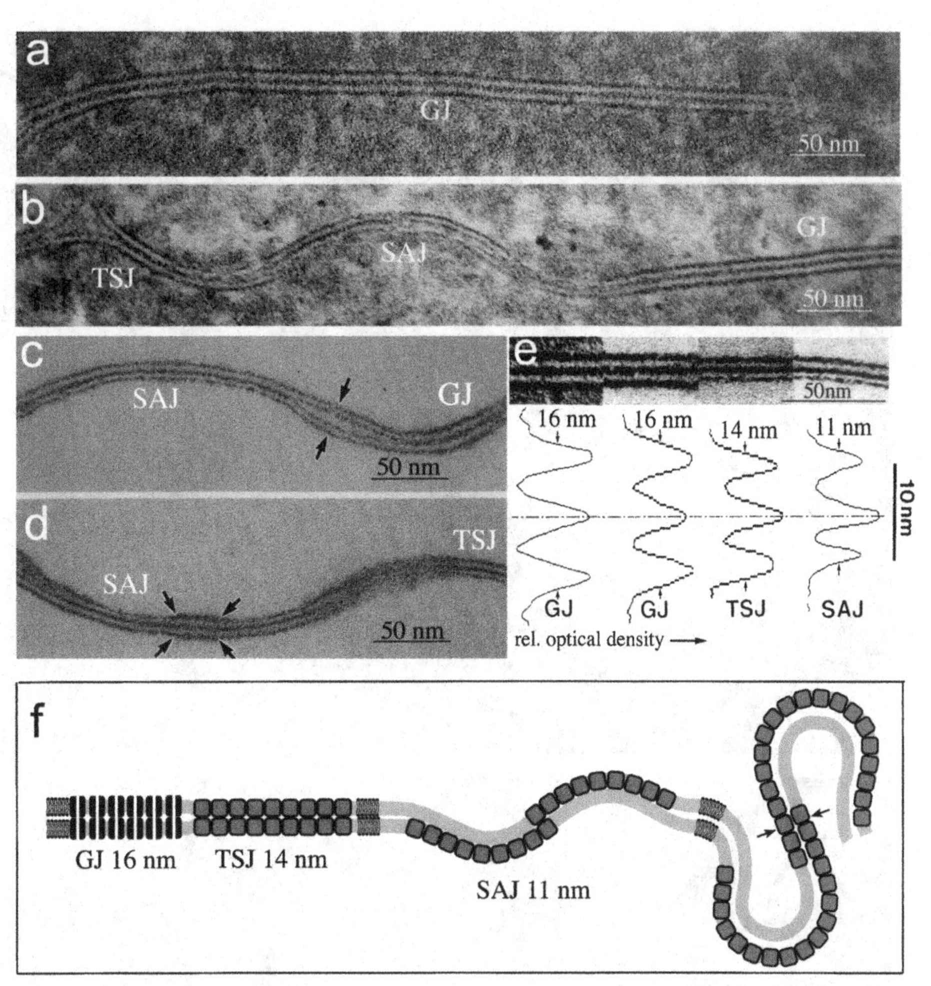

Figure 4.24. High-resolution transmission electron micrographs showing the different types of bovine fiber junctions. (a) Gap junction (GJ) from preserved intact embryonic nucleus. Note uniformity of stain and thickness. (b) Different junctional types (thin symmetric junction [TSJ], square array junction [SAJ], and GJ) from preserved intact inner cortical fibers. (c) Isolated membranes revealing an SAJ, nonjunctional plasma membranes (arrows), and a GJ. (d) Isolated membranes showing an SAJ, a region of abrupt change in membrane thickness (arrows), and a TSJ. (e) High-magnification comparison of (left to right) a GJ from intact tissue and a GJ, a TSJ, and an SAJ from isolated membranes. Compare and contrast the 14 nm–thick symmetric paired membranes of the TSJ with the 11 nm–thick asymmetric paired membranes of the SAJ. Optical density scans of calibrated images display the quantitative differences and give the average thickness of each type of junction. (f) Schematic diagram giving a molecular interpretation of the different fiber junctional types. The small squares represent MIP/aquaporin0. The TSJs are characterized by adherent layers of square arrays of aquaporin0 in register. Low- and high-amplitude undulating membranes are characterized by square arrays of aquaporin0 out of register (SAJ), and occasional short stretches of TSJs (arrows).

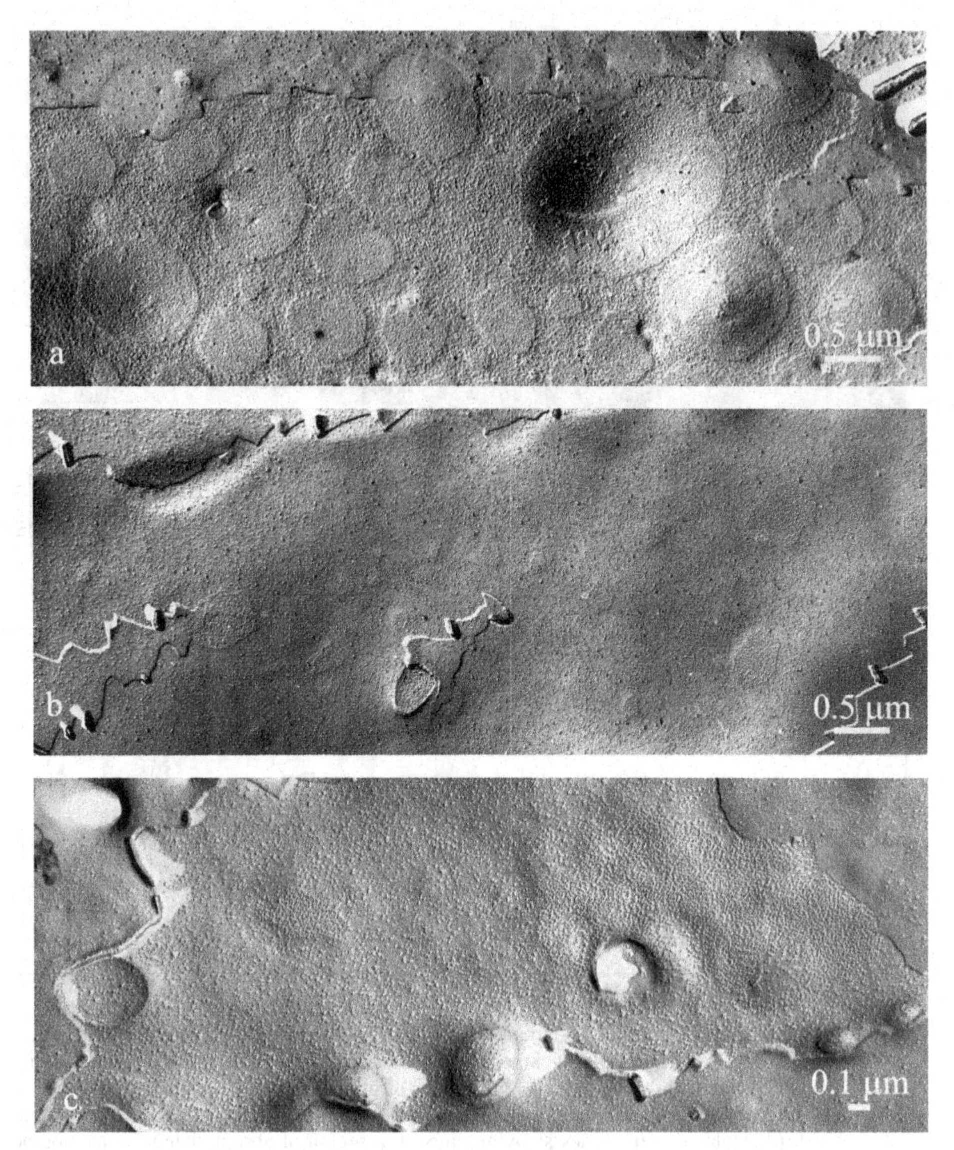

Figure 4.25. Transmission electron micrographs of freeze-etch–replicated adult cortical fiber middle segments from (a) rat, (b) frog, and (c) monkey lenses. Note the variation in the density of fiber gap junctions between these comparable fibers from different species. From Kuszak, 1995, courtesy of Academic Press, London, UK.

The fact that the character of the fiber membrane varies as a function of maturity and age has led to speculation that this is the result of a redistribution of MIP/aquaporin0 channels within the plane of the membrane. The functional significance of such a dramatic remodeling of the fiber membrane is unknown.

4.5.2.1. Cell-to-Cell Fusion of Fibers

Closely apposed regions of lateral membrane from neighboring fibers are frequently fused for a variable distance along the fiber length (Fig. 4.29; Kuszak et al., 1985; Shestapolov

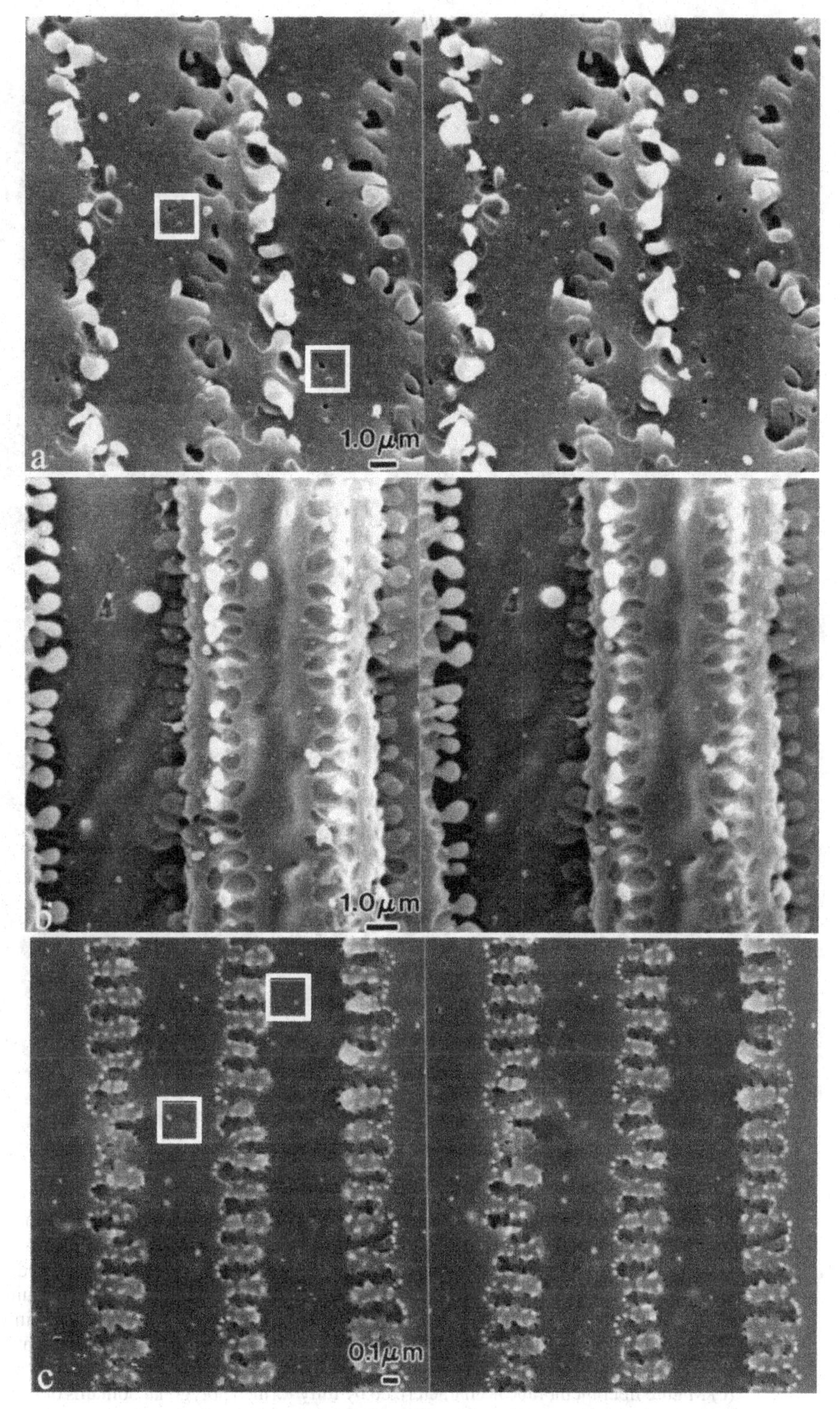

Figure 4.26. Scanning electron micrograph stereopairs showing different types and distributions of cortical fiber interdigitations (lateral edge processes) from (a) newborn human, (b) bovine, and (c) fish lenses. Note the complementarity between flaps and imprints of adjacent fibers. Examples of small balls and sockets are shown within white squares. From Kuszak, Brown and Peterson, 1996, courtesy of John Wiley and Sons, New York, NY.

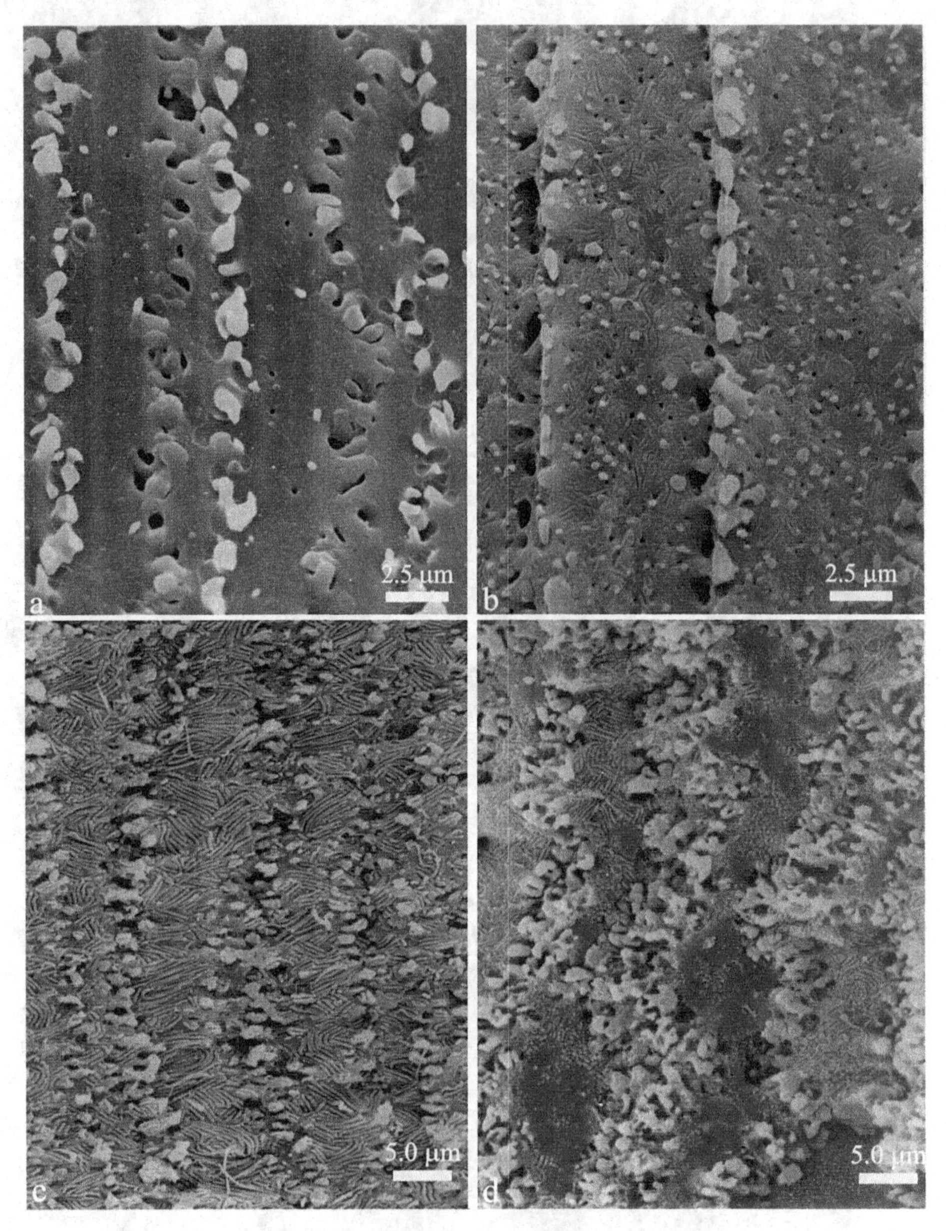

Figure 4.27. Scanning electron micrographs showing the surface alterations in primate fiber membrane as it transitions through periods of development, growth, and aging. (a) Newly formed fibers have smooth lateral membrane with lateral interdigitations arising from the angles formed between broad and narrow faces and between narrow faces aligned along their length. (b) As fully elongated fibers continue the maturation process by eliminating most of their internal organelles, aquaporin0, the main intrinsic membrane protein (MIP 26K), is redistributed within the plane of the membrane-forming polygonal domains of furrowed and ridged membrane (the undulating membrane seen by thin-section analysis). (c) Mature fiber membrane is characterized by fully formed polygonal domains of furrowed and ridged membrane and by lateral interdigitations. (d) As fibers age, compaction causes folds along the long axis, resulting in some polygonal domains of furrowed and ridged membrane becoming alternately internalized into or protruding from adjacent fibers. From Kuszak et al., 2000, courtesy of W. B. Saunders Co., Philadelphia, PA.

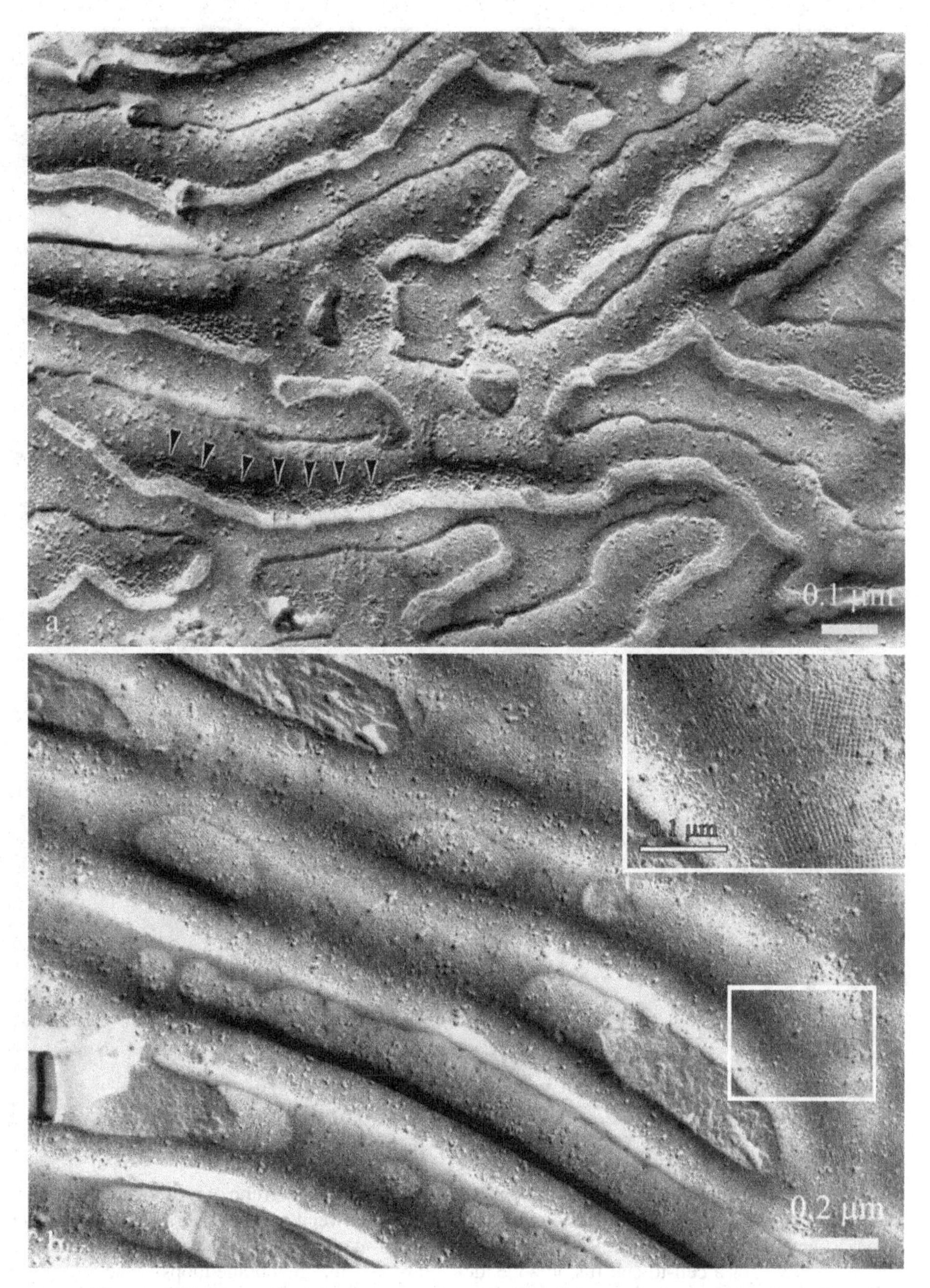

Figure 4.28. Transmission electron micrographs of freeze-etch–replicated adult nuclear fiber middle segments from monkey lenses. (a) In p-face exposure, patches of square array membrane consisting of aquaporin0 (arrowheads) are seen alternating with protein-free areas within the furrows of fiber membrane. (b) In e-face exposure, patches of square array membrane are seen in p-face exposure alternating with protein-free areas on the crests of ridged fiber membrane (higher magnification of representative square arrays shown in inset). (a) From Kuszak and Brown, 1993, courtesy of W. B. Saunders Co., Philadelphia, PA.

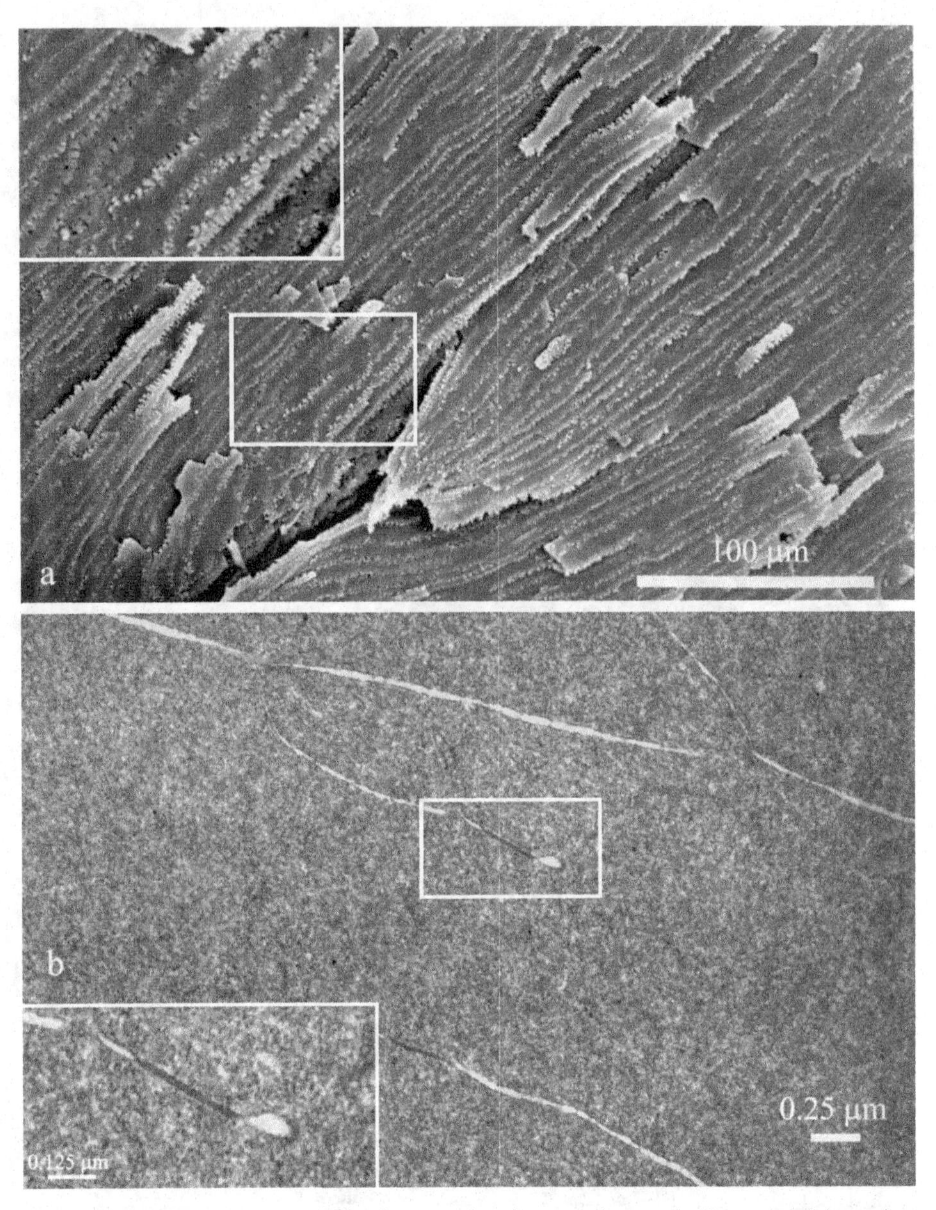

Figure 4.29. (a) Low-magnification scanning electron micrograph of a young adult monkey lens showing numerous cell-to-cell fusion zones between cortical fibers as they approach an anterior suture branch. (b) Transmission electron micrograph showing cytoplasmic continuity, or a cell-to-cell fusion zone, defined by plasma membrane loops at the ends of or proximal to gap junctions between rat elongating fibers in thin sections. White boxed areas shown at higher magnification in insets. From Kuszak, 1995, courtesy of Academic Press, London, UK.

and Bassnett, 2000b). Fusion zones are frequently seen to occur between the anterior segments or posterior segments of fiber cells as they approach their sutural locations. The frequent cell-to-cell fusion zones near sutures are thought to constitute a means for fibers to change their direction or curvature to conform to the precise modeling of sutures required for proper lens function. Cell-to-cell fusion zones also provide large patent pathways for intercellular transport between fibers for substances too large to pass through fiber gap junctions.

4.5.2.2. Compaction of Aged Fibers in Different Developmental Regions

Because the lens grows throughout life, it is necessary to compact the fiber mass so that the lens does not outgrow its place within the eye. Compaction, although not evident in young humans or animals, is particularly evident in aging humans.

The age-related changes in fiber compaction within the different developmental regions are best appreciated by relating equatorial thin-section TEM images, where all fibers are cut in cross section at their midpoints (Fig. 4.30), to SEM analyses of lenses split along their anteroposterior axis to reveal the end-to-end arrangement of nuclear fibers in growth shells and radial cell columns (Fig. 4.31).

In thick and thin sections taken through the lens equatorial plane, the reduction in fiber width as a result of fiber compaction is most evident at the transition from the cortex to the adult nucleus (Figs. 4.30a and b). This view also reveals that fiber compaction occurs in the juvenile and fetal nuclei, although to a lesser extent in the fetal nucleus (Taylor et al., 1996; Al-Ghoul and Costello, 1997). Comparable thick and thin sections through the embryonic nucleus suggest that age-related compaction in the oldest region of lenses is insignificant (Figs. 4.30c and d).

However, significant age-related compaction in the fetal and embryonic nuclei is readily apparent in SEM images (Fig. 4.31; Al-Ghoul et al., 2001). Low-magnification overviews (Figs. 4.31a and b) show that a young human lens has a thicker embryonic nucleus and a less acute fetal nuclear fiber angle (more spherical shape) than an older lens. The high-magnification views demonstrate that the fibers from the older lens have accordionlike folds that decrease the overall thickness of the embryonic nucleus along the optic axis (Figs. 4.31c and d). The SEM images also reveal the complex pattern of furrowed membranes that correspond exactly to the undulating membranes visible in TEM thin sections or freeze-etch replicas (Figs. 4.22 and 4.28). Finally, the knobs that appear on the fibers in the SEM images correspond to the collection of edge processes that are seen in TEM images (Figs. 4.22 and 4.30) and are exposed when adjacent fibers are split apart during the preparation of SEM specimens. An important conclusion is that a full appreciation of the complex topology of the fibers is only possible when both SEM and TEM techniques are employed.

4.6. Summary

The lens is a prime example of how an organ modifies its structure through terminal differentiation to accomplish its function. It produces fibers of defined shape and size and arranges these cells into radial cell columns and growth shells throughout a lifetime. It uses an abundance of specialized cytoplasmic crystallin proteins, cytoskeletal elements, intercellular contacts (gap junctions and square array membranes), and cell-to-cell fusion to produce a structure that will remain viable over many decades. The crystallin proteins

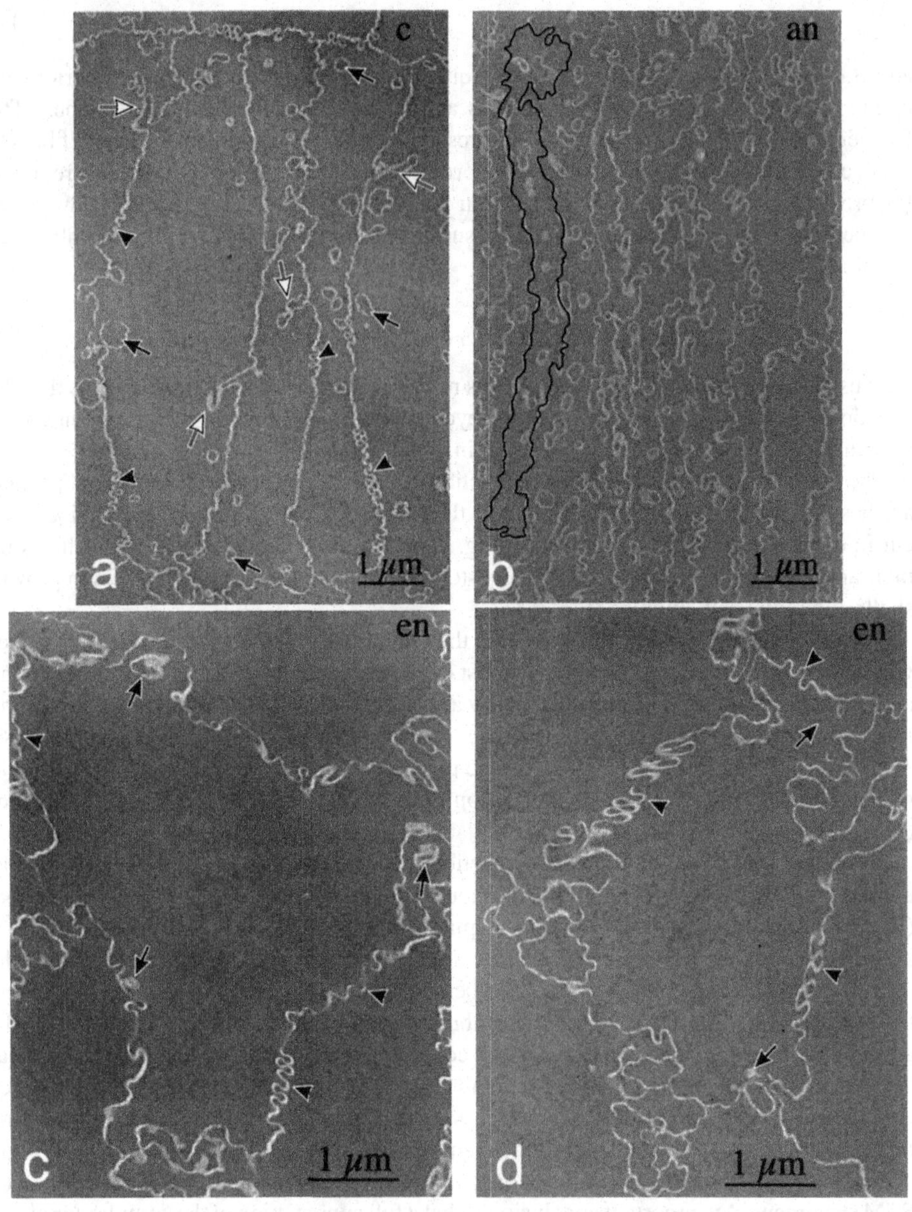

Figure 4.30. Transmission electron micrographs of human fibers. (a) Inner cortex showing smooth cytoplasm of five fibers in a radial cell column. Note the undulating membranes (arrowheads) and circular profiles (black arrows). It is likely, based on the marked intercellular projections (white arrows), that the circular profiles are derived from sections through the intercellular projections and do not represent vesicles or cellular breakdown products. (b) Adult nucleus revealing a very complex pattern of membranes. The increase in complexity can be accounted for by a significant compaction of fibers in which each fiber decreases its cross-sectional area by a factor of three. The interdigitations also become more complex, and more circular profiles appear within the fiber cytoplasm. These profiles are most likely derived from intercellular projections. (c) Embryonic nucleus from a young lens (22 years old). Note large irregular cell with undulating membranes (arrowheads) and cytoplasmic profiles (black arrows). (d) Embryonic nucleus from aged lens (67 years old). The fibers in panels c and d are similar because both have smooth cytoplasm and complex membrane interfaces. These fibers are influenced by compaction, which is more clearly demonstrated by SEM. (a) From Taylor et al., 1996, courtesy of ARVO. (b)–(d) From Al-Ghoul et al., 2001, courtesy of Academic Press, London, UK.

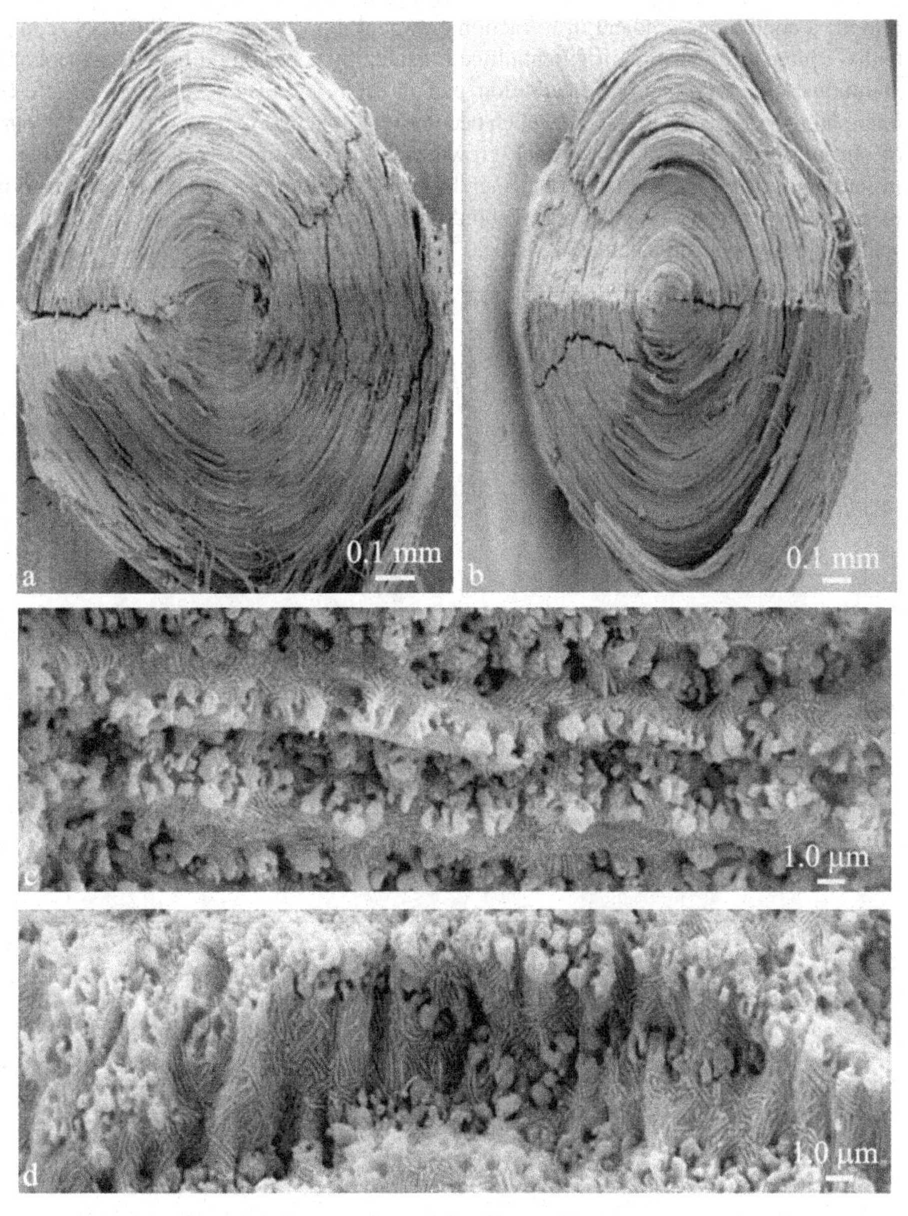

Figure 4.31. Scanning electron micrographs of human lenses. (a) Lens nucleus from a young adult (20 years old) split to reveal the embryonic nucleus and fetal nucleus. (b) Lens nucleus of a mature adult (71 years old) split to reveal a smaller embryonic nucleus and a fetal nucleus that shows a more acute angle of fibers. (c) Higher magnification of embryonic nuclear fibers in panel a showing furrowed membranes. (d) Higher magnification of embryonic nuclear fibers in panel b showing furrowed membranes and pronounced accordion-like folds that represent age-related compaction.

provide a continuous gradient of refraction between fibers. The extracellular space, a potential source of diffraction, is essentially eliminated by conjoining fibers via square array membranes and gap junctions. In addition, gap junctions and fusion zones provide intercellular pathways between the fibers, which become positioned farther away from their source of nutrition with the addition of each growth shell. But, ultimately, age-related changes in lens crystallins, cytoskeleton, and membrane render the lens incapable of preserving its necessary structure-function regimen indefinitely, leading to common lens pathologies, such as presbyopia and cataract.

5
Lens Crystallins

Melinda K. Duncan, Ales Cvekl, Marc Kantorow, and Joram Piatigorsky

5.1. Introduction

Since Kepler and Descartes first investigated the optics of the eye, the central role of the lens in light refraction has been appreciated. The lens must be extremely dense to refract light in the aqueous media in which it is suspended. The necessary density is achieved by the presence of the crystallins, proteins that accumulate to concentrations of 450 mg/ml or higher in the lens fiber cell cytoplasm (Fagerholm et al., 1981; Huizinga et al., 1989; Siezen et al., 1988). Since most proteins would aggregate and strongly scatter light long before accumulating to these high concentrations, the crystallins are believed to have a number of special properties that allow for the creation of the short range order necessary for lens transparency (Tardieu and Delaye, 1988). In the past 50 years, our understanding of the molecular nature of crystallins has increased exponentially, and now much is known about the structure, function and evolutionary origin of these proteins. Before the advent of molecular biology, proteins would be designated as crystallins if their concentration in the lens was sufficient to create a major peak on a size exclusion column, a band on a SDS-PAGE gel, or a spot on a two-dimensional protein gel. Practically, this working definition designated a protein as a crystallin if its concentration in the lens reached about 5% of the total water soluble protein (de Jong et al., 1994). This functional definition is somewhat difficult to sustain, however, since some crystallins do not meet the 5% cutoff in all of the vertebrates in which they are found. For instance, while βA4-crystallin protein is abundant in the bovine and human lens (Lampi et al., 1997; Slingsby and Bateman, 1990), this protein is not detectable on two-dimensional gels of chicken lens proteins (L. L. David, unpublished data), even though the βA4-crystallin gene is expressed at low levels (Duncan et al., 1995). Also, while all known γ-crystallin genes of the rat are highly transcriptionally active (vanLeen et al., 1987) and produce significant amounts of protein (Lampi et al., 2002), the expression level of some gamma crystallins is very low in the human lens (Lampi et al., 2002), and the presence of frameshift mutations in their coding sequence has shown that they are on the way to becoming pseudogenes (Meakin et al., 1987). Further, many species have taxon-specific crystallins expressed at high levels in their lenses, and these perform metabolic functions, and have different names, in nonlens tissues (Piatigorsky and Wistow, 1989).

While crystallins were first identified as proteins involved in light refraction, in recent years other functions of these proteins have been identified. The α-crystallins are now known to share significant sequence similarity with the small heat-shock proteins and are believed to protect lens proteins from aggregation caused by age-related damage (Horwitz, 1993). Further, the defective lens differentiation observed in mice harboring γ-crystallin mutations

(Graw, 1997) as well as αA/αB-crystallin double knockout mice (Brady and Wawrousek, 1997) argues for a direct role of at least some crystallins in developmental processes as well (Andley et al., 1998; Andley et al., 2000; Boyle and Takemoto, 2000).

The relative expression level of crystallin genes changes during development and results in quite large differences in crystallin composition in different portions of the lens. An example of this is the preponderance of δ-crystallin in the nucleus and β-crystallins in the cortex of the adult chicken lens (Ostrer and Piatigorsky, 1980; Thomson et al., 1978). This difference in crystallin composition results from changes in the relative amounts of crystallin gene transcription that occur in lens fibers born during the pre- and posthatching period (Hejtmancik et al., 1985). Since the efficiency of the vertebrate lens is dependent on a smooth refractive index gradient that corrects for spherical aberration (Land, 1988), it appears likely that differential transcriptional regulation of crystallin genes during development is important for lens function. In addition, since crystallins are often used as markers for lens differentiation, information on the control of their expression is essential to forward our understanding of the molecular mechanisms responsible for lens development.

Here we review the structure, function, and expression of crystallins during development.

5.2. Structure and Function of Crystallins

5.2.1. α-Crystallin

α-Crystallin, a member of the small heat-shock family of proteins, is an aggregate of two approximately 20 kDa polypeptides, αA and αB (de Jong et al., 1993; Groenen et al., 1994; Ingolia and Craig, 1982; Klemenz et al., 1991; Sax and Piatigorsky, 1994). These subunits most likely arose from a gene duplication event and share about 57% amino-acid identity. In humans, the αA-crystallin gene is localized on chromosome 21 and encodes a 173–amino acid polypeptide, and the αB gene is localized on chromosome 11 and encodes a 175–amino acid polypeptide. The two subunits form a soluble aggregate that has a molecular weight ranging from 300 kDa to over 1 thousand kDa, with an average of 800 kDa (Horwitz, 2003; Siezen et al., 1978; Veretout et al., 1989). When isolated from human, rat, and bovine lenses, the ratio of αA to αB is approximately 3:1 (Delcour and Papaconstantinou, 1974; Lampi et al., 2002; van Kamp et al., 1974).

Although the exact structure of α-crystallin has not been determined, it is known that α-crystallin consists of about 50% β-sheet and 10–15% α-helix (Horwitz et al., 1998; Koretz et al., 1998). Recently it has been shown by cryoelectron microscopy that aggregates of αB-crystallin contain approximately 32 subunits that form a globule with a hollow cavity (Haley et al., 1998; Horwitz et al., 1999; Smulders et al., 1998). The subunits making up this cavity freely exchange (van den Oetelaar et al., 1990) in a temperature-dependent manner (Bova et al., 1997).

Developmentally, α-crystallin is the first crystallin synthesized in the mammalian lens, where it is first detected in the lens placode (Haynes et al., 1996; Robinson and Overbeek, 1996; Zwaan, 1983). In chickens, α-crystallin appears after δ-crystallin in the lens vesicle (Ikeda and Zwaan, 1966, 1967). In the adult lens, α-crystallin transcription is mainly confined to the fiber cells (Sax and Piatigorsky, 1994), but significant amounts of both αA and αB mRNA can be detected in the lens epithelium (Robinson and Overbeek, 1996). In the mature mammalian lens, α-crystallin accounts for about 20–30% of water-soluble protein (Lampi et al., 1998; Lampi et al., 2002; Mehta and Lerman, 1972; Ueda et al., 2002).

α-Crystallin was first thought to be restricted to the lens, where it was assumed to have a completely refractive function. This view of α-crystallin changed when it was discovered by two independent groups that αB-crystallin was expressed in numerous tissues in addition to the lens (Bhat and Nagineni, 1989; Dubin et al., 1989), most notably in the heart, where it makes up as much as 3–5% of total cell protein (Benjamin et al., 1997). Although αA-crystallin expression is highly lens preferred, it can be detected in trace amounts in the spleen, thymus (Kato et al., 1991), and retina (Deretic et al., 1994). α-Crystallins are notably expressed in many other ocular tissues, including the Müller's cells of the retina (Deretic et al., 1994; Lewis et al., 1988; Moscona et al., 1985), the retinal photoreceptors (Deretic et al., 1994), the retinal pigmented epithelium (Nishikawa et al., 1994; Robinson and Overbeek, 1996), the corneal endothelium (Flugel et al., 1993), the ciliary muscle, and the trabecular meshwork (Siegner et al., 1996).

The α-crystallins have been implicated in numerous nonrefractive pathways, including those for stress response, phosphorylation, and cell protection (Jakob et al., 1993). In 1992, Joseph Horwitz found that α-crystallin is a molecular chaperone that binds unfolded or denatured proteins, thereby suppressing nonspecific irreversible protein aggregation (Horwitz, 1992). This discovery was later confirmed by Jakob et al. (1993), who demonstrated that, in addition to αB-crystallins, other small heat-shock proteins were capable of acting as molecular chaperones. Since this pioneering work, researchers have found evidence indicating that the binding of nonnative proteins to α-crystallin and other small heat-shock proteins creates a reservoir of unfolded proteins that can later interact with other chaperones to restore the unfolded proteins to a native state in an ATP-dependent process (Ehrnsperger et al., 1997; Lee et al., 1997; Wang and Spector, 2000). Consistent with the chaperone function of α-crystallin, it has been shown that both α-crystallin expression (Klemenz et al., 1991) and phosphorylation (Wang et al., 2000) are induced by physiological stress. Further, αB-crystallin is involved in numerous diseases outside of the eye (Head and Goldman, 2000; Welsh and Gaestel, 1998).

α-Crystallins have properties consistent with other nonrefractive functions. They participate in both cAMP-dependent (Chiesa et al., 1987; Spector et al., 1985; Voorter et al., 1986) and non-cAMP-dependent phosphorylation events (Kantorow et al., 1995; Kantorow and Piatigorsky, 1994), where they may be involved in signal transduction pathways. They also can bind cytoskeletal elements (Bloemendal et al., 1984; Del Vecchio et al., 1984; FitzGerald and Graham, 1991; Head and Goldman, 2000; Wisniewski and Goldman, 1998) and are associated with membrane binding (Cobb and Petrash, 2000; Ramaekers et al., 1980). α-Crystallins can also translocate to the nucleus (Bhat et al., 1999), where they may play a role in controlling lens cell differentiation (Boyle and Takemoto, 2000). α-Crystallins also exhibit antiapoptotic activity, with αA-crystallin having a greater protective effect than αB-crystallin (Andley et al., 2000).

Targeted disruption of the mouse αA-crystallin gene induces cataract (Brady et al., 1997), demonstrating its importance in lens function. The lens fiber cells of αA-crystallin null mice exhibit dense inclusion bodies consisting mainly of αB-crystallin, suggesting that high concentrations of αB-crystallin may be unstable in the absence of αA-crystallin *in vivo* and that αA-crystallin may have a solubilizing effect on αB-crystallin (Brady et al., 1997). In contrast, targeted disruption of the αB-crystallin gene and its linked relative the HSB2 gene (Iwaki et al., 1997) result in mice with relatively normal lenses. However, mice homozygous for targeted deletions of the αA-crystallin, αB-crystallin, and HSB2 genes have severe disruptions in lens morphology (Brady and Wawrousek, 1997). Studies of lens epithelial cells from the αA-crystallin knockout and αB-crystallin knockout mouse indicate that α-crystallin subunits also play major roles in regulating lens epithelial cell division and chromosomal stability (Andley et al., 1998; Andley et al., 2000).

A point mutation in the human αA-crystallin gene is linked to cataract (Litt et al., 1998), and a point mutation in the human αB-crystallin gene is linked to desmin-related myopathy (Vicart et al., 1998). These point mutations disrupt chaperone function (Kumar et al., 1999; Perng et al., 1999; Shroff et al., 2000), suggesting that α-crystallin chaperone function is important for the maintenance of both lens and nonlens tissues. Other diseases associated with αB-crystallin include Alexander's disease (Iwaki et al., 1989), Creutzfeldt-Jacob disease (Iwaki et al., 1992; Renkawek et al., 1992), Huntington's disease (Iwaki et al., 1992), multiple sclerosis, and a multitude of other disorders (Head and Goldman, 2000; Iwaki et al., 1992; Van Noort et al., 1998). The study of α-crystallin structure and function both in lens and nonlens tissues is one of the fastest growing and most exciting areas of crystallin research.

5.2.2. β/γ-Crystallins

Native β-crystallin is found in solutions of water-soluble vertebrate lens proteins as a polydisperse mixture of hetero-octomers (βHigh) and heterotetramers (βLow), along with lower amounts of dimers (Bindels et al., 1981). These β-crystallin macromolecules are formed from the seven known subunits (βA1, βA2, βA3, βA4, βB1, βB2, βB3), which range in size from 22 to 35 kDa, depending on the species (Asselbergs et al., 1979; Lampi et al., 1998; Ueda et al., 2002; Wistow et al., 1991). Cloning of the cDNAs revealed that these seven polypeptides are encoded by six genes, with the βA1- and βA3-crystallin subunits formed from alternate translational initiation from a common mRNA (Peterson and Piatigorsky, 1986). Orthologs of all six β-crystallin genes have been cloned from a number of vertebrates, including human, mouse, rat, cow, and chicken (Duncan et al., 1996a; Graw et al., 1999; Lampi et al., 1997; Lampi et al., 2002; Quax-Jeuken et al., 1984; vanRens et al., 1991). Since direct orthologs of several of these genes have also been reported in frogs and fish (Lu et al., 1996b; Wistow, 1995), it has been proposed that all six β-crystallin genes are found in all vertebrates.

In mammals, native γ-crystallin is found as a mixture of 22- to 25-kDa monomers consisting of the members γS (previously known as βS), γA, γB, γC, γD, γE, and γF (Lubsen et al., 1988). In mice and rats, all seven of these genes are functional, while in humans, only the γA-, γC-, γD-, and γS-crystallin genes contribute significantly to the lens proteins (Meakin et al., 1987; Zarina et al., 1992). Although birds generally lack γ-crystallins, it has been controversial whether bird lenses have γS-crystallin (Reton et al., 1984; vanRens et al., 1991). However, proteomic analysis of the chicken lens did not reveal γS-crystallin as a major lens protein (Wilmarth et al., under review). Proteins with sequence similarity to the mammalian γ-crystallins have been shown to be major contributors to the water-soluble proteins of fish, shark, reptile, and amphibian lenses (Chang et al., 1991; Chiou et al., 1987; Chuang et al., 1997; Lu et al., 1996a; Pan et al., 1994).

In the early 1980s, sequence analysis of the β- and γ-crystallins made it apparent that they have a common evolutionary ancestor (Driessen et al., 1981). The common structural feature of the β/γ-crystallins is a distinctive type of antiparallel β-sheet called the Greek key motif (Bax et al., 1990; Blundell et al., 1981). It has been proposed that the ancestor of the Greek key–containing proteins comprised a single motif that duplicated into two subtypes (A and B). A β/γ-crystallin domain formed when A- and B-type Greek key motifs fused. The complete β/γ-crystallin proteins were created when two β/γ-crystallin domains were linked together with an interdomain spacer (Wistow, 1990). In γ-crystallin, the two β/γ-crystallin domains interact intramolecularly, resulting in the monomeric native structure (Blundell et al., 1981). The two β/γ-crystallin domains of βB2-crystallin interact with their opposite number on another βB2-crystallin molecule (Bax et al., 1990). This results in the ability of βB2-crystallin to form stable dimers. In contrast, the β/γ-crystallin domains of βB1-crystallin interact intramolecularly, leaving several hydrophobic patches exposed to mediate dimerization (Van Montfort et al., 2003). While β-crystallin dimers further oligomerize into βHigh- and βLow-crystallin, the structure of these polydisperse molecules is not well characterized.

In all vertebrate lenses examined, the N- and C-terminal extensions of the β-crystallins are cleaved during the course of normal aging and development (David et al., 1993; Lampi et al., 1997; Ueda et al., 2002). This proteolysis is accelerated during the formation of cataract in rodent lenses treated with agents that disrupt calcium homeostasis (David et al., 1994). The mechanism responsible for the posttranslational proteolytic processing of human β-crystallins is still under investigation, but proteases of the calpain family are likely to be involved in rodents (David et al., 1994). While the function of crystallin proteolysis during lens maturation is still unknown, the observation that the processed β-crystallins tend to be found in the insoluble fraction of the lens nuclear proteins (Lampi et al., 1998; Ueda et al., 2002) suggests that the cleavage of the hydrophilic arms aids in the exclusion of water from the lens nucleus during lens maturation.

Unlike the α-crystallins, no nonrefractive physiological functions have been directly ascribed to the β/γ-crystallins. However, it is likely that nonrefractive functions do exist. First, the β/γ-crystallins are evolutionarily related to a number of proteins involved in stress response, including protein S, a spore coat protein of the colonial prokaryote *Myxococcus xanthus*, and spherulin 3a, an encystment-specific protein of the slime mold *Physarum polycephalum* (Bagby et al., 1994; Wistow, 1990). More recently, sequence similarity has been noted for the EP37s, a group of proteins expressed in the integument and digestive system of newts (Ogawa et al., 1997), and AIM1, a gene often disrupted in malignant melanoma of humans (Ray et al., 1997). Second, like protein S, β-crystallin can apparently bind calcium (Sharma et al., 1989), and several papers have reported that some β-crystallins are phosphorylated (Kantorow et al., 1997; Kleiman et al., 1988; Voorter et al., 1989). Further, it has recently been reported that β-crystallin is expressed in the neuronal cell line N1E-115 and that this protein translocates from the perinuclear zone to the cytoplasm in response to heat stress and cold shock (Coop et al., 1998).

The cataract in eye lens obsolescence (Elo) mice results from a frameshift mutation that destroys the fourth Greek key of γE-crystallin (Cartier et al., 1992). Developmental analysis of this cataract has shown that the earliest detectable morphological alteration is an impairment of primary fiber elongation at 12.5 days post coitum (dpc). Subsequently, a high proportion of these defective fibers undergo apoptosis or necrosis, which leads to a severe disruption in lens morphology (Oda et al., 1980). Further, both the $Cat2^{nop}$ mutation (caused by insertion or deletion of nucleotides from γB-crystallin) and the $Cat2^{ns}$ mutation (caused by the deletion of the 3′ end of γE-crystallin) have defects in secondary fiber elongation and nuclear degradation (Graw, 1997). Finally, both the expressed sequence tag (EST) effort of the human genome project and traditional approaches have found that βB2-, βB3-, βA3/A1-, and βA2-crystallin are expressed outside of the lens in mammals and birds (Head et al., 1991; Head et al., 1995; Magabo et al., 2000) and that some γ-crystallins are expressed outside of the lens in amphibians (Smolich et al., 1994) and mice (Jones et al., 1999; Sinha et al., 1998). In fact, some β- and γ-crystallins have been identified as major components of the vertebrate tooth proteome (Thyagarajan and Kulkarni, 2002). In aggregate, these observations strongly suggest that β/γ-crystallins have nonrefractive functions that remain to be identified.

Mutations in β/γ-crystallins have been associated with a number of congenital autosomal dominant cataract phenotypes in both humans and rodents. In humans, a mutation that results in the loss of the 51 C-terminal amino acids of βB2-crystallin results in the cerulean blue dot (Litt et al., 1997; Vanita et al., 2001) and Coppock-like cataracts (Gill et al., 2000). Mutations in the βA3/A1-crystallin gene have been linked to autosomal dominant cataract in two different human families (Bateman et al., 2000; Burdon et al., 2004; Kannabiran et al., 1998; Qi et al., 2004). The variable zonular pulverulent cataract is caused by mutation of the γC-crystallin gene (Ren et al., 2000). Finally, the aculeiform cataract has been associated with a mutation in the γD-crystallin gene (Heon et al., 1999; Stephan et al., 1999).

In mice, the Philly cataract is caused by an in-frame deletion of 12 basepairs from the βB2-crystallin gene that results in the deletion of four amino acids from the C-terminal Greek key motif (Chambers and Russell, 1991). Recently, it has been reported that mice homozygous for this mutation are subfertile, apparently due to defects in egg production and sperm production/function (Robinson et al., 2003). This may suggest a function for the βB2-crystallin protein detected in the adult rat testis (Magabo et al., 2000). In addition, a number of independent autosomal dominant cataract phenotypes have been mapped to the γ-crystallin locus, and potentially causative mutations have been found in the γA- (Klopp et al., 1998), γB- (Graw, 1997), γC- (Ren et al., 2000), γD- (Smith et al., 2000), and γE-crystallin (Klopp et al., 2001) genes. Finally, mutation of the γS-crystallin gene leads to the mouse Opj cataract, which exhibits defects in cortical fiber structure (Wistow et al., 2000b). Interestingly, the cataracts associated with γB-crystallin mutations have been attributed to the aberrant folding of the abnormal protein into amyloid-like structures which translocate into lens cell nuclei and disrupt their function (Sandilands et al., 2002).

5.2.3. Enzyme-Crystallins

Crystallins are very diverse proteins that are defined more for their prevalence in the lens than as members of any particular group of proteins. The first inkling of their diversity came when δ-crystallin was found to be a major protein in the chicken lens (Rabaey, 1962). Subsequently, it was found that δ-crystallin was confined to the lenses of almost all birds and reptiles (for a review, see Piatigorsky, 1984). Further studies showed that many species had specific crystallins in their lenses that were not found in the lenses of other species. Surprisingly, these so-called taxon-specific crystallins are similar or identical to metabolic enzymes. For example, ε-crystallin in ducks is lactate dehydrogenase B4, τ-crystallin in turtles is α-enolase, and δ-crystallin in chickens is argininosuccinate lyase (Wistow et al., 1987; Wistow and Piatigorsky, 1987). In many cases, the enzyme that is expressed at low levels outside of the lens and the enzyme-crystallin that is expressed at high concentration in the lens are encoded by the same gene; in other cases, one or more gene duplications have taken place, and the daughter gene (or genes) has specialized for lens expression, such as in the example of argininosuccinate lyase and δ-crystallin (Piatigorsky et al., 1988). The dual utilization of a gene as an enzyme for metabolism in many tissues and as a structural protein for refraction in the lens has been called "gene sharing" (Piatigorsky et al., 1988). The implications of gene sharing and the numerous examples of its use among the lens crystallins have been reviewed extensively and are beyond the scope of this chapter (for reviews, see de Jong et al., 1994; Piatigorsky, 1998; Wistow and Piatigorsky, 1988). Two of the most important conclusions from gene sharing and crystallin gene expression are that gene duplication is not necessary for innovating a new function for an old protein and that a protein does not need to abandon its original expression pattern and role in order to acquire a new function (Piatigorsky and Wistow, 1991). Indeed, changes in the regulation of gene expression appear to have been the critical evolutionary events that led to the recruitment of enzymes as lens crystallins.

5.2.4. Invertebrate Crystallins

In addition to compound eyes, whose gross anatomy is very different from that of vertebrate eyes, many invertebrates have complex eyes that look remarkably similar to vertebrate eyes, with a retina, pigmented epithelium, lens, and cornea. Although not proven, the prevailing view is that the eyes of vertebrates and invertebrates evolved divergently and have a common ancestral root (Gehring and Ikeo, 1999; Neumann and Nuesslein-Volhard, 2000; Pineda et al., 2000). This idea is based on similarities between their developmental pathways, especially the pivotal importance of Pax 6 (Gehring and Ikeo, 1999), as well as the common use of rhodopsin as the visual pigment (Yokohama, 2000). Despite the presumed evolutionary relationship among all eyes, early immunological tests indicated that the crystallins of invertebrates are different from those of vertebrates (for discussion and references, see Tomarev and Piatigorsky, 1996). None of the ubiquitous vertebrate crystallins (α- and β/γ-crystallins) have been found yet in invertebrate lenses. Instead, in the molluscs, as in some vertebrates, various enzymes have been recruited as lens enzyme-crystallins. In the few other invertebrate phyla that have been examined, the lens crystallins appear as novel proteins. We review here the crystallins in cellular lenses of invertebrates that have been characterized at the gene level. A more comprehensive discussion can be found elsewhere (Tomarev and Piatigorsky, 1996).

5.2.4.1. Mollusca

S-Crystallins. The most studied invertebrate crystallins are from cephalopods (squid and octopus). Initial characterization of the squid crystallins revealed a heterogeneous group of dimeric, water-soluble proteins (called S-crystallins) with high (~12%) methionine content (Chiou and Bunn, 1981; Chiou, 1984; Siezen and Shaw, 1982). The major S-crystallins have subunit molecular masses ranging from about 30 kDa (principally) to 22 kDa; the minor S-crystallins have additional subunits of 35–40 kDa. S-crystallins have basic isoelectric points and unblocked N-terminal ends, similar to the vertebrate γ-crystallins. Subsequent analyses of N-terminal amino acid (Wistow and Piatigorsky, 1987) and cDNA (Tomarev and Zinovieva, 1988) sequences showed that S-crystallins were related to the detoxification enzyme glutathione S-transferase (GST).

The most abundant S-crystallins have very little if any GST activity (Chiou et al., 1995; Tang et al., 1994; Tomarev et al., 1991). This appears to be due both to site-specific base pair changes in the coding regions as well as to the addition of exon 4 encoding a central peptide of variable length (Tomarev et al., 1995; Tomarev et al., 1992) in genes duplicated from the GST ancestor. However, at least one S-crystallin gene lacking exon 4 (SL11/Lops4) encodes a protein with some enzymatic activity, although much less than that of GST expressed in the digestive gland (Tomarev et al., 1995). Thus, the gene structure of SL11/Lops4 resembles that of the authentic GST protein expressed in the cephalopod digestive gland (Tomarev et al., 1993), which has high enzymatic activity (Tang et al., 1994; Tomarev et al., 1995).

Together, these data have led to a model for the recruitment of S-crystallins (Tang et al., 1994; Tomarev et al., 1995; Tomarev and Piatigorsky, 1996). In brief, this model proposes that an ancestral GST gene underwent gene duplication. One of the GST daughter genes duplicated in cephalopods and gave rise to the SL11/Lops4 gene, still encoding an active enzyme that was highly expressed in the lens. The S11/Lops4 gene in turn gave rise to a family of lens-specific genes. All the S-crystallins but SL11/Lops4 were inactivated during evolution by gradual drift in sequence and by the acquisition of the variable central peptide by exon shuffling. The single-copy, authentic GST gene encoding the enzyme is expressed primarily in the digestive gland of cephalopods (Tomarev et al., 1993), while the multiple S-crystallin genes are expressed in the lens (Tomarev et al., 1991) and cornea (Cuthbertson et al., 1992). Thus, the S-crystallins of cephalopods, unlike some of the enzyme-crystallins of vertebrates, appear to be an extreme case of gene duplication followed by specialization of the daughter genes for expression in the lens, where they have a structural role. In view of their abundance in the cephalopod lens, it is possible that one or more S-crystallins with limited activity have an enzymatic role, either during development or, more likely, for detoxification. GST is known to have a detoxification function in the mammalian lens (Jimenez-Asensio and Garland, 2000).

Crystallographic studies have shown that the overall structure of GST from the squid digestive gland is similar to that of the vertebrate enzymes (Ji et al., 1995). Nonetheless some structural differences were noted. Squid GST has a unique dimer interface with a relatively open active site, correlating with its high enzymatic activity. It is also relatively inefficient at catalyzing the addition of GSH to enones and epoxides. It has thus been suggested that squid GST forms a class of enzymes, called sigma, separate from the class alpha, mu, and pi enzymes of vertebrates.

Cephalopod eye development and lens structure appear unique in the animal kingdom (for description and references, see West et al., 1995). The lens has an anterior component and a posterior component, both of which consist of platelike shells with fiber extensions at the margins. The posterior lens primordium develops first from cellular processes extending from a middle group (group 2) of ectodermal, lentigenic cells. Later in development, an anterior group (group 1) of lentigenic cells in the ectoderm extend processes that are laid down in a circumferential fashion to form the anterior lens cap. The cornea of cephalopods comes from epithelial tissue that is pulled over the eyeball by muscle contraction (Arnold, 1984). The developmental expression of S-crystallins has been followed by immunolocalization in the squid (West et al., 1994). S-crystallins were observed first early in embryogenesis in the middle, group 2 lentigenic cells, the posterior lens primordium, and the processes connecting the lentigenic cells with the lens primordium. Two days later, S-crystallins were present in the anterior, group 1 lentigenic cells, the anterior lens primordium, and their connecting processes. The lentigenic cells of the ectoderm still stained positively for S-crystallins in the adult squid. Since the antibody did not discriminate among the S-crystallin family members, the extent of developmental regulation among the different S-crystallin genes is not known. Nothing is known yet about the developmental regulation of crystallin genes in the cornea. Indeed, it is not even known whether the S-crystallins that have been reported in the cornea (Cuthbertson et al., 1992) are products of the same genes as those that are expressed in the lens.

Ω-Crystallin. Another enzyme-crystallin in molluscs is Ω-crystallin. It is found as a minor crystallin in octopus and, in even lesser amounts, in squid (Chiou, 1988; Zinovieva et al., 1993). However, Ω-crystallin is the only crystallin present in the muscle-derived lens of the squid light organ, where it is called L-crystallin (Montgomery and McFall-Ngai, 1992), and in the eye lens of the scallop (Piatigorsky et al., 2000). Scallops have, amazingly, some 60–70 eyes along the mantle on each of their two shells. These eyes are highly developed, with two different functional layers of retinal photoreceptors, a cellular lens, an overlying cornea, and an argenteum, which is a mirrorlike layer beneath the retina that reflects an image upon the outer retina (Barber et al., 1967).

Ω-Crystallin was derived from aldehyde dehydrogenase (ALDH) class 1/2 (Zinovieva et al., 1993). The η-crystallin of elephant shrews is also an ALDH1/2 (Graham et al., 1996; Wistow and Kim, 1991). Thus, Ω-crystallin is the only lens crystallin that is represented in both vertebrates and invertebrates. Unlike the η-crystallin of elephant shrews, which has retinaldehyde dehydrogenase activity (Graham et al., 1996), no enzyme activity has ever been detected for Ω-crystallin (Montgomery and McFall-Ngai, 1992; Piatigorsky et al., 2000; Zinovieva et al., 1993). In general, the ALDH1/2 family members, including the Ω-crystallins of cephalopods, are tetrameric proteins. By contrast, scallop Ω-crystallin is a dimer and does not bind NAD(P) (Piatigorsky et al., 2000). Thus, as a group, Ω-crystallins are inactive ALDH-derived enzyme-crystallins that are highly specialized for crystallin function in the lens of the eye or light organ.

The developmental appearance of Ω/L-crystallin has been studied in the squid light organ. Ω/L-crystallin begins to accumulate 10 days after the hatching of the juvenile squid in a few cells of the ciliated duct adjacent to the symbiont-containing tissue as well as in the developing lens (Weis et al., 1993). The development of the squid light organ has been described elsewhere (Montgomery and McFall-Ngai, 1992). Subsequently, Ω/L-crystallin makes up approximately 70% of the total protein of the adult muscle-derived light organ lens.

5.2.4.2. Cnidaria

The crystallin genes of the cubomedusan jellyfish have also been cloned. Surprisingly, these ancient animals have complex, well-developed eyes, called "ocelli," which begin to develop as the polyps begin their metamorphosis into cubomedusae (Kostrouch et al., 1998). Cubomedusae have three families of monomeric crystallins, J1-, J2-, and J3-crystallin, with molecular masses of 35, 20, and 19 kDa, respectively (Piatigorsky et al., 1989). There are three intronless J1-crystallin genes (J1A, J1B, and J1C; Piatigorsky et al., 1993). While the coding sequences of the J1-crystallin genes are almost identical, their 5′ and 3′ untranslated sequences as well as their 5′ flanking sequences are entirely different. The J1-crystallins appear homologous to ADP-ribosylglycohydrolases (Z. Kozmik and J. Piatigorsky, unpublished data). J3-crystallin, encoded in a single-copy gene, shows homology to saposins, a conserved family of multifunctional proteins that bridge lysosomal hydrolases to membrane lipids and activate enzyme activity (Piatigorsky et al., 2001). Both J1- and J3-crystallins are expressed outside of the lens (especially in the statocyst), consistent with their having non-refractive as well crystallin functions.

5.2.4.3. Arthropoda

Drosocrystallin. In contrast to the complex eyes discussed above, compound eyes of arthropods lack cellular lenses. Compound eyes comprise hundreds of ommatidia, each with its own secreted, acellular cornea and an underlying cone containing a few cells. The acellular cornea is often called a lens even though it is structurally very different from the cellular lens of vertebrates or of ocelli. Some arthropods, such as *Drosophila*, have both compound eyes and ocelli. Further discussion of compound eyes can be found elsewhere (Land, 1988; Nilsson, 1990).

At least 14 different proteins have been identified in the cone of *Drosophila* (for discussion and references, see Tomarev and Piatigorsky, 1996). By contrast, 3 proteins with molecular masses of 52, 47, and 45 kDa make up the acellular cornea (lens; Komori et al., 1992). The most abundant of the 3 proteins is glycosylated and is called "drosocrystallin."

Drosocrystallin was cloned recently from *Drosophila* (Janssens and Gehring, 1999). A sequence relationship was found between drosocrystallin and some cuticular proteins. It is thus likely that drosocrystallin was recruited for its crystallin-like role in the acellular lens from a cuticular protein by a gene-sharing strategy (to extend this concept for crystallin evolution to the compound eye of insects). Antibody (Komori et al., 1992) and *in situ* hybridization (Janssens and Gehring, 1999) studies showed that drosocrystallin is synthesized during eye formation between 48 and 96 hours into the development of the pupa. They also indicated that drosocrystallin is synthesized in the primary pigment cells and cone cells of the compound eye during a narrow window of developmental time and secreted to the surface lens. Interestingly, drosocrystallin mRNA is also found in the corneagenous and retinula cells of the ocellus situated between the compound eyes of *Drosophila* (Janssens and Gehring, 1999). It is not known whether drosocrystallin accumulates in the ocellar lens of the fly.

5.3. Control of Crystallin Gene Expression

5.3.1. Introduction

The developmental expression of crystallin genes is as complex as their diversity. Each crystallin exhibits its unique "signature" pattern of expression in the embryonic, neonatal, and adult lens. What are these levels of complexity? First, lens epithelial cells express a different spectrum of crystallins than fiber cells, making crystallins prime markers of lens cell differentiation. Further, the amount of expression of each crystallin changes from the early embryo well into adulthood, probably as a way of establishing the refractive index gradient necessary for the function of an efficient lens. Another level of complexity derives from the fact that homologous crystallins have distinct qualitative and quantitative expression patterns in different species – patterns that may account for species-specific refractive properties. Finally, the expression of taxon-specific crystallins changes the overall quantitative distributions of common crystallins, presumably further modifying the optical properties of the lens.

How do lens cells achieve that unique temporal and spatial pattern of expression of their characteristic proteins? The prevailing view is that both the abundance and cellular distribution of all crystallin peptides are controlled principally at the level of gene transcription. If this is correct, lens cells require a very complex and delicately balanced transcriptional regulatory mechanism to ensure that specific crystallin mRNAs are made at the right time and place and in the right amounts. Here we deal primarily with this regulation at the level of individual transcription factors and promoters. However, the expression and function of these factors clearly must be controlled by cell-signaling cascades that are presumably involved in the withdrawal of equatorial epithelial cells from the cell cycle at the onset of fiber cell elongation as well as the detachment of lens fiber cells from the overlying epithelium and underlying capsule during the creation of the lens sutures and the creation of the organelle/nuclear free zone. Indeed, much evidence suggests that that cell cycle regulatory molecules, such as p57(KIP2), and fibroblast growth factors (FGFs) and their receptors play dominant roles in this process (de Iongh et al., 1997; Lang, 1999; Zhang et al., 1997; Zhang et al., 1998).

5.3.2. Transcriptional Regulation of Crystallin Gene Expression: General Aspects

Studies of prototype tissue-specific genes whose expression is characteristic for terminally differentiated cells (e.g., β^{A}-globin in red blood cells, albumin in the liver, and glucagon in the pancreas), combined with studies on the regulation of viral gene expression, have revealed the general mechanisms controlling the transcription of genes. Proximal elements, usually found in about 70 bp of DNA, include TATA boxes (TATAWAW; W = A or T) at position −30, a TFIIB recognition element (SSSCGCC; S = C or G) around position −35, initiator elements (Inr, CTCANTCT; N = any base) located around +1 with respect to the start site of transcription, and a downstream core promoter element (RGWCGTG; R = A or G; W = A or T) around position +30. Functionally, TATA and Inr elements specify the transcriptional start site by organizing the assembly of the basal transcriptional machinery. All crystallin genes discussed here contain TATA elements (Cvekl and Piatigorsky, 1996), while other core promoter sequences are less conserved in these crystallin genes. Basal (or core) promoters position the transcriptional start site and control the direction of transcription by controlling the assembly of the preinitiation complex containing RNA polymerase II and the basal transcriptional machinery. In transient transfections, basal crystallin promoter fragments often yield activities similar to the promoter-less parental reporter gene vector. Minimal promoters consist of the basal promoter plus additional position-dependent *cis*-regulatory elements that confer activities in transient transfections significantly higher than the basal promoter. While the advantage of transient transfections is their sensitivity to the presence of additional 5′ or 3′ sequences containing at least one position-dependent promoter element, in many instances this activity is not distinctly tissue-specific. Thus, we define a minimal lens-specific promoter as the smallest DNA fragment capable of driving reporter gene expression in the lens of transgenic mice.

Enhancers are defined as *cis* regulatory elements able to activate transcription of minimal promoters independent of their general position within a gene locus. Functional dissection of enhancers frequently reveals that they are composed of multiple copies of the same binding site or overlapping binding sites of multiple factors.

Additional, position- and orientation-dependent regulatory elements may be needed to achieve desired quantitative and qualitative levels of expression, and these are referred to as "modulatory elements." In some instances, transcription factor binding to modulatory elements mediates the transcriptional consequences of signal transduction pathway activation and are thus called "inducible/responsive elements." Finally, expression of some genes is regulated over long distances by locus control regions (LCRs) that may control the function of many genes.

Eukaryotic DNA is assembled into chromatin, which inhibits transcription by restricting access to RNA polymerase and transcription factors interacting with the positively acting regulatory sites. Actively transcribed genes are characterized by altered or "remodeled" chromatin. There are different levels of chromatin remodeling, including displacement of nucleosomes by DNA-binding transcription factors, which results in an increased sensitivity of DNA in transcribed genes to nucleases (observed as DNaseI hypersensitive regions). In addition to the general repressor role of chromatin, there are *cis*-regulatory elements, often called "silencers," bound by sequence-specific transcriptional repressors. The most recent studies have shown that transcriptional activators employ multiple mechanisms to influence the chromatin structure allowing RNA polymerase II to transcribe the gene.

Crystallin promoters have been functionally characterized through experiments employing immortalized and primary cultured lens and nonlens cells as well as transgenic animals. While cultured cells cannot completely mimic the unique properties of the *in vivo* lens, they have proven to be useful for studying transcriptional control of individual crystallin promoters. Immortalized lens epithelial cell lines such as N/N1003a and αTN4-1 have been successfully used to study both α- and γ-crystallins promoters. Primary lens cultures such as chicken lens epithelial explants, chicken patched lens epithelial cells, and rat lens epithelial explants represent more complicated experimental systems whose major advantage is the simulation of lens fiber cell differentiation. The data generated in these cell culture systems are invaluable in the design of more expensive and time consuming experiments testing the function of crystallin promoters *in vivo* using transgenic mice.

Extensive study of crystallin promoters during the past 20 years has revealed their basic architecture (see Fig. 5.1 for a summary) and a small group of regulatory proteins controlling their function (see Fig. 5.2 for a summary); nevertheless, the full complexity of gene regulation for any single crystallin gene is not yet understood. In the remainder of this section, the minimal promoters controlling lens-specific expression of crystallin genes are discussed (Figure 5.11), along with the many known *cis*-acting regulatory elements that control crystallin gene expression during lens development (Fig. 5.2). Finally, we focus on the transcription factors that bind to functionally important promoter elements, correlate their expression patterns during lens formation with that of the cognate crystallin gene, and propose their molecular mechanisms of function. It is noteworthy that no exclusively lens specific transcription factor is known, consistent with lens-specific expression being driven by a unique combination of multiple transcription factors interacting with different crystallin gene regulatory elements.

5.3.3. Vertebrate Crystallins

5.3.3.1. Transcriptional Control of the α-Crystallin Genes

Transcriptional Control of the αA-crystallin Gene. In all vertebrates studied, the αA-crystallin gene is predominately expressed in the lens; however, low levels of αA-crystallin protein have been detected in the rat thymus, spleen, cerebellum, liver, and kidney (Kato et al., 1991). Here we will only address the control of αA-crystallin expression in the lens, since the control of extralenticular αA-crystallin expression has not been studied. During mouse lens development, αA-crystallin mRNA and protein are first detected in the lens pit at 10–10.5 days post coitum (dpc; Webster and Zwaan, 1984; Zwaan, 1983; Zwaan and Silver, 1983) and is found at relatively equal levels in the presumptive epithelial and fiber cells of the lens vesicle (Robinson and Overbeek, 1996). During chicken lens development, αA-crystallin expression is first detected in the lens vesicle (at hour 60 of chick development) after the onset of δ- and β-crystallin expression (Ikeda and Zwaan, 1966, 1967; Zwaan, 1968). In both rodents and chickens, αA-crystallin mRNA levels increase as the cells in the posterior portion of the lens vesicle elongate into primary fibers. Throughout the remainder of lens development, which extends well into adulthood, αA-crystallin mRNA levels up-regulate precisely at the time that secondary lens fiber cells are created from the transitional

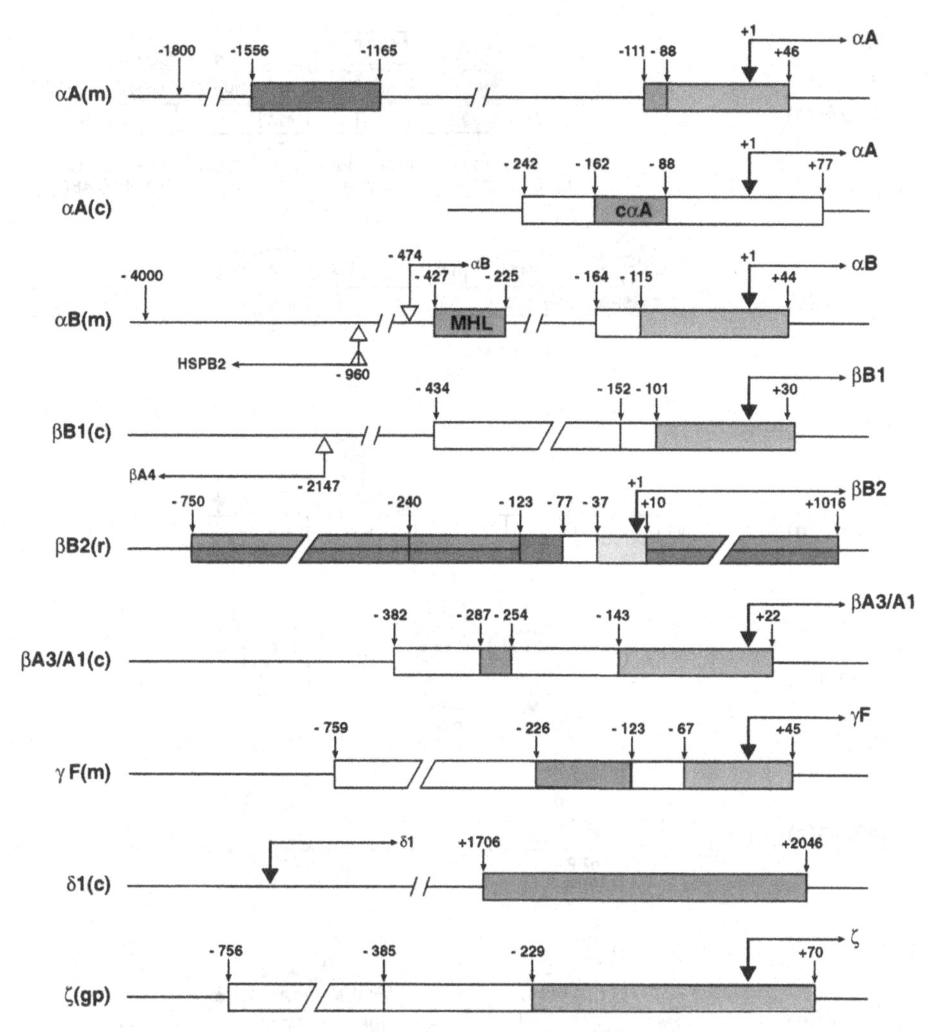

Figure 5.1. (See color plate VII.) Schematic representation of the mouse and chicken αA-, mouse αB-, chicken βB1-, rat βB2-, chicken βA3/A1-, mouse γF-, chicken δ1-, and guinea pig ζ-crystallin transcriptional control regions. Regulatory regions are shown in boxes, lens-specific minimal promoters are shown in green, enhancers are shown in orange, and negatively acting regions (repressors/silencers) are shown in red. Basal promoter of the rat βB2-crystallin gene determined in transient transfections is shown in yellow. Start sites of transcription in the lens (+1) are labeled by bold arrows. Alternate start sites of transcription in the mouse αB-crystallin gene, multiple start sites of HSB2 transcription (around −960) inside the αB-crystallin locus, and start site of the chicken βA4-crystallin transcription near the βB1-crystallin gene are labeled by open arrows. MHL: muscle, heart, lens enhancer of the αB-crystallin gene.

zone (Robinson and Overbeek, 1996). Thus, high-level expression of αA-crystallin can be considered diagnostic for lens fiber cells.

The human αA-crystallin gene is located on chromosome 21 (Hattori et al., 2000), and the mouse gene is found on a relatively short syntenic region of chromosome 17. Multiple sequence alignment has revealed that the mouse, human, hamster, and mole rat promoters are highly conserved (Sax and Piatigorsky, 1994), while the chicken αA-crystallin promoter

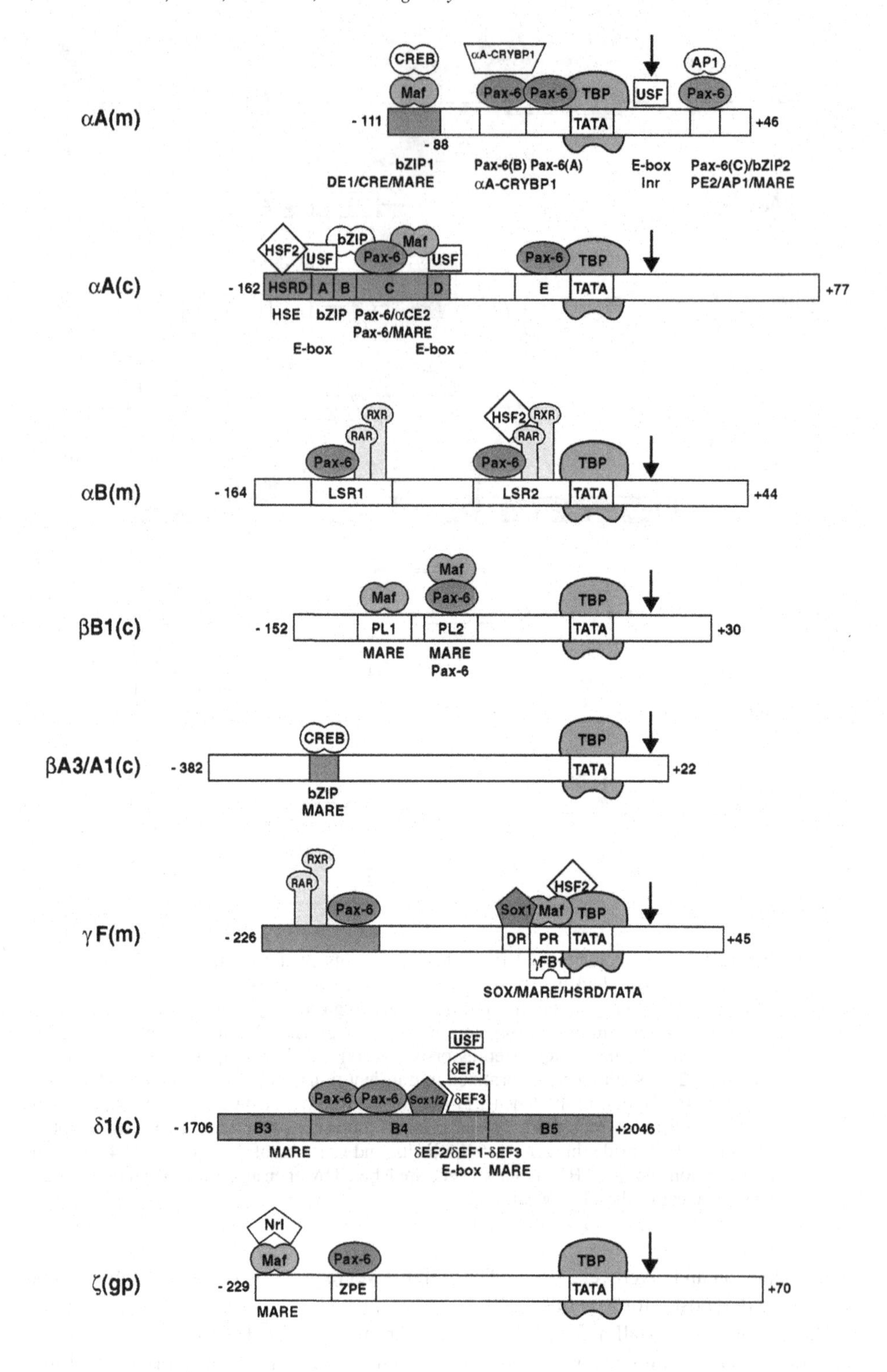
αA(m)
CREB
Maf
αA-CRYBP1
Pax-6
Pax-6
TBP
USF
AP1
Pax-6
- 111
TATA
+46
- 88
bZIP1
DE1/CRE/MARE
Pax-6(B) Pax-6(A)
αA-CRYBP1
E-box
Inr
Pax-6(C)/bZIP2
PE2/AP1/MARE
αA(c)
HSF2
USF
bZIP
Pax-6
Maf
USF
Pax-6
TBP
- 162
HSRD
A
B
C
D
E
TATA
+77
HSE
bZIP
Pax-6/αCE2
Pax-6/MARE
E-box
E-box
αB(m)
RXR
RAR
Pax-6
HSF2
RXR
RAR
Pax-6
TBP
- 164
LSR1
LSR2
TATA
+44
βB1(c)
Maf
Maf
Pax-6
TBP
- 152
PL1
PL2
TATA
+30
MARE
MARE
Pax-6
βA3/A1(c)
CREB
TBP
- 382
TATA
+22
bZIP
MARE
γF(m)
RXR
RAR
Pax-6
HSF2
Sox1
Maf
TBP
- 226
DR
PR
TATA
+45
γFB1
SOX/MARE/HSRD/TATA
δ1(c)
USF
δEF1
Pax-6
Pax-6
Sox1/2
δEF3
- 1706
B3
B4
B5
+2046
MARE
δEF2/δEF1-δEF3
E-box MARE
ζ(gp)
Nrl
Maf
Pax-6
TBP
- 229
ZPE
TATA
+70
MARE

is more divergent than expected. Both the mouse and chicken αA-crystallin promoters have been functionally characterized using transfection studies and transgenic mice. The majority of *in vivo* work in transgenic mice was done with the mouse αA-crystallin promoter. Thus, the chicken promoter will be discussed separately from the mouse promoter.

For the mouse αA-crystallin promoter, fragments ranging in size from −60/+46 to −1800/+46 have been studied using transgenic mice and transfection assays (Fig. 5.1). These investigations have determined that the minimal lens-specific region resides in fragment −88 to +46 (Wawrousek et al., 1990). Consequently, transcription factors interacting with this fragment of DNA are considered as essential for determining lens specificity. This minimal lens-specific promoter is activated by an element between −111 and −106 (Nakamura et al., 1989) in both transgenic mice and transfected cells, while −1556/−1165 functions as a repressor in transiently transfected lens epithelial cells (Sax et al., 1994). It should be noted that, though all functional fragments of the αA-crystallin promoter drive lens-specific expression in transgenic mice, all of the fragments tested to date produce levels of transgenic reporter mRNA at least 1000-fold lower than that of endogenous αA-crystallin. This suggests that one or more control elements or regions regulating αA-crystallin expression have yet to be identified.

Transient transfection studies conducted in the mouse transformed lens epithelial cell line αTN4-1, in the rabbit nontransformed cell line N/N1003A, and in chicken primary patched lens epithelial cells (PLEs) resulted in the identification of at least seven distinct regulatory regions – DE1/CRE/MARE, αA-CRYBP1/Pax-6 (site B), Pax-6 (site A), TATA box, PE1, Inr/E box, and PE2/AP1/MARE/Pax-6 (site C) – within the mouse −111 to +46 promoter fragment (Fig. 5.2; Cvekl et al., 1995a; Sax and Piatigorsky, 1994). *In vitro* and *in vivo* footprinting studies as well electrophoretic mobility shift assays (EMSAs) confirmed specific binding of nuclear proteins to those sites, further supporting their functional significance

Figure 5.2. (facing page) (See color plate VIII.) Schematic representation of the mouse and chicken αA-, mouse αB-, chicken βB1- and βA3/A1-, mouse γF-, chicken δ1-, and guinea pig ζ-crystallin regulatory sites and DNA-binding transcription factors. Promoter regions and/or enhancers are shown in open (orange) boxes. Transcription factors (AP1, αA-CRYBP1, CREB, δEF1, δEF3, γFB1, HSF2, Maf, Nrl, Pax6, RAR/RXR, Sox1, Sox1/2, TBP, and USF) are shown above the boxes, and the names of regulatory sites are given under the boxes. Commonly used abbreviations: bZIP, basic leucine zipper protein binding site; MARE, Maf factors binding site; E-box, basic helix-loop-helix transcription factor binding site; TATA, TBP/TFIID binding site; HSRD, heat shock transcription factor binding site. Start sites of transcription in the lens (+1) are labeled by bold arrows. In the mouse αA-crystallin promoter −111 to +46, shown are three Pax6 binding sites (A, B, and C). However, the functional significance of Pax6 binding to sites B and C remains to be determined. In the chicken αA-crystallin promoter −162 to +77, two Pax6 binding sites (C and E) and one L-Maf binding site (αCE2/MARE) are known. In the mouse αB-crystallin promoter −164 to +44, two regions of lens-specific binding proteins are labeled LSR1 and LSR2. Both LSR1 and LSR2 can interact with Pax6 and RAR/RXR. In the chicken βB1-crystallin promoter −152 to +30, regulatory regions PL1 and PL2 can interact with Maf and Pax6/Maf, respectively. In the chicken βA3/A1-crystallin promoter −382 to +22, an enhancer region containing the bZIP/MARE binding site is shown. DNaseI footprinting of this region revealed three potential Pax6 binding sites, at −5 to −14, −21 to −26, and −58 to −95. In the mouse γF-crystallin promoter −226 to +45, binding of RAR/RXR and Pax6 with the enhancer is shown, as is binding of Sox1, Maf, HSF2, γFB1, and TBP with regions DR and PR next to the TATA box. In the chicken δ1-crystallin enhancer −1706 to +2046, binding of Pax6, Sox1/2, and δEF3/δEF1/USF with the core enhancer region is shown. The δ1-crystallin promoter contains also a canonical TATA box (not shown). In the guinea pig ζ-crystallin promoter −229 to +70, Pax6 and Nrl/Maf binding sites are shown.

(Kantorow et al., 1993b). Specific transcription factors regulating the mouse αA-crystallin promoter were identified through a series of *in vivo* and *in vitro* experiments; these include tissue-restricted factors, the paired domain and homeodomain factor Pax 6, a large family of leucine-zipper – containing factors (e.g., c-Maf/MafB), and CREB/CREM/ATF. In addition, several other more widely expressed transcription factors (e.g., αA-CRYBP1, AP1, and USF) appear to also be involved in crystallin gene expression.

Knockout of the mouse *c-Maf* gene results in a lack of fiber cell differentiation coincident with a 90% reduction of endogenous αA-crystallin expression in *c-Maf* homozygous lenses (Kawauchi et al., 1999; Kim et al., 1999; Ring et al., 2000). *In vitro* studies confirmed binding of recombinant c-Maf protein to the regulatory site DE1/CRE (Ring et al., 2000) and c-Maf in combination with the transcriptional co-activators CBP/p300 strongly activate the mouse αA-crystallin promoter in co-transfections (Chen et al., 2002). In the lens, c-Maf was found at E10.5–11, when the posterior cells of the lens vesicle begin their differentiation (Ring et al., 2000), and later both in lens epithelium and fibers, with higher expression in lens fiber cells, which correlates well with the expression pattern of αA-crystallin. Details of the expression pattern of c-Maf are given in chapter 3. It is noteworthy that, in addition to the lens, c-Maf is highly expressed in embryonic mouse brain and kidney. However, deletion of the DE1 site (−111 to −89) or its mutagenesis had no effect on expression of this fragment in transgenic mice (Sax et al., 1993; Wawrousek et al., 1990). Several factors could contribute to this apparent discrepancy. First, the −88 to +46 promoter fragment may contain another c-Maf–binding site, with the region PE2 serving as a prime candidate due to its sequence similarity with the T-MARE (TRE-like Maf responsive element). Second, c-Maf may be important in the up-regulation of αA-crystallin expression in lens fiber cells but may not be essential to establish lens-specific expression. Third, c-Maf's major function may be to activate additional regulatory elements that lie outside of the minimal lens-specific region but play significant roles both in terms of the quantity of expression and temporal/spatial pattern of expression. These three models are not mutually exclusive, and ongoing studies will provide better insight into this issue.

It is also possible that inactivation of c-Maf has an indirect effect on the transcription of the αA-crystallin gene via other regulatory genes. The DE1/CRE region of the mouse αA-crystallin promoter contains consensus binding sites for the basic/leucine-zipper transcription factors, AP-1, CREB, CREM, and ATF. EMSA demonstrated that the DE1/CRE element binds to a number of lens nuclear factors, and immunological experiments suggested that CREB and/or CREM bind to this element (Cvekl et al., 1995a). Studies using cultured lens cells treated with 8-Bromo-cAMP and forskolin demonstrated that DE1 acts as a classical cAMP responsive element (CRE; Cvekl et al., 1995a). Since the temporal and spatial expression patterns of CREB and CREM in the lens are not known and no lens defects were reported in mice with corresponding null mutations (Hummler et al., 1994; Nantel et al., 1996), the functional significance of CREB/CREM binding remains uncertain.

In vitro experiments have also suggested that Pax6 is directly involved in mouse αA-crystallin gene regulation (Figure 5.2). Pax6, a paired and homeodomain-containing transcription factor, is essential for the early steps of lens induction (see chaps. 2 and 3). First, a Pax6-binding site (Pax6 site A) exists between the canonical TATA box and the αA-CRYBP-1–binding site (Cvekl et al., 1995a). In addition, Pax6 transactivated the mouse αA-crystallin promoter fused to the *cat* reporter gene in transiently cotransfected fibroblasts, and this transactivation was lost upon the introduction of a specific mutation (−45 to −40) into the Pax6-binding region. The functional significance of this interaction is less certain, since transgenic mice harboring the same mutation within the −111 to +46 background still express a CAT reporter gene (C. M. Sax, unpublished data). Additional *in vitro* experiments also suggest binding of recombinant Pax6 proteins to the αA-CRYBP1 region (Pax6 site B), 5′ to the DE1/CRE (−135 to −117, Pax6 site D), and PE2 (+15 to +34, Pax6 site C) regions (A. Cvekl and Z. Kozmik, unpublished data). While further work is necessary to elucidate the direct role of Pax6 in αA-crystallin expression in lens fibers and epithelium, it should be noted that if Pax6 is involved in the up-regulation of the αA-crystallin gene during fiber differentiation (when Pax6 mRNA and protein levels drop), then the lowered amount of protein must have sufficient activity to complex one or more binding sites in the promoter.

Interestingly, the *Pax6* gene is also critical for the development of the endocrine pancreas, where it regulates expression of tissue-specific genes such as glucagon, somatostatin, and insulin-I (Ritz-Laser et al., 1999; St-Onge et al., 1997). Activation of glucagon gene transcription is mediated by interaction of Pax6 and Cdx2 with the p300 coactivator (Hussain and Habener, 1999; Planque et al., 2001). Since p300 and its related CREB-binding protein, CBP, interact with CREB proteins, it is possible that similar synergistic interactions regulate the αA-crystallin promoter (Cvekl et al., 1995a).

While the αA-crystallin promoters contain a canonical TATA box, they also contain sequences similar to the Inr element. Both *in vitro* and *in vivo* footprinting studies indicated binding of proteins around the TATA box (Kantorow et al., 1993b), and binding of TBP, a DNA-binding subunit of TFIID, was shown *in vitro* (Sax et al., 1995). Unexpectedly, site-directed mutagenesis of the TATA box did not eliminate the lens-specific expression of promoter fragment −111 to +46 in transgenic mice lenses but resulted in multiple start sites of transcription, mainly between nucleotides −47 and +8 (Sax et al., 1995). Since Pax6 can physically interact with TBP/TFIID both *in vitro* and *in vivo*, it is possible that Pax6 recruits TFIID to this promoter in the presence of a mutated TATA (Cvekl et al., 1999).

Indirect data also suggest roles for two widely expressed transcription factors, αA-CRYBP1 and USF. αA-CRYBP1, also known as PRDII-BP1/AT-BP2/MBP-1/HIV-EP1, was the first transcription factor proposed to interact with the mouse αA-crystallin promoter (−66/+57) (Nakamura et al., 1990). This large (2,688–amino acid) zinc-finger–containing protein is expressed in many cell types (Brady et al., 1997; Nakamura et al., 1990), and anti–αA-CRYBP1 antibodies detect proteins of 50, 90, and 200 kDa in nuclear extracts prepared from αTN4-1 cells (Kantorow et al., 1993a). USF, a basic helix-loop-helix transcription factor, is a widely expressed protein that regulates the expression of many tissue-specific genes. A weak USF-binding site is found around the start site of transcription of the mouse αA-crystallin promoter (−7 to +5) and transient transfections of a promoter/reporter construct in the presence of specific oligonucleotide competitors (−15 to +15 region), and a consensus USF-binding site specifically reduced the wild-type promoter activities (Sax et al., 1997). The functional significance of αA-CRYBP1 and USF to αA-crystallin expression is not clear, since the expression pattern of neither transcription factor is known during lens development, and cotransfection assays with expression plasmids have not been reported.

Despite the sequence differences between the mouse and chicken αA-crystallin promoters, the chicken promoter also appears to be regulated by the specific factors Pax-6 and c-Maf/L-Maf and the general factors CREB/CREM and USF (Fig. 5.2). Initial transfection studies identified a fragment −162 to +77 as a minimal lens-specific promoter in cultured primary lens cells (Klement et al., 1989). Linker scanning analysis was performed between the position −162 and the TATA box of the intact promoter and revealed a series of *cis*-acting elements and potential protein-binding sites (Klement et al., 1993). A continuation of this analysis led to the discovery of two Pax-6–binding regions, sites C and E, and binding sites for USF and CREB/CREM proteins at positions different from that for the mouse αA-crystallin promoter (Cvekl et al., 1994).

The chicken αA-crystallin promoter region has been experimentally dissected into three regions (αCE1, αCE2, and αCE3) multimerized or studied individually or in combination with a heterologous β-actin basal promoter (Matsuo et al., 1991; Matsuo et al., 1992; Matsuo and Yasuda, 1992). Screening of chicken lens cDNA expression libraries with αCE2 yielded several clones encoding members of the *Maf* family of genes (Ogino and Yasuda, 1996).

One of them, L-Maf, was originally reported to be abundant in, and specific for the chick lens (Ogino and Yasuda, 1998), but is now known to be the chicken cognate of MafA which is expressed in a number of non-lens tissues including the retina, brain, dorsal root ganglia and pancreas (Benkhelifa et al., 1998; Kataoka et al., 2002; Lecoin et al., 2004). Overexpression of L-Maf/MafA transdifferentiates chick neural retina to lentoid bodies expressing αA-crystallin (Ogino and Yasuda, 1998) and can transactivate the chicken αA-crystallin promoter, although MafB and c-Maf were equally effective (Yoshida and Yasuda, 2002).

Transcriptional Control of the αB-Crystallin Gene. While αA-crystallin expression is highly lens preferred, αB-crystallin is also expressed in a number of extraocular sites, including the heart, the brain, skeletal muscle, and the lungs (Bhat and Nagineni, 1989; Dubin et al., 1989; Iwaki et al., 1989; Sax and Piatigorsky, 1994). In addition, αB-crystallin is expressed in many nonlens ocular cells, including those in the retinal pigment epithelium, optic nerve, extraocular muscle, iris, ciliary body, cornea, and trabecular meshwork (Robinson and Overbeek, 1996; Tamm et al., 1996). In fact, the first αB-crystallin expression detected during embryonic mouse development is in the cardiomyocytes of the early heart tube, at a time well before eye formation (Benjamin et al., 1997; Haynes et al., 1996). While αB-crystallin probably has important functions in these extralenticular sites (see above), here we will keep our discussion focused on the lens.

In the eye, endogenous mouse αB-crystallin mRNA is first detected at 9.5 dpc in the lens placode, preceding the expression of the αA-crystallin gene (Robinson and Overbeek, 1996). During subsequent stages of embryonic lens development, αB-crystallin expression remains highest in the epithelial cells; however, expression in the fiber cells increases between 12.5 and 15.5 dpc. Postnatally, expression of αB-crystallin in the mouse lens is more similar to the expression of αA-crystallin, in that both genes are expressed at low levels in the epithelium, and their levels increase in young cortical fiber cells (Robinson and Overbeek, 1996).

The human αB-crystallin gene is located on chromosome 11q22–q23, and the mouse gene is located on a syntenic region of chromosome 9 (Bova et al., 1999). Interestingly, the αB-crystallin gene (Fig. 5.1) is linked head to head with a related gene, *HSB2* (Iwaki et al., 1997). *HSB2* encodes a new member of the α-crystallin/small heat-shock protein family; however, this gene is not expressed in the lens. The proximal promoter sequences of αB-crystallin genes from humans, mice, and rats are similar, as expected for functional regulatory regions (Frederikse et al., 1994).

Studies of the mouse αB-crystallin promoter fused to the *LacZi* reporter gene determined that about 4 kb of the 5′-flanking sequence is sufficient to recapitulate the developmental expression pattern of the endogenous gene (Haynes et al., 1996). The minimal lens-specific αB-crystallin promoter reported to be active in transgenic lenses was the -115 to $+44$ region, while the -68 to $+44$ region was inactive in all tissues analyzed (Gopal-Srivastava et al., 1996). Interestingly, a −164 to +44 promoter fragment is active in lens fibers during embryogenesis and also postnatally in the corneal epithelium of transgenic mice (Gopal-Srivastava et al., 2000). This minimal lens-specific promoter is up-regulated by an upstream enhancer, MHL (−427 to −255). This enhancer is also responsible for αB-crystallin promoter activity in nonlens cells (Fig. 5.1).

The influence of the αB-crystallin enhancer on the αB-crystallin promoter is orientation-dependent in its natural context (Swamynathan and Piatigorsky, 2002). This explains in part why the enhancer does not promote significant expression of the neighbouring HSB2 gene in other tissues, including lens, skeletal muscle and heart. In addition, the existence of an insulator or silencer sequence appears to provide another functional barrier between the αB-crystallin enhancer and the HSB2 promoter.

While the mouse αB-crystallin promoter contains a canonical TATA box and possibly a cryptic TATA-rich region (−76 ATAATAAAA −68), the Inr is not present. *In vitro* footprinting indicated binding of proteins around the TATA box (Gopal-Srivastava et al., 1996). Site-directed mutagenesis of both the TATA box and the TATA-rich region in the context of the −661 to +44 genomic fragment reduced the promoter activity in lens cells. As with the αA-crystallin gene TATA-box mutations described earlier (Sax et al., 1995), mutations in the TATA regions of the αB-crystallin promoter did not eliminate the lens-specific expression in transgenic mouse lenses but did result in the abnormal start of transcription around nucleotide −50, which is weakly similar to the Inr consensus (Haynes et al., 1997). While it is clear that the elements responsible for the lens epithelial expression of αB-crystallin must reside between nucleotides −164 and approximately −4,000, the identities of these elements are not currently known.

The molecular basis of αB-crystallin expression in skeletal and cardiac muscle, lung, and brain has been investigated (Dubin et al., 1991; Gopal-Srivastava et al., 1995; Haynes et al., 1995; Haynes et al., 1996), but here we will focus on studies investigating the basis of lens-specific expression (Gopal-Srivastava et al., 1996; Gopal-Srivastava et al., 1998; Gopal-Srivastava et al., 2000; Gopal-Srivastava and Piatigorsky, 1994; Somasundaram and Bhat, 2000). DNaseI footprinting combined with electrophoretic mobility shift assays using lens and nonlens nuclear extracts and recombinant proteins revealed multiple binding sites for the tissue-restricted and developmentally controlled transcription factor Pax6 (Gopal-Srivastava et al., 1996), the retinoic acid (RA) nuclear receptors RARβ and RXRβ (Gopal-Srivastava et al., 1998), and members of the heat-shock transcription factor family (Somasundaram and Bhat, 2000). In addition, transient cotransfections of two lens-specific promoter fragments, −148 to +44 and −115 to +44, with cDNAs encoding Pax6, RARβ/RXRβ, and their combinations activated reporter gene expression in cultured lens cells. Treatment of cultured lens cells with RA induced endogenous αB-crystallin expression about threefold, confirming the role of RA nuclear receptors in the regulation of the αB-crystallin gene (Gopal-Srivastava et al., 1998). In contrast to αA-crystallin, the expression patterns of αB-crystallin (Robinson and Overbeek, 1996) and Pax6 (Walther and Gruss, 1991) overlap in the mouse eye both spatially and temporally, suggesting that Pax6 interactions with the αB-crystallin promoter are potentially more functionally significant. Since RA nuclear receptors are expressed at the onset of αB-crystallin transcription in the lens placode, they may cooperate with Pax6 to initiate the lens-specific expression of the αB-crystallin gene.

Heat-shock factors (HSF) present in lens cells but not in brain, heart, or liver extracts were shown to interact with the proximal (−54) pentamer motif (NGAAN; N = any base) found in the heat-shock element (HSE) of the rat αB-crystallin promoter, suggesting an additional mechanism responsible for regulated expression of the αB-crystallin in the lens

(Somasundaram and Bhat, 2000). These data, combined with the lack of promoter activity for −68 to +44 in transgenic mice, indicate that the promoter sequence specified as LSR2 (−78 to −46) contains multiple essential regulatory sites for expression of the αB-crystallin minimal promoter in lens fiber cells. *In vivo* experiments found reduced level of the endogenous αB-crystallin expression in c-*Maf* null lenses (Kim et al., 1999); however, no data are available on c-Maf binding to the αB-crystallin lens-specific minimal promoter. A candidate c-Maf binding site was noted as a part of lens-specific region LSR1 (Gopal-Srivastava and Piatigorsky, 1994), which is 5′ of the minimal lens-specific promoter (Fig. 5.2).

Finally, transgenic mouse experiments using the αB-crystallin promoter/enhancer region derived from the blind mole rat have revealed adaptive changes that have accompanied the subterranean evolution of this species (Hough et al., 2002). Blind mole rat eyes begin to form in the early embryo, but then degenerate into subcutaneous, non functional eyes, at least with respect to vision. In transgenic mice, the blind mole rat αB-crystallin promoter/enhancer:reporter transgene lost lens expression, maintained heart expression and increased (∼13-fold) skeletal muscle expression. The selective loss of lens expression was surprising since the αB-crystallin promoter sequences of the mole rat and mouse are quite similar. While the specific reasons for these adaptive changes in the αB-crystallin promoter/enhancer activity are not known yet, they illustrate the evolutionary refinement in gene expression that is associated with changes in function of the encoded protein.

5.3.3.2. Transcriptional Control of the β-Crystallin Genes

While six β-crystallin genes are known, generally the expression patterns and transcriptional control of these genes have only been investigated for βB1-, βB2-, and βA3/A1-crystallin.

βB1-Crystallin. Unlike the expression of most other crystallins, the expression of βB1-crystallin seems to be truly lens fiber cell specific. In mice, chickens, and frogs, βB1-crystallin gene expression is first detected in the elongating primary fibers of the lens vesicle (Altmann et al., 1997; Brahma, 1988; Duncan et al., 1996b). Further, throughout life, βB1-crystallin expression initiates precisely when secondary fibers first begin to elongate after leaving the transition zone (Brahma, 1988; Duncan et al., 1996b). Thus, βB1-crystallin is a useful marker for lens fiber cell differentiation.

In chickens and humans, the βB1-crystallin gene is linked head to head with the βA4-crystallin gene (Chen et al., 2001; Duncan et al., 1995), similar to the genomic arrangement described above for the αB-crystallin gene and *HSB2* (Iwaki et al., 1997). Comparison of the 5′-flanking sequences of the chicken, mouse, rat, and human βB1-crystallin genes has revealed a region of sequence conservation between −50 and +44 as well as other short regions of similarity (Chen et al., 2001; Roth et al., 1991). Since most work on the regulation of βB1-crystallin has been performed on the chicken promoter, we focus on that below.

A fragment consisting of −126 to +30 was determined to be the minimal promoter of chicken βB1-crystallin in transfected cells (Roth et al., 1991), while a fragment from −101 to +30 is the minimal promoter in transgenic mice (Fig. 5.1; Duncan et al., 1996b). The −126 to +30 fragment contains two elements, PL1 (−126 to −101) and PL2 (−96 to −76) (Roth et al., 1991), which appear to function synergistically in transgenic mice (Fig. 5.2; Duncan et al., 1996b). While PL1 and PL2 share the common core sequence TGATGA and bind some common factors, as assayed by EMSA, they appear to be functionally distinct; PL2 can confer lens-preferred activation to a heterologous promoter, while PL1 is a more general activator (Duncan et al., 1996b). The OL2 element, which resides at the 3′ end of the PL2 element, is also important for activity of the minimal promoter in transfections (Cui et al., 2004; Roth et al., 1991), although it has not been functionally tested in transgenic mice. In transfected lens and nonlens cells, sequences between −432 and −152 acted as a transcriptional repressor (Roth et al., 1991), while this same region was strongly activating in transgenic mice (Duncan et al., 1996b). Indeed, the −432 to +30 fragment of the chicken βB1-crystallin promoter was able to direct reporter gene expression that mimicked the spatial and temporal pattern as well as quantitative levels of the endogenous mouse βB1-crystallin gene (Duncan et al., 1996b; Taube et al., 2002).

Surprisingly, similar transfection experiments performed with a deletion series of mouse βB1-crystallin promoter fragments did not detect any promoter activity in fragments as long as −1800 to +45. However, this fragment was lens specific in transgenic mice even though its activity was approximately one-half that of the chicken −432 to +30 fragment in similar experiments (Chen et al., 2001).

The PL1 and PL2 elements of the chicken βB1-crystallin promoter both contain consensus MAREs (Maf responsive elements) that share sequence similarity with the αCE2 site of chicken αA-crystallin. Cotransfection analysis with L-Maf and a construct consisting of a multimerized PL2 element in front of the β-actin basal promoter showed that L-Maf can activate this element (Ogino and Yasuda, 1996, 1998). The relevance of this interaction is less certain, since L-Maf is expressed much earlier than βB1-crystallin, and L-Maf levels begin to down-regulate during embryogenesis at E16 in chickens (Ogino and Yasuda, 1996), the time at which βB1-crystallin (β35) mRNA levels increase (Hejtmancik et al., 1985). Both c-Maf and MafB also activate promoter activity via the PL1 and PL2 elements (Cui et al., 2004) and c-Maf may be the endogenous activator of the chicken βB1-crystallin gene into adulthood, since c-Maf levels increase in chicken lens fiber cells during late embryogenesis (Ogino and Yasuda, 1996). Further, the mouse βB1-crystallin promoter contains a MARE (Chen et al., 2001), and mice homozygous for a *c-Maf* deletion express much less βB1-crystallin in their lenses (Ring et al., 2000).

The full length chicken βB1-crystallin promoter is activated by co-transfection of Prox1, a homeodomain containing protein essential for lens fiber cell differentiation (Wigle et al., 1999). Notably, the OL2 element is highly similar to a consensus binding site for Prospero (Hassan et al., 1997), the Drosophila homolog of Prox1. Prox1 is able to weakly bind to the OL2 element in vitro and the Prox1 responsiveness of the chicken βB1-crystallin promoter appears to be at least partially controlled by this interaction (Cui et al., 2004).

Cotransfection studies of lens and nonlens cells demonstrated that Pax6 is a transcriptional repressor of chicken βB1-crystallin expression (Duncan et al., 1998). DNaseI footprinting assays and EMSAs demonstrated that this repression is likely to be mediated by direct Pax6 interaction with the PL2 element (Duncan et al., 1998) and Pax6 can largely block Maf and Prox1 mediated transactivation of the chicken βB1-crystallin promoter (Cui et al., 2004). These findings are consistent with the relative expression patterns of Pax6, c-Maf, Prox1 and βB1-crystallin. Pax6 protein levels are high in the lens placode and lens vesicle and decrease in developing fiber cells (Li et al., 1994). In the mature lens, Pax6 protein is predominately found in the lens epithelium, with much lower levels in the lens fibers (Duncan et al., 2000; Richardson et al., 1995). In contrast, Prox1 (Duncan et al., 2002) and c-Maf (Kim et al., 1999) expression upregulates during lens fiber cell differentiation. Since βB1-crystallin expression up-regulates approximately when Pax6 levels decrease, it can be hypothesized that Pax6 is important for controlling the fiber cell specificity of βB1-crystallin.

Transcriptional Control of βB2-Crystallin. In rodents, little to no βB2-crystallin expression is detected in mouse embryos, with expression initiating in new differentiating cortical fiber cells around birth (Carper et al., 1986). The minimal promoter of the rat βB2-crystallin gene functional in newborn rat lens epithelial explants treated with bFGF is −38/+10, while additional sequences upstream of this minimal promoter (−123/−77 and −750/−123) repress transcription in this system (Fig. 5.1; Dirks et al., 1996). However, inclusion of the first intron in the construct allows for this repression to be mostly overcome (Dirks et al., 1996). A consensus Pax6-binding site resides in the −77/+20 fragment and can interact with Pax6; however, neither mutation of this sequence nor cotransfection with a Pax6 expression vector affect promoter activity (Dirks et al., 1996). The minimal promoter of mouse βB2-crystallin functional in rabbit lens N/N1003a cells is −110/+30 (Chambers et al., 1995). While Pax6 interacts with the minimal mouse βB2-crystallin promoter fragments *in vitro* as well, this binding has not been shown to be functional (Chambers et al., 1995).

Mafs may play an important role in rat βB2-crystallin expression. c-Maf is able to activate the rat βB2-crystallin promoter in transfection assays and this activation is enhanced by the co-activators CBP/p300 (Chen et al., 2002). However, a dominant negative c-Maf fails to lower βB2-crystallin promoter activity, suggesting c-Maf is unlikely to be the endogenous activator of rat βB2-crystallin promoter (Doerwald et al., 2001). However, NRL, another member of the large Maf family, may be responsible, since the up-regulation of NRL in lenses coincides with the stimulation of rat βB2-crystallin expression.

Transcriptional Control of βA3/A1-Crystallin. In every species examined, the βA1- and βA3-crystallin peptides are produced by alternate translational initiation from a single βA3/A1 mRNA (Peterson and Piatigorsky, 1986). In the chicken, the only species in which βA3/A1-crystallin promoter function has been examined, the minimal promoter necessary for function in transfected chicken patched lens epithelial cells is −382/+22 (Fig. 5.1; McDermott et al., 1992). This fragment appears to contain a minimal promoter of −143 to +22, which can drive lens-specific expression in the lens of transgenic mice; a T-rich region; and an enhancer between −287 and −254. The enhancer includes a consensus binding sequence for basic leucine-zipper proteins such as AP1, CREB, and c-Maf (Fig. 5.2; McDermott et al., 1996; McDermott et al., 1997). Also, the βA3/A1-crystallin minimal promoter binds Pax6 at a nonconsensus site, and like in βB1-crystallin, cotransfection of Pax6 with the βA3/A1-crystallin promoter results in transcriptional repression (J. I. Haynes, unpublished data). This further suggests that Pax6 may function to prevent the expression of β-crystallins in the lens epithelium.

5.3.3.3. Transcriptional Control of the Mouse γF-Crystallin and Other γ-Crystallins

A cluster of six γ-crystallin genes is located on human chromosome 2q33–35; the mouse γ-crystallin gene cluster is located on a syntenic region of chromosome 1 (Shiloh et al., 1986; Skow et al., 1988). In rodents, all six γ-crystallins are induced in the embryonic lens fiber cells, and their expression drops after birth until only γB-crystallin transcripts are detectable in the adult rat lens (vanLeen et al., 1987). In humans, the six-gene cluster (γA, γB, γC, γD, γE, γF) is followed by a quarter gene fragment, γG. However, only the γC- and γD-crystallin genes produce abundant lens proteins (Lampi et al., 1998), while γE- and γF-crystallin are pseudogenes containing in-frame stop codons (Meakin et al., 1987). In rodents and humans, γ-crystallin gene expression is first detected in primary fibers as soon as they elongate to fill the lens vesicle, and this expression continues to be fiber cell specific during lens development (Treton et al., 1988; Treton et al., 1991). A detailed description of γ-crystallin expression has been published previously (vanLeen et al., 1987).

The 5′-flanking sequence of all the γ-crystallin genes have a highly conserved functional promoter region of about 90 bp upstream from the transcription start site (Peek et al., 1990). Since the majority of transcriptional work has been with the mouse γF-crystallin gene, we focus on that promoter.

Transgenic mouse experiments defined the fragment −67 to +45 as a minimal lens-specific γF-crystallin promoter active only in the central lens fibers, while an extended promoter fragment, −171 to +45, gave a broader pattern of fiber cell–specific expression (Goring et al., 1993). The −759 to +45 promoter fragment essentially recapitulated the expression of the endogenous γF-crystallin gene in the lens fibers (Goring et al., 1993). Studies of transiently transfected lens cells identified numerous candidate regulatory regions and a series of adjacent or overlapping *cis*-regulatory sites within the γF-crystallin minimal promoter fragment (Lok et al., 1989). Also identified were two enhancers, −392 to −278 and −226 to −121, that functioned both independently and cooperatively in transient transfection and transgenic mouse studies (Yu et al., 1990).

The earliest study of the transcription factors regulating γF-crystallin (Lok et al., 1989) demonstrated that −202 to −171 could interact with activator proteins. Subsequently, −208 to −190 was identified as a classic RA responsive element (RARE) that could be regulated by RA-activated nuclear receptors (Tini et al., 1995; Tini et al., 1993; Tini et al., 1994). In addition, the SRY proteins, Sox1 and Sox2, bind to positions −62/−51 and −37/−43 of the γF-crystallin promoter (Kamachi, 1996; Kamachi and Kondoh, 1993). Knockout of the *Sox1* gene demonstrated that it is essential for activation of γA- and γB-crystallin expression in the lens as well as the continued expression of the remaining γ-crystallins during development (Nishiguchi et al., 1998). A c-Maf binding site was also identified at −35/−47 (Ogino and Yasuda, 1998) and the γF-crystallin promoter is activated by cMaf in cooperation with the transcriptional co-activators CBP/p300 (Chen et al., 2002). Notably, studies of the c*Maf* knockout mouse have suggested that c-Maf is required for the expression of all γ-crystallin genes (Kawauchi et al., 1999; Kim et al., 1999). Six3 represses the γF-crystallin promoter through a responsive element found between positions −101 and −123, and Prox1 activates this promoter through a responsive element between positions −151 and −174 (Lengler et al., 2001).

Finally, sequences from −192 to −182 compose a canonical Pax6(5a)-binding site (Kozmik et al., 1997). Site-directed mutagenesis of this site repressed promoter activity in transiently transfected lens cells and also eliminated the RA inducibility (Kralova et al., 2002). Collectively, current data suggest that the mouse γF-crystallin gene is transcriptionally regulated by Sox1/2, members of the Maf family, RAR/RXR/ROR, and Pax6. How these transcription factors cooperate to impart fiber cell specificity on the γ-crystallin genes is still unknown.

In addition, rat γB- and γD-crystallin promoters were examined using the rat lens epithelial cell explant system (Klok et al., 1998; Peek et al., 1990). Despite the sequence similarity of all γ-crystallin promoters, subtle but functionally significant differences exist in two regulatory regions, the silencer region (a GC-rich region located between nucleotides −91 and −71 [numbering for the rat γD-crystallin promoter]) and the c-Maf binding site (T-MARE) described earlier (Fig. 5.2) which overlap with an imperfect AP-1 binding site (Klok et al., 1998). These may contribute to differential temporal expression of γ-crystallin genes. Indeed, activity of the γB-crystallin promoter, the last one to be expressed during development, could be attributed to the binding of a factor to the γD-crystallin silencer but not the γB-crystallin silencer. In addition, T-MARE sequences of the rat γB- and γD-crystallin promoters bind their common factor, presumably c-Maf, with different affinities. Finally, developmental silencing of the γD-crystallin promoter correlated with delayed demethylation of several CpG dinucleotides between promoter positions -10 and -50.

The γS-crystallin gene is not linked to the remainder of the γ-crystallin gene cluster; instead, it resides on human chromosome 3q25 (Kramer et al., 2000) and mouse chromosome 16 (Sinha et al., 1998). While little is known about the transcriptional regulation of γS-crystallin, its putative promoter contains sequences potentially able to bind homeodomain, Sox, and Maf proteins (Wistow et al., 2000a).

5.3.3.4. Transcriptional Control of the Avian δ1- and δ2-Crystallins

The δ1- and δ2-crystallin genes of chickens and ducks are linked head to tail, separated by 4–4.5 kb of intervening DNA (Hawkins et al., 1984; Nickerson et al., 1986; Nickerson et al., 1985; Piatigorsky et al., 1987). Both δ1- and δ2-crystallin can be considered enzyme-crystallins due to their similarity to arginosuccinate lyase (ASL), δ1-crystallin is enzymatically inactive, while δ2-crystallin is a functional enzyme (Barbosa et al., 1991; Kondoh et al., 1991; Piatigorsky and Horwitz, 1996; Piatigorsky et al., 1988). In chickens, δ1-crystallin protein is first detected in the lens placode before the onset of α- and β-crystallin expression, and it continues to be expressed in lens epithelial and fiber cells at levels approaching 50% of total soluble protein until late embryogenesis (Katoh and Yashida, 1973; Shinohara and Piatigorsky, 1976). At that point, δ1-crystallin expression levels drop in comparison to the β-crystallins, and they remain low throughout the posthatching period (Hejtmancik et al., 1985). Besides occurring in the lens, δ1-crystallin was found at appreciable levels in the cornea and retina (Li et al., 1993) and at lower levels in the brain and heart (Thomas et al., 1990).

Initial microinjection studies demonstrated that the chicken δ1-crystallin gene is expressed with lens specificity in mouse cells (Kondoh et al., 1983). In addition, mice transgenic for the δ1-crystallin locus expressed δ1-crystallin (Kondoh et al., 1987) in the lens even though δ1-crystallin is a taxon-specific crystallin only found in birds and reptiles (Hayashi et al., 1987). Transient transfections suggested that lens-specific expression of the chicken δ1-crystallin gene depended on an enhancer (B3–B5) located in the third intron (Hayashi et al., 1987). These initial observations on the evolutionarily conserved lens specificity of δ1-crystallin expression were later confirmed using the regulatory regions from δ1- and δ2-crystallin enhancers studied previously in transient transfections (Li et al., 1997; Takahashi et al., 1994).

Methylation of the chicken δ-crystallin genes as an additional transcriptional control mechanism was studied in expressing and nonexpressing cells (Sullivan et al., 1989; Sullivan et al., 1991). A series of CpG dinucleotides spread throughout the genomic locus are hypomethylated in lens cells expressing the δ-crystallins, while these sites are methylated in nonlens tissues that do not express these crystallins. Some of these CpG dinucleotides are also hypomethylated in nonlens tissues that express low levels of δ-crystallin message. Developmental studies showed that the hypomethylation of those critical CpG dinucleotides was not required for the onset and progression of δ-crystallin expression (Sullivan et al., 1991).

To a large degree, functional characterization of the δ1-crystallin promoter has centered on the lens-specific enhancer found in intron three. The enhancer was dissected into three fragments, B3, B4, and B5. While the B4 fragment by itself does not have any enhancer activity, it can participate in synergistic activation mediated by the flanking B3 and B5 regions. However, when B4 was multimerized into an array of 4 to 8 copies, it enhanced lens-specific expression of homologous or ehterologous promoters (Goto et al., 1990). Further functional characterization of B4 narrowed down the functional element to the DC5 region, which harbors a cluster of binding sites (δEF1, δEF2, and δEF3; Fig. 5.2) essential for transcriptional activators and repressors (Kamachi and Kondoh, 1993).

Studies on the δEF2 site suggested that the SRY-related proteins Sox1 and Sox2 could regulate crystallin gene expression (Kamachi, 1996). However, transient cotransfection studies found that both Sox1 and Sox2 could only activate reporter constructs harboring a multimerized δEF2 site in lens cells. This suggested that the binding of an additional lens-specific or preferred protein to an adjacent site was required to mediate Sox activation of δ-crystallin. Notably, cotransfection of Sox2 with MafA/L-Maf led to the cooperative activation of a construct containing the δ-crystallin enhancer (Shimada et al., 2003). A transcriptional repressor, δEF1, that binds to a site that overlaps δEF3 has been identified. δEF1 is a 124-kDa zinc-finger and homeodomain containing protein expressed in many tissues, including the lens (Funahashi et al., 1993; Sekido et al., 1997). Knockout of δEF1 resulted in multiple defects affecting the cranifacial areas and limb development and caused fused ribs, but unexpectedly it did not appear to lead to lens abnormalities (Takagi et al., 1998).

Two putative Pax6-binding sites were found 5′ of δEF2, and Pax6 activated the expression of a reporter construct driven by the δ1-crystallin promoter linked to the entire intron 3 enhancer in transiently transfected chicken fibroblasts (Cvekl et al., 1995b). Thus, a follow-up study focused on the determination of the precise temporal and spatial patterns of Pax6, Sox1/2/3, and δ1-crystallin expression in the developing chicken embryo (Kamachi et al., 1998). The earliest expression of Pax6 is in the head ectoderm in the region of lens-forming bias (Li et al., 1994). Shortly thereafter, the expression of Sox2/3 appears in an overlapping region (Kamachi et al., 1998). The cells that coexpress Pax6 and Sox2/3 give rise to the lens placode, which initiates δ-crystallin expression (Kamachi et al., 1998; Shinohara and Piatigorsky, 1976). The onset of Sox1 expression corresponds with the formation of the lens pit and is coincident with the elevation of δ1-crystallin expression. Extralenticular expression of δ-crystallin in Rathke's pouch and the spinal cord also coincides with expression of Pax6 and Sox2/3, consistent with the existence of an *in vivo* role for these factors in δ-crystallin expression (Kondoh et al., 2000; Sullivan et al., 1998). This is supported by the observation that Pax6 and Sox2 form a complex that binds to the δ-crystallin minimal enhancer and activates δ-crystallin transcription (Kamachi et al., 2001).

While a number of factors interacting with the B4 region of the intron 3 enhancer are known, much less is known about the factors regulating the B3 and B5 regions. Two putative L-Maf binding sites were found in the B3 and B5 regions near their borders with B4 (Ogino and Yasuda, 1998). Six copies of the putative L-Maf site cloned upstream of the heterologous β-actin minimal promoter strongly activated promoter activity in transfected lens cells. Since recombinant L-Maf binds to a single copy of this site *in vitro*, it is possible the L-Maf is involved in δ-crystallin gene regulation. MafB and c-Maf may also participate in the up-regulation of the δ-crystallin promoter (Yoshida and Yasuda, 2002).

Finally, RA activates δ1-crystallin gene expression in mouse teratocarcinoma stem cells stably transformed with the δ1-crystallin gene (Goto et al., 1988) and in cultured one-day-old chicken lens epithelial cells (Patek and Clayton, 1990). More recent experiments demonstrated that the complete intron 3 enhancer (B3–5) was required for activation of a δ1-crystallin reporter gene cotransfected with RARβ/RXRβ and treated with RA (Li et al., 1997). However, the precise nucleotides mediating the RA inducibility of δ-crystallin expression remain to be determined.

5.3.3.5. Transcriptional Control of Guinea Pig ζ-Crystallin

ζ-Crystallin is a taxon-specific crystallin expressed in the lens epithelium and cortical fiber cells of the guinea pig lens (Richardson et al., 1995). This gene is identical to NADPH quinone oxidoreductase, an enzyme constitutively expressed in the liver and other tissues (Rodokanaki et al., 1989). In most mammals, this gene is regulated from a TATA-less promoter. In guinea pigs, however, there is an additional TATA-containing promoter that is proximal to the TATA-less one and confers lens-specific expression (Gonzalez et al., 1994). Transgenic mice experiments found that fragments −756 to +70, −533 to +70, and −498 to +70 (+1 counted from the transcription start site controlled by the TATA-containing promoter) drove lens-specific expression, while two shorter fragments, −385 to +70 and −229 to +70, also directed expression to the brain (Lee et al., 1994; Sharon-Friling et al., 1998).

DNaseI footprinting identified a region between nucleotides −207 and −150, the ZPE element (Lee et al., 1994), that is important for lens-preferred promoter activity in transfected cells. Comparison of nucleotide sequences of the guinea pig ζ-crystallin gene with the corresponding region of the homologous NADPH quinone oxidoreductase gene from mice showed that ZPE is absent from the mouse promoter (Gonzalez et al., 1994). It has been proposed that the NADPH quinone oxidoreductase gene was recruited to serve as the ζ-crystallin gene after a ZPE-containing transposon was inserted between two 9-bp direct repeats (Lee et al., 1994). Pax6 binds to ZPE, and mutation of the Pax6 site in the context of the −756 to +70 guinea pig promoter led to the loss of lens-specific expression (Richardson et al., 1995). Further, a MARE resides adjacent to the Pax6 site (Sharon-Friling et al., 1998) and complexes with Nrl1, a member of the Maf subfamily of transcription factors (Sharon-Friling et al., 1998) *in vitro.*

5.3.4. Transcriptional Control of Invertebrate Crystallins

5.3.4.1. S-Crystallins

Sequence analysis of the squid SL20-1, SL11, and SL20-3 promoters revealed putative TATA boxes, which are present in all vertebrate crystallin promoters. In addition, these promoters have potential AP-1 sites, overlapping antioxidant responsive elements (AREs) and, in the case of SL20-3, a xenobiotic responsive element (Tomarev et al., 1994; Tomarev et al., 1992). AP-1 sites are common among vertebrate crystallin promoters (Piatigorsky and Zelenka, 1992), and antioxidant and xenobiotic responsive elements contribute to the activity of vertebrate GST and guinea pig ζ-crystallin promoters (for references, see Tomarev et al., 1992). It is also interesting that the SL11 promoter has a T-rich region, as do the vertebrate chicken βA3/A1-crystallin (McDermott et al., 1997) and α-enolase/τ-crystallin (Kim et al., 1991) promoters.

The SL20–1 and SL11 promoters are active in transfected rabbit lens epithelial cells and, to a lesser extent, in transfected monkey kidney cells and embryonic chicken fibroblasts

(Tomarev et al., 1994; Tomarev et al., 1992). Deletion (Tomarev et al., 1992) and site-directed mutagenesis (Tomarev et al., 1994) experiments indicated that the consensus AP-1 and overlapping ARE sites are necessary for the activity of these squid crystallin promoters in the transfected cells. Since the S-crystallin promoters are active in lens cells and fibroblasts of vertebrates, the importance of these *cis*-regulatory elements binding ubiquitous factors for lens specificity in cephalopods remains open.

Although Pax6 has been implicated in the activity of numerous crystallin promoters in vertebrates (Cvekl and Piatigorsky, 1996), it is unclear whether Pax6 regulates the promoter activity of S-crystallin genes. A squid Pax6 has been cloned and can induce ectopic eyes in *Drosophila,* as can Pax6 from other species (Tomarev et al., 1997). Pax6 mRNA first appears in the region of the rudimentary eye well before the first appearance of S-crystallins in the lentigenic cells at day 17 of development (West et al., 1995). Pax6 mRNA was not detected in the lentigenic cells that give rise to the posterior lens segment, but it was found in the anterior part of the lens segment, where two S-crystallin genes (Lops 7 and 12) are expressed. Moreover, Pax6 mRNA was observed in the fold of tissue that is posterior to the eye and gives rise to the cornea (Arnold, 1984), although it was not followed in the developing cornea (Tomarev et al., 1997).

5.3.4.2. Ω-Crystallin

The exon-intron structure of the scallop Ω-crystallin gene is similar to that of the mammalian *ALDH1/2* genes, with the exception that the scallop gene has an additional 5′ untranslated exon (Carosa et al., 2000). In contrast to the coding sequences and gene structure, however, the scallop ALDH/Ω-crystallin promoter region resembles that of the mouse and chicken αA-crystallin genes rather than the mammalian ALDH genes (Carosa et al., 2002). Gel shift experiments demonstrated binding of Pax6 and CREB to the Ω-crystallin promoter, and site-specific mutagenesis of the overlapping CREB/Jun and Pax6 sites abolished activity of the Ω-crystallin promoter in transfected cells. It appears as if convergent evolutionary adaptations underlie the preferential expression of crystallin genes in the lens of vertebrates and invertebrates.

5.3.4.3. J-Crystallins

The promoter regions of the known J-crystallin genes of jellyfish have putative TATA boxes and several CAAT sequences (Piatigorsky et al., 1993). There is also a CAAT sequence in the promoter region of the SL20–1 gene of the squid (Tomarev et al., 1992). The promoter regions of the J-crystallin genes have a RARE half-site that can bind histidine-tagged jRXR, an RA receptor that was cloned from, *Tripedalia cystophora*, a species of jellyfish (Kostrouch et al., 1998). The possibility that RA receptors regulate the jellyfish J-crystallin genes deserves further study, since these transcription factors activate the promoters of many vertebrate crystallin genes (Gopal-Srivastava et al., 1998; Li et al., 1997; Tini et al., 1995; Tini et al., 1993; Tini et al., 1994).

Compelling evidence exists for the role of Pax family members in crystallin gene expression in jellyfish (Kozmik et al., 2003). Unlike vertebrates, jellyfish have a protein called PaxB comprising a structural and functional Pax2/5/8-like DNA binding paired domain and octapeptide, and a Pax6-like homeodomain. Transfection experiments showed that PaxB strongly activates the J3-crystallin and weakly activates the J1A- and J1B-crystallin promoters in transfection tests. J3-crystallin promoter activation was accompanied by and dependent upon PaxB binding to cognate sequences at positions −66 to −34. Interestingly, the J3-crystallin promoter was weakly activated by Pax2, but not by Pax6. In addition to the interesting implications these results have for the evolution of eyes in general, they provide additional support for a critical role of Pax proteins for crystallin gene expression throughout evolution.

5.3.4.4. Drosocrystallin

Both antibody and hybridization tests indicated that drosocrystallin synthesis is eye specific in *Drosophila*. Transgenesis experiments in *Drosophila* showed that a 441-bp 5′-flanking sequence of the drosocrystallin gene fused to the *lacZ* reporter gene is sufficient to reproduce the endogenous expression pattern of drosocrystallin (Janssens and Gehring, 1999). This includes gene expression in the ocellus as well as the compound eye. Intron sequences also appear to have some regulatory capacity, as judged by their ability to drive expression in the imaginal discs. The drosocrystallin promoter was inactive in transgenic mice, however, when tested with the luciferase reporter gene (Carosa and Piatigorsky, unpublished data). Nothing is known yet about the transcription factors that might regulate drosocrystallin during fly development. Possibilities such as Pax6 and *prospero* have been considered elsewhere (Tomarev and Piatigorsky, 1996).

5.4. Lessons from Transcriptional Control of Diverse Crystallin Genes: A Common Regulatory Mechanism?

The common denominator among diverse crystallin promoters and/or enhancers is the presence of clusters of binding sites for a small group of DNA-binding transcription factors, including Pax6, c-Maf, Sox1/2, and RARβ/RXRβ. As shown earlier, neither Pax-6, c-Maf, Sox1/2, nor RARβ/RXRβ is a tissue-specific factor. Thus, the emerging model to explain lens-specific expression is based on the combinatorial principle used also by many other genes (Chen, 1999). As the expression of these transcription factors is quantitatively and spatially controlled during development, they presumably interact dynamically on specific arrays of positively and negatively acting *cis*-elements in crystallin promoters and enhancers. Promoter and enhancer site occupancy then would determine the level of crystallin gene expression in the lens epithelium and lens fibers.

Strong evidence for the combinatorial mechanism is that the same factors play regulatory roles in the transcription of different crystallin genes. As shown in Figure 5.2, many crystallin promoters and enhancers have at least one Pax6 and MARE/CRE/AP-1 site. It was proposed earlier that the unique environment of the lens (e.g., the redox state of the lens) may be advantageous for the function of specific transcription factors, and from an evolutionary perspective, genes recruited as crystallins were those genes employing transcription factors that worked "optimally" in the lens (Piatigorsky, 1993; Piatigorsky and Zelenka, 1992). Collectively, the combinatorial code employing about a dozen known or yet unknown tissue-restricted transcription factors and gene recruitment based on the transcriptional control by those genes are not mutually exclusive models.

In addition, to achieve transcription *in vivo*, numerous multiprotein complexes (e.g., TFIID, TFIIA, and chromatin-remodeling complexes) and factors (e.g., TFIIB), as well as the RNA polymerase II holoenzyme, have to be recruited to the vicinity of the start site of transcription. The DNA-binding transcription factors (Pax6, c-Maf, Sox1/2, RARβ/RXRβ, and TFIID) have to gain access to their cognate sites by displacing the nucleosomes and "opening" the chromatin for other transcription factors. Some DNA-binding factors may topologically alter the promoter and/or enhancer DNA through the mechanism of DNA bending and looping. Indeed, TFIID, Sox1/2/3, and possibly Pax6 can induce DNA bending (Connor et al., 1994; Gaston et al., 1992). In addition, some of the DNA-binding proteins (e.g., Pax6, c-Maf and RARβ/RXRβ) recruit transcriptional coactivators harboring the histone acetyltransferase (HAT) activity (e.g., p300 and CBP). HAT proteins modify histones by acetylating their tails and promote transcription of specific clusters of genes by remodeling the chromatin. It is noteworthy that lens fibers and nucleated erythrocytes derived from yolk sac are the only mouse embryonic cells expressing the histone H1° (Gjerset et al., 1982), and the prediction would be that there is a function for this histone in mouse embryonic lens. However, no abnormalities were found in histone H1° null lenses (Sirotkin et al., 1995).

While the research summarized in this chapter supports the global role of c-Maf, Pax6, RARβ/RXRβ, and Sox1/2, we do not know precisely how, where, and when all these factors interact *in vivo* with crystallin genes. The timing of the onset of the expression of crystallins, as well as the developmental control of the expression of those regulatory factors, is critical for understanding the complexity of crystallin-gene regulation (see Chapter 3). Pax6 is expressed in the head ectoderm/lens placode before the onset of αB- and αA-crystallins (Walther and Gruss, 1991), whereas c-Maf occurs later, at E10.5–11 (Ring et al., 2000). c-Maf is highly expressed in the differentiating lens fibers when the αA-crystallin gene and members of the β/γ-crystallin family of genes produce high amounts of transcripts. Later, c-Maf expression drops down in three-month-old rat lenses (Yoshida et al., 1997), but expression of some members of the β/γ-crystallin family is maintained after the cessation of c-Maf expression.

Similarly, Sox2 is initially expressed in the presumptive lens ectoderm from E8.5, and a high level of Sox2 is found during the formation of the lens pit; however, with the onset of Sox1 expression (at E11.5), Sox2 expression becomes down-regulated (Kamachi et al., 1998).

The RA-activated nuclear receptors RAR and RXR (present as α, β, and γ isoforms) form not only RAR/RXR heterodimers but also complexes with other nuclear receptors (e.g., T_3 and ROR). However, the developmental expression patterns of RAR and RXR in the lens are not well known. Using reporter genes with single or multiple copies of RAREs, it was possible to visualize retinoid signaling in the developing mouse lens (Balkan et al., 1992; Enwright and Grainger, 2000; Rossant et al., 1991). Retinoid signaling is active in the lens placode (at E9.5) and persists during the lens vesicle stage but is lost in the lens by E12.5. Interestingly, retinoid signaling in the lens is dependent on Pax6 (Enwright and Grainger, 2000), and consequently some phenotypes observed in RXR$\alpha^{-/-}$ and RAR$\gamma^{-/-}$ mice are similar to Peters' anomaly, a form of human ocular congenital defect due to mutations in the *PAX6* gene (Hanson et al., 1994). An RXR$\alpha^{-/-}$ phenotype resembles classic Peters' anomaly; the lens remains attached to the surface ectoderm and rotates ventrally (Kastner et al., 1994). When double mutants were examined, an increased severity of eye malformations was observed (Lohnes et al., 1994; Mendelsohn et al., 1994). It is not known whether expression of the αB- and γF-crystallins was affected in these mice. Thus, based on current data (summarized earlier), we propose here a model of coordinated crystallin gene expression as a sequence of events with at least four distinguishable phases.

First, as the head ectoderm is committed to become the lens placode, low-level expression of α-crystallins in the mouse (and δ-crystallin in the chicken) is initiated ("onset of crystallin transcription"), presumably as a result of transcriptional activation by Pax6 in combination with unknown factors (or perhaps Pax6 combined with Sox2/3 and RAR/RXR/ROR). As the lens placode becomes the lens pit and then the vesicle, these factors continue to cooperate to maintain crystallin expression in the presumptive lens epithelium. The repression of the β/γ-crystallin genes presumably occurs due to the absence of essential activators and to the presence of Pax6 acting as a repressor (Duncan et al., 1998). Then, when the cells in the posterior portion of the lens vesicle begin to elongate ("early embryonic fiber cell transcription"), c-Maf (or c-Maf and Sox1 together) becomes the prominent activator needed for high-level expression of the α-crystallins (or the β/γ-crystallins). While Pax6 protein expression is still detectable in secondary fiber cells (Duncan et al., 2000), its role in transcriptional control of α-/β-/γ-crystallins in fiber cells is not known. Similarly, RA receptors may terminate their engagement with αB-/γF-/δ1-crystallin promoters after E12.5 (Enwright and Grainger, 2000).

The next developmental switch in crystallin expression occurs around birth in rodents, coincident with the initiation of βB2- and γS-crystallin expression and the loss of γ-crystallin expression in secondary lens fiber cells ("postnatal switching phase"). A similar switch occurs in the chicken lens around E15, at which time the transcriptional activity of the δ1-crystallin gene down-regulates sharply and there is a coincident up-regulation of β-crystallin expression (Hejtmancik et al., 1985). While these changes in crystallin expression result in the crystallin composition of the lens nucleus differing from that of the cortex (and presumably are essential for the formation of the refractive index gradient), the molecular basis of these changes is unknown.

Finally, another switch occurs in the juvenile/adult lens and causes a difference in the crystallin composition of the newly formed cortical fibers that will reside behind the iris and thus outside of the optical axis throughout life. While this postnatal change in crystallin expression has been poorly described as yet and has no clear function, it is apparent that the ratio of crystallin messages made by the lens changes with age. A marked example is that expression of all γ-crystallin genes except γB-crystallin is lost from the adult rodent lens (Peek et al., 1990). Currently, the only hint of the molecular nature of this change is that c-Maf expression ceases in the lens after three months of age in the rat. We propose that changes in the relative expression levels of crystallin genes in lens fiber cells during development result from the changing levels and utilization of different transcription factors. It is possible that transcription factors with a broad expression pattern in the majority of cell types (e.g., CREB, CREM, AP-1, HSF, and USF) may participate in the maintained expression of crystallins in the postnatal lens without being essential for their developmentally regulated onset.

5.5. Current Questions

Early crystallin promoter studies were bolstered by the discovery that a short fragment of 5′-flanking sequence from the mouse αA-crystallin gene drove lens-specific expression in transfected cells (Chepelinsky et al., 1985) and transgenic mice (Overbeek et al., 1985). The strict evolutionary conservation of the mechanisms controlling lens-specific expression was then confirmed by the finding that chicken α-, β-, and δ-crystallin promoters also drove lens-specific expression in transgenic mice. In 1994–8, detailed study of the transcriptional regulation of αA- and δ1-crystallin made Pax6, Sox1/2/3, L-Maf/c-Maf/Nrl, and RAR/RXR leading candidates as transcription factors regulating the lens-specific expression of crystallin genes. Subsequently, binding sites for these transcription factors have been found in diverse crystallin promoters driving expression of both ubiquitous and taxon-specific crystallins. In 1998–2000, direct *in vivo* experiments confirmed that Sox1 and c-Maf are critical regulatory genes for the expression of many crystallin genes in the mouse lens.

While the minimal promoter necessary for lens-specific expression has been identified for many crystallins, in most, the sequences that control their precise temporal and spatial activity pattern during development are not known. Further, even when the known promoter sequences appear to recapitulate the developmental expression pattern of their genes, in only one case (chicken βB1-crystallin) have these sequences been shown to be sufficient to raise transgene expression to a level approaching that of an endogenous crystallin gene in the lens (Duncan et al., 1996b; Taube et al., 2002). Thus, at this time, the molecular mechanisms that allow a single copy gene to be expressed at levels approaching 50% of total cellular mRNA (chicken δ1-crystallin in embryonic lens fiber cells; Hejtmancik et al., 1985) are completely unknown. It can be hoped that the sequencing of the human, mouse, rat, and zebrafish genomes will allow for phylogenetic footprinting of the "ubiquitous" crystallin genes that may predict functionally conserved regulatory regions.

How the precise balance of transcription factors work together in the context of the complex, overlapping elements found in most crystallin minimal lens-specific promoter regions remains a key research area in crystallin gene expression. While c-Maf and Sox1 are expressed in lens fibers, other factors (e.g., Pax6) involved in crystallin gene expression are expressed primarily or preferentially in the lens epithelium. The investigation of this question should give great insight into both the mechanisms controlling tissue-specific gene expression during development and the biochemistry of the individual factors. It should be noted that, at this time, we cannot exclude the possibility that a truly lens specific transcriptional activator or coactivator exists.

There is accumulating evidence that growth factor–receptor (e.g., FGF family members and IGF-1; see chap. 11) and cell surface receptor-extracellular matrix interactions (e.g., integrin and lens capsule; see chap. 10) are involved in the regulation of lens fiber cell differentiation. In both cases, ligand binding to the receptor initiates a signal transduction cascade that changes the phosphorylation states of downstream signaling molecules. This signaling often changes the phosphorylation state of transcription factors, causing them to either translocate into the nucleus or change in function within the nucleus. Thus, integrin or growth factor signaling could cause a change in crystallin gene expression by influencing the function of preexisting transcription factors. Alternatively, this signaling could induce the expression of factors important for crystallin expression. While little direct evidence is available yet to support either possibility, it has been established that treatment of lens epithelial explants with bFGF (FGF-2) can induce crystallin gene expression (McAvoy et al., 1991). In addition, a recent study demonstrated that FGF-2 concentrations capable of promoting fiber cell differentiation in rat lens explants also activate a construct consisting of the mouse αA-crystallin –366 to +44 promoter fragment (Ueda et al., 2000). A better understanding of the downstream targets of signal transduction cascades in the lens, combined with information on how signal transduction influences factors known to be involved in crystallin transcription, is essential for a unified picture of the mechanisms regulating crystallin gene expression.

While most work on crystallin gene transcription has focused on the function of specific transcription factors, it is clear from studies in other systems that crystallin gene regulation is likely to be much more complex *in vivo* at the level of the endogenous gene. To address the role of chromatin in the transcriptional control of the mouse αA-crystallin promoter, two promoter fragments having 5′ ends at positions -1809 and -366 were introduced into stably transfected lens epithelial cells (Sax et al., 1996). The results showed substantial promoter activity of the −1809 to +46 fragment in stably transfected cells, while the same promoter fragment was less active in transient transfections. Thus, the role that chromatin organization and remodeling plays in crystallin gene expression is an important subject for future studies involving the identification of potential locus control regions (LCRs).

Finally, to understand both the evolution of lens function and crystallin gene expression, it is necessary to address the mechanisms by which crystallin genes have been recruited for high-level lens expression. Identification of common regulatory genes implicated in crystallin gene regulation may be used to explain some molecular events implicated in crystallin gene recruitment. We can experimentally test this hypothesis using transgenic "knock-in experiments." Three models may be particularly attractive in this respect. First, it is possible to knock the ZPE region from the guinea pig ζ-crystallin into the homologous mouse gene. Second, it is possible to knock LSR1/LSR2 from the mouse αB-crystallin promoter into the promoters of small heat-shock proteins, hsp27 and 29. Finally, chicken δ-crystallin/ASL lens-specific enhancer can be knocked in the mouse ASL gene locus.

5.6. Conclusion

During the past 20 years, our understanding of both the function of crystallin proteins and the control of their expression during lens development has increased greatly. While many unanswered questions remain, the rate of current progress in these fields suggests that many advances will be made in the next decade.

6

Lens Cell Membranes

Joerg Kistler, Reiner Eckert, and Paul Donaldson

6.1. Introduction

The transparency of the lens is closely linked to the structure and function of its cell membranes. To minimize light scattering, lens fiber cells are packed together in a tightly ordered array so that the space between the cells is smaller than the wavelength of light. Achievement of this configuration requires cell membranes to have sets of proteins able to facilitate the formation of junctions between cells, while its maintenance requires the effective regulation of cell volume. For a long time, much of the work on lens membranes was primarily devoted to the identification and biochemical characterization of intrinsic and peripheral membrane proteins and of the membrane lipids in which they are embedded. The main emphasis of this early work was on how these molecules were modified during lens aging and cataractogenesis, often without knowledge of their functions. It is only in the last 10 years that significant progress has been made on the important contributions cell membranes and their embedded proteins make to the physiology of the lens. The greatest quantum step in this context has without doubt been the discovery that the lens generates a circulating current (Robinson and Patterson, 1983). This current is believed to generate a circulating flux of ions and water that percolates through the lens carrying nutrients deeper into the lens and returning waste products to the lens surface more efficiently than is conceivable by passive diffusion alone (Mathias et al., 1997). While some aspects of this circulation have been demonstrated in experiments, others are still hypothetical and await further proof. More pieces of evidence in support of the lens circulation are continually being reported, but quite appropriately, to reflect a degree of uncertainty the literature generally refers to the lens circulation system as a model.

In concordance with these new developments, the present chapter focuses predominantly on the functional aspects of the lens cell membrane proteins and less on the details of protein and lipid composition and their modification during aging and cataractogenesis. Many of the membrane channels and transporters that are reviewed fit perfectly into the circulation model. An interesting aspect is that several newly identified membrane receptors have the potential to regulate the circulating fluxes. Further, there are the membrane receptors, which receive signals from growth factors, and the membrane proteins with adhesive properties, which together control the development and maintenance of the crystalline tissue architecture that is so crucial for lens transparency. Because of this emphasis on the relationship between membrane proteins and lens physiology, this chapter does not constitute an exhaustive review of lens membranes. In fact, it will ignore some portions of the excellent biochemical work that often forms the foundation for the functional experiments. We sincerely apologize to all whose important contributions may have gone unmentioned.

6.2. An Internal Circulation Is Generated by Spatial Differences in Membrane Proteins

While the transparent properties of the lens are a direct result of its structural adaptations, it is wrong to think of the lens as merely a passive optical element. Because of its size, the lens cannot rely on passive diffusion alone to transport nutrients to deeper lying cells or to transport waste products back to the surface. Furthermore, fiber cells deeper in the lens appear to lack the K^+ channels and Na^+/K^+ pumps necessary to generate the negative membrane potential used by other cells to counter Donnan forces so as to maintain their steady-state cell volume (Mathias et al., 1997). Thus, to maintain its structural organization and hence its transparency, the lens needs a specialized active transport system that delivers nutrients, removes waste products, and imposes the negative membrane potential required to maintain the steady-state volume of the fiber cells. The existence of such a system is suggested by the observation in all vertebrate lenses studied to date (Robinson and Patterson, 1983) of a standing flow of ionic current that is directed inward at the poles and outward at the equator (Fig. 6.1a). It has been suggested by Mathias et al. (1997) that this standing current is the basis for the generation of a unique internal microcirculation system that is responsible for maintaining fiber cell homeostasis and therefore lens transparency.

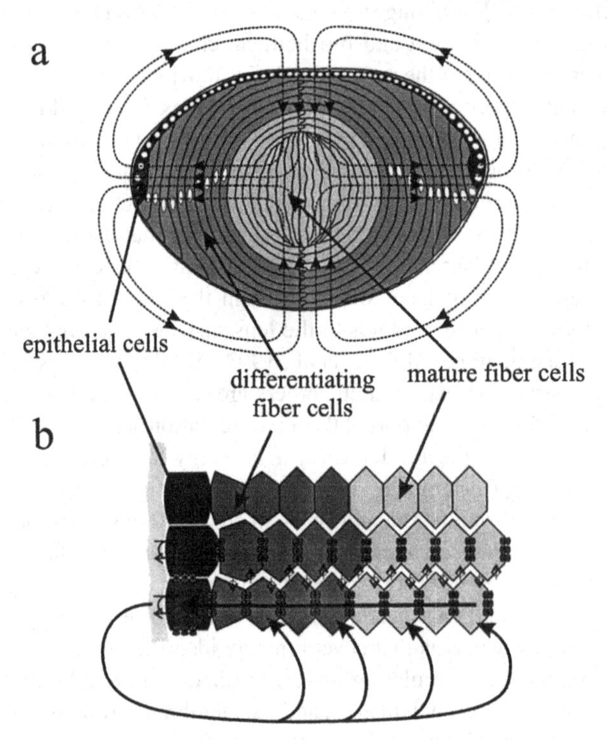

Figure 6.1. Structure and function of the mammalian lens. (a) Architecture of the lens showing the anterior epithelial monolayer, differentiating fiber cells, and mature fiber cells in the lens nucleus. Broken lines and arrows show the direction of the circulating currents measured by the vibrating probe (Robinson and Patterson, 1983). (b) Schematic cross section through the lens equator. Ions, solutes, and water are proposed to flow into the lens via the extracellular space, to cross fiber cell membranes, and to flow outwards via an intracellular pathway mediated by gap junction channels.

The circulating currents and theoretical work on modeling the transport properties of the lens have recently been reviewed (Mathias et al., 1997), and only a summary is provided here. Briefly, the current is carried primarily by Na^+ ions, which enter the lens via the extracellular clefts between fiber cells. Na^+ crosses the membranes of inner fiber cells and flows back toward the surface of the lens via an intracellular pathway mediated by gap junction channels. Finally, cells at the surface of the lens actively remove Na^+ to complete the loop (Fig. 6.1b). To preserve electroneutrality, the Na^+ flux is accompanied by a flux of Cl^- ions. These ionic fluxes in turn generate extracellular and intracellular fluid flows, which convect nutrients through the extracellular clefts toward the deeper lying fiber cells and remove metabolic waste products. The driving force for these fluxes is hypothesized to be the difference in the electromotive potential of surface cells and inner fiber cells. Fiber cell gap junctions are concentrated in the equatorial region, and their presence there is thought to direct the intracellular flow of Na^+ to the equatorial surface, where Na^+ is actively removed by Na^+/K^+ pumps.

An inherent feature of the model is the spatial localization of the different membrane transport proteins that generate the proposed circulating current (Fig. 6.2). Since the lens is

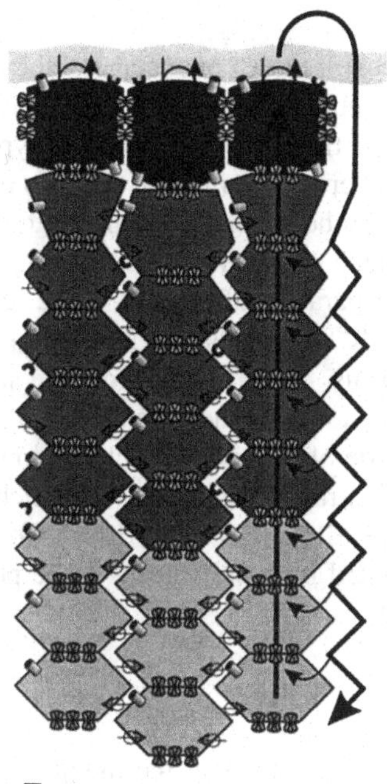

Figure 6.2. Spatial localization of the different classes of membrane proteins that play a role in lens transport. Differences in the distribution of pumps and channels are hypothesized to generate a circulating current that underpins the lens internal circulation. This brings nutrients to deeper regions of the lens, where transporters mediate their uptake into the fiber cells. Observed differences in receptor distribution throughout the lens raise the possibility that the magnitude of the circulating fluxes can be modulated.

continually adding new fiber cells at its equator, an area important for outward current flow, it is interesting to speculate that establishing and maintaining spatial differences in membrane transport proteins is an integral part of lens differentiation. To support this contention, the structural, functional, and molecular evidence that indicates the existence of spatial differences in membrane transport proteins between epithelial cells, newly differentiating fiber cells, and inner fiber cells is reviewed.

6.3. Membrane Conductances Vary between Lens Regions

Impedance analysis and ion substitution experiments have tentatively localized the majority of the K^+ conductance to cells near the surface of the lens and the Na^+ and Cl^- conductances to the inner fiber cells (Mathias et al., 1979). It has been calculated that the specific membrane resistance of the inner fiber cells is very high, indicating that under normal conditions the conductance of individual fiber cells is low. However, because the lens is made up of many fiber cells, their conductance generates a relatively large circulating ion flux. Further, because the surface and inner lens regions are electrically coupled via gap junction channels, this difference in electromotive potential generates the radially circulating current that powers the internal circulation while at the same time ensuring that the inner fiber cells have a negative membrane potential.

To determine the identity and spatial localization of these conductances, the patch clamp technique was initially used. More recently, molecular cloning techniques have been employed. Although we consider the membrane properties of three distinct cell types – epithelial cells, differentiating fiber cells, and mature fiber cells – most of the data accumulated concern the epithelial cells, as they have proven to be the most amenable to patch clamp analysis (Rae and Cooper, 1990). Attempts to patch clamp the inner fiber cells have been hampered by difficulties in obtaining a viable preparation of elongated fiber cells. This is of course not surprising, as these cells lack the K^+ channels and Na^+ pumps necessary for generating a negative membrane potential. Thus, once isolated from the surface cells, the inner fiber cells have no means of generating a membrane potential, and instead they depolarize. Their depolarization initiates a cascade of events that includes ion influx, cell swelling, and the Ca^{2+}-dependent activation of intracellular proteases, which ultimately leads to fiber cell globulization (Wang et al., 1997; Bhatnagar et al., 1997). To prevent this globulization, patch clamping of isolated elongated fiber cells needs to be performed in Ca^{2+}-free Ringers solutions (Eckert et al., 1998).

6.3.1. Potassium Conductance

The patch clamping of isolated epithelial cells has shown that the membrane properties of these cells are, as predicted, dominated by K^+ conductances. In subsequent studies, the molecular identity of the channel proteins underlying these conductances was revealed. This work has predominately been done by Rae and collegues (see Table 6.1 for references). They identified at least three major K^+ channel types: an inwardly rectifying K^+ channel, a delayed outwardly rectifying K^+ channel, and a large conductance Ca^{2+}-activated K^+ channel (Maxi-K). The relative expression of these different channel types varies in lenses from different species. An inwardly rectifying K^+ channel dominates the conductance properties of epithelial cells in rat, mouse, and rabbit lenses, while in the human, bovine, pig, frog, and chick lenses a delayed outwardly rectifying K^+ channel dominates (Table 6.1). It is believed that these two K^+ channel types are responsible for setting the membrane potential

Table 6.1. *Types of K^+ conductance in the lens epithelium*

Dominant channel type	Molecular identity	Species	Reference
Delayed outward rectifier (dOR)	$K_v2.1$ $K_v3.3$ $K_v9.1$ & 9.3*	Bovine, chick, human acute, human cultured, monkey	Cooper et al., 1990; Rae 1994; Rae & Shepard, 1998a, 2000; Shepard & Rae, 1999
Inward rectifier (IR)	$K_{ir}2.1$ (IRK1)	Chick, human, mouse, rabbit, rat, frog	Cooper et al., 1991, 1992; Rae & Shepard, 1998b
Maxi K (K_{Ca})	BKα BKβ	Chick, human, monkey, pig, rabbit	Rae et al., 1990; Rae & Shepard, 1998c

*Electrically silent channels

in the lens. Thus, lenses in which the inwardly rectifying K^+ channels dominate exhibit a higher potential than those in which the outwardly rectifying channels predominate. The Maxi-K channel has been identified in all species of lenses investigated to date, with the possible exception of the frog lens. The biophysical properties of the Maxi-K channel are such that it is inactive at normal resting membrane potentials. Activation of this type of channel can occur upon elevation of cytoplasmic Ca^{2+}. Thus, the recent finding (see below) that the lens contains an array of membrane receptors capable of mobilizing intracellular Ca^{2+} suggests that Maxi-K channels could play a role in the regulation of lens membrane potential, a notion that warrants further investigation.

A role for K^+ channels in fiber cells has not been as extensively investigated as in epithelial cells. It appears that isolated differentiating fiber cells are still capable of maintaining a negative membrane potential (Donaldson et al., 1995) and exhibit conductances that resemble the potassium conductances observed in epithelial cells (Fig. 6.3a). However, as the fiber cells differentiate and elongate, they become depolarized and their membrane properties become dominated by a large nonselective leak conductance (Fig. 6.3b). Thus, fiber cell differentiation appears to be associated with a loss of potassium conductance (Fig. 6.3c).

6.3.2. Sodium Conductance

Patch clamp experiments on epithelial cells have shown that, in addition to a K^+ conductance, epithelial cells often contain a Na^+ leak conductance, which may be due to a stretch-activated cation channel in amphibian lenses (Cooper et al., 1986) or a tetrodotoxin-sensitive Na^+ channel (Watsky et al., 1991). The relevance of these channels to lens physiology remains uncertain. While it is believed that inner fiber cells contain a Na^+ or cation conductance, the molecular nature of the channels that sustain this conductance is not presently known. Interestingly, in whole lens experiments, removal of extracellular Ca^{2+} induces an increase in the fiber cell cation conductance (Rae et al., 1992). Membrane currents of isolated elongated fiber cells bathed in zero Ca^{2+} to prevent fiber cell globulization also exhibit a dominant cation conductance (Eckert et al., 1998). One of these currents (Fig. 6.3b) exhibited a striking similarity to the so-called hemichannel currents recorded from single

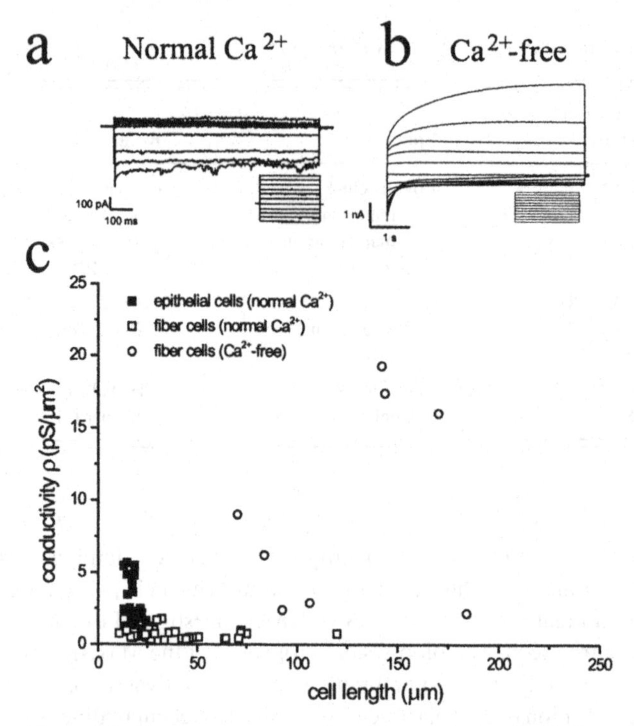

Figure 6.3. Conductance properties of acutely isolated cells from the rat lens. (a) Representative current trace recorded from a short (~60 μm) fiber cell bathed in normal Ringer's. (b) A large nonselective hemichannel-like conductance recorded from a long (~180 μm) fiber cell bathed in Ca^{2+}-free Ringer's. (c) A plot of specific conductivity (membrane conductance per area of membrane) versus cell length for isolated epithelial and fiber cells. This plot shows that, in normal Ringer's, the longer the cell, the lower the membrane conductance. However, incubation of longer fiber cells in Ca^{2+}-free Ringer's appears to induce a large increase in the specific conductivity, presumably via the activation of the hemichannel-like conductance.

oocytes expressing the gap junction protein connexin 46 (Ebihara and Steiner, 1993; Ebihara et al., 1995). In either cell system, these currents were only activated when Ca^{2+} was reduced and the cells were depolarized. Hence, hemichannels are unlikely to open in the lens under normal conditions, a prediction also consistent with electrical measurements of whole lenses, which characterize the fiber cells as electrically "tight" (Mathias et al., 1997). However, because lens depolarization is often observed early in cataractogenesis, it is conceivable that hemichannel currents become activated under pathologic conditions. This would lead to further depolarization of the fiber cells, causing an influx of ions and water and, ultimately, the cell swelling and globulization that is typically seen in cortical cataractogenesis (Bond et al., 1996).

6.3.3. Chloride Conductance

Patch clamp experiments have to date failed to identify a Cl^- selective current in lens epithelial cells (Mathias et al., 1997; Eckert, unpublished observations). More recently, transcripts for the Cl^- channel isoforms ClC2 and ClC3 where extracted from a cDNA library prepared from human lens epithelium (Shepard and Rae, 1998). In other expression systems, the activity of these Cl^- channel isoforms is increased by exposure to hypotonic solutions. Hence, in lens epithelial cells, these channels may be inactive under normal conditions and become activated only following changes in cell volume. Measurement of lens potential following anion substitution indicates that the fiber cells have a significant Cl^- conductance (Mathias et al., 1985). Consistent with this notion, Cl^- channels have been identified in membrane vesicles derived from bovine lens fiber cells (Zhang and Jacob 1994) and in isolated differentiating fiber cells (Zhang et al., 1994; Zhang and Jacob, 1996). In both preparations, Cl^- currents were blocked by a wide variety of Cl^- channel blockers, including tamoxifen and NPPB. Tamoxifen is a highly potent and selective blocker of p-glycoprotein, the multifunctional protein product of the multidrug resistance gene (*MDR1*). P-glycoprotein is also considered to be a regulator of volume-activated Cl^- channels. Interestingly, the addition of either tamoxifen or NPPB to organ-cultured lenses inhibits the ability of the lenses to volume regulate and causes lens opacification (Zhang et al., 1994; Zhang and Jacob, 1996; Tunstall et al., 1999). A role for Cl^- ions in lens volume regulation is more fully discussed later.

6.4. Lens Cells Are Connected by Gap Junction Channels

It has been long recognized that cells throughout the lens are extensively coupled via gap junctions (Goodenough, 1979; Goodenough et al., 1980; Mathias et al., 1981; Lo and Harding, 1986). Gap junctions connecting the fiber cells are particularly abundant and have been observed predominantly on the membrane broadsides (see chap. 4; see also Gruijters et al., 1987a, 1987b). This arrangement facilitates the convection of ions, solutes, and water in a radial direction, consistent with the outflow pathway of the lens circulation model. The first membrane protein correctly identified as a cell-cell channel-forming polypeptide in the lens is MP70, now known as connexin 50 (Kistler et al., 1985; Kistler et al., 1988; White et al., 1992). A closely related protein, connexin 46, was also identified, and together they form the gap junctions connecting fiber cells (Paul et al., 1991). The two isoforms assemble with each other to form heteromeric connexons (Jiang and Goodenough, 1996). A third isoform, connexin 43, was identified as the predominant gap junction protein in the lens epithelial cell membranes (Beyer et al., 1989). Other species isoforms were cloned from the lens as follows: chicken connexin 56 (Rup et al., 1993) and bovine connexin 44 (Gupta et al., 1994a) are homologous to rat connexin 46 (Paul et al., 1991); chicken connexin 45.6 (Jiang et al., 1994), ovine connexin 49 (Yang and Louis, 1996), and human connexin 50 (Church et al., 1995) correspond to mouse connexin 50 (White et al., 1992). No novel connexins have been found in the lens over and above the three isoforms initially identified.

Lens fiber cell connexins were isolated and confirmed to form channel structures by negative stain electron microscopy. This was possible because connexin 50 and 46 are soluble in mild detergents that leave other proteins embedded in the membrane (Kistler and Bullivant, 1988). Reconstitution of the solubilized channel structures with exogenous lipids resulted in the reassembly of gap junctions that were reminiscent of their native counterparts in lens fiber cell tissue (Kistler et al., 1994).

Lens connexins were expressed in *Xenopus* oocyte pairs to verify that they individually had the ability to form communicating channels (White et al., 1994). All three connexins formed channels that were sensitive to voltage and acidification. Patch clamping of acutely isolated epithelial cell pairs (Donaldson et al., 1994) and fiber cell pairs (Donaldson et al., 1995) established that channels with similar electrical gating properties indeed exist in the lens. Notably, the differential expression of connexins in epithelial and fiber cells results in significant changes in voltage dependence and unitary conductance, suggesting that altered gap junction permeabilities play a role in lens development. Whole lens experiments are also consistent, in that they were able to demonstrate pH-mediated uncoupling of fiber cells electrophysiologically (Baldo and Mathias, 1992; Emptage et al., 1992) and by dye transfer (Eckert et al., 1999) in the intact cortical tissue.

Phosphorylation of connexins appears to further regulate gap junction permeability in the lens. In lens cell cultures, activation of protein kinase C with a phorbol ester phoshorylated connexin 43 and significantly decreased dye transfer between epithelial cells (Reynhout et al., 1992). The situation is less clear for fiber cells. Coupling between fiber cells of cultured ovine lentoids expressing connexin 46 and 49 was unaffected by the phorbol ester treatment (Tenbroek et al., 1997). In contrast, the same treatment significantly reduced dye transfer between fiber cells in cultured chicken lentoids (Berthoud et al., 2000). It was independently confirmed that the fiber cell connexins were also phosphorylated in the whole lens and that protein kinase C was indeed involved (Jiang and Goodenough, 1998). Thus it appears possible that gap junction permeability is differentially regulated at different stages of lens development not only by a switch in connexin expression but also by protein kinase C–dependent connexin phosphorylation. However, there may be species variations with regard to the latter.

Lens casein kinase was also found to phosphorylate the connexin 50 homologs in chick and sheep (Cheng and Louis, 1999; Yin et al., 2000). The function of this phosphorylation event appears to be the destabilization and degradation of the connexin. Ser363 of connexin 45.6 was phosphorylated. In a mutant that had the serine replaced with alanine, this phosphorylation did not occur, and the result was a longer half-life than possessed by the wild-type connexin (Yin et al., 2000). Connexin 56 also seems to be affected in a similar way: One study detected two pools of connexin 56 that were phosphorylated but had different molecular weights (Berthoud et al., 1999). One form had a half-life of only a few hours, the other a half-life of days. These findings strongly suggest that gap junction stability in the lens is regulated by phosphorylation.

As another form of posttranslational processing, connexin 46 and 50 were both found to be cleaved upon fiber cell maturation at about the time when the cell nuclei were degraded (Fig. 6.4a; Kistler and Bullivant, 1987; Lin et al., 1997). This cleavage, accomplished by lens endogenous calpain, removes a significant portion of the cytoplasmic carboxyl tail and leaves the channel-forming portion in the membrane (Kistler et al., 1990; Lin et al., 1997). In the case of connexin 50, it has been possible to determine precisely the cleavage site in the molecule, allowing the construction of an appropriate truncation mutant for expression in *Xenopus* oocyte pairs (Fig. 6.4b). The functional consequences of the cleavage could thus be established (Fig. 6.4c), including the abolishment of the pH gating of the cell-cell channels (Lin et al., 1998). Similar efforts with connexin 46 have failed so far – the precise cleavage site remains unknown. However, it is reasonable to assume that its cleavage would have a similar effect on channel gating, as the same phenomenon was observed among other connexins. (Delmar et al., 2000).

The functional implications of this cleavage could be crucial for the lens. The lens nucleus is mildly acidic and has a pH around 6.5 (Mathias et al., 1991; Mathias et al., 1999), which is sufficient to close gap junction channels and thus uncouple the core region from the peripheral lens tissue. Thus, the lens has developed an elegant mechanism to keep the deeper-lying fiber cells communicating by simply abolishing the pH gating of the cell-cell channels through truncation of the connexins. That cell-to-cell communication in the lens

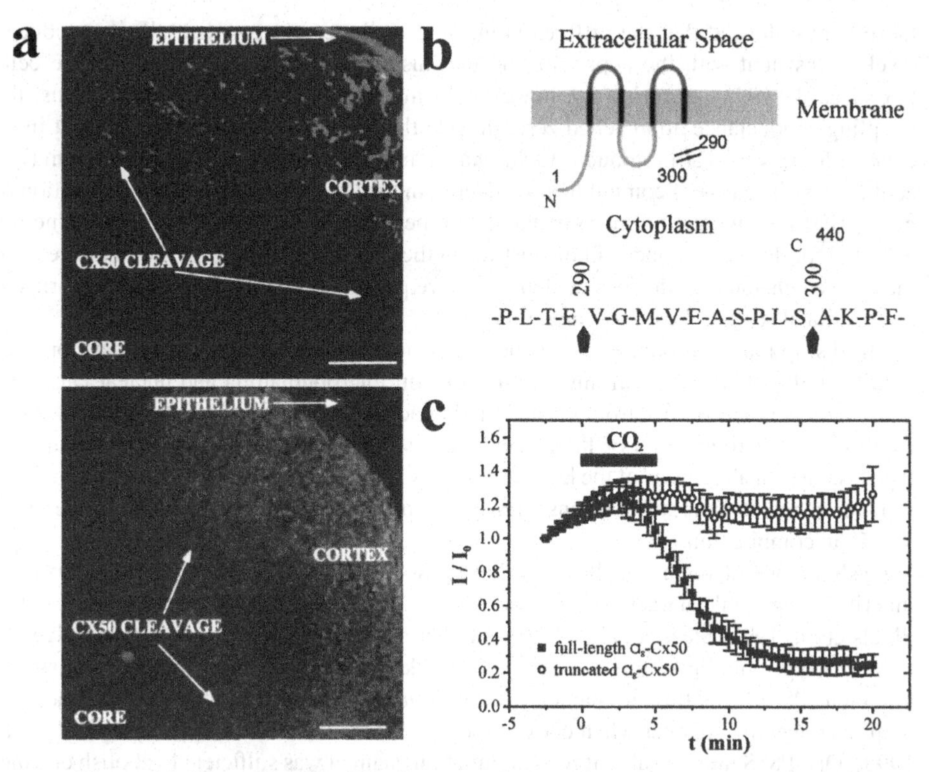

Figure 6.4. (See color plate IX.) Cleavage of connexins in the lens. (a) *Top panel:* axial section through the bow region of a mouse lens showing the coincidental loss of nuclei (red) and the carboxyl tail of connexin 50 (green). *Bottom panel:* equatorial section double-labeled with two different connexin 50 antibodies, one specific for the carboxyl tail (green) and the other specific for the cytoplasmic loop (red). This demonstrates that the membrane-embedded channel-forming portion of connexin 50 persists toward the center of the lens. Scale bar: 100 μm. (b) The Ca^{2+}-activated protease calpain cleaves the cytoplasmic tail of connexin 50 at amino acid residues 290 and 300. (c) Truncated recombinant connexin 50, which mimics the *in vivo* cleaved form, makes functional gap junction channels in *Xenopus* oocytes. Upon acidification of the oocytes by exposure of the bath solution to 100% CO_2, it becomes evident that the wild-type connexin 50 is pH-sensitive while the truncated version has lost pH-gating.

nucleus indeed occurs and is pH insensitive was verified in electrophysiological (Baldo and Mathias, 1992) and dye transfer experiments (Eckert et al., 2000).

The production of knockout mice for all three connexins expressed in the lens means that it is now possible to determine the contribution each isoform makes to lens transparency. In each case, the disruption of an individual connexin gene results in cataractogenesis. In the case of connexin 43, which is the predominant isoform forming the cell-cell channels in the epithelium (Beyer et al., 1989; Donaldson et al., 1994), the normally close cellular apposition is severely disrupted (Gao and Spray, 1998). Specifically, fiber cells are separated from the apical surfaces of the epithelial cells, and large vacuolar spaces are apparent between the fiber cells, most prominently in the deeper lens cortex. This damage phenotype is similar to those observed in cortical osmotic cataracts.

In contrast to disruption of the connexin 43 gene, disruption of the connexin 46 gene results in nuclear cataract (Gong et al., 1997). A detailed electrophysiological analysis of these lenses showed that the differentiating fiber cells remain coupled, albeit at reduced levels, consistent with the apparently normal distribution of connexin 50 in these cells (Gong et al., 1998). Surprisingly, however, in mature fiber cells deeper in the lens, the coupling conductance approached zero, despite the fact that connexin 50 is present in its cleaved form, which still produces functional channels in the *Xenopus* oocyte system (Lin et al., 1998). The loss of communication deeper in the lens correlates with the formation of nuclear cataract, indicating strongly that the proper functioning of gap junction channels is essential for the maintenance of homeostasis in the lens core region. It further indicates that the cleaved channels in the lens nucleus might require both fiber cell connexin isoforms to be functional (unlike the oocyte model system).

The disruption of the connexin 50 gene affects not only the lens but also eye development (White et al., 1998). The null mutant mice exhibit microphthalmia and nuclear cataracts. Lens mass is reduced 46% from normal, and opacities become evident shortly after birth. Analysis by microinjection of fluorescent tracers indicate that coupling between all cell types persists in all regions of the knockout lenses. The results suggest that both normal eye growth and maintenance of lens transparency depend on the unique properties of connexin 46 and 50 in combination.

Dysfunctions of connexin 46 or 50 have also been associated with certain forms of inherited congenital cataract in humans and mice (Mackay et al., 1997; Mackay et al., 1999; Shiels et al., 1998; Steele et al., 1998). The human cataracts are of the zonular pulverent (nuclear) type, similar to those observed in the knockout mice. A connexin 50 missense mutation (P88S) was further characterized in *Xenopus* oocyte pairs and shown to act in a dominant negative manner when coexpressed with normal human connexin 50 (Pal et al., 1999). One P88S mutant subunit per gap junction channel was sufficient to abolish channel function. These observations underline the importance of intercellular communication to the generation of the lens internal circulation.

6.5. Na^+ Pump Activity Is Greatest at the Lens Equator

The active extrusion of Na^+ by the Na^+ pump is critical to the generation of the circulating ion fluxes that drive the lens internal circulation. The circulation model predicts that most Na^+/K^+-ATPase activity is concentrated in the equatorial region of the lens. Gao et al. (2000) also measured the pump density and estimated that the total Na^+/K^+ pump activity per unit area of lens surface was about 20 times larger at the equator than at the anterior pole. This is consistent with an earlier study, in which the researchers separated the capsule, with the epithelium attached, from the bulk of the lens and then dissected a superficial anterior-equatorial cortex segment and a superficial posterior cortex segment (Alvarez et al., 1985). The principal Na^+/K^+-ATPase activity was found in the superficial anterior-equatorial cortex segment. By comparison, enzyme activity was 1.6 times less in the capsule epithelium and negligible in the posterior cortical segment. In older fiber cells deeper in the lens, pump protein was retained but had lost its functional ability (Delamere and Dean, 1993).

In addition to an increased activity of the Na^+/K^+-ATPase at the equator relative to the poles, a number of investigators have shown that there is also a change in the molecular composition of the Na^+ pump. There are at least three major isoforms of the Na^+/K^+-ATPase α subunit, and all three have been identified in the lens (Moseley et al., 1996; Garner and Kong, 1999; Tao et al., 1999). All three studies agree that these isoforms are differentially expressed but are inconsistent regarding the precise localization: In the first study, which used rat lenses, the epithelium contained predominantly $\alpha 1$ and $\alpha 2$ and less $\alpha 3$, while fiber cell membranes contained only $\alpha 1$ (Moseley et al., 1996). In the second study, which used bovine and human lenses, $\alpha 1$ and $\alpha 3$ were localized in the central epithelium, and $\alpha 2$ and $\alpha 3$ were found in cortical fiber cells (Garner and Kong, 1999). In the third study, abundant $\alpha 2$ and $\alpha 3$ were found in the epithelium of rabbit lenses, and minor amounts of $\alpha 3$ were detected in fiber cells form (Tao et al., 1999). Finally, in a recent report, $\alpha 2$ was the predominant isoform at the anterior pole of frog lenses, and $\alpha 1$ was predominant at the equator (Gao et al., 2000). While these inconsistencies may simply reflect species differences, the physiological relevance of the observed differential expression patterns is intriguing, especially in the light of recent reports that indicate that regulation of the Na^+ pump is isoform specific. Thus, the localization of a specific isoform at the equator may confer the ability to regulate lens circulation.

6.6. Water Flow across Lens Cell Membranes Is Enhanced by Aquaporins

The lens circulation model postulates that the circulating ion fluxes are accompanied by water flow. This necessitates that both epithelial and fiber cells have a significant water permeability. Consistent with this idea is the finding that each cell type expresses a distinct type of water channel or aquaporin (AQP) isoform. In the epithelial cell membranes, water permeability is enhanced by AQP1, while in the fiber cells, AQP0 (orginally called the main intrinsic polypeptide [MIP]) is responsible (Stamer et al., 1994; Patil et al., 1997; Lee et al., 1998; Hamann et al., 1998). When expressed in *Xenopus* oocytes, both aquaporin isoforms significantly increased the water permeability of the oocyte membrane, but AQP1 was about 40 times more effective as a water channel than AQP0/MIP (Chandy et al., 1997). In accordance with this, the water permeability of lens epithelial membranes is much greater than that of the fiber cell membranes (Fischbarg et al., 1999; Varadaraj et al., 1999). This difference may be crucial for the proper functioning of the lens circulation system, as it allows the epithelial membranes to cope with the large water flow arising from the cumulative entry into the large number of fiber cells.

MIP was not always recognized as a water channel and in fact has been attributed several functions over time. Initially isolated (Broekhuyse et al., 1976) and cloned (Gorin et al., 1984) from bovine lenses, MIP is the most prominent of all fiber cell membrane proteins and in some species constitutes up to 60% of total membrane protein. It is present throughout the lens cortex and nucleus and was first localized in the 11- to 13-nm "thin" fiber membrane junctions, which were for some time wrongly interpreted as communicating junctions (Bok et al., 1982; Sas et al., 1985). This mistake was recognized when it was demonstrated that MIP was unable to form communicating channels in the *Xenopus* paired oocyte expression system (Swenson et al., 1989). Later, MIP was demonstrated to form relatively large nonselective ion channels in planar lipid bilayers (Ehring et al., 1990; Shen et al., 1991; Modesto et al., 1996), a capability that could not be reproduced by expressing MIP in oocytes. Only more recently has MIP been recognized as the founding member of the MIP family of water and solute channels (Park and Saier, 1996; Heymann and Engel, 1999). When expressed in *Xenopus* oocytes, MIP increased membrane permeability for water and glycerol (Kushmerick et al., 1995; Mulders et al., 1995; Chandy et al., 1997). This water channel activity was confirmed also for the lens (Fig. 6.5a) on the basis that fiber cell membranes containing wild-type MIP had a significantly larger water

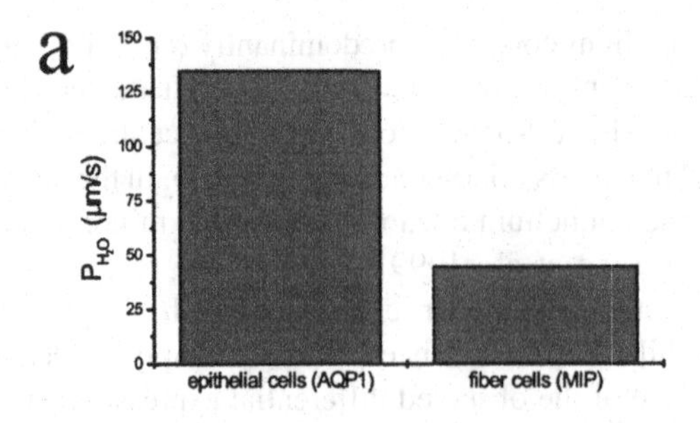

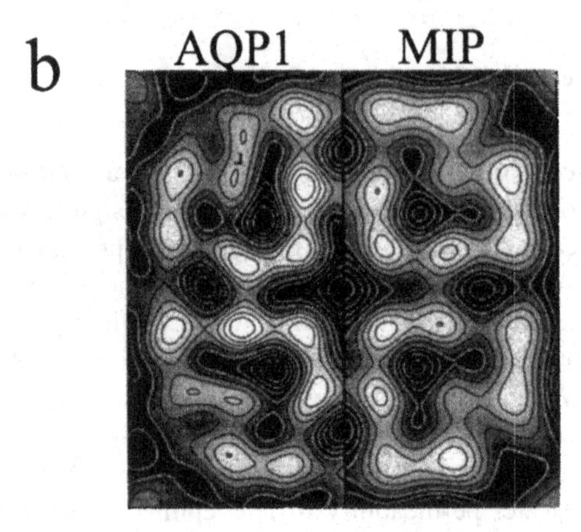

Figure 6.5. Water channels in the lens. (a) The water permeability (P_{H_2O}) of epithelial cells that express AQP1 exceeds that of membrane vesicles prepared from fiber cells containing MIP (Varadaraj et al., 1999). (b) A common structure for lens water channels. Projection maps for AQP1 and MIP show that these two membrane proteins have a similar structure (Hasler et al., 1998). The full width of the image corresponds to 6.4 nm.

permeability than fiber cell membranes containing a mutant form of MIP (Varadaraj et al., 1999).

MIP is unique among the aquaporins in that its water channel activity is pH dependent. When expressed in *Xenopus* oocytes, reduction of the pH to 6.5 increased the water transport activity of MIP 3.4 times (Nemeth-Cahalan and Hall, 2000). The prevailing pH in the lens nucleus is similar to this value due to the accumulation of lactate as the end product of glycolysis (Bassnett et al., 1987; Mathias et al., 1991; Mathias et al., 1999). Hence, it is possible that the water transport activity of MIP in the lens core is enhanced by the lower pH, which could be a mechanism facilitating the penetration of the circulating ion and water fluxes deeper into the lens. This feature of MIP could even represent a feedback mechanism whereby a decrease in intracellular pH could act as a signal of increased metabolic activity and thus could be a stimulus for greater fluid flow in regions of the lens with elevated metabolic activity.

Further support for MIP's role as a water channel is provided by its structural homology to the archetypical water channel AQP1. Following solubilization from fiber cell membranes with detergent, MIP was isolated as tetramers (Aerts et al., 1990; Konig et al., 1997; Hasler et al., 1998), which upon reconstitution with lipids crystallized into highly ordered square arrays (Hasler et al., 1998). Projection maps initially at 0.9 nm (Hasler et al., 1998) and then at 0.6 nm (Fotiadis et al., 2000) revealed several protein densities, which could be attributed to transmembrane helices, and displayed a configuration around a central pore similar to that observed for AQP1 (Fig. 6.5b). The agreement between structural and functional data makes a strong case that MIP indeed plays a crucial role in facilitating the flow of water that follows the circulating ion flux in the lens.

Finally, and somewhat ironically, an interesting aspect of the two-dimensional crystals of MIP is that most turned out to consist of two membrane sheets that faced each other with their extracellular sides. They were held together by tightly fitting tongue-and-groove interactions between the MIP molecules in the apposing membranes (Fotiadis et al., 2000). This feature of the *in vitro* reconstituted membranes is consistent with all earlier reports, which identified MIP as a junction-forming protein in the lens. Thus it is likely that MIP has dual functions: first as a water channel, second as a cell-to-cell adhesion protein. Consistent with this, MIP is present in junctional as well as nonjunctional portions of the fiber cell membranes (Fitzgerald et al., 1983; Paul and Goodenough, 1983). It appears possible that MIP directly contributes to the unusually tight packing of the fiber cells that is so important for lens transparency. Such a key role is supported by the observation that mutations of MIP cause cataracts in mice (Shiels and Bassnett, 1996).

6.7. Specialized Transporters Serve Nutrient Uptake

The circulating current creates a net flux of solute that generates an extracellular fluid flow that in turn convects nutrients toward the deeper-lying fiber cells, while the intracellular flow removes waste products from the intracellular compartment. For this model to be correct, the fiber cells need a mechanism for importing the nutrients that are being convected to them. Glucose is the principal fuel that the lens uses to support growth and homeostasis. In the lens, the epithelium and differentiating fiber cells are capable of oxidative phosphorylation, but the mature fiber cells, having lost their mitochondria, must rely solely on glycolysis for energy production. The lens is bathed by the aqueous humor, which contains glucose levels that mirror those in the plasma. Hence, lens cells near the periphery have access to an abundant supply of glucose; however, the supply of glucose to the deeper-lying fiber cells is likely to be limited by the tortuous nature of the extracellular space.

Transport studies to identify the site of glucose uptake in the lens are not entirely consistent with each other. One report identified the epithelium as the predominant site of glucose transport (Goodenough et al., 1980). Another study demonstrated that, in addition to the epithelium, fiber cells also have the ability to transport glucose. This was concluded from the observation that the capacity of glucose transport was comparable at both the anterior and posterior lens surfaces (Kern and Ho, 1973). Furthermore, approximately half of the total lens glucose transport remained following removal of the capsule and the adhering epithelial cell layer (Giles and Harris, 1959). In yet another study, glucose transporters were found predominantly in the lens nucleus and at lesser levels in the cortex (Lucas and Zigler, 1987, 1988; Kaulen et al., 1991). However, this localization was based on the binding of cytochalasin B to the transporters, and concerns remain about the specificity of the reagent. Also, the result is inconsistent with the observation that enzyme activities involved in the breakdown of glucose are greatest in the cortex and decrease toward the center of the lens (Zhang and Augusteyn, 1995). It appears that both the epithelial and cortical fiber cells are capable of glucose uptake.

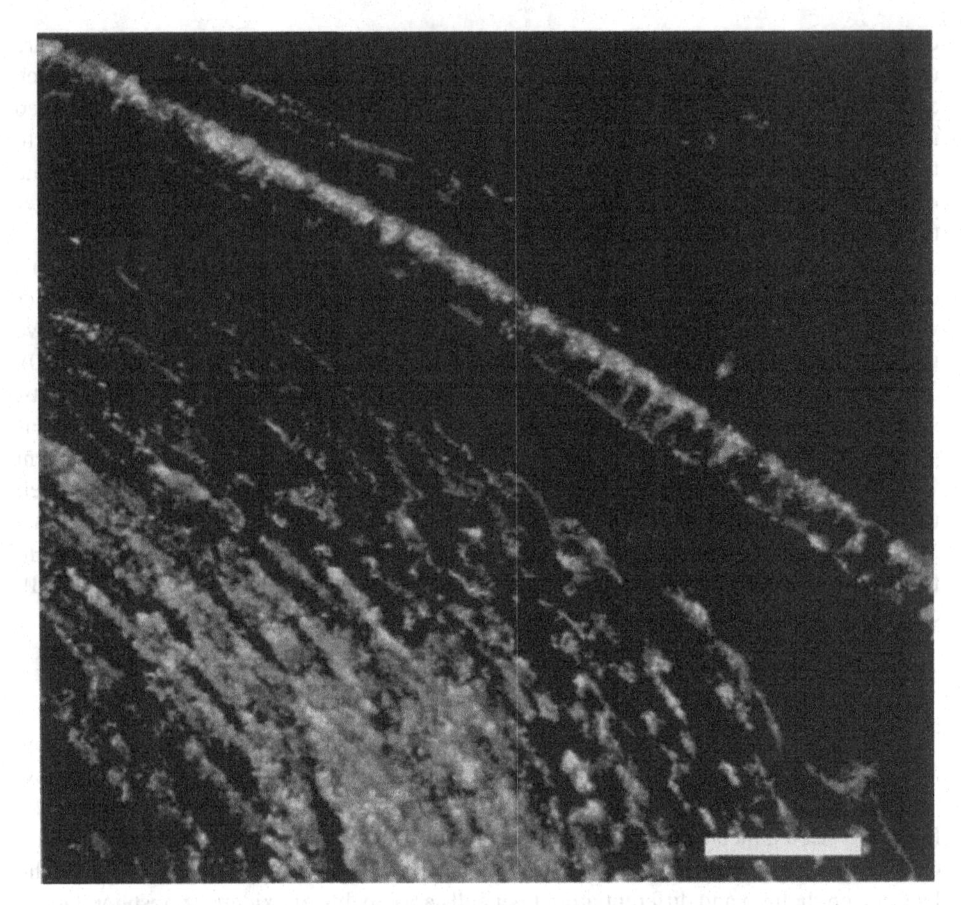

Figure 6.6. (See color plate X.) Differential expression of glucose transporters in the lens. An axial section of a rat lens showing the expression of GLUT1 (red) on the basolateral aspect of the epithelium and GLUT3 (green) in the differentiating fiber cells. Scale bar: 25 μm.

In other tissues, glucose uptake from the extracellular fluid is mediated by members of the glucose-facilitated transporter (GLUT) family (Gould and Holman, 1993). The identification and localization of glucose transporter isoforms in the lens has been controversial. For example, one study localized the glucose transporter GLUT1 in cortical fiber cells (Mantych et al., 1993), but another study reported a negative result for the same isoform (Kumagai et al., 1994). GLUT2, GLUT3, and GLUT4 were also investigated but could not be detected. A more recent study reported an absence of GLUT2 and GLUT4 in the rat lens but found strong evidence for the expression of GLUT1 and GLUT3 (Merriman-Smith et al., 1999). The latter two isoforms were reliably identified and localized at both transcript and protein levels. GLUT1 is predominantly expressed in the rat lens epithelium and differentiating cells at the equator, whereas GLUT3 expression is strongest in cortical fiber cells (Fig. 6.6). This differential expression is consistent with the different glucose environments these cells are exposed to. GLUT1 has a K_m appropriate for the abundant glucose supplies in the aqueous humor. GLUT3 has a lower K_m than GLUT1 and works most effectively in situations where glucose supplies are more restrictive, as is likely to be the case for deeper-lying fiber cells.

In addition to the transporters for glucose, which are the most extensively studied, the presence in the lens of transporters for two other substrates has been reported. The epithelium further expresses a sodium-dependent reduced glutathione transporter (Kannan et al., 1996). This transporter has the ability to take up glutathione into lens epithelial cells to a concentration sixfold that in the aqueous humor (Mackic et al., 1996). Thus it is conceivable that circulating glutathione in the aqueous is a major source of lens epithelial glutathione under physiological conditions. How this glutathione is conveyed to the fiber cells is presently unknown. The lens also contains a transporter for *myo*-inositol, for which further details are given below.

In conclusion, lens epithelial cell membranes constitute without doubt a major site of transport activity. This is consistent with the epithelium playing an important role in the maintenance of lens homeostasis. However, the strong presence of glucose transporters in cortical fiber cell membranes indicates that fiber cells also make an active contribution. This is entirely consistent with the lens circulation model, which postulates that the circulating ion and water fluxes carry nutrients deeper into the lens through extracellular pathways. It also lends strong support to the emerging view that fiber cells are not passive but make a greater contribution to overall lens function than has previously been acknowledged.

6.8. Changes in Membrane Channel- or Transporter-Activity May Result in Cataract

Assuming the lens internal circulation is integral to normal function, it stands to reason that any dysfunction in the components involved in this system is likely to have an impact on the crystalline tissue architecture. Because many membrane pumps and channels have distinct spatial distributions, one would expect that the dysfunction of any of these membrane proteins would produce a distinctive and localized damage phenotype. Localized tissue disruptions are indeed a common feature of many cataract types, and hence it is of interest to study the effect that the disruption of individual transporters or channels has on lens tissue integrity. This can be achieved experimentally in two ways: First, the function of a membrane transport or channel protein can be disrupted in normal lenses using specific inhibitors. Second, the normal expression pattern of a transporter or channel can be altered in transgenic animals. Insights into membrane protein function gained via both approaches are outlined below.

6.8.1. Inhibition of Cl^- Channels and Transporters

The application of various Cl^- channel inhibitors to organ-cultured normal rat lenses causes extensive localized tissue damage (Fig. 6.7). In an equatorial zone about 100 μm from the capsule, extracellular space is grossly dilated, resulting in massive tissue breakdown and liquefaction, whereas closer to the surface, the extracellular space remains tight but the fiber cells are swollen (Young et al., 2000). This damage phenotype is similar to the localized tissue disruption evident in the diabetic rat lens (Bond et al., 1996).

To understand how this localized damage is generated, it is first necessary to consider the role of Cl^- fluxes in the internal circulation model. Measurements of radial differences in membrane potential and the ionic concentration of Cl^- in the whole lens have been used to calculate the electrochemical gradient for Cl^- ion movement as a function of radial distance (Mathias and Rae, 1985). Such an analysis predicts that deeper in the lens Cl^- ions move from the extracellular space into fiber cells but nearer to the lens surface they move from the

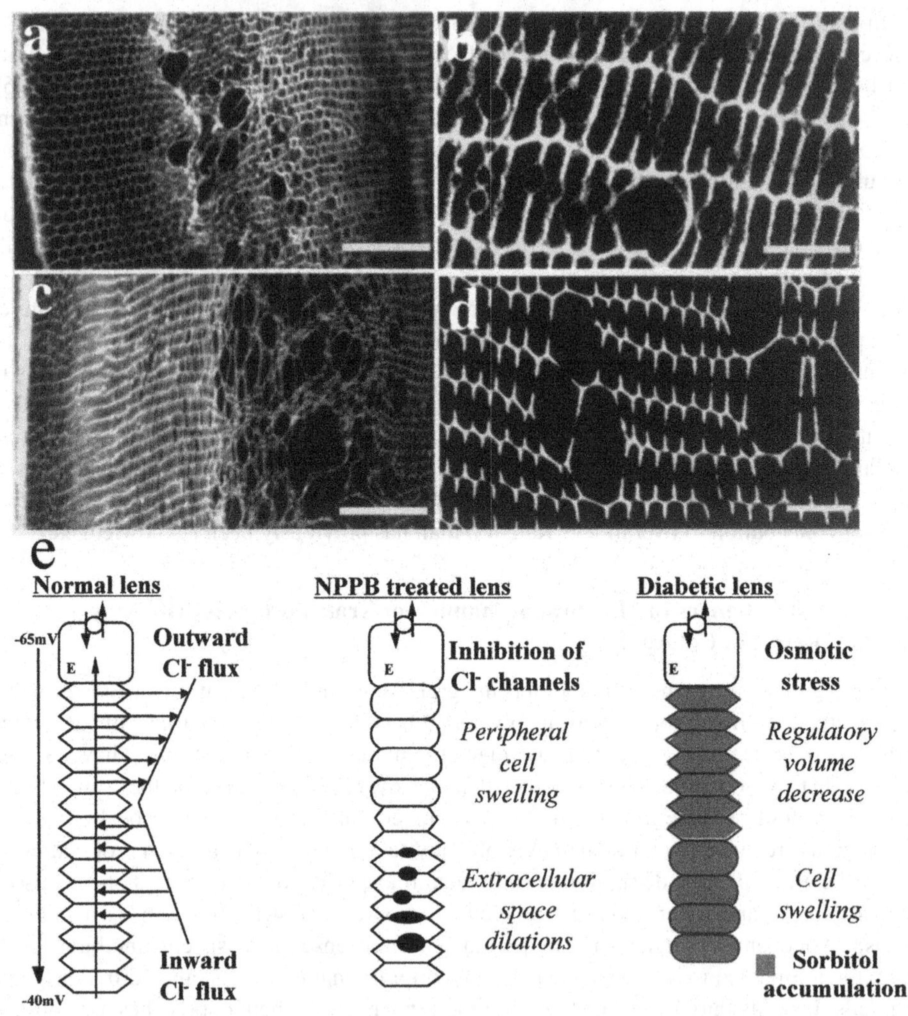

Figure 6.7. The involvement of Cl^- channels in the maintenance of steady-state lens volume. (a) An equatorial section from a rat lens incubated for 18 hours in the presence of the Cl^- channel blocker NPPB (100 μM) showing peripheral fiber cell swelling and an inner zone of extensive fiber cell damage. (b) By incubating lenses for only 6 hours, it can be determined that the dilation of the extracellular spaces is the precursor of the more extensive cellular breakdown in the inner zone. (c) Equatorial section through a diabetic rat lens showing disruptions that result in complete tissue breakdown at a later stage. (d) In the diabetic lens, these disruptions are caused by the uncontrolled swelling of fiber cells in the inner zone. (e) A unifying hypothesis. In the normal lens, radial differences in the membrane potential (−65 to −45 mV) of fiber cells relative to the extracellular space mean that the direction of the Cl^- flux (J_{Cl}) changes from an influx into deeper fiber cells to an efflux from the more peripheral fiber cells. Blocking Cl^- channels with NPPB is therefore expected to cause cell swelling in the lens periphery but extracellular space dilations in the deeper fiber cells. In the diabetic lens, accumulation of sorbitol in the fiber cells imposes osmotic stress, which in the peripheral fiber cells is compensated for by a regulatory volume decrease; in the inner zone, however, it is further aggravated by an influx of Cl^- ions.

cytoplasm of fiber cells to the extracellular space (Fig. 6.7e). Hence, one would predict that the inhibition of Cl^- channels in the inner cortex would block the uptake of Cl^- ions from the extracellular space by fiber cells. This would cause an accumulation of Cl^- ions and water in the tortuous extracellular space and lead to its dilation. Near the lens surface, the efflux of Cl^- ions from fiber cells would be blocked, causing intracellular accumulation of osmolytes and subsequent fiber cell swelling. The cellular changes observed in normal rat lenses treated with Cl^- channel inhibitors were precisely as predicted (Young et al., 2000). This agreement between predicted and observed histological changes lends strong support to the theory that active circulation occurs in the lens.

The above observations can now be used to explain the changes in tissue architecture observed in the diabetic rat lens. In this case, tissue in a similar zone about 100 μm from the equator is liquefied, but fiber cell swelling rather than extracellular space dilation is the initial cause (Bond et al., 1996). It has been shown that elevation of glucose leads to an accumulation in the fiber cells of the intracellular osmolyte sorbitol (Lee et al., 1995). This causes an osmotic insult, which in many cell types is compensated for by the opening of Cl^- and cation channels to release osmolytes and water (Lewis and Donaldson, 1990). The analysis of Cl^- fluxes in the lens shows that the peripheral fiber cells have a chloride equilibrium potential (E_{Cl}), which favors Cl^- efflux. Thus, these cells are able to undergo a regulatory volume decrease and appear normal. In deeper regions of the lens, E_{Cl} favors an influx rather than an efflux of Cl^- ions. In these cells, an opening of volume-regulated Cl^- channels will therefore cause an influx of Cl^- ions and further increase the rate of cell swelling. Thus, the knowledge now accumulated on the lens circulation system for the first time provides a satisfactory explanation for the localized cortical tissue damage observed in the early stages of cataractogenesis in the diabetic rat.

6.8.2. Genetic Ablation or Inactivation of Connexins and MIP

Connexin 43 knockout mice (Gao & Spray, 1998), connexin 46 knockout mice (Gong et al., 1997), connexin 50 knockout mice (Gong et al., 1998; White et al., 1998), and mice with inherited mutations of MIP (Shiels and Bassnett, 1996) all develop cataractous tissue changes with different degrees of severity. Such effects are expected given the key roles of these membrane components in the lens circulation system. The assessment of the damage phenotypes and physiological measurements of each of these mutants has been described earlier under the headings of each individual protein class.

*6.8.3. Overexpression of the Na^+/*Myo*-Inositol Cotransporter*

This transporter is normally expressed in the epithelium and appears involved in osmoregulation. It was found up-regulated upon exposure of the intact lens or of cultured lens epithelial cells to hypertonic stress (Morimura et al., 1997; Zhou et al., 1994; Cammarata and Chen, 1994). Increased expression of the Na^+-dependent *myo*-inositol transporter in transgenic mice was accompanied by increased *myo*-inositol levels in the lens. The mice developed cataract, the severity of which correlated with the level of transporter expression and accumulated *myo*-inositol (Cammarata et al., 1999; Jiang et al., 2000). Most affected were the differentiating fiber cells in the bow region and the subcapsular fiber cells in the central zone, which appeared swollen, indicating an osmotic basis for this cataract.

6.8.4. Overexpression of p-Glycoprotein

Overexpression of p-glycoprotein in transgenic mice also results in the development of an osmotic cataract. An initial swelling of fiber cells was typically observed, followed by tissue liquefaction in the lens cortex (Dunia et al., 1996). The relevance of this observation is still doubtful, as the expression of p-glycoprotein in the normal lens could not be demonstrated at the protein level. Yet, tamoxifen, a potent and selective inhibitor of p-glycoprotein, blocks Cl^- currents in isolated lens fiber cells, and in whole lenses it prevents volume regulation and causes lens opacification (Zhang et al., 1994). Further work is clearly required to better understand the role of p-glycoprotein, if any, in the normal lens.

6.9. Some Membrane Receptors Have the Potential to Regulate Lens Homeostasis

The realization that the lens possesses an active internal circulation system raises the question whether this circulation can be regulated, and if so, which receptors and signaling pathways are involved. In addition, as the lens has a highly organized cellular structure, one would expect that its development is under strict control from various growth factors, requiring further sets of specialized membrane receptors. Knowledge of the types of membrane receptors expressed in the lens has remained rudimentary for a long time but has recently expanded rapidly. Data are of two kinds: First, receptors have been identified by demonstrating an effect following treatment of cultured cells with pharmacological reagents, hormones, or growth factors (Wang et al., 1999; Richiert and Ireland, 1999; Churchill and Louis, 1997) or by observing developmental changes in transgenic lenses overexpressing selected growth factors (Srinivasan et al., 1998; Robinson et al., 1998). Second, receptors have been identified at the transcript or protein level, which has the advantage that receptor isoforms can be specified. The broad variety of receptor types that have come to light is surprising. It indicates either that the composition of the aqueous humor with regard to its content of signaling molecules is likely to be considerably more complex than previously assumed or that the lens releases a variety of autocrine signaling molecules (or both).

There is accumulating evidence that G-protein–linked receptors and their signaling pathways are active in cultured epithelial cells as well as in the whole lens. Beta-adrenergic receptors were identified by affinity labeling, and when activated, they increased cAMP in cultured epithelial cells (Ireland and Jacks, 1989; Ireland and Shanbom, 1991). Muscarinic acetylcholine receptors were demonstrated in cultured epithelial cells and play a role in the mobilization of intracellular Ca^{2+} (Williams et al., 1993; Gupta et al., 1994b). In the intact lens, acetylcholine reduces the translens current (Alvarez et al., 1995) and induces membrane potential oscillations throughout the tissue (Thomas et al., 1998). Purinergic receptors were also identified (Fig. 6.8). Transcripts for P2Y1 and P2Y2 were detected in cortical lens fiber cells but not in the epithelium (Merriman-Smith et al., 1998). However, a P2U receptor and α1A receptor were active in cultured lens epithelial cells as these cells mobilized Ca^{2+} from intracellular stores upon treatment with ATP and epinephrine, respectively (Churchill and Louis, 1997; Riach et al., 1995). This functional link of G-protein–coupled receptors with Ca^{2+} signaling is generating considerable interest because of the possible implications for cataractogenesis (Duncan and Wormstone, 1999, Duncan et al., 1994). For example, a prolonged increase of intracellular free Ca^{2+} could activate proteases such as calpain, which could cause irreversible damage to key cytoplasmic and membrane proteins. Alternatively,

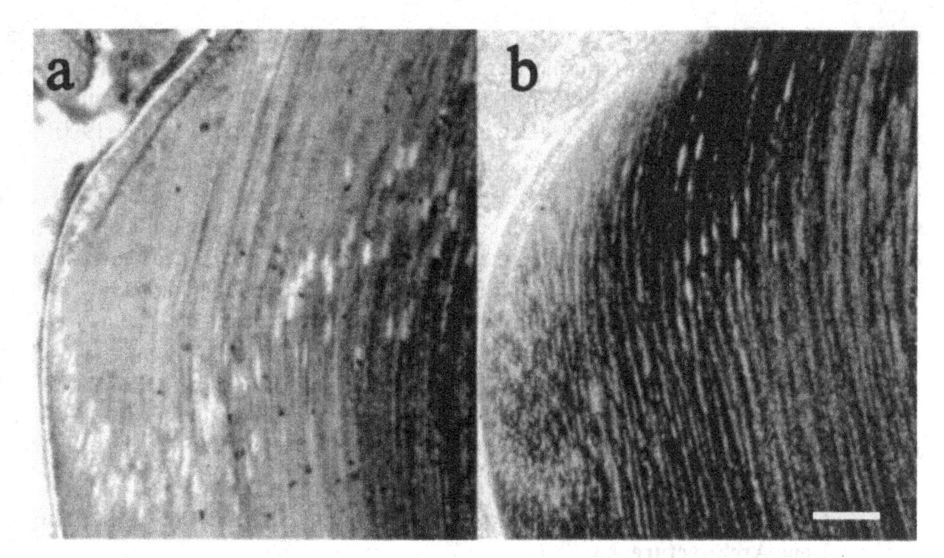

Figure 6.8. Differential expression of purinergic receptors in the rat lens. (a) P2Y1 is expressed in the inner cortex of the lens. (b) P2Y2 is expressed at an earlier stage of fiber cell differentiation. Both isoforms appear absent from the lens epithelium (Merriman-Smith et al., 1998). Scale bar: 100 μm.

Ca^{2+}-mediated signals could alter lens developmental and maintenance processes in a way that results in the formation of opacities.

6.10. Multiple Membrane Receptors May Control the Highly Organized Lens Tissue Architecture

It is obvious that the assembly of a highly ordered tissue such as the lens requires precise control by growth factors, growth factor receptors, and intracellular signaling pathways (see chap. 11). As expected, therefore, lens cells, predominantly the epithelial cells, express numerous types of growth factor receptors. Some of these have been demonstrated to play a role in specific developmental processes, others have merely been identified and localized, their functions so far unknown. The following is a brief summary of relevant growth factors, with an emphasis on inventory rather than the downstream effects these receptors have on activation.

The lens expresses multiple isoforms of the FGF (fibroblast growth factor) receptor family (Wang et al., 1999; Weng et al., 1997; de Iongh et al., 1997; McDevitt et al., 1997). Appropriate growth factors were found synthesized in the lens (Lovicu et al., 1997) as well as in ocular tissues surrounding the lens, consistent with signaling molecules from intralenticular and more distant sources regulating lens development. Lens epithelial cells were also shown to express types I and II TGF-beta (transforming growth factor) receptors (Obata et al., 1996; Richiert and Ireland, 1999). Activation of these receptors stimulate cell growth and results in increased secretion of fibronectin, indicating a likely role for TGF-beta in regulating the production of the lens capsule (Richiert and Ireland, 1999). The observation that overexpression of TGF-beta1 in transgenic mice results in the formation of anterior subcapsular cataract is also consistent with the presence of this receptor in the lens epithelium (Srinivasan et al., 1998). Epithelial cell proliferation is also stimulated by PDGF (platelet-derived growth factor), a result consistent with *in situ* hybridization experiments that identified PDGF-α receptors in lens epithelium (Potts et al., 1994; Potts et al., 1998; Reneker and Overbeek, 1996). Further, a receptor for IGF-1 (insulin-like growth factor-1) was detected in lens epithelial cells using affinity labeling or reverse transcription PCR (Palmade et al., 1994; Jacobs et al., 1992; Caldes et al., 1991; Bassnett and Beebe, 1990). The addition of IGF-1 to cultured epithelial cells increases the expression of integrins. Using a similar affinity-labeling procedure, an EGF (epidermal growth factor) receptor was detected that, when activated, enhances epithelial cell growth in a dose-dependent manner (Hongo et al., 1993). Finally, also in cultured lens epithelial cells, the pharmacological and molecular identification of prostaglandin receptors EP4 and FP was reported (Mukhopadhyay et al., 1999). In contrast to this emphasis on epithelial cells, several receptor isoforms for TNF (tumor necrosis factor) were detected in lens epithelial and fiber cells and are believed to play a role in the degeneration of the cell nuclei in differentiating fiber cells (Wride and Sanders, 1998). Obviously, the future challenge will be the integration of these observations into a unified model of lens development.

6.11. Proteins with Adhesive Properties Further Support the Crystalline Lens Architecture

In addition to the growth factor receptors, which are responsible for passing developmental signals across cell membranes, the lens expresses numerous intrinsic or peripherally associated membrane proteins with adhesive properties. In some cases, their inactivation or genetic ablation result in cataract, highlighting the importance of these proteins in aiding the proper development or maintenance of the crystalline lens architecture.

6.11.1. NCAM and Cadherins

The neural cell adhesion molecule NCAM was detected in both the lens epithelium and the differentiating fiber cells, consistent with the abundant presence of adherent junctions (Lo, 1988; Maisel and Atreya, 1990; Watanabe et al., 1992). NCAM expression decreases rapidly in cortical fiber cells and is further reduced in the lens nucleus. Two isoforms are differentially expressed: NCAM 140 and NCAM 120 are the predominant isoforms in epithelial and fiber cells, respectively (Katar et al., 1993). The presence of NCAM was found to be essential for lens differentiation (Watanabe et al., 1989).

The lens also differentially expresses N- and B-cadherin (Leong et al., 2000). B-cadherin was prefentially observed in fiber cells, while N-cadherin was found at similar levels in both epithelial and fiber cells. A series of elegant experiments using a chick embryo lens culture system showed that the addition of the N-cadherin function–blocking antibody, NCD-2, inhibited lens morphogenesis and differentiation (Ferreira-Cormwell et al., 2000). Specifically, changes were observed in the expression pattern of the molecular components of the cadherin-catenin complex and in their linkage with the actin cytoskeleton. Interestingly, the cells tried to compensate for the inactivation of N-cadherin by up-regulating another cadherin (or other cadherins) and plakoglobin, but this response was inadequate to restore tissue differentiation in this system, a strong indication that N-cadherin and its associated cytoskeleton play an important role in lens development.

6.11.2. Integrins

Information on the presence of heterodimeric integrins in the lens is still limited (see chap. 10). So far, $\beta 1$ and three α isoforms, $\alpha 3$, $\alpha 5$, and $\alpha 6$, have been detected in the lens

(Menko and Philip, 1995; Menko et al., 1998). $\beta1$ was found in both epithelial and fiber cell membranes. In contrast, the α isoforms are expressed differentially: $\alpha3$ and $\alpha5$ are predominantly located in the epithelium, and $\alpha6$ is primarily associated with the fiber cells. The combinations $\alpha3/\beta1$ and $\alpha5/\beta1$ have binding specificity for fibronectin, $\alpha6/\beta1$ for laminin. Consistent with these specificities, both fibronectin and laminin were detected in the lens (Kohno et al., 1987). Laminin was found exclusively in the lens capsule. This would indicate that $\alpha6/\beta1$ integrin might be responsible for the interaction between fiber cells and the posterior capsule. Fibronectin is more widely distributed in the lens, including the capsule, epithelium, and fiber cells. Interactions with fibronectin between the capsule and epithelial cells could thus be mediated primarily via the $\alpha3/\beta1$ and $\alpha5/\beta1$ heterodimers. Further consistent with the presence of integrins is the observation that cultured lens epithelial cells exhibit larger attachment "footprints" and longer survival when grown in the presence of extracellular matrix proteins (Oharazawa et al., 1999). More detailed immunolocalization using a wider collection of isoform-specific antibodies is now essential to complete the inventory of lens integrins and establish how they could be involved in the adhesion processes underpinning the development and maintenance of the lens.

6.11.3. MP20

MP20 is the next most abundant intrinsic protein of the fiber cell membranes after MIP. Initially designated MP17, it was shown to be lens specific and widely distributed in the fiber cell plasma membrane (Mulders et al., 1988; Voorter et al., 1989). The same protein, then named MP18, was demonstrated to be a phosphoprotein that binds calmodulin and has the ability to form membrane junctions (Louis et al., 1989; Galvan et al., 1989). Also of the identical protein, the genes coding for the human isoform MP19 (Church and Wang, 1993) and the rat isoform MP20 (Kumar et al., 1993) were isolated. Mutations to the MP20 gene have been shown to produce cataract in mice (Steele et al., 1997). Mice heterozygous or homozygous for the *To3* mutation, which substitutes a valine for glycine at position 15 in MP20, are born with total opacity of their lenses. The lenses are vacuolated and the fiber cells are completely disorganized. Cataract in the homozygotes is also accompanied by microphthalmia.

Despite its abundance and obvious importance to the maintenance of lens transparency, little is known about the function of MP20 in the lens. Sequence analysis suggested that the polypeptide chain traverses the membrane four times (4-TMS), which was later confirmed biochemically (Arneson and Louis, 1998). Furthermore, MP20 was identified as a distant relative of a wider gene family, including the peripheral myelin protein 22 (PMP22) and the epithelial membrane proteins 1, 2, and 3 (EMP-1, -2, and -3), which are all 4-TMS membrane proteins (Taylor et al., 1995; Lobsiger et al., 1996). A potential role for MP20 in mediating cell adhesion has emerged from a study (Gonen et al., 2000) that identified galectin-3 as a component of fiber cell membranes.

6.11.4. Galectin-3

Galectin-3 was identified as a 31-kD peripheral lens membrane protein on the basis that it could be removed from isolated crude fiber cell membranes by alkaline stripping (Gonen et al., 2000). Galectin-3 is a member of the galectin family and is known to function as a modulator of cell adhesion in a variety of tissues (Barondes et al., 1994; Kaltner and Stiersdorfer, 1998; Perillo et al., 1998). Galectin-3 can dimerize to become a divalent complex with two carbohydrate-binding sites, and by binding to membrane proteins, which contain β-galactoside, this complex has the ability to facilitate cell-cell contacts. In a search for binding partners in the lens, the junction forming MP20 was identified as a candidate ligand of galectin-3 (Gonen et al., 2000). While it has not yet been confirmed that MP20 is glycosylated, it appears possible that MP20 does not form membrane junctions by itself but rather forms them indirectly using galectin-3 as a linker.

6.11.5. GRIFIN

Galectin-related interfiber protein (GRIFIN) is a novel protein that was recently shown to be exclusively expressed in the lens (Ogden et al., 1998). Although closely related to the galectins, GRIFIN is significantly different in that it has retained the ability to form dimers but does not bind to β-galactoside sugars. GRIFIN is localized at the membrane interface between fiber cells. It is currently unclear whether GRIFIN is located inside or outside the membrane, and its function still remains an enigma. It would be interesting to know the relationship between galectin-3, GRIFIN, and MP20 and whether the distribution of GRIFIN changes during lens differentiation.

6.11.6. SPARC

Secreted protein acidic and rich in cystein (SPARC) is another matricellular protein that regulates cell adhesion and proliferation in a variety of tissues. SPARC is localized in the lens epithelium, and levels decrease abruptly in the equatorial bow region. It is absent in the lens capsule and differentiated fiber cells (Bassuk et al., 1999). It was discovered in the lens because disruption of the SPARC locus in mice resulted in cataractogenesis (Gilmour et al., 1998; Bassuk et al., 1999). Ligands of SPARC in the lens have not yet been identified, and therefore it is presently unclear how this polypeptide exerts its function. In the knockout mice lens, abnormalities become evident only four to six months after birth and are characterized predominantly by a severe disruption of equatorial epithelial and fiber cells, which appear swollen and vacuolated (Gilmour et al., 1998; Bassuk et al., 1999). Other changes include the incomplete elongation of fiber cells posteriorily and a consequent posterior displacement of the lens nucleus. Hence, SPARC is required for the normal differentiation of epithelial cells and the maturation of fiber cells.

6.12. Conclusion

There is a large and ever growing catalog of lens membrane proteins. Many of these proteins exhibit distribution patterns that are related to development. Some are likely to be intimately involved in controlling lens fiber cell differentiation, while others are involved in maintaining lens homeostasis and transparency. There is accumulating evidence that the spatial distribution of membrane transporters and channel proteins is critical to lens function. Thus, a knowledge of how the distribution of these proteins changes during lens development will be instrumental in deepening our understanding of lens transparency and the processes underlying cataractogenesis.

7

Lens Cell Cytoskeleton

Roy Quinlan and Alan Prescott

7.1. Introduction

The lens, like any other tissue, is dependent upon the cytoskeleton for its function. There are now a number of examples of mutated or overexpressed cytoskeletal proteins that have been shown to be the genetic basis of cataract, underlining the importance of the cytoskeleton to cell shape, intracellular organisation, and compartmentalisation in the lens. In other cell systems, the cytoskeleton enables the cell to maintain and diversify the internal complexities required for specialised cellular functions. In the lens, this requires specialised cell-cell interactions that allow adjacent plasma membranes to be closely apposed or interdigitated over most of their surface. Differentiation requires the programmed elongation of lens fiber cells and, by inference, the directed traffic of organelles, vesicles, and other cargoes in the highly elongated fiber cell. Finally, the lens is highly specialised in the maintenance and use of stable proteins, of which the cytoskeleton is but one example. Evolution has dictated the choice of the proteins and their structures to ensure that the lens efficiently refracts light onto the retina, and our task as cell biologists is to unravel this rich tapestry and to discover the contribution of the cytoskeleton to the function of this highly specialised tissue.

7.2. Major Components of the Lenticular Cytoskeleton

The lens, like all other tissues, possesses microtubules, microfilaments, and intermediate filaments, the three main cytoskeletal elements of most eukaryotic cells. These structural filaments in isolation are ineffective, as it is the linking, the attachment, and the transiently associating proteins that give the cytoskeleton functionality. The different elements of the cytoskeleton function as an integrated unit because of the numerous associated proteins that interlink, control, and regulate the different filament networks. Therefore, any complete definition of the cytoskeleton has to include not only the major structural proteins but all these other associated proteins. As will become apparent, the lens has adapted some of the typical cytoskeletal structures into specialised filaments that appear unique to the lens, such as the beaded filament. Therefore, the identification of cytoskeleton-associated proteins in the lens may also require some unique strategies. The lens has important and fascinating lessons for the cell biologist interested in the diversity of cytoskeletal function.

7.2.1. Intermediate Filament Proteins in the Lens

7.2.1.1. Intermediate Filaments Are Essential for Lens Transparency

Intermediate filament proteins are developmentally and differentially regulated in their expression, with each type of intermediate filament exhibiting a tight tissue-specific expression profile (Coulombe et al., 2000; Fuchs and Cleveland, 1998; Herrmann and Aebi, 2000; Quinlan et al., 1995; Quinlan et al., 1985). There are now over 60 different intermediate filament proteins that have been organised into six distinct classes based on gene structure and sequence similarities (Herrmann and Aebi, 2000; Quinlan et al., 1995). Their organisation can be correlated to particular physical characteristics of the constituent proteins, as the more robust intermediate filament proteins are found in those cells that are exposed to more demanding environmental stresses. The importance of their contribution to the physical integrity of the cell and the intracellular organisation is graphically illustrated by the initial discovery of the role of keratin filaments in the skin condition epidermolysis bullosa simplex (EBS). Here, a single mutation in the keratin 14 protein can cause traumatic skin blistering due to a breakdown in the epidermal barrier (Bonifas et al., 1991; Coulombe et al., 1991; Ishida-Yamamoto et al., 1991; for a recent review, see Irvine and McLean, 1999). The epidermis is a terminally differentiating system, and concomitant with its differentiation several specific keratins are expressed in the maturing layers of the epidermis, which explains the range of human diseases resulting from keratin mutations (Irvine and McLean, 1999). Like the suprabasal cells of the epidermis, the lens fibre cells also express their own specific intermediate filament proteins (Quinlan et al., 1996), and mutations in one of them, CP49, causes inherited cataract in humans (Conley et al., 2000; Jakobs et al., 2000). Mutations in the intermediate filament–associated proteins αA-crystallin (Litt et al., 1998; Pras et al., 2000) and αB-crystallin (Vicart et al., 1998) can also cause cataract. Thus, intermediate filaments appear to have an important role in lens transparency.

7.2.1.2. Lens-Specific Intermediate Filament Proteins

There are two lens-specific intermediate filament proteins: filensin (Merdes et al., 1991), formerly CP115 (FitzGerald, 1988), and CP49 (Hess et al., 1993), sometimes called phakinin (Merdes et al., 1993) or beaded filament structural protein 2 (Jakobs et al., 2000). The expression of each of these two proteins appears restricted to the lens fibres (Gounari et al., 1993; Hess et al., 1993; Merdes et al., 1993) and serves as an excellent marker for lens cell differentiation (Ireland et al., 2000). Although keratins (Kasper and Viebahn, 1992) and vimentin (Ellis et al., 1984; Kasper et al., 1988; Kasper and Viebahn, 1992; Ramaekers et al., 1982; Ramaekers et al., 1980) are also expressed in lens cells during lens development and differentiation, it is only filensin and CP49 that are found at all stages of lens fibre cell differentiation (FitzGerald, 1988; Ireland et al., 2000; Quinlan, 1991). Vimentin is lost from the later stages of lens fibre cell differentiation (Sandilands et al., 1995a), whilst keratins are only expressed very early in lens development (Kasper and Viebahn, 1992). The fact that immunoreactive fragments of both filensin and CP49 have been found in the lens nucleus (FitzGerald, 1988; Quinlan, 1991) also suggests that these proteins are expressed early in lens development, as this is the location of the embryonic primary fibre cells in the adult lens. This is indeed the case for the human lens, as shown in Figure 7.1.

In the chicken lens, IGF and insulin induce lens fibre cell differentiation (Beebe et al., 1987) and also induce CP49 expression (Le and Musil, 2001). Insulin acts through the ERK kinase cascade and can be inhibited by the MEK1/2 inhibitor, UO126. Interestingly, FGF

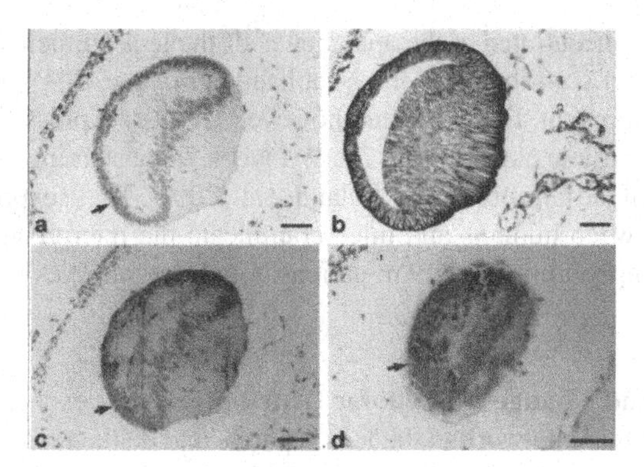

Figure 7.1. (See color plate XI.) Detection of filensin and CP49 within the embryonic lens vesicle. Panels (a–d) show sections of the primary lens vesicle from five-week-old human embryos stained with keratin antibody (a), vimentin antibody 3052 (b), CP49 antibody 2981 (c), and filensin rabbit polyclonal antibody fpa-R (d) using horseradish peroxidase and diaminobenzidine as the colour developer. The arrow in (a) indicates the trace keratin reactivity within the anterior lens cells. Both CP49 reactivity (c) and filensin reactivity (d) are confined to the lens vesicle and show no obvious cross-reactivity with the surrounding ectoderm. Filensin and CP49 reactivity is detected in the lens epithelium at this stage of development. At week 7 of development, there is intense CP49 reactivity, which is particularly strong at the posterior pole of the lens vesicle. There is also significant CP49 staining within the anterior lens epithelial cells. Filensin shows a similar distribution (see Fig. 2 in Ireland et al., 2000). Scale bars: 50 μm. We thank Michael Kasper (University of Dresden) for these images.

was also found to drive CP49 expression in this system, but it was not inhibited by UO126. These data therefore indicate that there are multiple signal transduction pathways that can drive the expression of CP49. Indeed, it is possible to uncouple cell differentiation from the expression of CP49 using cyclic AMP analogues (Ireland et al., 1997).

It is worth mentioning at this point that there are species differences in the expression of intermediate filament proteins in the lens. For instance, in birds and in particular the chicken, the *CP49* gene gives rise to two transcripts that are expressed in the lens, CP49 and CP49ins (Wallace et al., 1998). No equivalent to the CP49ins has yet been identified in mammalian lenses. Also in the chicken, the protein synemin was identified as a lens cytoskeletal component (Granger and Lazarides, 1984) that can coassemble with vimentin (Bilak et al., 1998). Equally, nestin, another intermediate filament protein that can coassemble with vimentin (Steinert et al., 1999), has been identified in the mammalian lens, and its expression pattern has been documented during lens development (Yang et al., 2000). Other intermediate filament proteins, such as desmin, are expressed in rodent lens epithelial cells in response to TGFß during anterior subcapsular formation (McAvoy et al., 1998; Lovicu et al., 2002), whilst in some strains of mice, GFAP is constitutively expressed in lens epithelial cells (Boyer et al., 1990; Hatfield et al., 1985), but these unexpected expression patterns have to be confirmed for other species. In conclusion, the expression pattern for intermediate filament proteins changes with respect to stage of development and differentiation and according to the species being studied.

The potential importance of such species differences is very relevant to CP49 and filensin, as it has been recently reported in the mouse that filensin and CP49 are expressed first in the more differentiated fibre cells, concentrated at the anterior ends of the fibres (Blankenship et al., 2001). At E14.5, nestin is concentrated at the apical ends of the lens epithelial cells, making it unlikely that this protein contributes to the establishment of the CP49-filensin network during mouse lens development. In the human, however, CP49 is expressed in all lens fibre cells at the time of lens vesicle closure (Fig. 7.1). At week 7, the staining appears concentrated at the posterior end of the lens fibre cells (Ireland et al., 2000). These expression pattern differences for CP49 between humans and mice complicate the interpretation of mouse models of cataract based upon human CP49 mutations (Conley et al., 2000; Jakobs et al., 2000).

Assembly Properties and Changes in Cellular Distribution. It is expected that CP49 and filensin will form a filament network in the lens separate and distinct from those of vimentin and keratin, despite common principles of assembly in each case (Coulombe et al., 2000; Fuchs and Cleveland, 1998; Herrmann and Aebi, 2000; Quinlan et al., 1995). *In vitro* assembly studies have shown that filensin and CP49 do not coassemble with keratins (Merdes et al., 1993) or with vimentin (Carter et al., 1995; Merdes et al., 1993). Indeed, filensin and CP49 need each other for *in vitro* assembly (Carter et al., 1995; Merdes et al., 1993), and in the lens immunofluorescence miscroscopy confirms that both proteins are always co-localised (Sandilands et al., 1995a). In the juvenile bovine lens, vimentin is lost during lens fibre cell differentiation (Ellis et al., 1984; Ramaekers et al., 1982; Ramaekers et al., 1980), whilst the CP49-filensin network is maintained (Sandilands et al., 1995a). A similar change is seen for the human lens with the apparent loss of vimentin at later stages of lens fibre differentiation (Fig. 7.2). This is not to say that the CP49-filensin network does

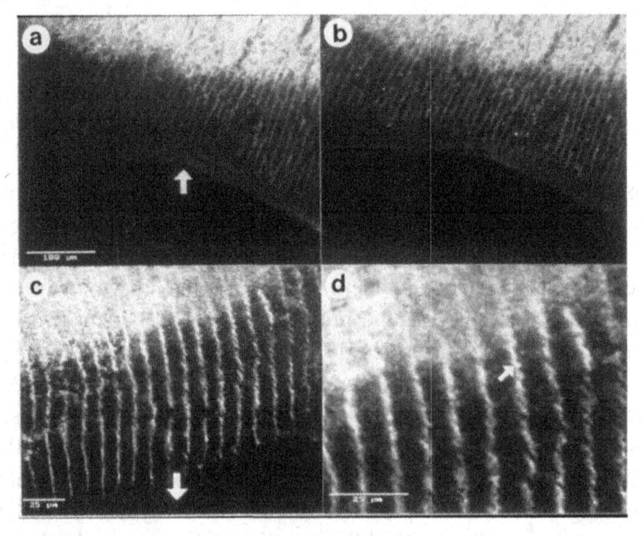

Figure 7.2. Filensin and CP49 undergo changes in their distribution within the lens during fibre cell ageing. Panels a and b show a confocal microscope image from a 12-year-old human lens double-labelled with filensin antibody 7B10 (a) and CP49 antibody 2981 (b). Both antibodies fail to react with the lens epithelium (arrow). Panels c and d show a confocal microscope image from a 2-year-old human lens single labelled with filensin antibody fpa-R. In c, the direction in which the edge of the section lies is indicated by an arrow. Panel d shows a higher magnification image from a region in c, demonstrating the localisation of the filensin antibody to the short sides of the lens fiber cells (arrow). We thank Aileen Sandilands (University of Dundee) for these images.

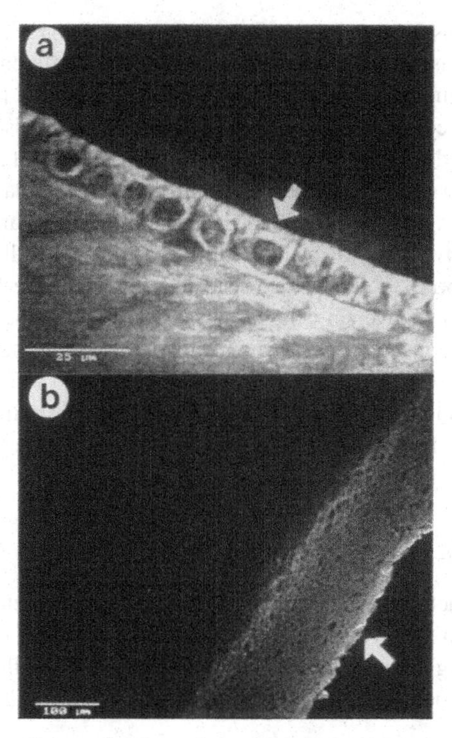

Figure 7.3. Vimentin reactivity is rapidly reduced during lens fiber cell maturation in the developed human lens. Panels a and b are single-labelled confocal microsope images of a section from a two-year-old human lens stained with vimentin monoclonal antibody V9. Panel a shows an area of the anterior lens epithelium (arrow), which is vimentin positive. Panel b shows a region of the posterior lens, demonstrating the abrupt reduction in vimentin reactivity in the fiber cells. The arrow indicates the outer edge of the section. Lens fiber cell age increases from right to left.

not change its distribution during lens fibre differentiation. Some 200–300 μm from the lens capsule in the juvenile bovine lens, the CP49-filensin network undergoes a dramatic redistribution as part of the differentiation programme (Sandilands et al., 1995a), and such a redistribution has been confirmed in the human lens (Fig. 7.3). The mechanism for this dramatic change in the subcellular distribution of CP49-filensin could be linked to the complex proteolytic processing that filensin undergoes during lens fibre cell differentiation (Sandilands et al., 1995b). Alternatively, phosphorylation could be involved, as seen with other intermediate filament proteins (Inagaki et al., 1996), for phosphorylation is at least partly responsible for the membrane association and insolubilisation of CP49 (Ireland et al., 1993). The redistribution of CP49-filensin coincides with the loss of nuclei and mitochondria in the differentiating lens fibre cells (Bassnett, 1997; Bassnett and Beebe, 1992; Kuwabara and Imaizumi, 1974; Sandilands et al., 1995a). There has not been a detailed enough characterisation of the beaded filament protein phosphorylation to link specific phosphorylation events with changes in filament distribution. Nevertheless, the discussion above details ample reasons to suggest that the filensin-CP49 network contributes fundamentally to lens fibre cell differentiation.

In the youngest fibre cells of adult or juvenile human lenses, filensin and CP49 appear in a predominantly membrane-associated distribution. It has been proposed that filensin and

CP49 compose a specialised fibre cell plasma membrane-associated cytoskeleton within the lens (Georgatos et al., 1994). Whilst this is probably true of the most recently differentiating fibre cells, the increased cytoplasmic staining observed in older fibres suggests that filensin and CP49 also compose a substantial cytoplasmic filament network (Sandilands et al., 1995b). Transgenic studies have revealed that the overexpression (Capetanaki et al., 1989) or the inappropriate expression of intermediate filaments within the lens (Dunia et al., 1990) leads to fibre cell plasma membrane disruption and the generation of cataract. In addition, such lenses show the ineffective removal of fibre cell organelles as well as the nucleus. The expression of mutated forms of CP49 also lead to cataract in the human (Conley et al., 2000; Jakobs et al., 2000). Conversely, the loss of vimentin does not induce cataracts in knockout mice (Colucci-Guyon et al., 1994), suggesting that an increased level or inappropriate expression of intermediate filament protein or the expression of mutated intermediate filament proteins is more damaging to the lens. Based on this, it is difficult to single out one specific function for intermediate filaments in the lens other than the generic structural role proposed for other tissues (Lazarides, 1980).

7.2.1.3. Intermediate Filament–Associated Proteins (IFAPs)

The points of attachment and the interactions available to the lenticular cytoskeleton will be key to any structural role. It is often the cytoskeletal-associated proteins that provide the physical links, and there is a growing interest in those proteins that link more than one element of the cytoskeleton, as these proteins are thought to be important in coordinating the cytoskeleton as a whole.

One protein that has been shown to associate with intermediate filaments in the lens is plectin (Weitzer and Wiche, 1987). The patterns for intermediate filament proteins and plectin overlap in lens fibre cells (Sandilands et al., 1995a), as does the pattern of a protein called intermediate filament–associated protein 300 (IFAP300; Lieska et al., 1991; Yang et al., 1985), which has recently been identified as plectin (Clubb et al., 2000). Interestingly, plectin introduces cross-links between intermediate filaments and microtubules (Svitkina et al., 1996) and also to actin (Andra et al., 1998). Plectin's role as a linker protein became clear when patients with both muscular dystrophy and Epidermolysis Bullosa Simplex were discovered to have plectin mutations (Smith et al., 1996), demonstrating how important these types of linker protein are to cytoskeletal function (Andra et al., 1997).

Plectin is also important for linking intermediate filament proteins to integrin complexes in hemidesmosomes. There are no hemidesmosomes in the lens, but the integrin complex $\alpha_6\beta_4$, which is found in hemidesmosomes, is expressed in the lens, at least in the chick, and is localised to the plasma membranes of the fibre cells (Walker and Menko, 1999; see also chap. 10). It is important to note that the expression of this $\alpha_6\beta_4$ integrin complex is not restricted to those basal membranes in contact with the posterior membrane (i.e., where there is no laminin), so the integrin complex here is certainly performing a different role in the lens than in other cell types (Walker and Menko, 1999). The location of this integrin complex would still permit signalling as well as anchorage of the cytoplasmic intermediate filaments via plectin. However, no cataract phenotype has so far been reported for patients carrying the mutant plectin or in plectin knockout mice (Andra et al., 1997), as might be expected for such an abundant plasma membrane component that has an important role in linking intermediate filaments, actin, and microtubule cytoskeletons. Possible explanations include the existence of tissue-specific splice variants (Fuchs et al., 1999) or perhaps compensation by other lens-specific proteins. This point has yet to be fully addressed, but plectin and the

$\alpha_6\beta_4$ integrin complex are strong candidates for the role of facilitating intermediate filament attachment to the plasma membrane.

The associated proteins are a major factor in determining how, for instance, intermediate filaments attach to other cytoskeletal elements, the plasma membrane, or other cellular components. Candidates include ankyrin (Georgatos and Marchesi, 1985) and spectrin (Macioce et al., 1999), both of which are present in the lens. Ankyrin and spectrin are better known as key components of the actin-based red blood cell cytoskeletal–plasma membrane complex. The other components of this complex, such as adducin (Kaiser et al., 1989), band 3 (Allen et al., 1987), band 4.1 (Granger and Lazarides, 1985), band 4.9 (Faquin et al., 1988), caldesmon (Bassnett et al., 1999), and tropomyosin and tropomodulin (Woo and Fowler, 1994), are also present in the lens. Indeed, electron microscopy studies show how the actin and intermediate filaments lie in close proximity to the plasma membranes of the lens fibre cells, concentrated along the shorter faces of these hexagonal cells (Lo et al., 1997). Proteins that enable cross-talk between these different cytoskeletal elements promote their integration, which is as functionally important to fibre cells as to other cell types.

The arrangement of intermediate filaments just below the plasma membrane resembles that of a filament mat (Franke et al., 1987). The protein plakoglobin, another IFAP, is heavily enriched in this region of the cell, and in fact the lens was used as a useful starting point for the purification of this protein from a non-keratinising tissue (Franke et al., 1987). Plakoglobin interacts with N-cadherin (A-CAM; Angst et al., 2001), the major cadherin in the lens (Volk and Geiger, 1986a) and a key component of the actin-containing adherens junctions at the lens plasma membrane, which is required for lens cell differentiation (Ferreira-Cornwell et al., 2000). In keratinising epithelia, plakoglobin links intermediate filaments to the membrane desmoglein-desmocollin complex via desmoplakin in the desmosome, but these have not been found so far in the lens (e.g., Koch et al., 1992; Ramaekers et al., 1980). Plakophilin 1 and 2 have been found in cultured lens cells (Heid et al., 1994; Mertens et al., 1999) though in the apparent absence of desmosomes, this suggests a novel type of adherens-type junction in the lens, as has recently been identified in the retina (Paffenholz et al., 1999). The attachment of intermediate filaments to the plasma membrane is therefore still an open question, and, as detailed above, there are many alternatives.

7.3. Microtubule Networks in the Lens

7.3.1. General Properties of Microtubules

Microtubules are essential cytoskeletal elements required for chromosome movement during meiosis and mitosis, for cytokinesis, for intracellular transport, and for the positioning of intracellular organelles. Microtubules are dynamic structures comprising α- and β-tubulin heterodimers that add only to the ends of the microtubules. The rate of this addition is different at the two ends of the microtubules, as these polymers have polarity, unlike intermediate filaments, for instance. Microtubule polarity and subunit dynamics are central to the functional properties of microtubules, and GTP hydrolysis establishes the dynamic properties of this filament system.

Microtubule organisation in cells is dictated by the location of microtubule-organising centres (MTOCs), such as the centrosome, which binds one specific end, termed the (−) end, whilst the distal end is designated the (+) end. Microtubule arrays, therefore, also have polarity, which means that the movement of organelles, vesicles, and other cargoes in cells can be vectorially organised. Specific motor proteins, such as kinesin, move to the (+) end, whilst other motor proteins, such as dynein, move to the (−) end.

Cells generally contain at least two microtubule populations, one a very dynamic population and the other a more stable population (Schulze and Kirschner, 1987). At present, the functional significance of the two populations is unclear, although the biochemical reasons for the increased stability are beginning to emerge (Infante et al., 2000). Nevertheless, the stable population can be distinguished from the dynamic using antibodies against post-translational modifications of tubulin which only occur when tubulin is assembled into microtubules. For instance, both acetylation and detyrosination are modifications that occur on the stable population of microtubules, and these markers can be used to define this subset in cells and tissues (Infante et al., 2000; Schulze and Kirschner, 1987).

In tissues, the organisation of microtubules can be quite different from that described for cells in culture (Mogensen et al., 2000). In particular, the MTOCs in cells from some tissues are peripherally located, in contrast to the perinuclear centrosomes described in most tissue culture cells (Joshi, 1994), and consequently microtubule networks originate in quite distinct non-periniuclear regions of the cell. Cultured cells can be induced to polarise the MTOCs (microtubule organisation reflects this polarity), with the MTOCs localised at the apical plasma membrane (Meads and Schroer, 1995). This arrangement of microtubules in polarised epithelia seems to depend, at least in part, upon the intermediate filaments (Ameen et al., 2001). In elongated cells like neurones, the microtubules are released into the axon (Baas, 1998) and organised to permit axonal function by facilitating vectorial transport within the axon (Miki et al., 2001; Zhao et al., 2001). What then is the situation in lens fibre cells?

7.3.2. Microtubules and Their Role in the Lens

The high energy required for microtubule dynamics has important implications for the lens. The normal process of differentiation removes mitochondria, and the resulting fibre cells then rely upon glycolysis to supply ATP. Perhaps as a consequence of this, microtubules are infrequently observed in denucleated fibre cells and have not been seen in the lens nucleus. Their contrbutions to lens function is therefore most significant in the lens epithelium and early fibre cells.

In the lens, cell differentiation involves a dramatic change in cell shape as well as the repositioning and elimination of a number of cell organelles (see Bassnett, 1992, 1997), and microtubules are likely to be involved in these numerous changes. Microtubules are present in differentiating fibre cells, as documented by electron microscopy studies (Beebe and Cerrelli, 1989; Kuwabara, 1968; Pearce and Zwaan, 1970), and in cultured lens epithelial cells, as demonstrated by indirect immunofluorescence labelling (Lonchampt et al., 1976). The motor proteins, kinesin and dynein, are present in the lens (Lo and Wen, 1999), suggesting that vectorial transport of cargoes occurs within lens fibre cells. Some of these cargoes will be other cytoskeletal proteins. For instance, the soluble precursors of intermediate filaments are transported in a kinesin-dependent manner on microtubules (Prahlad et al., 1998).

Despite these potential functions for microtubules in the lens, their role in lens fibre cell differentiation is disputed. Using a microtubule depolymerising drug, colchicine, initial research indicated that microtubules were necessary for the dramatic cell elongation that occurs during lens fibre cell differentiation in the chick embryo (Piatigorsky, 1975). The interpretation of this research was subsequently questioned (Beebe et al., 1979), and to this day the essential role of microtubules is open to debate. In other cell systems, we know that their role (see Neff et al., 1983) and the role of their associated motor proteins are vital to cell function (Miki et al., 2001; Zhao et al., 2001), as they are in the *Drosophila* eye, as shown by studies with *glued*, a cytoplasmic dynein (Fan and Ready, 1997). These data have altered the question from "Are microtubules important in the lens?" to "What do microtubules do in the lens?" The answer to this latter question is reflected in the organisation and distribution of microtubules and MTOCs.

7.3.3. MTOCs in the Lens

In lens epithelial cells, the centrosome is the major MTOC. The microtubules radiate from this structure, which is located close to the cell nucleus (Lonchampt et al., 1976; Prescott et al., 1991). In this respect, there is nothing unusual about the microtubule organisation in lens epithelial cells, as there is a stellate arrangement of the microtubules originating at the centrosome (Millar et al., 1997). This organisation, however, changes noticeably at the lens equator, when the more posterior epithelial cells and newly emerging fibre cells become aligned relative to one another. Coincidently, the microtubules also adopt a more regular organisation (Millar et al., 1997). The epithelial cell nuclei also undergo shape changes, and at this time there is significant changes in gene expression. It is tempting to speculate that microtubules help facilitate these rearrangements, but there is currently no evidence in support of this.

One surprising aspect of microtubule organisation in the lens is the apparent lack of a centrosome or another distinct microtubule-organising site during lens fibre cell differentiation. The epithelial cells in the post-germinative zone, closest to the first fibre cells, do not contain a γ-tubulin- or pericentrin-staining centrosome, but still have an intact microtubule array (Millar et al., 1997). This lack of a discrete MTOC structure is not unique. For example, the microtubule network in the supporting epithelial cells of the cochlea are not centrosome orientated (Mogensen et al., 2000), albeit the cells still contain centrosomes. The microtubules in nerve axons also lack γ-tubulin staining at their (−) ends (Baas, 1998), like those in lens fibre cells (Millar et al., 1997); however, a γ-tubulin–positive centrosome is still present in the cell body of these neurons. Other cell types lacking defined centrosomes include ciliated epithelial cells and retinal photoreceptor cells, but these have γ-tubulin associated with other MTOCs, such as basal bodies (Joshi, 1994). The surprise in the lens was the apparent disappearance of the centrosome from the lens fibre cells and the more posterior epithelial cells, at least as seen by the failure to stain these cells with the centrosome-specific markers γ-tubulin and pericentrin (Millar et al., 1997). Our observations suggest that the more mature fibre cells contacting the epithelium do regain a γ-tubulin- and pericentrin-positive MTOC located at their apical ends near the epithelium (unpublished results).

In these types of situation, a nucleation, release, and capture model of the centrosome has been proposed (Henderson et al., 1995), and evidence is accumulating in support of novel microtubule (−) end complexes involved in the stabilisation of microtubules (Mogensen et al., 2000). Perhaps the lens will be another example of such a system.

7.3.4. Microtubule Organisation in the Lens

In the lenses of mouse, rat, rabbit, and cow, the microtubule distribution in the lens epithelial cells (Fig. 7.4) is very similar to that previously reported for the frog lens (Prescott et al., 1991). Like the epithelial cells, the lens fibre cells also contain both stable microtubule

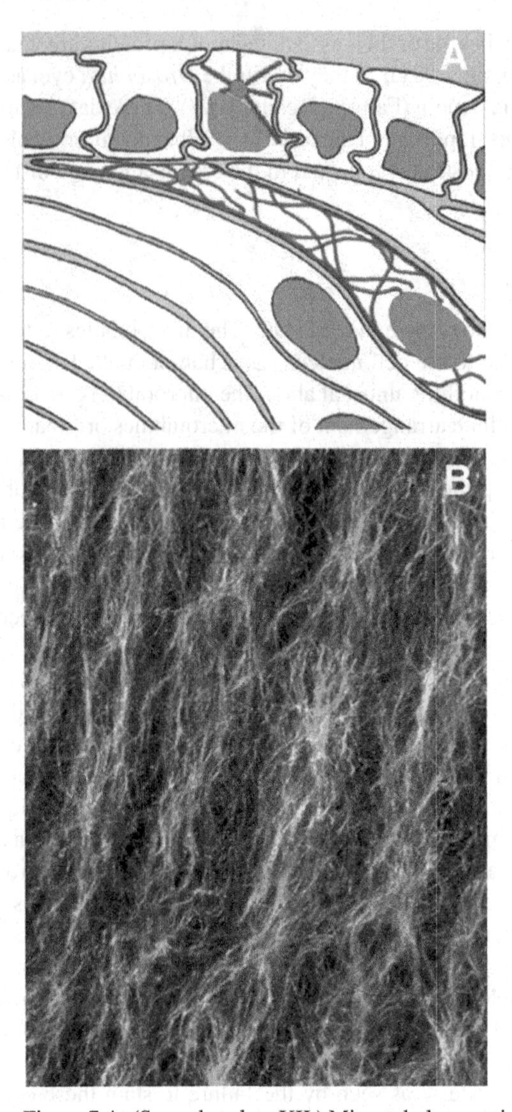

Figure 7.4. (See color plate XII.) Microtubule organisation in the lens. Microtubules (blue) are organised from a perinuclear centrosome (red) in the epithelium, while fiber cells have an MTOC at their apical end. This nucleates microtubules that run the length of the fibres. Microtubules are associated with the MTOC in the fibre cells; unmodified microtubules are in green and posttranslationally modified microtubules are in red.

populations (containing acetylated, detyrosinated tubulin) and dynamic microtubule populations. The stable microtubules, at least in the rat lens fibres, lie closer to the centre of the cell and run continuously from one end of the cell to the other. Other species show some differences from this organisation. For instance, the stable microtubules of the rabbit lens fibres are more helically arranged than those of the rat lens fibres.

The function of these two different microtubule populations has not been clearly defined, but some differences are beginning to emerge. For instance, kinesin preferentially associates with detyrosinated tubulin (Liao and Gundersen, 1998), and this population of microtubules is the more important, at least in fibroblasts, for intermediate filament redistributions (Gurland and Gundersen, 1995; Liao and Gundersen, 1998). Kinesin is also required for this process (Prahlad et al., 1998). The more dynamic population of microtubules is associated with the plasma membrane, whilst the stable population is more centrally localised in both epithelium and fibre cells. Earlier electron microscopy studies of embryonic chick lenses (Pearce and Zwaan, 1970) and human lenses (Kuwabara, 1968) showed that microtubules were predominantly orientated parallel to the long axis of the fibre cells. It is now clear that these represent the stable microtubules, whilst the dynamic microtubules are at an angle to the fibre cell long axis (Millar et al., 1997). These microtubules would not have been quite so obvious in the electron microscopy sections for a variety of reasons, not least the difficulty of rapidly fixing the tissue before microtubule disassembly occurred (Millar et al., 1997). As many of the microtubules are at the fibre cell plasma membrane, roles in membrane subcompartmentalisation (Gruijters et al., 1987), organelle positioning (Bassnett, 1995), and the redistribution of actin and intermediate filament networks (Sandilands et al., 1995a) are likely. Indeed, in the context of the other cytoskeletal elements, microtubules are important for the redistribution of intermediate filaments (Prahlad et al., 1998) and are required for the maintenance of adherens junctions (Waterman-Storer et al., 2000). Intermediate filaments also contribute to the establishment of cell polarity and microtubule organisation (Ameen et al., 2001), and the tumour-suppressor gene product APC (adenomatous polposis coli) provides a link between cell adhesion, microtubule stability, and cell differentiation (Aberle et al., 1996; Zumbrunn et al., 2001). There is thus extensive cross-talk between the microtubule, intermediate filament, and actin networks (Correia et al., 1999; Kaverina et al., 1998; Waterman-Storer et al., 2000), particularly in regard to common associated proteins (reviewed in Herrmann and Aebi, 2000) and the correlation between important events during lens development and differentiation and dramatic changes in the organisation of the cytoskeleton.

7.4. Actin in the Lens

Since the unequivocal identification of actin in cultured lens cells (Lonchampt et al., 1976) and the intact lens (Kibbelaar et al., 1980; Kibbelaar et al., 1979), the characterisation of the actin complex in the lens has progressed steadily. Although not yet tested experimentally, a role for actin in lens accommodation has been proposed (Rafferty and Goossens, 1978). However, there has been an accumulation of data correlating actin and its associated proteins with critical events during lens development and differentiation (see Beebe and Cerrelli, 1989; Ferreira-Cornwell et al., 2000; Lo et al., 2000).

During early development, the lens is formed from a vesicle that is pinched off from the surface head ectoderm (see chap. 1). It is hard to envisage how this process can be completed in the absence of a functional actin cytoskeleton given the fundamental role that the actin–adherens junction complex plays in morphogenic events during development. It is also important to remember that the lens continues to grow throughout life and that the differentiation process of the lens fibre cells requires movement of the fibre cells concomitantly across the posterior lens capsule and over the anterior lens epithelial cells. Of necessity, this means that the cell-cell junctions and cell-matrix junctions at the epithelial–fibre cell interface and at the fibre cell–capsule interface have to be dynamic, as do the associated cytoskeletal structures.

Defects in the actin machinery – defects in the structural arrangement and in the signalling pathways – are the cause of a wide spectrum of inherited human diseases, from haemolytic anemias (e.g., spectrin and ankyrin; see Bennett, 1990; Lux et al., 1990) to cancer (e.g., cadherin and APC; see Bienz and Clevers, 2000; Birchmeier, 1995). They are also the cause of eye diseases, such as Usher's 1B syndrome, in which myosin VIIa mutations cause syndromic (multiple phenotypic disease) deafness along with retinal degeneration (Weil et al., 1995). Cataract has been linked to both actin (Hales et al., 2000; Rafferty et al., 1993) and its regulatory proteins (Jin et al., 2000; Kluwe et al., 1995; Rao et al., 1997), adding urgency to discovering the precise roles of actin and its associated proteins in the lens.

7.4.1. Actin Expression and Function in the Lens

Actin is a highly conserved protein (Sheterline et al., 1995) found at all stages of lens development and cell differentiation (Lee et al., 2000; Ramaekers et al., 1981). It is present in the centre of the lens (Fig. 7.5), and so, like intermediate filaments but in contrast to microtubules, it is expected to play a role in all stages of lens cell differentiation. Actin

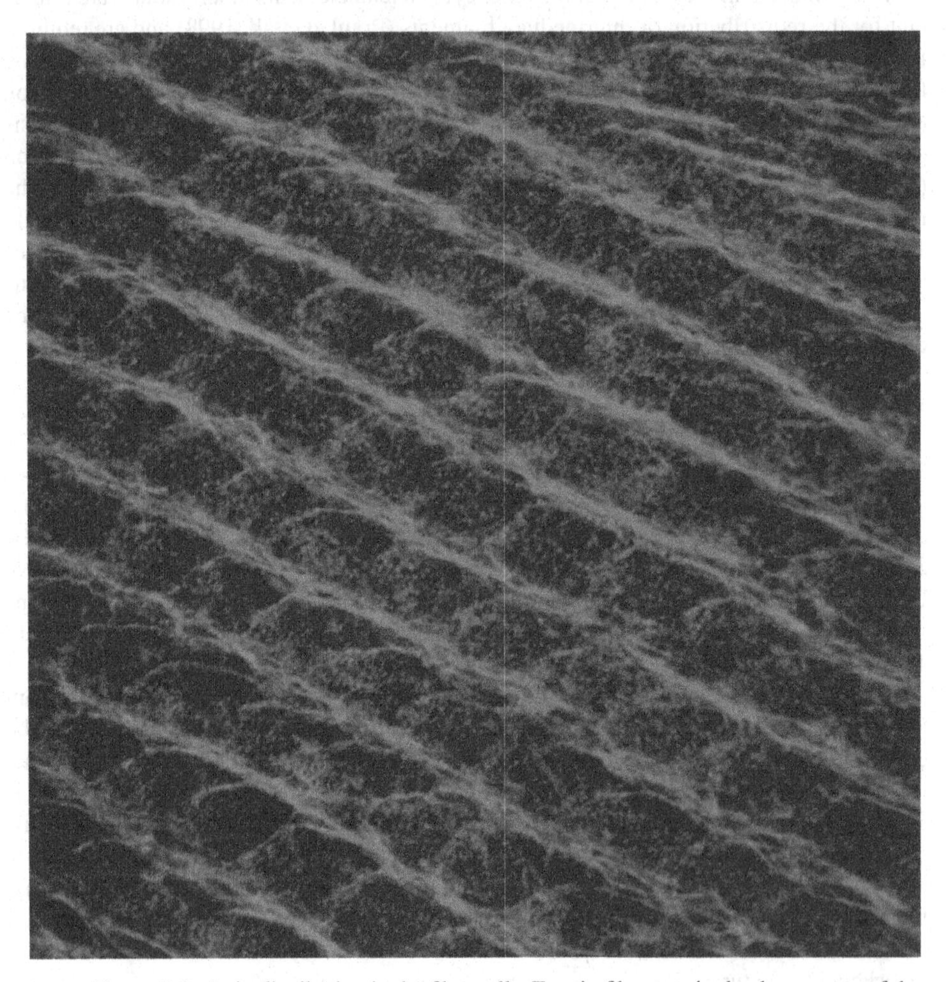

Figure 7.5. Actin distribution in the fiber cells. F-actin filaments in the deep cortex of the lens. Notice the staining is stronger at the short faces of the fiber cells, since there is more membrane associated with the ball-and-socket interdigitations.

exists as seven different isoforms in mammalian cells. In non-muscle cells, including lens cells (Schmitt et al., 1990), the complement of actin isoforms includes β- and γ-non-muscle actin. Only in some forms of cataract does this change, as with the expression of α-smooth muscle (α_{sm}) actin (Schmitt et al., 1990) in subcapsular cataract. In fact, it was proposed that α_{sm}-actin could be a marker for this cataract type (Hales et al., 1994). The exposure of isolated rat lenses to TGFβ induced the expression of α_{sm}-actin (Hales et al., 1994; Lovicu et al., 2002). Lens epithelial cells in culture also express α_{sm}-actin (Nagamoto et al., 2000), indicating the sensitivity of lens epithelial cells to altered growth environments and the importance of the cytoskeleton in the adaptive response of the lens and lens cells. Indeed, studies on *Drosophila* eye development demonstrate the importance of TGFβ family members for the eye, as ectopic expression of *decapentaplegic* (*dpp*) along the anterior margin of the eye disc induces new discs (Pignoni and Zipursky, 1997). Although the role of dpp in eye development is still being established (Chen et al., 1999), links have been proposed with the actin cytsoskeleton via band 4.1 proteins. This is because merlin (schwannomin; Claudio et al., 1997) is thought to mediate the effects of dpp on cell proliferation and differentiation in the eye (McCartney et al., 2000). These data collectively suggest that misexpression of TGFβ is at least an additional risk factor in cataractogenesis and that the regulation of actin is important during lens development and differentiation.

Support for this hypothesis came initially from drug studies. Cytochalasins have been used to explore actin function in the lens (Beebe and Cerrelli, 1989; Mousa and Trevithick, 1977), and these studies showed that lens fibre cell elongation could be prevented by these drugs. Interestingly, cytochalasin D prevented K^+ efflux, and this was highlighted as a reason for the prevention of cell elongation (Beebe and Cerrelli, 1989). The actin cytoskeleton is vital to establishing the plasma membrane compartments (Garner and Kong, 1999; Yeaman et al., 1999), and so it is easy now to understand how compromising the actin cytoskeleton may alter ion transport.

7.4.2. Actin Membrane Complexes in the Polarised Cells of the Lens

The actin–plasma membrane complex acts as a major contributor to cell polarity in epithelial cells by directing the sorting of adhesion complexes, pumps, and ion channels. Cell polarity is fundamental to cell function, and thus there are similar expectations of the lenticular actin cytoskeleton. The first comparative accounts of the actin cytoskeleton in mammalian lenses described a three-dimensional network (Lonchampt et al., 1976; Perry et al., 1981; Rafferty and Goossens, 1978), later described as a polygonal lattice (Rafferty and Scholz, 1984), that was concentrated at either side of the epithelial–fibre cell interface. The surfaces of this interface represent the apical ends of the epithelial and fibre cells, respectively. The actin concentrated in these regions is linked via adhesion junctions, as has recently been elegantly demonstrated (Lo et al., 2000). This study on the embryonic chick lens established the association of actin networks with zonulae adherens junctions throughout the plain of the epithelial and fiber cell apical domains. Indeed, it was proposed that during development the zonulae adherens form a continuous belt linking the epithelial cells and fibre cells in the equatorial region, where the epithelial cells are differentiated into fiber cells. Fasciae adherens were also seen between the apical ends of the epithelial and fibre cells (Lo et al., 2000). The fasciae adherens are not as robust as the zonulae adherens, for the epithelial–fibre cell interface is easily ruptured when the lens mass is removed, at least in the mouse (Liou, 1990) and cow (unpublished observation). Interestingly, this is not the case for the basal ends of the lens fibre cell, which are linked to the posterior lens capsule by another actin-enriched complex termed the basal membrane complex (BMC; Bassnett et al., 1999). Removal of the capsule from chick lenses ruptures the lens fibre cells at this point, leaving behind the cell-ECM complex. This interaction is calcium dependent and allows very effective purification of the complex (Bassnett et al., 1999; see also chap. 9).

The junctions themselves comprise a familiar list of actin-binding proteins and adherent junction components. These include α-actinin (Lo et al., 1997); α-, β-, and γ-catenin (Ferreira-Cornwell et al., 2000); N-cadherin (Volk and Geiger, 1986b; Watanabe et al., 1992); B-cadherin (Murphy-Erdosh et al., 1994); L-CAM (Thiery et al., 1984); band 4.1; paxillin (Beebe et al., 2001); plakoglobin (Franke et al., 1987); spectrin; and vinculin. The cell-cell adherens junctions themselves are considered unusual by some and have been given the name "cortex adhaerens" in recognition of the fact that contact extends over virtually the whole surface (Schmidt et al., 1994). These junctions are indeed unusual in as much as the $\alpha_6\beta_4$ integrin complex (Walker and Menko, 1999) and a β_1 complex (Bassnett et al., 1999) with α_6 (Menko and Philip, 1995) are localised to the same plasma membrane regions in the lens fibre cells. It is worth noting the potential of the different cadherins to localise to different subdomains in the plasma membrane (Leong et al., 2000). The cortex adhaerens is therefore not a homogenous mixture of adherent junction proteins. As has emerged recently, the expression of some of these named components changes dramatically during lens development and lens cell differentiation and also according to the subcellular location of the junctions (Bassnett et al., 1999; Beebe et al., 2001; Leong et al., 2000; Menko and Philip, 1995; Walker and Menko, 1999). This means that the adherens junctions probably perform very discrete roles depending on their local environment.

The most extensive studies on these changes in adherens junction components have been done in the chick (Bassnett et al., 1999; Beebe et al., 2001; Leong et al., 2000). As with the other elements of the cytoskeleton, there are dramatic changes in the expression of the adherens junction components. For instance, band 4.1 is not present in lens epithelial cells but is expressed in the newly differentiated fibre cells, where it appears concentrated along the four short faces of these hexagonal cells. This relative increase in expression begins to wane once the fibre cells have detached from the posterior capsule at a later stage in differentiation. This particular example illustrates the complexity of the changes involved, but it is difficult at this moment to relate such expression patterns to function. In another example, the actin-binding protein tropomodulin is concentrated at the apical and basal ends of the newly differentiated fibre cells (Lee et al., 2000) in an unusual structure that does not completely overlap the actin filaments in this region of the lens (Fischer et al., 2000). Tropomodulin, like tropomyosin, is not proteolysed during lens fibre cell differentiation, and this is thought to contribute to the stability of actin in the lens nucleus (Lee et al., 2000). Evidence from the bovine lens suggests that there are at least two distinct actin complexes associated with the plasma membrane, one containing tropomodulin and the other spectrin (Woo et al., 2000). The functional significance of this duality remains unknown, but the arrangement is interesting, as it is quite different from actin organisation in the erythrocyte plasma membrane. Such data do, however, define in more detail the finer points of lens fibre cell differentiation, which is important if we are to correlate structure with function.

From these studies, it should be clear that the adherens junctions and the associated actin network are dynamic, responding to changes in the local environment and to external signals. This point is made more poignant by the observed link between Pax6, N-cadherin expression, and eye defects. It has been shown that Pax6 levels determine the expression of N-cadherin in the lens placode and consequently the successful formation of a lens (van Raamsdonk and Tilghman, 2000). The *Sey* phenotype in the mouse appears to arise from inadequate adhesive properties due to a loss of cadherin function in these cells (Collinson et al., 2000; Quinn et al., 1996). The targeted expression of Pax6(5A) to the fibre cells induces cataract and up-regulates the level of integrins (Duncan et al., 2000), components of the cell-cell junctions in the fibre cells (Walker and Menko, 1999). The loss of Prox1 expression by a knockout strategy also results in the inappropriate expression of E-cadherin, the loss of cell polarisation, and the failure of the lens cells to elongate properly during lens vesicle closure (Wigle et al., 1999). These data highlight the importance of the cell junctions and the associated cytoskeletal machinery to lens development and illustrate the link between transcription factors, the cytoskeleton, and the processes of development and differentiation.

7.5. Conclusion

The lens cytoskeleton stands at the apex of all the important pathways that regulate lens cell development and differentiation. The cytoskeleton and associated structures are the verbs in the language used to make and maintain lens cells. They organise compartments, assist and even direct the transport and sorting of other cellular compartments, and provide the structural support required for lens accommodation and transparency. The cytoskeleton is a complex arrangement of filaments that provide a physical scaffold for facilitating protein interactions required in key cellular events, such as signal transduction and protein repair and maintenance. Medical and functional genetics have provided the evidence that the cytoskeleton is essential to the lens, and it is now up to us to fill in the details.

Note added in proof: Details of targeted (Alizadeh et al., 2002; Sandilands et al., 2003) and a natural CP49 knockout (Alizadeh et al., 2004) as well as a filensin knockout (Alizadeh et al., 2003) have been published. Some of these data demonstrate the great importance of the CP49 intermediate filament network to the optical properties of the lens (Sandilands et al., 2003) and also show the interdependence of the vimentin and the CP49-filensin networks (Sandilands et al., 2004). Details of the unique membrane complex on lens plasma membranes have also recently been published (Straub et al., 2003).

Part Three

Lens Development and Growth

8

Lens Cell Proliferation: The Cell Cycle

Anne E. Griep and Pumin Zhang

8.1. Introduction

Regulation of the cell division cycle is an essential process by which the cell monitors its growth and differentiation. Maintaining the proper controls on these cellular processes is essential not only during embryonic development but also throughout the lifetime of an animal. During embryonic development, in a temporally and topographically distinct manner, a wide variety of cells exhibit the capacity to become quiescent, to proliferate, and to irreversibly withdraw from the cell cycle and undergo terminal differentiation. Thus, both the entry of a cell into the cell cycle from a state of quiescence and the exit of the cell from active cycling must be precisely regulated if normal cell growth and differentiation are to be maintained. Furthermore, these two distinct types of cell cycle regulation must be coordinated with the regulation of differentiation. Over the past decades, much has been learned about the mechanisms that control cell cycle progression *in vitro*, primarily as it relates to cancer. Only in recent years has an understanding of how the cell cycle is controlled *in vivo* in normal development begun to emerge.

The ocular lens has served as a model system for unraveling the roles of cell cycle regulatory genes in a developmental context. A relatively simple tissue with a well-described blueprint of cell division and morphogenesis, the lens has been ideal for studying the coordination of both cell growth and differentiation *in vivo*. In this chapter, we first review the molecular mechanisms through which cell growth is thought to be controlled, as gleaned from the efforts of many to document cell cycle regulation in *in vitro* systems. Then we detail our current understanding of how the cell cycle is regulated in lens cells at the time of their differentiation into fiber cells (when they permanently withdraw from the cell cycle) and also how it is regulated in the cells of the epithelium. As it is impossible to cover the entire history of work on this topic, we have chosen to focus our attention on *in vivo* work, with an emphasis on the more recent mechanistic studies that involve genetic manipulation of gene function in the mouse.

8.2. Regulation of the Cell Cycle

This section reviews our current understanding of the mechanisms of cell cycle control as gleaned from extensive investigation of cell cycle control in tissue culture model systems. To a large degree, the findings in the cell cycle field have driven investigations in the lens field. Therefore, it is useful to review these insights before proceeding with a discussion of how cell cycle control is achieved in the lens.

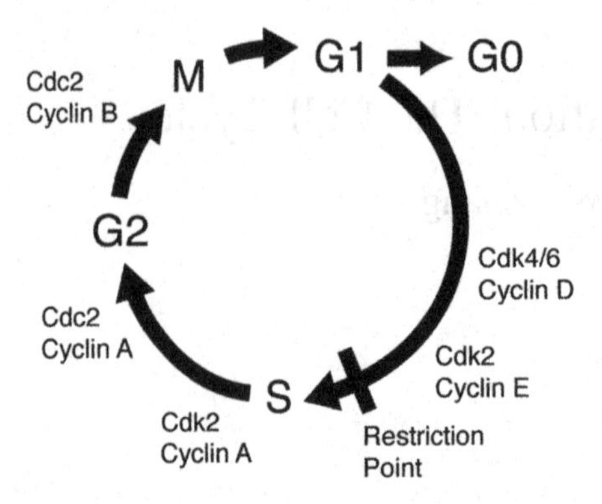

Figure 8.1. The cell cycle. The mammalian cell cycle is divided into four phases, G1, S, G2, and M, progression through which is catalyzed by cyclin-dependent kinases activated by different cyclins at different phases during the cycle. Cells can also enter a nondividing quiescent state (G0).

Cell division cycles are the fundamental means that an organism uses to grow. In most somatic cells, one cell cycle is divided into four phases: G1, S, G2, and M. Cells synthesize their genome in the S phase (DNA *s*ynthesis), then segregate the duplicated genome into two daughter cells in *m*itosis (the M phase). DNA synthesis and mitosis are separated by two *g*ap phases, G1 and G2 (Fig. 8.1). In addition, cells can enter a G0, or quiescent, state. Proliferation of eukaryotic cells is regulated primarily at two points in the cell cycle, in G1 prior to entry into the S phase and in G2 prior to entry into mitosis. Whether to commit to a round of cell division or exit the cell cycle is decided at a point in G1 referred to as the "restriction point" in mammalian cells (Draetta, 1994; Dulic et al., 1992; Ohtsubo and Roberts, 1993; Pardee, 1989; Quelle et al., 1993) or "START" in yeast (Koch and Nasmyth, 1994). In fibroblasts, passage through the restriction point depends critically on the signals that are received through mitogen-activated pathways, but once this point is passed, cells are committed to the S phase and the remainder of the cycle in a mitogen-independent manner (Pardee, 1989). In somatic tissues, passage through the restriction point is thought to be the primary event controlling cell proliferation.

Progression through the cell cycle is regulated by the action of two factors acting in concert, the cyclin-dependent kinases (Cdks) and the cyclins (see Table 8.1 for a summary). The Cdks are serine-threonine protein kinases that are activated by association with their cognate cyclin (see Fig. 8.1 and Table 8.1). Once activated by the cyclin, the Cdk is responsible for phosphorylating specific targets. There are five Cdks (Cdk1, 2, 3, 4, and 6) and four classes of cyclins (cyclin A, B, D, and E) that have been shown to regulate cell cycle progression. Other Cdks and their cyclin partners are involved in regulating other biological processes, such as transcription. Search of the near-complete human genome sequences identified three new cyclins but no new Cdks (Murray and Marks, 2001). Two of the new cyclins are likely to be involved in transcription regulation, and the third has weak homology to B-type cyclins (Murray and Marks, 2001). The levels of Cdks are generally constant throughout the cell cycle, while the levels of their cyclin partners fluctuate periodically during the cycle – hence their name, cyclins (Evans et al., 1983). Biochemical and genetic data from several systems have demonstrated that cyclins promote cell cycle transitions via their ability to associate with and activate their cognate Cdks (Coleman and Dunphy, 1994; Draetta, 1994; Hunter and Pines, 1994; King et al., 1994; Koch and Nasmyth, 1994; Meyerson et al., 1992; Nurse, 1994; Sherr, 1994). D-type and E-type cyclins function in the G1 phase of the

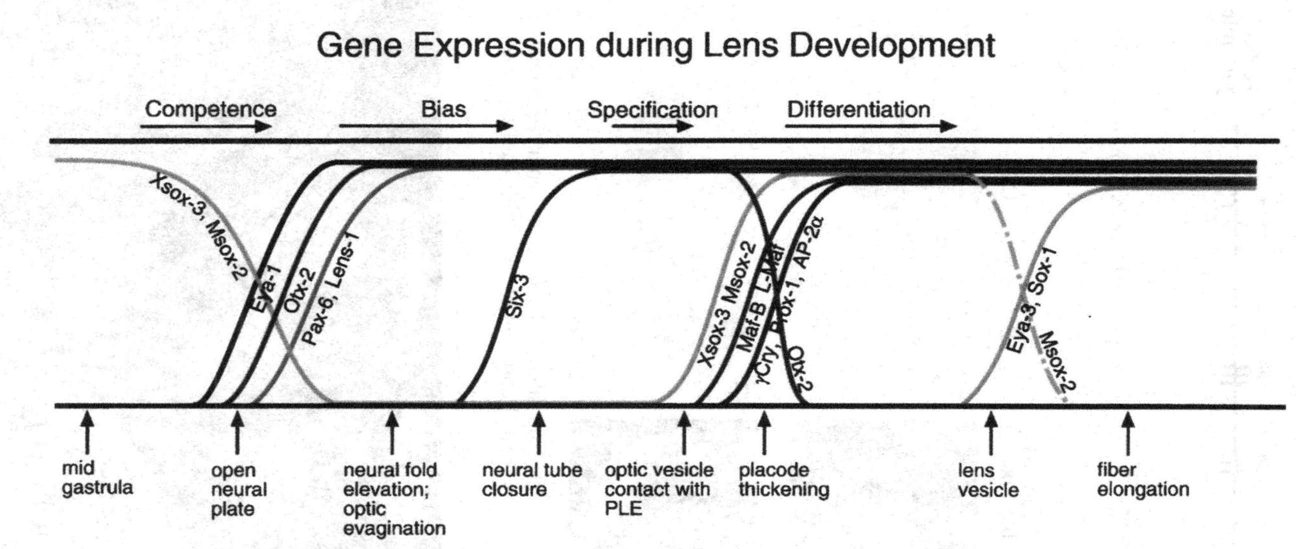

Plate I. Gene expression during lens determination. (Figure 2.2 in the text.)

Plates I-XX in this section are available for download in colour from www.cambridge.org/9780521184236

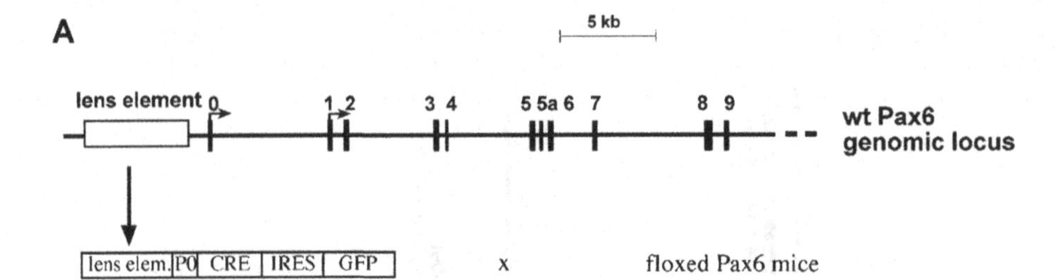

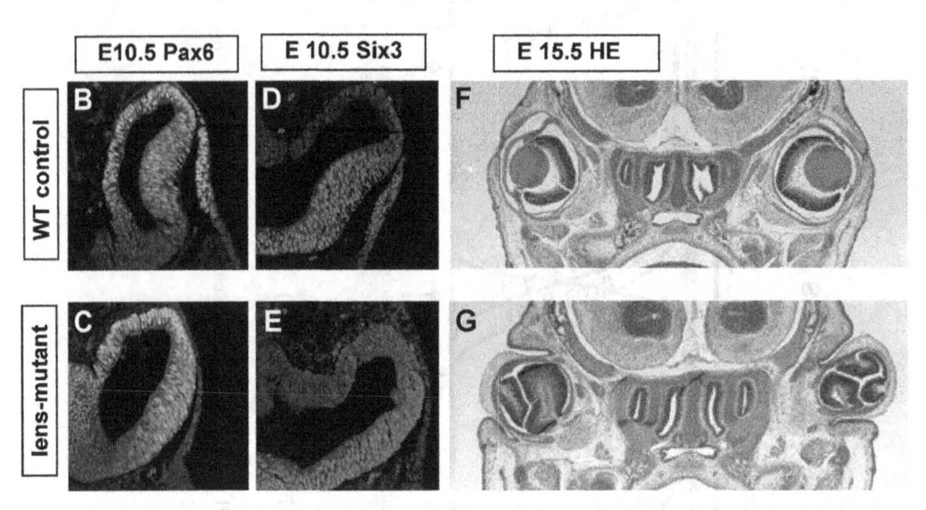

Plate II. The regulatory element responsible for Pax6 expression in the surface ectoderm and subsequently in the lens is located upstream of Exon 0 of the Pax6 genomic locus. (Figure 3.3 in the text.)

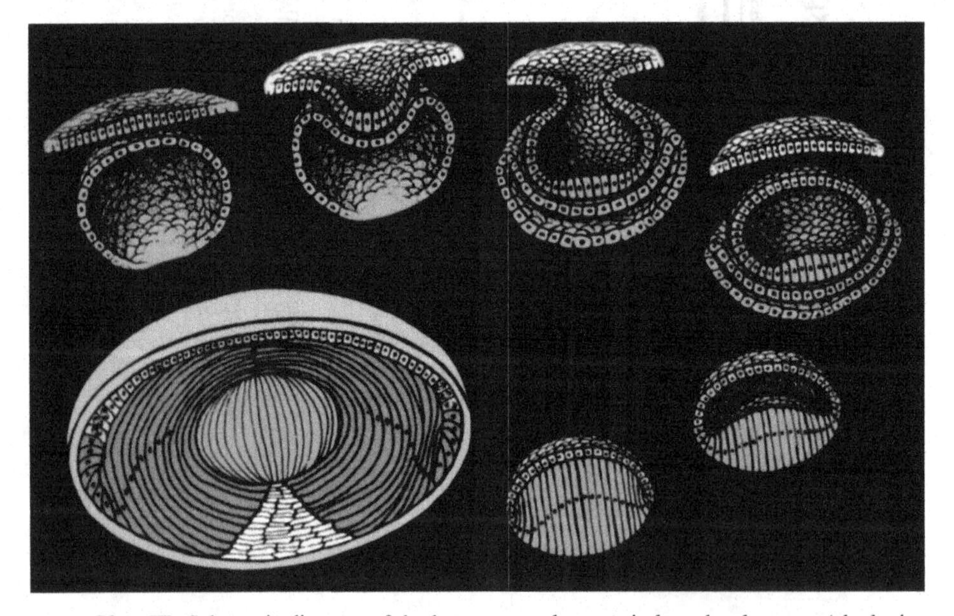

Plate III. Schematic diagram of the key structural events in lens development (clockwise from top left): The thickening of the lens placode (green). The invagination of the lens placode (green) toward the developing optic cup (orange). The inverted, or "inside out," lens vesicle (green). The elongation of posterior lens vesicle cells as they terminally differentiate into primary lens fibers (green). The obliteration of the lumen of the lens vesicle by fully elongated primary fibers (green). The continuous formation of secondary lens fibers (dark blue), resulting in successive growth shells being continuously added onto the primary fiber mass (the embryonic lens nucleus; green). (Figure 4.1 in the text.)

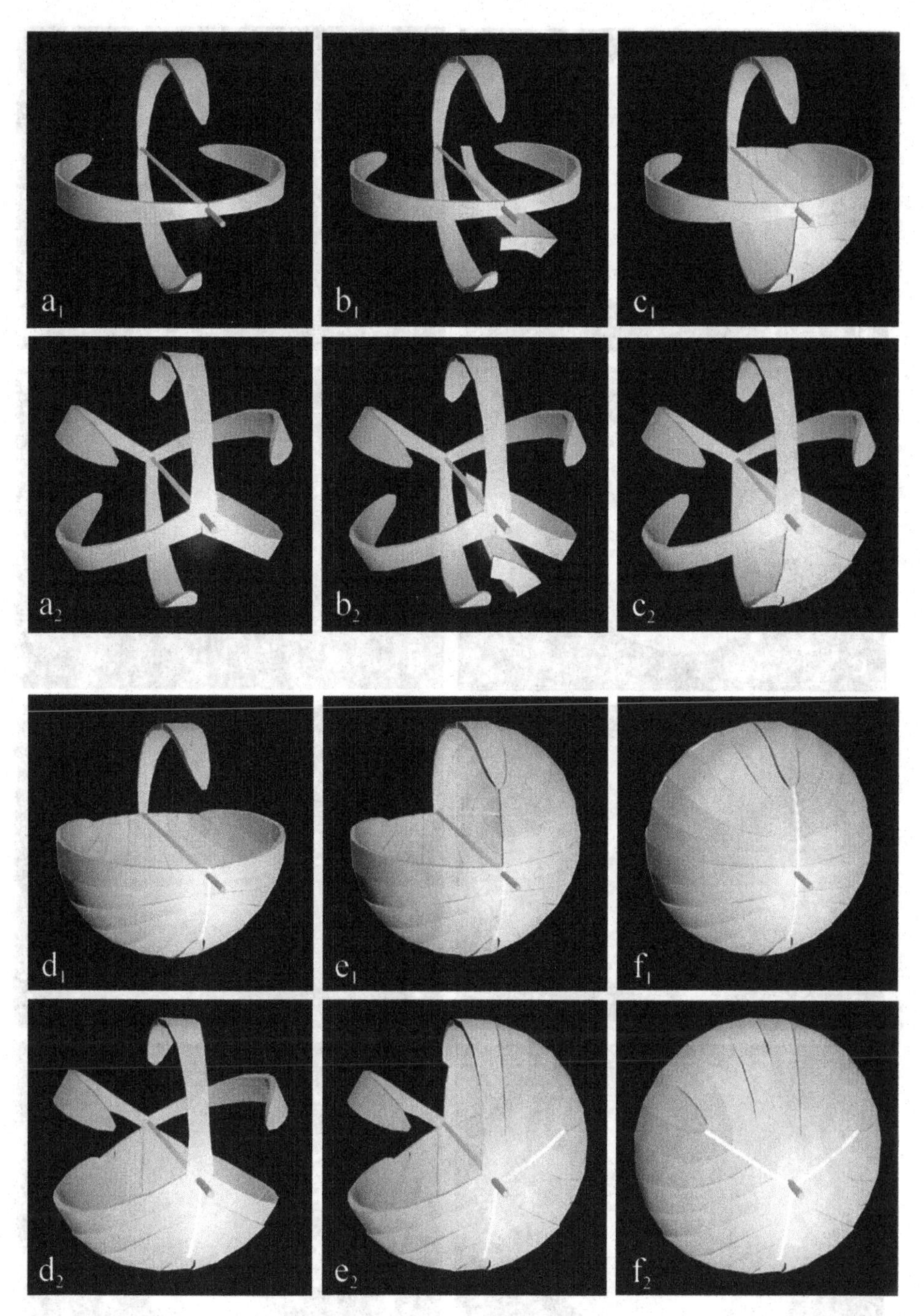

Plate IV. Scale 3D-CADs showing key structural elements in the formation of line suture lenses (a_1–f_1) and Y suture lenses (a_2–f_2). (Figure 4.7 in the text.)

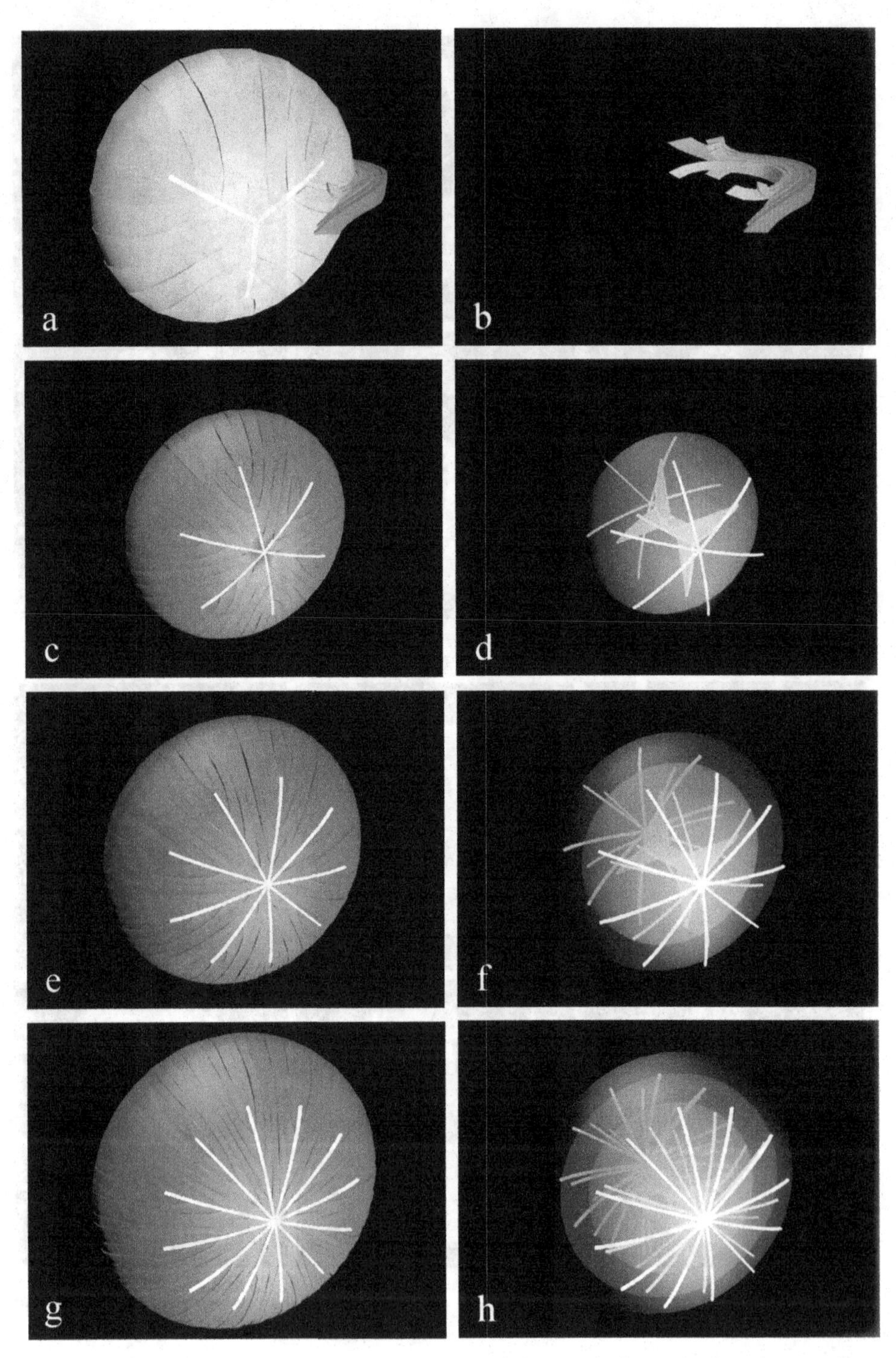

Plate V. Scale 3D-CADs showing key structural elements in the production of suture planes in primate lenses. (Figure 4.10 in the text.)

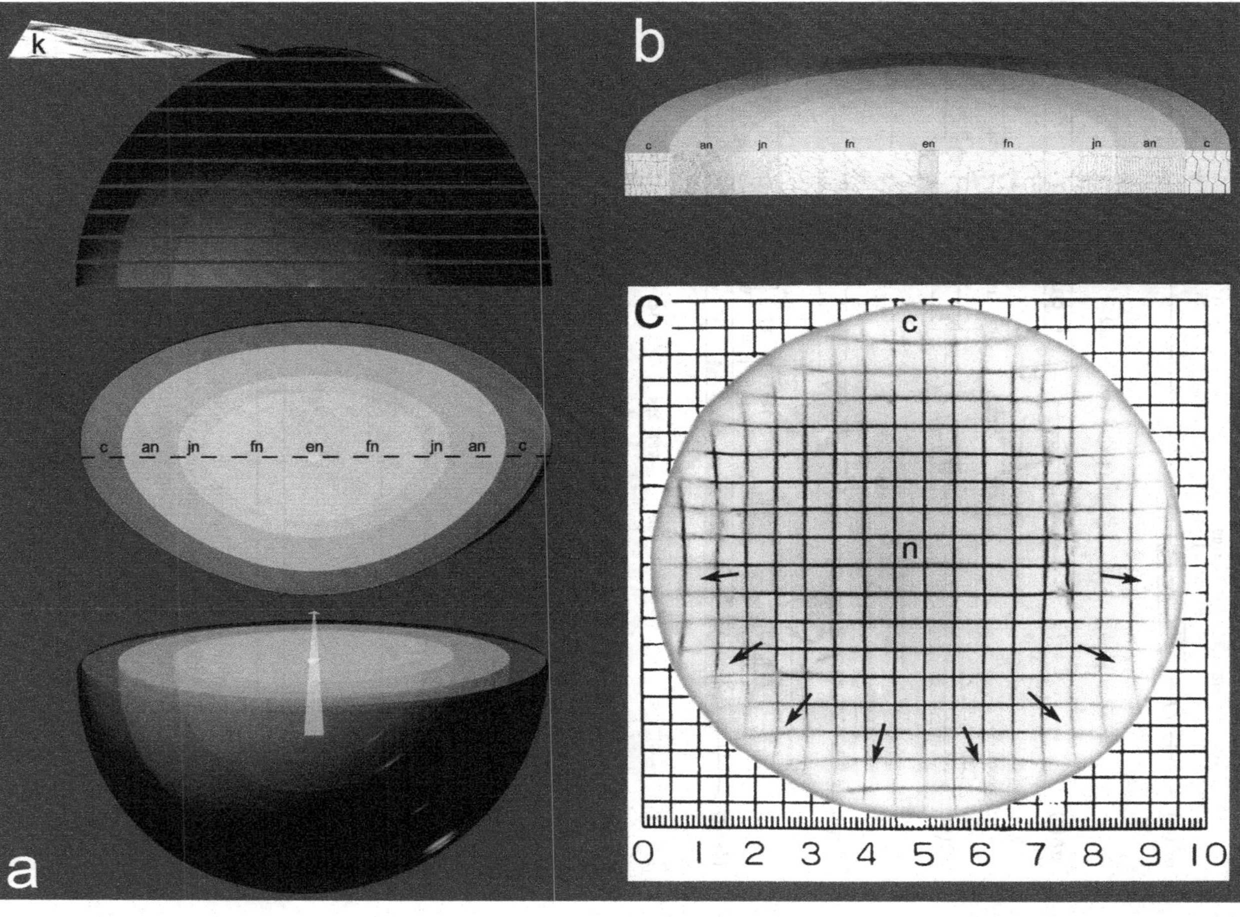

Plate VI. Dissection of fresh lenses for ultrastructural examination reveals internal organization in all developmental regions. (Figure 4.12 in the text.)

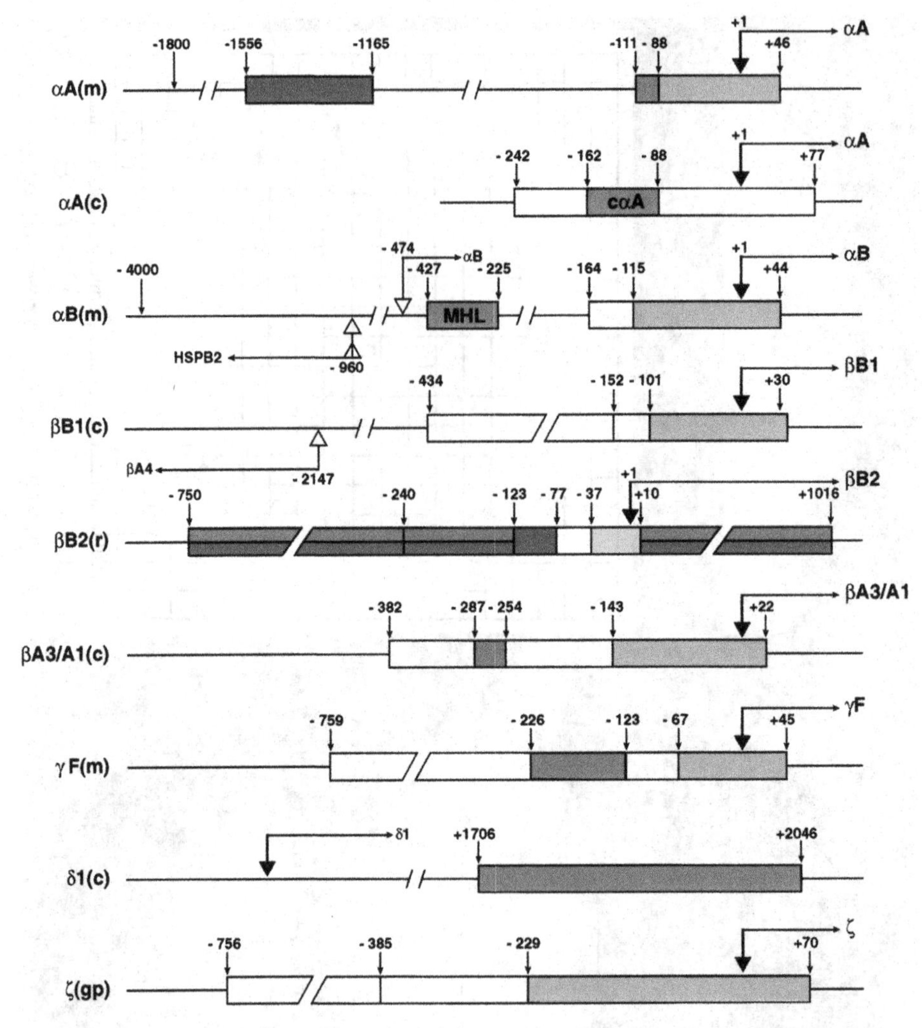

Plate VII. Schematic representation of the mouse and chicken αA-, mouse αB-, chicken βB1-, rat βB2-, chicken βA3/A1-, mouse γF-, chicken δ1-, and guinea pig ζ-crystallin transcriptional control regions. Regulatory regions are shown in boxes, lens-specific minimal promoters are shown in green, enhancers are shown in orange, and negatively acting regions (repressors/silencers) are shown in red. Basal promoter of the rat βB2-crystallin gene determined in transient transfections is shown in yellow. Start sites of transcription in the lens (+1) are labeled by bold arrows. Alternate start sites of transcription in the mouse αB-crystallin gene, multiple start sites of HSB2 transcription (around −960) inside the αB-crystallin locus, and start site of the chicken βA4-crystallin transcription near the βB1-crystallin gene are labeled by open arrows. MHL: muscle, heart, lens enhancer of the αB-crystallin gene. (Figure 5.1 in the text.)

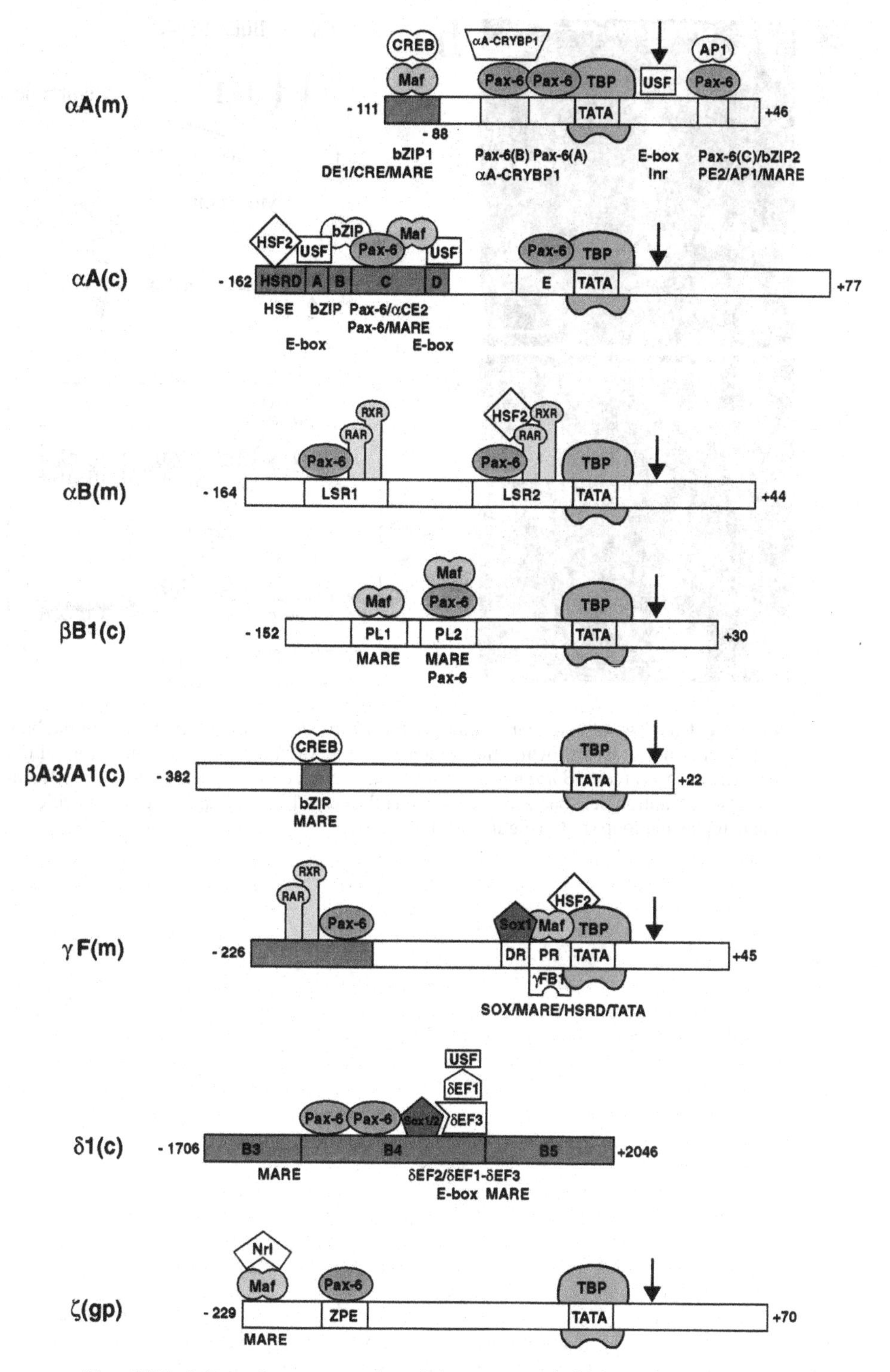

Plate VIII. Schematic representation of the mouse and chicken αA-, mouse αB-, chicken βB1- and βA3/A1-, mouse γF-, chicken δ1-, and guinea pig ζ-crystallin regulatory sites and DNA-binding transcription factors. (Figure 5.2 in the text.)

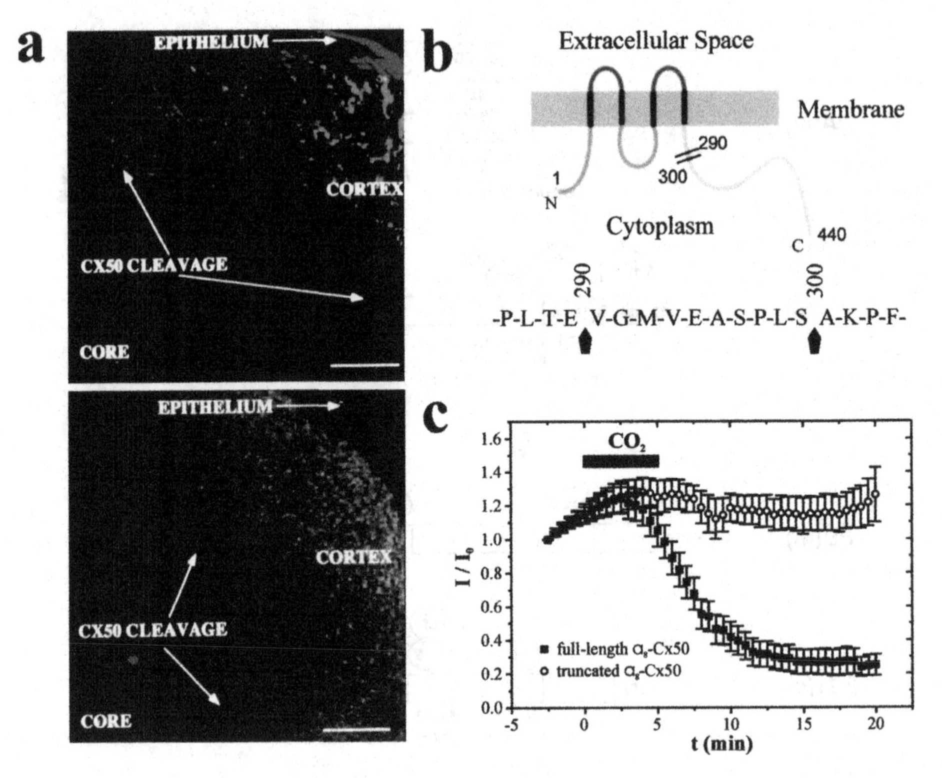

Plate IX. Cleavage of connexins in the lens. (a) *Top panel:* axial section through the bow region of a mouse lens showing the coincidental loss of nuclei (red) and the carboxyl tail of connexin 50 (green). *Bottom panel:* equatorial section double-labeled with two different connexin 50 antibodies, one specific for the carboxyl tail (green) and the other specific for the cytoplasmic loop (red). (Figure 6.4 in the text.)

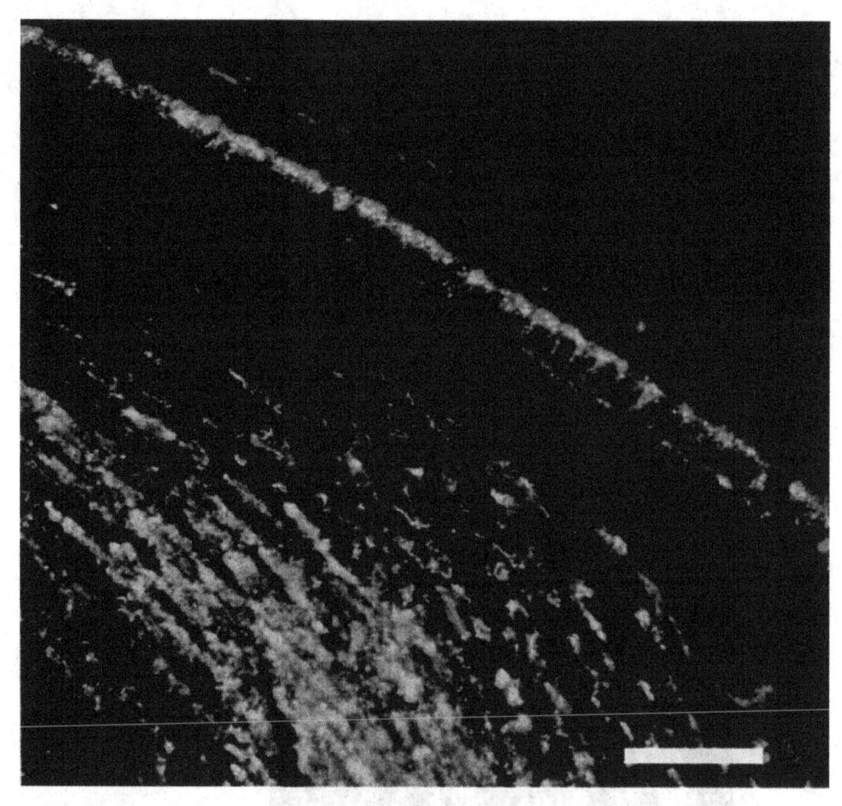

Plate X. Differential expression of glucose transporters in the lens. An axial section of a rat lens showing the expression of GLUT1 (red) on the basolateral aspect of the epithelium and GLUT3 (green) in the differentiating fiber cells. Scale bar: 25 μm. (Figure 6.6 in the text.)

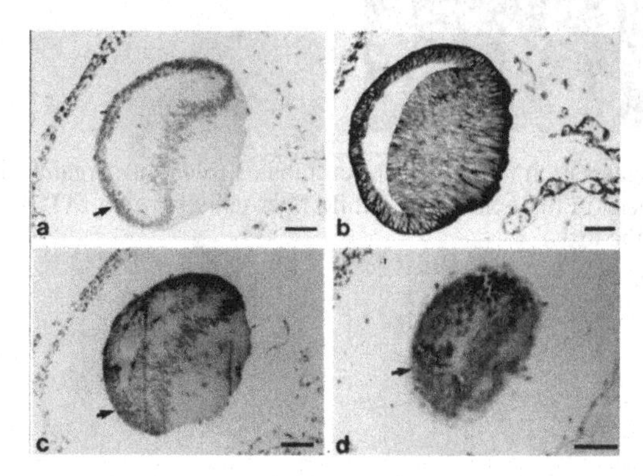

Plate XI. Detection of filensin and CP49 within the embryonic lens vesicle. (Figure 7.1 in the text.)

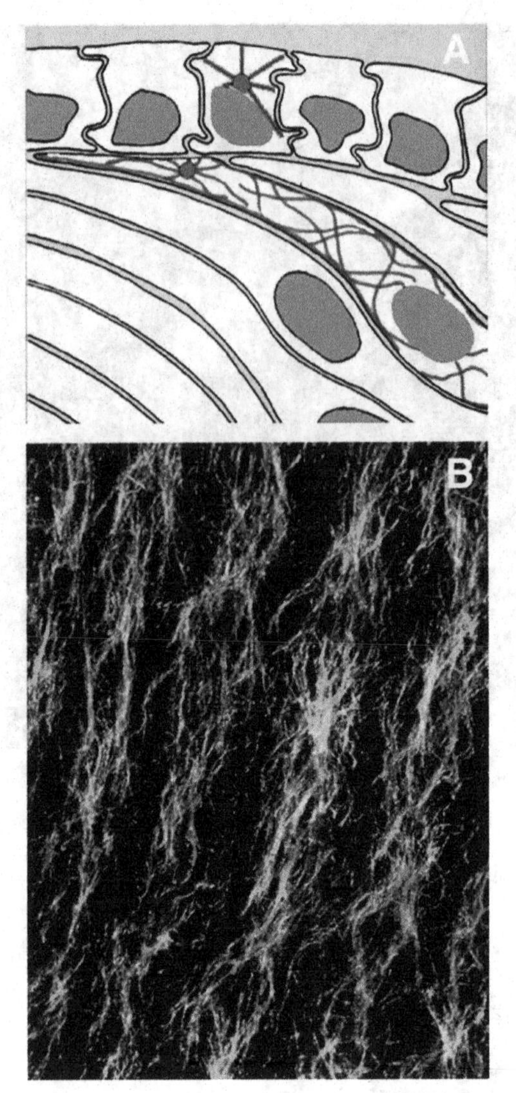

Plate XII. Microtubule organisation in the lens. Microtubules (blue) are organised from a perinuclear centrosome (red) in the epithelium, while fiber cells have an MTOC at their apical end. (Figure 7.4 in the text.)

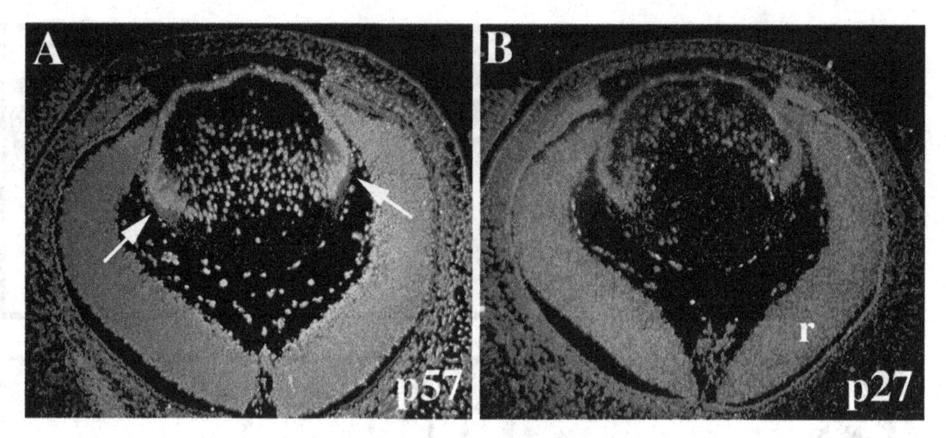

Plate XIII. Expression of Cdk inhibitor genes during lens development. (Figure 8.5 in the text.)

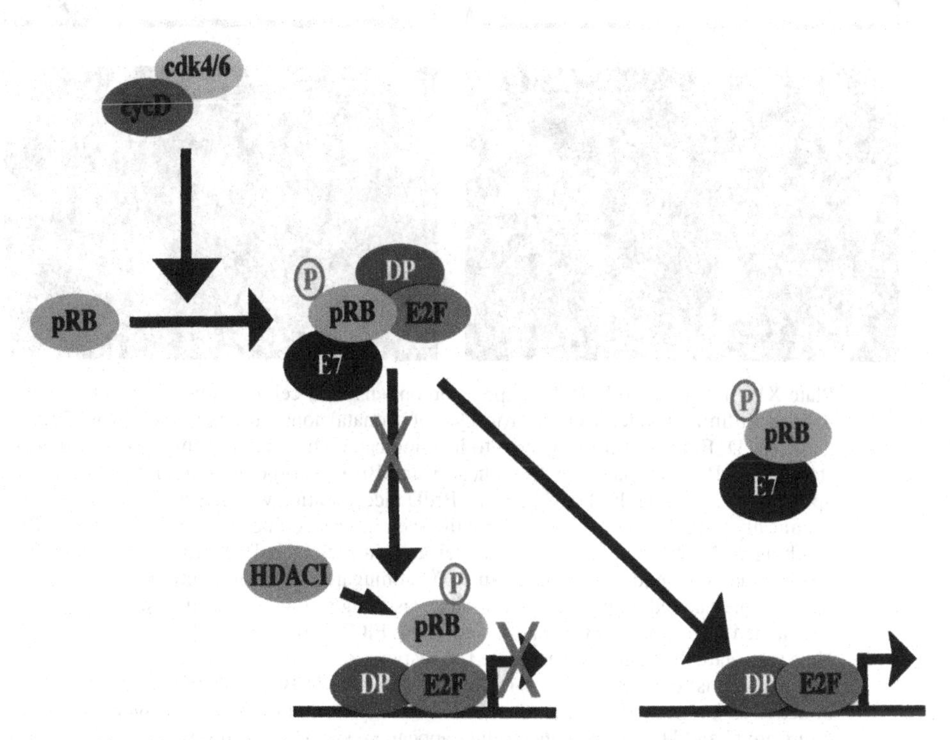

Plate XIV. Inactivation of pRb by PV-16 E7. (Figure 8.6 in the text.)

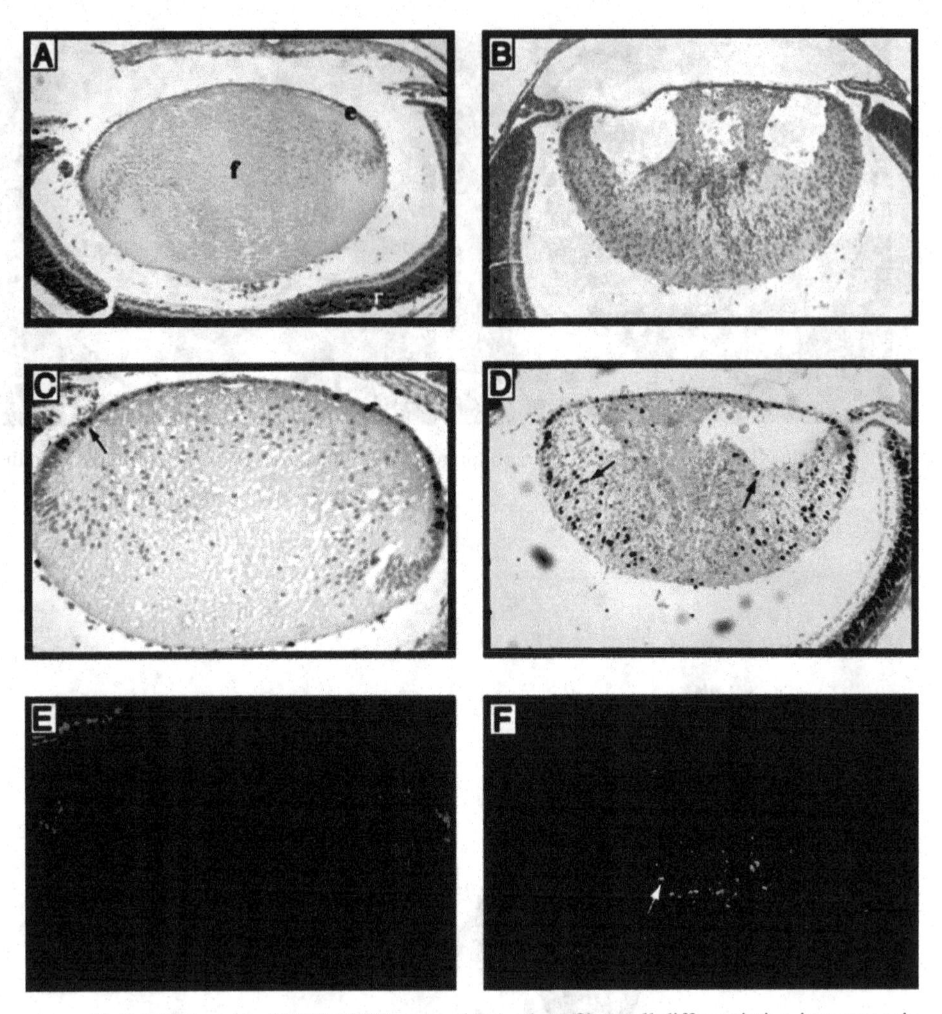

Plate XV. Impact of HPV-16 E7 expression on lens fiber cell differentiation in transgenic mice. Paraffin-embedded sections from eyes of neonatal nontransgenic (A, C, E) or E7 transgenic (B, D, F) mice were subjected to histological analysis using hematoxylin and eosin staining (A, B), analysis of DNA synthesis using BrdU incorporation (C, D), or analysis of apoptosis (E, F) using the TUNEL assay. BrdU incorporation was detected immmunohistochemically with diaminobenzidine, and the section was counterstained with hematoxylin, resulting in brown-stained positive nuclei and purple-stained negative nuclei. TUNEL-labeled nuclei were identified using an FITC-conjugated secondary antibody, and sections were counterstained with propidium iodide, resulting in green-stained positive nuclei and red-stained negative nuclei when viewed using FICT and Texas Red filters, respectively. Note the small, disorganized vacuolated appearance of the fiber cell compartment of the E7 transgenic lens (compare A and B). Note the presence of BrdU positive nuclei, indicative of active DNA synthesis, within the fiber cell compartment of the E7 transgenic lens only (compare C and D) and the concomitant appearance of TUNEL positive cells throughout the fiber cell compartment, indicative of apoptosis (compare E and F). e, lens epithelial cells; f, lens fiber cells; r, retina; black arrows (C and D), brown-staining, BrdU-positive nuclei; white arrow (F), green, TUNEL-positive apoptotic nuclei in fiber cell compartment. In all panels, the anterior of the lens is at the top. (Figure 8.7 in the text.)

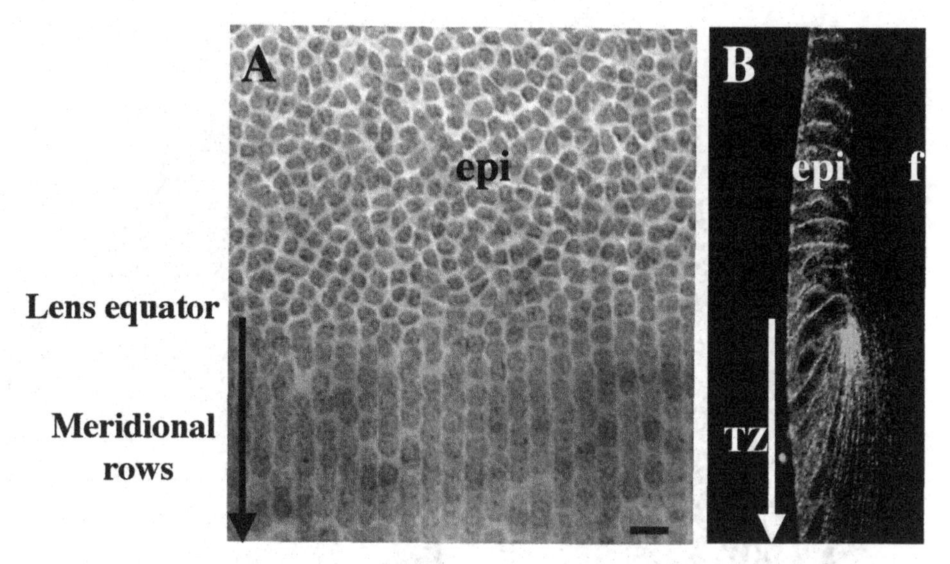

Plate XVI. A wholemount and a histological section of the lens epithelium of a weanling rat. (Figure 11.5 in the text.)

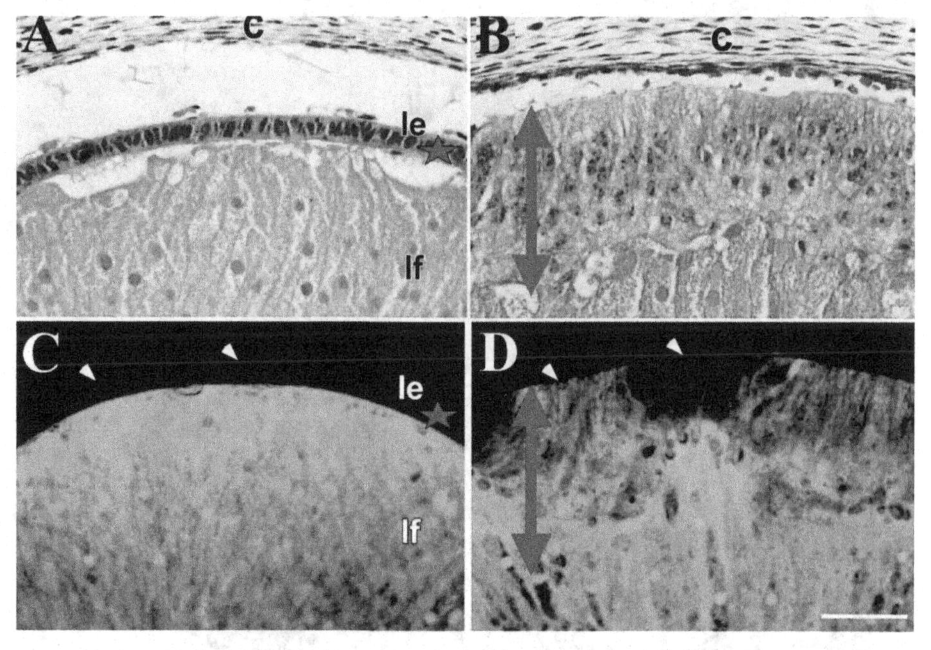

Plate XVII. Induction of fiber differentiation by FGF-4 in transgenic mice. (Figure 11.13 in the text.)

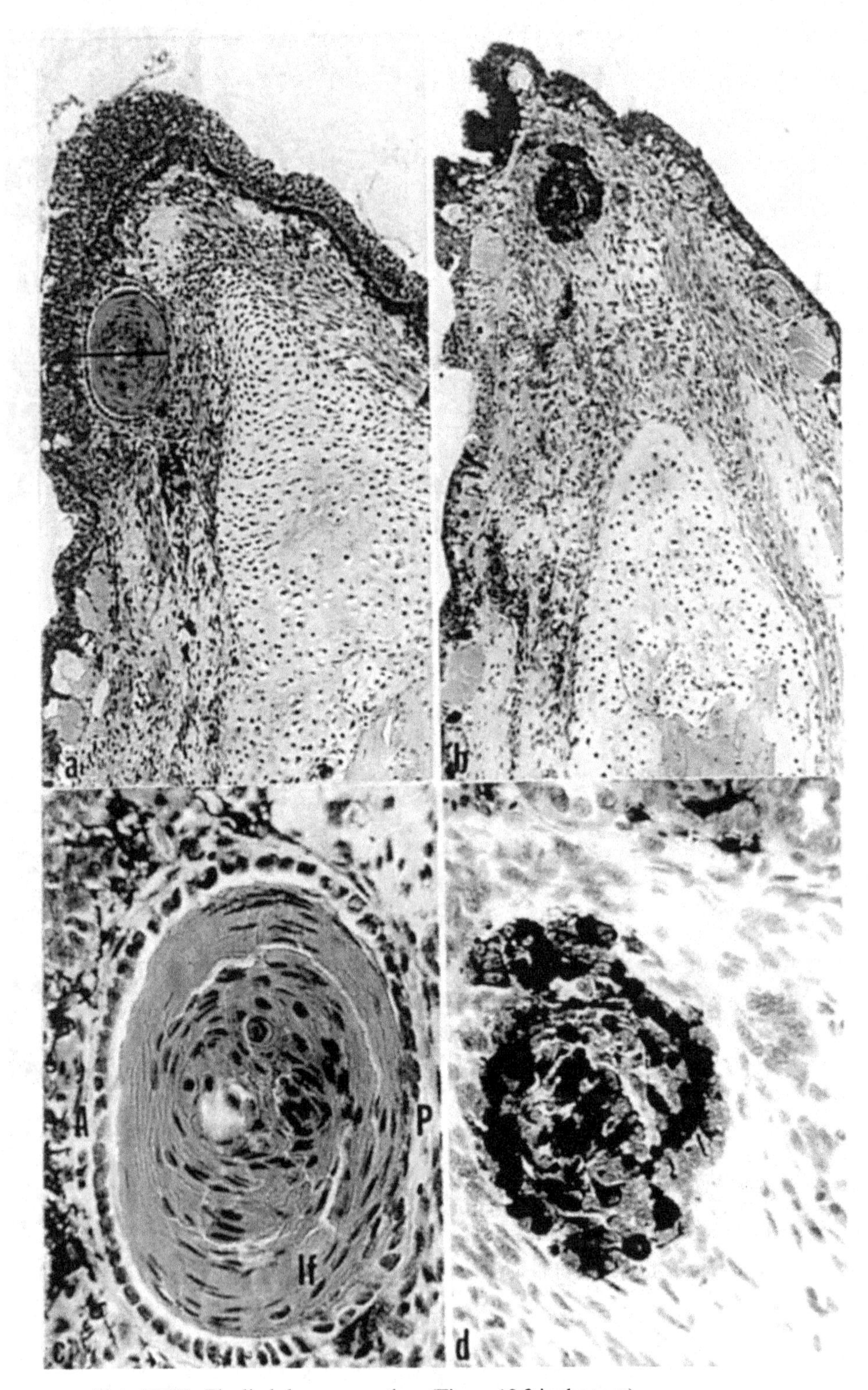

Plate XVIII. The limb-lens connection. (Figure 12.2 in the text.)

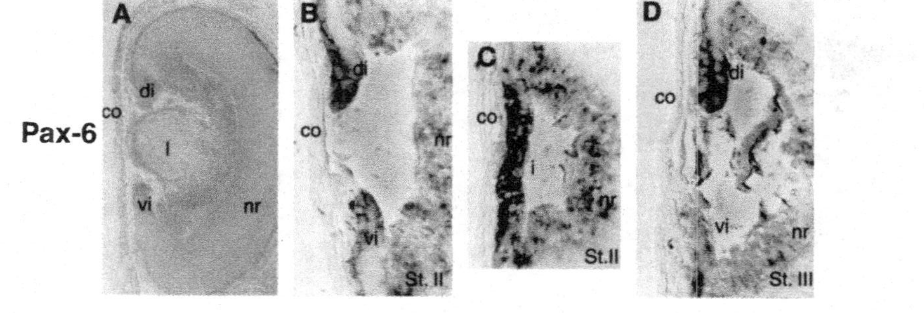

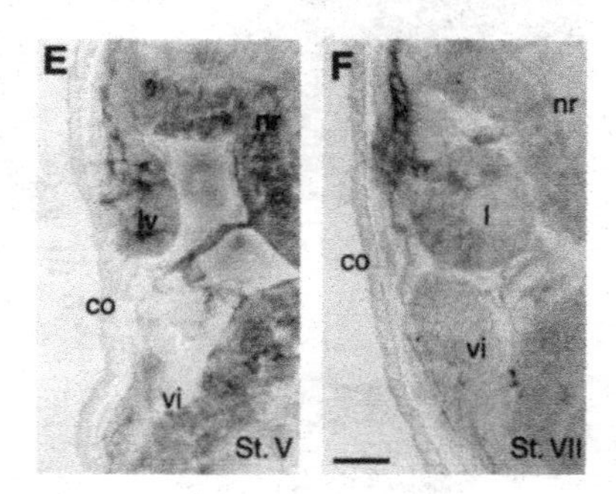

Plate XIX. Pax6 expression via *in situ* hybridization during the process of lens regeneration. (Figure 12.3 in the text.)

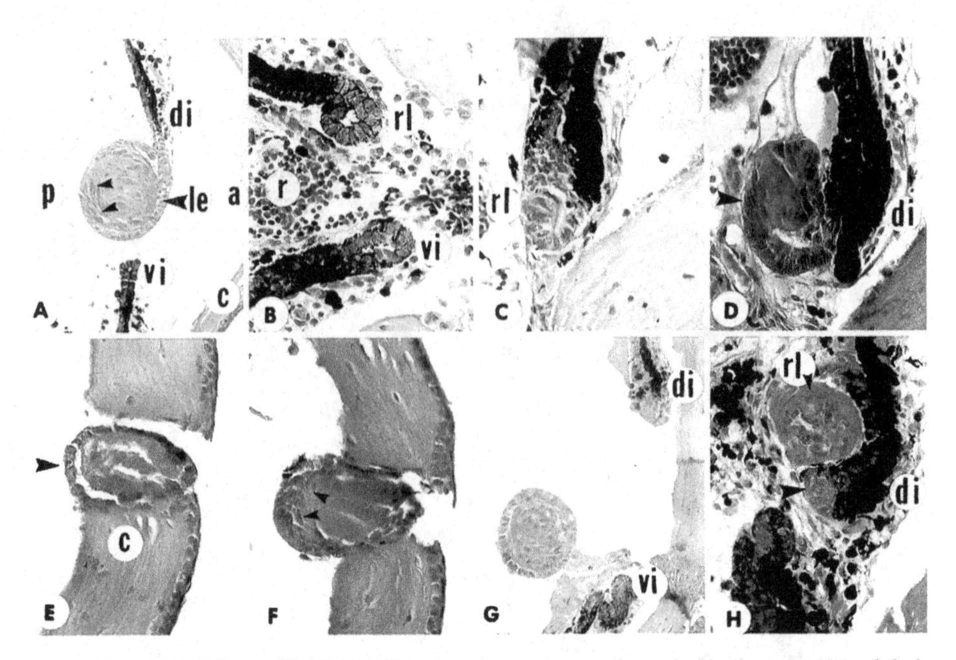

Plate XX. Effects of inhibiting RA signaling on lens regeneration in the newt *Notophthalmus viridecens*. (Figure 12.7 in the text.)

Table 8.1. *Cyclin-Dependent Kinases and Their Cyclin Partners*

Cyclin-dependent kinases	Cyclin partner	Place required in cell cycle
Cdc2 (Cdk1)	Cyclin A, B	G2, M
Cdk2	Cyclin E, A	Late G1/early S (E); S (A)
Cdk4	Cyclin D1, D2, D3	G1
Cdk6	Cyclin D1, D2, D3	G1

cell cycle (Baldin et al., 1993; Draetta, 1994; Dulic et al., 1992; Hunter and Pines, 1994; Koff et al., 1992; Matsushime et al., 1991; Meyerson et al., 1992; Sherr, 1994), and overexpression of cyclin Dl or cyclin E shortens G1, accelerating entry into the S phase (Ohtsubo et al., 1995; Quelle et al., 1993; Resnitsky et al., 1995; Resnitsky and Reed, 1995). D-type cyclins associate with Cdk4 and Cdk6 kinases, while cyclin E associates with Cdk2 (Sherr, 1993). In addition, a close homolog of Cdk2, Cdk3, is also thought to play a unique role in the G1-S transition (van den Heuvel and Harlow, 1993). However, Cdk3 is not essential in mice, as several laboratory strains encode a truncated form of the protein (Ye et al., 2001). Cyclin A binds Cdk2 and Cdc2 (Cdk1), another cyclin-dependant kinase that has a primary function in G2 and mitosis and is required for both the S phase and the G2-M transition (Girard et al., 1991; Pagano et al., 1992; Zindy et al., 1992), while cyclin B–Cdc2 complexes appear to be specific for control of mitotic entry.

What are the critical substrates for the cyclin-Cdk complexes in regulating cell cycle progression? One family of target substrates for these regulators is the retinoblastoma family of pocket proteins, pRb, p107, and p130 (Mulligan and Jacks, 1998; Nevins, 1998). The founding member of this family is pRb, the product of the retinoblastoma tumor susceptibility gene, whose mutational inactivation is the hallmark of the childhood retinal tumor retinoblastoma (Cavenee et al., 1983; Dryja et al., 1984; Friend et al., 1986; Fung et al., 1987; Lee et al., 1987). Since the cloning of *RB* in 1986 (Friend et al., 1986; Lee et al., 1987), there has been much effort to understand its function. These studies have led to the concept that pRb, the gene product of *RB*, is the master brake of the cell cycle, the molecular factor controlling the restriction point (Lundberg and Weinberg, 1999) in G1 phase of the cell cycle.

In cycling cells, pRb is unphosphorylated at the conclusion of mitosis. As the cell progresses through G1 (Fig. 8.2), pRb becomes phosphorylated by D-type cyclins in association with Cdk4 or 6 (Ewen et al., 1993; Hinds et al., 1992; Kato et al., 1993; Sherr, 1994; Weinberg, 1995). In this hypophosphorylated state, pRb binds to members of the E2F transcription factor family (preferentially E2Fs 1–3), masking the transactivation domain of the E2F but not interfering with the ability of E2F to bind to its DP partner (Dyson, 1998; Mulligan and Jacks, 1998; Nevins, 1998). Thus, when bound to E2Fs, the pRb-E2F-DP complex prevents the expression of target genes (Fig. 8.3). These E2F target genes encode enzymes for DNA synthesis and other cell cycle regulators such as cyclin E, DHFR (dihydrofolate reductase), cdc2, B-myb, c-myc, and N-myc (Dyson, 1998; Mulligan and Jacks, 1998; Nevins, 1998). Recent evidence has, however, modified the classical view that pRb suppresses E2F target genes only through masking E2F's activation domain (Dyson, 1998). It is recognized now that pRb also actively represses promoters of E2F target genes via its ability to recruit histone deacetylase (HDAC; Fig. 8.3; Brehm et al., 1998; Luo et al., 1998; Magnaghi-Jaulin et al., 1998). Histone deacetylation is thought to facilitate the formation of nucleosomes and therefore hinder access to promoters by transcription factors (Grunstein, 1997).

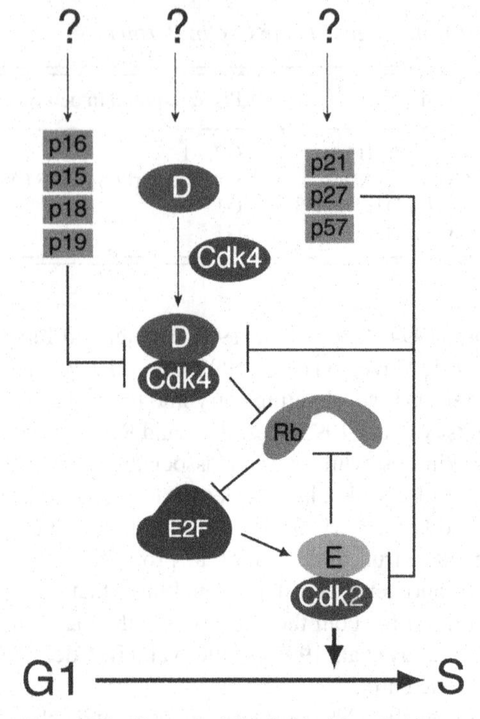

Figure 8.2. Control of G1-S transition. D-type cyclins, thought to be induced by mitogenic signals, complex with Cdk4 (or Cdk6) to carry out initial phosphorylation of pRb, leading to derepression of E2F. E2F activates the expression of cyclin E, which in turn activates Cdk2. Cdk2–cyclin E kinase further phosphorylates (hence inactivates) pRb, resulting in even more cyclin E expression. The activity of cyclin E–Cdk2 is necessary to drive the cell into the S phase. Counteracting the cyclins are two classes of cyclin-dependent kinase inhibitors, which are thought to be induced by negative growth signals. Induction of p15, p16, p18, p19, p21, or p57 will prevent phosphorylation of pRb by Cdks and keep the cell in G1. The signals that induce the cyclins or CKIs are yet to be identified.

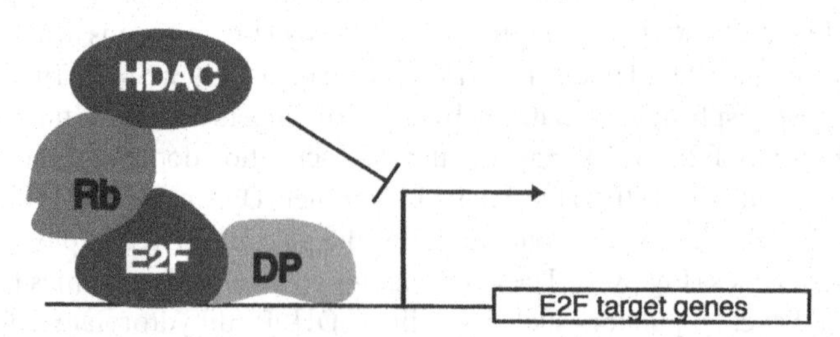

Figure 8.3. Suppression of the transcriptional activation activity of E2F by pRb. Binding of pRb to the E2F-DP complex inhibits transcription of E2F targets by masking E2F's transcriptional activation domain. Additionally, pRb can recruit HDAC (histone deacetylase) to the complex, which modifies local chromosomal structure in such a way that transcription is hindered.

As cells continue to progress through G1, pRb becomes further phosphorylated by cyclin-Cdk complexes (Fig. 8.2). This hyperphosphorylation of pRb leads to the dissociation of pRb from the pRb-E2F-DP complex, thus permitting either the derepression or activation of E2F targets (Dyson, 1998; Mulligan and Jacks, 1998; Nevins, 1998). Clearly, the hyperphosphorylation of pRb at the G1-S transition is important for that transition, and cyclin E is important in facilitating this event. However, there must be other roles for cyclin E in the G1-S transition because cyclin E is required for this transition in *Rb* null cells (Lukas et al., 1997; Ohtsubo et al., 1995). It is known that cyclin E is required for centrosome duplication (Hinchcliffe et al., 1999; Lacy et al., 1999; Matsumoto et al., 1999) and is involved in the up-regulation of histone transcription via phosphorylation of NPAT, nuclear protein mapped to the AT locus (Ma et al., 2000; Zhao et al., 2000). Thus, the evidence suggests that cyclin E–Cdk2 is the primary kinase involved in the G1-S transition (Dulic et al., 1992; Koff et al., 1992; Ohtsubo et al., 1995; Pagano et al., 1993; van den Heuvel and Harlow, 1993).

The roles of the pRb family members p107 and p130 are less well understood. The p130 protein is thought to play its role primarily in quiescent cells (i.e., cells that are in G0), whereas the p107 protein is thought to play its role primarily in the S phase (Dyson, 1998; Mulligan and Jacks, 1998). The list of targets for the pocket proteins is lengthy. One important family of targets, as mentioned above, is the E2F family of transcription factors, of which there are six members currently identified (Dyson, 1998; Nevins, 1998). Extensive *in vitro* analysis has led to the view that pRb preferentially binds to and regulates the activity primarily of E2Fs 1–3, p107 preferentially regulates E2F4, and p130 regulates E2Fs 4 and 5 (Bagchi et al., 1991; Cao et al., 1992; Chellappan et al., 1991; Chittenden et al., 1991; Devoto et al., 1992; Sardet et al., 1995; Shirodkar et al., 1992). The activities of pRb, p107, and p130 are regulated by their phosphorylation state, which in turn is regulated by cyclin-Cdk complexes in a cell cycle–dependent fashion.

In the S phase, cyclin A–Cdk2 is the primary cyclin complex, and it is thought to promote DNA synthesis by facilitating origin firing through phosphorylating protein substrates in the origin of replication complexes (Stillman, 1996). Inhibition of cyclin A function prevents DNA synthesis in cultured cells (Pagano et al., 1992). Inhibition of cyclin A's function through microinjection of antibodies into cultured human fibroblasts also resulted in inhibition of entry to mitosis (Pagano et al., 1992). Given cyclin A's function in DNA replication and entry to mitosis, it is not unexpected that mouse embryos deficient in cyclin A2 (an ubiquitously expressed form of cyclin A) die early in embryogenesis (Murphy et al., 1997). Deletion of cyclin A1 (germ cell–specific cyclin A) causes a block of spermatogenesis before the first meiotic division (Liu et al., 1998). In addition, cyclin A–Cdk2 inactivates E2F1 (and presumably E2F2 and 3 as well) in the S phase through phosphorylating their heterodimeric partner DP-1, therefore preventing E2Fs from binding DNA (Dynlacht et al., 1994; Krek et al., 1995). This function of cyclin A–Cdk2 cannot be accomplished by cyclin E–Cdk2 kinase activity (Dynlacht et al., 1994), and it is important for orderly S phase progression, as persistent E2F1 activity delays or arrests the S phase and leads to regrowth or apoptosis.

Cyclin B starts to accumulate in the S phase but is sequestered in the cytoplasm throughout S and G2 (Bailly et al., 1992; Ookata et al., 1992; Pines and Hunter, 1991). At the onset of mitosis, cyclin B is translocated into the nucleus, where it binds cdc2 and eventually leads to its activation. Activation of cdc2 marks the beginning of mitosis. Thus, cyclin B is called the mitotic cyclin and cdc2 the mitotic Cdk. Cyclin B–cdc2 is historically known as maturation-promoting factor (MPF), and it induces the maturation of *Xenopus* oocytes (Lohka et al., 1988; Masui and Markert, 1971; Newport and Kirschner, 1984). To prevent premature mitosis, cyclin B–cdc2 complex is held inactive by inhibitory phosphorylation on Thr-14 and Tyr-15 residues of cdc2 protein, which are abruptly removed by phosphatase cdc25 at the end of G2 (Kumagai and Dunphy, 1992). The activity of cdc2 triggers entry into prophase and eventually results in the assembly of spindles and metaphase plate in uncharacterized ways. To exit mitosis, cyclin B must be destroyed through an ubiquitin-mediated proteolysis pathway (see below). Ubiquitination of cyclin B is accomplished by the anaphase-promoting complex (APC; King et al., 1994). Destruction of cyclin B is important in reestablishing the G1 state in daughter cells.

A large body of evidence suggests that the critical functions of D-type cyclins in cell cycle control are to inactivate pRb by activating Cdk4 or 6 and to sequester p27^{KIP1} (Sherr, 1996), leading to activation of cyclin E. Indeed, when cells enter the cell cycle after growth factor exposure, D-type cyclins are expressed earlier than cyclin E and are expressed differentially in different cell types (Matsushime et al., 1991). Inhibition of the function of D-type cyclins has minimal effect on cell cycle progression in cells lacking pRb (Lukas et al., 1995; Medema et al., 1995). With cyclin E downstream of D-type cyclins, D-type cyclins become the sensors linking mitogenic stimuli and the cell cycle machinery (Pines, 1995; Sherr and Roberts, 1995). The intracellular signaling pathways that regulate D-type cyclin expression are multiple, mostly involving Ras and Myc (Leone et al., 1997).

Acting in opposition to cyclins are the cyclin-dependent kinase inhibitors (CKIs). Currently, two classes of CKIs have been identified in mammals, the p21^{CIP1} and p16^{INK4} families, which differ in structure, mechanism of inhibition, and specificity. The p16^{INK4} family (*in*hibitor of Cd*k4*) consists of ankyrin repeat proteins, including p15, p16, p18, and p19 (Harper and Elledge, 1996), that bind to and inhibit Cdk4 and Cdk6 kinases. The p21^{CIP1} family of CKIs, including p21^{CIP1} (p21), p27^{KIP1} (p27), and p57^{KIP2} (p57), inhibit all Cdks involved in the G1-S transition. The fact that there are so many genes encoding Cdk inhibitors reflects the challenge an organism faces in putting a brake on proliferation during development and tumorigenesis. Compared with cyclins, however, little is known about factors or extracellular stimuli that would induce expression of these negative cell cycle regulators.

Many key cell cycle regulators are also subject to ubiquitin-mediated proteolysis, making the processes they regulate irreversible. Ubiquitin-mediated proteolysis is a major protein degradation pathway in eukaryotic cells (Ciechanover et al., 2000). Proteins destined for degradation are tagged with the highly conserved protein ubiquitin. Multiple units of ubiquitin are added onto a protein substrate, producing polyubiquitin chains, which are recognized by the 26S proteasome – the protease that destroys the ubiquitin-tagged proteins (Voges et al., 1999). Polyubiquitination is the rate-limiting step in this process, which involves a cascade of ubiquitin transfer reactions requiring three components. First, an ubiquitin-activating enzyme (El) activates ubiquitin, that is transferred to the second component, the ubiquitin-conjugating enzyme (E2). The third component, E3, acts as an adapter between the E2 and the substrate by contacting both of them simultaneously. Specificity (as to which substrate to ubiquitinate) is largely provided by E3. Therefore, E3 is called the ubiquitin ligase. There are three major E3 classes: the HECT-domain proteins, the anaphase-promoting complex (APC), and the SKP1/CUL1/F-box (SCF) ubiquitin ligases (Harper and Elledge, 1999). In SCF-mediated ubiquitination, an F-box protein targets a substrate to the ligase. There are many different F-box proteins, reflecting the need of a cell to degrade a variety of proteins. F-box proteins usually recognize substrates that have been phosphorylated. Thus, phosphorylation provides a signal for degradation (Harper and Elledge, 1999; Montagnoli et al., 1999). A number of cell cycle regulators are degraded through the SCF pathway, including cyclins D and E, E2Fl, and p27, in which a F-box protein called SKP2 is required (Montagnoli et al., 1999; Nakayama et al., 2000). The C-terminal QT domain of p27 contains a critical threonine residue, which when phosphorylated by Cdk2–cyclin E leads to p27 degradation (Montagnoli et al., 1999).

Thus, at present considerable detail on the mechanisms of cell cycle regulation has been gleaned from *in vitro* studies. This body of knowledge has guided studies to determine how the cell cycle is controlled at the molecular level in the lens. After briefly reviewing the classical work on the patterns of cell proliferation in the developing lens, this chapter discusses how mechanisms of cell cycle control in the lens are similar to or different from the paradigm set forth by these *in vitro* analyses.

8.3. Cellular Proliferation in the Lens

This section reviews some of the seminal findings that have elucidated the patterns of cell proliferation in the developing vertebrate lens.

By day E11.0–E11.5 in mouse embryogenesis, the developing lens has entered the lens vesicle stage. Although work has demonstrated that all cells within this vesicle have the same developmental potential, the position of any such cell with respect to the developing optic cup dictates its fate in subsequent cellular differentiation stages of lens development. Cells in the posterior, in close proximity to the optic cup, are destined to become postmitotic and differentiate into the primary fiber cells, whereas cells in the anterior, directly opposite the optic cup, are destined to retain their epithelial character and proliferative capacity. After the initial differentiation of the primary fiber cells that fill the lens vesicle, the continued growth of the lens depends on the proliferation of lens epithelial cells and their subsequent differentiation into fiber cells. Because other chapters in this text discuss at length the early embryonic development of the lens (chaps. 1 and 2) and the nature of the environment that influences cells in the posterior as compared with those in the anterior (chaps. 2 and 11), this chapter focuses on the cell cycle machinery that permits the lens to maintain distinct populations of cells with differing proliferative characteristics from the lens vesicle stage forward.

One of the reasons that the lens is such an ideal *in vivo* model system for studying cell cycle control is that the program of differentiation is very well characterized at the histological level. This histological characterization includes the mapping of regions of high and low proliferative activity in the lens as a function of developmental age and placing this information in the context of cellular morphology and patterns of crystallin gene expression (Fig. 8.4). In the 1960s, a number of investigators, including Hanna and O'Brien (1961), Mikulicich and Young (1963), and Modak et al. (1968), used incorporation of tritiated thymidine into newly synthesized DNA and autoradiographic techniques to map the regions of the chick, rat, and mouse lens where cells are actively undergoing DNA synthesis. The measurements of DNA synthesis were combined with measurements of mitotic activity to determine cell cycle times for cells in different regions of the lens. These studies revealed that DNA synthesis and mitosis occur throughout the lens pit and the forming lens vesicle. However, as the vesicle stage proceeds to the beginnings of elongation of the primary fibers, regions of highest DNA synthesis become localized more anteriorly. Thus, DNA synthesis and mitosis become restricted to the anterior epithelium. As development proceeds in late embryos and especially postnatally, a distinct compartmentalization of regions with a high proliferative index and regions with substantially lower proliferative index arises within the epithelium. In young postnatal rats (birth to 35 days), the regions of highest proliferative

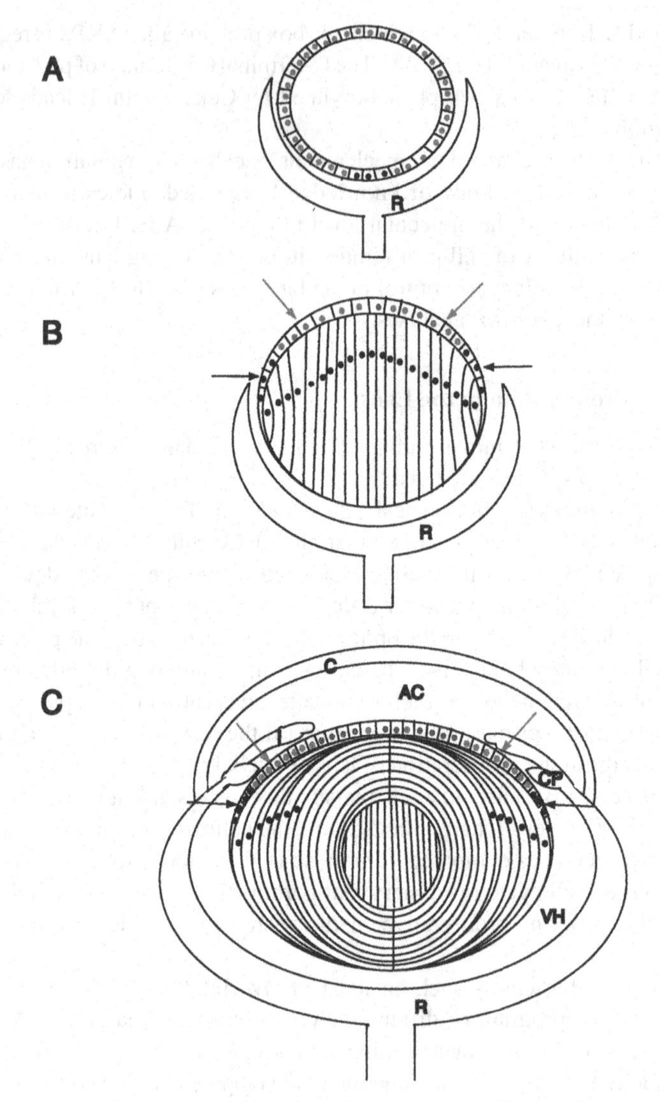

Figure 8.4. Schematic representation of regions of high and low proliferative activity in the mouse lens just after formation of the lens vesicle at day E11.0–11.5 (A), just after occlusion of the vesicle by elongating primary fiber cells at day E13.0–13.5 (B), and postnatally (C). During the early lens vesicle stage, virtually all cells in the lens vesicle have the same capacity to proliferate, but some in the extreme posterior may have already withdrawn from the cell cycle (A, dark nuclei). Soon thereafter proliferation becomes restricted to the undifferentiated anterior epithelium, as posterior cells withdraw from the cell cycle permanently and elongate to from the primary fibers (B, dark nuclei). As developmental age increases, the region of high proliferative activity becomes restricted to the proliferation or germinative zone (diagonal arrows), whereas withdrawal from the cell cycle occurs in newly differentiating secondary fiber cells in the transition zone (horizontal arrows). In the central epithelium, cells, although undifferentiated, become quiescent, the proportion of these increasing with developmental age. AC, anterior chamber; C, cornea; CP, ciliary process; I, iris; R, retina; VH, vitreous humor; diagonal arrows, cells in the germinative zone; horizontal arrows, cells in the elongation zone.

index become focused in narrow bands of approximately 50 cells located just anterior to the equator, a region known as the proliferation or germinative zone. The estimated cell cycle time is shortest in this region of high proliferative index and much longer in the central epithelium. With increasing age, cell cycle time increases throughout the epithelium.

In the 1970s, McAvoy (1978a) mapped the positions of high and low mitotic activity within the lens from embryonic and postnatal rats with respect to the aqueous and vitreous humors, cell morphology, and the expression pattern of crystallin genes. Consistent with earlier studies, he observed that, with increasing age of the animal, proliferating cells become progressively more localized to the germinative or proliferation zone and that these cells exhibit the shortest cell cycle time. McAvoy further established that α-crystallins first appear in the lens pit of the rat at day 12 in embryogenesis, β-crystallins are not detected until cells had stopped dividing and begun elongating, and γ-crystallins are not detected until even later in the elongation process. Much later, Zwaan determined that, in the mouse, α-crystallins first appear in the lens cup in actively cycling cells and do not lead to withdrawal from the cell cycle (Zwaan, 1983).

Many studies suggested that within the anterior of the eye, especially in the equatorial regions, proliferation factors were present that supported cell division in cells of the proliferation zone, whereas in the posterior of the eye, factors were present that supported elongation of cells in the elongation zone. Other studies compared the proliferative behavior of cells in the central epithelium to those in the proliferation zone. *In vivo*, the cycling cells in the central epithelium of the postnatal rodent become progressively fewer with increasing age. When cells from this region from 10- to 15-day-old rats are explanted into tissue culture and maintained in serum-free medium, there is a large boost in the number of proliferating cells. However, when cells from 3-day-old rats are placed in culture, typically there is very little boost in proliferation (McAvoy and McDonald, 1984). These results suggest that cells in the central epithelium withdraw from the cell cycle into a quiescent G0 phase and that the proportion of cells in quiescence increases greatly between 3 and 10 days of age. The results also suggest that the reduced capacity of cells in the anterior epithelium to proliferate may be due at least in part to diminishing environmental stimuli, such as the possible decrease in the availability of proliferation factors as these cells become further removed from their presumed source, the ciliary body. Alternatively, the factors inhibiting proliferation *in vivo* may be absent in this tissue culture system.

Collectively, these studies demonstrated that the pattern of growth and differentiation in the lens results in the compartmentalization of cells with differing proliferative capacities (Fig. 8.4). Cells in the central epithelium, which have a cuboidal morphology and lens epithelial gene expression patterns, exhibit substantial proliferative capacity in early embryos but lose this capacity *in vivo* as the animal ages. Cells in the proliferation zone, which have a cuboidal morphology and lens epithelial gene expression patterns, exhibit high mitotic activity. Cells in the transition zone (i.e., cells posterior to the lens equator) exhibit low mitotic activity and begin to show morphology and gene expression patterns consistent with early fiber cell differentiation. The cells that make up the bulk of the fiber cell compartment exhibit no mitotic activity, a highly elongated morphology and expression of fiber cell-specific markers such as β- and γ-crystallins. By mapping the mitotic activity within the lens, these studies, combined with the known morphology and pattern of activation of differentiation-specific crystallin gene expression, provided the reference for subsequent studies to address how at the molecular level cell proliferation and differentiation are regulated in the lens.

With this knowledge in hand, many investigators set out to identify growth factors that would stimulate lens epithelial cell proliferation *in vitro* using explants from chick and rat. The candidate proliferation factors included factors in embryo-derived serum (Hyatt and Beebe, 1993), EGF (Redden and Wilson-Dziedzic, 1983), PDGF (Brewitt and Clark, 1988), insulin-like growth factor (Redden and Dziedzic, 1982), and certain concentrations of fibroblast growth factors (McAvoy and Chamberlain, 1989). More recently, the activities of many of these factors in regulating lens epithelial cell growth *in vivo* have been examined. As the focus of this chapter is on the intracellular cell cycle machinery that regulates cell proliferation and differentiation, we direct your attention to chapter 11 for a full accounting of growth factor regulation of lens cell growth and differentiation.

8.4. Expression Patterns of Cell Cycle Regulatory Genes in the Developing Lens

As the first essential step toward understanding the mechanism of cell cycle regulation in the lens, many studies were performed to characterize expression patterns of cell cycle genes, primarily using the mouse and rat as model systems. In some instances, the analysis was carried out at the level of mRNA expression, while other studies sought to identify the proteins and their activities. From these efforts is emerging a good picture of the expression domains of the critical players. This section summarizes our current understanding of the expression patterns of cyclins, Cdks, CKIs, Rb family members, and E2Fs in the lens (see Table 8.2 for a summary).

In the developing mouse lens, cyclin A is expressed in the lens epithelial cells but not in the postmitotic fiber cells (Fromm and Overbeek, 1996; Hyde and Griep, 2002). Cyclin B is expressed in both epithelium and fiber cells (Gao et al., 1995). D-type cyclins are expressed throughout the lens, and the level of D2 is highest in the epithelium (Fromm and Overbeek, 1996; Gao and Zelenka, 1997; Geng et al., 1999; Zhang et al., 1998). Low levels of cyclin E message are detected (Lahoz et al., 1999). Although the localization of cyclin E within the lens has not been determined, it would not be unreasonable to predict that its expression is localized to the epithelium, as this is the compartment in which cycling cells are located. Several Cdks (Cdk2, Cdk4 and Cdc2) are detected at the mRNA and protein levels in the epithelium (Gao et al., 1995). Cdk4 is also present in the equatorial region, where cells stop dividing and start differentiation (Fromm and Overbeek, 1996). Cdc2 is found in the fiber cells (Gao et al., 1995; He et al., 1998). Cdk2-associated kinase activity is detected in epithelial cells but not in fiber cells (Gao et al., 1999). Other Cdks that are

Table 8.2. *Expression Domains of Cell Cycle Regulators in the Lens*

	Epithelium	Fibers
Cyclins	A,B,D,E*	B,D
Cdks	Cdk2, 4, Cdc2	Cdk4, Cdc2
CKIs	p57#, p16	p57, p27
Pocket proteins	pRb, p107, p130	pRb, p107
E2Fs	1, 2, 3, 4, 5	1, 3, 5

*Presumed to be epithelial specific due to very low expression level and function in cycling cells.
#Present in anterior epithelium at much lower level than in equatorial and transition zones.

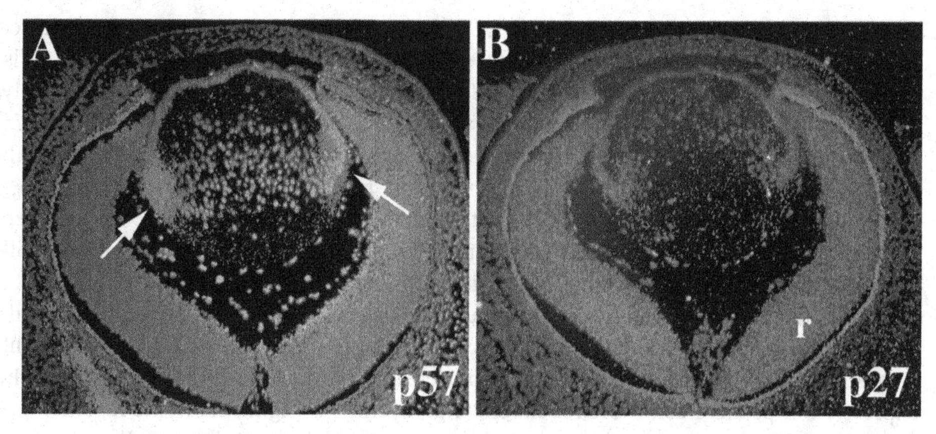

Figure 8.5. (See color plate XIII.) Expression of Cdk inhibitor genes during lens development. *In situ* hybridization was performed on transverse sections through the region of the eye of an E15.5 mouse embryo using antisense probes for *p57* (A) and *p27* (B). Arrows in A indicate cells in the equatorial zone of the lens, which express high levels of *p57*. *p27* is expressed at a low level throughout the lens and at a higher level in the retina (r).

not usually associated with cell cycle regulation, such as Cdk5, Cdk7, and Cdk8, are also expressed in the lens and presumably play different roles (Gao et al., 1999). The expression patterns of these positive cell cycle regulators are generally consistent with the fact that cell proliferation in the lens is restricted to the epithelial cells. However, *cyclin B* and *cdc2* are expressed in fiber cells, and furthermore cyclin B–cdc2 complexes with kinase activity have been detected in fiber cells at specific developmental time points. These data suggest that these factors, which are normally considered to be involved only in cell growth, may also play a role in cell differentiation (He et al., 1998).

Many CKIs are expressed in the lens. The expression patterns of these factors are consistent with a role in counteracting the positive effects of cyclin-Cdk complexes on cell cycle progression. The CKI p16 (Pan and Griep, unpublished data) is expressed in the epithelium. Two members of the p21 family, $p27^{KIP1}$ and $p57^{KIP2}$ (Zhang et al., 1998), are expressed in the lens in distinct but overlapping patterns (Fig. 8.5). The *p27* message seems to be present in all fiber cells in developing mouse lens. *p57* mRNA is found both in the epithelium and in fiber cells; however, interestingly, *p57* transcripts are most abundant in the equatorial region (Lovicu and McAvoy, 1999; Zhang et al., 1998). The founding member of this family of Cdk inhibitors, $p21^{CIP1}$, is not expressed in the lens at detectable levels (Zhang et al., 1998). The role of these CKIs presumably is to regulate the activity of various cyclin-Cdk complexes. Most notably, cyclin D–Cdk4 complexes would be expected to be inactivated in fiber cells. Indeed, Cdk4 activity is not found in rat fiber cells, and coimmunoprecipitation experiments have confirmed that p57 is bound to Cdk4 in these cells (Gao et al., 1999). Cdk4-p57 complexes are found in epithelial cells as well, suggesting a role for this CKI in maintaining normal cell cycling in the epithelium.

As noted, one important family of target proteins for cyclin-Cdk complexes is the pRb family. All three members of this family are expressed in the lens epithelium, and pRb and p107 are expressed in fiber cells (Rampalli et al., 1998; Pan and Griep, unpublished data). In the neonatal mouse lens, *Rb* transcripts are most abundant in the equatorial region (Pan and Griep, unpublished data), a pattern reminiscent of *p57* expression. Interestingly, p130 protein appears to be degraded through a ubiquitin-dependent process in fiber cells

(Rampalli et al., 1998). Members of the E2F family of transcription factors are expressed in the lens. In the epithelium, transcripts for *E2F*s 1–5 are found, while in the fiber cells, transcripts for *E2F*s 1, 3, and 5 are found (Pan, 1995; Rampalli et al., 1998). Presumably, one important role for the pocket proteins would be to regulate the activities of the E2F family members. This would imply that complexes of pocket proteins and E2F family members should be present. Indeed, complexes of E2Fs and pRb and p107 have been detected in lens epithelial and fiber cells, although the particular E2F member or members in these complexes have not been determined. Whether or not p130 is complexed with E2Fs in the epithelial cells also has not been determined.

How is the expression of cell cycle genes regulated in the developing lens? Certainly their expression must be controlled by the genetic program that directs lens development. At present, this genetic program is far from clear. Very little is known about how the expression of positive cell cycle regulators (e.g., cyclins) is regulated. Some regulators of CKI expression, which are transcription factors that regulate expression of cell cycle genes either directly or indirectly, may be involved. For example, the transcription factor Prox1 (see chap. 3) is required for the expression of both *p27* and *p57* in the lens (Wigle et al., 1999), and among other defects, lack of Prox1 leads to disrupted cell cycle regulation in the lens, similar to the disrupted regulation seen in mice lacking both *p27* and *p57* (Wigle et al., 1999). Whether Prox1 directly activates the transcription of these two CKIs or does so indirectly via its downstream target genes has yet to be determined. Another possible candidate is the c-*Maf* gene product (Ring et al., 2000; also see chap. 3). c-*Maf* deficiency results in lens phenotypes similar to the ones caused by *Prox1* inactivation, but the expression of *Prox1* in c-*Maf* null lenses is grossly normal, suggesting that c-*Maf* could be downstream of *Prox1*. How is expression of these transcription factors regulated? Presumably growth factor signaling pathways impact on their expression. This question is addressed in chapter 11.

With the comprehensive picture of the expression patterns of cell cycle factors in the lens, it is possible to formulate hypotheses as to how cell cycle control is achieved in the undifferentiated cells of the lens as well as in cells that are embarking on differentiation. The expression patterns suggest that when lens cells differentiate, factors such as pRb and p57 may play seminal roles. They further suggest that the components of cell cycle control set forth earlier in this chapter for undifferentiated proliferating populations of cells may be applicable, at least in part, to the control of cell proliferation in the lens epithelium. The application of technologies to manipulate gene function in the developing mouse lens, such as transgenic and knockout approaches, has provided avenues for testing these hypotheses.

8.5. Cell Cycle Regulation during Fiber Cell Differentiation

Differentiated lens fibers are postmitotic, highly elongated cells that lack subcellular organelles. Intuitively, it would seem that continued cell division would be incompatible with this terminal phenotype. Thus it would appear logical that cell cycle withdrawal and differentiation would be coupled mechanistically, at least at some level. How do lens fiber cells coordinate cell cycle withdrawal and differentiation? The first factors whose requirements for differentiation were tested specifically in this process were pRb and p57, in accordance with the data reviewed above (see sections 8.2 and 8.4). The experimental strategies used to address these hypotheses involved transgenic and gene disruption approaches. This section describes the results of genetic analysis of cell cycle gene function in lens fiber cell differentiation.

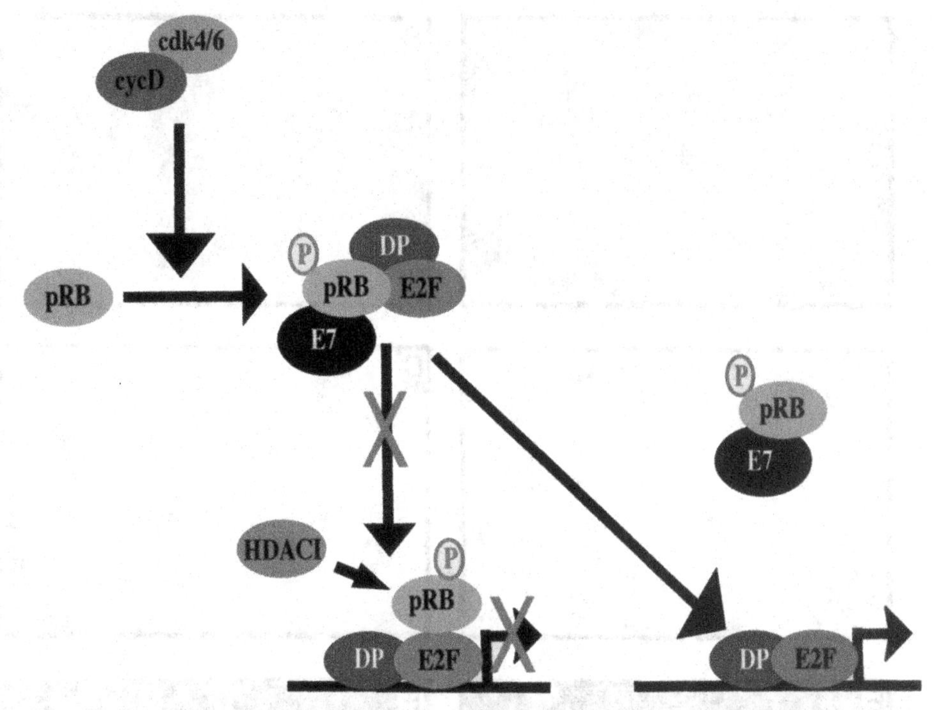

Figure 8.6. (See color plate XIV.) Inactivation of pRb by HPV-16 E7. Initial phosphorylation of pRb by cyclin D–cdk4/6 complexes normally generates functional, hypophosphorylated pRb that is able to bind to E2F and create a complex capable of repressing transcription of E2F target genes. The E7 oncoprotein from HPV-16 can bind to pRb, thus sequestering pRb from E2F. This leaves free E2F-DP complexes able to activate transcription of E2F targets.

The knowledge that pRb is an important regulator of the cell cycle *in vitro* prompted investigators to ask if pRb is required for cell cycle withdrawal during fiber cell differentiation. To this end, transgenic mice expressing the E7 oncoprotein from human papillomavirus type 16 (Pan and Griep, 1994) or a C-terminally truncated large T antigen from SV40 (Fromm et al., 1994) under the direction of the murine αA-crystallin promoter were generated. The αA-crystallin promoter would direct expression of linked genes to the lens cells at the time of their initial differentiation. As shown in Figure 8.6, the E7 oncoprotein has the capacity to bind to and inactivate the hypophosphorylated form of pRb (Dyson et al., 1989; Munger et al., 1989). The C-terminally truncated SV40 large T antigen similarly binds to pRb (DeCaprio et al., 1988). Mice expressing either oncoprotein display profound defects in the differentiation of lens fiber cells (Figure 8.7). Most notably, cells in the transition zone fail to withdraw from the cell cycle and fail to elongate. Instead, these abnormal cells showed a high propensity for apoptosis (Fromm et al., 1994; Pan and Griep, 1994). These defects likely are due to inactivation of pRb (or pRb family function) given that mutations in E7 that abrogate its ability to bind pRb also abrogate the phenotype (Pan and Griep, 1994). Consistent with this interpretation, germline mutation of *Rb* leads to a similar lens phenotype (Morgenbesser et al., 1994). Although activation of the expression of β- and γ-crystallins does occur in these models, immunohistochemistry suggests that the levels of these proteins are abnormally low (Morgenbesser et al., 1994). More recent evidence

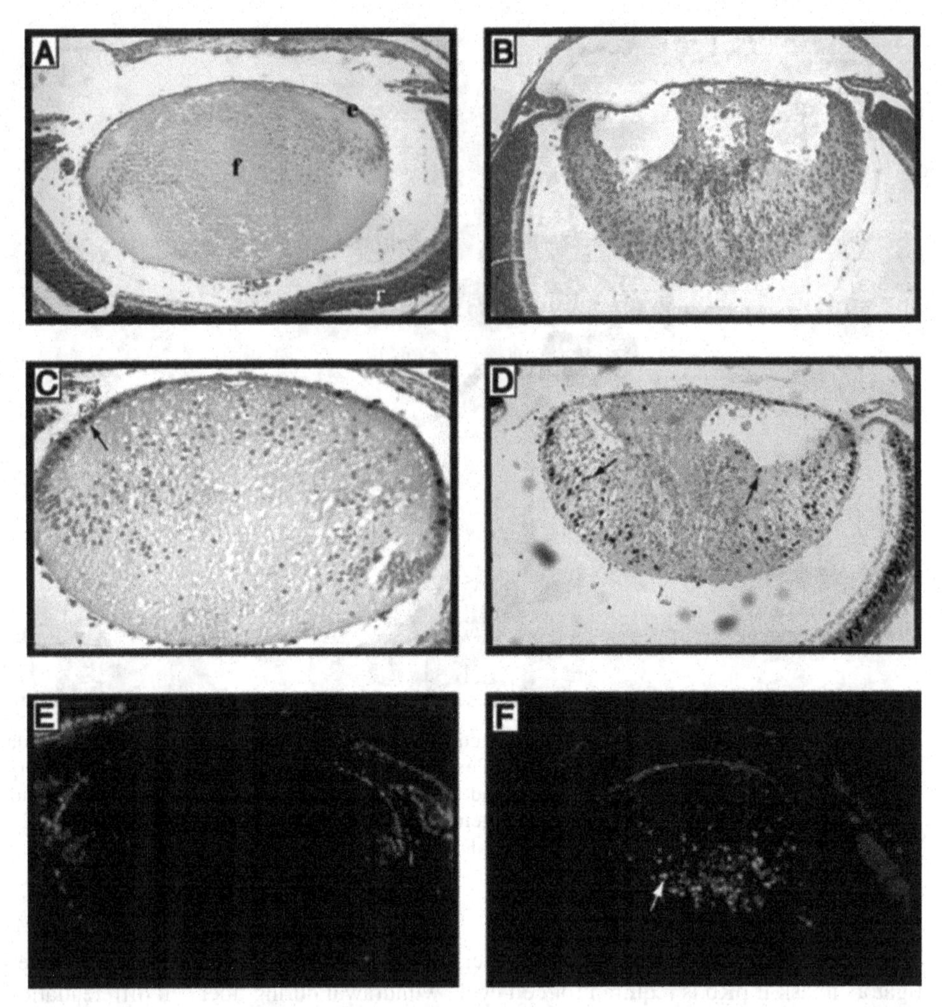

Figure 8.7. (See color plate XV.) Impact of HPV-16 E7 expression on lens fiber cell differentiation in transgenic mice. Paraffin-embedded sections from eyes of neonatal nontransgenic (A, C, E) or E7 transgenic (B, D, F) mice were subjected to histological analysis using hematoxylin and eosin staining (A, B), analysis of DNA synthesis using BrdU incorporation (C, D), or analysis of apoptosis (E, F) using the TUNEL assay. BrdU incorporation was detected immmunohistochemically with diaminobenzidine, and the section was counterstained with hematoxylin, resulting in brown-stained positive nuclei and purple-stained negative nuclei. TUNEL-labeled nuclei were identified using an FITC-conjugated secondary antibody, and sections were counterstained with propidium iodide, resulting in green-stained positive nuclei and red-stained negative nuclei when viewed using FICT and Texas Red filters, respectively. Note the small, disorganized vacuolated appearance of the fiber cell compartment of the E7 transgenic lens (compare A and B). Note the presence of BrdU positive nuclei, indicative of active DNA synthesis, within the fiber cell compartment of the E7 transgenic lens only (compare C and D) and the concomitant appearance of TUNEL positive cells throughout the fiber cell compartment, indicative of apoptosis (compare E and F). e, lens epithelial cells; f, lens fiber cells; r, retina; black arrows (C and D), brown-staining, BrdU-positive nuclei; white arrow (F), green, TUNEL-positive apoptotic nuclei in fiber cell compartment. In all panels, the anterior of the lens is at the top.

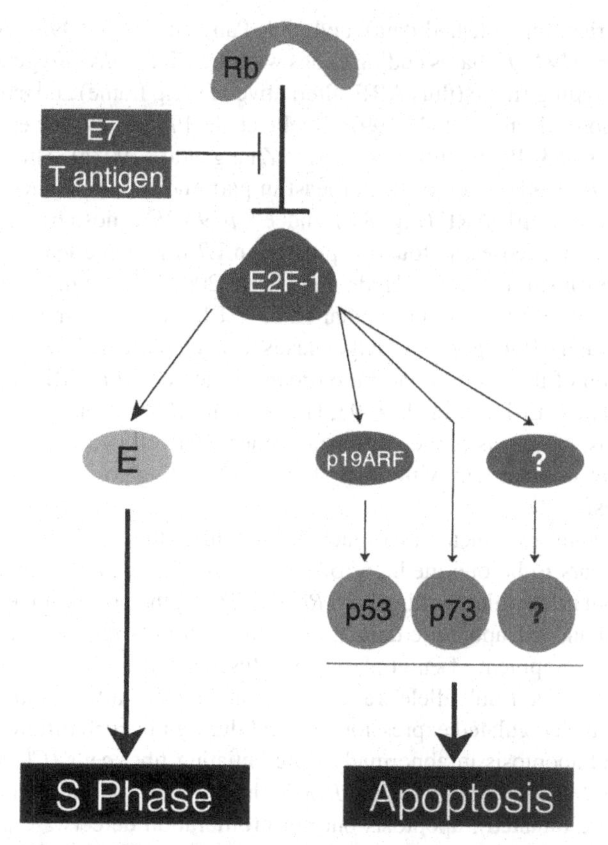

Figure 8.8. E2F-1 regulates both S phase entry and apoptosis. Once it is released from pRB suppression due to inactivation of pRB by Cdk phosphorylation or by binding of viral oncogene products such as E7 and T antigen, E2F-1 promotes S phase entry by activating cyclin E expression. Persistent activity of E2F-1 could lead to apoptosis through transcriptionally activating p19ARF, p73, or other unidentified factors.

suggests that there are reduced message levels for numerous β- and γ-crystallins in lenses from *Rb* null (Liu and Zacksenhaus, 2000) and E7 transgenic mice (Hyde, Potter, and Griep, unpublished data) as compared with nontransgenic control mice.

What is the molecular pathway through which *Rb*-deficient cells continue past the restriction point into the S phase and undergo apoptosis (see Fig. 8.8 for a summary)? As discussed in section 8.2, it is thought that a key role for pRb is to regulate the transcriptional activation activity of E2F transcription factor family members, leading to repression of E2F targets when pRb is bound to E2Fs. Cyclin E is known to be a key target of E2F1-dependent transcription activation. Thus, the prediction would be that inactivation of pRb results in continued expression of cyclin E in the abnormal lens fibers, which then drives G1-S progression, resulting in continued cell cycle progression. Indeed, cyclin E and a host of other cell cycle regulators whose expression is normally repressed during fiber cell differentiation continue to be expressed or are expressed in expanded patterns at the RNA level in the abnormal *Rb*-deficient fiber cells. These include *cyclins* and *Cdks* (Fromm and Overbeek, 1996), *E2Fs* 1–3 (McCaffrey et al., 1999; Hyde and Griep, 2002, and unpublished data),

Rb (H. Pan and A. E. Griep, unpublished data), and *p53* (Pan and Griep, 1995). Another target of E2F1 is *p19ARF. p19ARF* shares coding exons with the CKI *pl6INK4a* gene but is translated in a different reading frame (thus ARF, alternative reading frame) and expressed through a different promoter (Kamijo et al., 1997; Quelle et al., 1995). *p19ARF* encodes a polypeptide that stabilizes p53 (Pomerantz et al., 1998; Zhang et al., 1998a). Therefore, it is predicted that lack of *Rb* leads to apoptosis, at least in part due to stabilization of p53 by the increased expression of p19ARF (Fig. 8.8). Indeed, *p19ARF*, whose expression is undetectable in normal mouse lenses, is found expressed in E7 transgenic lenses, and its expression is entirely dependent on *E2F1* (Hyde and Griep, 2002). These data suggest a pathway that inactivation of pRb leads to up-regulation of E2F1, which in turn leads to the up-regulation of *cyclin E* and *p19ARF*. p19ARF then leads to an increase in the level of p53, which results in activation of the expression of the proapoptotic gene *Bax* (Miyashita and Reed, 1995) and the CKI *p21* (El-Diery et al., 1993; Harper et al., 1993; Xiong et al., 1993). In the context of the lens, apoptosis is the end result rather than cell cycle arrest. These data support a model that is consistent with other studies delineating common molecular pathways toward apoptosis.

Results from genetic analyses indicate that much but not all of this predicted model is correct. First, *E2F1* appears to be one mediator of the effects of loss of *Rb* function. In *Rb/E2F1* double knockout mice (mice lacking both *Rb* and *E2F1*), the proliferative defects are substantially rescued and the apoptotic defects are almost entirely rescued, at least at day 13.5 in embryonic development (Tsai et al., 1998). Rescue of the lens defects in the E7 transgenic mouse by the *E2F1* null allele was also substantial (McCaffrey et al., 1999). Consistent with these data, deregulated expression of *E2F1* during fiber cell differentiation leads to proliferation and apoptosis in abnormally differentiating fiber cells (Chen et al., 2000). Downstream of *E2F1* is *p19ARF*. In *Rb/pINK4a* double knockout mice (in which both *p16* and *p19ARF* are mutated), apoptosis but not proliferation defects are partially rescued (Pomerantz et al., 1998), although other studies document that the effect of a specific *p19ARF* knockout on apoptosis in the *Rb* null lens is minimal (Tsai et al., 2002). Downstream of *p19ARF* is *p53*. A *p53* null mutation rescues nearly completely the apoptotic defects of the *Rb* null mutation at day E13.5 (Morgenbesser et al., 1994). The same *p53* mutation provides near complete rescue of E7-induced apoptosis at day E13.5. The patterns and degree of rescue afforded by either the *E2F1* null allele or the *p53* null allele overlap, suggesting that these genetic factors lie in the same pathway. Finally, consistent with the up-regulation of *Bax* expression in E7 transgenic lenses, investigation of the effects of a *Bax* null mutation on E7-induced apoptosis has revealed that the *Bax* null mutation provides partial rescue to apoptosis in the same spatial regions as the *p53* null allele (Nguyen and Griep, unpublished data). Thus, initial analysis of the pathway regulated by *Rb* seems to support, at least in part, the existingmodels.

However, a more thorough analysis of mice phenotypes suggests that additional pathways may be activated by dysregulation of *Rb* function. For example, comparing the phenotypes of *Rb/E2F1* double null mice and *E7/E2F1* null mice shows that the rescue of E7-induced proliferation and apoptosis defects by the *E2F1* null mutation was not as great as the rescue of the *Rb* null–induced defects (Tsai et al., 1998; McCaffrey et al., 1999). More recent observations indicate that at least one of these other downstream targets is *E2F3* (Saavedra et al., 2002; Ziebold et al., 2001). Second, the *p19ARF* null mutation was at best only partially capable of rescuing the *Rb* null–induced defects, indicating that there must be other factors involved in regulating p53-dependent apoptosis. Third, while the *p53* null mutation nearly completely rescues *Rb* null–induced and E7-induced apoptosis at day E13.5, at later times in embryogenesis this allele only affords partial protection against E7-induced apoptosis. This result indicates that apoptosis occurs through both p53-dependent and p53-independent pathways that are temporally regulated during development (Pan and Griep, 1995). These *p53*-dependent and *p53*-independent pathways toward apoptosis are also spatially distinct (Pan and Griep, 1995). Interestingly, a second oncoprotein from HPV-16, E6, also was able to rescue E7-induced apoptosis when coexpressed with E7 in the lenses of transgenic mice. The rescue afforded by the expression of this oncoprotein was greater than the *p53* mutation alone, indicating that both p53-dependent and p53-independent apoptosis was blocked (Pan and Griep, 1994, 1995). The *p53*-dependent component of E6's activities is due presumably to E6's ability to target p53 for degradation via ubiquitin-dependent proteolysis (Huibregtse et al., 1991; Scheffner et al., 1990), and recent data support this hypothesis (M. M. Nguyen and A. E. Griep, unpublished data). Fourth, the model predicts that the rescue of apoptosis by the *E2F1* null mutation should be similar to that of the *p53* null mutation. However, the pattern of rescue by the *p53* null allele is more restricted than that of the *E2F1* null allele, suggesting that E2F1 affects both p53-dependent and p53-independent apoptosis (McCaffrey et al., 1999). Finally, the model predicts that rescue afforded by the *Bax* null mutation should be similar to that of the *p53* null mutation. However, rescue by the *Bax* null mutation, while similar, does not appear to be as complete (M. M. Nguyen and A. E. Griep, unpublished data). This suggests that *Bax* is a relevant downstream target of *p53* in mediating p53-dependent apoptosis in the lens; however, it is not the only one. Thus, although to the first approximation the data confirm that the predicted pathway is responsible for the proliferative and apoptotic defects of Rb inactivation, extensive analysis suggests that the story is not quite so simple.

There are several possible explanations for these apparent discrepancies between the actual data and the proposed model. First, the fact that E2F1 appears to affect apoptosis through pathways in addition to the p53 pathway comes from the recent work of Irwin et al. (2000). This group has shown that E2F1 can induce apoptosis through the *p73* (a homolog of *p53*) pathway (Irwin et al., 2000), which might also be involved in the apoptosis in the lens invoked by lack of *Rb* (Fig. 8.8). Second, in considering the extent to which the E7 transgenic lens system and the *Rb* null lens system are equivalent, it is important to recognize that E7 has the capacity to inactivate the pRb family members p107 and p130. At least, p107 is present in complexes with E2F in fiber cells (Rampalli et al., 1998). In the retina, it is clear that, in the absence of *Rb*, *p107* acts to partially suppress the effects of this mutation, as the *Rb/p107* null phenotype is more severe than the *Rb* null phenotype alone (Robanus-Maandag et al., 1998). By analogy, in the lens, inactivating both family members may lead to a more severe phenotype than inactivation of pRb alone. Alternatively, the differences between the effects of E7 and the *Rb* null mutation could be accounted for by differences in the ages of the embryos examined. Because the *Rb* null mutation leads to embryonic death shortly after day 13.5 in embryogenesis (Clarke et al., 1992; Jacks et al., 1992; Lee et al., 1992), it was not possible to determine the extent to which *E2F1* or *p53* null mutations would affect *Rb* null–induced proliferation and apoptosis at later time points. There is evidence that the *Rb* null mutation can lead to both p53-dependent and p53-independent apoptosis (Macleod et al., 1996). On the other hand, there is evidence to the contrary. When the early lethal effects of the *Rb* null mutation are rescued by expression of a *Rb* minigene, lens defects still persist. In this situation, the apoptosis in the lens appears to be entirely p53 dependent (Liu and Zacksenhaus, 2000). Further analysis will be required

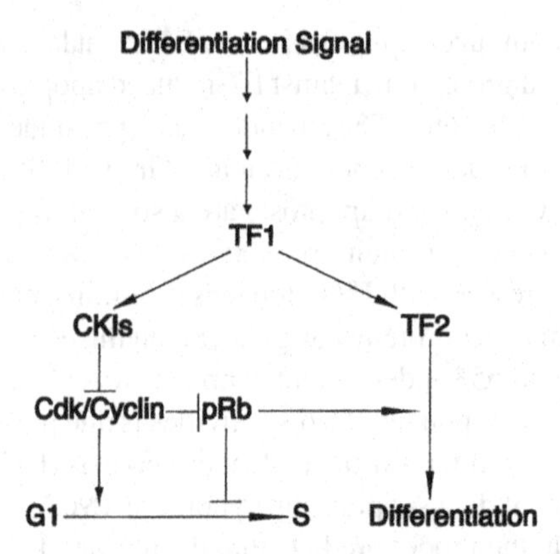

Figure 8.9. Cell cycle and differentiation. When cells receive differentiation signals, they activate expression of CKIs through the hypothesized transcription factor (TF1) to arrest the cell cycle. A second hypothesized transcription factor (TF2) activates the transcriptional program of differentiation, which may require the presence of active pRb. TF1 and TF2 could be the same transcription factor. In addition, it is possible that CKIs contribute directly to the differentiation process.

to determine if E7 is affecting p107 function and if this activity contributes to the overall phenotype.

In addition to defects in cell cycle regulation, *Rb* null, E7 transgenic, and SV40 truncT antigen transgenic mice show abnormal levels of differentiation specific marker proteins such as β- and γ-crystallins. This differentiation defect in *Rb* null lenses could be a result of deregulated cell cycle progression. In this model, the capacity to show morphological and biochemical traits of the differentiated fiber cell would be dependent on cell cycle withdrawal. In an alternative model, however, the effects of inactivating *Rb* on the expression of differentiation-specific genes could be attributable to the independent activity of pRb (Fig. 8.9). The latter alternative is possible given that pRb has been shown to interact with transcription factors that are critical regulators of differentiation. These factors include MyoD in skeletal muscle cells (Gu et al., 1993), c-Jun in keratinocytes (Nead et al., 1998), and CCAAT/enhancer-binding proteins (C/EBPs) in adipocytes (Chen et al., 1996). Additionally, this transcriptional coactivator function of pRb is separable from its ability to suppress E2F target genes (Sellers et al., 1998). It remains to be determined if pRb plays a direct role in the differentiation of lens fiber cells, and if so, by which means. However, it is known that pRb can bind to Pax6 *in vitro* and in embryonic chick lens (Cvekl et al., 1999), suggesting that it is possible that pRb-Pax6 interaction may be important for optimal Pax6-dependent transcriptional regulation. Alternatively, pRb could interact with c-Maf, whose absence in the lens is correlated with decreased mRNA levels of the differentiation-specific crystallins (Ring et al., 2000).

Given the fact that pRb is essential in lens development and is inactivated by G1 Cdks, what then keeps pRb in its hypophosphorylated active state once the lens fiber cells start to differentiate? The expression patterns of cell cycle genes indicate the presence of D-type cyclins in lens fiber cells, which could be balanced by the expression of a Cdk inhibitor. Indeed, if this balance is tipped off by forced expression of G1 cyclins and Cdks as transgenes in the lens, then overproliferation of lens fiber cells, presumably along with pRb inactivation, ensued (Lahoz et al., 1999). The effect of expression of these transgenes is dramatized if combined with the loss of the Cdk inhibitor $p57^{KIP2}$ (Lahoz et al., 1999). $p57^{KIP2}$ has been suggested to be the main effector that down-regulates Cdk activity to activate pRb, based on its high levels of expression in cells that are beginning to differentiate in the equator region of the lens (Lovicu and McAvoy, 1999; Matsuoka et al., 1995; Zhang et al., 1998). However, loss of $p57^{KIP2}$ in mice causes only a slight increase in the proliferation rates of these cells, much less severe than what has been observed in mice lacking *Rb*, suggesting that other mechanisms are complementing $p57^{KIP2}$ loss (Zhang et al., 1997). Analysis of mice doubly null for $p27^{KIP1}$ and $p57^{KIP2}$ indicates that $p27^{KIP1}$ plays a redundant role with $p57^{KIP2}$ in the development of the lens (Zhang et al., 1998). Loss of $p27^{KIP1}$ alone does not cause lens abnormalities, but it has a dramatic effect on the proliferation and differentiation of lens fiber cells when combined with $p57^{KIP2}$ deficiency (Zhang et al., 1998). The phenotypes of $p27^{KIP1}/p57^{KIP2}$ double null lenses bear strong similarities to those of *Rb* null lenses, indicating these two CKIs cooperatively activate pRb in the differentiation of lens fiber cells.

However, two significant differences exist between the phenotypes of the *Rb* null versus $p27^{KIP1}/p57^{KIP2}$ double null mutants. First, the extent of overproliferation as assessed by BrdU incorporation appears to be significantly greater in $p27^{KIP1}/p57^{KIP2}$ mutants than in *Rb* mutants. This may reflect the fact that these two CKIs function not only upstream of pRb by blocking Cyclin D–Cdk4 activity but also downstream of pRb by blocking cyclin E–Cdk2-mediated S-phase entry. Alternatively, the increase in Cdk activity due to CKI loss may result in inactivation of additional pRb family members such as p130 and p107, leading to a more severe proliferation defect than *Rb* loss alone. Thus, proliferation of lens fiber cells lacking pRb may be limited due to the action of p27 and p57 on Cdks other than Cdk4 and Cdk6. The second major difference is that the rates of apoptosis in CKI-deficient lenses are much lower than for *Rb*-deficient lenses and are similar to the rates seen in *Rb/p53* double mutant lenses. pRb is required to establish the transcriptional program that brings about differentiation of multiple cell types but has also been shown to inhibit apoptosis during myoblast differentiation and in other situations (Wang, 1997). Thus, low rates of apoptosis in $p27^{KIP1}/p57^{KIP2}$ double mutant lenses may reflect an antiapoptotic role for pRb. If the absence of $p27^{KIP1}$ and $p57^{KIP2}$ results in the inactivation of pRb to such an extent that it phenocopies the differentiation defect of *Rb* null mutant lenses, why the difference in apoptosis rates? There are several plausible explanations for this difference. First is that pRb could have an antiapoptotic function that is not regulated by Cdk phosphorylation and therefore would not be altered by CKI loss. Second is that, even in the absence of the CKIs, there may be residual pRb activity such that the apoptosis-inhibiting functions of pRb are largely intact. Even in the absence of CKIs, there is likely to be residual regulation of pRb if Cdk activity is still cyclical. In contrast, an *Rb* null cell would constitutively derepress all pRb-regulated genes such as *E2F1*, an apoptosis-inducing gene (Kowalik et al., 1995; Qin et al., 1994; Shan and Lee, 1994), and might display a more severe phenotype for this reason. Third, it is also possible that CKI mutant cells have higher Cdk activity levels, which prematurely inactivate E2F1 function (Dynlacht et al., 1994; Krek et al., 1994), thereby balancing the apoptotic-inducing consequences of inactivating pRb. The fact that the apoptosis rates of $p27^{KIP1}/p57^{KIP2}$ double mutants are similar to the rates observed in *Rb/p53* double mutant mice (Morgenbesser et al., 1994) is consistent with interfering with E2F1 function; apoptosis caused by *Rb* loss is partially mediated by *E2F1* (Tsai et al., 1998) and E2F1-mediated apoptosis is argued to be p53-dependent (DeGregori et al., 1997; Qin et al., 1994).

In sum, our knowledge of the mechanism through which cell cycle withdrawal is achieved during fiber cell differentiation has increased greatly over the past decade. This rapid expansion in our knowledge has resulted from *in vivo* genetic analysis of the function of components of the cell cycle machinery. Yet, despite our gains, many questions remain about the mechanism of cell cycle regulation in the lens and its coupling to differentiation.

8.6. Regulation of Proliferation in the Lens Epithelium

As discussed in section 8.3, the lens epithelium is a monolayer of cells that have proliferative capacity. Yet the cells within this layer differ in their proliferative characteristics, depending on their position relative to the ciliary process. In contrast to our rather extensive knowledge of how withdrawal from the cell cycle is achieved at the time of fiber cell differentiation, at present our knowledge of the mechanisms of cell cycle regulation in the epithelium is rather limited. We do, however, have a a basic understanding of the expression patterns and activities of many cell cycle genes, as discussed in section 8.4. The data indicate that in general the expected players are expressed and that the regulatory circuits are likely to be complex. As it was for the mechanisms of cell cycle withdrawal during differentiation, genetic studies are essential for determining the roles of specific factors or families of factors. Only recently, with the identification of molecular reagents that target the expression of transgenes to the epithelium, is progress beginning to be made on this important question. This section describes our current understanding, based on genetic studies, of cell cycle genes important for controlling cell proliferation in the epithelium.

Obvious candidates as regulators of the cell cycle in these cells are pRb and/or its family members p107 and p130. With some cells in the epithelium dividing only rarely, at least in postnatal animals, while others are actively cycling, it might be expected that the entire pocket protein family and all of its target E2F transcription factors would be important. Results from recent studies exploiting gene knockout and transgenic approaches that disrupt the function of pRb family members are consistent with this prediction. The first piece of evidence comes from analysis of the lens in mice rendered *Rb* deficient by gene-targeting strategies. Analysis of the lenses in these *Rb* null embryos failed to identify changes in the proliferative characteristics of the epithelial cells (Morgenbesser et al., 1994). However, because *Rb* null embryos die early in lens differentiation, it is possible that *Rb* is required in the epithelium at later stages of lens differentiation. Because of the strong desire of developmental biologists to study the role of *Rb* at later stages in embryogenesis, several groups have generated *Rb* null rescue chimeras or transgenic mice with *Rb* minigenes that permit analysis at later embryonic time points. Analysis of lens phenotypes in these animals also failed to identify alterations in epithelial cell proliferation characteristics that might suggest disruption of cell cycle control (Liegeois et al., 1996; Liu and Zacksenhaus, 2000; Maandag et al., 1994; Williams et al., 1994). Thus to date, a specific effect of Rb mutation in the lens epithelium has not been reported. Likewise, there appears to be no defects in the lens in *p107* or *p130* null mice (Cobrinik et al., 1996; Lee et al., 1996). It is possible that in the epithelium *Rb* family members may have redundant or compensatory functions that could be revealed by analysis of double or triple knockout mice. Due to early embryonic lethality, it was not possible to assess whether *Rb/p107* double null mice would have a phenotype in the lens epithelium (Lee et al., 1996). However, it was possible to examine lens phenotypes in *p107* null, *p130* null, and the double null mice. No phenotypes have been reported (Cobrinik et al., 1996), suggesting at least that the loss of function of these combinations of pocket proteins alone is not sufficient to dysregulate cell cycle control in the epithelium. Loss of all three *Rb* family members has recently been shown to be necessary for abrogating the G1 checkpoint in cultured mouse embryonic fibroblasts (Dannenberg et al., 2000; Sage et al., 2000), and pRb and p107/p130 appear to be required for regulation of different sets of genes (Hurford et al., 1997). Together, these data suggest that if the pocket proteins are essential regulators of lens epithelial cell growth, all of the pocket protein family members might be required.

If the pocket protein members are capable of providing redundant or compensatory functions in the epithelium, it is possible that transgenic expression in the epithelium of a DNA tumor virus oncoprotein that interferes with the function of all pocket proteins might be informative. It has been reported that transgenic mice in which the expression of HPV-16 E7 is regulated by the human keratin 14 develop cataracts (Herber et al., 1996). Recent analysis of the lens phenotype in these mice indicates that the transgenes are expressed postnatally in the epithelium and transition zone and that this expression is associated with increased numbers of proliferating cells throughout the epithelium (Nguyen et al., 2002). The capacity of E7 to induce high levels of proliferation in the lens epithelium in these mice is attributable to its capacity to regulate pocket protein activity, as mutations that abrogate E7's regulation of pRb and its family members also abrogate the lens phenotype (Nguyen et al., 2002). Thus, in contrast to the differentiating lens cell, where pRb function is necessary and sufficient for cell cycle withdrawal, it appears that all pRb family members may be required for maintaining cell cycle control in the epithelium.

Given the complex pattern of cell proliferation in the lens epithelium (see section 8.3), it is possible that multiple factors and/or pathways contribute to the regulation of cell cycle control in this layer. Interestingly, expression of the other HPV-16 oncogene, E6, in the lens epithelium in transgenic mice also leads to cataract formation (Song et al., 1999), and transgene expression was correlated with high levels of proliferation throughout the epithelium (Nguyen et al., 2002). The mechanism through which E6 induced high levels of proliferation appears to be its targeting of PDZ proteins such as the mouse homologs of *Drosophila* proteins Discs-large (DLG) and Scribble rather than its targeting of p53 (Song et al., 1999; Nguyen et al., 2002, Nguyen et al., 2003).

How is expression of cell cycle genes regulated in the lens epithelium? The answer is very much unknown at this point in time. It is known that certain growth factors such as PDGF stimulate the proliferation of cultured lens epithelial cells (Brewitt and Clark, 1988) and that overexpression of PDGF (Reneker and Overbeek, 1996) and IGF1 (Shirke et al., 2001) in transgenic mice lead to increased levels of proliferating cells in the epithelium. Also, preliminary investigations suggest that constitutive expression of activated H-*ras* in the epithelium can lead to heightened levels of proliferation in these cells (L. Reneker, personal communication). Given these data and the known connections between the ras-raf-MAPK pathway and the expression of cyclins (Lundberg and Weinberg, 1999), it would seem quite likely that a growth factor–mediated pathway would be operational in the lens. Initial studies are indeed now beginning to shed light on the important players in cell cycle control in the epithelium, but overall the mechanisms of cell cycle control in the epithelium remain to be elucidated.

8.7. Significance of Understanding Cell Cycle Control for Clinical Issues

The transparency of the lens depends on the formation and maintenance of its precise cellular structure. The formation of this structure in turn depends on the precise control of cell growth and differentiation throughout the life of the animal. When the lens structure breaks down, loss of transparency, or cataract, arises. As cataract is the leading cause of blindness worldwide, it is considered to be a major public health problem. The most common treatment for cataract today, surgical removal of the lens fiber mass while leaving the lens capsule and epithelium behind, leads to secondary cataracts in a high proportion of individuals. These secondary cataracts referred to as posterior subcapsular cataracts (PCOs), affect 50% of the population of cataract patients, children and adults alike. As the condition is caused by the proliferation and migration of the remaining lens epithelial cells posteriorly, with concomitant aberrant differentiation (Apple, 1992; Beebe, 1992; Kappelhof and Vrensen, 1992), understanding the factors that control cell proliferation is critical for our future ability to develop new strategies for effectively preventing, delaying, or treating cataract.

This chapter has focused on the many recent endeavors to understand the mechanism of cell cycle control, primarily through the analysis of animal models. Although the absolute changes in the rate at which the lens grows differ between humans and rodents, the general pattern of greater growth rate in embryos than in adults is conserved (Bron et al., 2000; van Heyningen, 1976). As discussed in this chapter and at length in other chapters in this text, it is thought that the proliferative and differentiative characteristics of the lens epithelial cells are influenced by their environment, but the exact nature of that environment is not understood. These uncertainties notwithstanding, the lens is most perfectly modeled *in vivo* by the intact lens itself. Therefore, the normal processes of growth and differentiation at various times in the lifespan of an animal, be it embryonic or postnatal, can provide models to elucidate how cataract may form. Animal models in which lens fibers are ablated but epithelial cells remain behind in an adult animal (Breitman et al., 1989; Landel et al., 1988; Pan and Griep, 1994) offer a way of mimicking the lens after cataract surgery. Using these models, it should be possible to learn more about the mechanisms that stimulate cell proliferation under abnormal conditions and compare them to how cell proliferation and differentiation are regulated during normal lens differentiation. From these comparisons may emerge new ideas for how to repress aberrant cell proliferation and its effects in the lens.

8.8. Key Questions for Future Investigation

As we have detailed in this chapter, a great deal has been learned about the mechanism through which the cell cycle is regulated in the lens, including withdrawal from the cell cycle at the time of fiber cell differentiation. Yet despite these advances, much remains to be learned. Among the many questions that remain to be answered about how cell cycle control is maintained, a few are appropriate to raise in the conclusion of this chapter. That lens cells are compartmentalized into domains with distinct proliferative activity has been known for decades. Yet very little is known at the mechanistic level about how this compartmentalization is achieved *in vivo*. This is especially true for the epithelium, which is subdivided into regions with distinct proliferative behavior that in addition is influenced by age. Ultimately, models of cell cycle control in the epithelium must account for how cells in the central epithelium are regulated differently than cells in the proliferation zone. What are the intracellular cell cycle components that are required and what is the function of each? What extracellular signals regulate the expression and activity of the cell cycle genes and through which intracellular signaling pathways are these signals mediated? And what are the direct targets of these factors? Although we have some initial clues as to some of these key regulators, much remains to determine if initial hypotheses are in fact correct.

Many of these same questions can also be asked of the differentiating fiber cells. Although we have a much clearer picture of what the key factors are for differentiating cells than for undifferentiated cells, many questions remain. What are the signal transduction pathways that regulate the expression and activity of key cell cycle genes? What are the relevant *in vivo* targets of key cell cycle factors such as pRb? What is the role of other pRb family members, if any, in lens differentiation? How are cell cycle withdrawal, fiber cell elongation, and differentiation-specific changes in gene expression coordinately regulated? What are the mechanisms through which aging affects cell cycle regulation? These are but a few of the many questions that need to answered if we are to understand how lens growth and differentiation are achieved and, ultimately, how these factors may impact cataract formation. The study of genetically manipulated animal models, in combination with advances in genomics and proteomics, should provide tools to answer these and many more questions about how the cell cycle is regulated in the lens.

9
Lens Fiber Differentiation

Steven Bassnett and David Beebe

9.1. Introduction

The lens consists of two morphologically distinct cell types, an unremarkable cuboidal epithelium that covers the anterior surface and concentric layers of fiber cells that account for the remainder, and vast majority, of the tissue volume (Fig. 9.1). The fiber cells are unique in the body. They have an enormous aspect ratio, being no more than a few micrometers wide but often exceeding a thousand micrometers in length. In cross-sectional profile, they appear as flattened hexagons, and their sharply angled membranes enclose a transparent cytoplasm that lacks the organelles found in typical cells. It is striking that these cells of remarkable shape and composition are derived from the more typical cells of the overlying epithelium.

In this chapter, we examine what is known (and, more often, what is not) about the process of terminal differentiation in the lens. We propose a staging system that allows one to discriminate critical periods in the maturation of a lens fiber cell. Using this system, we follow a hypothetical fiber cell through the differentiation program, from the time when it is an unspecialized epithelial cell near the lens equator to the cessation of protein synthesis that occurs when it is a mature fiber cell buried in the lens core. We include speculations on how the differentiation program might act to influence the shape and thus the optical properties of the lens as a whole. Finally, it seems evident that in some cataracts at least, the differentiation program has been interrupted or corrupted. We conclude, therefore, by examining the etiology and pathology of posterior subcapsular cataracts (PSCs) and secondary cataracts.

9.2. The Stages of Fiber Cell Differentiation

The lens grows by the steady addition of fiber cells at its periphery. All cells are retained within the lens. Consequently, fiber cells close to the surface of the lens are younger than the cells in the lens core. All lens fiber cells function as optical elements and share an elongated prismatic shape. Furthermore, the presence of abundant communicating junctions ensures that the lens functions as a syncytium with respect to small molecules. Faced with an apparently homogeneous system, researchers have had a tendency to attribute those properties of the lens that vary as a function of depth to the effects of aging alone. However, discrete changes occur during fiber cell differentiation. For example, cell adhesion complexes are extensively remodeled shortly after fibers detach from the capsule (Beebe et al., 2001). Nuclei, mitochondria, and the endoplasmic reticulum are abruptly degraded late in the process of fiber formation, after the fiber cells have fully elongated (Bassnett,

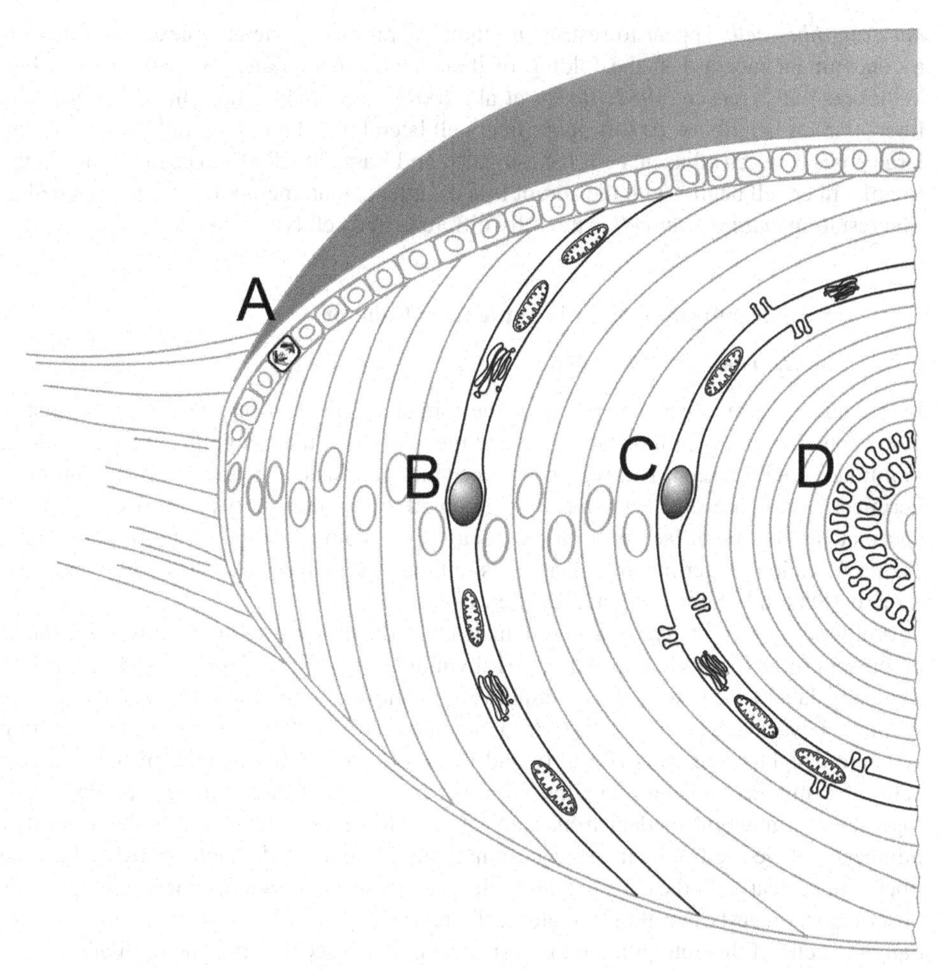

Figure 9.1. Stages in the differentiation of a fiber cell. (A) Mitotically active cell near the equatorial margin of the lens epithelium. (B) Elongating fiber cell. (C) Maturing fiber cell. (D) Mature fiber cell.

1992; Bassnett and Beebe, 1992; Bassnett, 1995, 1997; Bassnett and Mataic, 1997). In addition, the membranes of adjacent fiber cells partially fuse just before organelles are degraded (Shestopalov and Bassnett, 2000a). Therefore, during the process of terminal differentiation, fiber cells pass through specific programmed stages. Although these stages are not always marked by gross morphological alterations, they can be discerned by the presence or absence of characteristic biochemical markers.

We have divided the life of a fiber cell into four distinct stages (Fig. 9.1). *Fiber cell precursors* are produced by mitosis of cells in the germinative zone of the epithelium. Fiber cells withdraw from the cell cycle and elongate from columnar progenitor cells at the lens equator. *Elongating fiber cells* have relatively smooth membranes with distinct adherens junctions near their apical and basal tips. Elongating fibers make direct contact with the capsule and the epithelium at their basal and apical surfaces, respectively. When elongation is complete, fiber cells lose contact with the capsule and the epithelium. Distinct basal and apical adherens complexes can no longer be visualized by antibody staining. These *maturing fiber cells* appear to restructure their lateral membrane complexes, and there is a concomitant increase in the folding of their lateral membranes (Kuszak et al., 1980; Willekens and Vrensen, 1982; Beebe et al., 2001), presumably more firmly locking the fibers to their neighbors. At this point, fiber cell lateral membranes partially fuse, creating large pores between adjacent cells (Shestopalov and Bassnett, 2000a). The final stage in the life of a fiber cell begins with the abrupt loss of intracellular, membrane-bound organelles. The resulting *mature fiber cell* is the fully differentiated cell type of the lens.

9.3. Organization of Cells at the Lens Equator

9.3.1. The Germinative Zone

In the adult lens, mitotic activity is rare in the epithelial cells. It is only near the lens equator, at the boundary between the epithelium and the fiber cells, that a significant concentration of mitotic activity is seen. This region is termed the "germinative zone." The cells within the germinative zone are adjacent to the nonpigmented cells of the ciliary processes. It has been speculated that growth factors from the ciliary epithelium are responsible for stimulating the proliferation of germinative zone cells, although this hypothesis has not been tested directly (Schlotzer-Schrehardt and Dorfler, 1993).

Proliferating germinative zone cells can be labeled *in vivo* by intraperitoneal injection of precursors of DNA synthesis, such as ^{3}H-thymidine or 5-bromo-2′-deoxyuridine (BrdU). Typically, labeled cells from the germinative zone move into the early fiber cell population within a few days to a few weeks. In adult mice, at least 95% of these differentiating cells are daughter cell pairs (Rafferty and Rafferty, 1981). That is, both of the daughter cells of a mitotic division leave the germinative zone and differentiate into fiber cells together. Examination of data from Rafferty and Rafferty (1981) reveals that a similar number of labeled cells remain in the germinative zone after cell division as leave to become fibers. Since nearly all the cells that leave the germinative zone are daughter cell pairs, this observation suggests that those labeled cells remaining in the germinative zone are also daughter cells. If this interpretation is correct, it is reminiscent of the fate of dividing cells in the basal layer of the corneal epithelium. In that tissue, the daughter cells of a mitosis either both move out of the basal layer at the same time and differentiate together or both remain in the basal layer where they may divide again (Beebe and Masters, 1996). The factors that determine whether a pair of daughter cells remains in the germinative zone or leaves to begin the process of fiber cell differentiation are unknown.

From extant data it is not possible to determine whether there is a population of cells that resides in the germinative zone throughout the life of the lens or whether all germinative zone cells eventually differentiate into fibers and are replaced by cells from the adjacent peripheral epithelium. Further studies of the fates of dividing cells in the germinative zone are needed to clarify the pattern of mitosis and migration in this region of the lens and provide insight into the stability of this cell population.

9.3.2. The Transitional Zone

After leaving the germinative zone, cells move into the "transitional zone," a region in which cells are postmitotic but have not yet begun to form lens fibers. The transitional zone is about 6 or 7 cells wide in adult mice (Rafferty and Rafferty, 1981) and approximately 17 cells wide in adult rats (McAvoy, 1978a). In birds, the comparable region of postmitotic cells, the "annular pad" (see chap. 1), contains a much larger population of postmitotic cells. These cells form a thickened band of elongated epithelial cells that probably helps to cushion the lens during accommodation, since the ciliary apparatus of birds compresses the lens to increase its focal power.

The forces responsible for the movement of cells from the germinative zone into the transitional zone have not been examined directly in the intact lens. Because all lens cells are part of a continuous epithelial layer, one plausible view is that it is the increase in cell number resulting from mitosis in the germinative zone that causes cells to be displaced posteriorly. Consistent with this view, when cell division in the frog lens is blocked by removal of the pituitary gland, movement of cells through the transitional zone into the fiber mass stops (Hayden and Rothstein, 1979). Furthermore, restoration of lens epithelial cell division by repeated injection of growth hormone (somatotropin) or somatomedin-C (insulinlike growth factor-1 [IGF-1]) restores cell migration and fiber cell differentiation (Rothstein et al., 1980; Klein et al., 1989). However, it is possible that removal of the pituitary gland and/or supplementation with somatotropin or IGF-1 can influence fiber differentiation directly.

Other studies have suggested that factors that stimulate fiber cell differentiation may also cause cells to actively migrate toward the posterior of the lens. Treatment of rat lens epithelial cells with low concentrations of fibroblast growth factors (FGFs) stimulates cell proliferation, while higher concentrations cause cell migration, and still higher concentrations result in fiber cell differentiation (McAvoy and Chamberlain, 1989). This observation led to the hypothesis that mitosis in the germinative zone, migration into the transitional zone, and eventual differentiation into fibers is a response to a standing gradient of FGFs in the ocular fluids. Measurement of FGF levels in the aqueous humor that bathes the anterior epithelium and in the vitreous body adjacent to the lens fibers supports this view (Schulz et al., 1993). The presumed intermediate concentration of FGFs in the ocular fluids adjacent to the germinative and transitional zones could, therefore, stimulate migration from the germinative zone into the transitional zone.

Data from the frog suggesting that mitosis drives cell migration and data from the rat suggesting that cell migration may be regulated directly by growth factors in the ocular media have not been reconciled satisfactorily. It has not been possible to block cell proliferation in the rat lens *in vivo* (Klein et al., 1989), and to our knowledge the concentration of fiber differentiation factors has not been measured in the intraocular fluids of the frog eye after hypophysectomy.

The mechanism responsible for the cessation of cell division that occurs as cells move into the transitional zone is not well understood. As described above, *in vitro* experiments suggest that high concentrations of FGFs cause lens epithelial cells to stop dividing and to differentiate (McAvoy and Chamberlain, 1989). Lens epithelial cells also withdraw from the cell cycle and differentiate into fiber cells in transgenic animals that overexpress FGFs in the lens (Lovicu and Overbeek, 1998; Robinson et al., 1998). It is possible, therefore, that activating a "fiber cell differentiation program" is sufficient to assure that cells will exit from the cell cycle. Alternatively, FGFs or other fiber differentiation factors may block lens cell proliferation and trigger fiber cell differentiation independently.

Studies in chicken embryos have demonstrated an activity in the vitreous body that prevents lens cells from dividing in response to growth stimulatory factors present in the anterior chamber (Hyatt and Beebe, 1993). Similarly, when chicken embryo lenses are rotated so that the rapidly dividing epithelial cells are exposed to the vitreous humor, these cells withdraw from the cell cycle within nine hours (Zwaan and Kenyon, 1984). It is not evident from these studies whether the antimitotic activity present in the vitreous humor is caused by a fiber differentiation factor or another factor that assures that cells in the posterior of the lens do not proliferate.

Intracellular proteins that play critical roles in the regulation of cell proliferation elsewhere in the body have likewise been implicated in the cessation of cell proliferation in the lens. Studies in mice showed that the cyclin-dependent kinase inhibitor proteins $p27^{KIP1}$ and $p57^{KIP2}$ and the retinoblastoma protein (pRb) are needed for lens fiber cells to withdraw from the cell cycle (see chap. 8). Progression through the cell cycle requires the phosphorylation of cyclins by cyclin-dependent kinases. The cyclin-dependent kinase inhibitor protein $p57^{KIP2}$ inhibits this process, thereby blocking cell cycle progression. $p57^{KIP2}$ is expressed in cells of the transitional zone and the youngest lens fibers, where it is found bound to cyclins (Gao et al., 1999; Lovicu and McAvoy, 1999). In lenses that lack the gene for pRb, a well-known regulator of the cell cycle, $p57^{KIP2}$ mRNA levels are low and lens fiber cells do not withdraw from the cell cycle (Morgenbesser et al., 1994; Fromm and Overbeek, 1996). This finding suggests that pRb is required for the expression of $p57^{KIP2}$ and that at least one function of $p57^{KIP2}$ is to maintain lens fiber cells in the nonproliferating state by complexing with cyclins. Targeted deletion of both of the cyclin-dependent kinase inhibitors expressed in the lens, $p27^{KIP1}$ and $p57^{KIP2}$, leads to continued lens fiber cell proliferation, even in the presence of pRb, confirming that these molecules play an important role in the withdrawal of fiber cells from the cell cycle (Zhang et al., 1998). Lenses lacking both $p27^{KIP1}$ and $p57^{KIP2}$ do not synthesize detectable levels of crystallins, suggesting that the function of these molecules is also required, directly or indirectly, for normal gene expression during fiber cell differentiation (Zhang et al., 1998). Other than a requirement for pRb, the mechanisms that link fiber differentiation factors outside the lens cells to the increased expression of $p27^{KIP1}$ and $p57^{KIP2}$ within lens fiber cells have not been determined.

9.3.3. Studying Fiber Cell Differentiation in Vitro

The events that occur during fiber differentiation can be described in the intact lens, but studying the underlying mechanism(s) is more easily accomplished in cultured lens epithelial cells stimulated to differentiate into fiberlike cells *in vitro*. Epithelial explants (whole or partial sheets of lens epithelium) obtained from neonatal rats and chicken embryos have provided the most useful models for these studies.

Dissociated lens cells have also been used to study aspects of fiber cell differentiation *in vitro* (Menko et al., 1984; Blakely et al., 2000). However, these systems seem less relevant to the *in vivo* situation because the cells are no longer attached to their basal lamina, the lens capsule. Dissociated lens epithelial cells are also isolated from their neighbors and are usually cultured in serum-containing medium. Dissociation often alters the behavior of epithelial cells, and culture in serum-containing medium can be sufficient to trigger fiber cell differentiation, making it difficult to use such a system to study those factors that normally trigger this process *in vivo*.

Extended culture of dissociated lens epithelial cells may lead to the formation of "lentoid bodies," which are aggregates of fiberlike cells. The cells in lentoid bodies can be compared with nearby flattened epithelial cells as a means of studying the properties of "fiberlike" and "epithelial-like" cells in the same culture environment (Berthoud et al., 1999; Blakely et al., 2000; Ibaraki et al., 1996; Menko et al., 1984; Wride and Sanders, 1998).

In explanted chicken embryo lens epithelia, cells elongate and show increased rates of transcription and protein synthesis when cultured in media containing vitreous humor

(Beebe et al., 1980), fetal calf serum (Philpott and Coulombre, 1965; Piatigorsky et al., 1972; Piatigorsky et al., 1973a; Milstone and Piatigorsky, 1975; Milstone et al., 1976), or other differentiation factors (Beebe et al., 1987). These changes are evident within the first five hours in culture. Since differentiating fiber cells are exposed to vitreous humor *in vivo*, these responses are likely to replicate those occurring in the intact lens.

Cell elongation, increased RNA and protein synthesis, and specialization for crystallin synthesis have not been examined in the same detail in neonatal rat lens epithelial explants. However, from those studies that have been reported, it appears that these processes do not occur as rapidly in rat lens epithelial cells as they do in chicken lens epithelial explants treated in a similar fashion (Campbell and McAvoy, 1984; Peek et al., 1992). For example, in the rat system, cell migration begins at 1 hour; nucleolar swelling, a sign of increased protein synthesis, is seen at 16 hours; and increased α-crystallin mRNA accumulation is detected by 24 hours (Walton and McAvoy, 1984; Peek et al., 1992). It is possible that the slower rate of response seen is due to the fact that cultured rat lens epithelial cells have been studied only postnatally, while in chickens only embryonic lens epithelia have been examined. Fiber cell differentiation may occur more rapidly in tissues from embryos. Consistent with this possibility, a decrease in the rate and extent of fiber cell differentiation has been described in rat lens epithelial explants with increasing postnatal age (Richardson and McAvoy, 1990; Richardson et al., 1992).

There is another fundamental difference in the response of lens epithelial explants from chicken embryos and adult rat lenses to agents that stimulate fiber cell differentiation. Cell elongation in chicken embryo explants occurs perpendicular to the lens capsule, and the elongating cells maintain their attachments to the capsule and to their neighbors (Philpott and Coulombre, 1965; Piatigorsky et al., 1972). In rat lens epithelial explants, cells initially detach from both the capsule and their neighbors and migrate on the surface of the epithelial monolayer (McAvoy, 1988). They subsequently pile up and elongate perpendicular to the capsule (Walton and McAvoy, 1984). In spite of these differences, the fiberlike cells that eventually differentiate in these *in vitro* systems resemble authentic lens fibers in many ways. For example, the differentiating cells eventually degrade their organelles in a similar fashion to that observed *in vivo* (Piatigorsky et al., 1973a; McAvoy and Richardson, 1986).

9.4. The Initial Events in Lens Fiber Cell Differentiation

Because chicken embryo lens epithelial cells respond quickly to stimuli that cause fiber differentiation, this section concentrates on studies using this model. Where appropriate, examples from rodent lens epithelia are included.

Treatment of embryonic day 6 (E6) lens epithelial cells with vitreous humor, fetal calf serum, or insulin leads to a rapid increase in the methylation of the membrane phospholipid phosphatidylethanolamine (Zelenka et al., 1982). The addition of three methyl groups converts phosphatidylethanolamine to phosphatidylcholine. Remarkably, the synthesis of phosphatidylcholine by methylation reaches a peak within 6 seconds after addition of chicken embryo vitreous humor or fetal bovine serum. The newly synthesized phosphatidylcholine is degraded by 15 seconds after addition. Insulin, which also stimulates chicken embryo lens cells to elongate, exhibits a slower time course, with phosphatidylcholine being synthesized and degraded within one minute. The slower time course in response to insulin may be caused by the binding of insulin to the endogenous insulinlike growth factor receptor rather than the insulin receptor, since both receptors are present in the lens and insulin binds to the insulinlike growth factor receptor with lower affinity than does IGF-1 (Beebe et al., 1987; Alemany et al., 1990; Bassnett and Beebe, 1990). Inhibitors of phospholipid methylation block cell elongation with a dose-response relationship that parallels their ability to reduce phospholipid methylation (Zelenka et al., 1982). In these studies, it was not possible to determine whether the rate of phospholipid methylation and degradation remains elevated after stimulation or whether it rapidly returns to the same level as before stimulation.

Little more is known about the events that follow the synthesis and degradation of phosphatidylcholine in response to vitreous humor or other factors that stimulate fiber cell differentiation. The degradation of phosphatidylcholine by phospholipases releases lysophosphatidylcholine (Zelenka et al. 1982) and may eventually produce diacylglycerol, lysophosphatidic acid, and other lipid mediators that could activate protein kinase C or other downstream effectors. However, these possibilities have not been tested.

9.4.1. Fiber Cell Elongation

Elongation is the morphological hallmark of fiber cell differentiation. However, it has been difficult to study this process *in vivo* at more than a descriptive level. Therefore, most mechanistic studies have used epithelial explants.

Chicken embryo lens epithelial cells double in length during the first five hours of culture in medium supplemented with vitreous humor (Beebe et al., 1980), insulin (Piatigorsky, 1973; Piatigorsky et al., 1973b), IGF-1 (Beebe et al., 1987), or fetal calf serum (Philpott and Coulombre, 1965; Piatigorsky et al., 1972). After this initial burst, elongation continues at a slower rate for weeks (Piatigorsky et al., 1973b). The rapid initial phase of elongation can occur in the absence of protein synthesis (Piatigorsky et al., 1972).

Colchicine, a drug that depolymerizes microtubules, inhibits the initial phase of elongation, suggesting that microtubules play an essential role in this cell shape change (Piatigorsky et al., 1972). However, nocodazole, another microtubule-disrupting drug, depolymerizes microtubules without preventing lens cell elongation in epithelial explants (Beebe et al., 1979). Therefore, intact microtubules are not essential for fiber cell elongation. Additional studies have demonstrated that colchicine inhibits lens cell elongation by a mechanism distinct from its ability to depolymerize microtubules (Beebe et al., 1979). The cell components that interact with colchicine to prevent elongation are not known.

As lens fiber cells elongate, *in vivo* or *in vitro*, they maintain a similar cross-sectional area. This means that their volume increases in proportion to their length, a phenomenon that has been confirmed by three-dimensional reconstruction of elongating lens cells in tissue sections (Beebe et al., 1982). Cell elongation could, therefore, be due to an increase in cell length that is precisely accompanied by an increase in cell volume or to an increase in volume that provides the driving force for the increase in cell length. Measurements of ion fluxes in elongating lens cells provide support for the latter explanation.

Stimulation of cell elongation leads to a change in the flux of potassium ions across the plasma membrane (Beebe et al., 1986; Parmelee and Beebe, 1988). Potassium continues to enter the cell at the same rate as before stimulation by the action of the Na,K-ATPase, but the potassium efflux slows, leading to the accumulation of potassium and chloride ions in the cytoplasm (Fig. 9.2). The resulting increase in the osmolarity of the cytoplasm leads to an influx of water and a concomitant increase in cell volume (Parmelee and Beebe, 1988; Beebe et al., 1990). In support of these data, agents that increase the efflux of potassium or reduce potassium uptake prevent cell elongation (Parmelee and Beebe, 1988; Beebe and Cerrelli, 1989).

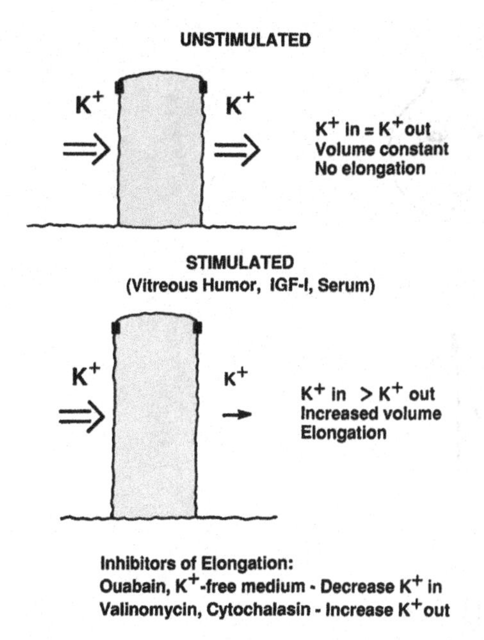

Figure 9.2. Lens cell elongation is driven by the accumulation of potassium ions.

The cells at the lens equator or in an epithelial explant are all increasing in volume at the same time. They are also attached to their neighbors by cell-cell adhesive junctions and to the lens capsule by basal adherens junctions (Bassnett et al., 1999). Therefore, the increase in cell volume triggered by fiber differentiation factors is translated into an increase in cell length (Fig. 9.3; Beebe et al., 1986).

While this hypothesis has not been tested explicitly, isolated chicken embryo lens epithelial cells do not elongate when cultured in fetal bovine serum or vitreous humor. Similarly, at early stages of elongation, cells from epithelial explants round up when isolated from their neighbors (unpublished observations of D. Beebe). Therefore, it seems plausible that a physiologically regulated increase in cell volume and the constraints imposed by cell-cell and cell-substrate adhesions could be sufficient to account for the morphological changes that characterize the early stages of lens fiber cell differentiation. Continued elongation is also associated with a proportional increase in cell volume. This volume increase is likely to be driven by protein accumulation.

9.4.2. Increased Protein Synthesis Early in Fiber Cell Differentiation

In chicken embryo lens epithelial cells stimulated to elongate with fetal bovine serum, the absolute rate of protein synthesis (molecules/second/cell) increases nearly twofold in the first 5 hours (Milstone and Piatigorsky, 1975; Beebe and Piatigorsky, 1976). During this period, the rate of synthesis of the major lens crystallin in the embryonic lens, δ-crystallin, increases by a similar amount. In the next 19 hours of culture, total protein synthesis continues at about the same rate as seen at 5 hours, but the rate of δ-crystallin synthesis nearly doubles (Milstone and Piatigorsky, 1975; Beebe and Piatigorsky, 1976). The specific increase in δ-crystallin synthesis is associated with an increase in the accumulation of δ-crystallin mRNA (Milstone et al., 1976). There is no evidence of an increase in the

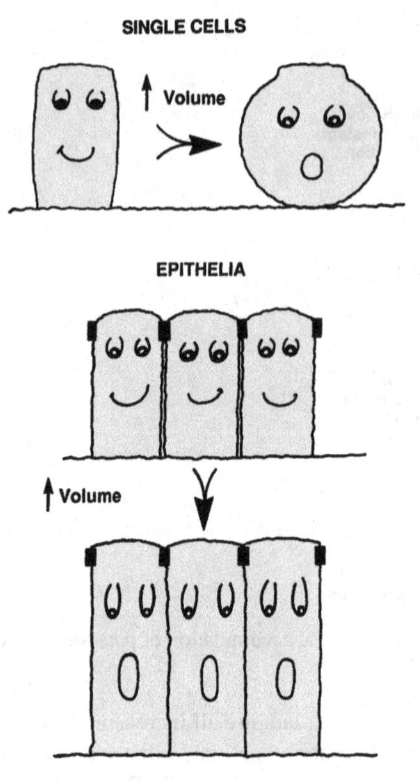

Figure 9.3. Effect of volume increase on single cells or cells in an epithelial monolayer.

rate of protein degradation during this period, suggesting that the increased rate of protein synthesis is translated into increases in protein accumulation (Milstone and Piatigorsky, 1975; Beebe and Piatigorsky, 1976). Similar increases in total protein synthesis are likely to occur early in the differentiation of rodent lens epithelial explants, although analyses have not been performed as explicitly as in chicken lens explants. The increase in the size of nucleoli reported by Walton and McAvoy (1984) 16 hours after exposure of rat lens epithelial explants to retina-conditioned medium suggests a general increase in protein synthesis. By 24 hours after exposure to FGF, α-crystallin begins to accumulate (Peek et al., 1992). Therefore, in these *in vitro* model systems, fiber cell differentiation appears to be first associated with an overall increase in protein synthesis and accumulation, then with specialization for the synthesis and accumulation of the crystallins.

9.4.3. Protein Synthesis and Accumulation in Later Stages of Fiber Cell Elongation

The accumulation of crystallin and noncrystallin polypeptides during lens fiber cell differentiation may be responsible for the continued elongation of lens fiber cells. Inhibition of protein synthesis does not prevent the initial doubling of cell length in chicken embryo explants, but blocks subsequent elongation (Piatigorsky et al., 1972). Recent studies in which crystallin gene expression was inhibited in mice by the targeted deletion of essential transcription factors also provide evidence that protein accumulation drives fiber cell elongation. Disruption of the transcription factor c-maf greatly reduces crystallin gene expression and fiber cell elongation (Kawauchi et al., 1999; Kim et al., 1999; Ring et al., 2000). Targeted deletion of the Sox1 or Prox1 transcription factors blocks fiber cell elongation at an early stage and prevents the accumulation of mRNAs for some of the γ-crystallins but not the α- and β-crystallins (Nishiguchi et al., 1998; Wigle et al., 1999). It is possible that the small amount of cell elongation that does occur when transcription factor genes have been knocked out reflects a low residual level of crystallin protein accumulation. Unfortunately, the amount of protein that accumulates in the lens cells of these genetically modified animals and the volumes of their shorter fiber cells have not been measured. Therefore, although these results are consistent with the idea that protein accumulation drives the later stages of lens fiber cell elongation, further study is needed to test this possibility.

It is interesting that the phenotype of the sox1 and prox1 knockout mice resembles that of the dominant Cat2^{elo} mutation, which results in truncation of γE-crystallin (Cartier et al., 1992). Apparently, normal expression of the γ-crystallin genes is essential for proper fiber cell elongation. It is not clear whether the inhibition of cell elongation that occurs when γ-crystallin synthesis is perturbed is due to a reduction in total protein accumulation or whether elongation depends in some specific way on having the proper complement of γ-crystallins.

The idea that fiber cell elongation is driven by an increase in cell volume is probably too simplistic to account for all aspects of fiber cell morphogenesis, but we know of no alternative mechanism that explains the coordinated increase in fiber cell length and volume. The lens fiber cell cytoskeleton may also contribute to the stabilization of the plasma membrane and to the development of the membrane specializations that are found in mature fiber cells (see below). However, the action of the cytoskeleton alone does not appear to be sufficient to account for fiber cell elongation.

Fiber cell elongation ceases when the apical and basal ends of the cells reach the sutures. Whether protein accumulation also slows or stops at this stage has not been determined. This is a critical stage in the life of a fiber cell, since suture formation involves separation of the fiber cell from the posterior capsule, establishment of connections with a fiber cell from the opposite hemisphere of the lens, and eventual burial of cells beneath layers of younger fibers. Little is known about changes in the synthetic or metabolic activity of fiber cells during this process.

9.4.4. Establishment of Regular Packing Order

As epithelial cells leave the transitional zone and begin to form fiber cells, a remarkable reorganization takes place. Instead of being randomly distributed within the cell layer, the early fiber cells line up in rows, or radial cell columns. This organization can be appreciated in whole mount preparations of the epithelium, where the aligned cells are referred to as "meridional rows" (see chap. 11). The precise alignment established in the meridional rows is maintained during the remainder of fiber cell differentiation.

Cross sections through epithelial cells show that the cell borders are irregular and that cells have variable numbers of neighbors. By contrast, fiber cells align with six neighboring cells in a precise packing geometry to form radial cell columns (Fig. 9.4). The cells in adjacent columns are interdigitated so that the cell nuclei in adjacent elongating fiber cells are displaced from each other by exactly one-half a cell thickness. Each fiber cell adheres to two fiber cells in the same radial column, one that is deeper in the lens and one that is more superficial. On each lateral border a fiber cell interacts with two cells in the adjacent radial cell column. This arrangement assures that fiber cells will have hexagonal cross sections.

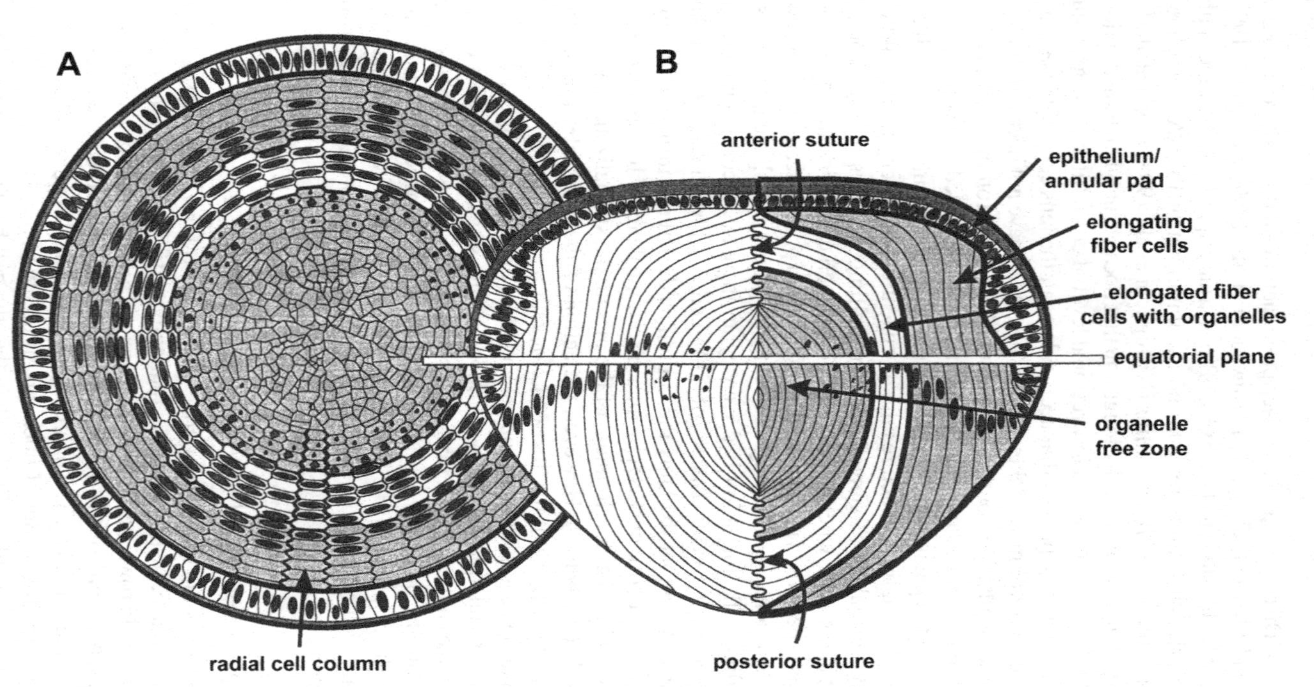

Figure 9.4. The tissue organization of the embryonic chicken lens viewed perpendicular (A) and parallel to the optic axis (B).

Hexagonal packing of solid objects leads to a geometric arrangement that minimizes extracellular space and surface area. Low extracellular space and minimal membrane surface area are important for reducing light scattering and therefore maximizing the transparency of lens fiber cells.

The mechanism responsible for the precise spatial organization of differentiating fiber cells is not known but may be related to the increase in cell volume that drives fiber cell elongation. As lens epithelial cells increase in volume during the early stages of fiber cell elongation, irregular cellular projections will become increasingly filled with cytoplasm. Assuming that that the cells do not lose contact with their neighbors and that the surfaces of adjacent cells can slide past each other, this volume increase will eventually result in the generation of hexagonal cross sections.

An analogy that may help to visualize this process is the inflation of a box of balloons of cylindrical shape. If the balloons are deflated or partially inflated, they can assume irregular shapes, perhaps intertwining with each other in various arrangements. As they are inflated, their shape would be increasingly restricted and more closely approximate a hexagon. If their surfaces were able to slide past each other, they would eventually displace all of the outside air and become a "solid" mass resembling a honeycomb. If one further assumed that the "necks" of these balloons were inserted through one side of the box, they would elongate and assume hexagonal cross sections as they were inflated. This is an approximate physical analogy to what happens to elongating fiber cells. They are all attached to the capsule at their basal ends and are increasing in cell volume. This transforms them from irregular solid shapes to elongated columns of hexagonal cross section.

It is not clear whether geometric constraints alone are sufficient to account for the precise packing of lens fiber cells during elongation. However, physical models such as the one proposed above have the advantage of being simple and amenable to simulation and testing. Furthermore, since biological systems must obey physical laws, asking first whether a simple physical model can account for the behavior of a biological system before invoking specialized adhesive or cytoskeletal mechanisms is both parsimonious and realistic.

9.5. The Elongating Fiber Cell

The most obvious physical manifestation of fiber cell differentiation is an enormous increase in cell length. Derived from epithelial precursors measuring less than 10 μm in length, fiber cells may ultimately extend for a centimeter or more around the lens equator. This increase in length is paralleled by an equally impressive increase in surface area, especially if one considers the complex lateral interdigitations of the plasma membrane that form during the elongation process.

The lens epithelial cells from which the fiber cells derive are polarized. Certain cellular components, for example, those involved in cell-substrate interactions, are localized specifically to the basal cell surface, subjacent to the lens capsule. Within the cytoplasm, organelles and other components are not evenly distributed along the polar axis. The Golgi apparatus, for example, is located on the apical side of the lens epithelial cell nucleus (Bassnett, 1995). From a physiological standpoint, the distribution of ion channels and pumps is especially important, and in this regard an apical domain and a basolateral domain are recognized. Tight junctions at the apicolateral border separate and define these membrane domains (Lo and Harding, 1983). Elongating fiber cells appear to inherit and retain the intrinsic polarity of their epithelial precursors, and basal, lateral, and apical membrane domains can be readily identified.

9.5.1. The Basal Membrane

Despite its relatively small area (less than 0.1% of the total plasma membrane), the fiber cell basal membrane and its associated cytoskeletal elements (here referred to collectively as the basal membrane complex [BMC]) are critical for lens function. The BMC anchors the fibers to the lens capsule. The attachment is strong enough to resist the forces generated during accommodation. Elongating fiber cells make up a migratory population of cells. During differentiation, their posterior tips move along the capsule from the equator to the posterior suture, often a journey of several millimeters. The fiber BMC is also responsible for secreting the capsule substrate upon which it migrates (Young and Ocumpaugh, 1966; Johnson and Beebe, 1984; Haddad and Bennett, 1988). At the suture, fiber cells dock with cells from the opposite hemisphere of the lens. The BMC ensures proper tracking of the fiber cells across the capsule, timely detachment from the capsule, and accurate docking at the suture. Collectively, the BMCs constitute the posterior face of the lens, a critical refractive surface. The radius of curvature of this surface plays an important role in determining the focal length of the lens.

9.5.1.1. Cell-Substrate Interaction

The fiber cell basal membrane domain is firmly attached to the lens capsule. The strength of this adhesive interaction can be demonstrated by mechanically stripping the capsule from the fibers. Under these circumstances, the basal portion of each fiber breaks away from the rest of the cell, remaining attached to the stripped capsule (Bassnett et al., 1999). Adhesion to the capsule is mediated by integrins located in the basal membrane (see chap. 10). Integrins are heterodimeric membrane proteins that bind to matrix components extracellularly and to the cytoskeleton intracellularly. In the fiber cells, $\alpha6/\beta1$ integrins predominate (Menko and Philip, 1995), and different isoforms of this molecule (with differing affinities for the cytoskeleton) are expressed during cell elongation (Walker and Menko, 1999). The interaction of integrins with extracellular matrix is dependent on the presence of divalent cations. Therefore, when fiber cells are incubated in Ca^{2+}- and Mg^{2+}-free solutions, their basal membranes become detached from the lens capsule. Further evidence for the importance of integrins in fiber cell adhesion has been provided by experiments in which a function-blocking antibody to $\beta1$ integrin caused fiber cells to detach from the capsule (Bassnett et al., 1999).

Although fiber cells are firmly anchored to the capsule, they also migrate across it. It is estimated that embryonic chicken fiber cells traverse the capsule at a rate of 120 μm/day (Bassnett et al., 1999). The BMC is highly enriched in actin, which is organized into a hexagonal filamentous latticework immediately beneath the basal membrane (Fig. 9.5). Spatially controlled polymerization of actin is at the origin of cell motility in many cell types and is likely implicated in fiber migration. Interestingly, it has been reported that small GTP-binding proteins, which are known to regulate the actin cytoskeleton, are enriched in elongating fiber cells (Deng et al., 2001). Although much is known about the movement of individual cells across a substratum, the coordinated migration of sheets of cells (such as epithelial cells during wound healing or fiber cells during normal differentiation) is poorly understood. Moreover, the migration of fiber cells in the living lens has not been observed directly in sufficient detail to allow a hypothesis to be proposed regarding mechanism. It is not known, for example, whether cells move by extension of lamellipodia, as seems to be the case for isolated motile cells.

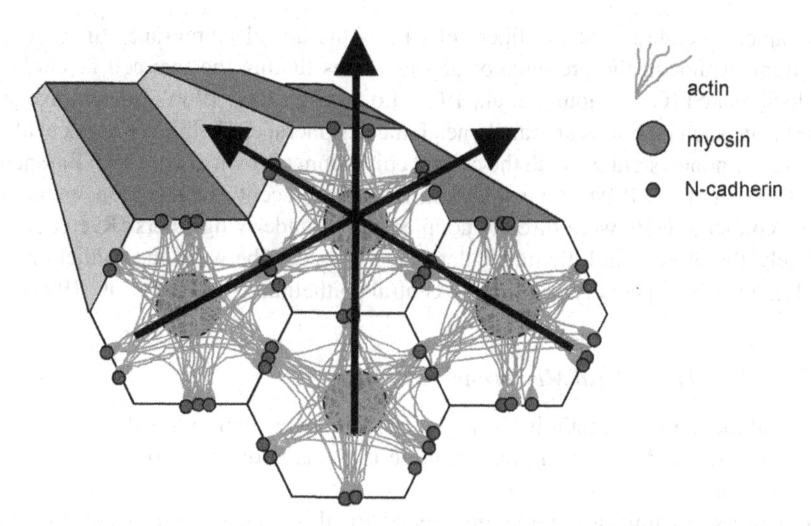

Figure 9.5. Arrangement of contractile and adhesive elements in the basal membrane complex (BMC) of the embryonic chicken lens. The f-actin bundles in the BMC of one cell are aligned with those in the next. Enrichment of N-cadherin at the points of membrane association of actin filaments may stabilize the latticework and ensure that contractile tone generated by actin-myosin interaction is transmitted across the posterior surface of the lens (along the axes indicated by the arrows).

9.5.1.2. Structure of the Basal Membrane Complex

In the chicken lens, the footprints of the fiber cells form perfect hexagons on the lens capsule (Fig. 9.5; Bassnett et al., 1999). Adjacent hexagons appear to be joined by the cell-cell adhesion protein N-cadherin, which is particularly concentrated at the midpoint of each face of the hexagon. Stress fiberlike cables of actin extend beneath the basal membrane from points of insertion in the N-cadherin rich foci. The actin-capping protein tropomodulin is enriched in the BMC (Lee et al., 2000). Myosin II is associated with the actin fibers, along with two other contractile proteins, myosin light-chain kinase and caldesmon. Together these components make up a regular latticework of structural elements cradling the posterior face of the lens. Because contractile proteins are present, it is possible that the contractile tone of this latticework plays a role in determining the radius of curvature of the posterior lens face. Interestingly, as the tips of the fiber cells approach the posterior suture, the BMC becomes less well organized.

9.5.2. The Apical Membrane

The functional organization of the apical membrane has received less attention than that of the basal membrane; however, it is evident that the apical membrane differs structurally from the basal membrane. The apical tips of the fibers are not packed in the precise, regular formation that characterizes the posterior tips (Kuszak et al., 1995). Although an apical actin web is observed, a precisely oriented latticework of contractile proteins is absent. The apical membrane is apposed to a cellular layer (the lens epithelium) rather than the acellular capsule. The anterior tips of the fibers appear to migrate at a slower rate than the posterior tips because their arrival at the anterior suture is somewhat delayed with respect to the posterior suture.

The apical membrane of the fiber cells faces the apical membranes of the overlying epithelium. Although the presence of gap junctions linking the two cell layers has been well documented (Goodenough et al., 1980; Lo and Harding, 1986), quantitative analysis suggests that such junctions are rare beneath the peripheral epithelium (Kuszak et al., 1995) and virtually nonexistent beneath the central epithelium (Brown et al., 1990; Bassnett et al., 1994; Prescott et al., 1994). One study, utilizing fluorescent dye injection, estimated that 10% of epithelial cells were directly coupled to the underlying fibers (Rae et al., 1996). It is likely, therefore, that little molecular transport occurs between the epithelium and the underlying fibers, especially beneath the central epithelium (Bassnett et al., 1994).

9.5.3. The Lateral Membrane

The lateral membrane domain is by far the largest plasma membrane domain of the elongating fiber cell, and there is a great increase in the area of this surface during fiber cell differentiation.

The increase in lipid accumulation needed for this process is provided by increased synthesis of most phospholipids and selective decreases in their degradation (Zelenka, 1978, 1983).

It is likely that the lateral membrane is composed of subdomains. In cross section, the elongating fiber cells have a flattened hexagonal appearance, with two broad faces and four narrow faces (see Fig. 9.4A). In the mammalian lens, gap junction plaques are prominent features of the broad faces, at least in the midportion of the fibers (Gruijters et al., 1987). Conversely, some membrane specializations (most notably those concerned with intercellular adhesion) are preferentially associated with the short faces. N-cadherin, vinculin, and paxillin (all components of fascia-type adherens junctions) are concentrated along the short sides of the fibers (Lo, 1988; Lagunowich and Grunwald, 1989; Lo et al., 2000; Beebe et al., 2001).

The fact that the apical, basal, and lateral membranes differ in composition implies the presence of mechanisms capable of targeting the delivery of newly synthesized components to these widely separated domains (and of subsequently preventing mixing of components in the plane of the membrane). It has recently been demonstrated that the microtubule orientation in elongating fiber cells is polarized (Lo and Wen, 1999), with the minus ends of the microtubules directed toward the apical membrane. In conjunction with molecular motor proteins such as kinesin and dynein (both of which are present in the lens), polarized cytoskeletal elements are likely to mediate the delivery of components to specific membrane domains in the elongating fiber cell.

As the posterior tips of the fiber cells approach the sutures, the BMC is disassembled. The basal membrane ultimately detaches from the capsule (by a mechanism that is not understood) and sutures with the basal membrane of fiber cells originating from the opposite hemisphere. A few days later, the anterior tips of the fibers reach the anterior suture. The sutured fiber cells then enter the maturing fiber cell population.

9.6. The Maturing Fiber Cell

9.6.1. Suture Patterns

Conceptually, the simplest suture pattern would have the ends of fiber cells meeting at the anterior and posterior poles of the lens. This type of suture, called an "umbilical suture," is present in the lenses of birds, reptiles, and teleosts (Priolo et al., 1999a). In mammals, the ends of the fiber cells meet at planes. This can be a simple "line suture," as seen in rabbits, or the more complex "Y" suture seen in most other mammals. In humans and other primates, a Y suture is present before birth, but as the lens grows the ends of the Y branch to form, by adulthood, a more complex "star suture" (Kuszak, 1995a; see also chap. 4).

The positioning and orientation of the suture planes constitute one of the most interesting puzzles in eye development. In animals with a Y suture, the Y is upright with respect to the dorsoventral axis of the eye in the anterior half of the lens and inverted in the posterior. Similarly, the anterior and posterior line sutures in rabbits are orthogonal to each other. The factors responsible for the formation of a particular number of suture planes and for the precise positioning of the sutures with respect to the axes of the eye are not known. The details of suture morphogenesis are discussed at greater length in chapter 4.

9.6.2. Suture Formation

Several important events occur during suture formation:

1. New basal-basal and apical-apical contacts are made between fiber cells from opposite sides of the lens.
2. The basal ends of the fiber cells detach from the capsule and the apical ends lose contact with the lens epithelium.
3. The cells become progressively isolated from diffusible nutrients and growth factors present in the ocular media.

Little is known about the mechanisms responsible for these events and the impact they may have on other aspects of fiber cell maturation and function.

The nature of the adhesive interactions that link the ends of fiber cells together at the sutures is largely unknown. Tensin, a molecule present in the basal membrane complex that links the fiber cell to the posterior capsule, is not detected in the posterior suture (Beebe et al., 2001). This suggests that at least some of the adhesive interactions that hold the ends of the fibers together after suture formation are different from those that attach the fiber cells to the capsule. To our knowledge, no cell-cell adhesion molecules appear at the sutures that are not already present in the basal membrane complex or along the lateral ends of the fibers.

Vinculin, a protein that is part of the complex that links transmembrane adhesive molecules, such as integrins and cadherins, to the cytoskeleton, is abundant in the basal membrane complex, but barely detectable along the lateral membranes of elongating fibers (Beebe et al., 2001). Soon after suture formation, vinculin becomes abundant along the lateral membranes of the fiber cells and especially prominent at the sutures. This increase in vinculin protein is accompanied by an increase in the amount of vinculin mRNA (Beebe et al., 2001). Presumably, vinculin is part of the integrin-mediated adhesion complex that links the basal ends of the fibers to the capsule (Menko and Philip, 1995; Bassnett et al., 1999; Walker and Menko, 1999). We assume that, at the sutures, vinculin associates with a cell-cell adhesion molecule that connects the apical and basal ends of fiber cells. This hypothetical molecule could be N-cadherin (Beebe et al., 2001), one or more of the integrins (Menko and Philip 1995), or another cell-cell adhesion molecule.

The attachment of epithelial cells to their basal lamina provides physical stability and is the source of signals that regulate cell differentiation and survival (Howlett and Bissell, 1993; Coucouvanis and Martin, 1995; Boudreau et al., 1996; Pullan et al., 1996). Therefore, the separation of lens fibers from their basal lamina during suture formation might be expected to trigger a series of biochemical changes. However, other than the increase in vinculin mRNA and protein described above, no changes in gene expression or protein accumulation at the time of suture formation have been described. A search for transcripts that are expressed only after suture formation in the chicken embryo lens has identified two novel cDNAs (Vasiliev and Beebe, in preparation). *In situ* hybridization has revealed that these transcripts accumulate soon after fiber cells detach from the capsule. The function of the proteins encoded by these transcripts is under study. However, the existence of these mRNAs shows that separation from the capsule is associated with the synthesis of new gene products. The proteins encoded by them may be important for the subsequent maturation of lens fiber cells.

9.6.3. Fiber Cell Maturation after Suture Formation

In fiber cells at early stages of elongation, the lateral plasma membranes are relatively smooth. However, as fiber cells reach the sutures, the membranes become progressively more interdigitated, forming interlocking "ball-and-socket" junctions (Kuszak et al., 1980; Willekens and Vrensen, 1982). Ball-and-socket junctions may stabilize the lateral membranes of the fiber cells and assure that the cells remain tightly connected during accommodation. The mechanisms responsible for the formation of these unusual membrane specializations have not been identified.

In chicken embryo lenses, there is an abrupt increase in paxillin accumulation along the lateral membranes of fiber cells just before they degrade their organelles (Beebe et al. 2001). The mRNA for paxillin also increases at this time. Paxillin is a protein that is usually found associated with cell-cell and cell-substrate adherens junctions (Turner, 1998). In cultured cells, changes in paxillin phosphorylation are associated with alterations in cell motility, but the role of this protein in epithelial tissues is not well understood. Interestingly, paxillin accumulation in chicken embryo lenses occurs at about the same time as an increase in cell-cell membrane fusions that open up new communication channels between adjacent fiber cells.

9.6.4. Cell-Cell Fusion

It has long been appreciated that the internal electrical resistance of the lens is negligible, reflecting the presence of gap junctions that couple neighboring fiber cells (Duncan, 1969). In some species, gap junctions account for more than 50% of the plasma membrane surface area (Kuszak et al., 1978). Gap junction channels are permeable to ions and other small molecules but restrict the passage of proteins and other macromolecules.

Recent evidence suggests that a pathway that permits the cell-cell diffusion of macromolecules may also link lens fiber cells just before they degrade their organelles. Evidence for the existence of this pathway has come from the ectopic expression of green fluorescent protein (GFP). If plasmids encoding GFP are injected into the lumen of the chicken lens vesicle early in development, scattered primary fiber cells express the protein (Shestopalov and Bassnett, 2000a). GFP is restricted to the cytoplasm of a few fiber cells until approximately day 8 of development (E8). After E8, GFP fluorescence is detected throughout the core of the lens, having apparently diffused from the cytoplasm of cells in which it was originally expressed (Shestopalov and Bassnett, 2000a). Further evidence for the existence of a novel cell-cell diffusion pathway in the lens core comes from experiments in which fluorescent dextran molecules (MW = 40 kD) were injected into the cytoplasm of cells situated at different depths in the embryonic chicken lens. Following injection into superficial fiber cells, the dextran was invariably retained within the cytoplasm of the injected cells. However, injection into deeper fiber cells resulted in diffusion of the fluorescent dextran into many neighboring cells (Shestopalov and Bassnett, 2000a). These data suggest that during differentiation a protein-permeable pathway between fiber cells becomes patent.

Remarkably, GFP-tagged membrane proteins exhibit similar behavior. For example, expression of a GFP-tagged construct encoding the major lens intrinsic protein (MIP) in superficial fiber cells results in uniform incorporation of fluorescent protein in the membranes of the injected cells. However, expression of GFP-MIP in core fiber cells never results in single-cell labeling. Rather, the membranes of a group of cells (>50) became fluorescent. Because membrane proteins are only able to diffuse laterally in the plane of a single continuous membrane, these observations indicate that the membranes of adjacent core fiber cells are fused in some fashion. Furthermore, the data suggest that membrane fusions occur on or about E8 in the chicken lens. This is four days before the organelles are degraded in the central fiber cells.

Three-dimensional reconstruction of stacks of two-photon microscope images revealed the presence of fusion pores between cells in the core of the embryonic chicken lens. Such fusions represent conduits for the diffusion of cytoplasmic and membrane proteins in the lens core. Fiber cell membrane fusions have been described in the lenses of several species (Kuszak et al., 1985; Kuszak et al., 1989; Kuszak et al., 1996) and appear to be particularly numerous once cells have reached the sutures.

Because cell fusions are difficult to visualize in the lens, a careful morphometric analysis has not yet been possible. Consequently, it is not known, for example, whether fusions, once formed, are permanent structures. Assuming that fusions are a universal feature of the vertebrate lens, many questions arise regarding their functional significance. In the embryonic chicken lens, the core syncytium includes a layer of nucleated fiber cells. This raises the possibility that newly synthesized proteins could diffuse into the metabolically inert core of the lens. An optical role has also been postulated. By allowing the intermingling of cytoplasmic components, fusions could equalize the refractive index in adjacent cells and thereby minimize internal light scattering within the tissue.

A few days after the fiber cells are incorporated into the core syncytium, they abruptly degrade their organelles and enter the population of mature fiber cells.

9.7. The Mature Fiber Cell

9.7.1. Morphological Characteristics

In the developing chicken lens, mature fiber cells are present after E12. At this stage, organelle degradation commences in fiber cells in the lens core (see below). From E12 onwards, the mature fiber cell population encompasses a central core of disorganized primary fibers and an increasing number of organized, morphologically homogeneous secondary fiber cells (Fig. 9.6). The primary fiber cells vary in both length and thickness. Short regions of the cells may be dilated, and these distended zones are often joined by thin cellular connections (Shestopalov and Bassnett, 2000b). This variation in cross-sectional size and shape is reflected in an irregular polygonal packing arrangement (Fig. 9.6D). In contrast, secondary fiber cells are much more uniform in shape and are packed in a regular hexagonal arrangement, reflecting the hexagonal cross sections of individual cells (Figs. 9.6A, B).

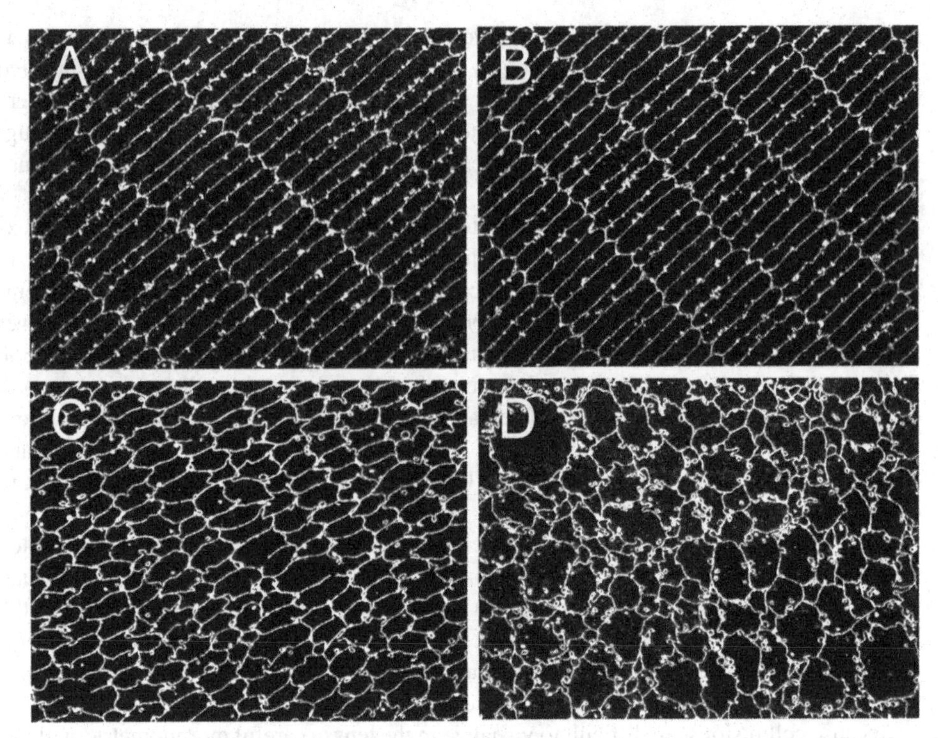

Figure 9.6. The cross-sectional appearance of fiber cells located at different depths in the adult (one-year-old) chicken lens. The cells in A are located at a depth of approximately 0.5 radius, those in B at 0.3 radius, those in C at 0.1 radius, and those in D are primary fiber cells located in the center of the lens. Despite their different appearance and packing organization, the cells shown in A–D are all "mature" fiber cells. The disorganized appearance of the primary fiber cells shown in D is not a function of aging. The appearance of these cells has not altered significantly from the time they first differentiated on day 4 of embryonic development.

Although the mature fiber cell population contains both organized and disorganized cells, the former do not give rise to the latter. The primary fibers are the oldest cells in the lens, but their relatively disorganized appearance is not the result of aging. Their irregular morphology was apparent at the time they first elongated to fill the lumen of the lens vesicle.

When examined by electron microscopy, mature fiber cells have an unusual appearance (Fig. 9.7). The membrane-bound organelles that once filled their cytoplasm are absent (Kuwabara, 1975). Mature fiber cells are located directly on the optical axis of the lens. Because organelles are large enough to scatter light, their absence from the cytoplasm of mature fibers is believed to assure the clarity of the tissue. In certain pathological conditions, however, organelles persist in mature fiber cells. For example, the failure of fibers to properly degrade their nuclei is a common feature of human congenital cataract (Zimmerman and Font, 1966).

The filamentous cytoskeleton, while present in mature fiber cells, is reduced in extent and complexity compared with that of the superficial cells. By E16, in the chicken lens, the intermediate filament protein vimentin has already been degraded in the central fiber cells (Ellis et al., 1984). In mammalian lenses, vimentin persists for a longer period, but eventually following the disappearance of the organelles, the vimentin filament network is degraded (Sandilands et al., 1995a). The lens-specific intermediate filament proteins filensin and

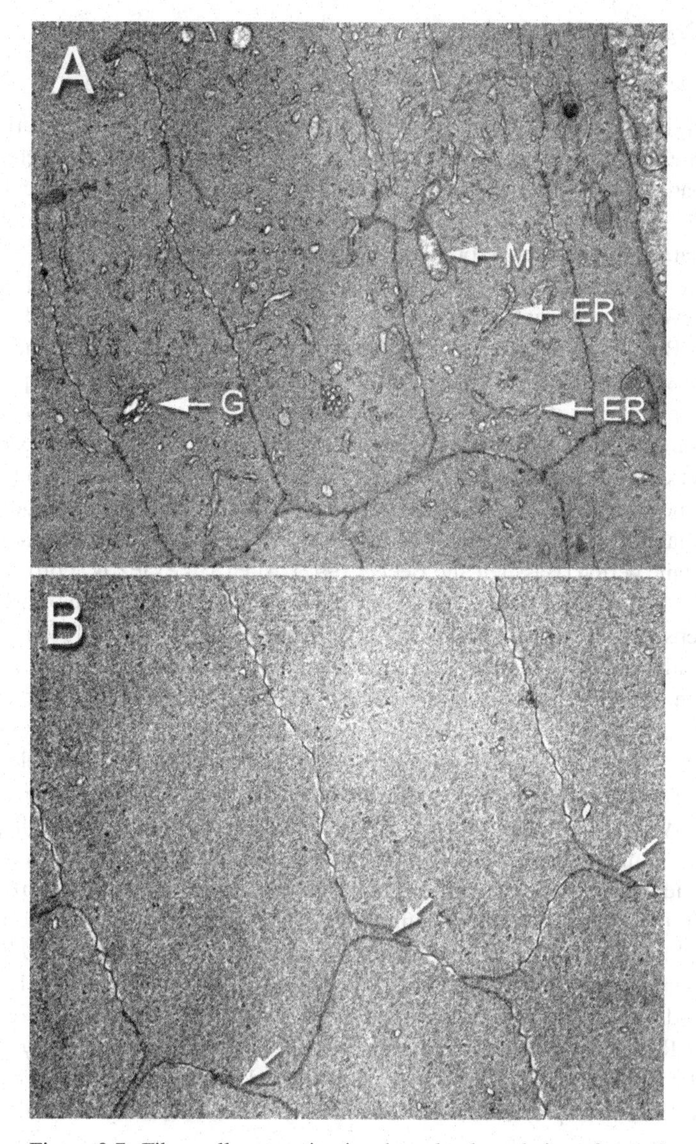

Figure 9.7. Fiber cell maturation involves the degradation of cytoplasmic organelles. (A) The superficial fibers of an E15 chicken lens (viewed here by transmission electron microscopy) contain a usual complement of organelles. G, Golgi apparatus; ER, endoplasmic reticulum; M, mitochondria. (B) Organelles are absent, however, from fibers located further into the lens. Note the fingerlike projections (arrows) that interlock neighboring fibers.

CP49 are present, but both are extensively proteolysed (Prescott et al., 1996). Microtubules, prominent constituents of elongating fiber cells, are absent from mature fibers (Kuwabara, 1975; Bradley et al., 1979), whereas the submembrane actin cytoskeleton persists (Kibbelaar et al., 1980; Ozaki et al., 1985; Bassnett and Beebe, 1992). In the absence of nuclei and other organelles, the principal ultrastructural feature of mature lens fiber cells is an elaborately folded plasma membrane that connects adjacent cells through a series of ball-and-socket interdigitations.

9.7.2. Organelle Degradation

9.7.2.1. Organelle Breakdown Is a Rapid, Coordinated Event

It has been known since the work of Rabl (1899) that fiber cell nuclei are degraded during differentiation. However, only recently, with the introduction of confocal imaging techniques, has the precise spatial and temporal pattern of nuclear loss been appreciated. Confocal imaging has revealed that the loss of nuclei coincides with the disappearance of other cytoplasmic organelles and that, in any given cell, organelle degradation is surprisingly rapid.

In the developing chicken lens, all cells initially contain a full complement of organelles (Fig. 9.7A). However, on or about E12, organelles are degraded in fiber cells located near the center of the lens (Bassnett and Beebe, 1992), resulting in the formation of an organelle-free zone (OFZ; Fig. 9.8). The OFZ initially encompasses only the primary fiber cell population, but as development proceeds it expands to include more and more secondary fiber cells.

During embryonic development, the OFZ increases in diameter at a rate of approximately 20 cell widths per day. Interestingly, newly differentiated fiber cells are added to the lens surface at about the same rate. The distance from the surface of the lens to the border of the OFZ thus remains constant, despite the considerable increase in lens size that occurs during the embryonic period (Bassnett and Beebe, 1992). At E18, the OFZ is approximately 2,600 μm wide in the equatorial dimension. Between E18 and E19, the wave of organelle degradation sweeps outward across a shell of >10,000 fiber cells (Bassnett, 1995). In each of these cells, rapid organelle degradation takes place.

Using organelle-specific vital dyes or antibodies raised against organelle-resident proteins, researchers have examined the fate of the organelle systems in fiber cells bordering the OFZ. At the border of the OFZ, mitochondria, which often have a highly elongated morphology in the superficial fiber cells, suddenly fragment. The disintegrating mitochondria lose the ability to accumulate the vital dye rhodamine 123, indicating a collapse of the mitochondrial membrane potential, $\Delta\Psi$ (Bassnett, 1992; Bassnett and Beebe, 1992). Histochemical detection of succinic dehyrogenase activity indicates that this enzyme, a component of the inner mitochondrial enzyme, is inactivated in parallel with the loss of $\Delta\Psi$ (Bassnett and Beebe, 1992), and immunofluorescence studies indicate the sudden loss of the mitochondrial marker proteins BAP37 and prohibitin (Dahm et al., 1998a). The fiber cell ER (visualized by staining with the lipophilic dye $DiOC_6$ or by immunofluorescence detection of the ER protein, protein disulfide isomerase) is also degraded rapidly in cells bordering the OFZ (Bassnett, 1995). It has been estimated that the degradation of mitochondria and the ER is completed in 2–4 hours in any given cell (Bassnett and Beebe, 1992).

In contrast to the sudden disappearance of mitochondria and the ER, the breakdown of the nucleus is a relatively lengthy process (although functional inactivation may be rapid). In some species, residual nuclear structures can be detected in cells lying deep within the OFZ (Bassnett, 1997). The relative position of these cells indicates that weeks or months have elapsed since organelle loss was initiated.

The first indication that the denucleation process has begun is observed in cells just outside the OFZ and consists of a change in nuclear shape, from an elongated to a more spherical form. The multiple nucleoli that are present in elongating fiber cells fuse into a single, large nucleolus with a spokelike configuration (Kuwabara and Imaizumi, 1974). As the nucleus rounds up, nuclear DNA begins to condense at the nuclear margin, and holes appear in the nuclear lamina (Bassnett and Mataic, 1997; Dahm et al., 1998a). The nuclear envelope subsequently disintegrates into a chain of membrane vesicles, and the

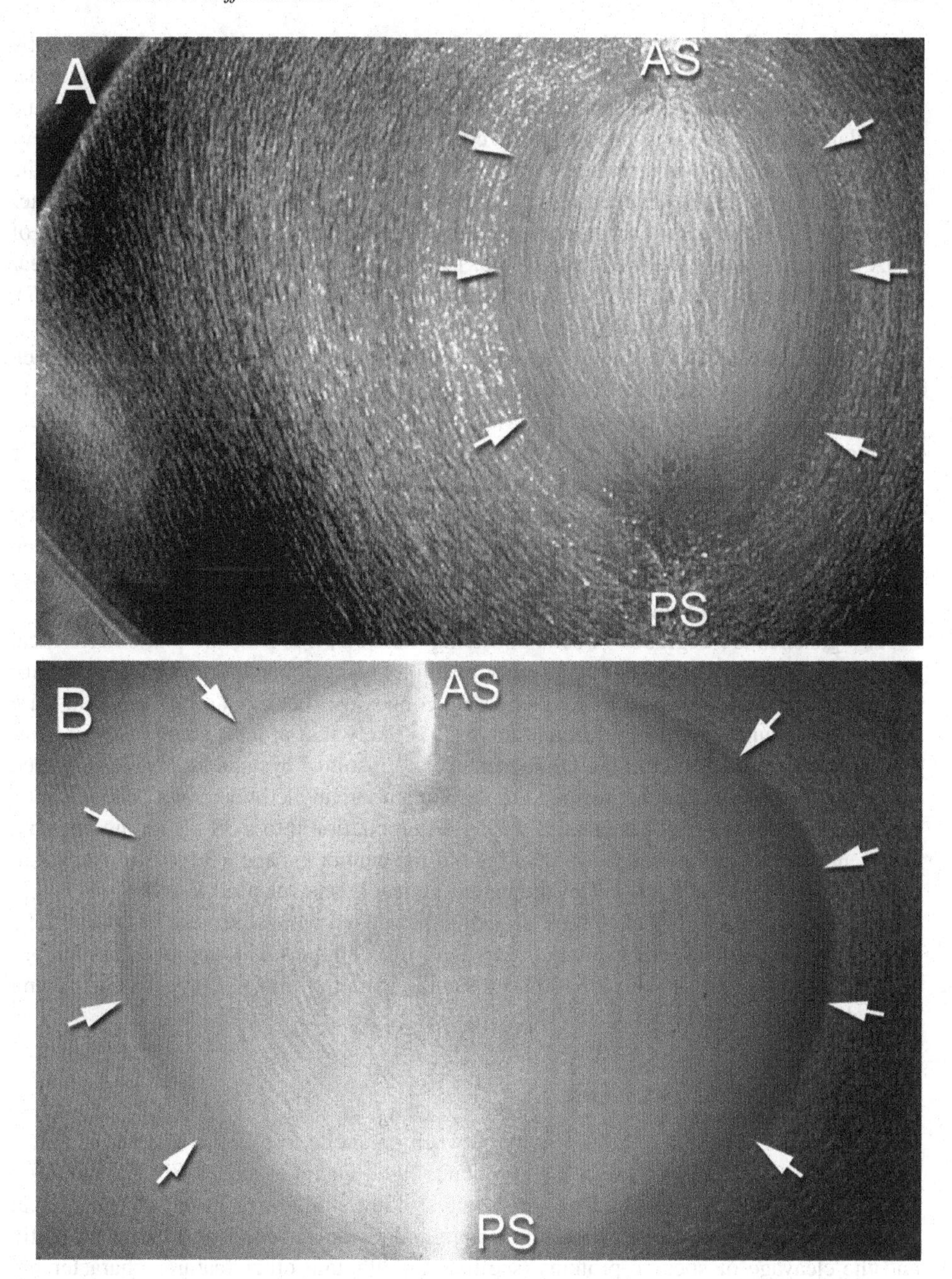

Figure 9.8. Formation of the organelle-free zone (OFZ). (A) Confocal images of a living slice prepared from an E12 lens and stained with a mitochondrial probe, rhodamine 123. The staining pattern reveals a core of primary fiber cells, which have lost their mitochondria. The resulting organelle-free zone (arrows in A and B) expands during embryonic development and by E16 has become sharply defined (B). As, anterior suture; PS, posterior suture.

nuclear contents become indistinguishable from the cytoplasm at the electron microscopy (EM) level (Kuwabara and Imaizumi, 1974). Immunofluorescence studies indicate that during this phase, cytoplasmic components, such as CP49, freely invade the nuclear volume (Sandilands et al., 1995a). A residual body, composed of condensed chromatin and putative nucleolar material, persists in the cytoplasm for an extended period after the loss of the nuclear envelope. In the chicken lens, final breakdown of the residual body occurs after a delay of 2–3 days and is accompanied by the sudden appearance of a large number of DNA strand scissions (Bassnett and Mataic, 1997). In the primate lens, a residual nuclear structure, perhaps consisting of nucleolar material, persists indefinitely (Bassnett, 1997). The condition of the nuclear DNA can be assessed by TUNEL (*t*erminal deoxynucleotidyl transferase-mediated d*U*TP-biotin *n*ick-*e*nd *l*abeling) assay (Gavrieli et al., 1992), which labels 3′-OH ends of nicked DNA. The onset of fiber DNA degradation occurs quite suddenly at the time of or shortly after mitochondrial and ER breakdown. At this stage, the nuclear chromatin is in the process of condensing and is strongly labeled by the TUNEL assay (Bassnett, 1997; Bassnett and Mataic, 1997; Dahm et al., 1998a; Ishizaki et al., 1998; Wride and Sanders, 1998). The appearance of TUNEL-positive nuclear remnants is paralleled by the release of nucleosome-sized fragments of DNA into the fiber cell cytoplasm. When resolved by agarose gel electrophoresis, the cleaved fragments of DNA form an evenly spaced ladder (Appleby and Modak, 1977).

As organelles are degraded in cells bordering the OFZ, it is possible that other critical maturation events occur concomitantly. For example, many abundant membrane and cytoplasmic proteins of the lens (among them MIP, connexins, spectrin, and β-crystallins) are known to be truncated during differentiation. It is likely that at least some of these truncations occur at the border of the OFZ, perhaps as a result of bystander damage sustained during the proteolysis that is presumed to accompany organelle breakdown. For example, connexin 50 (connexin $\alpha8$) is cleaved during differentiation into a 38-kD amino terminal fragment that remains embedded in the fiber plasma membrane and a 32-kD fragment that is derived from the carboxyl tail of the parent molecule and released into the cytoplasm. Proteolysis destroys a critical epitope, allowing the cells in which cleavage of connexin 50 has occurred to be identified by immunocytochemical techniques. This approach has demonstrated that proteolysis of connexin 50 occurs at the border of the OFZ, in cells undergoing organelle destruction (Lin et al. 1997).

9.7.2.2. *Role of Caspases*

Several authors have noted the similarity between organelle loss in the lens and classical apoptosis (Bassnett, 1997; Bassnett and Mataic, 1997; Dahm et al., 1998b; Dahm, 1999; Wride et al., 1999). Certainly, there are some striking similarities in the morphological features of the two processes, including the marginalization and condensation of chromatin and the cleavage of specific proteins (see below), although other features characteristic of apoptosis, such as membrane blebbing, are not observed during organelle loss. Recent findings also hint at conservation at the biochemical level.

A family of cysteine proteases has been shown to play a central role in apoptosis. These proteases, which are related to the interleukin-1 β-converting enzyme, are known as caspases (Alnemri et al., 1996). Caspases operate in a cascade, where proteolytic activation of upstream regulatory members of the family (e.g., caspase-8, -9, and -10) results in activation of effector caspases (e.g., caspase-3, -6, and -7). The effector caspases, in turn, cleave critical substrate molecules within the apoptotic cell, including poly(ADP-ribose) polymerase

(PARP), lamin B, DNA fragmentation factor (DFF45), and spectrin. The central role of caspases in apoptotic cell death is amply demonstrated by the observation that, in many instances, synthetic caspase inhibitors can prevent cells from dying in response to stimuli that would otherwise prove lethal.

PARP is a ubiquitous nuclear enzyme that binds to either double- or single-stranded DNA breaks and catalyzes the transfer of an ADP-ribose moiety from NAD^+ to itself and other nuclear acceptor proteins. The binding of PARP to broken DNA ends triggers a 500-fold stimulation in enzymatic activity. Following exposure of cells to environmental or endogenous genotoxic agents, PARP serves to trigger the base excision repair pathway. Caspase-mediated cleavage of PARP during apoptosis is thus thought to incapacitate the DNA damage surveillance network and eliminate fruitless attempts to repair the rapidly fragmenting DNA.

Like most eukaryotic cells, lens fibers possess extensive DNA repair machinery. Chicken fiber cells respond to x-radiation by an increase in nuclear ADP-ribosylation (Counis et al., 1985), presumably mediated by PARP. Interestingly, the capacity of lens fiber cells to repair damaged DNA decreases during fiber cell differentiation *in vitro*.

The catalytic and DNA-binding domains of PARP are separated by the caspase target sequence DEVD. Caspase-mediated cleavage of PARP generates 80-kDa and 20-kDa products corresponding to the catalytic and DNA-binding domains, respectively. The latter may play a direct role in the later stages of apoptosis by binding to DNA breaks and thereby preventing access by other repair enzymes.

Full-sized PARP ($\sim$120 kDa) is observed in the developing rat lens but is progressively cleaved to an $\sim$80-kDa fragment (Ishizaki et al., 1998). In the outer fibers of the adult rat lens, the 80-kDa fragment is the predominant form. Similarly, in the chicken lens, cleaved PARP is only detected late in embryonic development, after the onset of organelle degradation (Wride et al., 1999).

Although there is no direct evidence that PARP is cleaved by caspase during fiber cell differentiation *in vivo, in vitro* experiments support this notion. Lens epithelial cells cultured under appropriate conditions undergo many of the morphological transformations observed during fiber differentiation *in vivo,* including progressive cleavage of PARP. The addition of the pan-caspase inhibitor zVAD-fmk to cultures of rat or chicken lens cells inhibited PARP cleavage (Ishizaki et al., 1998; Wride et al., 1999). Taken together, these observations suggest that PARP is degraded during fiber cell differentiation by the action of one or more caspases.

Immunofluorescence studies have indicated that the nuclear lamina is dismantled during organelle loss (Bassnett and Mataic, 1997; Dahm et al., 1998a) and that lamin B, a prominent component of the nuclear lamina, is ultimately degraded (Bassnett and Mataic, 1997). Lamin B is thought to be specifically susceptible to digestion by caspase-3 or caspase-7 (Slee et al., 2000).

Immuncytochemical studies indicate that DFF45 is expressed in the outer fiber cells and that cleavage to a 30-kD product is first observed at E12 in the chicken lens (Wride et al., 1999). This corresponds to the stage at which organelle loss is initiated.

Thus, at least three classic caspase substrates, PARP, lamin B, and DFF45, appear to be proteolytically cleaved at the border of the OFZ. All are well-characterized caspase-3 substrates but may be cleaved by other members of the caspase family. In contrast, the cleavage of alpha II-spectrin appears to be mediated specifically by caspase-3 (Janicke et al., 1998). Caspase-cleaved spectrin has recently been identified in the developing chicken lens at E12 (Lee et al., 2001), the stage at which organelle degradation commences in core fiber cells.

Caspase-3 is itself cleaved, and thereby activated, in apoptotic cells. Caspase-3 has been localized in the embryonic chicken lens by immunocytochemistry, and a 17-kD fragment has been identified by Western blotting of lens lysates from >E12 embryos (Wride et al., 1999). Caspase-3 activity has not yet been measured directly in the developing lens. However, in lentoid cultures from rat or chicken, treatment with the pan-caspase inhibitors Z-VAD-FMK and Boc-D-FMK leads to a significant reduction in the number of TUNEL-labeled nuclei (Ishizaki et al., 1998; Wride et al., 1999). Interestingly, in this *in vitro* setting, the caspase-3–specific inhibitor Z-DEVD-FMK is ineffective (Wride et al., 1999), perhaps suggesting the presence of other effector caspases that might substitute for caspase-3.

Thus, a strong circumstantial case has emerged that implicates caspases, particularly caspase-3, in organelle degradation: Caspase-3 activation coincides with the onset of organelle breakdown; four well-characterized caspase substrates (PARP, lamin B, DFF45, and spectrin) are cleaved during organelle degradation; and, at least *in vitro*, caspase inhibitors block certain aspects of organelle degradation. Given that effector caspases appear to be activated during organelle loss and that the expected apoptotic substrates are cleaved, it is pertinent to ask why lens fiber cells do not undergo classic apoptosis rather than organelle degradation. Interestingly, the nuclear events preceding organelle loss closely resemble those observed during apoptosis, but the membrane events do not. It is not currently understood how the nuclear and membrane elements of the apoptotic program have been dissociated from each other during lens organelle loss.

9.7.2.3. Role of Nucleases

The disintegration of the fiber cell nucleus is accompanied by the progressive cleavage of DNA. Strand scissions produce multiple 3′-OH ends in the DNA (Bassnett and Mataic, 1997) and ultimately release nucleosome-sized fragments into the cytoplasm (Appleby and Modak, 1977). The identity of the nuclease (or nucleases) responsible has not been determined, but several candidate enzymes have been detected.

The endonuclease responsible for DNA laddering during apoptosis is DNA fragmentation factor (DFF; Liu et al. 1997). In its inactive form, DFF is a heterodimer composed of a 45-kDa inhibitory chaperone subunit (DFF45) and a 40-kDa latent endonuclease subunit (DFF40). The DFF complex resides in the nucleus (Samejima and Earnshaw, 2000). Caspase-3 specifically cuts DFF45, resulting in the dissociation of the cleaved DFF45 from DFF40. The released endonuclease forms homo-oligomers that are the enzymatically active form of DFF40 (also known as CAD or CPAN). Studies on the cleavage preferences of DFF40 on naked DNA have demonstrated that this nuclease has a pH optimum of 7.5, requires Mg^{2+} (but not Ca^{2+}), and is inhibited by Zn^{2+}. DFF40 attacks the chromatin in the internucleosomal linker, producing blunt ends or ends with 1-base 5′-overhangs possessing 5′-phosphate and 3′-hydroxyl groups (Widlak et al., 2000). Prolonged incubation of DNA with DFF40 results in the generation of oligonucleosomal DNA ladders. Immunocytochemical studies on the lens have demonstrated the presence of DFF45, although expression is apparently restricted to the cytoplasm (rather than the nuclei) of epithelium and outer fiber cells (Wride et al., 1999). Significantly, immunoblotting experiments detected the 30-kDa cleaved form of DFF45 in the embryonic chicken lens only at >E12, suggesting that, *in vivo*, the activation of DFF may coincide with the initiation of organelle loss. It is tempting to speculate that DFF activation may result from proteolysis by caspase-3, especially as this enzyme is also known to be activated in the lens at E12 (Wride et al., 1999). However, incubation of lens cultures with a variety of caspase inhibitors failed to prevent cleavage of DFF, indicating that, at least *in vitro*, DFF activation occurs via a caspase-independent pathway (Wride et al., 1999).

DNases may be grouped into three functional categories: Mg^{2+}-dependent endonucleases (of which DFF40 is the best-studied example), Ca^{2+}/Mg^{2+}-dependent endonucleases, and cation-independent endonucleases (Counis and Torriglia, 2000). As a group, nucleases are poorly characterized at the molecular level. In only a few cases are primary sequences available, and rarely has a physiological role in chromatin breakdown been demonstrated unequivocally. The best-characterized Ca^{2+}/Mg^{2+}-dependent endonuclease is DNase I. This enzyme was first purified from bovine pancreas and has since been cloned and crystallized for structural studies. The cleavage of DNA by DNase I generates strand breaks with free 3′-OH ends, and this is the characteristic cleavage pattern observed during fiber cell maturation (Bassnett and Mataic, 1997; Ishizaki et al., 1998; Wride and Sanders, 1998). However, RT-PCR analysis of lens cDNA has failed to detect any transcripts for DNase I (Hess and FitzGerald, 1996). Therefore, if a Ca^{2+}/Mg^{2+}-dependent nuclease is involved in fiber denucleation, it is unlikely to be DNase I itself. A DNase I–like 30-kDa protein has been identified in the lens by DNase activity gels and immunoblots (Counis et al., 1991; Arruti et al., 1995; Torriglia et al., 1995). A 40-kDa DNase, inhibited by high concentrations of divalent ions but with an absolute requirement for Ca^{2+} and Mg^{2+}, has also been demonstrated in both epithelial and fiber cells (Counis et al., 1991).

The Ca^{2+}/Mg^{2+}-independent nucleases include DNase II–like enzymes. DNase II is a ubiquitously expressed enzyme (Counis and Torriglia, 2000) and has been implicated in DNA fragmentation during apoptosis (Eastman, 1994; Belmokhtar et al., 2000). Four DNase II–like proteins have been identified in the lens by immunoblot analysis (Torriglia et al., 1995), and evidence for a role in fiber cell denucleation was provided by experiments in which antibodies to DNase II inhibited DNA degradation in isolated lens cell nuclei. Immunocytochemical localization of DNase II–like nucleases in fiber cells indicated that this enzyme is highly concentrated in the fiber cell nuclei (Torriglia et al., 1995; Counis et al., 1998). It should be noted, however, that DNase II digestion of DNA generates 3′-phosphate ends and that there is little evidence for accumulation of this kind of DNA damage during fiber differentiation *in vivo* (Bassnett, 1997; Bassnett and Mataic, 1997).

9.7.2.4. What Triggers Organelle Degradation?

Perhaps the most striking aspect of organelle degradation in the lens is that it appears to occur simultaneously in a thin shell of fiber cells. It is not known what serves to trigger organelle loss in this particular cell layer. In principal, the trigger could be spatial or temporal in nature (or both). The border of the OFZ is located approximately 800 μm beneath the lens surface in the embryonic chicken lens (Bassnett and Beebe, 1992) and 100–200 μm below the surface in the adult monkey lens (Bassnett, 1992).

Because the distance between the lens surface and the border of the OFZ remains relatively constant throughout chicken lens embryonic development (see Fig. 9.9), it has been suggested that organelle degradation may be triggered by a gradient of a diffusible substance (Bassnett and Mataic, 1997). As the volume of the lens increases, many standing gradients will be established because of the avascularity of the tissue and the relatively slow diffusion-limited exchange of metabolites with the surrounding humors of the eye. For example, glycolysis in the bulk of the tissue results in the establishment of a lactic acid gradient (Bassnett et al., 1987) and therefore a gradient of intracellular pH (Bassnett et al., 1987; Mathias et al., 1991). Any metabolite diffusing from the surface and consumed by the lens

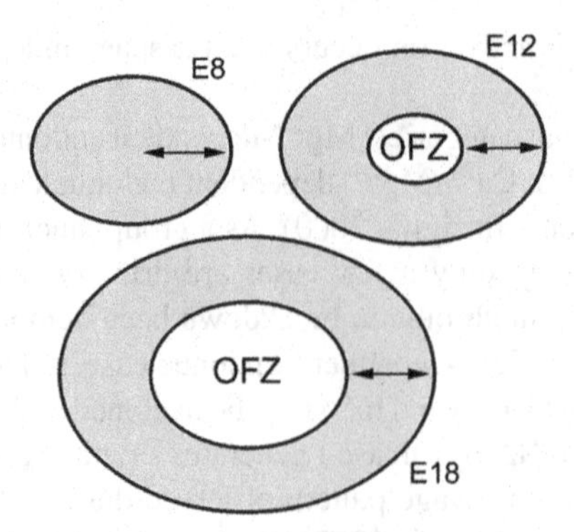

Figure 9.9. The relative sizes of the lens and the organelle-free zone (OFZ) during embryonic development. The images are drawn to scale from data presented in Bassnett and Beebe (1992). The arrows indicate the constant thickness of the layer of organelle-containing fiber cells.

will likely be found at lowest concentration in the lens core. Of particular relevance may be intracellular oxygen. Oxygen is a permeable species that will be consumed in mitochondria located in the outer layers. As the lens increases in volume, oxygen concentration in the lens core is expected to fall. If the oxygen reaches a sufficiently low concentration, the integrity of the mitochondria would be compromised, perhaps enough to trigger an apoptoticlike cascade, including release of cytochrome C and activation of the caspase pathway. This is an attractive hypothesis, but there is currently little experimental evidence to support it. Oxygen tension in the eye is low, and the noninvasive measurement of intracellular oxygen gradients in the lens is technically challenging.

Lens fiber cells express TNFα and TNF receptors (TNFR1 and TNFR2), and it has been proposed that this proinflammatory cytokine may play a role in triggering lens fiber organelle degradation (Wride and Sanders, 1998). Of the 17 or more members of the TNF superfamily, TNF is probably the most potent inducer of apoptosis, acting largely through the type 1 receptor (Rath and Aggarwal, 1999). *In vitro*, the number of TUNEL-positive cells in differentiating lens cell cultures increases following application of TNFα or agonistic antibodies to TNFR1 and TNFR2. Conversely, neutralizing antibody to TNFα caused a significant reduction in TUNEL labeling (Wride and Sanders, 1998).

9.7.3. Cessation of Transcription and Translation

9.7.3.1. Transcription and Translation at the Border of the OFZ

Presumably, with the loss of the fiber cell nucleus, transcription stops and any residual translation must depend on the stability of extant mRNAs and the durability of the protein synthetic machinery. It is important, however, to determine precisely when transcription ceases. It is possible, for example, that transcription is terminated long before organelles are degraded. Morphological data suggest that transcription may stop in cells immediately adjacent to the OFZ. Coilin and fibrillarin are markers for distinct nuclear compartments (the coiled body and nucleolar compartments, respectively) that show characteristic distributions in transcriptionally active cells. In the lens, both of these markers disappear shortly before nuclear condensation. This has been interpreted as signifying transcriptional shutdown (Dahm et al., 1998a).

The transcriptional and translational competence of differentiating fiber cells has been assessed directly by testing the ability of a fiber cell to synthesize the autofluorescent protein GFP from a microinjected plasmid template (Shestopalov and Bassnett, 1999). Twenty-four hours after plasmid injections into superficial regions of the lens, strongly fluorescent fiber cells were observed. As injections were made at progressively greater depths, a point was reached where injections never resulted in GFP synthesis. The last GFP-synthesizing fiber was estimated to lie immediately adjacent to the OFZ. From these functional measures and correlative morphological analyses, it appears that fiber cells are transcriptionally and translationally competent until shortly before organelle degradation.

9.7.3.2. Stability of RNA in Mature Fiber Cells

Because they lack an ER, mature fiber cells presumably are no longer able to synthesize membrane proteins or secreted proteins. However, it is conceivable that residual synthesis of cytoplasmic proteins (e.g., crystallins) might continue in these cells. In this regard, it is interesting to note that certain mRNAs are unusually long lived in lens fibers. For example, Northern blot analysis has demonstrated the persistence of full-length δ-crystallin mRNA transcripts in the core region of lenses from three-month-old chickens (Treton et al., 1982). Thus, in comparison with most mRNAs (which have half-lives measured in minutes or hours), δ-crystallin mRNA in lens fibers is remarkably stable. The stability of this transcript may also reflect intrinsically low RNAse activity in the core of the lens or the presence of an endogenous ribonuclease inhibitor (Ortwerth and Byrnes, 1971, 1972).

Despite the persistence of specific mRNAs, it appears unlikely that significant protein synthesis occurs in cells within the OFZ. In the majority of these cells, ribosomes are completely absent (Kuwabara, 1975), and most attempts to metabolically label nascent proteins have proved unsuccessful (Wannemacher and Spector, 1968; Bagchi et al., 1981; Russell et al., 1996). Proteins that are highly expressed in differentiated cells are often translated from unusually stable mRNAs, and this is thought to account, in part, for their abundance. Thus, a half-life of >15 hours has been reported for myosin (Buckingham et al., 1974), immunoglobulin (Cowan and Milstein, 1974), and ovalbumin (Palmiter and Carey, 1974) mRNAs. By this argument, the stabilization of crystallin mRNA templates may be part of the mechanism by which fiber cells are able to synthesize these proteins in such abundance. The hypothesis that specific mRNAs are stabilized to allow their translation in anucleated core fiber cells appears to be less likely.

9.8. How Is the Process of Fiber Cell Differentiation Related to the Overall Shape of the Lens?

Lenses depend for their function on having a precise shape and refractive index. The shape of the lens is essentially the shape of the fiber cell mass, which is directly related to the shape of the fiber cells themselves. It is remarkable, therefore, that the shape of the lens may change dramatically during development and that the lenses of different species vary greatly in shape, from nearly spherical to distinctly oblate, with different anterior and posterior surface curvatures. To our knowledge, no studies have related the shape of the lens to the properties of individual fiber cells. What little is known about lens shape is that it is determined by the actions of factors outside the lens.

The first study to clearly demonstrate the effect of the extralenticular environment on lens morphogenesis was performed by Jane Coulombre and Chris Coulombre (1963). They showed that if the lens were rotated 180° so that the epithelium faced the retina, the lens repolarized, forming a new fiber mass from the epithelial cells and reforming an epithelium over what had previously been the basal ends of the fiber cells. In later experiments, these authors showed that when the original lens was removed and two lenses were implanted into the eye in its place, the two implanted lenses grew as if they were a single lens, irrespective of their orientation at the time of implantation (Coulombre and Coulombre, 1969). This "compound" lens eventually approximated the size and shape of the single lens from the contralateral, unoperated eye. This elegant experiment showed that the shape of the lens was not an intrinsic property of the tissue itself but was imposed by factors in the ocular environment. It would be interesting to extend this experiment to see whether lenses transplanted to the eye of a different species would assume a shape characteristic of the host or the donor.

9.9. Lens Pathology: Cataracts Caused by Abnormal Fiber Cell Differentiation

9.9.1. Posterior Subcapsular Cataracts

There are three major types of human age-related cataracts: nuclear, cortical, and posterior subcapsular. Nuclear and cortical cataracts occur in the fully differentiated cells of the lens nucleus and cortex, respectively. Typical posterior subcapsular cataracts (PSCs) are due to an accumulation of swollen cells at the posterior sutures, just beneath the posterior capsule. The morphology of PSCs suggests that they are caused by the abnormal differentiation of lens fiber cells (Streeten and Eshaghian, 1978; Eshaghian and Streeten, 1980).

Examination of the posterior regions of lenses with PSCs shows one or more "streams" of cells apparently migrating on the inner surface of the posterior capsule and extending from the lens equator to the plaque of swollen cells at the posterior pole. Examination of the cells in these streams shows that, as they migrate toward the posterior pole, they come to resemble swollen lens fiber cells (Eshaghian and Streeten, 1980). The swollen cells, along with cellular debris from the death of some of the migrating cells, accumulate at the posterior pole, causing the light scattering that is responsible for the cataract (Eshaghian and Streeten, 1980; Eshagian, 1982).

The causes of PSCs are not well understood. These cataracts occur with lower frequency than nuclear or cortical cataracts in most populations (approximately 10% of cataract surgeries). There is an increased risk for developing PSCs in patients on steroid therapy and in individuals with hereditary retinal degenerations, like retinitis pigmentosa. PSCs are one of the early signs of neurofibromatosis type 2 (NF2), a hereditary disease caused by mutations in the merlin tumor suppressor gene (Kaiser-Kupfer et al., 1989; Lim et al., 2000). However, there is little information to suggest why steroid therapy or retinal degeneration should lead to the formation of a PSC, and the mechanism of PSC formation in NF2 is not known. Research in this area has been hampered by the lack of an animal model that develops PSCs resembling those seen in human patients.

The morphology of the cells responsible for PSC formation suggests that this type of cataract is caused by a failure of some of the cells in the lens to differentiate properly. If a proportion of lens cells failed to elongate at the equator, they would remain attached to the posterior capsule and be carried to the posterior pole of the lens by the movement of the basal ends of adjacent fiber cells. Based on the mechanism of cell elongation proposed above, one way that cells might fail to elongate would be if they lost their adhesive contacts with their neighbors. This would cause them to round up as they increased in volume rather than elongate. Such a view is consistent with the emerging role of the merlin tumor suppressor gene as a regulator of cell migration and, possibly, in connecting the actin cytoskeleton with the plasma membrane (Lim et al., 2000; Reed and Gutmann, 2001).

9.9.2. Secondary Cataracts

Modern cataract surgery is performed by peeling off the central portion of the lens epithelium, breaking up and removing the lens fiber mass, and implanting a clear plastic lens into the remaining "capsular bag." Epithelial cells from the lens equator sometimes migrate on the denuded posterior capsule beneath the lens implant. There they may differentiate into small clusters of lens fiber cells (Elschnig's pearls) that resemble the lentoid bodies that often form in primary cultures of lens epithelial cells. They may also form plaques of fibrous tissue. These plaques contain myofibroblast-like cells embedded in an abundant extracellular matrix. Both Elschnig's pearls and fibrous plaques scatter light and degrade the visual image, forming secondary cataracts. Secondary cataracts are also often referred to as "posterior capsule opacification" (PCO) or "after cataracts."

The cells in Elschnig's pearls closely resemble lens fiber cells (Kappelhof et al., 1986; Sveinsson, 1993; Marcantonio and Vrensen, 1999). It is probable that these fiberlike cells are derived from equatorial epithelial cells that have migrated onto the posterior capsule, responded to the normal environment there, and begun the process of fiber cell differentiation. The abnormal size and shape of these lentoids are likely to be due to the abnormal physical environment in which they are differentiating. They do not have a "template" of preexisting fiber cells on which to elongate, possibly explaining their swollen shape, and the size of the tiny lenslike structures that are formed are disproportionately small for the adult eye. We know of no studies that indicate the degree to which the biochemical characteristics of these fiberlike cells resemble authentic lens fiber cells.

In additions to forming Elschnig's pearls on the posterior capsule, lens fiber cells frequently differentiate near the lens equator. Here they often surround the haptics that position the IOL in the capsular bag. The aggregate of fiberlike cells that forms in this region is called Soemmering's ring. Because it is positioned outside of the optic axis, Soemmering's ring does not degrade the visual image. In young animals that have had experimental cataract surgery, the differentiation of fiber cells at the lens equator can re-form a fiber mass that has optical qualities comparable to the original (Gwon et al., 1989; Gwon et al., 1990; Gwon et al., 1992).

In contrast to the similarity of the cells in Elschnig's pearls to lens fiber cells, the fibrous plaques found in secondary cataracts result from the transdifferentiation of epithelial cells into myofibroblast-like cells (Novotny and Pau, 1984; Schmitt-Graff et al., 1990; Hales et al., 1994; Marcantonio and Vrensen, 1999; Nagamoto et al., 2000). These cells are contractile and express α-smooth muscle actin, collagen type I, and a variety of other markers of fibrotic connective tissues (Hales et al., 1994; Hales et al., 1995; Kurosaka et al., 1995; Lee and Joo, 1999; Marcantonio and Vrensen, 1999; Nagamoto et al., 2000; Wunderlich et al., 2000). In experimental studies, lens epithelial cells can be stimulated to acquire many of the properties of myofibroblasts by treatment with the cytokine TGFβ (Hales et al., 1994; Kurosaka et al., 1995; Lee and Joo, 1999). However, it has not yet been demonstrated that signaling through this pathway is responsible for myofibroblast differentiation during secondary cataract formation *in vivo*.

9.10. Concluding Remarks

The transparency and refractive properties of the lens depend directly on cellular features that arise during fiber cell differentiation. Therefore, a detailed knowledge of the differentiation process is central to an understanding of the fundamental optical properties of the lens. It also provides the basic information necessary to understand the etiology of certain types of cataract.

In this chapter we divided fiber cell differentiation into four stages that are identified by major physical landmarks common to all lenses. We hope that these stages provide a useful conceptual framework for the design and interpretation of future studies.

The formation of lens fiber cells is one of the most extreme examples of cell differentiation in nature, culminating in a degree of specialization scarcely equaled by any other cell type. However, as this chapter illustrates, there are many areas that remain to be explored before we can claim to have a full understanding of the events that underlie the differentiation process.

10

Role of Matrix and Cell Adhesion Molecules in Lens Differentiation

A. Sue Menko and Janice L. Walker

Epigenetic signals resulting from either cell-matrix or cell-cell interactions are critical to the regulation of cell differentiation and development. This is particularly true in the complex differentiation process that enables a lens epithelial cell to become a differentiated lens fiber cell and in the developmental events that direct a region of head ectoderm to invaginate, pinch off, and begin to form the lens. In this chapter we discuss the role of cell adhesion molecules in lens differentiation and development.

10.1. Extracellular Matrix

The molecular organization of the basement membrane can profoundly influence cellular behavior by providing information that can affect the genetic program of a cell. Its major components include proteins such as laminin, collagen type IV, fibronectin, and proteoglycans. These extracellular matrix (ECM) proteins direct differentiation-specific gene expression in most cell and tissue types (Bissell and Barcellos-Hoff, 1987; Bissell et al., 1982; Streuli et al., 1991). Their ability to orchestrate both cell differentiation and tissue development requires interaction with cell surface receptors (e.g., the integrins), by which they initiate specific intracellular signaling pathways (Giancotti and Ruoslahti, 1999). The expression and distribution of ECM proteins in the lens is well characterized, both throughout development and in the mature lens. The knowledge that has been gained, in combination with mutational and inhibitor studies, provides insight into the role of ECM molecules in the process of lens development.

10.1.1. Lens Determination

Proteins that compose the ECM are involved in lens development as early as the lens determination stage, when the optic vesicle interacts with the region of head ectoderm destined to become the lens. The correct spatiotemporal expression of glycosaminoglycans (GAGs) and proteoglycans appears to be critical to this early developmental step. Cross-linking and polymerization of sulfated GAGs present in the intercellular space between the lens primordia and the optic vesicle is thought to promote the attachment of the presumptive lens to the optic vesicle (Webster et al., 1983). Alteration of GAG composition, as occurs in aphakia (Zwaan and Webster, 1984), trisomy 1 (Smith, 1989), or myelencephalic bleb (*my*) mice (Center and Polizotto, 1992), disrupts the contact between the optic vesicle and the lens and results in abnormal lens development. Changes in the distribution of GAGs in the intercellular space between the lens vesicle and overlying cornea also occur in the early stages of lens development. Chondroitin sulfate is expressed in this matrix compartment before separation of the lens vesicle from the region of the head ectoderm, while heparan sulfate appears in the intercellular matrix between the lens and the cornea after separation (Mizuno et al., 1995). Therefore, GAGs play an important role in maintaining the boundaries required for separation of developing eye structures as well as for proper lens morphogenesis.

The syndecans, a family of transmembrane receptors, are unique proteoglycans expressed by the lens. Their cytoplasmic domain interacts with the actin cytoskeleton, and their extracellular domain binds ECM components such as collagen and fibronectin. Syndecans play a role in modulating cellular response to fibroblast growth factor (FGF; Horowitz et al., 2002; Kirsch et al., 2002). The expression of syndecans is spatiotemporally regulated with lens development. Syndecan 1 is expressed in the region of head ectoderm that will become the lens placode (Trautman et al., 1991). Once the head ectoderm differentiates into the lens placode, syndecan 1 expression is lost. In contrast, syndecan 3 is expressed by the epithelial cells of the lens placode but is absent in head ectoderm before it is specified as a lens (Gould et al., 1995). Syndecan 3 persists in the lens as it differentiates and is expressed by the lens fiber cells.

The ECM proteins laminin, fibronectin (Haloui et al., 1988; Parmigiani and McAvoy, 1984), and collagen type IV (Hilfer and Randolph, 1993) are also localized to the region between the optic vesicle and the head ectoderm at the time when they are closely associated. Laminin and fibronectin (Parmigiani and McAvoy, 1984), along with type IV collagen and heparan sulfate proteoglycan (Haloui et al., 1988), become even more strongly associated with the basal surfaces of the lens epithelial cells as the lens placode invaginates to form the lens pit. It has been suggested that these ECM components regulate the coordinate invagination of the lens pit and the optic cup.

10.1.2. Lens Development

From the time the lens vesicle pinches off the head ectoderm, it is surrounded by a basement membrane capsule. Throughout their differentiation from epithelial to fiber cells, lens cells remain in contact with this capsule, leaving only when they reach the lens suture. The makeup of the capsule has been determined for a number of species. Components of this complex basement membrane include laminin, type IV collagen, heparan sulfate proteoglycan, fibronectin, tenascin, and SPARC (Cammarata et al., 1986; Fitch et al., 1983; Menko et al., 1998; Parmigiani and McAvoy, 1991; Sawhney, 1995). Most are distributed throughout the thickness of the capsule (Cammarata et al., 1986). Many of these capsule proteins are expressed by multiple cell types in the developing eye (Dong and Chung, 1991). Therefore, it is possible that the matrix proteins that compose the lens capsule could be synthesized by other ocular tissues as well as by the lens.

10.1.2.1. Fibronectin

Even though fibronectin is only a minor component of the epithelial cell basement membrane, it plays an important role in initiating the signaling pathways required for cell proliferation (Bourdoulous et al., 1998; Kuwada and Li, 2000). While most studies have shown fibronectin expression in the lens capsule, details as to its spatiotemporal distribution during development and in the mature lens are conflicting. This may be due to differences between species, different specificities of the antibodies used, or differences of antibody accessibility to fibronectin in the capsule. Immunolocalization analysis has shown that fibronectin expression in the embryonic rat lens capsule is patchy and weak (Parmigiani and McAvoy, 1991). These studies demonstrate that although fibronectin is present in the lens capsule early in rat development, it becomes an increasingly minor component as the lens develops. In the embryonic chick lens, fibronectin is compartmentalized within the capsule, is associated primarily with cells of the anterior epithelium, and extends into the most anterior region of the equatorial epithelium, in association with the germinative zone (Menko et al., 1998). In these lenses, fibronectin was not found in regions of the capsule associated with postproliferative cells. In contrast, in the mature mouse lens, fibronectin is expressed in both anterior and posterior zones of the capsule, although in the anterior capsule, fibronectin is limited to the side of the capsule facing the aqueous humor (Duncan et al., 2000). These differences in fibronectin expression and distribution in the lens capsule in different species and different stages of development may reflect its role as a dynamic signaling molecule. It is also possible that we will find that these discrepancies may reflect variations in fibronectin isoform expression.

10.1.2.2. Collagen Type IV

Collagen type IV is found throughout the lens capsule at the time the lens forms, but its distribution within the capsule is regulated in conjunction with the state of lens development. Expression of collagen type IV decreases in the posterior lens capsule with increased developmental age, while it remains high in the anterior capsule (Fitch et al., 1983). An examination of the different collagen IV family members in lens development shows that collagen IV subunits alpha1(IV), alpha2(IV), alpha5(IV), and alpha6(IV) are expressed early in development. These collagens form fibrillar and elastic protomers. Later in development, collagen alpha3(IV) and alpha4(IV) were detected in the more cross linked collagen protomers. It is thought that the more elastic collagen fibers facilitate the rapid growth phase of the lens whereas the stronger, more cross linked forms of the matrix may provide the strength to withstand the mechanical stresses to which older lenses are subjected (Kelley et al., 2002). Specific mutations that lead to altered expression of collagen type IV in the lens indicate that the regulated expression of collagen type IV is required not only for normal lens development but for homeostasis. Patients with Alport's syndrome carry a mutation in the collagen type IV gene that leads to lens capsule abnormalities. These lenses are characterized by a marked thinning of the anterior lens capsule, which leads to weak and ruptured capsules and the formation of cataracts (Colville and Savige, 1997; Olitsky et al., 1999; Takei et al., 2001). Lens capsule abnormalities and the development of cataracts are also found in mice carrying the Cat Fraser mutation, here thought to result from the increased synthesis of collagen type IV by lens epithelial cells (Haloui et al., 1988; Haloui et al., 1990).

10.1.2.3. Laminin

Laminins are heterotrimeric molecules that make up a large family of proteins. In many tissues, these basement membrane proteins have been shown to signal cells to withdraw from the cell cycle and initiate their differentiation (Kubota et al., 1988; Streuli et al., 1991; Von and Ocalan, 1989). Laminin has been shown by many studies to be a major component of the lens capsule (Cammarata et al., 1986; Cammarata and Spiro, 1985; Parmigiani and McAvoy, 1991; Walker and Menko, 1999). In the developing lens, it is present throughout

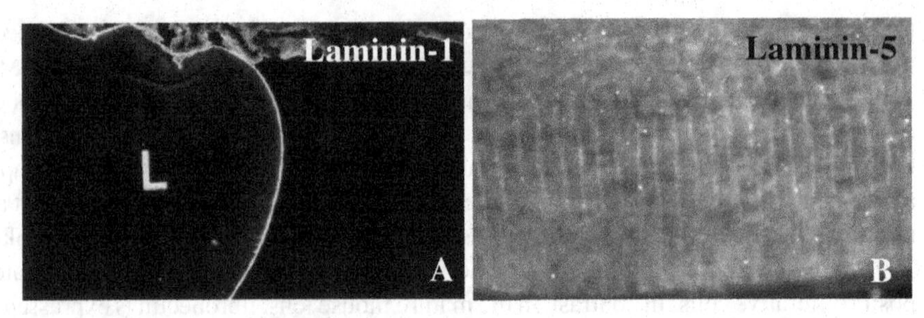

Figure 10.1. Distribution of laminins in the lens. Cryostat sections were cut from E10 chick embryo lenses (A) and adult mouse lenses (B). Sections were immunostained with antibodies to laminin (A) or laminin-5 (B). Staining for laminin, most likely to reflect laminin-1 distribution, is detected only in the lens capsule (A), while laminin-5 localizes to cell–cell interfaces along the posterior aspects of lens fiber cells (B).

all regions of the capsule (Fig. 10.1A). Most of the early studies of laminin in the lens were performed with antibodies that did not distinguish the different laminin family members. More recently it has been shown that the laminin-1 chain, a component of laminin-1, is expressed by the lens (Falk et al., 1999). Another laminin family member, laminin-5, has a pattern of distribution distinct from that of laminin-1. This laminin is typically associated with hemidesmosomes, adhesion complexes not found in the lens, and is involved in directing cell migration. Laminin-5 is found along the anterior and posterior tips of developing lens fiber cells, where they contact the anterior epithelium and posterior capsule, respectively, and it extends laterally for a short distance between the fiber cells (Fig. 10.1B; Menko, unpublished manuscript). This localization pattern suggests that it may mediate migration of these cells as they move toward the central suture. Laminin-5 is unique among ECM components in the lens, being the only matrix protein to have a noncapsular localization. The expression and distribution of other members of the laminin family of proteins in the lens capsule remain to be determined. However, the presence of bilateral cataracts in individuals with a congenital form of muscular dystrophy associated with defects in the laminin family member merosin (Reed et al., 1999; Topaloglu et al., 1997) suggests that this laminin family member is not only present in the lens capsule but important for its homeostasis.

10.1.2.4. Proteoglycans

Proteoglycans are also a significant component of the lens capsule. These matrix molecules typically bind to other components of the ECM such as collagen and fibronectin. Their importance in lens development and differentiation is shown by studies in which sulfated proteoglycan synthesis in the lens is altered by treatment with β-D xyloside. This results in the failure of primary lens fiber cells to elongate, the persistence of the lens cavity, and the degeneration of the anterior epithelium (Alonso et al., 1996). The role of proteoglycans in lens development and differentiation may in part be due to their ability to bind FGF and modulate its activity. In support of this, FGF and heparan sulfate proteoglycans have been demonstrated to co-localize in the lens capsule (Lovicu and McAvoy, 1993). This association is particularly important to the role of proteoglycans in lens development, since FGF has been reported to promote both lens proliferation and differentiation.

10.1.2.5. Tenascin

Tenascin is a secreted glycoprotein that exhibits both adhesive and anti-adhesive properties. This matrix protein can bind to cell surface receptors of the integrin family ($\alpha V\beta 3$) and the IgG superfamily. Within the matrix, it binds proteoglycans and other ECM proteins. It is has been implicated in the regulation of cell proliferation, cell migration, and apoptosis and the modulation of cell-matrix interactions. Although tenascin is found throughout the capsule in the lens vesicle stage (Tucker, 1991), it is more concentrated in the posterior capsule. Later in embryonic development, tenascin becomes restricted to the posterior capsule, where it is in contact with the differentiating fiber cells and extends into the most posterior regions of the equatorial zone, partially overlapping with fibronectin (Menko et al., 1998). As cells in the developing lens leave the equatorial epithelium and enter the cortical fiber zone, they establish extensive intercellular junctions and dramatically change their relationship with the lens capsule, becoming less adherent to the capsule (Atreya et al., 1989; Lagunowich and Grunwald, 1991). Tenascin would allow for the plasticity necessary for these changes in lens cell morphogenesis that occur as the cells move from strong cell-substrate interactions to strong cell-cell interactions. The adhesive and antiadhesive properties of tenascin also have clear functional implications for the cells associated with the posterior capsule. Tenascin has been shown to mediate cell sliding or sheet movement. In this capacity, it could allow differentiating lens fiber cells to maintain tight intercellular contact as they migrate along the posterior capsule toward the central suture.

10.1.2.6. SPARC

SPARC/osteonectin/BM40 (*s*ecreted *p*rotein *a*cidic and *r*ich in *c*ysteine) is a secreted Ca^{2+}-binding matricellular glycoprotein that is implicated in tissue repair, matrix remodeling, morphogenesis, antiproliferation, and modulation of cytokine activity (Brekken and Sage, 2000; Yan and Sage, 1999). SPARC binds to ECM components such as collagen and vitronectin and has antiadhesive properties, reducing focal contacts and cell-cell and cell-matrix adhesion (Brekken and Sage, 2000). There is as yet no known receptor for SPARC. SPARC is known as a matricellular protein because of its proposed dual roles within the matrix and within the cell.

In the lens, SPARC is found in both the lens capsule and the lens epithelium. Its expression is highest in the peripheral epithelium and ends in the equatorial region, where lens fiber cell differentiation begins (Bassuk et al., 1999; Gilbert et al., 1999; Kantorow et al., 1998; Kantorow et al., 2000; Maillard et al., 1992; Sawhney, 1995; Yan et al., 2000). The absence of SPARC leads to the formation of cataracts (Bassuk et al., 1999; Gilmour et al., 1998; Norose et al., 1998). In the lenses of SPARC null mice, the posterior lens fiber cells fail to fully elongate and degenerate, and the posterior lens capsule ruptures (Gilmour et al., 1998), but there are no major changes in the expression of the lens crystallins (Bassuk et al., 1999). The anterior lens capsule is thicker in SPARC null mice, yet no changes were detected in the ECM composition of the capsule (Bassuk et al., 1999). In precataractous SPARC null mice, filopodial extensions of both epithelial and fiber cells into the lens capsule were detected (Norose et al., 2000). This failure to maintain the normal cell–basement membrane interaction indicates that there are major changes in cell-matrix interaction involving integrin receptors that precede the onset of cataract in SPARC null mice. Interestingly, the small Rho GTPase family member cdc42, which can be activated by integrins, signals filopodial extension (Price et al., 1998). Examination of integrin function in SPARC null mice may elucidate mechanisms involved in SPARC's role in regulating lens differentiation and homeostasis.

10.1.3. Lens Capsule Matrix Proteins as Substrates for Lens Cell Growth and Differentiation in Culture

The lens capsule has long been known for its properties as a support for cell growth in culture. The ability of lens epithelial cell explants to differentiate in culture is thought to be due to their associated capsular matrix. When lens epithelial cells are isolated as single cells before plating in cell culture, the matrix on which they are plated is important for directing their attachment, migration, and differentiation. Studies have shown that type IV collagen is the best substrate for promotion of lens cell adhesion and that fibronectin is the best substratum for directing lens cell migration (Olivero and Furcht, 1993). While lens cells isolated from different stages of lens development can migrate on laminin, their ability to migrate on fibronectin decreases with developmental age (Parmigiani and McAvoy, 1991). Other studies have shown that whereas type IV collagen is a good substrate for lens cell attachment, it is a poor substrate for directing lens cell differentiation (Menko, unpublished manuscript). Although fibronectin can support lens cell differentiation, the best substrate for promotion of lens cell differentiation is laminin (Menko, unpublished manuscript). Interestingly, under most plating conditions, lens cells quickly remodel their substrate to provide themselves with one permissive for differentiation. For example, growth of lens epithelial explants in the presence of vitreous stimulated synthesis of a basement membrane containing matrix proteins such as laminin and heparan sulfate proteoglycan (Lovicu et al., 1995).

10.1.4. Matrix and Lens Disease

Typically, collagen type I, well known as a matrix protein produced by mesenchymal cells, is not detected in the capsule of normal lenses. However, it is expressed in cataractous lenses (Azuma and Hara, 1998; Hales et al., 1995). The expression of collagen type I by epithelial cells is correlated with their transformation into mesenchymal cells. When lens epithelial cells are plated in culture within collagen type I gels, they undergo an epithelial to mesenchymal transition. These cells lose their polarity and alter their matrix synthesis as well as their repertoire of integrin receptors (Zuk and Hay, 1994). Such lens-derived mesenchymal cells stop producing laminin and collagen type IV and in their place secrete collagen type I. These results demonstrate the importance of matrix composition in the control of lens cell behavior and suggest that the growth of lens epithelial cells in collagen gels is a good model system for studying cataract formation.

10.2. Integrin Receptors

Integrins make up a family of transmembrane receptors that consist of noncovalently bound α and β subunits (Hynes, 1992). Their principal ligands are ECM proteins, from which they receive signals that elicit changes in gene expression necessary for directing cell proliferation, migration, differentiation, and survival. Integrin receptors can exhibit distinct ECM ligand specificities and signaling functions. At their cytoplasmic domain, they interact with a complex of signaling and cytoskeletal proteins known as a "focal adhesion complex" (Kelley et al., 2002). These molecules regulate integrin receptor coupling to the actin cytoskeleton and their activation of downstream signaling effectors.

Integrins are bidirectional signaling receptors, participating in both outside-in and inside-out signaling pathways (Hynes, 1992). In their role as outside-in signaling receptors, engagement by their matrix ligand results in the formation of the focal adhesion complex and initiation of a specific signaling cascade. As inside-out signaling receptors, integrins receive signals from within the cell that cause changes in their conformation, activating them for binding their ECM ligands (Hynes, 1992).

Evidence of integrin function in eye development comes from studies in which Arg-Gly-Asp (RGD) peptides (which inhibit integrin adhesion to specific RGD-containing matrix ligands) or anti-integrin antibodies were injected into preoptic regions of the chick embryo. These treatments caused disruption of the invagination of both the optic cup and the lens vesicle (Svennevik and Linser, 1993). The ability to block lens cell differentiation in culture with a function-blocking antibody to $\beta1$ integrin suggests that a $\beta1$ integrin heterodimer is required for normal lens cell differentiation (Menko, unpublished manuscript).

10.2.1. $\beta1$ Integrin

A large repertoire of integrins are expressed in the developing lens, and their expression and distribution are regulated with the state of lens cell differentiation. It is likely that each integrin has a unique function in regulating lens cell differentiation. $\beta1$ integrins, which serve as heterodimeric partners for many different α integrins, are found in many different regions of the embryonic lens (Menko and Philp, 1995). They are found along the basal surfaces of lens epithelial and fiber cells, where they contact their matrix ligands in the capsule (Figs. 10.2A, B), and along the epithelial-fiber interface, where they are likely to be involved in cell-cell interactions (Fig. 10.2B). Most surprisingly, $\beta1$ integrins also are found all along the cell-cell interfaces of differentiating lens fiber cells, both in the equatorial epithelium (Fig. 10.2B) and in the cortical region of the developing lens (Figs. 10.2C, D).

10.2.2. $\alpha1\beta1$ and $\alpha2\beta1$ Integrins

Embryonic lenses express both $\alpha1\beta1$ and $\alpha2\beta1$ integrins. Although each can function as a laminin receptor, these integrins have distinct binding preferences for different collagen family members. Cells expressing $\alpha1\beta1$ integrin prefer to bind collagen type IV, whereas cells expressing $\alpha2\beta1$ prefer binding to collagen type I (Dickeson et al., 1999). The pattern of expression of $\alpha1$ integrin in the embryonic lens appears to parallel the regionalization of collagen type IV to the anterior lens capsule. $\alpha1\beta1$ integrin expression is limited to the embryonic lens epithelial cells and is highest in the equatorial zone (Walker et al., 2002c). In contrast, $\alpha2\beta1$ integrin is present in both epithelial and fiber cells of the embryonic lens (Wu and Santoro, 1994). This integrin is not detected in the adult lens. Because collagen type I is unlikely to be present in significant amounts in the normal embryonic lens, laminin is the most probable ligand for $\alpha2\beta1$ integrin in this tissue. Both $\alpha1\beta1$ and $\alpha2\beta1$ have been implicated in the regulation of cell proliferation events (Pozzi et al., 1998; Wary et al., 1996; Wary et al., 1998; Zutter et al., 1999). $\alpha2\beta1$ Integrin has also been shown to play a role in cell migration as well as in cell differentiation and morphogenesis (Zutter et al., 1999). In the embryonic lens epithelium, the cells must first proliferate and then withdraw from the cell cycle as they initiate their differentiation. During lens development, proliferation becomes restricted to the germinative zone, a region within the equatorial epithelium in which both $\alpha1$ and $\alpha2$ integrins are highly expressed. Consequently, $\alpha1\beta1$ and $\alpha2\beta1$ integrins may function in the regulation of lens cell proliferation. The persistence

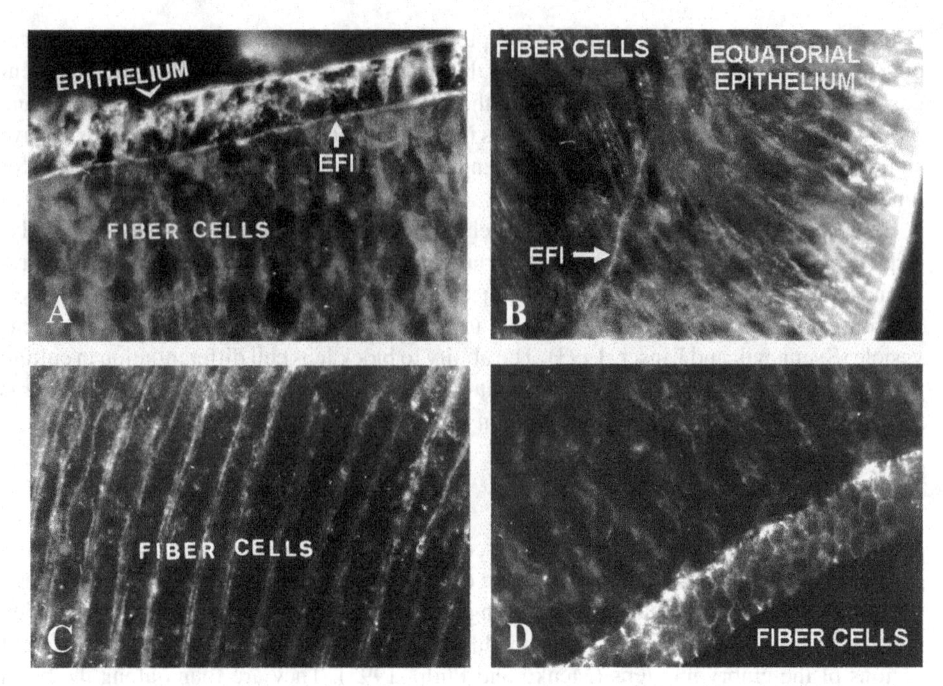

Figure 10.2. Distribution of $\beta1$ integrin in the lens. Cryostat sections of E10 chicken embryo lenses were immunostained with an antibody to the $\beta1$ integrin subunit. (A) In the lens epithelium, $\beta1$ integrin localizes to the basal surfaces where the cell interacts with the lens capsule (arrowhead), the epithelial–fiber interface (EFI, arrow), and along the lateral cell borders. (B) In the equatorial zone, $\beta1$ integrin is found at cell–cell borders and along the cell–capsule interface. (C) In differentiating lens fiber cells, $\beta1$ integrin is found along their lateral borders. (D) A tangential section through the posterior region of the lens clearly shows $\beta1$ integrin labeling the lateral membranes of the hexagonally packed fiber cells.

of $\alpha2\beta1$ in differentiating lens fiber cells may indicate another role for $\alpha2\beta1$ integrin in the regulation of early lens fiber differentiation.

10.2.3. $\alpha5\beta1$ Integrin

As discussed above, the ECM protein fibronectin has a unique distribution in the chick embryo lens. It is expressed very early in lens development in regions where the optic vesicle meets the head ectoderm to become the lens. Later in development, fibronectin is found in the lens capsule, and in the chick embryo it is associated with the lens epithelium, persisting into the region of the lens capsule in contact with the germinative zone. The principal cell surface receptor for fibronectin is $\alpha5\beta1$ integrin (Hynes, 1992). This integrin typically functions in the regulation of cell attachment, migration, and proliferation. In the embryonic lens, the pattern of distribution of $\alpha5$ integrin is similar to that of its ligand fibronectin (Menko, unpublished manuscript). $\alpha5$ integrin expression is high in both the lens anterior epithelial cells and in the equatorial epithelium, within which is located the germinative zone. Biochemical analysis has demonstrated that a low amount of $\alpha5$ is also expressed by differentiating lens fiber cells. The distinct similarities in the spatiotemporal distribution of fibronectin and $\alpha5\beta1$ integrin in the embryonic lens suggests that they play a role in regulating the proliferative events in the embryonic lens.

10.2.4. $\alpha3\beta1$ Integrin

$\alpha3\beta1$ Integrin is a high affinity receptor for laminin-5 and laminin-10/11 and a low affinity receptor for laminin-1. Numerous functions have been attributed to $\alpha3\beta1$ integrin, including organization of basement membranes, regulation of cell cycle, and determination of cell polarity (De Arcangelis et al., 1999; Gonzales et al., 1999; Hodivala-Dilke et al., 1998; Menko et al., 2001). It has been suggested that this integrin can also play an important role in the transdominant inhibition of function of other integrin receptors, such as $\alpha5\beta1$ (DiPersio et al., 1997). $\alpha3\beta1$ integrin expression occurs most predominately in lens epithelial cells, decreasing with lens fiber cell differentiation (Menko and Philp, 1995). No dramatic lens phenotype has been observed in the $\alpha3$ integrin knockout mouse (Menko, unpublished manuscript) or in the $\alpha6$ integrin (another laminin receptor) knockout mouse (Georges-Labouesse et al., 1996), but when $\alpha3/\alpha6$ integrin double knockout mice were created, lens defects were detected (De Arcangelis et al., 1999). These lenses developed breaches in the anterior lens capsule, which allowed the contents of the lens to spill out. This lens defect is most likely a consequence of the failure to organize and/or maintain a normal capsular basement membrane.

10.2.5. $\alpha6\beta1$ and $\alpha6\beta4$ Integrins

$\alpha6$ integrin forms heterodimers with both $\beta1$ and $\beta4$ integrins. $\alpha6\beta1$ integrin is principally a receptor for laminin-1, while $\alpha6\beta4$ serves as a receptor for laminin-5. Both $\alpha6\beta1$ and $\alpha6\beta4$ are expressed in the embryonic lens throughout lens cell differentiation (Menko and Philip, 1995; Walker and Menko, 1999). The activation of this integrin subunit is correlated with its linkage to the cytoskeleton (Shaw et al., 1990). In the embryonic lens, $\alpha6$ integrin insolubility in Triton X-100, correlated with linkage to the cytoskeleton, was highest in regions actively undergoing lens fiber cell differentiation, the equatorial epithelium and the cortical fiber cells (Walker and Menko, 1999). These data suggest that $\alpha6$ may modulate lens cell differentiation through changes in its linkage to the cytoskeleton. There are two well-characterized cytoplasmic variants of $\alpha6$, $\alpha6$A and $\alpha6$B, that have distinct signaling functions and differentiation-specific expression patterns (Jiang and Grabel, 1995; Shaw et al., 1993; Tamura et al., 1991; Wei et al., 1998). As lens cells differentiate from epithelial to fiber cells, a switch in isoform expression from $\alpha6$B to $\alpha6$A occurs (Fig. 10.3, Walker and Menko, 1999). This could provide a mechanism by which $\alpha6$ integrin can alter its signaling response to laminin present in the lens capsule. Immunolocalization studies demonstrate that, in addition to its role as a receptor for laminin in the capsule, $\alpha6$ integrin is the heterodimeric partner for $\beta1$ integrin all along the cell-cell interfaces of differentiating lens fiber cells (Fig. 10.4A; Walker and Menko, 1999). Much of this region is devoid of laminin, suggesting a potential laminin-independent function for $\alpha6$ in lens fiber cell differentiation. Interestingly, there is a high level of the uncleaved form of $\alpha6$ integrin expressed at the cell surface of cortical lens fiber cells (Walker et al., 2002a). While not as well characterized as the cleaved, disufide-bonded form of $\alpha6$ integrin, uncleaved $\alpha6$ integrin has similar ligand-binding properties but altered signaling function (Delwel et al., 1996; Delwel et al., 1997). These results have suggested a third type of signaling for $\alpha6\beta1$ integrin in the lens, inside-in signaling, where both activation of the integrin receptor and downstream signaling by this activated $\alpha6\beta1$ integrin occur within the cell (Walker and Menko, 1999).

Localization of $\alpha6\beta4$ integrin in the embryonic lens also is unusual, though coincident with that of its matrix ligand, laminin-5. $\beta4$ Integrin is found along the anterior and posterior

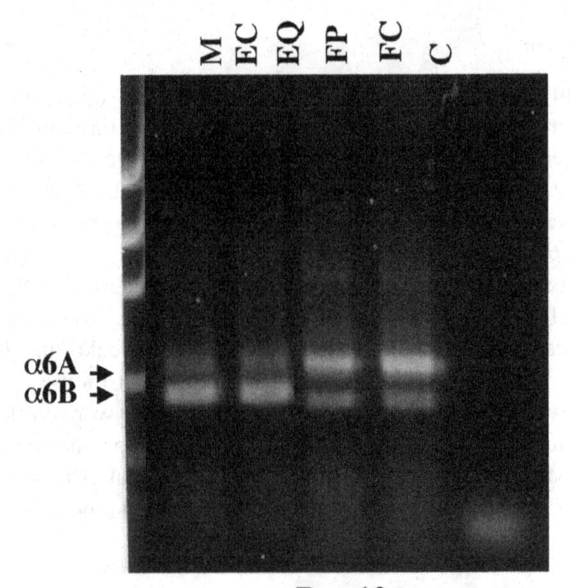

Figure 10.3. A switch in integrin isoform predominance from α6B to α6A, with lens cell differentiation. RT-PCR was performed using RNA extracted from microdissected regions of embryonic day 10 chick lenses and a primer set that recognizes both isoforms of α6 integrin, generating a 482-bp product for α6A and a 377-bp product for α6B. α6B integrin is the predominant α6 integrin isoform in the two epithelial regions of the day 10 lens (EC and EQ). Fiber cell differentiation is accompanied by a switch in isoform predominance to α6A. EC, central epithelium; EQ, equatorial epithelium; FP, cortical fiber cells; FC, nuclear fiber cells.

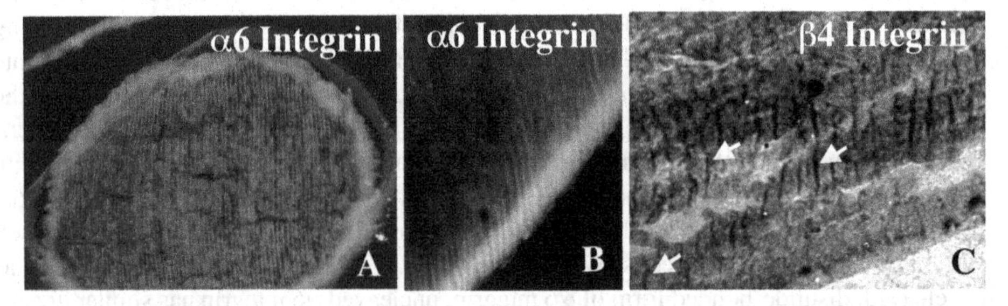

Figure 10.4. Distribution of α6 and β4 integrins in the lens. Cryostat sections were cut from E10 chick embryo lenses (B) or adult mouse lenses (A, C). Sections were immunostained with antibodies to α6 integrin (A, B) or β4 integrin (C). In a section cut through the cortical fiber zone near the equator of the lens, α6 integrin is found all along the cell–cell borders (A). In the more differentiated regions of the E10 chick embryo lens, α6 integrin was found at cell–cell borders but was restricted to a region near the posterior capsule (B), similar to the distribution of β4 integrin (arrows, C).

tips of developing lens fiber cells, extending laterally for a short distance between the fiber cells (Fig. 10.4C; Walker and Menko, unpublished manuscript). In this region of the lens, intense staining for $\alpha 6$ integrin is also detected (Fig. 10.4B). Interestingly, the intermediate filament protein vimentin was similarly localized in the embryonic lens (Walker and Menko, unpublished manuscript; Sandilands et al., 1995). The co-localization pattern for $\alpha 6$ integrin, $\beta 4$ integrin, and laminin-5 and the close association of vimentin suggest a novel function for $\alpha 6\beta 4$ in the lens, possibly the formation of a unique junction in lens fiber cells and the enablement of their coordinate migration along the anterior epithelium and the posterior capsule.

Studies of the signaling role of $\alpha 6$ integrin in the embryonic lens have shown that this integrin is present in differentiation-specific signaling complexes with the activated IGF-1 receptor (Walker et al., 2002c). In addition, signaling proteins typical of growth factor receptor signaling pathways are recruited to $\alpha 6$ integrin signaling complexes as lens cells differentiate, including Shc (Walker et al., 2002a), Grb2 (Walker et al., 2002a), and activated ERK (Walker et al., 2002c). These results suggest that $\alpha 6$ integrin signaling in the developing lens involves recruitment of growth factor receptors and their downstream effectors.

10.2.6. αV Integrin Receptors

αv integrin forms heterodimers with multiple β subunits (Hynes, 1992). At least one of these receptors, $\alpha v\beta 3$, is expressed in the embryonic lens. Both the αv and $\beta 3$ integrin subunits are associated with lens cells throughout their differentiation from epithelial to fiber cells (Walker et al., 2002c). $\alpha v\beta 3$ and $\alpha 6\beta 1/\beta 4$ are the only integrins that persist into the most terminally differentiated regions of the embryonic lens. The αv integrin heterodimers are receptors for multiple ligands, including vitronectin, tenascin, fibronectin and thrombospondin (Hynes, 1992), and regulate such events as migration and apoptosis. As a receptor for tenascin in the lens capsule, $\alpha v\beta 3$ integrin is likely to be involved in regulating the migration of lens fiber cells. This possibility is supported by the presence of large αv-containing focal contacts in lens cells grown in culture (Fig. 10.5A; Menko, unpublished manuscript).

10.2.7. Integrins and Lens Disease

When lens cells are plated within a collagen type I gel, they undergo an epithelial-mesenchymal transformation that alters their integrin and matrix expression. Under these conditions, lens cells lose polarity, becoming bipolar and spindle shaped. These lens cells down-regulate the differentiation-specific $\alpha 6$ integrin and its ligand laminin while up-regulating the mesenchymal cell–specific $\alpha 5$ integrin and its ligand fibronectin (Zuk and Hay, 1994). Cataract formation in transgenic mice that overexpress PAX6(5a) is accompanied by the elevated expression of $\alpha 5$ integrin in the lens fiber cells (Duncan et al., 2000). This increased expression of $\alpha 5$ integrin was correlated with the finding of potential PAX6(5A)-binding sites in the human $\alpha 5$ promoter. The effects on lens cells after plating in collagen gels are reminiscent of changes in lens epithelial cells during postcapsular opacification following cataract surgery. This secondary cataract forms following cataract surgery when the remaining anterior epithelial cells migrate inappropriately along the posterior capsule. The differentiation program of these cells is altered, as evidenced by changes in gene expression, including an altered repertoire of integrins and induction of both collagen type I and α-smooth actin. Blocking the function of the newly expressed or activated integrins

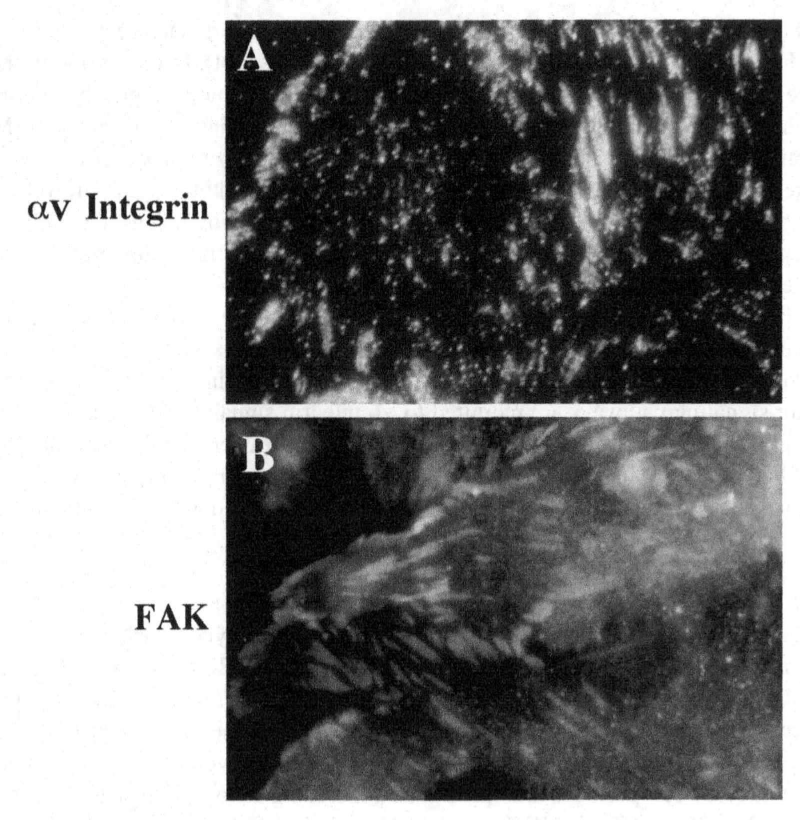

Figure 10.5. Focal adhesion plaques in chick embryo lens cultures contain both αv integrin and FAK. Chick embryo lens cells were plated in culture and immunostained using antibodies to αv integrin or FAK (focal adhesion kinase). Both were found to be components of large focal adhesion plaques in these well-spread lens epithelial cells.

is likely to lead to a lowered incidence of postcapsular opacification. Since $\alpha3\beta1$ integrin has been proposed as a transdominant inhibitor of $\alpha5\beta1$ integrin function (Hodivala-Dilke et al., 1998), these results raise the possibility that $\alpha3\beta1$ repression of $\alpha5\beta1$ function may be critical to the maintenance of cell polarity in the lens epithelium.

10.2.8. Integrin-Associated Focal Adhesion Complexes

Upon integrin engagement by ligand, focal adhesion kinase (FAK), a nonreceptor tyrosine kinase, is activated and recruited to integrin complexes, along with cytoskeletal proteins characteristic of the focal adhesion complex (FAC). These FAC proteins include talin, α-actinin, paxillin, and vinculin. Embryonic lens cells express all of these FAC proteins (Menko et al., 1998). However, their expression and distribution is regulated with the state of lens cell differentiation. Talin, α-actinin, and FAK are more highly expressed in lens epithelial cells than in lens fiber cells, while vinculin expression is up-regulated with lens fiber cell differentiation (Menko et al., 1998). Vinculin is found in both cell-matrix adhesion complexes (FACs) and cell-cell adhesion complexes in the lens. The increased expression of vinculin observed in lens fiber cells appears to be related to its association with fiber cell membranes just after fiber cells detach from the lens capsule (Beebe et al., 2001). The ability of lens cells to organize FACs is best visualized when lens cells are grown in culture and immunostained for a FAC protein such as FAK (Fig. 10.5B; Menko, unpublished manuscript). The FACs are typically linked to the cytoskeleton. Therefore, it is not surprising that many of the FAC proteins expressed by the embryonic lens change their linkage to the actin cytoskeleton as the lens cells differentiate. One such example is paxillin. Although its expression changes little with embryonic lens cell differentiation, it becomes highly linked to the cytoskeleton in the lens fiber cells (Menko et al., 1998). Paxillin's increased linkage to the cytoskeleton is correlated with increased localization to lateral fiber cell membranes at a time just before fiber cell nuclei and organelles are degraded (Beebe et al., 2001). In contrast to its expression in the embryonic lens, paxillin is down-regulated in the fiber cell region of the adult mouse lens (Duncan et al., 2000). However, constitutive expression of PAX6(5a) as a transgene in the mouse lens results in increased expression of paxillin, along with $\alpha5$ integrin, in the dysgenic fiber cells (Duncan et al., 2000). Many of the molecular components of FACs, including actin, paxillin, and FAK, have also been identified as components of the unique adhesion structures that form between differentiating lens fiber cells and the posterior capsule, known as basal membrane complexes (BMCs; Bassnett et al., 1999). Formation of the BMCs, which remain associated with the posterior lens capsule following mechanical stripping of the capsule from the lens, is dependent on $\beta1$ integrin function.

10.3. Cadherins

Cadherins make up a family of integral membrane proteins that mediate homophilic calcium-dependent cell-cell adhesion. They are involved in the cell-sorting events that regulate early mophogenetic developmental processes and in the regulation of cell differentiation (Takeichi, 1991, 1995). Their ability to form functional junctions (adherens junctions) is dependent on their interaction with the actin cytoskeleton, which is mediated by β-catenin or γ-catenin (plakoglobin), both of which associate with the cadherin cytoplasmic domain (Aberle et al., 1996; Gumbiner and McCrea, 1993). These molecules, in turn, associate with α-catenin, which binds to the actin cytoskeleton either directly or through its association with α-actinin. Enhanced tyrosine phosphorylation of these catenins has been shown to result in the blocking of cadherin linkage to the cytoskeleton, thereby inhibiting cadherin function (Takeda et al., 1995; Volberg et al., 1991).

In the mouse embryo lens, E-cadherin is expressed exclusively by lens epithelial cells, disappearing as these cells begin their differentiation (Wigle et al., 1999; Xu et al., 2002). In contrast, expression of N-cadherin, which also is found in lens epithelial cells, is maintained until the fiber cells mature (Xu et al., 2002). While E-cadherin is not expressed in the chick, two other cadherin family members have been identified in the chick embryo lens, with distinctive patterns of expression, N-cadherin (Atreya et al., 1989; Lagunowich and Grunwald, 1989; Volk and Geiger, 1986a, 1986b) and B-cadherin (the chicken homolog of P-cadherin; Leong et al., 2000). In differentiating chick lens cell cultures, the temporal localization of N-cadherin to cell-cell interfaces precedes that of B-cadherin (Leong et al., 2000). α-Catenin, β-catenin, and γ-catenin, all members of the cadherin complexes, are localized to lens cell-cell interfaces (Ferreira-Cornwell et al., 2000). Although N-cadherin is expressed in most regions of the chick embryonic lens, it is reported to be lost from the nuclear fiber region of the lens with age (Atreya et al., 1989; Beebe et al., 2001). This loss may be related to the turnover of N-cadherin through the regulated action of a

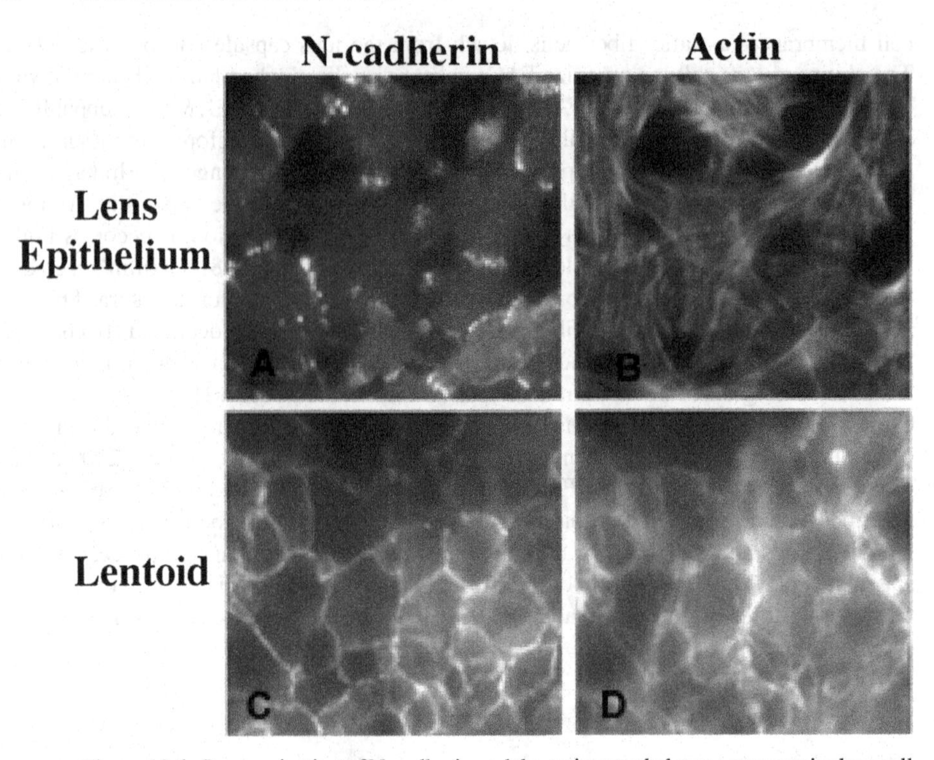

Figure 10.6. Reorganization of N-cadherin and the actin cytoskeleton accompanies lens cell differentiation *in vitro*. Cell cultures were established from day 10 chick embryo lenses. The panels represent undifferentiated lens epithelial cells (A, B) and differentiating lens fiber cells (C, D) stained either with an antibody to N-cadherin (A, C) or with phalloidin to detect polymerized actin (B, D). In lens epithelial cells, N-cadherin is organized in adhesion plaque-like structures limited to areas of cell–cell contact, and actin was found in stress fibers that ended at the N-cadherin adhesion plaques. As lens cells differentiated, N-cadherin became concentrated at cell–cell borders. This was paralleled by the reorganization of actin filaments to a cortical distribution.

calcium-activated protease (Maisel and Atreya, 1990) or an increase in its phosphorylation state (Lee et al., 1997; Volberg et al., 1991). In the embryo, the state of tyrosine phosphorylation of N-cadherin is regulated in a stage-specific manner (Lagunowich and Grunwald, 1991). In lens cultures, the formation of cell-cell junctions is prevented by the constitutive expression of the protein tyrosine kinase v-Src (Menko and Boettiger, 1988).

Primary lens epithelial cells in culture first form a well-spread monolayer in which N-cadherin is found in short cell-cell adhesion structures perpendicular to the membrane interface, evocative of a zipper (Fig. 10.6A; Ferreira-Cornwell et al., 2000). The actin cytoskeleton in these cells is organized as stress fibers (Fig. 10.6B). The initiation of lens fiber cell differentiation in culture is characterized by a compaction of the epithelial cell monolayer. This is accompanied by localization of N-cadherin all along cell-cell interfaces (Fig. 10.5C) and reorganization of actin stress fibers to a cortical distribution (Fig. 10.5D; Ferreira-Cornwell et al., 2000). Src is a negative regulator of cadherin junction assembly in the lens. The suppression of Src kinase activity in lens cultures rapidly induces the formation of stable N-cadherin junctions (Walker et al., 2002b). Induction of the cell cycle inhibitors p27 and p57, withdrawal of lens epithelial cells from the cell cycle, and induction of lens cell differentiation follow cadherin junction assembly (Walker et al., 2002b).

In lens fiber cells, N-cadherin can be found localized specifically to lateral regions of the basal membrane complex, which anchors fiber cells to the capsule and facilitates cell migration (Bassnett et al., 1999; see also chap. 9). It is in these N-cadherin–rich regions that F-actin bundles are seen to insert into regions of the lateral fiber cell membrane. This N-cadherin–actin complex forms a hexagonal lattice along the posterior face of the lens (Bassnett et al., 1999). When chick lens cells are grown in culture in the presence of the N-cadherin function–blocking antibody NCD-2, the formation of normal N-cadherin cell-cell junctions is blocked, the cells fail to form a tightly packed monolayer, and differentiation, both morphological and biochemical, is inhibited (Ferreira-Cornwell et al., 2000). Normally, both N-cadherin and B-cadherin linkage to the actin cytoskeleton increases with the chick lens cell differentiation state (Ferreira-Cornwell et al., 2000; Leong et al., 2000). In NCD-2–treated lens cultures, N-cadherin fails to become linked to the actin cytoskeleton and remains dispersed, and filamentous actin is distributed as stress fibers, both features of undifferentiated lens epithelial cells. These effects were paralleled by the suppression of expression of components of the cadherin-catenin complex, β-catenin and α-catenin, suggesting the importance of these proteins in regulating the function of N-cadherin junctions in the differentiating lens.

The formation of N-cadherin junctions is necessary for the formation of functional gap junctions, mediators of intercellular communication, between lens fiber cells (Frenzel and Johnson, 1996). While N-cadherin does not co-localize with the lens gap junctional protein connexin 56 in lens fiber cells formed in culture (Berthoud et al., 1999), temporal localization of B-cadherin closely parallels that of connexin 56 (Leong et al., 2000).

10.4. Other Lens Cell Adhesion Molecules

N-CAM is a calcium-independent homotypic cell-cell adhesion molecule that functions in cell adhesion, cell signaling, and neurite outgrowth (Beggs et al., 1994; Beggs et al., 1997; Kolkova et al., 2000; Maness et al., 1996; Walsh and Doherty, 1997). It is expressed by lens epithelial cells, and its expression decreases with lens fiber cell differentiation (Watanabe et al., 1989). N-CAM expression in the lens is also isoform specific. Whereas N-CAM 140 is the predominant isoform expressed in lens epithelial cells, N-CAM 120 is found principally in lens fiber cells (Katar et al., 1993). Glycosylation of N-CAM in the lens is regulated with developmental age. In embryonic lens epithelial cells, N-CAM has less sialic acid than in epithelial cells of the adult lens. Increased quantity of sialic acid in N-CAM is correlated with decreased adhesion (Watanabe et al., 1992). When N-CAM function is blocked in lens epithelial cell explants, the cells appear thinner and contain less mature gap junctions (Watanabe et al., 1989). These results suggest a role for N-CAM in lens differentiation, possibly through the regulation of fiber cell gap junctions or adherens junctions.

Galectin-3, a member of the galactin family of lectins, is a secreted β-galactoside–binding protein with adhesive properties whose expression is developmentally regulated (Perillo et al., 1998). It localizes to the plasma membrane of lens fiber cells, where it is associated as a peripheral protein. In the lens fiber cells, the junction-forming intrinsic fiber cell membrane protein MP20 has been identified as a candidate ligand for galectin-3 (Gonen et al., 2000). GRIFIN, a protein related to the galectins but missing β-galactoside–binding activity, is expressed exclusively in the lens, where it is localized to the cell-cell interfaces of lens fiber cells (Ogden et al., 1998).

10.5. Summary

For more than a century now, scientists have been trying to unravel how cells differentiate into distinct cell types. The process of differentiation is controlled by an elaborate plan and is influenced by intrinsic as well as extrinsic factors. Clearly, there is epigenetic control of lens cell differentiation, since environmental factors can infuence the outcome of both lens cell differentiation and development.

In this chapter we have reviewed matrix molecules and adhesion receptors that are implicated in the process of regulating the differentiation of a lens epithelial cell into a lens fiber cell. They are all expressed in distinct spatiotemporal patterns. Some of these molecules probably regulate lens cell proliferation while others appear to be involved in signaling the initiation of lens cell differentiation or the distinct morphogenetic changes that accompany the formation of the lens fiber cells. It is evident that we can no longer consider just one set of molecules to be the critical regulators of lens cell differentiation and development; rather we must determine how the complex interplay of molecules determines the differentiation state of cells in the lens.

11

Growth Factors in Lens Development

Richard A. Lang and John W. McAvoy

How the lens develops and acquires its distinctive morphology and growth patterns has been a major research focus for developmental biologists. Growth factors are known to play key roles in influencing cell behavior and cell fates during development. In recent years researchers have identified some of the growth factor families involved in regulating the processes of lens induction, morphogenesis, and growth. The aim of this chapter is to review the current state of knowledge in this key area of lens research.

The lens develops from head ectoderm that is associated with an evagination of the developing brain: the optic vesicle (Fig. 11.1). Soon after these two tissues become associated, the presumptive lens ectoderm grows and thickens to form the lens placode. Subsequent invagination of the placode forms the lens pit, which later closes to form the lens vesicle. Cells in the posterior segment of the lens vesicle, next to the optic cup, elongate to form the primary fibers, whereas cells in the anterior segment of the vesicle differentiate into epithelial cells. These divergent fates of embryonic lens cells give the lens its distinctive polarity. From this stage onwards, the lens grows by continued proliferation of epithelial cells and differentiation of fiber cells. Proliferation initially occurs throughout the lens epithelial compartment but during development becomes progressively restricted to a band of cells above the equator, known as the germinative zone. Progeny of divisions that shift below the equator enter the transitional zone and elongate to give rise to secondary fibers. These growth patterns ensure that lens polarity is maintained as new fibers continue to differentiate and are added to the fiber mass throughout life. This is crucial for the maintenance of the ordered cellular architecture that contributes to the transparency and optical properties of the lens.

Throughout its morphogenesis and differentiation, the developing lens is closely associated with the optic vesicle/cup. At early stages, this involves a close physical association between these two tissues; for example, cellular processes have been described that connect them (Mann, 1950; McAvoy, 1980), and extracellular matrix accumulates between them (Parmigiani and McAvoy, 1984). At later stages, the tissues move apart as vitreous forms between them (Fig. 11.1F). There is evidence that at both early (induction and morphogenesis) and late (differentiation and growth) stages, growth factors mediate and orchestrate at least some of the interactions that occur between the developing lens and retina.

11.1. Lens Induction and Morphogenesis

Embryologists at the turn of the century hypothesized that, because of their close spatial association, the optic vesicle was the tissue that provided lens induction signals (Spemann, 1901). This was supported by early experiments with amphibians. For example, in many

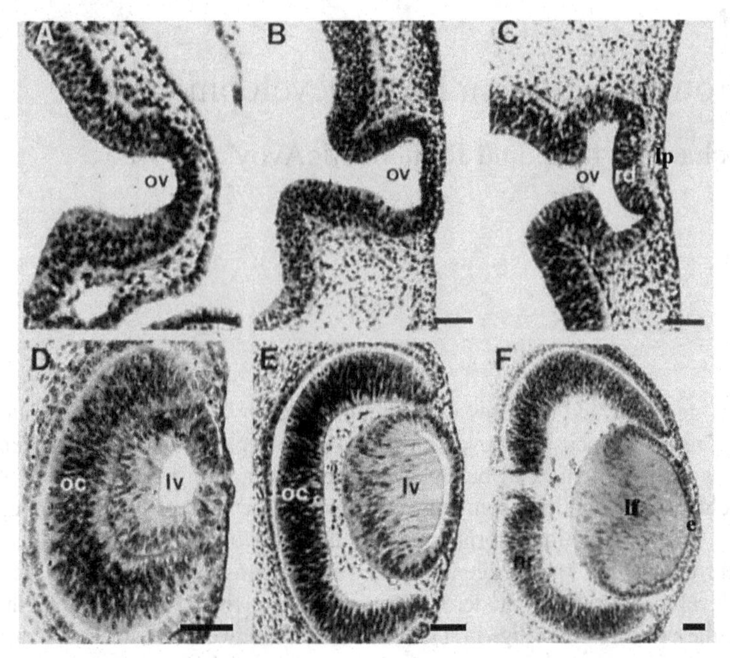

Figure 11.1. (A–F) Sections through the developing eyes of E10–14 rat embryos stained with haematoxylin and phloxine. (A) At E10, bilateral outgrowths from the developing brain form the optic vesicles (ov). (B) In early E11 embryos, the optic vesicle is closely associated with a region of head ectoderm that is destined to form a lens. (C) In the late E11 embryo, both the ectoderm and the neuroectoderm are thickened along the region of close proximity, forming the lens placode (lp) and retinal disc (rd), respectively. (D) Invagination of the lens placode and the optic vesicle at E12 leads to the formation of the lens pit/vesicle (lv) and the optic cup (oc), respectively. (E) By E13, the lens vesicle has completely closed and detached from the optic cup (oc). The posterior lens vesicle cells elongate to form primary lens fiber cells, leading to narrowing of the vesicle lumen. (F) By E14, the lens vesicle lumen has disappeared, and the primary lens fibers (lf) are in contact with the anterior lens vesicle cells, which form the epithelium (e). Vitreous humor and hyaloid vasculature forms between the developing lens and retina. The inner layer of the optic cup will form the neural retina (nr). Scale bars: (A), 50 μm; (B, C), 75 μm; (D–F), 100 μm. (Adapted from de Iongh and McAvoy, 1993.)

species, removal of the optic vesicle primordium resulted in no lens formation. Later experiments, in which other species developed lenses in the absence of the optic vesicle (Mencl, 1903), indicated a more complex situation. Embryological manipulations performed by a number of groups, including the Grainger Laboratory (Grainger et al., 1992), have indicated that lens induction is a multistage process that probably involves multiple types of inductive signaling. For example, it is possible to define stages of lens competence and "lens-forming bias" that are clearly distinct. The idea that lens induction is a multistage process is beginning to receive support from molecular genetic analysis. For example, there are two phases of *Pax6* expression in the lens lineage (see below). Gene expression in these phases is distinctly regulated, but both are required for lens development. The embryological manipulations that have led to the multistage model of lens induction are dealt with in detail in chapter 2 of this volume.

In the mouse, the optic vesicle is essential for lens induction. This is illustrated by explant experiments in which recombination of optic vesicle and presumptive lens ectoderm indicates that the optic vesicle is only dispensable for lens development after approximately 9.5 days of embryonic development (E9.5; Furuta and Hogan, 1998). Induced mutations in genes that are expressed exclusively in the neural component of the eye also illustrate this point. For example, *Lhx2* null mice form optic vesicles but do not form lenses (Porter et al., 1997), while *Rx* null mice do not form optic vesicles and also do not form lenses (Mathers et al., 1997). These and other observations have motivated those interested in lens induction in the mouse to focus their efforts on understanding the mediators of the optic vesicle–presumptive lens interaction.

11.1.1. Transcription Factors Involved in Lens Induction

A number of transcription factors have activities that are critical for normal eye development. In many cases, transcription factor gene expression has been used as a molecular indicator of the activity of a growth factor signaling pathway that lies genetically upstream. From this point of view, to understand how growth factor signaling pathways are integrated during lens development, we must first describe how the involved transcription factors participate.

11.1.1.1. Pax6

Pax6 is a transcription factor that contains both a paired domain and a homeodomain (Walther and Gruss, 1991). The phenotypes of both mice and humans with null mutations in the *Pax6* gene indicate that this transcription factor has a critical role in eye development. Heterozygosity in either species leads to microphthalmia, iris and optic nerve hypoplasia (Hill et al., 1991; Hogan et al., 1986; Jordan et al., 1992), and in some circumstances the persistent lens stalk that manifests as Peters' anomaly (Hanson et al., 1994). *Pax6* homozygosity leads to major craniofacial defects, including the absence of eyes (Hill et al., 1991; Hogan et al., 1986). In the developmental context, *Pax6*$^{Sey/Sey}$ mice display an arrest of eye development after the optic vesicles have formed. This manifests as an absence of formation of the lens placode and a failure of morphological patterning in the optic vesicle (Grindley et al., 1995; Hogan et al., 1986).

A number of experiments have indicated that Pax6 is necessary and sufficient for development of the lens. In explant culture, recombination of wild-type presumptive lens epithelium and optic vesicle leads to development of the lens (Fujiwara et al., 1994). If the optic vesicle in such a recombination experiment is derived from *Pax6*$^{Sey/Sey}$ mice, lens development also occurs. By contrast, if presumptive lens epithelium is taken from *Pax6*$^{Sey/Sey}$ mice, lens development is prevented. This indicates an autonomous requirement for Pax6 in the presumptive lens region. Generation of wild-type/*Pax6*$^{Sey/Sey}$ chimeric mice leads to a similar conclusion, as *Pax6*$^{Sey/Sey}$ cells do not contribute to developing lens structures either at the placodal stage (Collinson et al., 2000) or as the lens matures (Quinn et al., 1996). Finally, conditional deletion of *Pax6* using a Cre recombinase-based strategy also indicates that Pax6 is required for lens development (Ashery-Padan et al., 2000).

In *Drosophila,* the Pax6 homolog *eyeless* has the remarkable ability to induce ectopic eyes when misexpressed (Halder et al., 1995). An ectopic eye formation response has also been observed in vertebrates when Pax6 is misexpressed. Specifically, in *Xenopus laevis,* injection of Pax6-encoding RNA early in development resulted in the formation of eyes with all the major features of the mature structure, including retinal neurons and lens (Chow et al., 1999). A modification in the experimental protocol led to a dramatic increase in the frequency with which isolated ectopic lenses would form (Altmann et al., 1997). Many

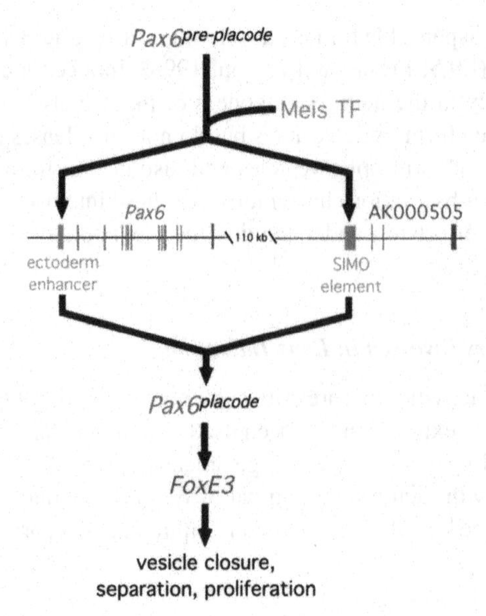

Figure 11.2. A model for the genetic pathways regulating lens induction. The arrows indicate genetic interactions, except in the case of the Meis transcription factor–enhancer physical complex. At the apex of the hierarchy is the preplacodal phase of *Pax6* expression ($Pax6^{pre\text{-}placode}$). It is understood that the later phase of *Pax6* expression in the lens placode ($Pax6^{placode}$) is dependent upon earlier activity of Pax6. $Pax6^{placode}$ is apparently regulated by at least two enhancers, represented by thick gray bars underneath the bifurcated arrow points on the schematic of the *Pax6* gene (exons, thin black vertical bars). One is the so-called ectoderm enhancer (abbreviated EE), and the other is the SIMO element, located in the final intron of the adjacent gene, AK000505, which is transcribed in the opposite direction to *Pax6*. The Meis transcription factors are likely to regulate the expression of *Pax6* directly by binding to the ectoderm enhancer and possibly to the SIMO element. The similar phenotypes (reduced lens lineage proliferation and lens vesicle closure and separation failure) of the *dysgenetic lens* (*FoxE3*) and $Pax6^{\Delta EE/\Delta EE}$ (homozygous targeted deletion of the ectoderm enhancer) mutant mice suggest that *FoxE3* might participate in this pathway. Lack of expression of the *FoxE3* gene in *Small eye* (*Pax6* mutant) homozygous mice and ectoderm enhancer deletion mice indicates that *FoxE3* occupies a downstream position.

of these lenses had all the features of a normal lens, including an epithelial cell layer, differentiated fiber cells, and the normal polarized morphology. This indicates that, in the context of the *Xenopus* embryo, Pax6 is sufficient for lens formation.

A simple analysis of *Pax6* gene expression demonstrates that there are multiple phases during development of the lens; this analysis has also established the basis for a genetic pathway for lens induction (Treisman and Lang, 2002). The $Pax6^{Sey\text{-}1Neu}$ allele is a point mutation that prevents the formation of a protein product but allows gene transcription (Grindley et al., 1997). Assessment of *Pax6* expression throughout eye development indicates that the early head ectoderm phase of *Pax6* expression is unaffected by the absence of functional transcription factor but that the later, placodal phase of expression does not occur (Grindley et al., 1995). This indicates that there are two phases of *Pax6* expression and that the later phase (corresponding to the lens placode) is dependent upon functional Pax6 in the head ectoderm (Fig. 11.2).

At least two transcriptional enhancers have been identified that mediate *Pax6* expression in the presumptive lens ectoderm. The first of these has been designated the "ectoderm enhancer" (EE), as it directs reporter gene expression to the ectoderm that gives rise to the lens, the lacrimal gland epithelium, and the corneal epithelium (Kammandel et al., 1999; Williams et al., 1998; Xu et al., 1999). Deletion of the EE by gene targeting results in mice with small eyes, small lenses, and a persistent lens stalk (Dimanlig et al., 2001). The presence of lenses in these animals, as well as a lower than normal level of Pax6 immunoreactivity in the lens placode, indicates that a second *Pax6* enhancer must also have activity in the lens lineage (Dimanlig et al., 2001). Indeed, more recent analysis has identified the so-called SIMO element, a *Pax6* enhancer that has activity in the lens lineage and is positioned 140 kb downstream (Kleinjan et al., 2001). SIMO was discovered through analysis of aniridia patients, in whom a translocation separated the SIMO element from the Pax6-coding region. When we combine (1) evidence that Pax6 is necessary and sufficient for lens induction with (2) analysis showing two distinct phases of Pax6 expression and (3) evidence for the involvement of the EE and SIMO enhancers, we can propose a model (Fig. 11.2) describing some features of the genetic regulation of lens induction.

11.1.1.2. Meis1

The mammalian Meis family transcription factors are TALE-class homeodomain transcription factors (Burglin, 1997) and homologs of the homothorax gene of *Drosophila melanogaster* that has been implicated in eye development (Bessa et al., 2002). Recent analysis has implied that Meis1 and Meis2 have important roles in regulating lens induction in vertebrates (Zhang et al., 2002). Specifically, Meis-binding sites have been identified in the ectoderm enhancer of *Pax6*, and, furthermore, various experimental outcomes support the case that *in vivo* Meis transcription factors regulate *Pax6* expression through this element. The experimental evidence includes (1) immunoidentification of Meis in a complex with EE probes in mobility shift assays, (2) transgenic mice showing that the activity of EE is dependent upon Meis-binding sites, (3) a genetic interaction between a Meis2 transgene and the *Pax6*$^{Sey1\text{-}Neu}$ allele, and (4) the demonstration that suppressor forms of Meis1 can down-regulate *Pax6* expression when transiently expressed in the lens placode. Given the existence of two enhancers regulating *Pax6* expression in the lens placode (Fig. 11.2), the latter experimental outcome requires that Meis transcription factors affect both the EE and SIMO elements. The identification of Meis-binding sites in the SIMO element is of great interest, but further experimentation will be required to precisely define the molecular genetics of this interaction. In particular, it will be very interesting to further analyze the eye phenotypes that arise in mice null or conditionally targeted for the Meis genes (Zhang et al., 2002). Since expression of the Meis genes is independent of *Pax6*, they are best incorporated into the model for genetic regulation of lens induction as an input upstream of both the EE and SIMO elements (Fig. 11.2).

11.1.1.3. FoxE3

FoxE3 is a *forkhead* family transcription factor that begins to be expressed in the mouse in a small region of the midbrain and the presumptive lens ectoderm at approximately E8.75 (Brownell et al., 2000). Analysis has indicated that *FoxE3* is the gene mutated in the dysgenetic lens (dyl) mouse, which shows defective lens development (Blixt et al., 2000; Brownell et al., 2000). The phenotype apparent in the dyl mouse includes development of a small lens, probably due to reduced proliferation, and the lack of separation of lens vesicle and surface ectoderm (a Peters' anomaly–like feature). It has been shown that in gene-targeted mice missing the ectoderm enhancer (Dimanlig et al., 2001), *FoxE3* expression in the lens placode and lens pit is dramatically down-regulated. This suggests that *FoxE3* is genetically downstream of the enhancer (Fig. 11.2; Dimanlig et al., 2001) and is consistent with earlier observations indicating that in *small-eye* homozygous mice *FoxE3* expression is absent (Brownell et al., 2000).

11.1.2. Bmp7 is a Lens Inducer

The bone morphogenetic proteins (Bmps) are a large family of secreted signaling molecules that function by binding cell surface receptors and eliciting cellular responses through a serine-threonine kinase-driven pathway (Massague, 1998). Bmps have been implicated in many different developmental processes, including development of the eye. Indeed, the first signaling molecule implicated in lens induction was Bmp7.

In homozygous *Bmp7* null mice, there is a variably penetrant eye phenotype that ranges from mild microphthalmia to anophthalmia (Dudley et al., 1995). An analysis of lens placode development in homozygous embryos indicates that Bmp7 is required (Wawersik et al., 1999). Specifically, *Pax6* expression was not maintained in the lens placode in the absence of Bmp7. Given the timing of *Pax6* expression loss, this suggests that Bmp7 functions upstream of both the EE and SIMO enhancers in allowing *Pax6* expression at the placodal phase (Wawersik et al., 1999). This information can be incorporated into the genetic model of lens induction (Fig. 11.3; Treisman and Lang, 2002).

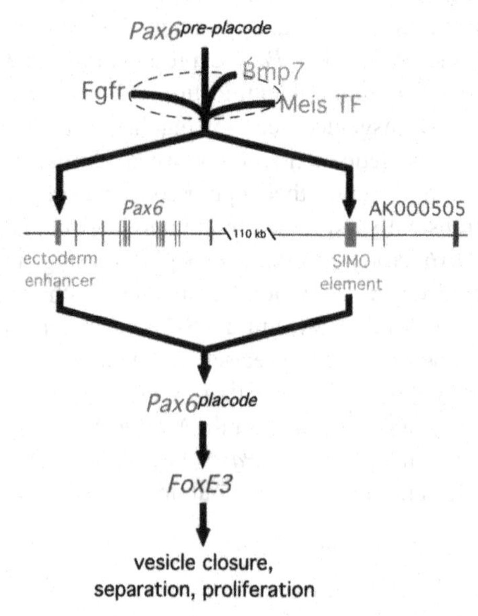

Figure 11.3. Fgf receptor and Bmp7 signaling cooperate in lens induction. Since both placodal Pax6 and *FoxE3* levels are reduced in mouse embryos that express a dominant-negative Fgf receptor in the lens lineage, it is likely that both *Pax6*^{placode} and *FoxE3* lie downstream of Fgf receptor activity. Given that even lower levels of Pax6 and *FoxE3* can be recorded in embryos that express the dominant-negative Fgf receptor *and* have half the normal level of Bmp7, we might suggest that Fgf receptor and *Bmp7* signaling cooperate to maintain the placodal phase of *Pax6* expression. Previous analysis has shown that the early phase of *Pax6* expression is unaffected in the *Bmp7* null mice and thus FGF and *Bmp7* signaling must converge on the pathway downstream of *Pax6*$^{pre\text{-}placode}$.

11.1.3. FGF Receptor Signaling Plays a Role in Lens Induction and Cooperates with Bmp7

It is frequently the case that developmental events require the coordinated action of distinct signaling pathways. This and the observation that FGF signaling regulates lens polarization (section 11.2) provided the motivation to examine the question of whether FGF signaling is involved in lens induction. Expression of an inhibitory, dominant-negative form of FGFR1IIIc in the lens placode using the *Pax6* ectoderm enhancer resulted in transgenic mice (designated *Tfr7*) with distinctive defects in lens and eye development (Faber et al., 2001). The *Tfr7* mice showed a delay in the thickening and invagination of the lens placode, a persistent lens stalk, a small lens pit and vesicle, and ultimately microphthalmia. An examination of Pax6 levels showed that immunoreactivity in the lens placode was lower than in wild-type mice. This suggested that, like Bmp7, FGF receptor signaling was important for lens induction and was genetically upstream of the EE and SIMO regulatory elements (Fig. 11.3; Faber et al., 2001).

These data also implied that Bmp7 and FGF receptor signaling might cooperate in lens induction. To test this possibility, *Bmp7* heterozygote and *Tfr7* mice were crossed through two generations, and the resulting phenotype was examined in animals of compound genotype. Interestingly, the severity of the lens phenotype was dramatically increased in *Tfr7/Tfr7,Bmp7*$^{+/-}$ embryos compared with *Bmp7*$^{+/-}$ or *Tfr7/Tfr7* embryos. This implied a genetic interaction between the two loci and suggested Bmp7, FGF receptor signaling cooperation (Faber et al., 2001). To test this possibility at stages of lens induction, an assessment of Pax6 levels was also performed in wild-type, *Tfr7/Tfr7*, and *Bmp7*$^{+/-}$*,Tfr7/Tfr7* embryos at E9.5. This showed that while *Tfr7/Tfr7* embryos had lower levels than wild-type embryos, the lowest levels were apparent in *Tfr7/Tfr7,Bmp7*$^{+/-}$embryos. This indicated a strong genetic interaction between the *Bmp7* and *Tfr7* loci and provided further evidence for cooperation between Bmp7 and FGF receptor signaling in the regulation of *Pax6* expression (Fig. 11.3; Faber et al., 2001). The progressively lower level of *FoxE3* expression in wild-type, *Tfr7/Tfr7*, and *Tfr7/Tfr7,Bmp7*$^{+/-}$ mice confirmed the proposed genetic relationship of FGF receptor and Bmp7 signaling to *FoxE3* (Faber et al., 2001).

11.1.4. Bmp4 Involvement in Lens Induction and Development

Another member of the Bmp family, Bmp4, has been implicated in early development of the lens. Some *Bmp4* null embryos survive until E10.5, and this is just sufficient to examine the consequences of Bmp4 absence for the first stages of eye development (Furuta and Hogan, 1998). Interestingly, absence of Bmp4 results in the lack of lens formation. Tissue explantation techniques have also been employed to extend the period in which analysis can be performed beyond E10.5 (Furuta and Hogan, 1998), and this strategy has been used to show that lens formation in *Bmp4* null eye primordia could be rescued if the primordia were cultured in the presence of recombinant Bmp4 (Furuta and Hogan, 1998). Furthermore, lens formation from *Bmp4* null presumptive lens ectoderm was rescued if the ectoderm was explanted adjacent to wild-type optic vesicle but not rescued if it was explanted only with Bmp4. These data have been interpreted to suggest that lens induction is a result of presumptive lens ectoderm responding to Bmp4 and at least one other signal derived from the optic vesicle (Furuta and Hogan, 1998). Alternatively, it is possible that Bmp4 has an essential role in the development of the optic vesicle and that in the absence of this factor optic vesicle–derived lens induction signals are not produced.

Interestingly, the absence of Bmp4 does not affect the expression of *Pax6* in the lens lineage (Furuta and Hogan, 1998). This is not what might be expected of a signaling molecule that is a direct lens inducer given the necessity (Ashery-Padan et al., 2000; Collinson et al., 2000; Fujiwara et al., 1994) and sufficiency of *Pax6* for lens induction (Altmann et al., 1997; Chow et al., 1999). As a result of this observation, we must propose that Bmp4 is genetically downstream of *Pax6*^{placode} or participates in lens induction in a parallel pathway.

It is likely that Bmp4 functions in lens induction in a pathway that involves the transcriptional regulator Sox2. Sox2 is an HMG box transcription factor related to the sex-determining factor SRY (Kamachi et al., 1995). Sox2 and family members Sox1 and Sox3 have been implicated in lens development through their expression patterns and through their regulation of crystallin genes (Kamachi et al., 1995; Kamachi et al., 1998). In particular, Sox2 is known to regulate δ-crystallin expression in the chick in a complex with Pax6 (Kamachi et al., 2001). Thus, the observation that *Sox2* expression in the lens lineage is not up-regulated in the usual way in *Bmp4* null embryos suggests the appealing model that Bmp4 stimulates *Sox2* expression in preparation for crystallin gene regulation by a Sox2-Pax6 complex (Fig. 11.4). When combined with the observation that *Sox2* is

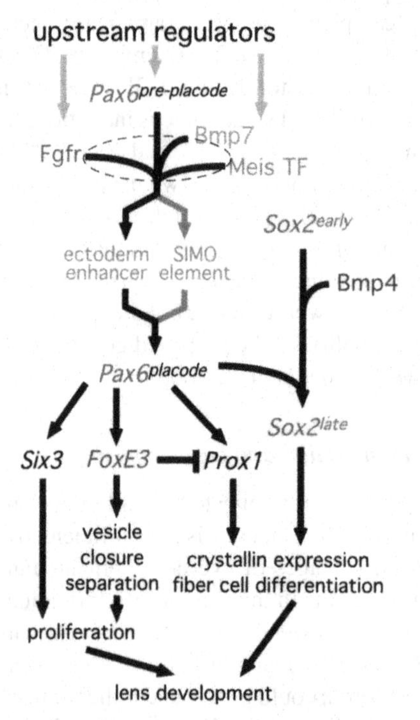

Figure 11.4. Multiple elements can be incorporated into a genetic pathway describing lens induction. The homeodomain transcription factor *Six3* lies genetically downstream of *Pax6*^{placode}, as mice that do not express placodal *Pax6* also do not express placodal *Six3*. This is also true for *Prox1*. Since *FoxE3*$^{dyl/dyl}$ mice show an expansion of the *Prox1* expression domain in the lens epithelium, this suggests that FoxE3 normally suppresses *Prox1* at later stages of lens development. While we currently do not understand the genetic relationship between *FoxE3* and *Sox2*, it is clear from previous analyses that *Sox2* lies downstream of *Pax6*^{placode}. Since *Sox2* (but not *Pax6*) expression is diminished in the *Bmp4* null mice, Bmp4 signaling must contribute to the pathway between *Pax6*^{placode} and *Sox2*.

also not up-regulated in the usual way in *Pax6*$^{Sey/Sey}$ embryos, we can suggest that *Sox2* is downstream of *Pax6*$^{pre\text{-}placode}$ and participates with Bmp4 in the proposed parallel pathway (Fig. 11.4). It is likely that currently unpublished work examining the requirement for different Bmp receptors in lens induction will help refine our understanding of Bmp4 involvement in lens induction.

11.1.5. Retinoic Acid

The study of retinoic acid (RA) involvement in any developmental process has some inherent difficulties because this signaling agent is not a gene product (one researcher recently stated that "we work on retinoic acid signaling, but we don't like it"). Despite this, there is emerging evidence that RA and its related molecules have an important role to play in lens development.

RA receptors (RARs) reside in the nucleus and are ligand-inducible transcriptional regulators (Petkovich et al., 1987). RARs belong to a superfamily of nuclear receptors and are involved in a wide range of developmental processes. It now appears that there are two families of retinoid receptors. Members of the RAR family (RARs α, ß, and γ and their isoforms) are all activated by both all-*trans* RA and 9-*cis* RA. In contrast, members of the retinoid X receptor (RXR) family (RXRs α, ß, and γ) are activated only by 9-*cis* RA. In addition to the ligand and its receptors, there are also cellular retinoic acid–binding proteins (CRABPs) that bind RA and can modulate the action of endogenous retinoids by sequestering RA, thus preventing it from activating the RARs (Perez-Castro et al., 1993).

A role for retinoid signaling in late lens development is supported by evidence that expression of some crystallin genes is regulated by retinoid receptors. For example, RARs and RXRs function additively with Pax6 to activate transcription of the mouse αB-crystallin gene (Gopal-Srivastava et al., 1998). Consistent with this is the observation that transgenic mice expressing a RA-binding protein at high levels in the lens show defects in the development of lens fiber cells. However, there is also strong evidence that RA signaling has an important role in lens induction.

When early mouse embryos are exposed to retinol-binding protein antisense oligonucleotides, RA signaling is effectively inhibited according to readout from a RA-sensitive lacZ reporter (Bavik et al., 1996). The inhibition of RA signaling results in defects in eye formation, specifically a failure of the lens placode to invaginate and form the lens pit (Bavik et al., 1996). This suggests that RA signaling has a critical early role in lens induction. Supporting this idea is the observation that the RA-responsive lacZ reporter is expressed in the neuroepithelium of the optic pit at E8.5 and in the lens lineage from E8.75 (Enwright and Grainger, 2000), at stages when lens induction signals are believed to be critical. Furthermore, *Pax6*$^{Sey/Sey}$ embryos, in which Pax6 activity is absent, show diminished expression of the RA-sensitive reporter in the lens placode (Enwright and Grainger, 2000), suggesting that by some means Pax6 regulates the activity of this pathway. It has thus been suggested (Enwright and Grainger, 2000) that some aspects of the Pax6 deficiency phenotypes are a result of diminished RA signaling. This notion is supported by the observation that RAR/RXR null mice have defects in lens development (Kastner et al., 1994) and that exposure of mouse embryos to 13-*cis* RA produces a Peters' anomaly – like defect characteristic of *Pax6* heterozygosity (Cook and Sulik, 1988).

Further implicating RA as an important lens induction signal is evidence that two RA responsive transcription factor genes are implicated in lens induction. The Meis transcription factors have already been mentioned in the context of *Pax6* regulation, but in the limb it is clear that *Meis1* and *Meis2* are downstream of RA signaling and are critical proximal limb determinants (Capdevila et al., 1999; Mercader et al., 1999; Mercader et al., 2000). Given the possibility of conserved pathway function, it is worth investigating whether in lens induction RA signaling regulates the transcription of Meis genes. Similarly, the AP-2α gene is RA responsive and when mutated results in distinctive defects in early development of the eye, including the lens (Nottoli et al., 1998; West-Mays et al., 1999). With this information available, there is every opportunity to devise how RA regulates AP-2α and Meis transcription factors in the context of lens induction.

11.1.6. Other Elements of Lens Development Genetic Pathways

While there are still many genetic relationships that are not completely understood, several additional elements can be placed in the model of genetic regulation of early lens development. Six3 is a homeodomain transcription factor related to *Drosophila sine oculis* (Oliver et al., 1995). Analysis of *Six3* expression in embryos with a conditional deletion of *Pax6* in the lens placode (Ashery-Padan et al., 2000) indicates that *Six3* lies genetically downstream of $Pax6^{placode}$ (Fig. 11.4). This is also true for Prox1 (Ashery-Padan et al., 2000), a homeobox transcription factor that is related to *Drosophila prospero* (Tomarev et al., 1998) and is required for lens fiber cell development (Wigle et al., 1999). Interestingly, the $FoxE3^{dyl/dyl}$ mice show an expansion of the *Prox1* expression domain in the lens epithelium (Blixt et al., 2000), suggesting that FoxE3 plays a role in suppressing *Prox1* at later stages of lens development (Fig. 11.4).

11.1.7. Future Challenges in the Analysis of Lens Induction Mechanisms

Several challenges face researchers interested in understanding lens induction. Despite the relatively complex genetic pathways that can be proposed to describe lens induction, we currently know remarkably little about how the various components orchestrate the cell-cell interactions that are presumably crucial for development of the lens. For example, the *Bmp7* gene is expressed in the presumptive lens ectoderm (Wawersik et al., 1999), and it is therefore tempting to suggest that it may signal to the presumptive retina. Thus far, however, there is no evidence for this type of signaling event. So too we currently have no direct evidence that an Fgf ligand is the mediator of an optic vesicle to presumptive lens signal in lens induction.

A further challenge will be to understand how the various signaling pathways involved in lens induction are integrated. There is strong evidence for involvement of Fgf and Bmp signaling pathways and circumstantial evidence for the involvement of RA signaling. Wnt signaling pathways have also been implicated. Modulation of Wnt signaling through misexpression of the Wnt receptor Xfz3 in *Xenopus laevis* results in the formation of lenses in the context of ectopic eyes (Rasmussen et al., 2001), and deletion of the Wnt signaling coreceptor Lrp6 results in a microphthalmia (Pinson et al., 2000) that appears to have its origin in the inductive phases of eye development (Stump et al., 2003). Thus, there are as many as four distinct pathways that must be integrated in a finely tuned way if the eye is to develop normally and reproducibly. An analysis of pathway integration will require all of the tools from the developmental geneticist's toolkit.

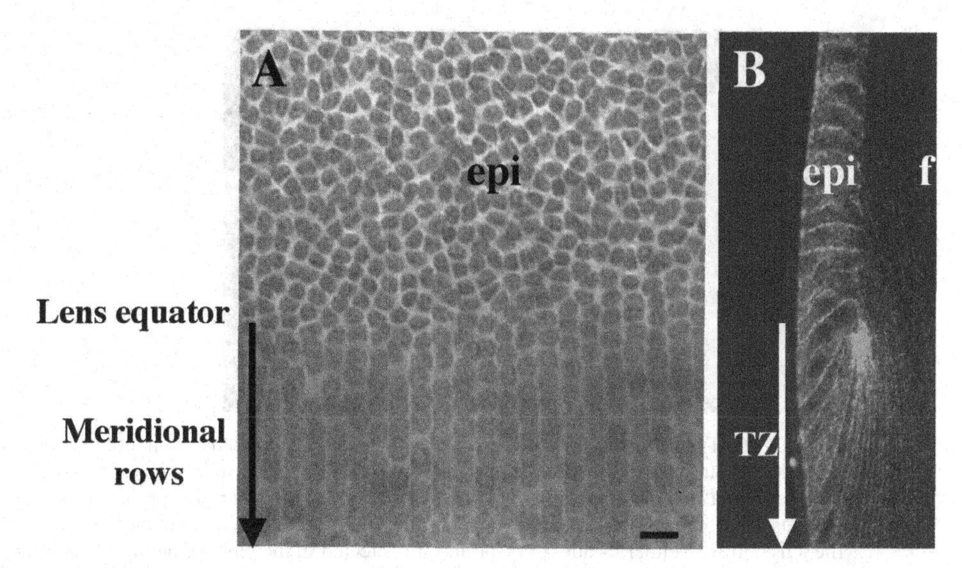

Figure 11.5. (See color plate XVI.) (A) Wholemount of the lens epithelium of a weanling rat, stained with hematoxylin. The equatorial region is shown. Anterior to the equator, the lens epithelial (epi) cells exist as a monolayer and show cobblestone-like packing. Below the equator, the cells are aligned in very distinct meridional rows. (B) The histological section of the weanling rat lens is stained with an antibody for ß-catenin to show cell boundaries. Cells above the equator are columnar in shape, whereas below the equator in the transitional zone (TZ) the cells are elongated. The arrows indicate equivalent regions known as meridional rows in wholemounts (A) and the transitional zone in sections (B). Together the wholemount and section show that the elongating fiber cells are highly aligned. Scale bar: 11 μm.

11.2. Lens Differentiation and Growth

The later stages of lens morphogenesis are characterized by the differentiation of two forms of lens cells from the lens vesicle. Differentiation of epithelial and fiber cells involves acquisition of distinctive morphological and molecular characteristics required for the function of cells in the two lens compartments. The epithelial cells are cuboidal and possess strong intercellular adhesion and communication properties. They are firmly attached to the lens capsule and form an epithelial sheet that covers the anterior surface of the fiber mass. The fibers are highly elongated cells. During differentiation, each cell develops a similar hexagonal shape, with four short sides and two long sides, and this results in the fibers assuming a highly ordered packing arrangement (see chap. 4 of this volume). This is readily evident in wholemount preparations that include the meridional rows (Fig. 11.5).

One of the best general markers for lens cell differentiation is the presence of crystallins. These proteins are abundant in the lens and progressively accumulate during morphogenesis and growth. In mammals, α-crystallin appears during formation of the lens pit, and as lens morphogenesis progresses, it is produced by all lens vesicle cells and their progeny, epithelial and fiber cells (McAvoy, 1978b). ß-Crystallin and γ-crystallin first appear in the posterior cells of the lens vesicle that form the primary fibers, but unlike α-crystallin, their expression remains restricted to the fiber cell compartment throughout life (Fig. 11.6; McAvoy, 1978a). Differences in crystallin composition are only one feature that distinguishes epithelial and fiber cells at the molecular level. Epithelial cells and fiber cells express their own distinctive repertoires of communication, adhesion, and cytoskeletal molecules, and these determine

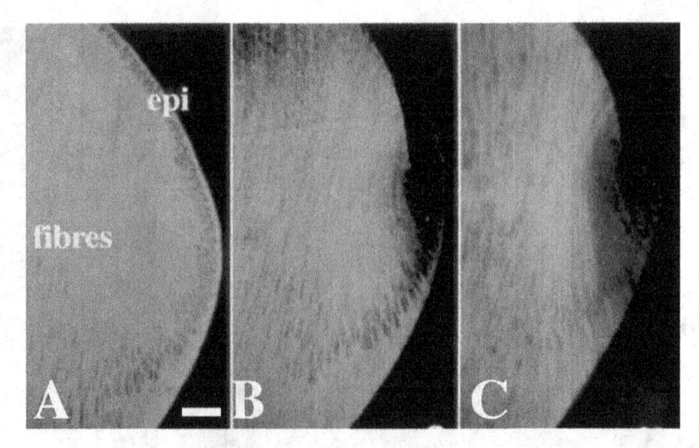

Figure 11.6. Typical distribution of crystallins in the neonatal rat lens. Immunohistochemistry carried out on serial sections shows α-crystallin (A) is detected in both lens epithelial (epi) and fiber cells. ß-crystallin (B) is first detected early in fiber elongation just below the lens equator (see Fig. 11.5), and γ-crystallin (C) is first detected in young fiber cells in the lens cortex. Neither ß- nor γ-crystallins are detected in the lens epithelium. Scale bar: 50 μm. (Adapted from McAvoy, 1978a.)

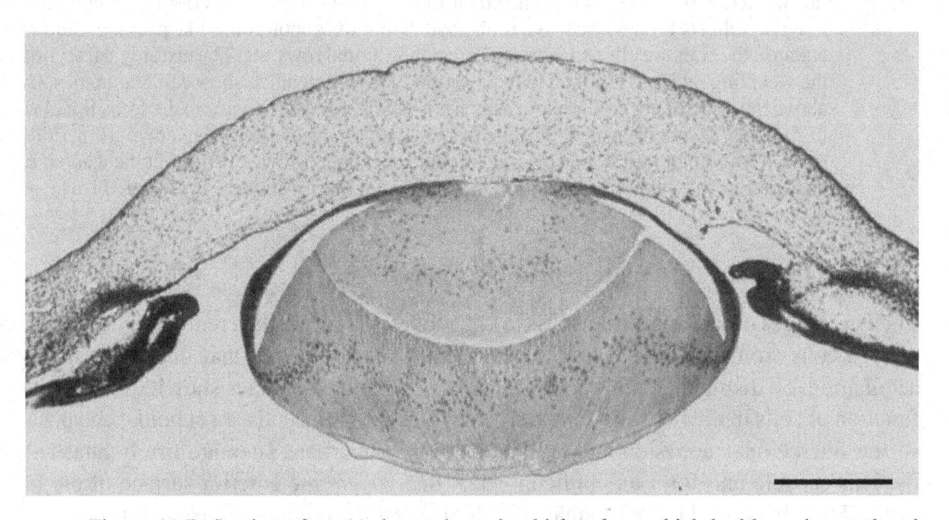

Figure 11.7. Section of an 11-day embryonic chicken lens which had been inverted and left *in situ* for five days. The polarity of the lens has become reversed. A new fiber mass has developed posteriorly from the original lens epithelium, and a new epithelial layer has formed anteriorly. Scale bar: 1 mm.

their different structures and functions (see chaps. 6 and 7 of this volume). How the two forms of lens cells differentiate to bring about lens polarity has been a major focus of research.

The importance of the ocular environment for determining and maintaining lens polarity was demonstrated in the classic lens inversion experiments of Coulombre and Coulombre (1963). They turned the chicken lens upside down so that the epithelial cells, which normally faced the aqueous and cornea, faced the vitreous and neural retina instead (Fig. 11.7). In this new environment, the epithelial cells elongated and differentiated into a new fiber mass, whereas a new epithelial sheet grew over the surface of the lens facing the cornea. Similar lens inversion experiments using mouse eyes confirmed that this phenomenon also occurred in mammals (Yamamoto, 1976). These experiments showed that the vitreous environment promotes the differentiation of fiber cells and that the aqueous environment promotes epithelial differentiation and growth.

The ocular media have been shown to constitute a rich source of growth and regulatory factors (McAvoy and Chamberlain, 1990). The lens itself expresses members of major growth factor families and a variety of growth factor receptors and molecules involved in a range of signaling pathways. Studies over the last 20 years have mostly concentrated on identifying the factor(s) that controls fiber differentiation. The influence of members of major growth factor families has been investigated, and as a result of promising leads, members of the insulin/IGF, FGF, and TGFß growth factor families have received a lot of attention. This work will be evaluated in this chapter, and other factors that have been shown to influence the behavior of lens cells will also be reviewed.

11.2.1. Regulation of Fiber Differentiation

11.2.1.1. Insulin/IGF

The first factor reported to influence lens fiber differentiation was insulin. Piatigorsky (1973) showed that insulin stimulated explanted lens epithelial cells from six-day-old embryonic chicks to elongate. Beebe et al. (1980) reported that a factor from the vitreous, lentropin, stimulated chick lens epithelial explants to elongate and accumulate δ-crystallin, a crystallin that is markedly up-regulated during fiber differentiation in chicks. Subsequently, it was shown that lentropin is functionally and immunologically related to IGF-1 (Beebe et al., 1987). More recent studies have confirmed the ability of insulin/IGF-1 to induce chick lens epithelial explants to elongate and accumulate fiber differentiation markers, including CP49, a component of the fiber-specific beaded filaments (Le and Musil, 2001).

The influence of insulin and IGF-1 on the differentiation of epithelial explants from neonatal rats has also been studied. In contrast to their influence in chicks, insulin and IGF-1 each have only a slight stimulatory effect on the fiber differentiation response. However, in the presence of FGF, insulin or IGF can synergistically enhance fiber differentiation (Chamberlain et al., 1991). Further analysis of this phenomenon has shown that members of the insulin/IGF family of growth factors can maintain fiber differentiation once initiated by FGF (Klok et al., 1998; Leenders et al., 1997). Interestingly, an exposure of 15 minutes is sufficient to initiate this process, which IGF can then maintain. Members of the insulin/IGF growth factor family are present in the ocular media (Arnold et al., 1993), and because of their biological activity *in vitro*, they are likely to have a role in regulating the process of fiber differentiation *in vivo*. Their role may be less prominent in mammals than in chicks, as the fiber differentiation effects of insulin/IGFs (on their own) appear to be much less pronounced in the former. This is supported by studies with transgenic mice. When IGF-1 was overexpressed in the lens, from the αA-crystallin promoter there was no induction of premature differentiation in the lens epithelial compartment. Instead there was an apparent expansion of the germinative and transitional zones toward the posterior lens pole (Shirke et al., 2001), indicating a role for IGF-1 in regulating proliferation *in vivo* (see section 2.2.1).

11.2.1.2. FGF

Researchers over a number of years have developed a strong case that members of the FGF family of growth factors play a central role in regulating fiber differentiation (see

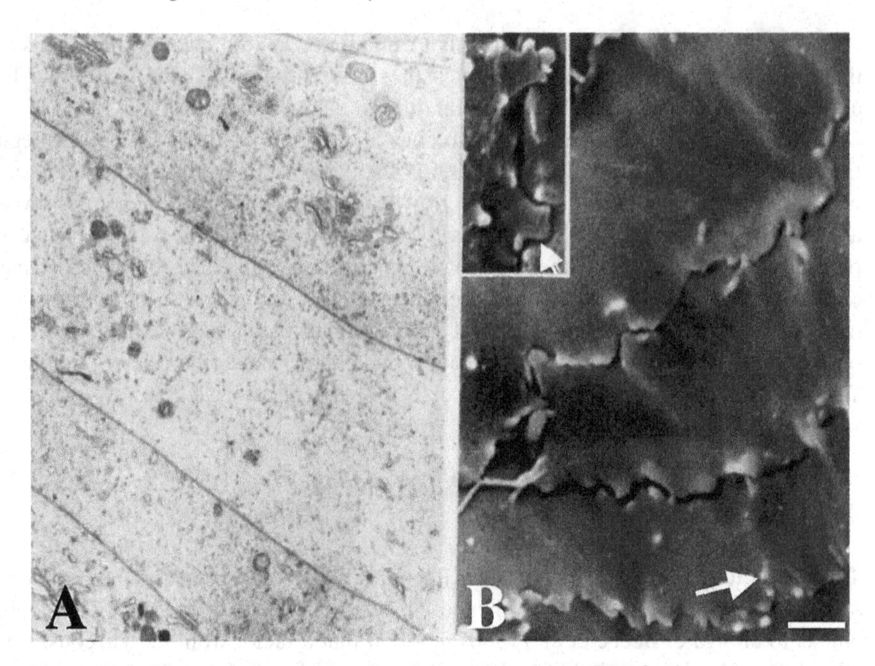

Figure 11.8. Electron micrographs of explants cultured with FGF for four days. Transmission electron microscopy (A) and scanning electron microscopy (B) show parallel packing of flattened fiber cells, with interlocking processes at the lateral margins; for example, tonguelike flaps overlap complementary imprints in neighboring cells (arrows, B and inset). Scale bar: (A), 0.5 μm; (B), 1 μm; (B) inset, 0.4 μm. (Adapted from Lovicu and McAvoy, 1989.)

Chamberlain and McAvoy, 1997). This case was initially based on evidence from *in vitro* studies but has received support from a variety of *in vivo* models designed to test the hypothesis that FGF is a key regulator of this process.

***In Vitro* Studies.** The development of a rat lens epithelial explant system was central to the identification of members of the FGF family as potential regulators of fiber differentiation. In this culture system, FGF induces lens epithelial cells to undergo morphological and molecular changes characteristic of fiber differentiation (Lovicu and McAvoy, 1989). The explants become multilayered as epithelial cells migrate and elongate into fibers. Structural specializations include loss of cytoplasmic organelles, formation of specialized cell-cell junctions, and denucleation (Fig. 11.8). These and other FGF-induced changes are characteristic of lens fiber differentiation *in situ* (Fig. 11.9).

A Role for FGF *in Vivo*. For FGF to be a regulator of fiber differentiation *in vivo*, it must be present in the lens environment, and lens cells must express the appropriate receptors *in situ*. Currently there are 23 known members of the FGF family (Bansal, 2002). FGF-1 and FGF-2 were the first FGFs to be identified, and their distribution in the eye has been studied extensively. For most of the more recent additions to the family, detailed distribution studies have not yet been carried out. Using immunohistochemical and biochemical methods, it has been shown that, in addition to being present in lens cells and the lens capsule, FGF-1 and FGF-2 are present in the ocular media that bathe the lens and tissues near the lens,

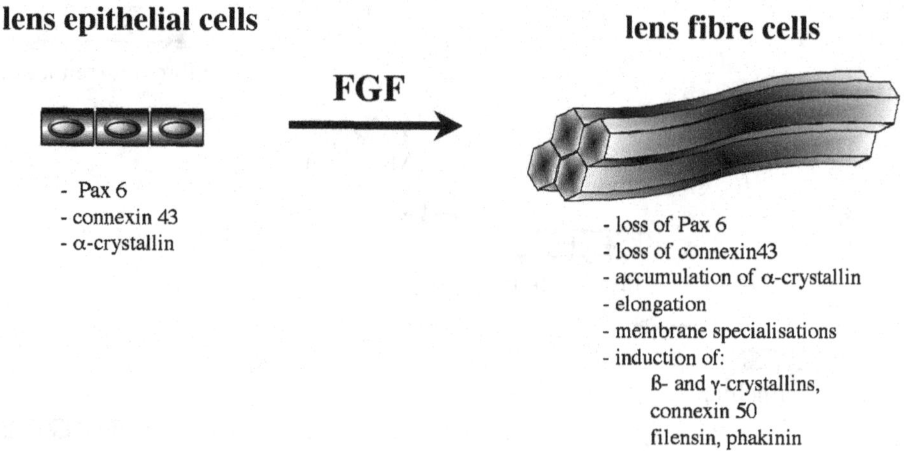

lens epithelial cells

FGF

lens fibre cells

- Pax 6
- connexin 43
- α-crystallin

- loss of Pax 6
- loss of connexin43
- accumulation of α-crystallin
- elongation
- membrane specialisations
- induction of:
  ß- and γ-crystallins,
  connexin 50
  filensin, phakinin

Figure 11.9. Diagram illustrating FGF induction of fiber differentiation, including cell elongation and fiber-specific changes in expression of molecular markers.

including the retina, cornea, ciliary body, and iris (Chamberlain and McAvoy, 1997). *In situ* hybridization studies showed that FGF-1 and FGF-2 mRNAs have a ubiquitous distribution in eye tissues similar to that of their proteins (Lovicu et al., 1997). The distribution of other FGFs has been studied, and so far FGF-12 is the only other family member reportedly expressed in the lens (Hartung et al., 1997). Other FGFs, including 3, 5, 8, 9, 11–13, 15, and 19, are also expressed in the retina at various stages of its development (Govindarajan and Overbeek, 2001; Vogel-Hopker et al., 2000; Xie et al., 1999; Zhao et al., 2001). Thus, a number of FGFs are potentially available to lens cells and could have roles in regulating fiber differentiation during development. However, the wide distribution of FGFs in both anterior and posterior segments of the eye raises the question of how fiber differentiation is spatially restricted to the posterior segment. Normally the only cells that undergo fiber differentiation are those below the lens equator, and this is crucial for maintaining lens polarity during the lifelong process of lens growth.

FGF Induces Different Responses at Different Concentrations. A possible mechanism whereby FGF may influence lens polarity and growth patterns emerged from dose-response studies. Analysis of explants cultured with FGF identified three distinct cellular responses: proliferation, migration, and fiber differentiation. Interestingly, these responses were induced sequentially as the FGF concentration was increased (Fig. 11.10; McAvoy and Chamberlain, 1989). Therefore, the concentration of FGF has the potential to influence the nature of the response of lens epithelial cells both qualitatively and quantitatively.

Investigations into the mechanisms that underlie these dose-dependent responses have focused on identifying the relevant signaling pathways. So far, mitogen-activated protein kinase (MAPK) signaling has received the most attention. It has been shown that FGF induces a dose-dependent activation of ERK-1/2 in lens epithelial cells (Lovicu and McAvoy, 2001). Blocking experiments have shown that activation of ERK is required for FGF-induced lens cell proliferation and fiber differentiation. However, inhibition of ERK signaling can block only the morphological changes associated with FGF-induced lens fiber differentiation and not the synthesis of some of the molecular differentiation markers, such as ß-crystallin.

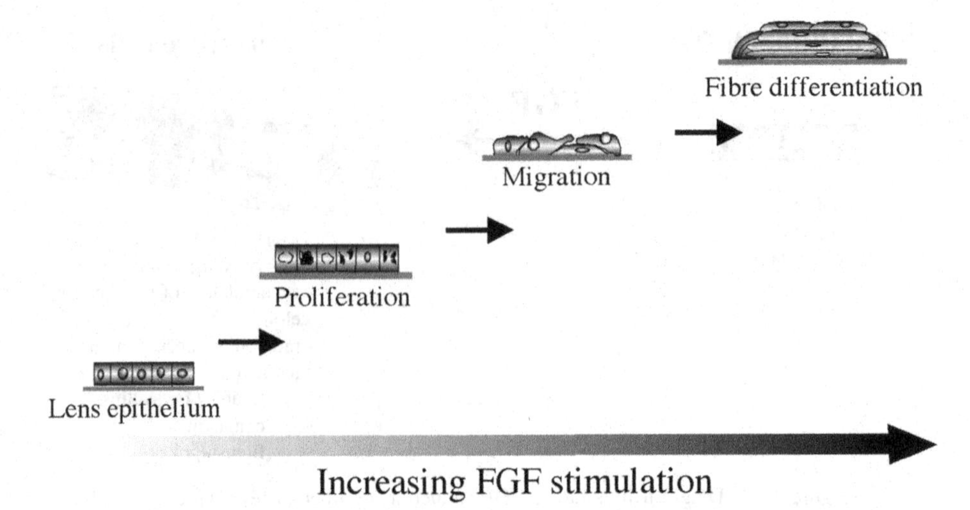

Figure 11.10. Diagram illustrating the three different responses to FGF observed in lens epithelial explants *in vitro* and their sequential stimulation by increasing concentrations. (Adapted from Chamberlain and McAvoy, 1997.)

Other studies have also reported an association between ERK signaling and FGF-mediated mammalian (Chow et al., 1995; Govindarajan and Overbeek, 2001) and avian lens fiber differentiation (Le and Musil, 2001). In addition, the involvement of ERK signaling in fiber differentiation is consistent with the immunohistochemical localization of the phosphorylated (active) forms of ERK-1/2 in the transitional zone, where cells elongate and undergo the earliest morphological changes associated with fiber differentiation (Lovicu and McAvoy, 2001). Taken together, these data indicate that FGF-induced ERK signaling is important for proliferation and the regulation of early morphological events associated with fiber differentiation. However, other FGF-induced signaling pathways required for the process of lens fiber differentiation and maturation remain to be identified.

The FGF Gradient Hypothesis. Given the dose-dependent response to FGF and the highly ordered anteroposterior sequence of proliferation, migration (or displacement), and fiber differentiation in the lens, it has been proposed that these polarized patterns of lens cell behavior are determined by an FGF gradient (Fig. 11.11; Chamberlain and McAvoy, 1997). The ability of growth factors to induce different cellular responses at different concentrations has also been shown in other systems and may involve a general mechanism that underlies the establishment of growth and differentiation patterns within tissues (e.g., see Lillien and Wancio, 1998; Pizette et al., 1996).

In the lens, a number of different approaches have been used to test the FGF gradient hypothesis, such as the analysis of FGF distribution, the analysis of FGF receptor expression, and the generation of transgenic mice with altered patterns of FGF distribution and receptor expression.

Differential Distribution of FGF in the Ocular Media. *In vivo*, the anteroposterior patterns of lens cell behavior correlate with the distribution of the ocular media. The anterior region of the lens is bathed by aqueous, while the region of the lens below the equator (where fibers differentiate) is bathed by vitreous. To test the hypothesis that the ocular

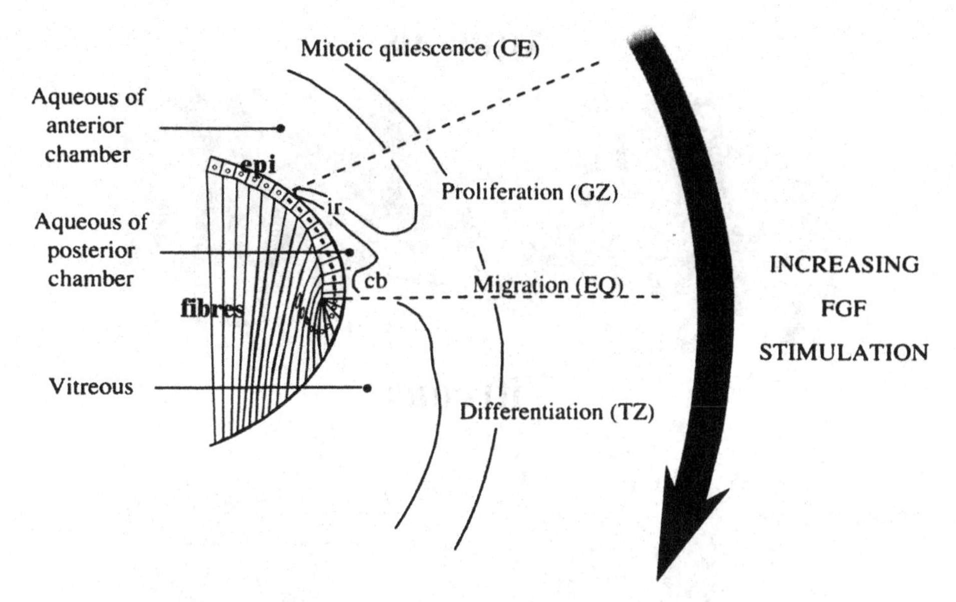

Figure 11.11. Diagram indicating how anteroposterior patterns of lens cell behavior may be determined by a gradient of FGF stimulation. In the mature lens, the epithelium (epi) can be divided into two main zones, the central epithelium (CE) and the germinative zone (GZ), which extends to the equator (EQ). Immediately posterior to the equator is the transitional zone (TZ). These zones coincide with compartments defined by the anatomy of the eye: Central epithelial cells are exposed to aqueous of the anterior chamber; germinative zone cells are exposed to aqueous of the posterior chamber, which is demarcated by the ciliary body (cb) and iris (ir); the transitional zone extends into the compartment that contains the vitreous. The cellular behaviors indicated are observed both *in vivo* and in lens epithelial explants cultured with increasing concentrations of FGF. (Adapted from Chamberlain and McAvoy, 1997.)

media influence the differentiated fate of lens cells, epithelial explants were cultured in either aqueous or vitreous. In explants exposed to vitreous, the cells became elongated and multilayered and acquired morphological and molecular markers of fiber cells. In contrast, the cells in explants cultured with aqueous maintained an epithelial phenotype (Fig. 11.12; Lovicu et al., 1995).

Thus, vitreous induced a fiber differentiation response but aqueous did not. Analysis by SDS-PAGE and Western blotting showed that FGF-1 and FGF-2 are present in both media but that more of these FGF isoforms are recovered from vitreous than from aqueous. By fractionating vitreous and testing fractions by FGF ELISA (using antibodies specific for FGF-1 and FGF-2) and biological assay, it was shown that a large proportion of the fiber-differentiating activity of vitreous is FGF associated (Schulz et al., 1993). However, several fractions, containing about 16% of the total activity, appeared to have virtually no FGF. This observation indicates that factors other than FGF-1 and FGF-2 are present in the ocular media and have fiber-differentiating activity. Such factors may be unrelated to FGFs but may also have the ability to activate signaling pathways involved in the differentiation process. Alternatively, these molecules may be other members of the FGF family. Of the ones that have been studied, FGFs 3, 5, 8, 9, 12, and 15 have been detected in the eye. Given that FGFs 3, 8, and 9 (among other FGFs; see Lovicu and Overbeek, 1998) have been shown to induce premature fiber differentiation in the lens epithelium in transgenic

Aqueous

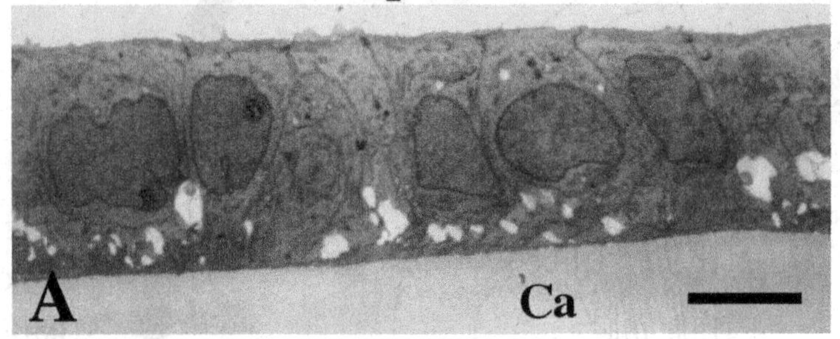

Vitreous

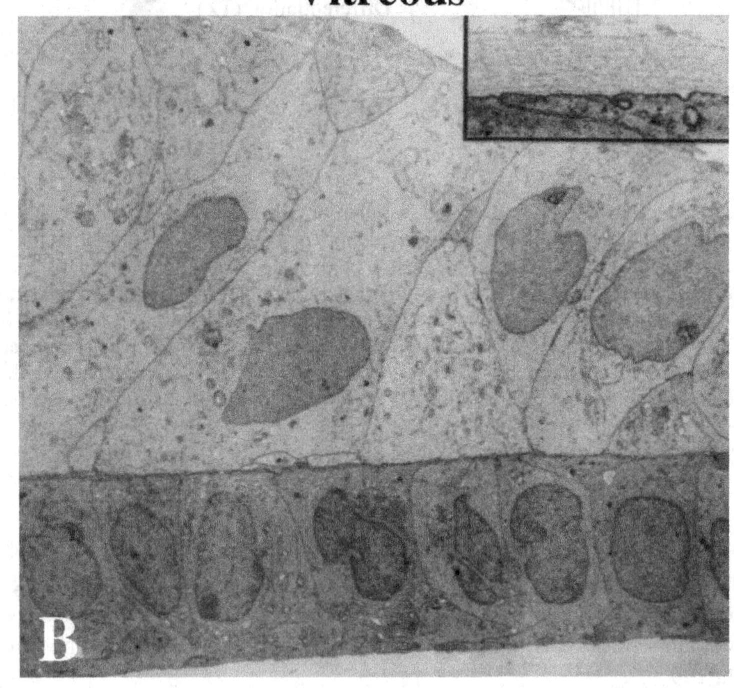

Figure 11.12. Transmission electron micrographs of neonatal rat lens epithelial explants cultured with ocular media for five days. (A) In the presence of aqueous, cells remained in a monolayer on the lens capsule (Ca) and retained a cuboidal epithelial morphology. (B) In the presence of vitreous, explants thickened, and many cells showed early fiber elongation characteristic of cells in the transitional zone (see Fig. 11.5B). The inset in B also shows that a capsulelike matrix forms on the exposed surface of the elongating cells. Fresh bovine aqueous and vitreous were diluted with an equal volume of culture medium before use. Scale bar: 5 μm. (Adapted from Lovicu et al., 1995.)

mice (see below), it is possible that a number of FGFs may work together to maintain a fiber-differentiating environment in the vitreous.

A higher concentration of FGF in the vitreous than in the aqueous may help to ensure that fiber differentiation is restricted to the posterior compartment; however, the ability of the vitreous to promote fiber differentiation may not be due solely to a relatively high concentration of FGF. Other factors present in the vitreous may positively modulate the effects of FGF. Possible modulators include extracellular molecules such as heparan sulphate (HS) and heparan sulphate proteoglycan (HSPG), which are known to stabilize FGF, and cell surface HSPGs, which appear to be involved in presenting the ligand to the high-affinity tyrosine kinase receptors (Ornitz and Itoh, 2001). There is very little information available about such molecules in the aqueous and vitreous and how they could influence FGF activity and bioavailability. However, some anteroposterior differences in HSPGs have been reported in the lens capsule. For example, there appears to be a higher HSPG:core protein ratio in the posterior capsule than in the anterior capsule (Schulz et al., 1997). It is not known if this influences FGF activity in relation to lens cells, but it is interesting that greater FGF biological activity and reactivity resides in the posterior than in the anterior capsule (de Iongh and McAvoy, 1992; Lovicu and McAvoy, 1993; Schulz et al., 1993). Clearly, much more needs to be done to investigate the complex interactions between FGF and extracellular matrix molecules such as HS and HSPG in the ocular media and capsule in order to understand how they influence patterns of lens cell behavior.

Differential Distribution of FGF Receptors. Cellular responses to FGF are mediated via a family of four receptor genes (*FGFR1–4*; Bansal, 2002; Ornitz and Itoh, 2001). Lens cells have been shown to express three of the *FGFR* genes: *FGFR1*, *FGFR2*, and *FGFR3*. Each receptor has its own unique pattern of expression, but overall there is a substantial increase in receptor expression below the lens equator (McAvoy et al., 1999). Such an anteroposterior gradient in receptor expression could contribute to an anteroposterior gradient of FGF signaling in the lens.

Transgenic and Knockout Studies. Studies with transgenic mice provide strong support for the "FGF gradient hypothesis." Through use of the α-crystallin promoter, transgenic mice were generated that express high levels of various FGFs specifically in the lens. In these transgenic mice, overexpression of FGF-1, FGF-3, FGF-4, FGF-5, FGF-7, FGF-8, or FGF-9 induced the anterior epithelial cells to undergo premature fiber differentiation (Fig. 11.13; Lovicu and Overbeek, 1998; Robinson et al., 1995a; Robinson et al., 1998). They lost their typical cuboidal epithelial morphology, became elongated, and accumulated ß-crystallin, which is normally restricted to the fiber mass. In most cases the lens lost its characteristic cellular polarity. In studies using this transgenic model, FGF-2 is the only member of the FGF family reported so far that did not induce differentiation in the epithelium (Stolen et al., 1997). Furthermore, in studies using a similar transgenic approach, the process of lens fiber differentiation was impaired by the overexpression of either a signaling-defective, inhibitory FGFR (Chow et al., 1995; Robinson et al., 1995b; Stolen and Griep, 2000) or a specific secreted FGF receptor (Govindarajan and Overbeek, 2001).

Results from these transgenic studies lend strong support to the hypothesis that in the normal situation an anteroposterior FGF gradient is involved in regulating fiber differentiation and determining lens polarity. For example, the simplest interpretation of the FGF overexpression studies is that the normal ocular FGF gradient is destroyed. Hence, the anterior epithelial cells are exposed to higher than normal levels of FGF and consequently undergo an inappropriate fiber differentiation response. The transgenic studies also show that most of the FGFs tested exhibit a capacity to induce fiber differentiation. This is consistent with the observation that lens cells express three of the four FGF receptor genes, thereby gaining

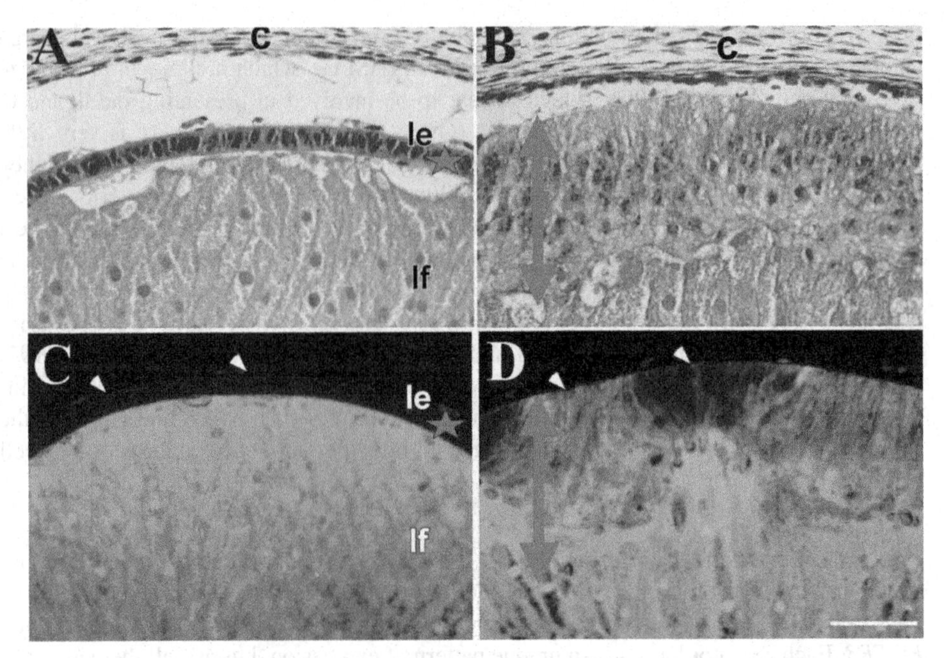

Figure 11.13. (See color plate XVII.)Induction of fiber differentiation by FGF-4 in transgenic mice. Sections of embryonic day 15.5 nontransgenic (A, C) and transgenic (B, D) eyes, either stained with haematoxylin and eosin (A, B) or immunolabeled for ß-crystallin expression (C, D). In transgenic mice, anterior lens epithelial cells undergo extensive elongation (B, double arrow), differentiating into fibers, as shown by the expression of ß-crystallin (D, double arrow). In nontransgenic mice, the lens maintains its distinct polarity with a monolayer of cuboidal epithelial cells (A, le, asterisk), overlying a full complement of fiber cells (lf), reactive for ß-crystallin. Characteristically, ß-crystallin is not expressed in the anterior lens epithelium (C, le, asterisk). The small arrowheads in C and D define the anterior surface of the lens. Scale bar: 60 μm. (Adapted from Lovicu and Overbeek, 1998.)

the capacity to bind a wide range of FGF family members. However, it is not clear yet which of the 23 currently known FGFs are involved in this process in the normal lens.

In this context it is also important to note that, although FGF-1 and FGF-2 are present in the ocular environment and appear to provide a major component of the fiber-differentiating activity of vitreous humor, studies with adult mice made homozygous null for both FGF-1 and FGF-2 show no defect in lens or eye development (Miller et al., 2000). This result clearly shows that neither FGF-1 nor FGF-2, alone or in combination, is necessary for lens fiber differentiation. The same applies to FGF-9, as mice made homozygous null for the FGF-9 gene show no lens phenotype (unpublished data), although a retinal phenotype was reported (Zhao et al., 2001). As more FGFs are tested for their activity in lens fiber differentiation, we may identify an FGF or FGFs that are critical for lens fiber differentiation. However, perhaps it is more likely that redundancy exists in the system and that several FGFs are responsible for fiber differentiation. In this scenario, the absence of one or a couple of FGFs would have little or no effect, as the remaining FGFs would compensate and maintain normal function. The definitive test for the critical role of FGFs in fiber differentiation awaits the results from experiments in which all FGF signaling has been blocked, such as in mice with lenses that are null for all FGF receptors.

11.2.1.3. TGFβ Superfamily

Bmps. In a previous section, we summarized how Bmp signaling participates in lens induction. Recent evidence indicates that Bmp signaling pathways continue to play a role as the lens matures. A number of observations have suggested that Bmp signaling is important for fiber cell maturation. Inhibition of Bmp activity with recombinant noggin (Zimmerman et al., 1996) in explants of the mouse eye (Faber et al., 2002) or in noggin retrovirus-infected chick eyes (Belecky-Adams et al., 1997) has resulted in reduced elongation of lens fiber cells. A chick lens epithelial explant assay was also used to show that fiber cell elongation activity present in vitreous could be inhibited by noggin (Belecky-Adams et al., 1997) and furthermore that vitreous depleted of noggin-binding factors could have its fiber cell elongation activity restored by the addition of Bmps (Belecky-Adams et al., 1997). Transgenic mice expressing an inhibitory, dominant-negative form of the Bmp receptor Alk6 also showed reduced fiber cell elongation, although this was restricted to the ventronasal quadrant of the lens vesicle (Faber et al., 2002). Finally, both mouse (Faber et al., 2002) and chick (Belecky-Adams et al., 1997) lens vesicles showed the anticipated nuclear pattern of immunoreactivity with antibodies (Korchynskyi et al., 1999) to the active, phosphorylated form of the Bmp signaling proteins Smad1, 5, and 8 (Massague, 1998). Combined, these data indicate that the Bmp signaling machinery is active in the lens at stages when fiber cell elongation is initiated and suggest that Bmp pathways may be evolutionarily conserved and important for the regulation of lens polarity.

TGFß. Members of the TGFß growth family and their receptors are expressed in the lens, and *in vitro* studies show that TGFß induces pathological changes in the lens epithelium (de Iongh et al., 2001a). However, recent studies with transgenic mice also indicate other important functions for TGFß in fiber differentiation (de Iongh et al., 2001b). Expression of a mutant TGFß type II receptor transgene under the control of the αA-crystallin promoter results in severe degeneration of fiber cells in the inner cortex. In these "dominant-negative" TGFß receptor mice, lens fiber differentiation is disrupted in the inner cortex at about the stage of early nuclear condensation in the lens bow. These fibers first become swollen, then disintegrate. Thus, inhibition of TGFß receptor signaling disturbs fiber cell maturation and/or maintenance, suggesting an important role for TGFß signaling in coordinating events in fiber differentiation. At this time, however, it is not clear which specific event or events this signaling pathway regulates.

11.2.1.4. Other Growth Factors in Fiber Differentiation

PDGF. Members of this family of growth factors are expressed near the lens (Reneker and Overbeek, 1996). Transgenic mice that overexpressed PDGF-A specifically in the lens displayed visible ocular abnormalities early in development (E15), including lenticular defects such as lens enlargement and the appearance of cataracts (Reneker and Overbeek, 1996). The lens epithelium was observed to be multilayered, consisting of a surface epithelial monolayer and an underlying subepithelium. A considerable increase in DNA synthesis was detected in the epithelium of these transgenic lenses compared with the wild type (see later). The transgenic lenses also exhibited areas of cells in both the subepithelial and surface epithelial layers that expressed fiber-specific ß-crystallin. Hence, the overexpression of PDGF-A caused some cells to undergo at least some aspects of fiber cell differentiation. However, it was not clear in this transgenic study if the phenotype was due to the influence of PDGF on fiber differentiation or if it was a result of secondary effects induced by disturbing patterns of PDGF signaling. Results from explant studies support the latter possibility. In epithelial explants, PDGF is unable to induce fiber differentiation on its own but can potentiate the fiber-differentiating activity of FGF (Kok et al., 2002). It is noteworthy that PDGFR-α expression is lost early in the transitional zone in the process of fiber differentiation (Reneker and Overbeek, 1996); therefore, any influence that PDGF may have on fiber differentiation is likely to be restricted to early stages of the process.

EGF. Members of the EGF family of growth factors and their receptors have been detected in the lenses of a number of species (Weng et al., 1997). Experimental studies have also indicated that EGF signaling may be involved in regulating proliferation and differentiation of lens cells in various species, including birds and mammals. EGF has been shown to have an effect on proliferation in organ cultures as well as cell cultures (Reddan and Wilson-Dziedzic, 1983), and the appearance of lentoids is more frequent in long-term cultures treated with EGF (Ibaraki et al, 1995). Whether EGF directly influences differentiation is still not clear, as cell cultures invariably include serum. However, recent studies with chick lenses indicate that EGFR signaling occurs in fresh annular pad cells, and cultures of these cells respond to TGF-α by up-regulating filensin, an early marker for fiber differentiation (Ireland and Mrock, 2000). This up-regulation was further increased when cells were costimulated with cAMP analogs. Taken together, these results indicate that EGFR signaling may have a role in regulating aspects of the fiber differentiation response.

11.2.1.5. Overview

Thus, the picture that is emerging from growth factor signaling studies is that the process of lens fiber differentiation and maturation may depend on a combination of growth factor–induced signaling pathways. Evidence from *in vitro* and *in vivo* studies indicate that key roles may be played by FGFs and the TGFß superfamily. Other factors that can influence the behavior of lens cells *in vitro* may also have roles in different aspects of this process, and there are indications that their involvement may occur in a temporal sequence. For example, if PDGF plays a role, it is likely to have its effects early in the process, simply because fiber differentiation involves the loss of PDGF receptors. In this context, it is important to note that, so far, the only factors shown to be sufficient for initiating fiber differentiation when overexpressed are the FGFs. When members of the Bmp (Hung et al., 2002), TGFß (Srinivasan et al., 1998), and IGF (Shirke et al., 2001) families are overexpressed, they fail to elicit premature fiber differentiation of lens epithelial cells into fiber cells, although they are clearly functional, as they have other effects. This in no way excludes members of these growth factor families from having key roles in fiber differentiation; rather it indicates that their roles are likely to be downstream of initiation of the process. The observation that FGF is sufficient to initiate fiber differentiation from epithelial cells *in vivo* and *in vitro* indicates that FGF signaling is at the top of a hierarchy of growth factor signaling, at least for secondary fiber differentiation. The FGF gradient hypothesis proposes that fiber differentiation is initiated by FGF once cells shift below the lens equator and enter the vitreous environment. Propagation and progression of the process may then depend on a cascade of signaling events brought about by other exogenous (and probably also endogenous) growth factors, such as members of the TGFß superfamily (Fig. 11.16). An important goal for future research will be to further test this hypothesis and provide more detailed information on the growth factor signaling pathways involved and their complex interactions.

11.2.2. Factors Involved in Epithelial Differentiation, Growth, and Maintenance

Whilst a lot of effort has been spent in identifying factors that control fiber differentiation, little attention has been given to the other major lens polarity question: what factors control the formation, growth, and maintenance of the epithelial monolayer? Indeed, the Coulombres' observation (1963) that a new epithelial layer formed anteriorly in the inverted chicken lens provides strong evidence that epithelial-promoting factors predominate in the anterior environment (Fig. 11.7). In addition to FGF, a number of other growth factors that are present in the anterior segment and have been shown to influence the behavior of lens epithelial cells, particularly proliferation, have been studied in various degrees of detail. RA and its naturally occurring retinoid analogs make up another class of compounds that regulate gene expression in lens cells and appear to have important roles in the maintenance of the epithelial monolayer.

11.2.2.1. Mitogenic Factors

PDGF. Platelet-derived growth factor (PDGF) stimulates proliferation of cultured bovine lens epithelial cells (Wunderlich and Knorr, 1994) and chick (Hyatt and Beebe, 1993; Potts et al., 1994; Potts et al., 1998) and rat (Kok et al., 2002) lens epithelial explants. During murine embryonic development, the expression of PDGF-A and its receptor, PDGFR-α, correlates with the proliferative activity of the lens epithelium (Reneker and Overbeek, 1996). During postnatal development, the expression of PDGF-A becomes restricted to the iris epithelium and ciliary body, which are in close apposition to the lens epithelial cells of the germinative zone. Expression of PDGFR-α is restricted to the lens epithelium at all ages and becomes localized to the epithelial cells of the germinative zone during postnatal development. Hence, both the ligand and its receptor are localized to the region where lens epithelial cells proliferate. In line with a mitogenic role for PDGF, transgenic mice that overexpressed PDGF-A specifically in the lens displayed a considerable increase in DNA synthesis in the epithelium compared with the wild type. The lens epithelium became multilayered and consisted of a surface epithelial monolayer and an underlying subepithelium (Reneker and Overbeek, 1996).

Mice that, through gene targeting, lack PDGFR-α have apparently normal lenses (Soriano, 1997), and as reported by Potts et al. (1998), they show normal levels of cell proliferation. Potts et al. argue that, whilst the results from various studies show that PDGF can regulate proliferation in lens epithelial cells under various conditions, the lack of phenotype in *PDGFR*-$\alpha^{-/-}$ mice indicates that PDGF is probably not involved in maintaining lens cell proliferation *in vivo*. However, another interpretation is that this result may reflect redundancy in the system; for example, another growth factor, such as FGF, could compensate for the absence of PDGF-α signaling. FGF has been shown to induce a proliferative response in lens epithelial cells (McAvoy and Chamberlain, 1989), and it is expressed in the lens and in tissues near the lens, with particularly high levels of expression near the lens equator and in the neighboring ciliary body and iris (de Iongh and McAvoy, 1993; Lovicu et al., 1997). Therefore, bearing in mind the localization of PDGF and FGF to this region, it is likely that lens cells in the germinative zone are exposed to both PDGF and FGF *in vivo* (and very likely to other mitogens; see below). Since the combination of these growth factors *in vitro* (Kok et al., 2002) had an additive effect on lens epithelial proliferation, both may contribute to the formation and maintenance of the germinative zone in this region *in vivo*.

Insulin/IGF. The insulin/IGF family has also been studied in a variety of experimental systems. Insulin and IGF have been shown to stimulate proliferation of cultured lens cells (Reddan and Wilson-Dziedzic, 1983). IGF-1 and IGF-2 induced DNA synthesis in lens epithelial explants, and combining them with FGF synergistically enhanced their effect (Liu et al., 1996). IGF-1 and insulin receptors are expressed in mammalian lens epithelial cells, and IGFs and various binding proteins have been detected in the ocular media of various species (Chamberlain and McAvoy, 1997). Transgenic mice in which IGF-1 was overexpressed from the αA-crystallin promoter showed no premature differentiation of lens epithelial cells but rather an apparent expansion of the epithelial compartment toward the posterior pole (Shirke et al., 2001). This was accompanied by increased proliferation in the germinative zone and its expansion posteriorly. This study also showed that, in the wild type, IGF-1 is expressed in the epithelial cells at the lens equator. Taken together with the presence of IGF-1 receptors in the lens epithelium, these studies indicate that IGFs, as well as other mitogenic factors (see above), may influence spatial patterns of cell proliferation in the lens. A local domain of endogenous IGF-1 stimulation at the lens equator may provide a cue that defines the extent of the germinative zone and the location of the transitional zone.

Other Factors. Other factors that may be involved in regulating cell proliferation in the lens include the EGF/TGFα and HGF growth factor families. EGF and EGF receptors have been detected in the lens of a number of species (Weng et al., 1997), and the effect of EGF on proliferation has been shown in organ explants as well as cell cultures (Reddan and Wilson-Dziedzic, 1983). Expression of HGF and its receptor c-met has been reported in lens epithelial cells of a number of species (Fleming et al., 1998; Weng et al., 1997), and among its other effects, it has been shown to stimulate DNA synthesis (Wormstone et al., 2000).

11.2.2.2. Inhibitory Factors

TGFβ. In addition to the mitogenic factors described above, other growth factors may act as negative regulators of lens growth. For example, the ability of TGFß to inhibit the growth of various cell types is now well established (Akhurst and Derynck, 2001). Moreover, TGFß has been shown to inhibit serum-stimulated cell proliferation in passaged bovine lens epithelial cells. Antibody-blocking experiments also indicate that TGFß in aqueous can inhibit lens cell proliferation (Kurosaka and Nagamoto, 1994). *In situ* hybridization studies showed that TGFß1 and TGFß2 but not TGFß3 mRNA is expressed in the lens during embryonic and postnatal development (Gordon-Thomson et al., 1998) and furthermore that proteins for all three TGFß isoforms are localized in the lens (Gordon-Thomson et al., 1998; Pelton et al., 1991). Both TGFß type I and type II receptors are required for TGFß signaling, but only the type I receptor is expressed in the epithelium throughout development. However, postnatally the type II receptor also appears in the epithelium, coincident with the development of the competence of lens epithelial cells to respond to TGFß (de Iongh et al., 2001a). As TGFß is potentially available to lens cells *in situ*, suppression of cell proliferation by TGFß may have a role in maintaining mitotic quiescence in the central lens epithelium postnatally.

11.2.2.3. Epithelial Differentiation and Maintenance

Besides growth regulation, there is evidence that growth and other regulatory factors play key roles in the differentiation and maintenance of the lens epithelial phenotype.

Retinoic Acid. As mentioned earlier in the discussion of lens induction, RA and its naturally occurring analogs appear to play important roles in regulating aspects of lens development. The importance of retinoid signaling has been established by applying exogenous RA (Hyatt et al., 1996a; Hyatt et al., 1996b), by deleting various combinations of RAR and RXR genes (Lohnes et al., 1993; Lohnes et al., 1994), and by overexpressing CRABP I (Perez-Castro et al., 1993). In lens cell cultures, RA has been shown to induce the expression of αB-, δ1- and γF-crystallin genes (Gopal-Srivastava et al., 1998). In addition, RA has been shown to inhibit the epithelial-mesenchymal transition that lens cells undergo when cultured in type I collagen gels (Mattern et al., 1993). In this study, RA treatment induced the deposition of a basement membrane–like material, a result consistent with evidence that RA induces production of extracellular matrix components in lens cells, including type IV procollagen (Sawhney et al., 1997). Combined with experiments showing that RA is present in lens epithelial cells and the ocular media (Wakabayashi et al., 1994), these studies provide strong evidence that RA signaling influences many aspects of lens biology, including having an important role in maintaining the epithelial phenotype. This is supported by studies on the RA-inducible transcription factor AP-2α, which is expressed in the lens epithelium but not the fibers (Ohtaka-Maruyama et al., 1998a). AP-2α null mice develop an abnormal lens that adheres to the cornea and lacks a typical lens epithelium (West-Mays et al., 1999).

Wnt. Members of the Wnt growth factor family have been shown to be involved in the regulation of epithelial differentiation in a number of organ systems (see, e.g., Brisken et al., 2000; Itaranta et al., 2002; Miller and Sassoon, 1998; Tebar et al., 2001). Studies on the lens identified the expression of a number of Wnts and their Frizzled receptors in the chick and rodent lens epithelium (Jasoni et al., 1999; Stump et al., 2003). Wnt isoforms 5a, 5b, 7a, 7b, 8a, 8b and Frizzled receptors 1, 2, 3, 4 and 6 were detected in the postnatal rodent lens and showed similar patterns of expression (Stump et al., 2003). Expression was generally present throughout the lens epithelium and extended into the transitional zone, where fiber elongation occurs, but was lost in the more mature fibers of the lens cortex. Coreceptors, LDL-related proteins (LRP) 5 and 6, which are required for Wnt signaling through the ß-catenin pathway, were also expressed in the lens. In addition, expression of other molecules that are known to be involved in Wnt signaling and its regulation, including Dishevelleds, Dickkopfs, and secreted Frizzled-related proteins, were detected in the lens (Leimeister et al., 1998; Stump, 2003). Analysis of mice with a null mutation of *Lrp6* (Pinson et al., 2000) provided more direct evidence of a role for Wnt/β-catenin signaling in differentiation of the lens (Stump et al., 2003). These mice had small eyes and aberrant lenses, characterized by an incompletely formed anterior epithelium, resulting in extrusion of the lens fibers into the overlying corneal stroma. Taken together, these results indicate a role for Wnt signaling in regulating the differentiation and formation of the epithelial sheet.

LEDGF. Lens epithelium–derived growth factor (LEDGF) has been recently identified and characterized (Shinohara et al., 2002). It was cloned from a lens epithelium cDNA library and is a member of the hepatoma-derived growth factor family. It is inducible in lens cells under stress and appears to have a variety of functions related to its ability to promote resistance of lens epithelial cells to various kinds of stresses and enhance cell survival. Not only lens cells but corneal epithelium, photoreceptor cells, and RPE show increased survival when exposed to LEDGF. Thus, it appears that LEDGF may be an important survival or maintenance factor in the lens epithelium as well as other ocular tissues.

11.2.2.4. Factors That Induce Aberrant Growth and Differentiation of the Epithelium Cause Cataract

TGFß. One of the growth factors that is abundant in the lens and the ocular media that bathe it is transforming growth factor beta (TGFß) (Gordon-Thomson et al., 1998). The effects of TGFß on lens cells were first studied in rats using epithelial explants and whole lens cultures. TGFß was found to induce lens epithelial cells to undergo a pathway of differentiation that is distinctly different from that induced by FGF. In whole lens cultures, TGFß induces the formation of punctate opacities (Hales et al., 1995). These opacities correspond to plaques of spindle-shaped cells that contain α-smooth muscle actin and desmin and to accumulations of extracellular matrix that contain collagen types I and III, fibronectin, and tenascin (Hales et al., 1995; Lovicu et al., 2002). Apoptotic cell death and localized capsule wrinkling are also induced by TGFß in explants and whole lenses (Hales et al., 1995; Liu et al., 1994; Maruno et al., 2002). Similarly, overexpression of a constitutively active form of TGFß in transgenic mice under the control of the αA-crystallin promoter also results in the development of anterior subcapsular plaques that grow progressively during postnatal life (Fig. 11.14; Lovicu et al., 2002; Srinivasan et al.,

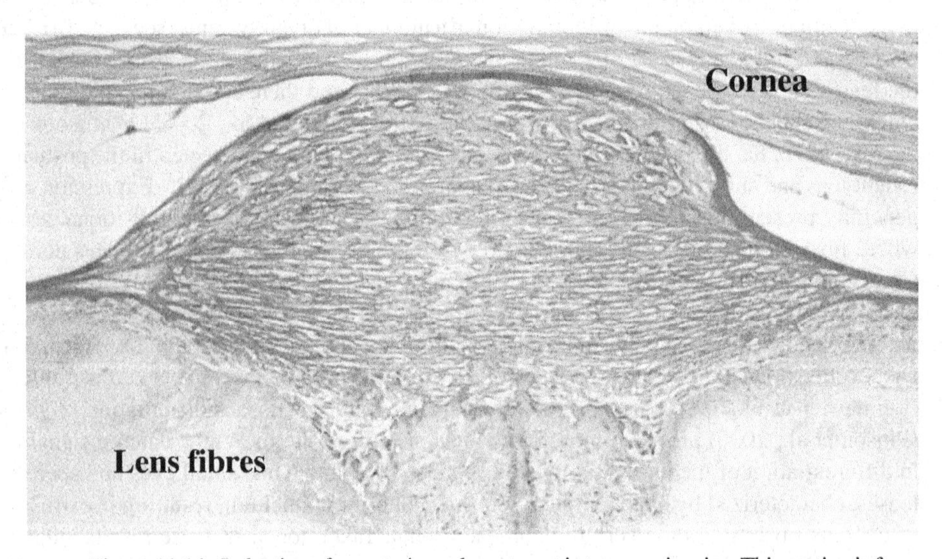

Figure 11.14. Induction of an anterior polar cataract in transgenic mice. This section is from the lens of a 140-day-old transgenic mouse overexpressing TGFß1 in the lens. Hypertrophy of the epithelium at the anterior pole has resulted in the formation of a large plaque consisting of laminations of extracellular matrix. Spindle-shaped cells lie between the laminae. Stained with periodic acid-Schiff reagent and haematoxylin. Scale bar: 40 μm. (Adapted from Lovicu et al., 2002.)

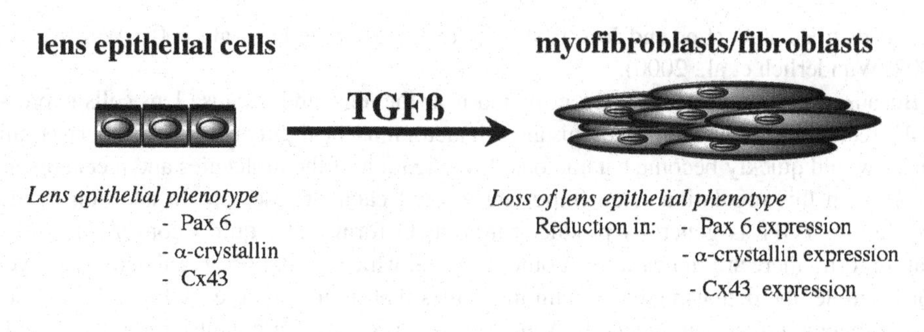

Figure 11.15. Diagram illustrating the epithelial-mesenchymal transition induced by TGFß. (Adapted from McAvoy et al., 2000.)

1998). As with the lenses cultured with TGFß, the cells in the transgenic plaques express markers of myofibroblastic/fibroblastic cells. These studies clearly show that TGFß disrupts the normal lens epithelial architecture and induces an epithelial-mesenchymal transition. In addition to induction of new patterns of gene expression, there is also evidence that TGFß down-regulates normal epithelial markers such as Pax-6, α-crystallin, and connexin 43 (Lovicu et al., 2002). Thus one of the effects of TGFß may be to inhibit the signaling pathways induced by growth factors involved in the maintenance of the lens epithelial phenotype. As a result, cells lose their normal cell-cell and cell-matrix associations. They also lose their normal polarity and monolayered arrangement as they undergo a marked phenotypic change (Fig. 11.15).

Some or all of these TGFß-induced changes, as shown in the various animal models, are typically found in subcapsular cataracts in humans. Following ocular trauma or eye surgery or the onset of other conditions (e.g., atopic dermatitis and retinitis pigmentosa), anterior subcapsular cataracts (ASCs) can develop (Sasaki et al., 1998). These feature the development of subcapsular fibrotic plaques that obscure vision. Similar fibrotic plaques, as well as capsular wrinkles, also frequently arise in the lens bag after cataract surgery and lead to posterior capsular opacification (PCO; Wormstone, 2002). This condition, also known as "secondary cataract" and "aftercataract," arises as a result of the aberrant growth and differentiation of epithelial cells that are invariably left behind after cataract surgery. In PCO and ASC, the opaque plaques are formed by abnormal accumulations of lens cells and extracellular matrix (Novotny and Pau, 1984; Pau et al., 1985; Wormstone, 2002). Immunolabeling studies of these cataracts have revealed the presence of cytoskeletal and extracellular matrix proteins not normally expressed by human lens cells. α-Smooth muscle actin and types I and III collagen have been immunolocalized in the plaques in ASC (Hatae et al., 1993; Schmitt-Graff et al., 1990; Wunderlich et al., 2000) and PCO (Frezzotti, 1990; Ishibashi et al., 1994; Saika et al., 1998; Wunderlich et al., 2000). Tenascin and fibronectin have also been localized in ASC and PCO (Saika et al., 1998; Wunderlich et al., 2000). In these cataracts, it also appears that an initial TGFß insult induces other factors, such as connective tissue growth factor and TGFß-inducible gene-H3, and other autocrine signaling events, including endogenous TGFß signaling, that may be central to the progressive fibrosis that leads to cataract (Lee and Joo, 1999; Lee et al., 2000; Saika et al., 2001; Wormstone, 2002; Wunderlich et al., 2000).

Because TGFß is normally present in and near the lens, and because lens cells express TGFß receptors, bioavailability of this growth factor must be tightly regulated; otherwise all lenses would quickly become cataractous. Investigation of the molecules and mechanisms involved in this regulation is an important area of cataract research. In addition to the dogma that TGFß is generally produced in a latent form that requires conversion to the mature form, there are indications of other levels at which it might be regulated in the eye. For example, the ocular media contain molecules that inhibit mature TGFß and block its cataractogenic effects on lens cells. Vitreous, in particular, is a potent inhibitor of TGFß (Schulz et al., 1996). Vitreous contains the serum protein α2-macroglobulin, which has been shown to protect lens cells from the cataractogenic effects of TGFß. At another level, the responsiveness of lens cells to TGFß may be modulated by various factors. For example, studies with ovariectomized rats have shown that estrogen can protect the lens from TGFß-induced cataractous changes (Hales et al., 1997). This is consistent with a trend reported in human epidemiological studies, namely, that female hormones may help prevent or slow the development of some forms of cataract (Younan et al., 2002).

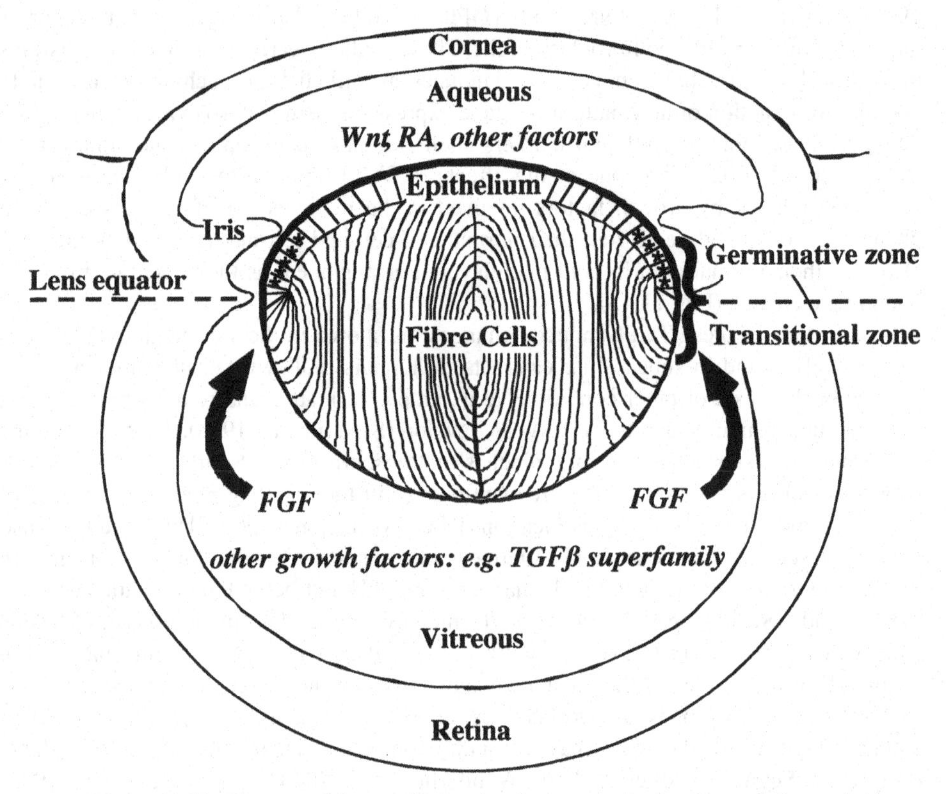

Figure 11.16. Diagram of the eye illustrating the proposed role of growth factors in determining lens polarity. In the anterior segment, factors involved in promoting formation and maintenance of the lens epithelium may include Wnts and retinoic acid. Posteriorly, lens cells may receive a strong enough FGF stimulus to initiate fiber differentiation when they enter the vitreous. Other factors, notably members of the TGFß superfamily, may promote the fiber differentiation response.

HGF. This growth factor is frequently up-regulated during wound healing and has a major influence on cell spreading (it is also known as "scatter factor") and invasive growth in metastasis (Trusolino and Comoglio, 2002). Consistent with this general role, HGF has been reported to be substantially up-regulated during the development of PCO (Wormstone, 2002). Its abundance in human capsular bag cultures and its ability to stimulate lens cell proliferation and migration indicate that it may contribute to the development of PCO (Wormstone et al., 2000).

11.3. Overview

The polarity of the lens is likely to depend on the influence of factors that inhibit or stimulate epithelial cell proliferation or differentiation. Growth factors that are known to be present in the anterior segment include FGF, TGFß, insulin/IGF, EGF/TGFα, PDGF, HGF, and Wnt. RA and its naturally occurring retinoid analogs make up another class of regulatory compounds that are present. Among these factors, FGF, insulin/IGF, EGF/TGFα, PDGF, and HGF have been shown to be mitogenic for cultured lens epithelial cells. This is consistent with their roles in other cellular systems, and it appears likely that at least some of these ligands function as lens mitogenic factors *in vivo* and that they probably work in concert. On the other hand, TGFß appears to be an inhibitor of the proliferation of lens cells and may have a role in maintenance of a mitotically quiescent central epithelium.

In addition to its role in normal lens biology, TGFß can induce pathological changes in the lens. TGFß has a major disruptive effect on the lens epithelial sheet and leads to the formation of plaques of aberrant cells that are characteristically found in subcapsular cataracts. Besides a direct effect, it is likely that TGFß destabilizes lens cells by down-regulating the molecules and mechanisms that are involved in the formation and maintenance of the normal epithelial phenotype. Currently little is known about the identity of such factors; however, emerging evidence indicates important roles for Wnt and RA. As with fiber differentiation, it is likely that the situation will be complex and that additional factors will be involved in activating complex signaling cascades that regulate gene expression and lens epithelial cell behavior (Fig. 11.16).

Clearly, there is much to do to develop a good understanding of the molecular interactions that determine the lens epithelial phenotype. Some momentum for this research may be gained by the growing realization that it is critical for understanding the molecular basis of cataracts involving aberrant epithelial growth and differentiation. These include subcapsular cataracts as well as posterior capsular opacification, which is a common sequel to cataract surgery and has become a major drain on health care budgets worldwide.

12
Lens Regeneration
Katia Del Rio-Tsonis and Goro Eguchi

12.1. Introduction

One of the most remarkable processes in nature is the process of replacing or regenerating damaged tissue. Some salamander species possess the capacity to regenerate a variety of tissues and organs as adult organisms. Other higher vertebrate species also possess regenerative abilities, but these are limited to early embryonic stages and/or tissues that can undergo renewal (Tsonis, 2000, 2001). Lens regeneration in the adult urodele amphibian represents one of these unique processes in which major cellular events such as dedifferentiation and transdifferentiation regulate tissue replacement. Dedifferentiation involves terminally differentiated cells reentering the cell cycle and losing the typical characteristics of their origin, whereas transdifferentiation allows a cell to change its identity and become a completely different cell type. During lens regeneration, the cells that undergo this transformation are the pigment epithelial cells (PECs) of the dorsal iris. This cell-type conversion is not usually observed in terminally differentiated cells that have followed a developmental path and had been determined in phenotype and function. Cancer cells share similarities with the PECs that undergo the regenerative process. In the former, during oncogenesis, the original phenotype is destabilized and the cells divide, resulting in uncontrolled growth, eventual invasion to other organs/tissues, and the production of tumors. During lens regeneration, there must be a mechanism or program that destabilizes the cell phenotype but at the same time carefully directs these cells to divide, reorganize, and redifferentiate to new cell types that will be responsible for replacing the lost parts. If we come to understand how these cellular mechanisms are turned on and off, we will have at hand the tools to induce this amazing capacity in higher vertebrates, including humans.

12.2. General Background on the Process of Lens Regeneration

Scientists have been perplexed at the intricate process of lens regeneration in urodele amphibians since it was first discovered in the late 18th century by Bonnet (1781) and later investigated by Collucci (1891), who studied the regenerative capacity of the eye as a whole. Wolff (1895) was the first to actually show that the lens itself could be regenerated from the dorsal marginal iris. For this reason, lens regeneration is referred to as "Wolffian regeneration." A detailed histological analysis of the process of lens regeneration was first done by Sato (1940), who methodically assigned 10 critical stages (Sato stages) of tissue change to this process in *Triton taeniatus* and *Diemyctylus (Triturus, Cynops) pyrrhogaster*. Later, Stone and Steinitz (1953) and Yamada (1967) reported stages for adult *Notophthalmus viridecens*. In addition, Reyer (1954, 1962) reported lens regeneration stages for larvae.

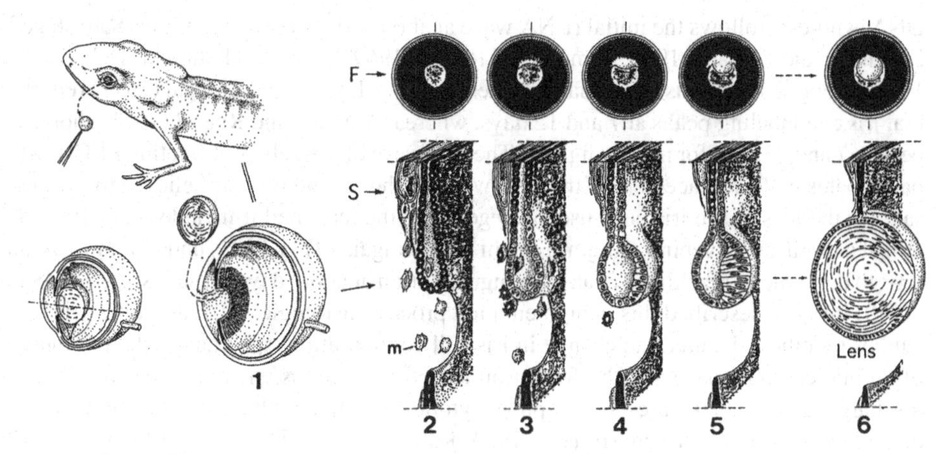

Figure 12.1. Lens regeneration in the newt. Sequential steps from lens removal (1) to completion of lens regeneration (6). Frontal view of the eye at different stages of regeneration is indicated by F. The vertical lines indicate the dorsoventral plane of sectioning through the iris corresponding to the sections shown below and indicated by S. F, frontal; S, sagittal; m, macrophage. (From Eguchi, 1998.)

The major histological events of the lens regenerative process in the newt will be described first, followed by the cellular and molecular events that take place during the different stages.

12.2.1. Histology of the Regenerating Eye

Upon removal of the newt lens (Fig. 12.1, step 1), the pupillary margin of the dorsal iris swells, and the iris epithelium divides into inner and outer regions. A small vesicle soon forms at the tip of the dorsal iris due to the active proliferation of the PECs (Sato stages I and II, 4–8 days). The cells in this vesicle depigment as the cells divide (Sato stage III, 8–10 days; Fig. 12.1, step 2). The vesicle grows in size to form a vesicular lens rudiment corresponding to the lens vesicle in normal development (Sato stage IV, 9–15 days; Fig. 12.1, step 3). Then the inner part of the vesicle thickens as the cells in this region elongate to eventually differentiate into primary lens fibers (Sato stages V and VI, 12–16 days; Fig. 12.1, steps 4 and 5). Primary lens fibers continue to form on the inside of this lens vesicle, while secondary lens fibers form in the outer layers (Sato stage VIII and beyond). The growth of the lens is directed by the continuous differentiation of secondary lens fibers from lens epithelial cells. Finally, around 25 days postlentectomy (Sato stage X; Fig. 12.1, step 6) a fully regenerated lens is formed, with a lens epithelial layer surrounding a mass of lens fibers. At a later stage (Sato stage XI), the lens detaches from the dorsal iris and stands as an independent entity (Reyer, 1977).

12.2.2. Early Cellular and Molecular Events of Lens Regeneration

Among the first events to take place upon lentectomy are decondensation of chromatin and nucleolar activation of the dorsal iris PECs. These events are early signs that a cell is becoming ready for cell division and active transcription. Nucleolar activation, initiated with the enlargement of nucleoli and the increase in the amount of granule content (Eguchi, 1963, 1964; Karasaki, 1964; Dumont et al., 1970), is followed by active rRNA synthesis (Yamada and Karasaki, 1963; Reese et al., 1969). Along with RNA synthesis, cistrons for rRNA sequences are significantly amplified during the first days of this process (Collins, 1972).

DNA synthesis follows the initial rRNA wave at about 4 days postlentectomy (Sato stage I; Eisenberg and Yamada, 1966; Yamada and Roesel, 1969; Eguchi and Shingai, 1971; Reyer, 1971), along with mitosis (Yamada and Roesel, 1971). Eguchi and Shingai (1971) reported that iris cell labeling peaks at 7 and 12 days, whereas Yamada and Roesel (1971) showed a peak at 7 and 15 days for mitotic figures. The first wave of actively proliferating PEC mostly participates in the replacement of the lens, whereas the second wave is believed to compensate for the loss of the iris cells used to regenerate the lens, and it includes cells from the dorsal as well as the ventral margin (Eguchi and Shingai, 1971). PECs undergo mitosis and start to depigment, reaching a peak of depigmentation at about day 13 (Sato stage IV). Sato (1940) initially described this depigmentation process histologically. Later, ultrastructural studies by others detailed the change in iris cell configuration, specifically the production of an intercellular space and the formation of cell projections where melanosomes await discharge, as well as the presence of phagocytotic cells (macrophages), which aid the process of cytoplasmic shedding (Eguchi, 1963; Karasaki, 1964; Dumont and Yamada, 1972; Yamada and Dumont, 1972; Reyer, 1990a , 1990b). For a detailed account of melanosome discharge and cytoplasmic shedding, consult Yamada's (1977) review. Along with these changes, dedifferentiating PECs lose intercellular communication mediated via gap junctions, and their cell-cell adhesion decreases (Eguchi, 1976; Kodama and Eguchi, 1994). Also, the increased levels of alkaline phosphatase in the dorsal margin of the iris are reflective of the disorganization and dedifferentiation taking place there (Eguchi and Ishikawa, 1963). Interestingly, during this time, there is an increase in the number of cytoplasmic organelles (Eguchi, 1964; Karasaki, 1964; Yamada, 1967; Dumont and Yamada, 1972).

12.2.3. Differentiation and Morphogenesis of the Regenerating Lens

As the dorsal iris cells divide and depigment, they form a lens vesicle that eventually thickens between 12 and 15 days postlentectomy (Sato stage V). An inner layer of cells is then formed, and these cells stop dividing and enter the differentiation mode, soon to become lens fiber cells (Eisenberg and Yamada, 1966; Yamada, 1966, 1967; Reyer, 1971). It is at this stage that β- and γ-crystallin proteins are first detected (McDevitt and Brahma, 1982). Just before this stage (Sato stage IV), the cells in the vesicle express α-, β-, and γ-crystallin mRNA (Mitashov et al., 1992; Mizuno et al., 2002). As these cells become primary fibers, they continue to synthesize β- and γ-crystallin proteins (Sato stages VI and VII), and α-crystallin protein is first detected (Takata et al., 1964, 1966; McDevitt and Brahma, 1982). The lens epithelial cells left in the outer layer continue to proliferate and begin to express α- and β-crystallin mRNA at Sato stage VI (Mizuno et al., 2002). Crystallin proteins are not evident until later stages in the epithelium, when α- and β- crystallin proteins are detected at Sato stage VIII (Takata et al., 1964, 1966; McDevitt and Brahma, 1982, 1990). As the epithelial cells reach the equatorial zone, they cease to divide and begin elongating to form secondary lens fibers that surround the primary lens fibers (Eguchi, 1964; Dumont and Yamada, 1972; Yamada and Dumont, 1972; Reyer, 1977, 1982; Yamada, 1977, 1982; Yamada and McDevitt, 1984). The secondary lens fibers express αA-, βB1-, and γ-crystallin genes at Sato stage VIII (Mizuno et al., 2002). No crystallin mRNAs are detected in terminally differentiated primary fibers (McDevitt and Brahma, 1982; Mizuno et al., 2002). By day 25 (Sato stage X), the lens has completely formed, and dividing cells are observed only in the lens epithelium. The differentiation of the lens vesicle to form a complete lens mimics the normal lens development process, where lens fibers differentiate from lens epithelial cells and even express crystallin genes similarly (McDevitt and Brahma, 1982, 1990; Mizuno et al.,

2002). In this sense, similar events might take place during development and regeneration of the lens even though inductive interactions such as those seen during lens development (between surface ectoderm and optic vesicle) are not necessary for lens regeneration.

12.2.4. Extracellular Matrix of the Regenerating Lens

A lens capsule is formed from the basal lamina of the iris epithelium. Before lentectomy, the basal lamina of the iris epithelium is composed of fibronectin, heparan sulfate proteoglycan (HSPG), and nidogen-entactin (Elgert and Zalik, 1989; Ortiz et al., 1992). Upon lentectomy and up to about day 15, the expression of fibronectin and nidogen-entactin decreases in the iris. During the differentiation phase of lens regeneration, the signal for these two extracellular matrix molecules in the iris epithelium and new capsule increases. HSPG was detected only after completion of regeneration in the iris and the lens capsule. On the other hand, laminin was found only in the new lens capsule during the differentiation stage of lens regeneration, at around 20 days (Ortiz et al., 1992). It has also been shown that the dorsal iris has high activity for enzymes such as hyaluronidase, which may promote the remodelisg of extracellular matrix components (Kulyk et al., 1987). It is clear from these studies that a significant change in the composition of the extracellular matrix is important for the cell type conversion that takes place during lens regeneration.

12.3. Problems Involved in the Study of Lens Regeneration

In the past one hundred years or so, there has been a significant effort to answer the basic questions about the process of lens regeneration. The main questions in the field of lens regeneration still stand: How can a terminally differentiated group of cells (iris PECs) transdiffrentiate and give rise to a completely functional lens? What is the mechanism? What molecules are involved or responsible? Can this ability be induced in other species?

The classic approaches to addressing these questions depended largely on the available technology and therefore concentrated on the cellular level, which was studied mostly by means of labeling and transplantation experiments. The early focus was on finding out, among other things, which cells are responsible for the regenerative capabilities, if there was a retinal factor involved in the process, and if there was a lens factor involved. The effort now is geared more toward discovering the molecular mechanism involved in the regenerative process by trying to identify the molecular masterminds of the process.

The field of lens regeneration is in need of a revolution at the molecular level that will allow it to elucidate the molecular mechanisms by which iris cells destabilize, reenter the cell cycle, dedifferentiate, and finally redifferentiate into the final product, the lens. These cells undergo changes at many levels, including a phenotypic change, a change in cell-cell communication, a change in cell–extracellular matrix interaction, and changes at the physiological and biochemical levels. All of these eventually must be carefully directed and orchestrated.

In the modern approaches, *in vitro* and *in vivo* models have been selected to study and understand cell conversion. *In vitro* models have been crucial for understanding cell type conversion. Several *in vitro* models used to study this phenomenon have been described (Eguchi, 1998). One of these includes the newt iris/retinal PEC model, in which iris/retinal PECs are isolated and grown under optimal conditions to transdifferentiate to lens cells (Eguchi, 1967, 1976, 1979; Zalik and Meza, 1968; Eguchi et al., 1974; Horstman and Zalik, 1974). Another *in vitro* model includes chick and human retinal PECs, which can actually be directed to transdifferentiate to either the lens phenotype or the retinal phenotype (Eguchi and Okada, 1973; Eguchi, 1976; Itoh, 1976; Yasuda, 1979; Masuda and Eguchi, 1982; Itoh and Eguchi, 1986a, 1986b; Kosaka et al., 1998). With these models, molecular switches for cell proliferation, dedifferentiation, and eventual differentiation can be carefully dissected. Some progress has been achieved, but there are still many pieces of the puzzle to be solved.

As for the *in vivo* model, the recent focus is on the elucidation of the factors that can induce or inhibit the process of lens regeneration in the newt eye. Approaches so far have concentrated on identifying important molecules involved in the process that are differentially expressed in the dorsal iris versus the ventral iris. Once a crucial molecule or factor is identified, functional studies to knock out or overexpress its function can be utilized. Using this approach, factors such as FGF and their receptors have been identified as essential for the process of lens regeneration (Del Rio-Tsonis et al., 1997; Del Rio-Tsonis et al., 1998). There are limitations as to the type of molecules (i.e., transcriptional factors) that can be manipulated with the existing technology.

Recently, Hayashi et al. devised an *in vitro/in vivo* system that recapitulates the *in vivo* conditions and is suitable for efficient genetic manipulation (Ito et al., 1999; Hayashi et al., 2001; Hayashi et al., 2002). Isolated PECs from the dorsal or ventral iris are cultured for about two weeks. Then the cells are aggregated and implanted into a lentectomized eye. The implanted ventral aggregates fail to form lenses, but the dorsal aggregates are successful in forming a lens regenerate. More importantly, transfection of genes has been established with an efficiency as high as 80% (Hayashi et al., 2001). Using this system, researchers can bypass the problem of *in vivo* transgenesis and create ventral transgenic aggregates with key genes that may control transdifferentiation of the dorsal iris. These transgenic aggregates can be implanted in host eyes to assay for their lens-forming potential. Similarly, transcription/translation of genes can be inhibited *in vitro* by the use of morpholino (Heasman, 2002) or RNAi (Zhou et al., 2002; Scherr et al., 2003) technology, creating thus a knockout condition in the system.

In the next sections, we discuss both the classic approaches and the modern approaches used to study the process of lens regeneration.

12.4. Classic Approaches to the Problems

Scientists in the field of lens regeneration have concentrated on trying to understand the intrinsic process of lens regeneration and transdifferentiation. Some of the basic questions pondered are discussed in this section.

12.4.1. How Does the Lens Regenerate? Which Cells Are Responsible for Its Regenerative Capabilities?

Following the histological description set up by Collucci (1891) and Wolff (1895), investigators accepted the idea that the lens regenerated from the dorsal iris PEC but still questioned whether these were the only cells involved. To address this issue, a series of experimental approaches were set forward in the first half of the 20th century, including microsurgical transplantation. With this approach it was shown that grafts of dorsal iris placed into the optic cavity of lentectomized eyes gave rise to lens (Wachs, 1914; Sato, 1930, 1935; Mikami, 1941). The most convincing experimental approach was the introduction of regeneration competent dorsal iris into the cavity of a lentectomized eye from a species of salamander that lacks the regenerating capacity (Ikeda, 1934; Amano and Sato, 1940; Reyer, 1956).

Autoradiographic studies helped to clarify the origin of the regenerate as well. Labeled cells were followed to determine their fate. Even though both dorsal and ventral irises were labeled, only the dorsal labeled cells were found to give rise to the regenerating lens (Reyer, 1966a, 1971; Eisenberg and Yamada, 1966; Eguchi and Shingai, 1971). Zalik and Scott (1971) went even farther, transplanting dorsal irises labeled *in vitro* and then grafting them into a host lentectomized eye; indeed, these labeled cells gave rise to a regenerating lens. Ultrastructural studies performed in the 1960s confirmed the early work of Wolff, basically verifying that the iris PECs do undergo depigmentation and eventually give rise to a regenerate (Eguchi, 1963, 1964; Dumont and Yamada, 1972; Yamada and Dumont, 1972; Reyer, 1990a, 1990b).

12.4.2. Is There a Retinal Factor Involved in the Process?

One of the issues that still occupy scientists today is the contribution of the retina to lens regeneration. Knowing that the dorsal iris PECs of several salamander species, such as those belonging to the genara *Triturus* and *Salamandra*, are competent to give rise to a regenerating lens, researchers asked why these competent dorsal irises, if implanted, would give rise to lenses in certain circumstances and not in others. If these irises were implanted in the cavities of lentectomized eyes from a species of salamander that lacks the regenerating ability, such as *Amblystoma punctatum*, lenses would form. On the other hand, a competent dorsal iris, placed in the body cavity of an adult salamander or in a subcutaneous location would not form a lens, regardless of whether the salamander was competent to regenerate its lens or not (Ikeda, 1935, 1936; Stone, 1958a; Reyer, 1953, 1954).

Interestingly, though, if the neural retina was included in these transplants, then lenses would form (Stone, 1958a). So a series of experiments were performed to try to elucidate the role of the neural retina in lens regeneration. One experiment that pinpointed the contribution of the retina to lens regeneration involved removing the neural retina, choroid, and retinal pigment epithelium along with the lens. In this case, no lens regenerated. If only the neural retina and the lens were removed, then the lens would regenerate but only after the neural retina started to regenerate as well (Stone, 1958a). In other experiments, the physical block or separation of the neural retina and the iris prevented any lens formation (Stone, 1958b; Zalokar, 1944). *In vitro* studies directed by Eguchi (1967) and Yamada et al. (1973), the culturing of iris-corneal complexes or fragments of iris pigment epithelium from the dorsal margin with neural retina supported the growth and differentiation potential of the retinal factor.

Not only was it determined that there existed one or more retinal factors that promoted lens regeneration, but it was shown that this same influence was responsible for lens polarity – the formation of the lens fibers in the posterior region facing the neural retina and of lens epithelial cells in the anterior region facing the cornea (Reyer, 1948; Stone, 1954a, 1954b). It is tempting to speculate that this factor is the same factor that supported the formation of a perfect lens in a regenerating newt limb when dorsal iris explants were transplanted to the blastema (Fig. 12.2). Even the anteroposterior polarity of the lens was maintained, with the anterior (front) part of the lens facing the wound epithelium (Reyer et al., 1973; Ito et al., 1999). The molecules responsible for this activity might include FGFs and FGFRs, because these have been implicated in determining the lens polarity in the developing eye and because the wound epithelium of the regenerating limb is a rich source of FGFs (Caruelle et al., 1989; Chamberlain and McAvoy, 1989; de Iongh and McAvoy, 1993; Lovicu and McAvoy, 1993; Lovicu and Oberbeek, 1998; Robinson et al., 1995a; Robinson et al., 1995b; Robinson et al., 1998; Mullen et al., 1996; Zenjari et al., 1997; Stolen et al., 1997, Stolen and Griep, 2000).

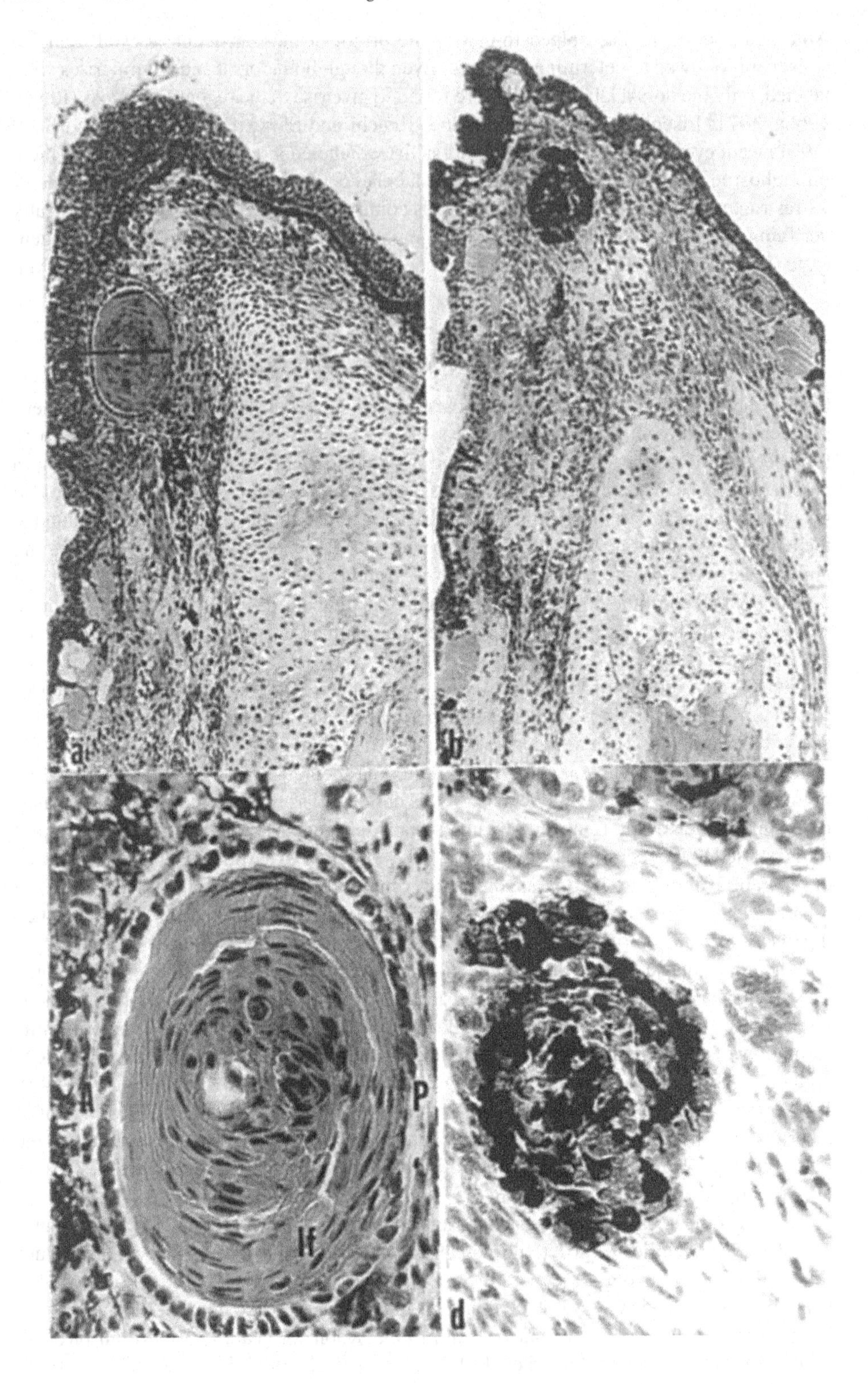
a
b
A
P
lf
c
d

12.4.3. Is There a Lens Factor Involved?

A controversial issue that to date remains unresolved is the type of influence the lens itself has. When a lens is removed and then placed back in the regenerating eye, it exerts a negative influence on the lens regeneration process, and its effect seems to depend on how close the lens is to the dorsal iris. If very close, no regeneration is observed; if placed at some distance from the dorsal iris, then regeneration initiates but is significantly inhibited (Wachs, 1914; Ikeda and Kojima, 1940; Ikeda and Amatatu, 1941; Eguchi, 1961; Dinnean, 1942; Uno, 1943; Stone, 1943; Zalokar, 1944; Takano et al., 1958; Reyer, 1961). To challenge the idea that the lens and its suspensory ligaments act as a physical barrier to the neural factor, dorsal iris transplants were inserted in the anterior chamber, and lenses were either left *in situ* or removed and then immediately replaced. In this case, no lens regenerates were observed (Wachs, 1914; Stone, 1943, 1952; Reyer, 1961). When dorsal iris transplants were inserted in the vitreous chamber through an incision in the ventral part of the eyecup without disturbing the host lens, regenerates did form, but the regenerates did not behave the same as similar implants where the host lens was removed (Reyer, 1966b). It was concluded that there is a physical barrier factor partially responsible but also that some other factor was influencing the formation of the regenerating lens. It was hypothesized that the lens secretes a substance that either competes with the neural factor or acts as an inhibitory substance specific to the regenerating lens (Zalokar, 1944; Stone and Vultee, 1949; Stone, 1953). Williams and Higginbotham (1975) performed experiments similar to those done by Stone (1966) in order to clarify the role of the intact lens on the lens regeneration process. In these experiments, the ventral iris was replaced by a donor dorsal iris, and the original lens was removed, so that two lenses were created, one from the donor dorsal iris and the other one from the host dorsal iris. When the host lens was removed 53–91 days after the original lenetectomy, it regenerated in the presence of the donor lens, but it was noted that in all cases there was a significant amount of space between the dorsal iris and the donor lens, allowing the neural factor to be available. Eguchi's (1961) experiments supported the idea that displaced lenses must be at a certain distance from the competent iris to permit regeneration.

12.5. Modern Approaches to Lens Regeneration

One approach to dissecting the molecular pathway involved in the process of lens regeneration has been to analyze molecules known to play a role during vertebrate lens development. Some of these molecules include the eye-determining gene *Pax6* (Ton et al., 1991; Grindley et al., 1995; Duncan et al., 2000) and *Prox1* (Oliver et al., 1993; Tomarev et al., 1996;

Figure 12.2. (facing page) (See color plate XVIII.) The limb-lens connection. Lens formation in the regenerating limb. (a) A perfectly formed lens after dissociated pigment epithelial cells from the dorsal iris were implanted into the blastema. The lens possesses a normal appearance and normal posteroanterior polarity (direction of the arrow), with the anterior (front) part facing the wound epithelium. In (c), the lens is magnified to show its normal appearance. The lens epithelium in the anterior (A) shows the characteristic cuboidal shape, while in the posterior (P), the cells become elongated and differentiate to lens fibers (lf). The anteroposterior polarity of lens does not coincide with the anteroposterior axis of the limb. (b) Failure to form a lens after transplantation of dissociated pigment epithelium cells from the ventral iris. The cells have not dedifferentiated and have remained pigmented. (d) magnification of the pigmented aggregate shown in (b). (Courtesy of Dr. M. Okamoto.)

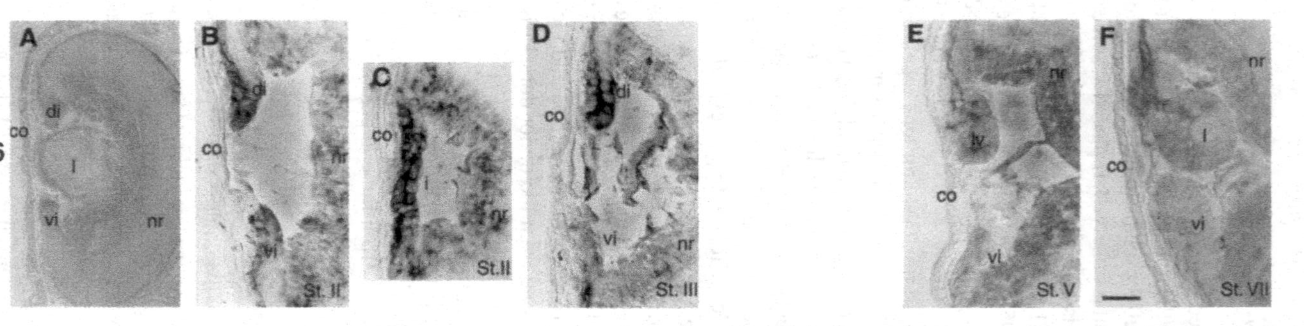

Figure 12.3. (See color plate XIX.) Pax-6 expression via *in situ* hybridization during the process of lens regeneration. (A) Intact eye of the newt *Cynops pyrrhogaster* showing no expression in the dorsal iris (di) or ventral iris (vi) of the eye. (B) Stage II lens regenerate (2–5 days postlentectomy) showing expression of Pax-6 in both the dorsal iris (di) and ventral iris (vi). (C) Section through the lateral part of the eye in B showing expression of Pax-6 throughout the entire margin of the iris (i). (D) Stage III lens regenerate (4–7 days postlentectomy) showing specific expression of Pax-6 in the dorsal iris (di). (E) Stage V lens regenerate (8–11 days postlentectomy) again showing dorsal iris (di)–specific expression and expression in the lens vesicle (lv). (F) Stage VII lens regenerate (13–15 days postlentectomy) showing expression of the dorsal iris (di) and the regenerating lens (l). co, cornea; di, dorsal iris; vi, ventral iris; i, iris; l, lens; lv, lens vesicle; nr, neural retina. Scale bars: 50 μm. (From Mizuno et al., 1999.)

Tomarev, 1997; Wigle et al., 1999), both early molecular markers for lens development. Even though the mechanisms for lens development and lens regeneration are different, it seems that the basic molecular regulators do participate in both.

12.5.1. Expression of Molecules Involved in Lens Development and Regeneration

12.5.1.1. Role of Homeobox-Containing Genes

During vertebrate lens development, the lens arises from the head ectoderm after receiving inductive signals from the optic vesicle. This region of the head ectoderm seems to be predisposed to become a lens, as it expresses *Pax6* even before optic vesicle induction (Li et al., 1994). Tissue recombination experiments in rats have shown that *Pax6* expression in the head ectoderm is essential for lens development (Fujiwara et al., 1994), and studies using conditional *Pax6* knockouts have shown that Pax6 is autonomously required for lens formation (Ashery-Padan et al., 2000). On the other hand, *Prox1* expression is apparent just before the optic vesicle induces the head ectoderm to become the lens placode, the structure that will develop into the lens vesicle (Tomarev et al., 1996). Recently, Wigle et al. (1999) demonstrated that *Prox1* knockout mice develop a hollow lens because the mutant lens cells fail to polarize and elongate properly. In addition, both Pax6 and Prox1 are transcription factors that regulate lens crystallin gene expression (Cvekl and Piatigorsky, 1996; Gopal-Srivastava et al., 1996; Tomarev et al., 1996; Duncan et al., 1998; Wigle et al., 1999; Lengler et al., 2001).

During lens regeneration in the newt, the lens is replaced by the PECs of the dorsal iris. These cells are derived from the neural ectoderm, not the head surface ectoderm, as for normal lens development. These cells will transdifferentiate into lens cells and replace the lost lens. Even though, in regeneration and development, the origin of the lenses differs, as does the process of their formation, *Pax6* and *Prox1* expression seems to be common to both. Soon after lentectomy, *Pax6* was found to be expressed in both the ventral and a dorsal iris PECs, but *Pax6* expression becomes restricted to the dorsal PECs at a stage where dedifferentiation is apparent (Fig. 12.3; Mizuno et al., 1999). This gene continues to be expressed in the lens vesicle and eventually becomes restricted to the lens epithelial layer of the regenerating lens (Del Rio-Tsonis et al., 1995; Mizuno et al., 1999). *Pax6* is expressed in the adult intact newt eye; however, its expression is not evident in the axolotl, a urodele unable to regenerate the lens (Del Rio-Tsonis et al., 1995). *Prox1* has been found to be specifically expressed and regulated in the pigment epithelium of the adult newt dorsal iris and in the dorsal iris during lens regeneration (Fig. 12.4; Del Rio-Tsonis et al., 1999; Mizuno et al., 1999). Such expression patterns suggest a role for these two genes in regeneration-competent PECs. Functional studies are currently being undertaken to clarify the role of these molecules.

12.5.1.2. Role of Fibroblast Growth Factors

Another set of molecules involved in lens development and lens regeneration consists of FGFs and their receptors. FGF signaling is essential for the early inductive events taking place during the lens placode stage (Faber et al., 2001). In addition, it has been proposed that FGFs determine the polarity of the lens (Lovicu and McAvoy, 1993; Schulz et al., 1993; Govindarajan and Overbeek, 2001). It has been shown that FGF-1 is present as a gradient in the vertebrate eyeball, with higher concentration needed for fiber differentiation in the

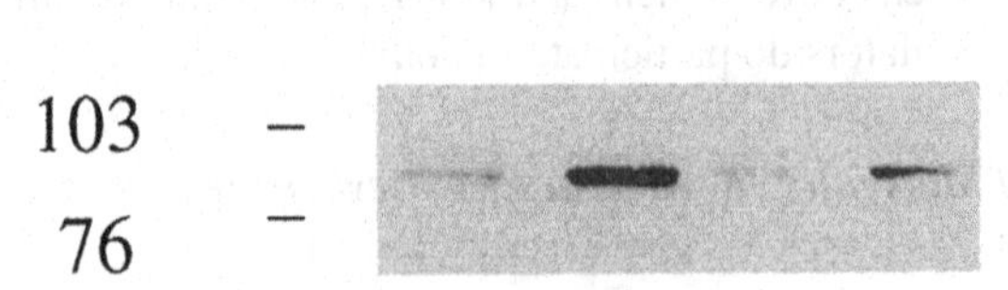

Figure 12.4. Regulation of Prox-1 protein during the process of lens regeneration in the newt *Notophthalmus viridecens*. Expression via Western blot analysis. Lane 1, extracts from ventral iris of the regenerating eye; lane 2, extracts isolated from the dorsal iris of the regenerating eye; lane 3, extracts from the ventral iris of the intact eye; lane 4, extracts from the dorsal intact eye. (From Del Rio-Tsonis et al., 1999.)

posterior chamber and lower concentration in the anterior, where the lens epithelial cells are proliferating (Chamberlain and McAvoy, 1989; McAvoy and Chamberlain, 1989; McAvoy et al., 1991; Le and Musil, 2001). A series of mice transgenic for FGFs and their receptors have been generated, and they show that when these molecules are misexpressed in the eye, the normal development and differentiation of the lens is disrupted (Robinson et al., 1995a; Robinson et al., 1995b, Robinson et al., 1998; Stolen et al., 1997; Stolen and Griep, 2000; Lovicu and Overbeek, 1998; Govindarajan and Overbeek, 2001).

In newt lens regeneration, FGF-1, FGFR-1, FGFR-2, and FGFR-3 have been shown to be expressed in the dedifferentiating cells of the regenerating lens vesicle as well as in the subsequent stages of lens fiber differentiation (Del Rio-Tsonis et al., 1997; Del Rio-Tsonis et al., 1998; McDevitt et al., 1997). However, only FGFR-1 product seems to be present specifically in the dorsal iris during dedifferentiation (Fig. 12.5). Functional studies in which the function of FGFR-1 was blocked using an inhibitor supported its role in regulating lens regeneration (Fig. 12.6). The results showed inhibition of lens regeneration and lens fiber differentiation (Del Rio-Tsonis et al., 1998). In addition, as in cases of FGF transgenic mice, exogenous FGFs were capable of inducing similar abnormalities in the regenerating lens (Del Rio-Tsonis et al., 1997). These abnormalities included vacuolated lens, double lens formation, and lenses with abnormal polarity.

Proteoglycans, specifically heparan sulphate proteoglycans, can bind to many FGF molecules at a time enhancing the ability of FGFs to bind to their receptors resulting in FGFR dimerization (Spivak-Kroizman et al., 1994). It is interesting to note that during lens regeneration a sequential loss of cell surface molecules, including proteoglycans, takes place (Zalik and Scott, 1973; Ortiz et al., 1992). This disappearance may affect the availability of FGFs for activating their receptors or their mode of action in such activation. Further research in this area is needed in order to establish clear relationships between key cell surface changes and other events taking place during the dedifferentiation process.

12.5.1.3. Role of Retinoids

Retinoids and their receptors are another set of important players in eye development and also in axis determination during many developmental processes. During eye development, retinoic acid is synthesized in the retina and controls the fate of the cells composing the neural retina (McCaffery et al., 1992; McCaffery et al., 1993; Hyatt et al., 1996a; Wagner et al., 2000). In mice lacking retinoid receptors, the ventral iris is not developed (Kastner et al., 1994). In addition to these roles in retina development, exogenous retinoic acid has been implicated in the induction of ectopic lens differentiation during eye development (Manns

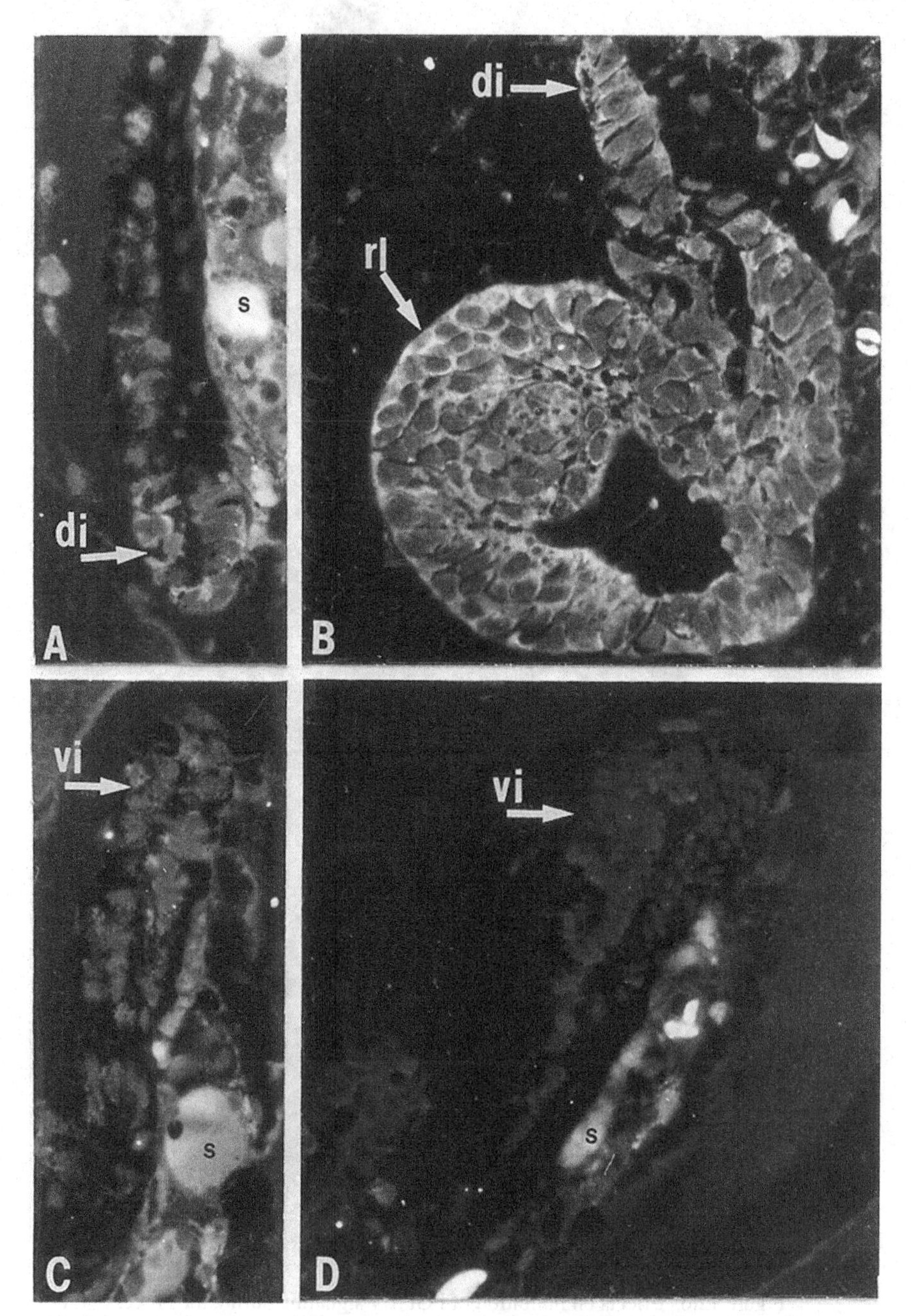

Figure 12.5. Expression of FGFR-1 protein during the process of lens regeneration in the newt *Notophthalmus viridecens*. After lentectomy, FGFR-1 protein was detected (A) at day 10 when the tip of the dorsal iris (di) undergoes dedifferentiation. (B) FGFR-1 protein was found to be present in the regenerating lens vesicle and differentiating fibers of the regenerating lens (rl) 15 days postlentectomy. (C) FGFR-1 was not detectable in the ventral iris (vi) at 10 days and (D) 15 days postlentectomy. The presence of FGFR-1 protein in the dorsal iris suggests that this molecule must play an important role in the process of lens regeneration. (From Del Rio-Tsonis et al., 1998.)

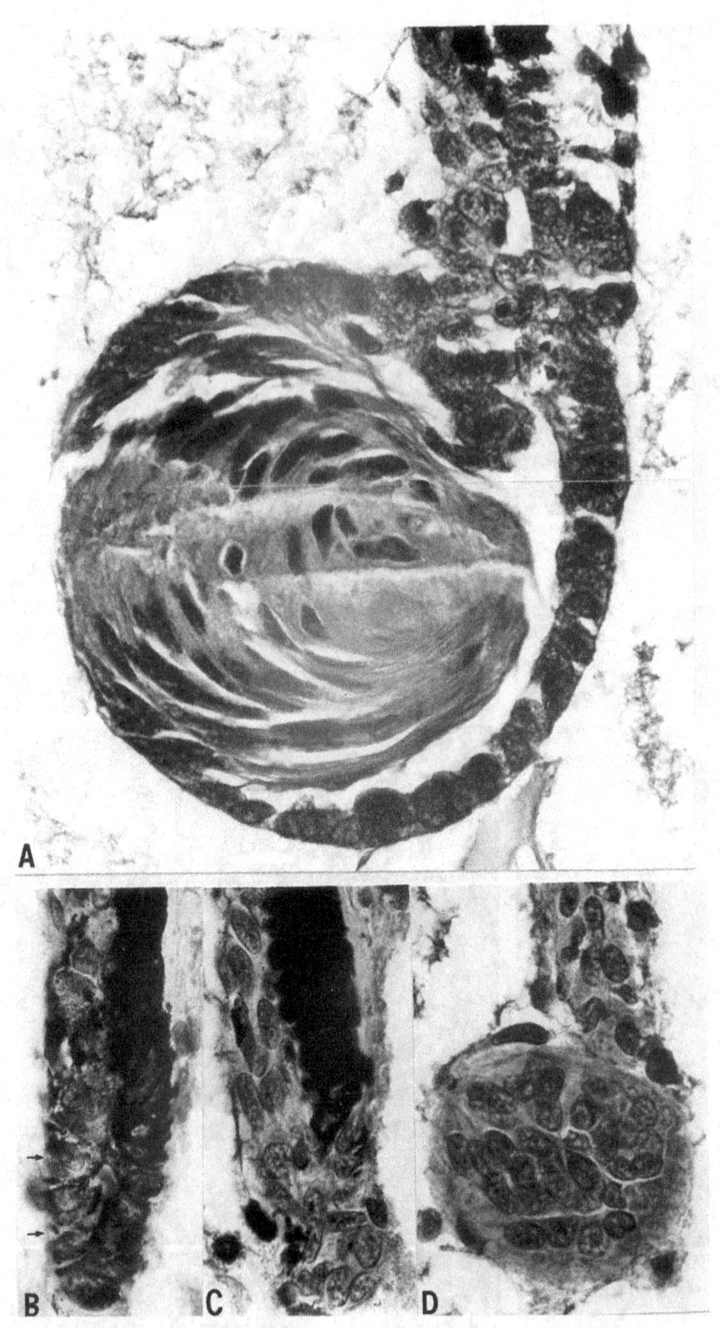

Figure 12.6. Effects on inhibition of FGFR-1 signaling during the process of lens regeneration in the newt *Notophthalmus viridecens*. To try to establish the role of FGFR-1 in lens regeneration, an FGFR-1–specific inhibitor was used. This inhibitor binds to FGFR-1 to produce a conformational change, which blocks the function of FGFR-1 in the signal transduction pathway. (A) Twenty day lens regenerate showing normal lens formation. (B–D) FGFR-1 inhibitor–treated eyes; lens regeneration did not advance beyond the dedifferentiation stage. No lens or fiber differentiation could be observed. (D) The most advanced stage of regeneration was a small lens vesicle, comparable to Sato stage IV, that did not show any clear primary fiber differentiation. (From Del Rio-Tsonis et al., 1998.)

and Fritzsch, 1991). Retinoid receptors also regulate expression of crystallins, including αB, E, and F crystallins (Gopal-Srivastava et al., 1998; Kralova et al., 2002).

Retinoic acid has also been shown to be important during newt lens regeneration. If the synthesis of retinoic acid is inhibited by disulfiram, a compound that can prevent its synthesis (Vallari and Pietruszko, 1982; Maden, 1997), or if the function of the retinoid receptors is impaired using a retinoic acid receptor antagonist, the process of lens regeneration can be dramatically affected (Fig. 12.7). In the majority of the cases, lens regeneration from the dorsal iris is inhibited (Figs. 12.7B, C). In a few cases, ectopic lens regeneration from places

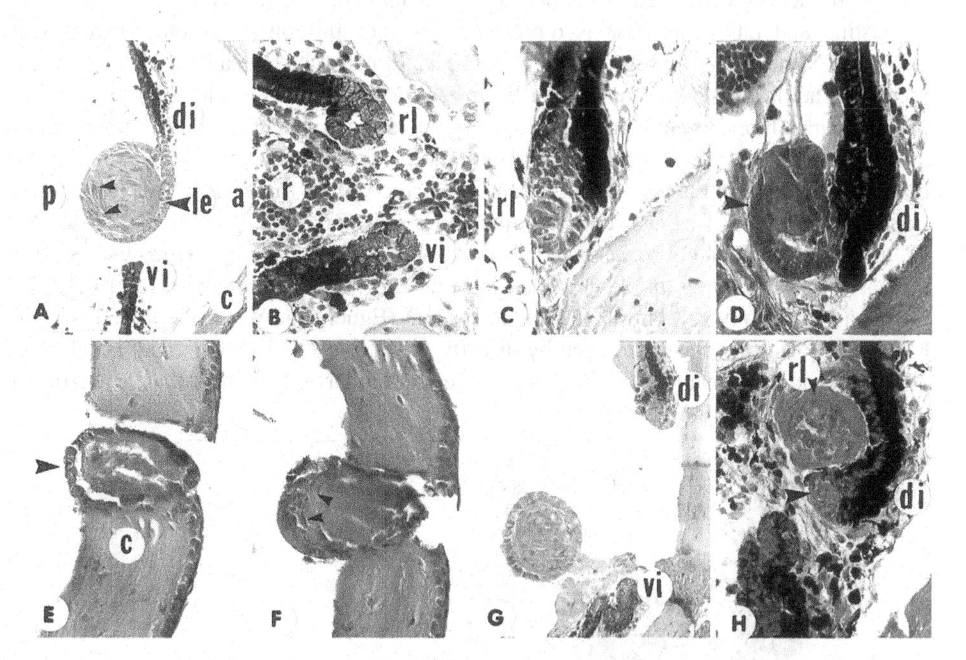

Figure 12.7. (See color plate XX.) Effects of inhibiting RA signaling on lens regeneration in the newt *Notophthalmus viridecens.* (A) A section through a normal lens (control) regenerated 20 days postlentectomy (10×). Note the lens regenerated from the dorsal iris (di). vi, ventral iris; c, cornea; le, lens epithelium (large arrowhead; the small arrowheads indicate the differentiation of lens fibers [elongated cells] in the posterior part of the lens); a, anterior part of the eye; p, posterior. (B) Lens regeneration in an eye treated with disulfiram. Note the inhibition of the regenerating lens (rl) at the tip of the dorsal iris. The tip of the ventral iris (vi) also shows some degree of dedifferentiation. Also, detachment of the retina (r) can be observed (20×). (C–H) Lens regeneration in eyes treated with RAR antagonist. (C) Inhibition of lens regeneration from the dorsal iris. The regenerating lens (rl) has not advanced beyond the initial stages of dedifferentiation (20×). (D) Retardation and abnormalities of the regenerating lens (arrowhead). The lens is oblong and positioned more dorsally and posteriorly than a normal regenerating lens (20×). di, dorsal iris. (E and F) A case of lens regeneration from the cornea. In E, note the continuity between the regenerating lens (arrowhead) and the cornea (c). In F, we can observe the differentiation of elongated primary lens fibers (arrowheads) in the posterior part of the lens, which is a dominant feature of lens morphology (20×). (G) A case of lens regeneration in the ventral part of the eye. The regenerating lens is clearly associated with the ventral iris (vi) and not the dorsal iris (di) (10×). (H) A case of double lens regeneration from the dorsal iris. Note the small lens vesicle (arrowhead) at the tip of the dorsal iris (di) (at the correct location) and the abnormal regenerating lens (rl) dorsal and posterior to the dorsal iris (20×). All eyes were examined 20 days postlentectomy. (From Tsonis et al., 2000.)

other than the normal site was observed (Figs. 12.7D–H). The most spectacular case was the regeneration of a lens from the cornea, an event only observed in premetamorphic frogs. These data show that inhibition of retinoid receptors is paramount for the normal course and distribution of lens regeneration (Tsonis et al., 2000; Tsonis et al., 2002).

12.5.2. Manipulating the Regenerative Process

One way to try to dissect the process of lens regeneration has been by manipulating the process by adding or depleting factors or chemicals believed to be important. One very interesting and intriguing case is represented by the addition of a potent carcinogen, N-methyl-N′-nitro-N-nitrosoguanidine (MNNG). Eguchi and Watanabe (1973) experimented by adding this potent carcinogen during the process of lens regeneration, and instead of producing tumors, its application led to the regeneration of an ectopic lens from the ventral iris (Fig. 12.8) This report represents the only case (excluding transplants) where a regenerate developed from the non-regeneration-competent ventral iris *in vivo*. This transformation of the ventral PECs was well maintained after a secondary lentectomy was performed 12 months later. In addition, isolated ventral irises were cultured *in vitro*, treated with the carcinogen, and then implanted back into the eye cavity of lentectomized newt eyes. Again, a number of ventral implants produced a lens (Eguchi and Watanabe, 1973). MNNG has been shown to act as a mutagen by directly modifying DNA (Singer and Grunberger, 1983). It is possible that this carcinogen modifies certain DNA sequences that switch on the

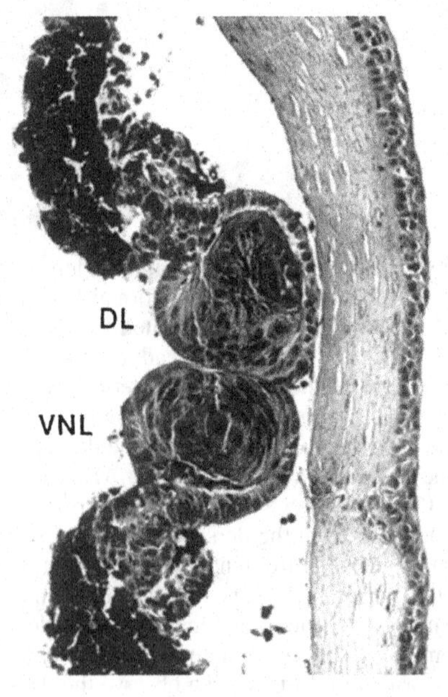

Figure 12.8. Induction of a second lens from the ventral iris using MNNG during the process of lens regeneration in the newt *Cynops pyrrhogaster*. A histological section of a typical case, in which a supernumerary lens regenerated from the ventronasal portion of the iris in addition to the dorsal lens regenerate. This case was administered with an MNNG crystal (ca.10 ng) 5 days postlentectomy and was fixed 35 days postlentectomy. One large lens regenerate (DL) was formed from the middorsal iris, the other (VNL) from the ventronasal iris (90×). (From Eguchi and Watanabe, 1973.)

regeneration program in the PECs of the ventral iris. This carcinogen has been shown to produce tumors in other systems (Sugimura et al., 1970; Seto et al., 1999). On the other hand, not all carcinogens have this effect. For example, nickel subsulfide, another potent carcinogen, has been tested in this system, with results ranging from inhibition of regeneration to development of tumors in the ventral iris, depending on the concentration (Okamoto, 1987, 1997). The fact that MNNG was able to reprogram the PECs of the ventral iris implies that the regeneration machinery was switched on and a series of genes activated or repressed. It is clear from these experiments that this regenerating system responds differently to the different stimuli, and the fact that the ventral iris can be induced to transdifferentiate *in vivo* under these extreme conditions indicates that this reaction can be induced in other species.

As previously mentioned, if we disrupt the signal transduction pathway of FGFR-1 or the action or synthesis of retinoids, the regeneration will be modified, even inhibited. On the other hand, if we add extra factors or chemicals, we can observe changes as well. For example, by ectopically placing FGF-1 and FGF-4 into lentectomized eyes, we can affect lens fiber differentiation and polarity (Del Rio-Tsonis et al., 1997).

12.5.3. Differences between Dorsal and Ventral Iris PECs

One important approach to understanding the restriction of the lens regenerative process is to look at the molecules that may be differentially expressed or present in the dorsal versus the ventral iris. As discussed earlier, a few of the molecules involved in lens development also seem to play a role in lens regeneration, and these seemed to be preferentially expressed in the dorsal PECs during regeneration (eg. Pax6, Prox1, FGFR-1).

Imokawa and Eguchi (1992) and Imokawa et al. (1992) generated a monoclonal antibody (2NI-36) that detects an antigen expressed in the ventral iris but not in the dorsal during lens regeneration. If this antibody is used to treat the ventral iris *in vitro*, and if this treated iris is then transplanted into the eye cavity, a regenerate forms. It has been speculated that 2NI-36 antigen is a glycoprotein that helps to stabilize the differentiated state of the iris cells and that these cells can destabilize and transdifferentiate if the antigen is lost (see also section 7.1.4).

12.6. Lens Regenerative Capacity of Vertebrates

12.6.1. Lens Regenerative Abilities in Amphibians

It is curious that a limited number of species have the gift of lens regeneration. Within vertebrates, the amphibians seemed to be among the fortunate ones. As adult organisms, some urodele amphibians have this amazing capability. Among the species that can replace the missing lens are members of the genera *Triturus, Pleurodeles, Salamandra, Salamandrina, Typhlotriton*, and *Eurycea* (Stone, 1967). Stone also found eight species unable to regenerate their lens, including members of the families *Ambystomidae* and *Plethodontidae* and even some species of *Eurycea*. The fact that two closely related species do not share the same regenerative capabilities raises the issue of evolutionary conservation and selective pressures to keep or lose such an ability. Anuran amphibians such as *Xenopus laevis* can regenerate their lens but only premetamorphosis. In addition, the mode of regeneration differs from that of the salamanders. In these animals, the lens regenerates from the inner cell layer of the outer cornea (Freeman, 1963).

12.6.2. Lens Regenerative Abilities in Other Vertebrates

Some species of adult fish can regenerate their lens via transdifferentiation of the dorsal iris, similarly to the preselected salamanders (Sato, 1961; Mitashov, 1966). Again, here we can pose the question of how evolutionary pressures selected completely different organisms from different classes of vertebrates to share regenerative abilities.

Within other classes of vertebrates, chicks have been reported to be able to replace their lenses upon removal during a small window during their development. It is not clear if these embryos actually regenerate their lenses via the transdifferentiation of the PECs of the iris or if regeneration results from the induction of the still competent ectoderm by the optic vesicle (van Deth, 1940; McKeehan, 1961; Génis-Gálvez, 1962; Niazi, 1967; Wedlock and McCallion, 1968). On the other hand, rabbits and cats have been reported to be able to regenerate their lenses upon removal, but only if the anterior and posterior capsular bags are left intact. The regenerates seem to be the end result of the growth of lens epithelial cells left in the capsular bags (Gwon et al., 1989; Gwon et al., 1990; Gwon et al., 1992; Gwon et al., 1993a; Gwon et al., 1993b).

12.7. Transdifferentiation of PECs as the Basis of Lens Regeneration

Although lens regeneration from the dorsal iris in the newt has convincingly been shown histologically (as previously mentioned), the ultimate proof of such cell type conversion has come from *in vitro* studies.

12.7.1. Iris PECs as the Actual Regeneration Cells

In experiments done in the 1970s, the dorsal and ventral iris PECs of newt eyes have been separately dissociated and cultured *in vitro*. After a long lag of more than two weeks, iris PECs started to grow vigorously. The actively growing cells gradually lost melanosomes and formed a typical monolayer sheet. From the progeny of non-pigmented epithelial cells in such monolayer cell sheets, three-dimensional transparent structures expressing lens-specific molecules developed within 40 days of culturing. No significant difference was observed between the dorsal and ventral PECs (Fig. 12.9; Eguchi et al., 1974). These findings had been confirmed by clonal analysis. More than 15% of the clonal colonies grown from singly dissociated dorsal or ventral iris PECs differentiated into lentoids with lens specificities (Abe and Eguchi, 1977). These experiments showed that the iris PECs can transdifferentiate *in vitro* and thus be the source of the lens epithelial cells that will reorganize the new forming lens.

12.7.2. Lens-Forming Potential of the Iris PE as a Whole

Application of MNNG during the process of lens regeneration in the newt (Eguchi and Watanabe, 1973) suggested that cell-cell and cell-substrate adhesion and communication were reduced throughout the iris pigment epithelium (PE); supernumerary lens regeneration was induced at different regions of the iris, the ventral iris in addition to the dorsal marginal region, where regeneration usually takes place (Fig. 12.8). These findings also suggest that normally the tissue architecture at the dorsal marginal PE must be unstable enough to react to micro environmental changes caused by lentectomy. Therefore, changes in the ventral iris PECs in cell culture conditions were also studied and compared with those in the dorsal iris

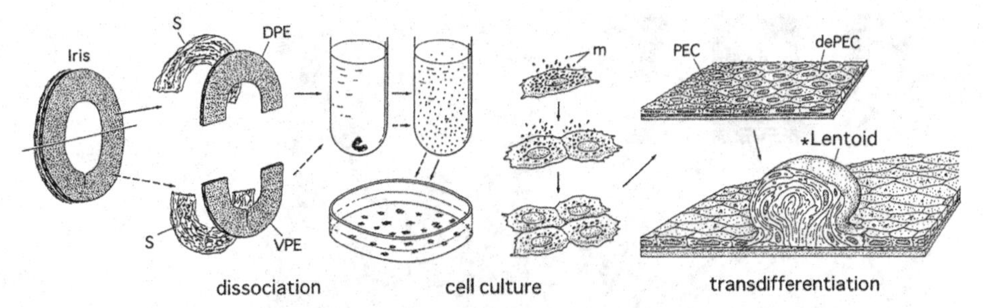

Figure 12.9. *In vitro* lens transdifferentiation of the newt iris PECs. The dorsal and ventral PE (DPE and VPE, respectively) are separated from the stromal tissue (S), dissociated, and cultured. Lentoids expressing lens specificities differentiate from populations of dedifferentiated PECs (dePEC) in cultures of both dorsal and ventral iris PECs. m, melanosome discharged from PECs. (From Eguchi, 1998.)

PECs (Eguchi et al., 1974). It has been clearly demonstrated that the iris PECs as a whole conserve the capacity to transdifferentiate to lens cells (Fig. 12.9) and that the iris PECs are the source of the cells that replace the lens (Eguchi and Shingai, 1971; Eguchi et al., 1974; Abe and Eguchi, 1977). In addition, it has been strongly suggested by these studies that the tissue architecture of the ventral iris PECs *in situ* must be much more stable than that of the dorsal iris PECs, but its destabilization can be induced by artificial dissociation and long-term culturing.

12.7.3. Lens Transdifferentiation in Higher Vertebrates PE

There is a possibility that PECs of chick embryos may also transdifferentiate to lens cells postlentectomy, since the lens is thought to be replaced either from the rudiment of the iris epithelium or from the still competent ectoderm during a restricted period of their development (van Deth, 1940; Génis-Gálvez, 1962). Recently, it has been established that iris PECs from a one-day-old chick can readily transdifferentiate to lens cells *in vitro*. This cell type can provide a very useful cell culture system for studying transdifferentiation from iris PECs to lens cells (Kosaka et al., 1998). In addition, well-differentiated retina PECs dissociated from eyes of older chick embryos have been cultured *in vitro* (Eguchi and Okada, 1973). Retina PECs of chick embryos, maintained in successive culturing, eventually develop into lentoids expressing lens-specific molecules. Direct evidence of lens transdifferentiation in chick embryo retinal PECs was established by clonal analysis, and the results of this study led to the establishment of an effective cell culture system that provides a pure population of cells at each step of lens transdifferentiation, particularly multipotent dedifferentiated PECs (Itoh and Eguchi, 1986a, 1986b). This cell culture system has facilitated the analysis of the cellular and molecular mechanisms of transdifferentiation. *In vitro* cell culture studies have been extended to other higher vertebrates, and lens transdifferentiation of iris or retina PECs in conditions of cell culture has been confirmed in the following: Japanese quail (Coturnix), mouse (Mus), bovine (Bos), and ape (Macaca) (Eguchi, unpublished research). Lens transdifferentiation has been confirmed even in human iris and retina PECs dissociated from adult (80-year-old and 22-year-old) and fetal (16-week-old) eyes (Eguchi, 1988, 1993, 1998; Tsonis et al., 2001). In the case of PECs from the 80-year-old donor, singly dissociated PECs were seeded at clonal cell density. Many of these single PECs grew vigorously, rapidly losing their phenotype and formed clonal colonies consisting of

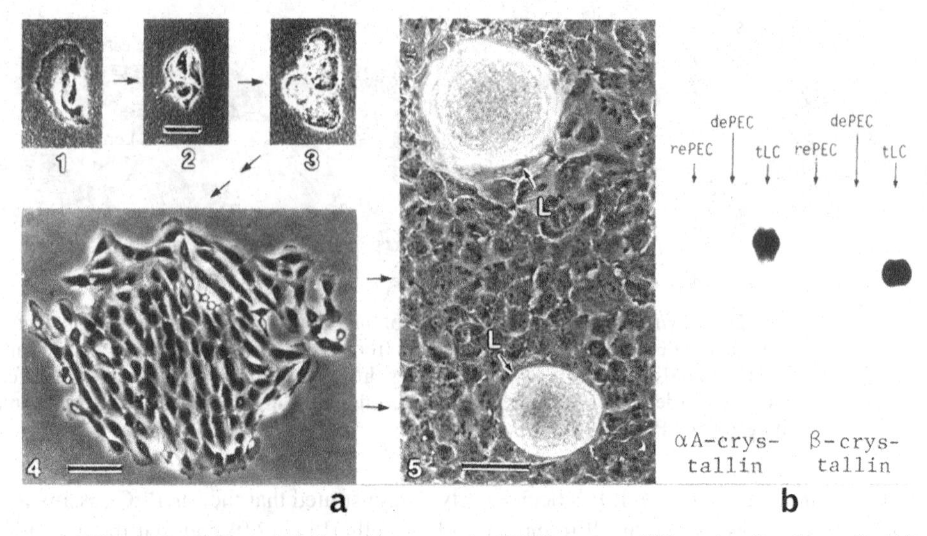

Figure 12.10. Potential of aged human retinal PECs to transdifferentiate into lens cells. (a) Clonal growth (l–4) and generation of lentoids (L) actively expressing lens specificities (5) from a single PEC derived from an 80-year-old donor (l–3: bar = 20 μm; 4: bar = 50 μm, 5: bar = 150 μm). (b) Expression of human lens crystallin genes αA- and β-crystallin by cultured cells differentiating into lentoids. Both αA- and β-crystallin gene products are obviously detected only in transdifferentiated lens cells generating lentoids (tLC) (shown in a5) but never in both dedifferentiated cells (dePEC) and redifferentiated cells (rePEC) – dePEC cells that reverted to the pigmented phenotype.

dedifferentiated cells (Fig. 12.10al–4). In many of these clonal colonies, differentiation of lentoids expressing lens-specific molecules was observed within 10 days (Fig. 12.10a and b; Eguchi, 1988, 1993, 1998). The same results were obtained from cell culture analysis of PECs of both a 22-year-old adult donor and a 16-week-old fetus. Thus it has been concluded that the potential for lens transdifferentiation of PECs is conserved even in humans throughout life.

12.7.4. Molecular Events and Factors Regulating Lens Transdifferentiation of PECs

12.7.4.1. Cell Type–Specific Gene Expression

Using the cell culture system of chick embryonic PECs (Itoh and Eguchi, 1986b), transcription patterns for the following genes have been analyzed: PEC-specific *MMP115* (Mochii et al., 1988a; Mochii et al., 1988b; Mochii et al., 1991) and *pP344* (Iio et al., 1994; Kobayashi, Agata, and Eguchi, unpublished research); the genes for lens cell–specific αA-, βA3/Al-, βB1-, βB2-, γ-, and δ-crystallin; and the chick *myc*, *erb*-B, and *ras* proto-oncogenes (Agata et al., 1993). Expression analysis clearly indicates that the PEC-specific genes are transiently repressed at the stage of dedifferentiation. In contrast, transcription of crystallin genes is activated upon induction of transdifferentiation of PECs into lens cells (Fig. 12.11a and b). Lens transdifferentiation of PECs proceeds through a stable intermediate state in which neither lens cell–specific nor PEC-specific genes are expressed.

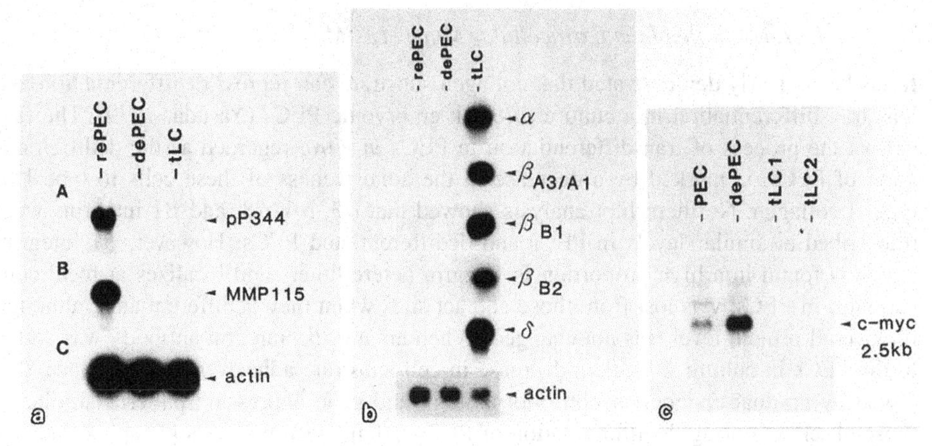

Figure 12.11. Results of gene expression analysis by Northern blotting during lens transdifferentiation of retinal PECs from older chick embryos. (a) Expression of the PEC specific gene pP344 and of the melanosomal matrix protein gene MMP115 is observed only in redifferentiated PECs (rePEC). Five micrograms of Poly(A)$^+$ mRNA were hybridized with pP344, MMP115, and β-actin probes. (b) Expression of lens cell–specific crystallin genes is observed only in transdifferentiated lens cells (tLC). Ten micrograms of poly(A)$^+$ mRNA from redifferentiated PECs (rePEC), dedifferentiated PECs (dePEC), and transdifferentiated PECs (tLC) were used. (c) High level of expression of the c-*myc* proto-oncogene was observed in dedifferentiated PECs. Five micrograms each of poly(A)$^+$ mRNA from well-differentiated PECs and dedifferentiated PECs and 0.5 μg of poly(A)$^+$ mRNA from transdifferentiated lens cells maintained in culture for either one (tLC1) or two (tLC2) weeks were used. (From Agata et al., 1993.)

Dedifferentiated PECs, which are vigorously growing, consume glucose instead of oxygen and lose the property of contact inhibition of growth (Eguchi and Itoh, 1982; Itoh and Eguchi, 1982) in addition to losing gap junctional cell-cell communication (Kodama and Eguchi, 1994). Such properties resemble those of neoplastic cells. Expression of c-*myc* increases significantly in dedifferentiated PECs (Fig. 12.11c), whereas other oncogenes are not activated (Agata et al., 1993), suggesting that increased expression of the c-*myc* oncogene is associated with the high growth potential of dedifferentiated PECs.

12.7.4.2. Growth Factors

Through careful testing of FGFs purified from bovine brain, it has been confirmed that both FGF-1 and FGF-2 dramatically enhance lens transdifferentiation of well-differentiated PECs from older chick embryos (Hyuga et al., 1993), strongly suggesting that FGFs must be among the essential factors regulating lens transdifferentiation. This possibility has been supported by studies on the regulatory functions of FGFs in lens regeneration in the newt *in vivo* (Figs. 12.5 and 12.6; Del Rio-Tsonis et al., 1997; Del Rio-Tsonis et al., 1998; McDevitt et al., 1997). It has recently been found that epidermal growth factor (EGF) is also essential for lens transdifferentiation of one-day-old chick iris PECs, which are much more stable than embryonic PECs. Transdifferentiation of PECs can be dramatically enhanced by EGF in the presence of FGF (Kosaka, Mochii, and Eguchi, unpublished research).

12.7.4.3. Role of the Extracellular Matrix (ECM)

It has been clearly demonstrated that collagen substrate can repress dedifferentiation and lens transdifferentiation in a culture of chick embryonic PECs (Yasuda, 1979). The first half of the process of transdifferentiation in PECs *in vitro*, regarded as the dedifferentiation of PECs, is marked by a decrease in the adhesiveness of these cells to type I or type II collagen. Northern blot analysis showed that $\alpha3$, $\alpha6$, $\alpha8$, and $\beta1$ integrins were transcribed at similar levels in PECs and dedifferentiated PECs. However, $\beta1$ integrin, which is found in a high proportion in integrin heterodimers and localizes at focal contact sites in PECs, was lost from those contact sites when they dedifferentiated, although the overall protein level was not changed. When an anti–$\beta1$ integrin antibody was added to the PECs in culture, a marked decrease in cell-substrate adhesiveness took place, followed by gradual changes in both morphology and gene expression patterns similar to those observed during dedifferentiation of PECs. $\beta1$ Integrin becomes phosphorylated at the onset of dedifferentiation, and this results in loss of its localization at the focal contact sites in the dedifferentiated PECs (Mazaki et al., 1996). In addition, the tyrosine residue of proteins such as paxillin, $p125^{FAK}$, tensin, and p130Cas, which are involved in signal transduction, was phosphorylated in differentiated PECs, whereas no tyrosine phosphorylation of tensin and p130Cas was confirmed in dedifferentiated PECs (Eguchi, unpublished research). Based on these findings, it is possible to postulate that integrin stimulation in differentiated PECs induces actin fiber organization and that mitosis is regulated through protein tyrosine phosphorylation and cytoskeletal changes. In contrast, mitosis of dedifferentiated PECs must be regulated by other signal transducers such as members of the Src kinase family, since integrin does not mediate cell-ECM adhesion in dedifferentiated PECs (Fig. 12.12). In conclusion, an appropriate distribution of $\beta1$ integrin might be essential for maintaining the differentiated state of PECs through cell-ECM adhesion. The transdifferentiation potential of PECs must be completely repressed by some unknown mechanism in the normal eye *in situ*. A search for molecules responsible for the stabilization of the tissue architecture of the PECs was initiated, and a molecule designated as 2NI-36 was

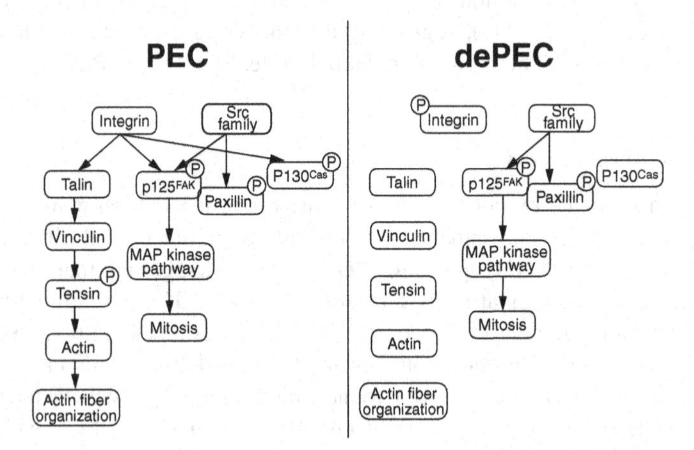

Figure 12.12. Illustration of the hypothesis that integrins control the organization of actin fiber, cytoskeletal complexes, and mitosis in PECs (left) but not in dedifferentiated PECs (right), where signal transduction might be controlled by a member of the Src kinase family. (From Eguchi, 1998.)

identified as a candidate (Eguchi, 1988; Imokawa and Eguchi, 1992; Imokawa et al., 1992). This molecule cannot be detected in newt iris and retina PECs that are actively growing *in vitro*, but it does become detectable when PECs begin to organize epithelial structures.

12.8. Future Prospects

The fact that pigment epithelial cells from anywhere in the eyecup are capable of transdifferentiating to lens cells in culture, without any species restrictions, renders this a potential system for clinical applications. Even PECs from aged humans have this capability (Eguchi, 1988, 1993, 1998; Tsonis et al., 2001). This suggests that if we can find the molecular switches that turn on this transdifferentiation *in vitro* as well as *in vivo*, we will have at hand the power to regenerate lenses in higher vertebrates, including humans.

It is important to apply state-of-the-art technology to elucidate these molecular switches. The use of techniques such as *in vitro* transfection of ventral iris cells to introduce potential transdifferentiation regulators and the subsequent implantation of these transgenic aggregated cells into host lentectomized newt eyes could elucidate the molecular mechanisms behind the regeneration process (Hayashi et al., 2001; Hayashi et al., 2002). The use of techniques, including morpholino (Heasman, 2002) and RNAi (Zhou et al., 2002; Scherr et al., 2003) technology, to block the activity of putative essential regeneration molecules should also be a priority. Pinpointing the molecules responsible for these molecular switches will contribute not only to the field of eye research but to the investigation of many other systems in need of regenerative capacities. On the other hand, the basic knowledge obtained from this research will have an impact on our understanding of cancer growth, since it seems that both regeneration and cancer growth share similar initial activities and must share similar regulators early in the game. In addition, because in regeneration the ultimate result is an organized tissue, we can learn how this regenerating, "provisionally destabilized" tissue organizes itself and avoids awry growth and lack of controlled differentiation to form a copy of the original missing part.

Bibliography

Abdelhak, S., Kalatzis, V., Heilig, R., Compain, S., Samson, D. V. C., Weil, D., Cruaud, C., Sahly, I., Leibovici, M., Bitner-Glindzicz, M. F. M., et al. (1997). A human homologue of the *Drosophila eyes absent* gene underlies branchio-oto-renal (BOR) syndrome and identifies a novel gene family. *Nat. Genet.* **15**, 157–64.

Abe, S. I. and Eguchi, G. (1977). An analysis of differentiative capacity of pigmented epithelial cells of adult newt iris in clonal cell culture. *Dev. Growth Differ.* **19**, 309–17.

Aberle, H., Schwartz, H. and Kemler, R. (1996). Cadherin-catenin complex: protein interactions and their implications for cadherin function. *J. Cell. Biochem.* **61**, 514–23.

Acampora, D., Avantaggiato, V., Tuorto, F., Barone, P., Perera, M., Choo, D., Wu, D., Corte, G. and Simeone, A. (1999a). Differential transcriptional control as the major molecular event in generating $Otx1^{-/-}$ and $Otx2^{-/-}$ divergent phenotypes. *Development* **126**, 1417–26.

Acampora, D., Gulisano, M. and Simeone, A. (1999b). *Otx* genes and the genetic control of brain morphogenesis. *Mol. Cell. Neurosc.* **13**, 1–8.

Acampora, D., Mazan, S., Avantaggiato, V., Barone, P., Tuorto, F., Lallemand, Y., Brulet, P. and Simeone, A. (1996). Epilepsy and brain abnormalities in mice lacking the *Otx1* gene. *Nat. Genet.* **14**, 218–22.

Acampora, D., Mazan, S., Lallemand, Y., Avantaggiato, V., Maury, M., Simeone, A. and Brulet, P. (1995). Forebrain and midbrain regions are deleted in $Otx2^{-/-}$ mutants due to a defective anterior neuroectoderm specification during gastrulation. *Development* **121**, 3279–90.

Adelmann, H. B. (1966). *Marcello Malpighi and the Evolution of Embryology*. Ithaca: Cornell University Press.

Aerts T., Xia, J. Z., Slegers, H., de Block, J. and Clauwaert, J. (1990). Hydrodynamic characterization of the major intrinsic protein from the bovine lens fibre membranes. *J. Biol. Chem.* **265**, 8675–80.

Agata, K., Kobayashi, H., Itoh, Y., Mochii, M., Sawada, K. and Eguchi, G. (1993). Genetic characterization of the multipotent dedifferentiated state of pigmented epithelial cells *in vitro*. *Development* **118**, 1025–30.

Akhurst, R. J. and Derynck, R. (2001). TGF-beta signaling in cancer: a double-edged sword. *Trends. Cell Biol.* **11**, S44–51.

Akimenko, M.-A., Ekker, M., Wegner, J., Lin, W. and Westerfield, M. (1994). Combinatorial expression of three zebrafish genes related to *Distal-less*: Part of a homeobox gene code for the head. *J. Neurosci.* **14**, 3475–86.

Albert, D. M. (1996a). Greek, Roman and Arabian ophthalmology. In D. M. Albert and D. D. Edwards (Eds.), *The History of Ophthalmology*. Cambridge, MA: Blackwell Science.

Albert, D. M. (1996b). The development of ophthalmic pathology. In D. M. Albert and D. D. Edwards (Eds.), *The History of Ophthalmology*. Cambridge, MA: Blackwell Science.

Albert, D. M. (1996c). Discovering the anatomy of the eye. In D. M. Albert and D. D. Edwards (Eds.), *The History of Ophthalmology*. Cambridge, MA: Blackwell Science.

Albert, D. M. and D. D. Edwards, eds. (1996). *The History of Ophthalmology*. Cambridge, Mass.: Blackwell Science.

Alcala, J. and Maisel, H. (1985). Biochemistry of lens plasma membrane and cytoskeleton. In *The Ocular Lens: Structure, Function and Pathology*, ed. H. Maisel. New York: Marcel Dekker, pp. 169–222.

Alemany, J., Girbau, M., Bassas, L. and de Pablo, F. (1990). Insulin receptors and insulin-like growth factor I receptors are functional during organogenesis of the lens. *Mol. Cell. Endocrinol.* **74**, 155–62.

Alexander, L. E. (1937). An experimental study of the role of optic cup and overlying ectoderm in lens formation in the chick embryo. *J. Exp. Zool.* **75**, 41–74.

Al-Ghoul, K. J. and Costello, M. J. (1997). Light microscopic variation of fiber cell size, shape and ordering in the equatorial plane of bovine and human lenses. *Mol. Vis.* **3**, 2. Available at http://www.emory.edu/molvis/v3/al-ghoul.

Al-Ghoul, K. J., Kirk, T., Kuszak, A. J., Zoltoski, R. K., Shiels, A. and Kuszak J. R. (2003). Lens structure in MIP-deficient mice. *Anat. Rec. Part A*, **273A**, 714–30.

Al-Ghoul, K. J., Nordgren, R. K., Kuszak, A. J., Freel, C. D., Costello, M. J. and Kuszak, J. R. (2001). Structural evidence of human nuclear fiber compaction as a function of aging and cataractogenesis. *Exp. Eye Res.* **72**, 199–214.

Alizadeh, A., Clark, J., Seeberger, T., Hess, J., Blankenship, T. and FitzGerald, P. G. (2003). Targeted deletion of the lens fiber cell-specific intermediate filament protein filensin. *Invest. Ophthalmol. Vis. Sci.* **44**, 5252–8.

Alizadeh, A., Clark, J., Seeberger, T., Hess, J., Blankenship, T. and FitzGerald, P. G. (2004). Characterization of a mutation in the lens-specific CP49 in the 129 strain of mouse. *Invest. Ophthalmol. Vis. Sci.* **45**, 884–91.

Alizadeh, A., Clark, J. I., Seeberger, T., Hess, J., Blankenship, T., Spicer, A. and FitzGerald, P. G. (2002). Targeted genomic deletion of the lens-specific intermediate filament protein CP49. *Invest. Ophthalmol. Vis. Sci.* **43**, 3722–7.

Allen, D. P., Low, P. S., Dola, A. and Maisel, H. (1987). Band 3 and ankyrin homologues are present in eye lens: evidence for all major erythrocyte membrane components in same non-erythroid cell. *Biochem. Biophys. Res. Commun.* **149**, 266–75.

Alnemri, E. S., Livingston, D. J., Nicholson, D. W., Salvesen, G., Thornberry, N. A., Wong, W. W. and Yuan, J. (1996). Human ICE/CED-3 protease nomenclature [letter]. *Cell* **87**, 171.

Alonso, M. I., Gato, A., Moro, J. A., Martin, C., Barbosa, M., Callejo, S. and Barbosa, E. (1996). Role of sulfated proteoglycans in early lens development. *Int. J. Dev. Biol.* Suppl 1, 249S–50S.

Altmann, C. R., Chow, R. L., Lang, R. A. and Hemmati-Brivanlou, A. (1997). Lens induction by Pax-6 in *Xenopus laevis*. *Dev. Biol.* **185**, 119–23.

Alvarez, L. J., Candia, O. A. and Grillone, L. R. (1985). Na^+/K^+ ATPase distribution in frog and bovine lenses. *Curr. Eye Res.* **4**, 143–52.

Alvarez, L. J., Candia, O. A. and Zamudio, A. C. (1995). Acetylcholine modulation of the short-circuit current across the rabbit lens. *Exp. Eye Res.* **61**, 129–40.

Amano, U. und Sato, J. (1940). Uber die xenoplastische Implantation der larvalen des Triturus pyrrhogaster in das entlinste Auge der Larven des *Hynobius nebulosus*. *Jpn. J. Med. Sci. I. Anat.* **8**, 75–81.

Amaya, E., Offield, M. F. and Grainger, R. M. (1998). Frog genetics: *Xenopus tropicalis* jumps into the future. *Trends Genet.* **14**, 253–5.

Ameen, N. A., Figueroa, Y. and Salas, P. J. (2001). Anomalous apical plasma membrane phenotype in CK8-deficient mice indicates a novel role for intermediate filaments in the polarization of simple epithelia. *J. Cell Sci.* **114**, 563–75.

Andley, U. P., Song, Z., Wawrousek, E. F. and Bassnett, S. (1998). The molecular chaperone alphaA-crystallin enhances lens epithelial cell growth and resistance to UVA stress. *J. Biol. Chem.* **273**, 31252–61.

Andley, U. P., Song, Z., Wawrousek, E. F., Fleming, T. P. and Bassnett, S. (2000). Differential protective activity of alpha A- and alpha B-crystallin in lens epithelial cells. *J. Biol. Chem.* **275**, 36823–31.

Andra, K., Lassmann, H., Bittner, R., Shorny, S., Fassler, R., Propst, F. and Wiche, G. (1997). Targeted inactivation of plectin reveals essential function in maintaining the integrity of skin, muscle, and heart cytoarchitecture. *Genes Dev.* **11**, 3143–56.

Andra, K., Nikolic, B., Stocher, M., Drenckhahn, D. and Wiche, G. (1998). Not just scaffolding: plectin regulates actin dynamics in cultured cells. *Genes Dev.* **12**, 3442–51.

Ang, S.-L., Conlon, R. A., Jin, O. and Rossant, J. (1994). Positive and negative signals from mesoderm regulate the expression of mouse *Otx2* in ectoderm explants. *Development* **120**, 2979–89.

Ang, S.-L., Jin, O., Rhinn, M., Daigle, N., Stevenson, L. and Rossant, J. (1996). A targeted mouse

Otx2 mutation leads to severe defects in gastrulation and formation of axial mesoderm and to deletion of rostral brain. *Development* **122**, 243–52.

Angst, B. D., Marcozzi, C. and Magee, A. I. (2001). The cadherin superfamily. *J. Cell Sci.* **114**, 625–6.

Apple, D. J. (1992). Posterior capsule opacification. *Surv. Ophthalmol.* **37**, 73–116.

Appleby, D. W. and Modak, S. P. (1977). DNA degradation in terminally differentiating lens fiber cells from chick embryos. *Proc. Natl. Acad. Sci. USA* **74**, 5579–83.

Arneson, M. L. and Louis, C. F. (1998). Structural arrangement of lens fiber cell plasma membrane protein MP20. *Exp. Eye Res.* **66**, 495–509.

Arnold, D. R., Moshayedi, P., Schoen, T. J., Jones, B. E., Chader, G. J. and Waldbillig, R. J. (1993). Distribution of IGF-I and -II, IGF binding proteins (IGFBPs) and IGFBP mRNA in ocular fluids and tissues: potential sites of synthesis of IGFBPs in aqueous and vitreous. *Exp. Eye Res.* **56**, 555–65.

Arnold, J. M. (1984). Closure of the squid cornea: a muscular basis for embryonic tissue movement. *J. Exp. Zool.* **232**, 187–95.

Arruti, C., Chaudun, E., De Maria, A., Courtois, Y. and Counis, M.-F. (1995). Characterization of eye-lens DNases: long term persistence of activity in post apoptotic lens fiber cells. *Cell Death Diff.* **2**, 47–56.

Ashery-Padan, R., Marquardt, T., Zhou, X. and Gruss, P. (2000). Pax6 activity in the lens primordium is required for lens formation and for correct placement of a single retina in the eye. *Genes Dev.* **14**, 2701–11.

Asselbergs, F. A., Koopmans, M., van Venrooij, W. J. and Bloemendal, H. (1979). Improved resolution of calf lens beta-crystallins. *Exp. Eye Res.* **28**, 223–8.

Atreya, P. L., Barnes, J., Katar, M., Alcala, J. and Maisel, H. (1989). N-cadherin of the human lens. *Curr. Eye Res.* **8**, 947–56.

Azuma, N. and Hara, T. (1998). Extracellular matrix of opacified anterior capsule after endocapsular cataract surgery. *Graefes Arch. Clin. Exp. Ophthalmol.* **236**, 531–6.

Azuma, N., Hirakiyama, A., Inoue, T., Asaka, A. and Yamada, M. (2000). Mutations of a human homologue of the *Drosophila eyes absent* gene (*EYA1*) detected in patients with congenital cataracts and ocular anterior segment anomalies. *Hum. Mol. Genet.* **9**, 363–6.

Baas, P. W. (1998). The role of motor proteins in establishing the microtubule arrays of axons and dendrites. *J. Chem. Neuroanat.* **14**, 175–80.

Baas, D., Bumsted, K. M., Martinez, J. A., Vaccarino, F. M., Wikler, K. C. and Barnstable, C. J. (2000). The subcellular localization of *Otx2* is cell-type specific and developmentally regulated in the mouse retina. *Mol. Brain Res.* **78**, 26–37.

Bagby, S., Harvey, T. S., Eagle, S. G., Inouye, S. and Ikura, M. (1994). Structural similarity of a developmentally related bacterial spore coat protein to $\beta\gamma$-crystallins of the vertebrate eye lens. *Proc. Natl. Acad. Sci. USA* **91**, 4308–12.

Bagchi, M., Alcala, J. R. and Maisel, H. (1981). Delta-crystallin synthesis by the adult chicken lens. *Exp. Eye Res.* **32**, 251–4.

Bagchi, S., Weinmann, R. and Raychaudhuri, P. (1991). The retinoblastoma protein copurifies with E2F-I, an E1A-regulated inhibitor of the transcription factor E2F. *Cell* **65**, 1063–72.

Bailly, E., Pines, J., Hunter, T. and Bornens, M. (1992). Cytoplasmic accumulation of cyclin B1 in human cells: association with a detergent-resistant compartment and with the centrosome. *J. Cell Sci.* **101**, 529–45.

Baldin, V., Lukas, J., Marcote, M. J., Pagano, M. and Draetta, G. (1993). Cyclin D1 is a nuclear protein required for cell cycle progression in G1. *Genes Dev.* **7**, 812–21.

Baldo, G. J. and Mathias, R. T. (1992). Spatial variations in membrane properties in the intact rat lens. *Biophys. J.* **63**, 518–29.

Balkan, W., Klintworth, G. K., Bock, C. B. and Linney, E. (1992). Transgenic mice expressing a constitutively active retinoic acid receptor in the lens exhibit ocular defects. *Dev. Biol.* **151**, 622–5.

Bansal, R. (2002). Fibroblast growth factors and their receptors in oligodendrocyte development: implications for demyelination and remyelination. *Dev. Neurosci.* **24**, 35–46.

Barabanov, V. M. (1977). Detection of δ-crystallins in the adenohypophysis of chick embryos. *Dokl. Akad. Nauk. USSR* **234**, 195–8.

Barabanov, V. M. and Fedtsova, N. G. (1982). The distribution of lens differentiation capacity in the head ectoderm of chick embryos. *Differentiation* **21**, 183–90.

Barber, V. C., Evan, E. M. and Land, M. F. (1967). The fine structure of the eye of the mollusk *Pecten maximus*. *Z. Zellforsch.* **76**, 295–312.

Barbosa, P., Wistow, G. J., Cialkowski, M., Piatigorsky, J. and O'Brien, W. E. (1991). Expression of duck lens delta-crystallin cDNAs in yeast and bacterial hosts: delta 2-crystallin is an active argininosuccinate lyase. *J. Biol. Chem.* **266**, 22319–22.

Barr, F. G. (1999). The role of chimeric paired box transcription factors in the pathogenesis of pediatric rhabdomysarcoma. *Cancer Res.* **59** (Suppl), 1711–5.

Barondes, S. H., Cooper. D. N. W., Gitt, M. A. and Leffler, H. (1994). Structure and function of a large family of animal lectins. *J. Biol. Chem.* **269**, 20807–10.

Barrett, J. C. and Preston, G. (1994). Apoptosis and cellular senescence: forms of irreversible growth arrest. In *Apoptosis II: The Molecular Basis of Apoptosis in Disease*, ed. L. D. Tomei and F. O. Cope. Plainview, New York: Cold Spring Harbor Laboratory Press, pp. 253–81.

Bassnett, S. (1992). Mitochondrial dynamics in differentiating fiber cells of the mammalian lens. *Curr. Eye Res.* **11**, 1227–32.

Bassnett, S. (1995). The fate of the Golgi apparatus and the endoplasmic reticulum during lens fiber cell differentiation. *Invest. Ophthalmol. Vis. Sci.* **36**, 1793–803.

Bassnett, S. (1997a). Chromatin degradation in differentiating fiber cells of the eye lens. *J. Cell Biol.* **137**, 37–49.

Bassnett, S. (1997b). Fiber cell denucleation in the primate lens. *Invest. Ophthalmol. Vis. Sci.* **38**, 1678–87.

Bassnett, S. and Beebe, D. C. (1990). Localization of insulin-like growth factor-1 binding sites in the embryonic chicken eye. *Invest. Ophthalmol. Vis. Sci.* **31**, 1637–43.

Bassnett, S. and Beebe, D. C. (1992). Coincident loss of mitochondria and nuclei during lens fiber cell differentiation. *Dev. Dyn.* **194**, 85–93.

Bassnett, S., Croghan, P. C. and Duncan, G. (1987). Diffusion of lactate and its role in determining intracellular pH in the lens of the eye. *Exp. Eye Res.* **44**, 143–7.

Bassnett, S., Kuszak, J. R., Reinisch, L., Brown, H. G. and Beebe, D. C. (1994). Intercellular communication between epithelial and fiber cells of the eye lens. *J. Cell Sci.*, **107**, 799–811.

Bassnett, S., Missey, H. and Vucemilo, I. (1999). Molecular architecture of the lens fiber cell basal membrane complex. *J. Cell Sci.* **112**, 2155–65.

Bassuk, J. A., Birkebak, T., Rothmier, J. D., Clark, J. M., Bradshaw, A., Muchowski, P. J., Howe, C. C., Clark, J. I. and Sage, E. H. (1999). Disruption of the SPARC locus in mice alters the differentiation of lenticular epithelial cells and leads to cataract formation. *Exp. Eye Res.* **68**, 321–31.

Bateman, J. B., Geyer, D. D., Flodman, P., Johannes, M., Sikela, J., Walter, N., Moreira, A. T., Clancy, K. and Spence, M. A. (2000). A new betaA1-crystallin splice junction mutation in autosomal dominant cataract. *Invest. Ophthalmol. Vis. Sci.* **41**, 3278–85.

Bavik, C., Ward, S. and Chambon, P. (1996). Developmental abnormalities in cultured mouse embryos deprived of retinoic acid by inhibition of yolk-sac retinol binding protein synthesis. *Proc. Natl. Acad. Sci. USA* **93**, 3110–4.

Bax, B., Lapatto, R., Nalini, V., Driessen, H., Lindley, P. F., Mahadevan, D., Blundell, T. L. and Slingsby, C. (1990). X-ray analysis of βB2-crystallin and evolution of oligomeric lens proteins. *Nature* **347**, 776–9.

Beebe, D., Snellings, K., Silver, M. and Van Wyk, J. (1986). Control of lens cell differentiation and ion fluxes by growth factors. *Prog. Clin. Biol. Res.* 365–9.

Beebe, D. C. (1992). The control of lens growth: relationship to secondary cataract. *Acta Ophthalmol. Suppl.* **205**, 53–7.

Beebe, D. C. and Cerrelli, S. (1989). Cytochalasin prevents cell elongation and increases potassium efflux from embryonic lens epithelial cells: implications for the mechanism of lens fiber cell elongation. *Lens Eye Toxic Res.* **6**, 589–601.

Beebe, D. C., Compart, P. J., Johnson, M. C., Feagans, D. E. and Feinberg, R. N. (1982). The mechanism of cell elongation during lens fiber cell differentiation. *Dev. Biol.* **92**, 54–9.

Beebe, D. C., Feagans, D. E., Blanchette-Mackie, E. J. and Nau, M. E. (1979). Lens epithelial cell elongation in the absence of microtubules: evidence for a new effect of colchicine. *Science* **206**, 836–8.

Beebe, D. C., Feagans, D. E. and Jebens, H. A. (1980). Lentropin: a factor in vitreous humor which promotes lens fiber cell differentiation. *Proc. Natl. Acad. Sci. USA* **77**, 490–3.

Beebe, D. C. and Masters, B. R. (1996). Cell lineage and the differentiation of corneal epithelial cells. *Invest. Ophthalmol. Vis. Sci.* **37**, 1815–25.

Beebe, D. C., Parmelee, J. T. and Belcher, K. S. (1990). Volume regulation in lens epithelial cells and differentiating lens fiber cells. *J. Cell. Physiol.* **143**, 455–9.

Beebe, D. C. and Piatigorsky, J. (1976). Differential synthesis of crystallin and noncrystallin polypeptides during lens fiber cell differentiation in vitro. *Exp. Eye Res.* **22**, 237–49.

Beebe, D. C., Silver, M. H., Belcher, K. S., Van Wyk, J. J., Svoboda, M. E. and Zelenka, P. S. (1987). Lentropin, a protein that controls lens fiber formation, is related functionally and immunologically to the insulin-like growth factors. *Proc. Natl. Acad. Sci. USA* **84**, 2327–30.

Beebe, D. C., Vasiliev, O., Guo, J., Shui, Y. B. and Bassnett, S. (2001). Changes in adhesion complexes define stages in the differentiation of lens fiber cells. *Invest. Ophthalmol. Vis. Sci.* **42**, 727–34.

Beggs, H. E., Baragona, S. C., Hemperly, J. J. and Maness, P. F. (1997). NCAM140 interacts with the focal adhesion kinase p125(fak) and the SRC-related tyrosine kinase p59 (fyn). *J. Biol. Chem.* **272**, 8310–9.

Beggs, H. E., Soriano, P. and Maness, P. F. (1994). NCAM-dependent neurite outgrowth is inhibited in neurons from Fyn-minus mice. *J. Cell Biol.* **127**, 825–33.

Belecky-Adams, T., Tomarev, S., Li, H. S., Ploder, L., McInnes, R. R., Sundin, O. and Adler, R. (1997). Pax-6, Prox 1, and Chx10 homeobox gene expression correlates with phenotypic fate of retinal precursor cells. *Invest. Ophthalmol. Vis. Sci.* **38**, 1293–303.

Belmokhtar, C. A., Torriglia, A., Counis, M. F., Courtois, Y., Jacquemin-Sablon, A. and Segal-Bendirdjian, E. (2000). Nuclear translocation of a leukocyte elastase inhibitor/elastase complex during staurosporine-induced apoptosis: role in the generation of nuclear L-DNase II activity. *Exp. Cell Res.* **254**, 99–109.

Benedetti, E. L., Dunia, I., Bentzel, C. J., Vermorken, A. J. M., Kibbelaar, M. and Bloemendal, H. (1976). A portrait of plasma membrane specializations in lens epithelium and fibers. *Biochim. Biophys. Acta* **457**, 353–84.

Benjamin, I. J., Shelton, J., Garry, D. J. and Richardson, J. A. (1997). Temporospatial expression of the small HSP/alpha B-crystallin in cardiac and skeletal muscle during mouse development. *Dev. Dyn.* **208**, 75–84.

Benkhelifa, S., Provot, S., Lecoq, O., Pouponnot, C., Calothy, G. and Felder-Schmittbuhl, M. P. (1998). MafA, a novel member of the maf proto-oncogene family, displays developmental regulation and mitogenic capacity in avian neuroretina cells. *Oncogene* **17**, 247–54.

Bennett, V. (1990). Spectrin: a structural mediator between diverse plasma membrane proteins and the cytoplasm. *Curr. Opin. Cell Biol.* **2**, 51–6.

Berthoud, V. M., Bassnett, S. and Beyer, E. C. (1999). Cultured chicken embryo lens cells resemble differentiating fiber cells in vivo and contain two kinetic pools of connexin56. *Exp. Eye Res.* **68**, 475–84.

Berthoud, V. M., Westphale, E. M., Grigoryeva, A. and Beyer, E. C. (2000). PKC isoenzymes in the chicken lens and TPA-induced effects on intercellular communication. *Invest. Ophthalmol. Vis. Sci.* **41**, 350–8.

Bessa, J., Gebelein, B., Pichaud, F., Casares, F. and Mann, R. S. (2002). Combinatorial control of *Drosophila* eye development by Eyeless, Homothorax, and Teashirt. *Genes Dev.* **16**, 2415–27.

Beyer, E. C., Kistler, J., Paul, D. L. and Goodenough, D. A. (1989). Antisera directed against connexin43 peptides react with a 43-kD protein localized to gap junctions in myocardium and other tissues. *J. Cell Biol.* **108**, 595–605.

Bhat, S. P., Hale, I. L., Matsumoto, B. and Elghanayan, D. (1999). Ectopic expression of alpha B-crystallin in Chinese hamster ovary cells suggests a nuclear role for this protein. *Eur. J. Cell. Biol.* **78**, 143–50.

Bhat, S. P. and Nagineni, C. N. (1989). Alpha B subunit of lens-specific protein alpha-crystallin is present in other ocular and non-ocular tissues. *Biochem. Biophys. Res. Commun.* **158**, 319–25.

Bhatnagar, A., Dhir, P., Wang, L. F., Ansari, N. H., Lo, W. K. and Srivastava, S. K. (1997). Alterations in the light transmission through single lens fibers during calcium-mediated disintegrative globulization. *Invest. Ophthalmol. Vis. Sci.* **38**, 586–92.

Bienz, M. and Clevers, H. (2000). Linking colorectal cancer to Wnt signaling. *Cell* **103**, 311–20.

Bilak, S. R., Sernett, S. W., Bilak, M. M., Bellin, R. M., Stromer, M. H., Huiatt, T. W. and Robson, R. M. (1998). Properties of the novel intermediate filament protein synemin and its identification in mammalian muscle. *Arch. Biochem. Biophys.* **355**, 63–76.

Bindels, J. G., Koppers, A. and Hoenders, H. J. (1981). Structural aspects of bovine β-crystallins: physical characterization including dissociation-association behavior. *Exp. Eye Res.* **33**, 333–43.

Birchmeier, W. (1995). E-cadherin as a tumor (invasion) suppressor gene. *Bioessays* **17**, 97–9.

Bishop, K. M., Goudreau, G. and O'Leary, D. D. (2000). Regulation of area identity in the mammalian neocortex by *Emx2* and *Pax6*. *Science* **288**, 344–9.

Bissell, M. J. and Barcellos-Hoff, M. H. (1987). The influence of extracellular matrix on gene expression: is structure the message? *J. Cell Sci.* **8** (Suppl), 327–43.

Bissell, M. J., Hall, H. G. and Parry, G. (1982). How does extracellular matrix direct gene expression? *J. Theor. Biol.* **99**, 31–68.

Blakely, E. A., Bjornstad, K. A., Chang, P. Y., McNamara, M. P., Chang, E., Aragon, G., Lin, S. P., Lui, G. and Polansky, J. R. (2000). Growth and differentiation of human lens epithelial cells in vitro on matrix. *Invest. Ophthalmol. Vis. Sci.* **41**, 3898–907.

Blank, V. and Andrews, N. C. (1997). The Maf transcription factors: regulators of differentiation. *Trends Biochem. Sci.* **22**, 437–41.

Blankenship, T. N., Hess, J. F. and FitzGerald, P. G. (2001). Development- and differentiation-dependent reorganization of intermediate filaments in fiber cells. *Invest. Ophthalmol. Visual Sci.* **42**, 735–42.

Blixt, Å., Mahlapuu, M., Aitola, M., Pelto-Huikko, M., Enerbäck, S. and Carlsson, P. (2000). A forkhead gene, *FoxE3*, is essential for lens epithelial proliferation and closure of the lens vesicle. *Genes Dev.* **14**, 245–54.

Bloemendal, H., Berbers, G. A., De Jong, W. W., Ramaekers, F. C., Vermorken, A. J., Dunia, I. and Benedetti, E. L. (1984). Interaction of crystallins with the cytoskeletal–plasma membrane complex of the bovine lens. *CIBA Found. Symp.* **106**, 177–90.

Blundell, T. L., Lindley, P. F., Miller, L., Moss, D. S., Slingsby, C., Tickle, I. J., Turnell, W. G. and Wistow, G. J. (1981). *Nature* **289**, 771–7.

Bok, D., Dockstader, J. and Horwitz, J. (1982). Immunocytochemical localization of the lens main intrinsic polypeptide (MIP26) in communicating junctions. *J. Cell Bio.* **92**, 213–20.

Bond, J., Green, C., Donaldson, P. J. and Kistler, J. (1996). Liquefaction of cortical tissue in diabetic and galactosemic rat lenses defined by confocal laser scanning microscopy. *Invest. Ophthalmol. Vis. Sci.* **37**, 1557–65.

Bonifas, J. M., Rothman, A. L. and Epstein, E. H., Jr. (1991). Epidermolysis bullosa simplex: evidence in two families for keratin gene abnormalities [see comments]. *Science* **254**, 1202–5.

Bonini, N. M., Bui, Q. T., Grayboard, G. L. and Warrick, J. M. (1997). The *Drosophila eyes absent* gene directs ectopic eye formation in a pathway conserved between flies and vertebrates. *Development* **124**, 4819–26.

Bonini, N. M., Leiserson, W. M. and Benzer, S. (1993). The eyes absent gene: genetic control of cell survival and differentiation in the developing *Drosophila* eye. *Cell* **72**, 379–95.

Bonnet, C. (1781). *Sur les reproductions des salamanders: Oeuvres d'Histoire Naturelle et de philosophie*, vol. 2. Neuchatel: Chez S. Fauche, pp. 175–9.

Borsani, G., DeGrandi, A., Ballabio, A., Bulfone, A., Bernard, L., Banfi, S., Gattuso, C., Mariani, M., Dixon, M., Donnai, D., et al. (1999). *EYA4*, a novel vertebrate gene related to *Drosophila eyes absent*. *Hum. Mol. Genet.* **8**, 11–23.

Bosher, J. M., Totty, N. F., Hsuan, J. J., Williams, T. and Hurst, H. C. (1996). A family of AP-2 proteins regulates c-*erb*B-2 expression in mammary carcinoma. *Oncogene* **13**, 1701–7.

Boudreau, N., Werb, Z. and Bissell, M. J. (1996). Suppression of apoptosis by basement membrane requires three-dimensional tissue organization and withdrawal from the cell cycle. *Proc. Natl. Acad. Sci. USA* **93**, 3509–13.

Bourdoulous, S., Orend, G., MacKenna, D. A., Pasqualini, R. and Ruoslahti, E. (1998). Fibronectin matrix regulates activation of RHO and CDC42 GTPases and cell cycle progression. *J. Cell Biol.* **143**, 267–76.

Bova, M. P., Ding, L. L., Horwitz, J. and Fung, B. K. (1997). Subunit exchange of alphaA-crystallin. *J. Biol. Chem.* **272**, 29511–17.

Bova, M. P., Yaron, O., Huang, Q., Ding, L., Haley, D. A., Stewart, P. L. and Horwitz, J. (1999). Mutation R120G in alphaB-crystallin, which is linked to a desmin-related myopathy, results in an irregular structure and defective chaperone-like function. *Proc. Natl. Acad. Sci. USA* **96**, 6137–42.

Bovolenta, P., Mallamaci, A., Briata, P., Corte, G. and Boncinelli, E. (1997). Implication of OTX2 in pigment epithelium determination and neural retina differentiation. *J. Neurosc.* **17**, 4243–52.

Boyadijiev, S. A. and Jabs, E. W. (2000). Online Mendelian Inheritance in Man (OMIM) as a knowledgebase for human developmental disorders. *Clin. Genet.* **57**, 253–66.

Boyd, J. and Barrett, J. C. (1990). Tumor supressor genes: possible functions in the negative regulation of cell proliferation. *Mol. Carcinog.* **3**, 325–9.

Boyer, S., Maunoury, R., Gomes, D., de Nechaud, B., Hill, A. M. and Dupouey, P. (1990). Expression of glial fibrillary acidic protein and vimentin in mouse lens epithelial cells during development in vivo and during proliferation and differentiation in vitro: comparison with the developmental appearance of GFAP in the mouse central nervous system. *J. Neurosci. Res.* **27**, 55–64.

Boyle, D. L. and Takemoto, L. (2000). A possible role for alpha-crystallins in lens epithelial cell differentiation. *Mol. Vis.* **6**, 63–71.

Bradley, R. H., Ireland, M. and Maisel, H. (1979). The cytoskeleton of chick lens cells. *Exp. Eye Res.* **28**, 441–53.

Brady, J. P., Garland, D., Duglas-Tabor, Y., Robison, W. G. Jr., Groome, A. and Wawrousek, E. F. (1997). Targeted disruption of the mouse alpha A-crystallin gene induces cataract and cytoplasmic inclusion bodies containing the small heat shock protein alpha B-crystallin. *Proc. Natl. Acad. Sci. USA* **94**, 884–9.

Brady, J. P. and Wawrousek, E. F. (1997). Targeted disruption of the mouse αB-crystallin gene. *Invest. Opthalmol. Vis. Sci.* **38**, S935.

Brahma, S. K. (1988). Ontogeny of βB1-crystallin polypeptide during chicken lens development. *Exp. Eye Res.* **47**, 507–10.

Brehm, A., Miska, E. A., McCance, D. J., Reid, J. L., Bannister, A. J. and Kouzarides, T. (1998). Retinoblastoma protein recruits histone deacetylase to repress transcription. *Nature* **391**, 597–601.

Breitman, M. L., Bruce, D. M., Giddens, E., Clapoff, S., Goring, D., Tsui, L. C., Klintworth, G. K. and Bernstein, A. (1989). Analysis of lens cell fate and eye morphogenesis in transgenic mice ablated for cells of the lens lineage. *Development* **106**, 457–63.

Brekken, R. A. and Sage, E. H. (2000). SPARC, a matricellular protein: at the crossroads of cell-matrix. *Matrix Biol.* **19**, 569–80.

Brewitt, B. and Clark, J. (1988). Growth and transparency in the lens, an epithelial tissue, stimulated by pulses of PDGF. *Science* **242**, 777–9.

Brisken, C., Heineman, A., Chavarria, T., Elenbaas, B., Tan, J., Dey, S. K., McMahon, J. A., McMahon, A. P. and Weinberg, R. A. (2000). Essential function of Wnt-4 in mammary gland development downstream of progesterone signaling. *Genes Dev.* **14**, 650–4.

Broekhuyse, R. M., Kuhlmann, E. D. and Stols, A. L. (1976). Lens membranes. II. Isolation and characterization of the main intrinsic polypeptide (MIP) of bovine lens fiber membranes. *Exp. Eye Res.* **23**, 365–71.

Bron, A. J., Vrensen, G. F. J. M., Koretz, J., Maraini, G. and Harding, J. J. (2000). The aging lens. *Ophthalmologica* **214**, 86–104.

Brown, H. G., Ireland, M., Pappas, G. D. and Kuszak, J. R. (1990). Ultrastructural, biochemical and immunological evidence of receptor-mediated endocytosis in the crystalline lens. *Invest. Ophthalmol. Vis. Sci.* **31**, 2579–92.

Brown, H. G., Pappas, G. D., Ireland, M. E. and Kuszak, J. R. (1990). Ultrastructural, biochemical and immunologic evidence of receptor-mediated endocytosis in the crystalline lens. *Invest. Ophthalmol. Vis. Sci.* **31**, 2579–92.

Brownell, I., Dirksen, M. and Jamrich, M. (2000). Forkhead Foxe3 maps to the *dysgenetic lens* locus and is citical in lens development and differentiation. *Genesis* **27**, 81–93.

Buckingham, M. E., Caput, D., Cohen, A., Whalen, R. G. and Gros, F. (1974). The synthesis and stability of cytoplasmic messenger RNA during myoblast differentiation in culture. *Proc. Natl. Acad. Sci. USA* **71**, 1466–70.

Budtz, P. E. (1994). Epidermal homeostasis: A new model that includes apoptosis. In *Apoptosis II: The Molecular Basis of Apoptosis in Disease*, ed. L. D. Tomei and F. O. Cope. Plainview, New York: Cold Springs Harbor Laboratory Press, pp. 165–83.

Burdon, K. P., Wirth, M. G., Mackey, D. A., Russell-Eggitt, I. M., Craig, J. E., Elder, J. E., Dickinson, J. L. and Sale, M. M. (2004). Investigation of crystallin genes in familial cataract, and report of two disease associated mutations. *Br. J. Ophthalmol.* **88**, 79–83.

Burglin, T. R. (1997). Analysis of TALE superclass homeobox genes (MEIS, PBC, KNOX, Iroquois, TGIF) reveals a novel domain conserved between plants and animals. *Nucleic Acids Res.* **25**, 4173–80.

Caldes, T., Alemany, J., Robcis, H. L. and de Pablo, F. (1991). Expression of insulin-like growth factor I in developing lens is compartmentalized. *J. Biol. Chem.* **266**, 20786–90.

Call, M. K., Grogg, M. W., Del Rio-Tsonis, K. and Tsonis, P. A. (2004). Lens regeneration in mice: implications in cataracts. *Exp. Eye Res.* **78(2)**, 297–9.

Callaerts, P., Halder, G. and Gehring, W. J. (1997). *PAX-6* in development and evolution. *Annu. Rev. Neurosci.* **20**, 483–532.

Cammarata, P. R., Cantu-Crouch, D., Oakford, L. and Morrill, A. (1986). Macromolecular organization of bovine lens capsule. *Tissue Cell* **18**, 83–97.

Cammarata, P. R. and Chen, H. Q. (1994). Osmoregulatory alterations in myo-inositol uptake by bovine lens epithelial cells: a hypertonicity-induced protein enhances myo-inositol transport. *Invest. Ophthalmol. Vis. Sci.* **35**, 1223–35.

Cammarata, P. R. and Spiro, R. G. (1985). Identification of noncollagenous components of calf lens capsule: evaluation of their adhesion-promoting activity. *J. Cell Physiol.* **125**, 393–402.

Cammarata, P. R., Zhou, C., Chen, G., Singh, I., Reeves, R. E., Kuszak, J. R. and Robinson, M. L. (1999). A transgenic animal model of osmotic cataract Part I: over-expression of bovine Na^+/myo-inositol cotransporter in lens fibers. *Invest. Ophthalmol. Vis. Sci.* **40**, 1727–37.

Campbell, M. T. and McAvoy, J. W. (1984). Onset of fibre differentiation in cultured rat lens epithelium under the influence of neural retina-conditioned medium. *Exp. Eye Res.* **39**, 83–94.

Cao, L., Faha, B., Denbski, M., Tsai, L.-H., Harlow, E. and Dyson, N. (1992). Independent binding of the retinoblastoma protein and p107 to the transcription factor E2F. *Nature* **355**, 176–9.

Capdevila, J., Tsukui, T., Rodriquez Esteban, C., Zappavigna, V. and Izpisua Belmonte, J. C. (1999). Control of vertebrate limb outgrowth by the proximal factor Meis2 and distal antagonism of BMPs by Gremlin. *Mol. Cell* **4**, 839–49.

Capetanaki, Y., Smith, S. and Heath, J. P. (1989). Overexpression of the vimentin gene in transgenic mice inhibits normal lens cell differentiation. *J. Cell Biol.* **109**, 1653–64.

Carosa, E., Kozmik, Z., Horwitz, J., Robison, G. and Piatigorsky, J. (2000). Aldehyde dehydrogenase/O-crystallin of the scallop lens. *Invest. Ophthalmol. Vis. Sci.* **41** (Suppl), S585.

Carosa, E., Kozmik, Z., Rall, J. E. and Piatigorsky, J. (2002). Structure and expression of the scallop Omega-crystallin gene. Evidence for convergent evolution of promoter sequences. *J. Biol. Chem.* **277**, 656–64.

Carper, D., Smith-Gill, S. J. and Kinoshita, J. H. (1986). Immunocytochemical localization of the 27Kbeta-crystallin polypeptide in the mouse lens during development using a specific monoclonal antibody: implications for cataract formation in the Philly mouse. *Dev. Biol.* **113**, 104–9.

Carter, J. M., Hutcheson, A. M. and Quinlan, R. A. (1995). In vitro studies on the assembly properties of the lens beaded filament proteins: co-assembly with α-crystallin but not with vimentin. *Exp. Eye Res.* **60**, 181–92.

Cartier, M., Breitman, M. L. and Tsui, L.-C. (1992). A frameshift mutation in the γE-crystallin gene of the Elo mouse. *Nat. Genet.* **2**, 42–5.

Caruelle, D., Groux-Muscatelli, B., Gaudric, A., Sestier, C., Coscas, G., Caruelle, J. P. and Barritault, D. (1989). Immunological study of acidic fibroblast growth factor (aFGF) distribution in the eye. *J. Cell Biochem.* **39**, 117–28.

Cavenee, W. K., Dryja, T. P., Philips, R. A., Benedict, W. F., Godbout, R., Gallie, B. L., Murphee, A. L., Strong, L. C. and White, R. L. (1983). Expression of recessive alleles by chromsomal mechanism in retinoblastoma. *Nature* **305**, 779–84.

Center, E. M. and Polizotto, R. S. (1992). Etiology of the developing eye in myelencephalic blebs (my) mice. *Histol. Histopathol.* **7**, 231–6.

Chalcroft, J. P. and Bullivant, S. (1970). An interpretation of liver cell membrane and junction structure based on observation of freeze-fracture replicas of both sides of the fracture. *J. Cell Biol.* **47**, 49–60.

Chamberlain, C. G. and McAvoy, J. W. (1989). Induction of lens fibre differentiation by acidic and basic fibroblast growth factor (FGF). *Growth Factors.* **1**, 125–34.

Chamberlain, C. G. and McAvoy, J. W. (1997). Fiber differentiation and polarity in the mammalian lens: a key role for FGF. *Prog. Ret. Eye Res.* **16**, 443–78.

Chamberlain, C. G., McAvoy, J. W. and Richardson, N. A. (1991). The effects of insulin and basic fibroblast growth factor on fiber differentiation in rat lens epithelial explants. *Growth Factors* **4**, 183–8.

Chambers, C., Cvekl, A., Sax, C. M. and Russell, P. (1995). Sequence, initial functional analysis and protein-DNA binding sites of the mouse beta B2-crystallin–encoding gene. *Gene* **166**, 287–92.

Chambers, C. and Russell, P. (1991). Deletion mutation in an eye lens β-crystallin: an animal model for inherited cataracts. *J. Biol. Chem.* **266**, 6742–6.

Chandy, G., Zampighi, G. A., Kreman, M. and Hall, J. E. (1997). Comparison of the water transporting properties of MIP and AQP1. *J. Membr. Biol.* **159**, 29–39.

Chang, T., Lin, C. L., Chen, P. H. and Chang, W. C. (1991). Gamma-crystallin genes in carp: cloning and characterization. *Biochim. Biophys. Acta* **1090**, 261–4.

Chazaud, C., Oulad-Abdelghani, M., Bouillet, P., Decimo, D., Chambon, P. and Dolle, P. (1996). AP-2.2, a novel gene related to AP-2, is expressed in the forebrain, limbs and face during mouse embryogenesis. *Mech. Dev.* **54**, 83–94.

Chellappan, S. P., Hiebert, S., Mudryj, M., Horowitz, J. M. and Nevins, J. R. (1991). The E2F transcription factor is a cellular target for the RB protein. *Cell* **65**, 1053–61.

Chen, L. (1999). Combinatorial gene regulation by eukaryotic transcription factors. *Curr. Opin. Struct. Biol.* **9**, 48–55.

Chen, P. L., Riley, D. J., Chen, Y. and Lee, W. H. (1996). Retinoblastoma protein positively regulates terminal adipocyte differentiation through direct interaction with C/EBPs. *Genes Dev.* **10**, 2794–804.

Chen, Q., Dowhan, D. H., Liang, D., Moore, D. D. and Overbeek, P. A. (2002). CREB-binding protein/p300 co-activation of crystallin gene expression. *J. Biol. Chem.* **277**, 24081–9.

Chen, Q., Hung, F.-C., Fromm, L. and Overbeek, P. A. (2000). Induction of cell cycle entry and cell death in postmitotic lens fiber cells by overexpression of E2F1 or E2F2. *Invest. Ophthalmol. Vis. Sci.* **41**, 4223–31.

Chen, R., Amoui, M., Zhang, Z. H. and Mardon, G. (1997a). Dachshund and eyes absent proteins form a complex and function synergistically to induce ectopic eye development in *Drosophila*. *Cell* **91**, 893–903.

Chen, R., Halder, G., Zhang, Z. and Mardon, G. (1999). Signaling by the TGF-beta homolog decapentaplegic functions reiteratively within the network of genes controlling retinal cell fate determination in *Drosophila*. *Development* **126**, 935–43.

Chen, S. M., Wang, Q. L., Nie, Z. Q., Sun, H., Lennon, G., Copeland, N. G., Gilbert, D. J., Jenkins, N. A. and Zack, D. J. (1997b). Crx, a novel Otx-like paired-homeodomain protein, binds to and transactivates photoreceptor cell–specific genes. *Neuron* **19**, 1017–30.

Chen, W. V., Fielding Hejtmancik, J., Piatigorsky, J. and Duncan, M. K. (2001). Functional conservation of the βB1-crystallin promoter during evolution. *Biochim. Biophys. Acta* **1519**, 30–8.

Cheng, H. L. and Louis, C. F. (1999). Endogenous casein kinase I catalyzes the phosphorylation of the lens fiber cell connexin49. *Eur. J. Biochem.* **263**, 276–86.

Chepelinsky, A. B., King, C. R., Zelenka, P. S. and Piatigorsky, J. (1985). Lens-specific expression of the chloramphenicol acetyltransferase gene promoted by 5′ flanking sequences of the murine alpha A-crystallin gene in explanted chicken lens epithelia. *Proc. Natl. Acad. Sci. USA* **82**, 2334–8.

Cheyette, B. N., Green, P. J., Martin, K., Garren, H., Hartenstein, V. and Zipursky, S. L. (1994). The *Drosophila sine oculis* locus encodes a homeodomain-containing protein required for the development of the entire visual system. *Neuron* **12**, 977–96.

Chiesa, R., Gawinowicz-Kolks, M. A. and Spector, A. (1987). The phosphorylation of the primary gene products of alpha-crystallin. *J. Biol. Chem.* **262**, 1438–41.

Chiou, S.-H. (1984). Physicochemical characterization of a crystallin from the squid lens and its comparison with vertebrate lens crystallins. *J. Biochem.* **95**, 75–82.

Chiou, S.-H. (1988). A novel crystallin from octopus lens. *FEBS Lett.* **241**, 261–4.

Chiou, S.-H. and Bunn, H. F. (1981). Characterization of a new crystallin from the squid lens and its biochemical comparison with the bovine ß-crystallin. *Invest. Ophthalmol. Vis. Sci.* **20** (Suppl), 138.

Chiou, S. H., Chang, W. P., Lo, C. H. and Chen, S. W. (1987). Sequence comparison of gamma-crystallins from the reptilian and other vertebrate species. *FEBS Lett.* **221**, 134–8.

Chiou, S.-H., Yu, C.-W., Lin, C.-W., Pan, F.-M., Lu, S.-F., Lee, H.-J. and Chang, G.-G. (1995). Octopus S-crystallins with endogenous glutathione S-transferase (GST) activity: sequence comparison and evolutionary relationships with authentic GST enzymes. *Biochem. J.* **309**, 793–800.

Chittenden, T., Livingston, D. M. and Kaelin, W. J. (1991). The T/E1A-binding domain of the retinoblastoma product can interact selectively with a sequence-specific DNA-binding protein. *Cell* **65**, 1073–82.

Chow, R. L., Altmann, C. R., Lang, R. A. and Hemmati-Brivanlou, A. (1999). Pax6 induces ectopic eyes in a vertebrate. *Development* **126**, 4213–22.

Chow, R. L., Roux, G. D., Roghani, M., Palmer, M. A., Rifkin, D. B., Moscatelli, D. A. and Lang, R. A. (1995). FGF suppresses apoptosis and induces differentiation of fiber cells in the mouse lens. *Development* **121**, 4383–93.

Chuang, M.-H., Pan, F.-M. and Chiou, S.-H. (1997). Sequence characterization of γ-crystallins from lip shark (*Chiloscyllium colax*): existence of two cDNAs encoding γ-crystallins of mammalian and teleostean classes. *J. Protein Chem.* **16**, 299–307.

Church, R. L. and Wang, J. H. (1993). The human lens fiber-cell intrinsic membrane protein MP19 gene: isolation and sequence analysis. *Curr. Eye Res.* **12**, 1057–65.

Church, R. L., Wang, J. H. and Steele, E. (1995). The human lens intrinsic membrane protein MP70 (Cx50) gene: clonal analysis and chromosome mapping. *Curr. Eye Res.* **14**, 215–21.

Churchill, G. C. and Louis, C. F. (1997). Stimulation of P2U purinergic or α1A adrenergic receptors mobilizes Ca^{2+} in lens cells. *Invest. Ophthalmol. Vis. Sci.* **38**, 855–65.

Ciechanover, A., Orian, A. and Schwartz, A. L. (2000). Ubiquitin-mediated proteolysis: biological regulation via destruction. *Bioessays* **22**, 442–51.

Clarke, A. R., Maandag, E. R., van Roon, M., van der Lugt, N. M., van der Valk, M., Hooper, M. L., Berns, A. and te Riele, H. (1992). Requirement for a functional Rb-1 gene in murine development. *Nature* **359**, 328–30.

Claudio, J. O., Veneziale, R. W., Menko, A. S. and Rouleau, G. A. (1997). Expression of schwannomin in lens and Schwann cells. *Neuroreport* **8**, 2025–30.

Clubb, B. H., Chou, Y. H., Herrmann, H., Svitkina, T. M., Borisy, G. G. and Goldman, R. D. (2000). The 300-kDa intermediate filament-associated protein (IFAP300) is a hamster plectin ortholog. *Biochem. Biophys. Res. Commun.* **273**, 183–7.

Cobb, B. A. and Petrash, J. M. (2000). Characterization of alpha-crystallin–plasma membrane binding. *J. Biol. Chem.* **275**, 6664–72.

Cobrinik, D., Lee, M., Hannon, G., Mulligan, G., Bronson, R., Dyson, N., Harlow, E., Beach, D., Weinberg, R. and Jacks, T. (1996). Shared role of the pRB-related p130 and p107 proteins in limb development. *Genes Dev.* **10**, 1633–44.

Coleman, T. R. and Dunphy, W. G. (1994). Cdc2 regulatory factors. *Curr. Opin. Cell Biol.* **6**, 877–82.

Collignon, J., Sockanathan, S., Hacker, A., Cohen-Tannoudji, M., Norris, D., Rasta, S., Stevanovich, M., Goodfellow, P. N. and Lovell-Badge, R. (1996). A comparison of the properties of *Sox-3* with *Sry* and two related genes, *Sox-1* and *Sox-2*. *Development* **122**, 509–20.

Collins, J. M. (1972). Amplification of ribosomal ribonucleic acid cistrons in the regenerating lens of *Triturus*. *Biochemistry* **11**, 1259–63.

Collinson, J. M., Hill, R. E. and West, J. D. (2000). Different roles for *Pax6* in the optic vesicle and facial epithelium mediate early morphogenesis of the murine eye. *Development* **127**, 945–56.

Collinson, J. M., Quinn, J. C., Buchanan, M. A., Kaufman, M. H., Wedden, S. E., West, J. D. and Hill, R. E. (2001). Primary defects in the lens underlie complex anterior segment abnormalities of the Pax6 heterozygous eye. *Proc. Natl. Acad. Sci. USA* **98**, 9688–93.

Colluci, V. L. (1891). Sulla rigenerazione parziale dell'occhio nei Tritoni-Istogenesi e sviluppo: Studio sperimentale. *Mem. R. Acad. Sci. Ist. Bologna, Ser.* **51**, 593–629.

Colucci-Guyon, E., Portier, M.-M., Dunia, I., Paulin, D., Pournin, S. and Babinet, C. (1994). Mice lacking vimentin develop and reproduce without an obvious phenotype. *Cell* **79**, 679–94.

Colville, D. J. and Savige, J. (1997). Alport syndrome: a review of the ocular manifestations. *Ophthalmic Genet.* **18**, 161–73.

Conley, Y. P., Erturk, D., Keverline, A., Mah, T. S., Keravala, A., Barnes, L. R., Bruchis, A., Hess, J. F., FitzGerald, P. G., Weeks, D. E. et al. (2000). A juvenile-onset, progressive cataract locus on chromosome 3q21–q22 is associated with a missense mutation in the beaded filament structural protein-2. *Am. J. Hum. Genet.* **66**, 1426–31.

Connor, F., Cary, P. D., Read, C. M., Preston, N. S., Driscoll, P. C., Denny, P., Crane-Robinson, C. and Ashworth, A. (1994). DNA binding and bending properties of the post-meiotically expressed Sry-related protein Sox-5. *Nucleic Acids Res.* **22**, 3339–46.

Cook, C. S. and Sulik, K. K. (1988). Keratolenticular dysgenesis (Peters' anomaly) as a result of acute embryonic insult during gastrulation. *J. Pediatr. Ophthalmol. Strabismus* **25**, 60–6.

Coop, A., Wiesmann, K. E. and Crabbe, M. J. (1998). Translocation of β-crystallin in neural cells in response to stress. *FEBS Lett.* **431**, 319–21.

Cooper, K., Gates, P., Rae, J. L. and Dewey, J. (1990). Electrophysiology of cultured human lens epithelial cells. *J. Membr. Biol.* **117**, 285–98.

Cooper, K., Rae, J. L. and Dewey, J. (1991). Inwardly rectifying potassium current in mammalian lens epithelial cells. *Am. J. Physiol.* **261**, C115–23.

Cooper, K., Watsky, M. and Rae, J. (1992). Potassium currents from isolated frog lens epithelial cells. *Exp. Eye Res.* **55**, 861–8.

Cooper, K. E., Tang, J. M., Rae, J. L. and Eisenberg, R. S. (1986). A cation channel in frog lens epithelia responsive to pressure and calcium. *J. Membr. Biol.* **93**, 259–69.

Cordes, S. P. and Barsh, G. S. (1994). The mouse segmentation gene *kr* encodes a novel basic domain–leucine zipper transcription factor. *Cell* **79**, 1025–34.

Cornell, R. A., Musci, T. J. and Kimelman, D. (1995). FGF is a prospective competence factor for early activin-type signals in *Xenopus* mesoderm induction. *Development* **121**, 2429–37.

Correia, I., Chu, D., Chou, Y. H., Goldman, R. D. and Matsudaira, P. (1999). Integrating the actin and vimentin cytoskeletons: adhesion-dependent formation of fimbrin-vimentin complexes in macrophages. *J. Cell Biol.* **146**, 831–42.

Costello, M. J., Al-Ghoul, K. J., Oliver, T. N., Lane, C. W., Wodnicka, M. and Wodnicki, P. (1993). Polymorphism of fiber cell junctions in mammalian lens. In *Proceedings of the 51st Annual Meeting of the Microscopy Society of America*, ed. G. W. Bailey. San Francisco: San Francisco Press, pp. 200–1.

Costello, M. J., McIntosh, T. J. and Robertson, J. D. (1985). Membrane specializations in mammalian lens fiber cells: distribution of square arrays. *Curr. Eye Res.* **4**, 1183–201.

Costello, M. J., McIntosh, T. J. and Robertson, J. D. (1989). Distribution of gap junctions and square array junctions in the mammalian lens. *Invest. Ophthalmol. Vis. Sci.* **30**, 975–89.

Coucouvanis, E. and Martin, G. R. (1995). Signals for death and survival: a two-step mechanism for cavitation in the vertebrate embryo. *Cell* **83**, 279–87.

Coulombre, J. L. and Coulombre, A. J. (1963). Lens development: Fiber elongation and lens orientation. *Science* **142**, 1489–90.

Coulombre, J. L. and Coulombre, A. J. (1969). Lens development. IV. Size, shape, and orientation. *Invest. Ophthalmol.* **8**, 251–7.

Coulombe, P. A., Bousquet, O., Ma, L., Yamada, S. and Wirtz, D. (2000). The "ins" and "outs" of intermediate filament organization. *Trends Cell Biol.* **10**, 420–8.

Coulombe, P. A., Hutton, M. E., Letai, A., Hebert, A., Paller, A. S. and Fuchs, E. (1991). Point mutations in human keratin 14 genes of epidermolysis bullosa simplex patients: genetic and functional analyses. *Cell* **66**, 1301–11.

Counis, M. F., Chaudun, E., Arruti, C., Oliver, L., Sanwal, M., Courtois, Y. and Torriglia, A. (1998). Analysis of nuclear degradation during lens cell differentiation. *Cell Death Diff.* **5**, 251–61.

Counis, M. F., Chaudun, E., Courtois, Y. and Allinquant, B. (1991). DNAase activities in embryonic chicken lens: in epithelial cells or in differentiating fibers where chromatin is progressively cleaved. *Biol. Cell* **72**, 231–8.

Counis, M. F., Chaudun, E., Courtois, Y. and Skidmore, C. J. (1985). Nuclear ADP-ribosylation in the chick lens during embryonic development. *Biochem. Biophys. Res. Commun.* **126**, 859–66.

Counis, M. F. and Torriglia, A. (2000). DNases and apoptosis. *Biochem. Cell. Biol.* **78**, 405–14.

Cowan, N. J. and Milstein, C. (1974). Stability of cytoplasmic ribonucleic acid in a mouse myeloma: estimation of the half-life of the messenger RNA coding for an immunoglobulin light chain. *J. Mol. Biol.* **82**, 469–81.

Cui, W., Tomarev, S. I., Piatigorsky, J., Chepelinsky, A. B. and Duncan, M. K. (2004). Mafs, Prox1 and Pax6 can regulate chicken beta B1-cystallin gene expression. *J. Biol. Chem.* (in press).

Curran, K. L. and Grainger, R. M. (2000). Expression of activated MAP kinase in *Xenopus laevis* embryos: evaluating the roles of FGF and other signaling pathways in early induction and patterning. *Dev. Biol.* **228**, 41–56.

Cuthbertson, R. A., Tomarev, S. I. and Piatigorsky, J. (1992). Taxon-specific recruitment of enzymes as major soluble proteins in the corneal epithelium of three mammals, chicken, and squid. *Proc. Natl. Acad. Sci. USA* **89**, 4004–8.

Cvekl, A., Kashanchi, F., Brady, J. N. and Piatigorsky, J. (1999). Pax-6 interactions with TATA-box–binding protein and retinoblastoma protein. *Invest. Opthal. Vis. Sci.* **10**, 1343–50.

Cvekl, A., Kashanchi, F., Sax, C. M., Brady, J. N. and Piatigorsky, J. (1995a). Transcriptional regulation of the mouse αA-crystallin gene: activation dependent on a cyclic AMP-responsive element (DE1/CRE) and a Pax-6–binding site. *Mol. Cell. Biol.* **15**, 653–60.

Cvekl, A. and Piatigorsky, J. (1996). Lens development and crystallin gene expression: many roles for Pax-6. *Bioessays* **18**, 621–30.

Cvekl, A., Sax, C. M., Bresnick, E. H. and Piatigorsky, J. (1994). A complex array of positive and negative elements regulates the chicken αA-crystallin gene: involvement of Pax-6, USF, CREB and/or CREM, and AP-1 proteins. *Mol. Cell. Biol.* **14**, 7363–76.

Cvekl, A., Sax, C. M., Li, X., McDermott, J. B. and Piatigorsky, J. (1995b). Pax-6 and lens-specific transcription of the chicken δ1-crystallin gene. *Proc. Natl. Acad. Sci. USA* **92**, 4681–5.

Czerny, T. and Busslinger, M. (1995). DNA-binding and transactivation properties of Pax-6: three amino acids in the paired domain are responsible for the different sequence recognition of Pax-6 and BSAP (Pax-5). *Mol. Cell. Biol.* **15**, 2858–71.

Czerny, T., Halder, G., Kloter, U., Souabni, A., Gehring, W. J. and Busslinger, M. (1999). *Twin of eyeless*, a second *Pax-6* gene of *Drosophila*, acts upstream of *eyeless* in the control of eye development. *Mol. Cell* **3**, 297–307.

Dahl, E., Koseki, H. and Balling, R. (1997). *Pax* genes and organogenesis. *Bioessays* **19**, 755–65.

Dahm, R. (1999). Lens fibre cell differentiation: a link with apoptosis? *Ophthalmic Res.* **31**, 163–83.

Dahm, R., Gribbon, C., Quinlan, R. A. and Prescott, A. R. (1998a). Changes in the nucleolar and coiled body compartments precede lamina and chromatin reorganization during fibre cell denucleation in the bovine lens. *Eur. J. Cell Biol.* **75**, 237–46.

Dahm, R., Gribbon, C., Quinlan, R. A. and Prescott, A. R. (1998b). Susceptibility of lens epithelial and fibre cells at different stages of differentiation to apoptosis. *Biochem. Soc. Trans.* **26**, S349.

Danchakoff, V. (1926). Lens ectoderm and optic vesicles in allantois grafts. *Carnegie Contrib. Embryol.* **18**, 63–78.

Dannenberg, J.-H., van Rossum, A., Schuijff, L. and te Riele, H. (2000). Ablation of the retinoblastoma gene family deregulates G1 control causing immortalization and increased cell turnover under growth-restricting conditions. *Genes Dev.* **14**, 3051–64.

David, L. L., Azuma, M. and Shearer, T. R. (1994). Cataract and the acceleration of calpain-induced β-crystallin insolubilization occuring during normal maturation of rat lens. *Invest. Ophthalmol. Vis. Sci.* **35**, 785–93.

David, L. L., Shearer, T. R. and Shih, M. (1993). Sequence analysis of lens β-crystallins suggests involvement of calpain in cataract formation. *J. Biol. Chem.* **268**, 1937–40.

Davis, R. J., Shen, W., Sandler, Y. I., Amoui, M., Purcell, P., Maas, R., Ou, C. N., Vogel, H., Beaudet, A. L. and Mardon, G. (2001). Dach1 mutant mice bear no gross abnormalities in eye, limb, and brain development and exhibit postnatal lethality. *Mol. Cell. Biol.* **21**, 1484–90.

De Arcangelis, A., Mark, M., Kreidberg, J., Sorokin, L. and Georges-Labouesse, E. (1999). Synergistic activities of alpha3 and alpha6 integrins are required during apical ectodermal ridge formation and organogenesis in the mouse. *Development* **126**, 3957–68.

DeCaprio, J. A., Ludlow, J. W., Figge, J., Shew, J.-Y., Huang, C.-M., Lee, W.-H., Marsilio, E., Paucha, E. and Livingston, D. M. (1988). SV40 large tumor antigen forms a specific complex with the product of the retinoblastoma susceptibility gene. *Cell* **54**, 275–83.

DeGregori, J., Leone, G., Miron, A., Jakoi, L. and Nevins, J. R. (1997). Distinct roles for E2F proteins in cell growth control and apoptosis. *Proc. Natl. Acad. Sci. USA* **94**, 7245–50.

de Iongh, R. U., Gordon-Thomson, C., Chamberlain, C. G., Hales, A. M. and McAvoy, J. W. (2001a). Tgfbeta receptor expression in lens: implications for differentiation and cataractogenesis. *Exp. Eye Res.* **72**, 649–59.

de Iongh, R. U., Lovicu, F. J., Chamberlain, C. G. and McAvoy, J. W. (1997). Differential expression of fibroblast growth factor receptors during rat lens morphogenesis and growth. *Invest. Ophthalmol. Vis. Sci.* **38**, 1688–99.

de Iongh, R. U., Lovicu, F. J., Overbeek, P. A., Schneider, M. D., Joya, J., Hardeman, E. D. and McAvoy, J. W. (2001b). Requirement for TGFbeta receptor signaling during terminal lens fiber differentiation. *Development* **128**, 3995–4010.

de Iongh, R. and McAvoy, J. W. (1992). Distribution of acidic and basic fibroblast growth factors (FGF) in the foetal rat eye: implications for lens development. *Growth Factors* **6**, 159–77.

de Iongh, R. and McAvoy, J. W. (1993). Spatio-temporal distribution of acidic and basic FGF indicates a role for FGF in rat lens morphogenesis. *Dev. Dyn.* **198**, 190–202.

de Jong, W. W., Leunissen, J. A. and Voorter, C. E. (1993). Evolution of the alpha-crystallin/small heat-shock protein family. *Mol. Biol. Evol.* **10**, 103–26.

de Jong, W. W., Lubsen, N. H. and Kraft, H. J. (1994). Molecular evolution of the eye lens. *Prog. Ret. Eye Res.* **13**, 391–442.

Delamere, N. A. and Dean, W. L. (1993). Distribution of lens sodium-potassium-adenosine triphosphatase. *Invest. Ophthalmol. Vis. Sci.* **34**, 2159–63.

Delcour, J. and Papaconstantinou, J. (1974). A change in the stoichiometry of assembly of bovine lens alpha-crystallin subunits in relation to cellular differentiation. *Biochem. Biophys. Res. Commun.* **57**, 134–41.

Delmar, M., Stergiopoulous, K., Homma, N., Calero, G., Morley, G., Ek-Vitorin, J. F. and Taffet, S. M. (2000). A molecular model for the chemical regulation of connexin43 channels, the "ball-and-chain" hypothesis *Curr. Top. Memb. Trans.* **49**, 223–69.

Del Rio-Tsonis, K., Jung, J. C., Chiu, I. M. and Tsonis, P. A. (1997). Conservation of fibroblast growth factor function in lens regeneration. *Proc. Natl. Acad. Sci. USA* **94**, 13701–6.

Del Rio-Tsonis, K., Tomarev, S. I. and Tsonis, P. A. (1999). Regulation of Prox-1 during lens regeneration. *Invest. Ophthalmol. Vis. Sci.* **40**, 2039–45.

Del Rio-Tsonis, K., Trombley, M. T., McMahon, G. and Tsonis, P. A. (1998). Regulation of lens regeneration by fibroblast growth factor receptor 1. *Dev. Dyn.* **213**, 140–6.

Del Rio-Tsonis, K., Washabaugh, C. H. and Tsonis, P. A. (1995). Expression of Pax-6 during urodele eye development and lens regeneration. *Proc. Natl. Acad. Sci. USA* **92**, 5092–6.

Del Vecchio, P. J., MacElroy, K. S., Rosser, M. P. and Church, R. L. (1984). Association of alpha-crystallin with actin in cultured lens cells. *Curr. Eye Res.* **3**, 1213–9.

Delwel, G. O., Hogervorst, F. and Sonnenborg, A. (1996). Cleavage of the alpha6 subunit is essential for activation of the alpha6Abeta1 integrin by phorbol 12-myristate 13-acetate. *J. Biol. Chem.* **271**, 7293–6.

Delwel, G. O., Kuikman, I., van der Schors, R. C., de Melker, A. A. and Sonnenberg, A. (1997). Identification of the cleavage sites in the alpha6A integrin subunit: structural requirements for cleavage and functional analysis of the uncleaved alpha6Abeta1 integrin. *Biochem. J.* **324**, 263–72.

Deng, P., Maddala, R. and Rao, P. (2001). Expression and distribution of Rho and Rac GTPases and their effector proteins in lens tissue. *Invest. Ophthalmol. Vis. Sci.* **42**, S289.

Deretic, D., Aebersold, R. H., Morrison, H. D. and Papermaster, D. S. (1994). Alpha A- and alpha B-crystallin in the retina: association with the post-Golgi compartment of frog retinal photoreceptors. *J. Biol. Chem.* **269**, 16853–61.

Desplan, C. (1997). Eye development: governed by a dictator or a junta? *Cell* **91**, 861–4.

Devoto, S. H., Mudryj, M., Pines, J., Hunter, T. and Nevins, J. R. (1992). A cyclin A-protein kinase complex possesses sequence-specific Dna binding activity: p33 cdk2 is a component of the E2F-cyclin A complex. *Cell* **68**, 167–76.

Dickeson, S. K., Mathis, N. L., Rahman, M., Bergelson, J. M. and Santoro, S. A. (1999). Determinants of ligand binding specificity of the alpha(1)beta(1) and alpha(2)beta(1) integrins. *J. Biol. Chem.* **274**, 32182–91.

Dimanlig, P. V., Faber, S. C., Auerbach, W., Makarenkova, H. P. and Lang, R. A. (2001). The upstream ectoderm enhancer in Pax6 has an important role in lens induction. *Development* **128**, 4415–24.

Dinnean, F. L. (1942). Lens regeneration from the dorsal iris and its inhibition by lens reimplantation in *Triturus torosus* larvae. *J. Exp. Zool.* **90**, 461–78.

DiPersio, C. M., Hodivala-Dilke, K. M., Jaenisch, R., Kreidberg, J. A. and Hynes, R. O. (1997). Alpha3beta1 integrin is required for normal development of the epidermal basement membrane. *J. Cell Biol.* **137**, 729–42.

Dirks, R. P. H., Kraft, H. J., Genessen, S. T. V., Klok, E. J., Pfundt, R., Schoenmakers, J. G. G. and Lubsen, N. H. (1996). The cooperation between two silencers creates an enhancer element that controls both the lens-preferred and the differentiation stage–specific expression of the rat beta B2-crystallin gene. *Eur. J. Biochem.* **239**, 23–32.

Dirksen, M.-L., Morasso, M. I., Sargent, T. D. and Jamrich, M. (1994). Differential expression of a *Distal-less* homeobox gene *Xdll-2* in ectodermal cell lineages. *Mech. Dev.* **46**, 63–70.

Doerwald, L., Nijveen, H., Civil, A., van Genesen, S. T. and Lubsen, N. H. (2001). Regulatory elements in the rat betaB2-crystallin promoter. *Exp. Eye Res.* **73**, 703–10.

Dohrmann, C., Gruss, P. and Lemaire, L. (2000). Pax genes and the differentiation of hormone-producing endocrine cells in the pancreas. *Mech. Dev.* **92**, 47–54.

Donaldson, P. J., Dong, Y., Roos, M., Green, C., Goodenough, D. A. and Kistler, J. (1995). Changes in lens connexin expression lead to increased gap junctional voltage dependence and conductance. *Am. J. Physiol.* **269**, C590–600.

Donaldson, P. J., Roos, M., Evans, C., Beyer, E. and Kistler, J. (1994). Electrical properties of mammalian lens epithelial gap junction channels. *Invest. Ophthalmol. Vis. Sci.* **35**, 3422–8.

Dong, L. J. and Chung, A. E. (1991). The expression of the genes for entactin, laminin A, laminin B1 and laminin B2 in murine lens morphogenesis and eye development. *Differentiation* **48**, 157–72.

Draetta, G. F. (1994). Mammalian G1 cyclins. *Curr. Opin. Cell Biol.* **6**, 842–6.

Driessen, H. P. C., Herbrink, P., Bloemendal, H. and DeJong, W. W. (1981). Primary structure of β-crystallin Bp chain: internal duplication and homology with γ-crystallin. *Eur. J. Biochem.* **121**, 83–91.

Dryja, T. P., Cavenee, W., White, R., Rapaport, J. M., Peterson, R., Albert, D. M. and Bruns, G. A. P. (1984). Homozygosity of chromosome 13 in retinoblastoma. *New Eng. J. Med.* **310**, 550–3.

Dubin, R. A., Gopal-Srivastava, R., Wawrousek, E. F. and Piatigorsky, J. (1991). Expression of the murine alpha B-crystallin gene in lens and skeletal muscle: identification of a muscle-preferred enhancer. *Mol. Cell. Biol.* **11**, 4340–9.

Dubin, R. A., Wawrousek, E. F. and Piatigorsky, J. (1989). Expression of the murine alpha B-crystallin gene is not restricted to the lens. *Mol. Cell. Biol.* **9**, 1083–91.

Dudley, A. T., Lyons, K. M. and Robertson, E. J. (1995). A requirement for bone morphogenetic protein-7 during development of the mammalian kidney and eye. *Genes Dev.* **9**, 2795–807.

Duke-Elder, S. (1958). System of Ophthalmology, vol. 1: *The eye in evolution*, ed. S. Duke-Elder. London: Kimpton.

Duke-Elder, S. and Cook, C. (1963). Normal and abnormal development. In *System of Ophthalmology*, vol. 3. St. Louis: C. V. Mosby, pp. 309–24.

Dulic, V., Lees, E. and Reed, S. I. (1992). Association of human cyclin E with a periodic G1-S phase protein kinase. *Science* **257**, 1958–61.

Dumont, J. N. and Yamada, T. (1972). Dedifferentitation of iris epithelial cells. *Dev. Biol.* **29**, 385–401.

Dumont, J. N., Yamada, T. and Cone, M. V. (1970). Alteration of nucleolar ultrastructure in iris epithelial cells during initiation of Wolffian lens regeneration. *J. Exp. Zool.* **174**, 187–204.

Duncan, G. (1969). The site of the ion restricting membranes in the toad lens. *Exp. Eye Res.* **8**, 406–12.

Duncan, G., Williams, M. R. and Riach, R. A. (1994). Calcium, cell signalling and cataract. *Prog. Ret. Eye Res.* **13**, 623–52.

Duncan, G. and Wormstone, I. M. (1999). Calcium cell signalling and cataract, role of the endoplasmic reticulum. *Eye* **13**, 480–3.

Duncan, M. K., Banerjee-Basu, S., McDermott, J. B. and Piatigorsky, J. (1996a). Sequence and expression of chicken βA2- and βB3-crystallins. *Exp. Eye Res.* **62**, 111–19.

Duncan, M. K., Cui, W., Oh, D.-J. and Tomarav, S. I. (2002). Prox1 is differentially localized during lens development. *Mech. Dev.* **112**, 195–8.

Duncan, M. K., Cvekl, A., Li, X. and Piatigorsky, J. (2000). Truncated forms of Pax-6 disrupt lens morphology in transgenic mice. *Invest. Ophthalmol. Vis. Sci.* **41**, 464–73.

Duncan, M. K., Haynes, J. I., II, Cvekl, A. and Piatigorsky, J. (1998). Dual roles for Pax-6: a transcriptional repressor of lens fiber cell–specific β-crystallin genes. *Mol. Cell. Biol.* **18**, 5579–86.

Duncan, M. K., Haynes, J. I., II and Piatigorsky, J. (1995). The chicken βA4- and βB1-crystallin–encoding genes are tightly linked. *Gene* **162**, 189–96.

Duncan, M. K., Kozmik, Z., Cveklova, K., Piatigorsky, J. and Cvekl, A. (2000). Overexpression of Pax-6 (5a) in lens fiber cells results in cataract and upregulation of $\alpha5\beta1$ integrin expression. *J. Cell Sci.* **113**, 3173–85.

Duncan, M. K., Li, X., Ogino, H., Yasuda, K. and Piatigorsky, J. (1996b). Developmental regulation of the chicken βB1-crystallin promoter in transgenic mice. *Mech. Devel.* **57**, 79–89.
Dunia, I., Pieper, F., Manenti, S., van der Kemp, A., Devilliers, G., Benedetti, E. L. and Bloemendal, H. (1990). Plasma membrane-cytoskeleton damage in eye lenses of transgenic mice expressing desmin. *Eur. J. Cell Biol.* **53**, 59–74.
Dunia, I., Smit, J. J. M., van der Valk, M. A., Bloemendal, H., Borst, P. and Benedetti, E. L. (1996). Human multidrug resistance 3-P-glycoprotein expression in transgenic mice induces lens membrane alterations leading to cataract. *J. Cell Biol.* **132**, 701–16.
Dynlacht, B., Flores, O., Lees, J. and Harlow, E. (1994). Differential regulation of E2F transactivation by cyclin/cdk2 complexes. *Genes Dev.* **8**, 1772–86.
Dyson, N. (1998). The regulation of E2F by pRB-family proteins. *Genes Dev.* **12**, 2245–62.
Dyson, N., Howley, P., Munger, K. and Harlow, E. (1989). The human papilloma virus-16 E7 oncoprotein is able to bind to the retinoblastoma gene product. *Science* **243**, 934–7.
Eastman, A. (1994). Deoxyribosenuclease II in apoptosis and the significance of intracellular acidification. *Cell Death Differ.* **1**, 7–9.
Eberhard, D., Jimenez, G., Heavey, B. and Busslinger, M. (2000). Transcriptional repression by Pax5 (BSAP) through interaction with corepressors of the Groucho family. *EMBO J.* **19**, 2292–303.
Ebihara, L., Berthoud, V. M. and Beyer, E. C. (1995). Distinct behavior of connexin56 and connexin46 gap junction channels can be predicted from the behavior of their hemi-gap-junctional channels. *Biophys. J.* **68**, 1796–803.
Ebihara, L. and Steiner, E. (1993). Properties of a nonjunctional current expressed from a rat connexin46 cDNA in *Xenopus* oocytes. *J. Gen. Physiol.* **102**, 59–74.
Eckert, R., Adams, B., Kistler, J. and Donaldson, P. J. (1999). Quantitative determination of gap junctional permeability in the lens cortex. *J. Membr. Biol.* **169**, 91–102.
Eckert, R., Donaldson, P. J., Goldie, K. and Kistler, J. (1998). A distinct membrane current in rat lens fiber cells isolated under calcium-free conditions. *Invest. Ophthalmol. Vis. Sci.* **39**, 1280–5.
Eckert, R., Donaldson, P., Lin, J. S., Bond, J., Green, C., Merriman-Smith, R., Tunstall, M. and Kistler, J. (2000). Gating of gap junction channels and hemichannels in the lens: a role in cataract? *Curr. Topics Membr.* **49**, 343–56.
Ede, D. A. and Kelly, W. A. (1964). Developmental abnormalities in the head region of the *talpid*[3] mutant of the fowl. *J. Embryol. Exp. Morphol.* **12**, 161–82.
Edwards, D. D. (1996). Ophthalmology before Hippocrates. In *The History of Ophthalmology*, ed. D. M. Albert and D. D. Edwards. Cambridge, Mass.: Blackwell Science.
Eggert, T., Hauck, B., Hildebrandt, N., Gehring, W. J. and Walldorf, U. (1998). Isolation of a *Drosophila* homolog of the vertebrate homeobox gene Rx and its possible role in brain and eye development. *Proc. Natl. Acad. Sci. USA* **95**, 2343–8.
Eguchi, G. (1961). The inhibitory effect of the injured and displaced lens on the lens-formation in *Triturus* larvae. *Embryologia* **6**, 13–35.
Eguchi, G. (1963). Electron microscopic studies on lens regeneration. I. Mechanisms of depigmentation of the iris. *Embryologia* **8**, 45–62.
Eguchi, G. (1964). Electron microscopic studies on lens regeneration. II. Formation and growth of lens vesicle and differentiation of lens fibers. *Embryologia* **8**, 247–87.
Eguchi, G. (1967). *In vitro* analyses of Wolffian lens regeneration: differentiation of the regenerating lens rudiment of the newt, *Triturus pyrrhogaster*. *Embryologia* **9**, 246–66.
Eguchi, G. (1976). "Transdifferentiation" of vertebrate cells in *in vitro* cell culture. In *Embryogenesis in Mammals* (Ciba Foundation Symposium 40). Amsterdam: Elsevier, pp. 241–58.
Eguchi, G. (1979). "Transdifferentiation" in pigmented epithelial cells of the vertebrate eye *in vitro*. In *Mechanisms of Cell Change*, ed. J. D. Ebert and T. S. Okada. New York: Wiley, pp. 273–91.
Eguchi, G. (1986). Instability in cell commitment of vertebrate pigmented epithelial cells and their transdifferentiation into lens cells. *Curr. Top. Dev. Biol.* **20**, 21–37.
Eguchi, G. (1988). Cellular and molecular background of Wolffian lens regeneration. In *Regulatory Mechnisms in Developmental Process*, ed. G. Eguchi, T. S. Okada, and L. Saxen. Amsterdam: Elsevier, pp. 147–58.
Eguchi, G. (1993). Lens transdifferentiation in the vertebrate retinal pigmented epithelial cell. In *Progress in Retinal Research 12*, ed. N. N. Osborne and J. Chader. London: Pergamon, pp. 205–30.

Eguchi, G. (1998). Transdiffrentiation as the basis of eye lens regeneration. In *Cellular and Molecular Basis of Regeneration: From Invertebrates to Humans*, ed. P. Ferreti and J. Géraudie. Chichester, England: Wiley, pp. 207–28.

Eguchi, G., Abe, S. I. and Watanabe, K. (1974). Differentiation of lens-like structures from newt iris epithelial cells *in vitro*. *Proc. Natl. Acad. Sci. USA* **71**, 5052–6.

Eguchi, G. and Ishikawa, M. (1963). Alkaline phosphatase in the dorsal and ventral halves of the iris during early stages of lens regeneration in the newt. *Embryologia* **7**, 296–305.

Eguchi, G. and Itoh, Y. (1982). Regeneration of the lens as a phenomenon of cellular transdifferentiation: regulability of the differentiated state of the vertebrate pigmented epithelial cells. *Trans. Ophthalmol. Soc. UK* **102**, 374–8.

Eguchi, G. and Okada, T. S. (1973). Differentiation of lens tissue from the progeny of chick retinal pigment cells cultured *in vitro*: a demonstration of a switch of cell types in clonal cell culture. *Proc. Natl. Acad. Sci. USA*. **70**, 495–9.

Eguchi, G. and Shingai, R. (1971). Cellular analysis on localization of lens-forming potency in the newt iris epithelium. *Dev. Growth Differ.* **13**, 337–49.

Eguchi, G. and Watanabe, K. (1973). Elicitation of lens formation from the ventral iris epithelium of the newt by a carcinogen, MNNG. *J. Embryol. Exp. Morphol.* **30**, 63–71.

Ehring, G. R., Zampighi, G., Horvitz, J., Bok, D. and Hall, J. E. (1990). Properties of channels reconstituted from the major intrinsic protein of lens fiber membranes. *J. Gen. Physiol.* **96**, 631–64.

Ehrnsperger, M., Graber, S., Gaestel, M. and Buchner, J. (1997). Binding of non-native protein to Hsp25 during heat shock creates a reservoir of folding intermediates for reactivation. *Embo. J.* **16**, 221–9.

Eichmann, A., Grapin-Botton, A., Kelly, L., Graf, T., Le Douarin, N. M. and Sieweke, M. (1997). The expression pattern of the *mafB/kr* gene in birds and mice reveals that the *kreisler* phenotype does not represent a null mutant. *Mech. Dev.* **65**, 111–22.

Eisenberg, S. and Yamada, T. (1966). A study of DNA synthesis during the transformation of the iris into lens in the lentectomized newt. *J. Exp. Zool.* **162**, 353–68.

El-Diery, W. S., Tokino, T., Velculescu, V. E., Levy, D. B., Parsons, R., Trent, J. M., Lin, D., Mercer, W. E., Kinzler, J. W. and Vogelstein, B. (1993). WAF1, a potential mediator of p53 tumor suppression. *Cell* **75**, 817–25.

Elgert, K. L. and Zalik, S. E. (1989). Fibronectin distribution during cell type conversion in newt lens regeneration. *Anat. Embryol. (Berl)* **180**, 131–42.

Ellis, M., Alousi, S., Lawniczak, J., Maisel, H. and Welsh, M. (1984). Studies on lens vimentin. *Exp. Eye Res.* **38**, 195–202.

Emptage, N. J., Duncan, G. and Croghan, P. C. (1992). Internal acidification modulates membrane and junctional resistance in the isolated lens of the frog *Rana pipiens*. *Exp. Eye Res.* **54**, 33–9.

Enersen, O. D. (2003). "Robert Remak." Retrieved February 3, 2004, from Who Named it? Web site: http://www.whonamedit.com/doctor.cfm/1180.html

Enwright, J. F., III and Grainger, R. M. (2000). Altered retinoid signalling in the heads of Small eye mouse embryos. *Dev. Biol.* **221**, 10–22.

Epstein, J., Cai, J., Glaser, T., Jepeal, L. and Maas, R. (1994a). Identification of a *Pax* paired domain recognition sequence and evidence for DNA-dependent conformational changes. *J. Biol. Chem.* **269**, 8355–61.

Epstein, J. A., Glaser, T., Cai, J., Jepeal, L., Walton, D. S. and Maas, R. L. (1994b). Two independent and interactive DNA-binding subdomains of the Pax6 paired domain are regulated by alternative splicing. *Genes Dev.* **8**, 2022–34.

Eshaghian, J. and Streeten, B. W. (1980). Human posterior subcapsular cataract: an ultrastructural study of the posteriorly migrating cells. *Arch. Ophthalmol.* **98**, 134–43.

Eshagian, J. (1982). Human posterior subcapsular cataracts. *Trans. Ophthalmol. Soc. UK* **102**, 364–8.

Evans, T., Rosenthal, E. T., Youngblom, J., Distel, D. and Hunt, T. (1983). Cyclin: a protein specified by maternal mRNA in sea urchin eggs that is destroyed at each cleavage division. *Cell* **33**, 389–96.

Eves, H. (1978). Geometry: mensuration formulas. In *CRC Handbook of Mathematical Sciences*, 5th ed., ed. W. H. Beyer. Boca Raton, Fla.: CRC Press, pp. 188–210.

Ewen, M. E., Sluss, H. K., Sherr, C. J., Matsushime, H., Kato, J. and Livingston, D. M. (1993). Functional interactions of the retinoblastoma protein with mammalian D-type cyclins. *Cell* **73**, 487–97.

Faber, S. C., Dimanlig, P., Makarenkova, H. P., Shirke, S., Ko, K. and Lang, R. A. (2001). Fgf receptor signaling plays a role in lens induction. *Development* **128**, 4425–38.

Faber, S. C., Robinson, M. L., Makarenkova, H. P. and Lang, R. A. (2002). Bmp signaling is required for development of primary lens fiber cells. *Development* **129**, 3727–37.

Fagerholm, P. P. and Philipson, B. T. (1981). Human lens epithelium in normal and cataractous lenses. *Invest. Ophthalmol. Vis. Sci.* **21**, 408–14.

Fagerholm, P. P., Philipson, B. T. and Lindstrom, B. (1981). Normal human lens: the distribution of proteins. *Exp. Eye Res.* **33**, 615–20.

Falk, M., Ferletta, M., Forsberg, E. and Ekblom, P. (1999). Restricted distribution of laminin alpha1 chain in normal adult mouse tissues. *Matrix Biol.* **18**, 557–68.

Fan, S. S. and Ready, D. F. (1997). Glued participates in distinct microtubule-based activities in *Drosophila* eye development. *Development* **124**, 1497–507.

Faquin, W. C., Husain, A., Hung, J. and Branton, D. (1988). An immunoreactive form of erythrocyte protein 4.9 is present in non-erythroid cells. *Eur. J. Cell Biol.* **46**, 168–75.

Fedtsova, N. G. and Barabanov, V. M. (1978). Lenticular and adenohypophyseal differentiation in the oral region ectoderm of chick embryos in tissue culture. *Ontogenez* **9**, 609–15.

Fernald, R. D. (2000). Evolution of eyes. *Curr. Opin. Neurobiol.* **10**, 444–50.

Ferreira-Cornwell, M. C., Veneziale, R. W., Grunwald, G. B. and Menko, A. S. (2000). N-cadherin function is required for differentiation-dependent cytoskeletal reorganization in lens cells in vitro. *Exp. Cell Res.* **256**, 237–47.

Finkelstein, R. and Boncinelli, E. (1994). From fly head to mammalian forebrain: the story of otd and Otx. *Trends Genet.* **10**, 310–15.

Fischbarg, J., Diecke, F. P. J., Kuang, K., Yu, B., Kang, F., Iserovich, P., Li, Y., Rosskothen, H. and Koniarrek, J. P. (1999). Transport of fluid by lens epithelium. *Am. J. Physiol.* **276**, C548–57.

Fischer, R. S., Lee, A. and Fowler, V. M. (2000). Tropomodulin and tropomyosin mediate lens cell actin cytoskeleton reorganization in vitro. *Invest. Ophthalmol. Vis. Sci.* **41**, 166–74.

Fitch, J. M., Mayne, R. and Linsenmayer, T. F. (1983). Developmental acquisition of basement membrane heterogeneity: type IV collagen in the avian lens capsule. *J. Cell Biol.* **97**, 940–3.

FitzGerald, P. G. (1988). Immunochemical characterization of a Mr 115 lens fiber cell-specific extrinsic membrane protein. *Curr. Eye Res.* **7**, 1243–53.

Fitzgerald, P. G., Bok, D. and Horwitz, J. (1983). Immunocytochemical localization of the main intrinsic polypeptide (MIP) in ultrathin frozen sections of rat lens. *J. Cell Biol.* **97**, 1491–9.

FitzGerald, P. G. and Graham, D. (1991). Ultrastructural localization of alpha A-crystallin to the bovine lens fiber cell cytoskeleton. *Curr. Eye Res.* **10**, 417–36.

Fleming, T. P., Song, Z. and Andley, U. P. (1998). Expression of growth control and differentiation genes in human lens epithelial cells with extended life span. *Invest. Ophthalmol. Vis. Sci.* **39**, 1387–98.

Flugel, C., Liebe, S., Voorter, C., Bloemendal, H. and Lutjen-Drecoll, E. (1993). Distribution of alpha B-crystallin in the anterior segment of primate and bovine eyes. *Curr. Eye Res.* **12**, 871–6.

Fotiadis, D., Hasler, L., Müller, D. J., Stahlberg, H., Kistler, J. and Engel, A. (2000). Surface tongue-and-groove contours on lens MIP facilitate cell-to-cell adherence. *J. Mol. Biol.* **300**, 779–89.

Franke, W. W., Kapprell, H. P. and Cowin, P. (1987). Plakoglobin is a component of the filamentous subplasmalemmal coat of lens cells. *Eur. J. Cell Biol.* **43**, 301–15.

Frederikse, P. H., Dubin, R. A., Haynes, J. I., 2nd and Piatigorsky, J. (1994). Structure and alternate tissue-preferred transcription initiation of the mouse alpha B-crystallin/small heat shock protein gene. *Nucleic Acids Res.* **22**, 5686–94.

Freel, C. D., Gilliland, K. O., Lane, C. W., Giblin, F. J. and Costello, M. J. (2002). Fourier analysis of cytoplasmic texture in nuclear fiber cells from transparent and cataractous human and animal lenses. *Exp. Eye Res.* **74**, 689–702.

Freeman, G. (1963). Lens regeneration from the cornea in *Xenopus laevis*. *J. Exp. Zool.* **154**, 39–66.

Frenzel, E. M. and Johnson, R. G. (1996). Gap junction formation between cultured embryonic lens cells is inhibited by antibody to N-cadherin. *Dev. Biol.* **179**, 1–16.

Freund, C., Horsford, D. J. and McInnes, R. R. (1996). Transcription factor genes and the developing eye: a genetic perspective. *Hum. Mol. Genet.* **5**, 1471–88.
Frezzotti, R. (1990). Pathogenesis of posterior capsular opacification. Pt. II.
Frezzotti, R. Caporossi, A., Mastrangelo, D., Hadjistilianou, T., Tosi, P., Cintrorino, M., and Minacci, C. (1990). Pathogenesis of posterior capsular opacification. Part II: Histopathological and in vitro culture findings. *J. Cataract. Refract. Surg.* **16**, 353–60
Friend, S. H., Bernards, R., Rogeli, S., Weinberg, R. A., Rapaport, J. M., Albert, D. M. and Dryja, T. P. (1986). A human DNA segment with properties of the gene that predispose to retinoblastoma and osteosarcoma. *Nature* **323**, 643–6.
Fromm, L. and Overbeek, P. A. (1996). Regulation of cyclin and cyclin-dependent kinase gene expression during lens differentiation requires the retinoblastoma protein. *Oncogene* **12**, 69–75.
Fromm, L., Shawlot, W., Gunning, K., Butel, J. S. and Overbeek, P. A. (1994). The retinoblastoma protein-binding region of simian virus 40 large T antigen alters cell cycle regulation in lenses of transgenic mice. *Mol. Cell Biol.* **14**, 6743–54.
Fuchs, E. and Cleveland, D. W. (1998). A structural scaffolding of intermediate filaments in health and disease. *Science* **279**, 514–9.
Fuchs, P., Zorer, M., Rezniczek, G. A., Spazierer, D., Oehler, S., Castanon, M. J., Hauptmann, R. and Wiche, G. (1999). Unusual 5′ transcript complexity of plectin isoforms: novel tissue-specific exons modulate actin binding activity. *Hum. Mol. Genet.* **8**, 2461–72.
Fujiwara, M., Uchida, T., Osumi-Yamashita, N. and Eto, K. (1994). Uchida rat (rSey): A new mutant with craniofacial abnormalities resembling those of the mouse Sey mutant. *Differentiation* **57**, 31–8.
Funahashi, J., Sekido, R., Murai, K., Kamachi, Y. and Kondoh, H. (1993). Delta-crystallin enhancer binding protein delta EF1 is a zinc finger–homeodomain protein implicated in postgastrulation embryogenesis. *Development* **119**, 433–46.
Fung, Y.-K. T., Murphee, A. L., T'Ang, A., Qian, J., Hinrichs, S. H. and Benedict, W. F. (1987). Structural evidence for the authenticity of the human retinoblastoma gene. *Science* **236**, 1657–61.
Furukawa, T., Morrow, E. M. and Cepko, C. L. (1997). *Crx*, a novel *otx*-like homeobox gene, shows photoreceptor-specific expression and regulates photoreceptor differentiation. *Cell* **91**, 531–41.
Furuta, Y. and Hogan, B. L. M. (1998). BMP4 is essential for lens induction in the mouse embryo. *Genes Dev.* **12**, 3764–75.
Gabella, G. (1979). Hypertonic smooth muscle. III. Increase in number and size of gap junctions. *Cell Tiss. Res.* **201**, 263–76.
Gallardo, M. E., Lopez-Rios, J., Fernaud-Espinosa, I., Granadino, B., Sanz, R., Ramos, C., Ayuso, C., Seller, M. J., Brunner, H. G., Bovolenta, P., et al. (1999). Genomic cloning and characterization of the human homeobox gene *SIX6* reveals a cluster of *SIX* genes in chromosome 14 and associates *SIX6* hemizygosity with bilateral anophthalmia and pituitary anomalies. *Genomics* **61**, 82–91.
Galvan, A., Lampe, P. D., Hur, K. C., Howard, J. B., Eccleston, E. D., Arneson, M. and Louis, C. F. (1989). Structural organization of the lens fiber cell plasma membrane protein MP18. *J. Biol. Chem.* **264**, 19974–8.
Gao, C. Y., Bassnett, S. and Zelenka, P. S. (1995). Cyclin B, p34cdc2, and H1-kinase activity in terminally differentiating lens fiber cells. *Dev. Biol.* **169**, 185–94.
Gao, C. Y. Rampalli, A. M., Cai, H., He, H. and Zelenka, P. S. (1999). Changes in cyclin dependent kinase expression and activity accompanying lens fiber cell differentiation. *Exp. Eye Res.* **69**, 695–703.
Gao, C. Y. and Zelenka, P. S. (1997). Cyclins, cyclin-dependent kinases and differentiation. *Bioessays* **19**, 307–15.
Gao, J., Sun, X., Yatsula, V., Wymore, R. S. and Mathias, R. T. (2000). Isoform-specific function and distribution of Na/K pumps in the frog lens epithelium. *J. Memb. Biol.* **178**, 89–101.
Gao, Y. and Spray, D. C. (1998). Structural changes in lenses of mice lacking the gap junction protein connexin43. *Invest. Ophthalmol. Vis. Sci.* **39**, 1198–209.
Garland, D. L., Duglas-Tabor, Y., Jimenez-Asensio, J., Datiles, M. B. and Magno, B. (1996). The nucleus of the human lens: demonstration of a highly characteristic protein pattern by two-dimensional electrophoresis and introduction of a new method of lens dissection. *Exp. Eye Res.* **62**, 285–91.
Garner, M. H. and Kong, Y. (1999). Lens epithelium and fiber Na,K-ATPases: distribution and localization by immunocytochemistry. *Invest. Ophthalmol. Vis. Sci.* **40**, 2291–8.

Gaston, K., Bell, A., Busby, S. and Fried, M. (1992). A comparison of the DNA bending activities of the DNA binding proteins CRP and TFIID. *Nucleic Acids Res.* **20**, 3391–6.

Gavrieli, Y., Sherman, Y. and Ben-Sasson, S. A. (1992). Identification of programmed cell death in situ via specific labeling of nuclear DNA fragmentation. *J. Cell Biol.* **119**, 493–501.

Gehring, W. J., Affolter, M. and Burglin, T. (1994). Homeodomain proteins. *Ann. Rev. Biochem.* **63**, 487–526.

Gehring, W. J. and Ikeo, K. (1999). *Pax 6*: mastering eye morphogenesis and eye evolution. *Trends Genet.* **15**, 371–7.

Geng, Y., Whoriskey, W., Park, M. Y., Bronson, R. T., Medema, R. H., Li, T., Weinberg, R. A. and Sicinski, P. (1999). Rescue of cyclin D1 deficiency by knockin cyclin E. *Cell* **97**, 767–77.

Génis-Gálvez, J. M. (1962). The results of the total and partial removal of the lens primordium in the chick embryo: contribution to the study of lens regeneration. *An. Desarr.* **10**, 249–67.

Georgatos, S. D., Gounari, F. and Remington, S. (1994). The beaded intermediate filaments and their potential functions in eye lens. *Bioessays* **16**, 413–18.

Georgatos, S. D. and Marchesi, V. T. (1985). The binding of vimentin to human erythrocyte membranes: a model system for the study of intermediate filament–membrane interactions. *J. Cell Biol.* **100**, 1955–61.

Georges-Labouesse, E., Messaddeq, N., Yehia, G., Cadalbert, L., Dierich, A. and Le Meur, M. (1996). Absence of integrin alpha 6 leads to epidermolysis bullosa and neonatal death in mice. *Nat. Genet.* **13**, 370–3.

Giancotti, F. G. and Ruoslahti, E. (1999). Integrin signaling. *Science* **285**, 1028–32.

Gilbert, R. E., Cox, A. J., Kelly, D. J., Wilkinson-Berka, J. L., Sage, E. H., Jerums, G. and Cooper, M. E. (1999). Localization of secreted protein acidic and rich in cysteine (SPARC) expression in the rat eye. *Connect. Tissue Res.* **40**, 295–303.

Giles, K. M. and Harris, J. E. (1959). The accumulation of 14C from uniformly labelled glucose by the normal and diabetic rabbit lens. *Am. J. Ophthalmol.* **48**, 508–17.

Gill, D., Klose, R., Munier, F. L., McFadden, M., Priston, M., Billingsley, G., Ducrey, N., Schorderet, D. F. and Heon, E. (2000). Genetic heterogeneity of the Coppock-like cataract: a mutation in CRYBB2 on chromosome 22q11.2. *Invest. Ophthalmol. Vis. Sci.* **41**, 159–65.

Gilliland, K. O., Freel, C. D., Lane, C. W., Fowler, W. C. and Costello, M. J. (2001). Multilameliar bodies as potential scattering particles in human age-related nuclear cataracts. *Mol. Vis.* **7**, 120–30.

Gilmour, D. T., Lyon, G. J., Carlton, M. B. L., Sanes, J. R., Cunningham, J. M., Anderson, J. R., Hogan, B. L. M., Evans, M. J. and Colledge, W. H. (1998). Mice deficient for the secreted glycoprotein SPARC/osteonectin/BM40 develop normally but show severe age-onset cataract formation and disruption of the lens. *EMBO J.* **17**, 1860–70.

Girard, F., Strausfeld, U., Fernandez, A. and Lamb, N. J. (1991). Cyclin A is required for the onset of DNA replication in mammalian fibroblasts. *Cell* **67**, 1169–79.

Gjerset, R., Gorka, C., Hasthorpe, S., Lawrence, J. J. and Eisen, H. (1982). Developmental and hormonal regulation of protein H1 degrees in rodents. *Proc. Natl. Acad. Sci. USA* **79**, 2333–7.

Glasgow, E. and Tomarev, S. I. (1998). Restricted expression of the homeobox gene prox 1 in developing zebrafish. *Mech. Dev.* **76**, 175–8.

Gonen, T., Donaldson, P. J. and Kistler, J. (2000). Galectin-3 is associated with the plasma membrane of lens fiber cells. *Invest. Ophthalmol. Vis. Sci.* **41**, 199–203.

Gong, X., Baldo, G. J., Kumar, N. M., Gilula, N. B. and Mathias, R. T. (1998). Gap junctional coupling in lenses lacking α3 connexin. *Proc. Natl. Acad. Sci. USA* **95**, 15303–8.

Gong, X., Li, E., Klier, G., Huang, Q., Wu, Y., Lei, H., Kumar, N. M., Horwitz, J. and Gilula, N. B. (1997). Disruption of alpha3 connexin gene leads to proteolysis and cataractogenesis in mice. *Cell* **91**, 833–43.

Gonzales, M., Haan, K., Baker, S. E., Fitchmun, M., Todorov, I., Weitzman, S. and Jones, J. C. (1999). A cell signal pathway involving laminin-5, alpha3beta1 integrin, and mitogen-activated protein kinase can regulate epithelial cell proliferation. *Mol. Biol. Cell* **10**, 259–70.

Gonzalez, P., Hernandez-Calzadilla, C., Rao, P. V., Rodriguez, I. R., Zigler, J. S., Jr. and Borras, T. (1994). Comparative analysis of the zeta-crystallin/quinone reductase gene in guinea pig and mouse. *Mol. Biol. Evol.* **11**, 305–15.

Goodenough, D. A. (1979). Lens gap junctions: a structural hypothesis for non-regulated low-resistance intercellular pathways. *Invest. Ophthalmol. Vis. Sci.* **11**, 1104–22.

Goodenough, D. A., Dick, J. S. B. and Lyons, J. E. (1980). Lens metabolic cooperation: a study of mouse lens transport and permeability visualized with freeze-substitution autoradiography and electron microscopy. *J. Cell Biol.* **86**, 576–89.

Gopal-Srivastava, R., Cvekl, A. and Piatigorsky, J. (1996). Pax-6 and αB-crystallin/small heat shock protein gene regulation in the murine lens: interaction with the lens-specific regions, LSR1 and LSR2. *J. Biol. Chem.* **271**, 23029–36.

Gopal-Srivastava, R., Cvekl, A. and Piatigorsky, J. (1998). Involvement of retinoic acid/retinoid receptors in the regulation of murine αB-crystallin/small heat shock protein gene expression in the lens. *J. Biol. Chem.* **273**, 17954–61.

Gopal-Srivastava, R., Haynes, J. I. and Piatigorksy, J. (1995). Regulation of the murine αB-crystallin/small heat shock protein gene in cardiac muscle. *Mol. Cell. Biol.* **15**, 7081–90.

Gopal-Srivastava, R., Kays, W. T. and Piatigorsky, J. (2000). Enhancer-independent promoter activity of the mouse alphaB-crystallin/small heat shock protein gene in the lens and cornea of transgenic mice. *Mech. Dev.* **92**, 125–34.

Gopal-Srivastava, R. and Piatigorsky, J. (1994). Identification of a lens-specific regulatory region (LSR) of the murine alpha B-crystallin gene. *Nucleic Acids Res.* **22**, 1281–6.

Gordon-Thomson, C., de Iongh, R. U., Hales, A. M., Chamberlain, C. G. and McAvoy, J. W. (1998). Differential cataractogenic potency of TGF-beta1, -beta2, and -beta3 and their expression in the postnatal rat eye. *Invest. Ophthalmol. Vis. Sci.* **39**, 1399–409.

Gorin, M. B., Yancey, S. B., Cline, J., Revel, J. P. and Horwitz, J. (1984). The major intrinsic protein (MIP) of the bovine lens fiber membrane: characterization and structure based on cDNA cloning. *Cell* **39**, 49–59.

Goring, D. R., Bryce, D. M., Tsui, L. C., Breitman, M. L. and Liu, Q. (1993). Developmental regulation and cell type–specific expression of the murine gamma F-crystallin gene is mediated through a lens-specific element containing the gamma F-1 binding site. *Dev. Dyn.* **196**, 143–52.

Gorthy, W. C. and Anderson, J. W. (1980). Special characteristics of the polar regions of the rat lens: morphology and phosphatase histochemistry. *Invest. Ophthalmol. Vis. Sci.* **19**, 1038–52.

Goto, K., Hayashi, S., Shirayoshi, Y., Takeichi, M. and Kondoh, H. (1988). Exogenous delta-crystallin gene expression as probe for differentiation of teratocarcinoma stem cells. *Cell Diff.* **24**, 139–47.

Goto, K., Okada, T. S. and Kondoh, H. (1990). Functional cooperation of lens-specific and nonspecific elements in the delta 1-crystallin enhancer. *Mol. Cell. Biol.* **10**, 958–64.

Götz, M., Stoykova, A. and Gruss, P. (1998). *Pax6* controls radial glia differentiation in the cerebral cortex. *Neuron* **21**, 1031–44.

Gould, G. W. and Holman, G. D. (1993). The glucose transporter family: structure function and tissue specific expression. *Biochem. J.* **295**, 329–41.

Gould, S. E., Upholt, W. B. and Kosher, R. A. (1995). Characterization of chicken syndecan-3 as a heparan sulfate proteoglycan and its expression during embryogenesis. *Dev. Biol.* **168**, 438–51.

Gounari, F., Merdes, A., Quinlan, R., Hess, J., FitzGerald, P. G., Ouzounis, C. A. and Georgatos, S. D. (1993). Bovine filensin possesses primary and secondary structure similarity to intermediate filament proteins. *J. Cell Biol.* **121**, 847–53.

Govindarajan, V. and Overbeek, P. A. (2001). Secreted FGFR3, but not FGFR1, inhibits lens fiber differentiation. *Development* **128**, 1617–27.

Graham, C., Hodin, J. and Wistow, G. (1996). A retinaldehyde dehydrogenase as a structural protein in a mammalian eye lens: gene recruitment of η-crystallin. *J. Biol. Chem.* **271**, 15623–8.

Grainger, R. M. (1992). Embryonic lens induction: shedding light on vertebrate tissue determination. *T.I.G.* **8**, 349–55.

Grainger, R. M. (1996). New perspectives on embryonic lens induction. *Sem. in Cell Devel. Biol.* **7**, 149–55.

Grainger, R. M. and Gurdon, J. B. (1989). Loss of competence in amphibian induction can take place in single nondividing cells. *Proc. Natl. Acad. Sci. USA* **86**, 1900–4.

Grainger, R. M., Henry, J. J. and Henderson, R. A. (1988). Reinvestigation of the role of the optic vesicle in embryonic lens induction. *Development* **102**, 517–26.

Grainger, R. M., Henry, J. J., Saha, M. S. and Servetnick, M. (1992). Recent progress on the mechanisms of embryonic lens formation. *Eye*, 117–22.

Grainger, R. M., Mannion, J. E., Cook, T. L., Jr. and Zygar, C. A. (1997). Defining intermediate stages in cell determination: acquisition of a lens-forming bias in head ectoderm during lens determination. *Dev. Genetics* **20**, 246–57.

Granger, B. L. and Lazarides, E. (1984). Expression of the intermediate-filament-associated protein synemin in chicken lens cells. *Mol. Cell. Biol.* **4**, 1943–50.
Granger, B. L. and Lazarides, E. (1985). Appearance of new variants of membrane skeletal protein 4.1 during terminal differentiation of avian erythroid and lenticular cells. *Nature* **313**, 238–41.
Graw, J. (1997). The crystallins: genes, proteins and diseases. *Biol. Chem.* **378**, 1331–48.
Graw, J. (2000). Mouse mutants for eye development. *Results Probl. Cell Differ.* **31**, 219–56.
Graw, J., Jung, M., Loster, J., Klopp, N., Soewarto, D., Fella, C., Fuchs, H., Reis, A., Wolf, E., Balling, R. et al. (1999). Mutation in the betaA3/A1-crystallin encoding gene Cryba1 causes a dominant cataract in the mouse. *Genomics* **62**, 67–73.
Grindley, J. C., Davidson, D. R. and Hill, R. E. (1995). The role of Pax-6 in eye and nasal development. *Development* **121**, 1433–42.
Grindley, J. C., Hargett, L. K., Hill, R. E., Ross, A. and Hogan, B. L. (1997). Disruption of PAX6 function in mice homozygous for the Pax6Sey-1Neu mutation produces abnormalities in the early development and regionalization of the diencephalon. *Mech. Dev.* **64**, 111–26.
Groenen, P. J., Merck, K. B., de Jong, W. W. and Bloemendal, H. (1994). Structure and modifications of the junior chaperone alpha-crystallin: from lens transparency to molecular pathology. *Eur. J. Biochem.* **225**, 1–19.
Grondona, J. M., Kastner, P., Gansmuller, A., Decimo, D., Chambon, P. and Mark, M. (1996). Retinal dysplasia and degeneration in RARβ2/RARγ2 compound mutant mice. *Development* **122**, 2173–88.
Gruijters, W. T. M., Kistler, J. and Bullivant, S. (1987a). Formation, distribution, and dissociation of intercellular junctions in the lens. *J. Cell Sci.* **88**, 351–59.
Gruijters, W. T. M., Kistler, J., Bullivant, S. and Goodenough, D. A. (1987b). Immunolocalization of MP70 in lens fiber 16–17nm intercellular junctions. *J. Cell Biol.* **104**, 565–72.
Grunstein, M. (1997). Histone acetylation in chromatin structure and transcription. *Nature* **389**, 349–52.
Gu, W., Schneider, J. W., Condorelli, G., Kaushal, S., Mahdavi, V. and Nadal-Ginard, B. (1993). Interaction of myogenic factors and the retinoblastoma protein mediates muscle commitment and differentiation. *Cell* **72**, 309–24.
Gumbiner, B. M. and McCrea, P. D. (1993). Catenins as mediators of the cytoplasmic functions of cadherins [review]. *J. Cell Sci.* **17** (Suppl), 155–8.
Gupta, V. K., Berthoud, V. M., Atal, N., Jarillo, J. A., Barrio, L. C. and Beyer, E. C. (1994a). Bovine connexin44, a lens gap junction protein: molecular cloning, immunologic characterization, and functional expression. *Invest. Ophthalmol. Vis. Sci.* **35**, 3747–58.
Gupta, N., Drance, S. M., McAllister, R., Prasad, S., Rootman, J. and Cynader, M. S. (1994b). Localization of M3 muscarinic receptor subtype and mRNA in the human eye. *Ophthalmic Res.* **26**, 207–13.
Gurland, G. and Gundersen, G. G. (1995). Stable, detyrosinated microtubules function to localize vimentin intermediate filaments in fibroblasts. *J. Cell Biol.* **131**, 1275–90.
Gwon, A., Enomoto, H., Horowitz, J. and Garner, M. H. (1989). Induction of de novo synthesis of crystalline lenses in aphakic rabbits. *Exp. Eye Res.* **49**, 913–26.
Gwon, A., Gruber, L. J. and Mantras, C. (1993a). Restoring lens capsule integrity enhances lens regeneration in New Zealand albino rabbits and cats. *J. Cataract Refract. Surg.* **19**, 735–46.
Gwon, A., Gruber, L. J., Mantras, C. and Cunanan, C. (1993b). Lens regeneration in New Zealand albino rabbits after endocapsular cataract extraction. *Invest. Ophthalmol. Vis. Sci.* **34**, 2124–9.
Gwon, A. E., Gruber, L. J. and Mundwiler, K. E. (1990). A histologic study of lens regeneration in aphakic rabbits. *Invest. Ophthalmol. Vis. Sci.* **31**, 540–7.
Gwon, A. E., Jones, R. L., Gruber, L. J. and Mantras, C. (1992). Lens regeneration in juvenile and adult rabbits measured by image analysis. *Invest. Ophthalmol. Vis. Sci.* **33**, 2279–83.
Haddad, A. and Bennett, G. (1988). Synthesis of lens capsule and plasma membrane glycoproteins by lens epithelial cells and fibers in the rat. *Am. J. Anat.* **183**, 212–25.
Hai, T. W., Liu, F., Coukos, W. J. and Green, M. R. (1989). Transcription factor ATF cDNA clones: an extensive family of leucine zipper proteins able to selectively form DNA-binding heterodimers. *Genes Dev.* **3**, 2083–90.
Halder, G., Callaerts, P., Flister, S., Walldorf, U., Kloter, U. and Gehring, W. J. (1998). Eyeless initiates the expression of both *sine oculis* and *eyes absent* during *Drosophila* compound eye development. *Development* **125**, 2181–91.

Halder, G., Callaerts, P. and Gehring, W. J. (1995). Induction of ectopic eyes by targeted expression of the *eyeless* gene in *Drosophila. Science* **267**, 1788–92.
Hales, A. M., Chamberlain, C. G. and McAvoy, J. W. (1995). Cataract induction in lenses cultured with transforming growth factor-beta. *Invest. Ophthalmol. Vis. Sci.* **36**, 1709–13.
Hales, A. M., Chamberlain, C. G. and McAvoy, J. W. (1997). Estrogen protects lenses against cataract induced by transforming growth factor-B (TGFB). *J. Exp. Med.* **185**, 273–80.
Hales, A. M., Chamberlain, C. G. and McAvoy, J. W. (2000). Susceptibility to TGFbeta2-induced cataract increases with aging in the rat. *Invest. Ophthalmol. Vis. Sci.* **41**, 3544–51.
Hales, A. M., Schulz, M. W., Chamberlain, C. G. and McAvoy, J. W. (1994). TGF-beta-1 induces lens cells to accumulate alpha-smooth muscle actin, a marker for subcapsular cataracts. *Curr. Eye Res.* **13**, 885–90.
Haley, D. A., Horwitz, J. and Stewart, P. L. (1998). The small heat-shock protein, alphaB-crystallin, has a variable quaternary structure. *J. Mol. Biol.* **277**, 27–35.
Haloui, Z., Jeanny, J. C., Jonet, L., Courtois, Y. and Laurent, M. (1988). Immunochemical analysis of extracellular matrix during embryonic lens development of the Cat Fraser mouse. *Exp. Eye Res.* **46**, 463–74.
Haloui, Z., Pujol, J. P., Galera, P., Courtois, Y. and Laurent, M. (1990). Analysis of lens protein synthesis in a cataractous mutant mouse: the Cat Fraser. *Exp. Eye Res.* **51**, 487–94.
Hamann, S., Zeuthen, T., La Cour, M., Nagelhus, E. A., Ottersen, O. P., Agre, P. and Nielsen, S. (1998). Aquaporins in complex tissues: distribution of aquaporins 1–5 in human and rat eye. *Am. J. Physiol.* **274**, C1332–45.
Hamburger, V. and Hamilton, H. L. (1951). A series of normal stages in the development of the chick embryo. *J. Morph.* **88**, 49–92.
Hammond, K. L., Hanson, I. M., Brown, A. G., Lettice, L. A. and Hill, R. E. (1998). Mammalian and *Drosophila dachshund* genes are related to the *Ski* proto-oncogene and are expressed in eye and limb. *Mech. Dev.* **74**, 121–31.
Hanna, C. and O'Brien, J. E. (1961). Cell production and migration in the epithelial layer of the lens. *Arch. Ophthalmol.* **66**, 103–7.
Hanson, I. M., Fletcher, J. M., Jordan, T., Brown, A., Taylor, D., Adams, R. J., Punnett, H. H. and van Heyningen, V. (1994). Mutations at the PAX6 locus are found in heterogeneous anterior seqment malformations including Peters' anomaly. *Nat. Genet.* **6**, 168–73.
Harding, C. V., Reddan, J. R., Unakar, N. J. and Bagchi, M. (1971). The control of cell division in the ocular lens. *Int. Rev. Cytol.* **31**, 215–300.
Harocopos, G. J., Kolker, A. E. and Beebe, D. C. (1996). Is apoptosis associated with cataract formation in humans? *Invest. Ophthalmol. Vis. Sci.* **37**, s651.
Harper, J. W., Adami, G. R., Wei, N., Keyomarsi, K. and Elledge, S. J. (1993). The p21 Cdk-interacting protein Cip1 is a potent inhibitor of G1 cyclin-dependent kinase. *Cell* **75**, 805–16.
Harper, J. W. and Elledge, S. J. (1996). Cdk inhibitors in development and cancer. *Curr. Opin. Genet. Dev.* **6**, 56–64.
Harper, J. W. and Elledge, S. J. (1999). Skipping into the E2F1-destruction pathway. *Nat. Cell. Biol.* **1**, E5–7.
Hartung, H., Feldman, B., Lovec, H., Coulier, F., Birnbaum, D. and Goldfarb, M. (1997). Murine FGF-12 and FGF-13: expression in embryonic nervous system, connective tissue and heart. *Mech. Dev.* **64**, 31–9.
Hasler, L., Walz, T., Tittmann, P., Gross, H., Kistler, J. and Engel, A. (1998). Purified lens major intrinsic protein (MIP) forms highly ordered tetragonal two-dimensional arrays by reconstitution. *J. Mol. Biol.* **279**, 855–64.
Hassan, B., Li, L., Bremer, K. A., Chang, W., Pinsonneault, J. and Vaessin, H. (1997). Prospero is a panneural transcription factor that modulates homeodomain protein activity. *Proc. Natl. Acad, Sci. USA* **94**, 10991–6.
Hatae, T., Ishibashi, T., Yoshitomo, F. and Shibata, Y. (1993). Immunocytochemistry of type I–IV collagen in human anterior subcapsular cataracts. *Graefes Arch. Clin. Exp. Ophthalmol.* **231**, 586–90.
Hatfield, J. S., Skoff, R. P., Maisel, H., Eng, L. and Bigner, D. D. (1985). The lens epithelium contains glial fibrillary acidic protein (GFAP). *J. Neuroimmunol.* **8**, 347–57.
Hattori, M., Fujiyama, A., Taylor, T. D., Watanabe, H., Yada, T., Park, H. S., Toyoda, A., Ishii, K.,

Totoki, Y., Choi, D. K., et al. (2000). The DNA sequence of human chromosome 21: the chromosome 21 mapping and sequencing consortium [see comments]. *Nature* **405**, 311–9.

Hauck, B., Gehring, W. J. and Walldorf, U. (1999). Functional analysis of an eye specific enhancer of the *eyeless* gene in *Drosophila*. *Proc. Natl. Acad. Sci. USA* **96**, 564–9.

Hawkins, J. W., Nickerson, J. M., Sullivan, M. A. and Piatigorsky, J. (1984). The chicken delta-crystallin gene family: two genes of similar structure in close chromosomal approximation. *J. Biol. Chem.* **259**, 9821–5.

Hayashi, S., Goto, K., Okada, T. S. and Kondoh, H. (1987). Lens-specific enhancer in the third intron regulates expression of the chicken delta 1-crystallin gene. *Genes Dev.* **1**, 818–28.

Hayashi, T., Mizuno, N., Owaribe, K., Kuroiwa, A. and Okamoto, M. (2002). Regulated lens regeneration from isolated pigmented epithelial cells of newt iris in culture in response to FGF2/4. *Differentiation* **70**, 101–8.

Hayashi, T., Yamagishi, A., Kuroiwa, A., Mizuno, N., Kondoh, H. and Okamoto, M. (2001). Highly efficient transfection system for functional gene analysis in adult amphibian lens regeneration. *Dev. Growth Differ.* **43**, 361–70.

Hayden, J. H. and Rothstein, H. (1979). Complete elimination of mitosis and DNA synthesis in the lens of the hypophysectomized frog: effects on cell migration and fiber growth. *Differentiation* **15**, 153–60.

Haynes, J. I., Duncan, M. K. and Piatigorsky, J. (1996). Spatial and temporal activity of the αB-crystallin/small heat shock protein gene promoter in transgenic mice. *Dev. Dyn.* **207**, 75–88.

Haynes, J. I., 2nd, Gopal-Srivastava, R., Frederikse, P. H. and Piatigorsky, J. (1995). Differential use of the regulatory elements of the alpha B-crystallin enhancer in cultured murine lung (MLg), lens (alpha TN4-1) and muscle (C2C12) cells. *Gene* **155**, 151–8.

Haynes, J. I., 2nd, Gopal-Srivastava, R. and Piatigorsky, J. (1997). Alpha B-crystallin TATA sequence mutations: lens-preference for the proximal TATA box and the distal TATA-like sequence in transgenic mice. *Biochem. Biophys. Res. Commun.* **241**, 407–13.

He, H., Gao, C., Vrensen, G. and Zelenka, P. S. (1998). Transient activation of cyclin B/cdc2 during terminal differentiation of lens fiber cells. *Dev. Dyn.* **211**, 26–34.

Head, M. W. and Goldman, J. E. (2000). Small heat shock proteins, the cytoskeleton, and inclusion body formation. *Neuropathol. Appl. Neurobiol.* **26**, 304–12.

Head, M. W., Peter, A. and Clayton, R. M. (1991). Extralenticular expression of members of the β-crystallin gene family in the chick and comparison with δ-crystallin during differentiation and transdifferentiation. *Differentiation* **48**, 147–56.

Head, M. W., Sedowofia, K. and Clayton, R. M. (1995). Beta B2-crystallin in the mammalian retina. *Exp. Eye Res.* **61**, 423–8.

Heanue, T. A., Reshef, R., Davis, R. J., Mardon, G., Oliver, G., Tomarev, S., Lassar, A. B. and Tabin, C. J. (1999). Synergistic regulation of vertebrate muscle development by *Dach2*, *Eya2*, and *Six1*, homologs of genes required for *Drosophila* eye formation. *Genes Dev.* **13**, 3231–43.

Heasman, J. 2002. Morpholino oligos: making sense of antisense? *Dev. Biol.* **243**, 209–14.

Heid, H. W., Schmidt, A., Zimbelmann, R., Schafer, S., Winter Simanowski, S., Stumpp, S., Keith, M., Figge, U., Schnolzer, M. and Franke, W. W. (1994). Cell type-specific desmosomal plaque proteins of the plakoglobin family: plakophilin 1 (band 6 protein). *Differentiation* **58**, 113–31.

Hejtmancik, J. F., Beebe, D. C., Ostrer, H. and Piatigorsky, J. (1985). δ- and β-crystallin mRNA levels in the embryonic and posthatched chicken lens: temporal and spatial changes during development. *Dev. Biol.* **109**, 72–81.

Henderson, C. G., Tucker, J. B., Mogensen, M. M., Mackie, J. B., Chaplin, M. A., Slepecky, N. B. and Leckie, L. M. (1995). Three microtubule-organizing centres collaborate in a mouse cochlear epithelial cell during supracellularly coordinated control of microtubule positioning. *J. Cell Sci.* **108**, 37–50.

Henry, J. J. and Grainger, R. M. (1987). Inductive interactions in the spatial and temporal restriction of lens-forming potential in embryonic ectoderm of *Xenopus laevis*. *Dev. Biol.* **124**, 200–14.

Henry, J. J. and Grainger, R. M. (1990). Early tissue interactions leading to embryonic lens formation in *Xenopus laevis*. *Dev. Biol.* **141**, 149–63.

Heon, E., Priston, M., Schorderet, D. F., Billingsley, G. D., Girard, P. O., Lubsen, N. and Munier, F. L. (1999). The gamma-crystallins and human cataracts: a puzzle made clearer. *Am. J. Hum. Genet.* **65**, 1261–7.

Herber, R., Liem, A., Pitot, H. and Lambert, P. F. (1996). Squamous epithelial hyperplasia and carcinoma in mice transgenic for the human papillomavirus type 16 E7 oncogene. *J. Virol.* **70**, 1873–81.

Herrmann, H. and Aebi, U. (2000). Intermediate filaments and their associates: multi-talented structural elements specifying cytoarchitecture and cytodynamics. *Curr. Opin. Cell Biol.* **12**, 79–90.

Hess, J. F., Casselman, J. T. and FitzGerald, P. G. (1993). cDNA analysis of the 49 kDa lens fiber cell cytoskeletal protein: a new, lens-specific member of the intermediate filament family? *Curr. Eye Res.* **12**, 77–88.

Hess, J. F. and FitzGerald, P. (1996). Lack of DNase I mRNA sequences in murine lenses. *Mol. Vis.* **2**, 8.

Hettmann, T., Barton, K. and Leiden, J. M. (2000). Microphthalmia due to p53-mediated apoptosis of anterior lens epithelial cells in mice lacking the CREB-2 transcription factor. *Dev. Biol.* **222**, 110–23.

Heymann, J. B. and Engel, A. (1999). Aquaporins: phylogeny, structure, and physiology of water channels. *News Physiol. Sci.* **14**, 187–93.

Hilfer, S. R. and Randolph, G. J. (1993). Immunolocalization of basal lamina components during development of chick otic and optic primordia. *Anat. Rec.* **235**, 443–52.

Hill, R., Favor, J., Hogan, B., Ton, C., Saunders, G., Hanson, I. M., Prosser, J., Jordan, T., Hastie, N. and van Heyningen, V. (1991). Mouse *Small-eye* results from mutations in a paired-like homeobox-containing gene. *Nature* **354**, 522–5.

Hinchcliffe, E. H., Li, C., Thompson, E. A., Maller, J. L. and Sluder, G. (1999). Requirement of Cdk2–cyclin E activity for repeated centrosome reproduction in *Xenopus* egg extracts. *Science* **283**, 851–4.

Hinds, P. W., Mittnacht, S., Dulic, V., Arnold, A., Reed, S. I. and Weinberg, R. A. (1992). Regulation of retinoblastoma protein functions by ectopic expression of human cyclins. *Cell* **70**, 993–1006.

Hirsch, N. and Grainger, R. M. (2000). Induction of the lens. *Results Probl. Cell Differ.* **31**, 51–68.

Hirschberg, J. (1982). Antiquity, Vol. X in The History of Ophthalmology (F. C. Blodi, Trans.). Bonn: Wayenborgh.

Hobert, O. and Westphal, H. (2000). Functions of LIM-homeobox genes. *Trends Genet.* **16**, 75–83.

Hodivala-Dilke, K. M., DiPersio, C. M., Kreidberg, J. A. and Hynes, R. O. (1998). Novel roles for alpha3beta1 integrin as a regulator of cytoskeletal assembly and as a trans-dominant inhibitor of integrin receptor function in mouse keratinocytes. *J. Cell Biol.* **142**, 1357–69.

Hoenders, H. J. and Bloemendal, H. (1981). Aging of lens proteins. In *Molecular and Cellular Biology of the Eye Lens*, ed. H. Bloemendal. New York: Wiley, pp. 279–326.

Hogan, B. L., Hirst, E. M., Horsburgh, G. and Hetherington, C. M. (1988). *Small eye* (*Sey*): a mouse model for the genetic analysis of craniofacial abnormalities. *Development* **103** (Suppl), 115–19.

Hogan, B. L., Horsburgh, G., Cohen, J., Hetherington, C. M., Fisher, G. and Lyon, M. F. (1986). *Small eyes* (*Sey*): a homozygous lethal mutation on chromosome 2 which affects the differentiation of both lens and nasal placodes in the mouse. *J. Embryol. Exp. Morphol.* **97**, 95–110.

Hongo, M., Itoi, M., Yamamura, Y. and Imanishi, J. (1993). Distribution of epidermal growth factor receptors in rabbit lens epithelial cells. *Invest. Ophthalmol. Vis. Sci.* **34**, 401–4.

Horowitz, A., Tkachenko, E. and Simons, M. (2002). Fibroblast growth factor–specific modulation of cellular response by syndecan-4. *J. Cell Biol.* **157**, 715–25.

Horstman, L. P. and Zalik, S. E. (1974). Growth of newt iris epithelial cells in vitro: a study of the cell cycle. *Exp. Cell Res.* **84**, 1–14.

Horwitz, J. (1992). Alpha-crystallin can function as a molecular chaperone. *Proc. Natl. Acad. Sci. USA* **89**, 10449–53.

Horwitz, J. (1993). The function of alpha-crystallin. *Invest. Ophthalmol. Vis. Sci.* **34**, 10–22.

Horwitz, J. (2003). Alpha crystallin. *Exp. Eye Res.* **76**, 145–53.

Horwitz, J., Bova, M. P., Ding, L. L., Haley, D. A. and Stewart, P. L. (1999). Lens alpha-crystallin: function and structure. *Eye* **13**, 403–8.

Horwitz, J., Huang, Q. L., Ding, L. and Bova, M. P. (1998). Lens alpha-crystallin: chaperone-like properties. *Methods Enzymol.* **290**, 365–83.

Hough, R. B., Avivi, A., Davis, J., Joel, A., Nevo, E. and Piatigorsky, J. (2002). Adaptive evolution of small heat shock protein/alpha B-crystallin promoter activity of the blind subterranean mole rat, Spalax ehrenbergi. *Proc. Natl. Acad. Sci. USA* **99**, 8145–50.

Howlett, A. R. and Bissell, M. J. (1993). The influence of tissue microenvironment (stroma and extracellular matrix) on the development and function of mammary epithelium. *Epithelial Cell Biol.* **2**, 79–89.

Huibregtse, J. M., Scheffner, M. and Howley, P. M. (1991). A cellular protein mediates association of p53 with the E6 oncoprotein of human papillomavirus types 16 and 18. *EMBO J.* **10**, 4129–35.

Huizinga, A., Bot, A. C. C., deMul, F. F. M., Vrensen, G. F. J. M. and Greve, J. (1989). Local variation in absolute water content of human and rabbit eye lenses measured by Raman microspectroscopy. *Exp. Eye Res.* **48**, 487–96.

Hummler, E., Cole, T. J., Blendy, J. A., Ganss, R., Aguzzi, A., Schmid, W., Beermann, F. and Schutz, G. (1994). Targeted mutation of the CREB gene: compensation within the CREB/ATF family of transcription factors. *Proc. Natl. Acad. Sci. USA* **91**, 5647–51.

Hung, F. C., Zhao, S., Chen, Q. and Overbeek, P. A. (2002). Retinal ablation and altered lens differentiation induced by ocular overexpression of BMP7. *Vision Res.* **42**, 427–38.

Hunter, T. and Pines, J. (1994). Cyclins and cancer. II. Cyclin D and CDK inhibitors come of age. *Cell* **79**, 573–82.

Hurford, R. K. J., Cobrinik, D., Lee, M.-H. and Dyson, N. (1997). pRB and p107/p130 are required for the regulated expression of different sets of E2F responsive genes. *Genes Dev.* **11**, 1447–63.

Hussain, M. A. and Habener, J. F. (1999). Glucagon gene transcription activation mediated by synergistic interactions of pax-6 and cdx-2 with the p300 co-activator. *J. Biol. Chem.* **274**, 28950–7.

Hyatt, G. A., and Beebe, D. C. (1993). Regulation of lens cell growth and polarity by an embryo-specific growth factor and by inhibitors of lens cell proliferation and differentiation. *Development* **117**, 701–9.

Hyatt, G. A., Schmitt, E. A., Fadool, J. M. and Dowling, J. E. (1996a). Retinoic acid alters photoreceptor development in vivo. *Proc. Natl. Acad. Sci. USA* **93**, 13298–303.

Hyatt, G. A., Schmitt, E. A., Marsh-Armstrong, N., McCaffery, P., Drager, U. C. and Dowling, J. E. (1996b). Retinoic acid establishes ventral retinal characteristics. *Development* **122**, 195–204.

Hyde, R. K. and Griep, A. E. (2002). Unique roles for *E2f1* in the mouse lens in the absence of functional pRB proteins. *Invest. Ophthalmol. Vis. Sci.* **43**, 1509–16.

Hynes, R. O. (1992). Integrins: versatility, modulation, and signaling in cell adhesion [review]. *Cell* **69**, 11–25.

Hyuga, M., Kodama, R. and Eguchi, G. (1993). Basic fibroblast growth factor as one of the essential factors regulating lens transdifferentiation of pigmented epithelial cells. *Int. J. Dev. Biol.* **37**, 319–26.

Ibaraki, N., Lin, L. R. and Reddy, V. N. (1995). Effects of growth factors on proliferation and differentiation in human lens epithelial cells in early subculture. *Invest. Ophthalmol. Vis. Sci.* **36**, 2304–12.

Ibaraki, N., Lin, L. R. and Reddy, V. N. (1996). A study of growth factor receptors in human lens epithelial cells and their relationship to fiber differentiation. *Exp. Eye Res.* **63**, 683–92.

IHGS Consortium. (2001). Initial sequencing and analysis of the human genome. *Nature* **409**, 860–921.

Iio, A., Mochii, M., Agata, K., Kodama, R. and Eguchi, G. (1994). Expression of the retinal pigmented epithelial cell–specific pP344 gene during development of the chicken eye and identification of its product. *Dev. Growth Differ.* **36**, 155–64.

Ikeda, Y. (1934). Beitrage zur Analyse der Wolffschen Linsenregeneration durch xenoplastische Implantation der Iris in das entlinste Auge bei *Triton* und *Hynobius*. *Arb. Anat. Inst. Sendai* **16**, 69–82.

Ikeda, Y. (1935). Uber die Regeneration von Augenbechern an verschiedenen Korperstellen durch isolierte Irisstucke. *Arb. Anat. Inst. Sendai* **17**, 11–54.

Ikeda, Y. (1936). Neue Versuche zur Analyse deer Wolffschen Linsenregeneration. *Arb. Anat. Inst. Sendai* **18**, 1–16.

Ikeda, Y. and Amatatu, H. (1941). Uber den Unterschied der Erhaltungsmoglich-keit der Linse bei zwei Urodelenarten (*Triturus pyrrhogaster* and *Hynobius nebulosus*), die sich bezuglich der Fahigkeit zur Wolffschen Linsenregeneration voneinander wesentlich verschieden verhalten. *Jpn. J. Med. Sci. I. Anat.* **8**, 205–26.

Ikeda, Y. and Kojima, T. (1940). Zur Frage der paralysierenden Wirkung der Linse auf die auslosenden Faktoren fur die Wolffsche Linsenregeneration. *Jpn. J. Med. Sci. I. Anat.* **8**, 51–73.

Ikeda, A. and Zwaan, J. (1966). Immunofluorescence studies on induction and differentiation of the chicken eye lens. *Invest. Ophthalmol.* **5**, 402–12.
Ikeda, A. and Zwaan, J. (1967). The changing cellular localization of alpha-crystallin in the lens of the chicken embryo, studied by immunofluorescence. *Dev. Biol.* **15**, 348–67.
Imokawa, Y. and Eguchi, G. (1992). Expression and distribution of regeneration responsive molecule during normal development of the newt, *Cynops pyrrhogaster*. *Int. J. Dev. Biol.* **36**, 407–12.
Imokawa, Y., Ono, S., Takeuchi, T. and Eguchi, G. (1992). Analysis of a unique molecule responsible for regeneration and stabilization of differentiated state of tissue cells. *Int. J. Dev. Biol.* **36**, 399–405.
Inagaki, M., Matsuoka, Y., Tsujimura, K., Ando, S., Tokui, T., Takahashi, T. and Inagaki, N. (1996). Dynamic property of intermediate filaments: regulation by phosphorylation. *Bioessays* **18**, 481–7.
Infante, A. S., Stein, M. S., Zhai, Y., Borisy, G. G. and Gundersen, G. G. (2000). Detyrosinated (Glu) microtubules are stabilized by an ATP-sensitive plus-end cap. *J. Cell Sci.* **113**, 3907–19.
Ingolia, T. D. and Craig, E. A. (1982). Four small *Drosophila* heat shock proteins are related to each other and to mammalian alpha-crystallin. *Proc. Natl. Acad. Sci. USA* **79**, 2360–4.
Inoue, T., Nakamura, S. and Osumi, N. (2000). Fate mapping of the mouse prosencephalic neural plate. *Dev. Biol.* **219**, 373–83.
Ireland, M. and Mrock, L. (2000). Differentiation of chick lens epithelial cells: involvement of the epidermal growth factor receptor and endogenous ligand. *Invest. Ophthalmol. Vis. Sci.* **41**, 183–90.
Ireland, M. E., Goebel, D. J., Maisel, H., Kiner, D. and Poosch, M. S. (1997). Quantification and regulation of mRNAs encoding beaded filament proteins in the chick lens. *Curr. Eye Res.* **16**, 838–46.
Ireland, M. E. and Jacks, L. A. (1989). Initial characterization of lens beta-adrenergic receptors. *Invest. Ophthalmol. Vis. Sci.* **30**, 2190–4.
Ireland, M. E., Klettner, C. and Nunlee, W. (1993). Cyclic AMP-mediated phosphorylation and insolubilization of a 49-kDa cytoskeletal marker protein of lens fiber terminal differentiation. *Exp. Eye Res.* **56**, 453–61.
Ireland, M. E. and Shanbom, S. (1991). Lens beta-adrenergic receptors: functional coupling to adenylate cyclase and photoaffinity labeling. *Invest. Ophthalmol. Vis. Sci.* **32**, 541–8.
Ireland, M. E., Wallace, P., Sandilands, A., Poosch, M., Kasper, M., Graw, J., Liu, A., Maisel, H., Prescott, A. R., Hutcheson, A. M., et al. (2000). Up-regulation of novel intermediate filament proteins in primary fiber cells: an indicator of all vertebrate lens fiber differentiation? *Anat. Rec.* **258**, 25–33.
Irvine, A. D. and McLean, W. H. I. (1999). Human keratin diseases: increasing spectrum of disease and subtlety of phenotype-genotype correlation. *Brit. J. Dermatol.* **140**, 815–28.
Irwin, M., Marin, M. C., Phillips, A. C., Seelan, R. S., Smith, D. I., Liu, W., Flores, E. R., Tsai, K. Y., Jacks, T., Vousden, K. H., et al. (2000). Role for the p53 homologue p73 in E2F-1–induced apoptosis. *Nature* **407**, 645–8.
Isaacs, H. V. (1997). New perspectives on the role of the fibroblast growth factor family in amphibian development. *Cell Mol. Life Sci.* **53**, 350–61.
Ishibashi, T., Hatae, T. and Inomata, H. (1994). Collagen types in human posterior capsule opacification. *J. Cat. Refract. Surg.* **20**, 643–6.
Ishibashi, S. and Yasuda, K. (2001). Distinct roles of *maf* genes during *Xenopus* lens development. *Mech. Dev.* **101**, 155–66.
Ishida-Yamamoto, A., McGrath, J. A., Chapman, S. J., Leigh, I. M., Lane, E. B. and Eady, R. A. (1991). Epidermolysis bullosa simplex (Dowling-Meara type) is a genetic disease characterized by an abnormal keratin-filament network involving keratins K5 and K14. *J. Invest. Dermatol.* **97**, 959–68.
Ishizaki, Y., Jacobson, M. D. and Raff, M. C. (1998). A role for caspases in lens fiber differentiation. *J. Cell Biol.* **140**, 153–8.
Ishizaki, Y., Voyvodic, J. T., Burne, J. F. and Raff, M. C. (1993). Control of lens epithelial survival. *J. Cell Biol.* **121**, 899–908.
Itaranta, P., Lin, Y., Perasaari, J., Roel, G., Destree, O. and Vainio, S. (2002). Wnt-6 is expressed in the ureter bud and induces kidney tubule development in vitro. *Genesis* **32**, 259–68.
Ito, M., Hayashi, T., Kuroiwa, A. and Okamoto, M. (1999). Lens formation by pigmented epithelial

cell reaggregate from dorsal iris implanted into limb blastema in the adult newt. *Dev. Growth Differ.* **41**, 429–40.
Itoh, Y. (1976). Enhancement of differentiation of lens and pigment cells by ascorbic acid in cultures of neural retinal cells of chick embryos. *Dev. Biol.* **54**, 157–62.
Itoh, Y. and Eguchi, G. (1982). Characterization of various cell states in transdifferentiation of pigmented epithelial cells. *Dev. Growth Differ.* **24**, 369 (B38).
Itoh, Y. and Eguchi, G. (1986a). *In vitro* analysis of cellular metaplasia from pigmented epithelial cells to lens phenotypes: a unique model system for studying cellular and molecular mechanisms of "transdifferentiation." *Dev. Biol.* **115**, 353–62.
Itoh, Y. and Eguchi, G. (1986b). Enhancement of expression of lens phenotype in cultures of pigmented epithelial cells by hyaluronidase in the presence of phenylthiourea. *Cell Differ.* **18**, 173–82.
Ivashkiv, L. B., Liou, H. C., Kara, C. J., Lamph, W. W., Verma, I. M. and Glimcher, L. H. (1990). mXBP/CRE-BP2 and c-Jun form a complex which binds to the cyclic AMP, but not to the 12-O-tetradecanoylphorbol-13-acetate response element. *Mol. Cell. Biol.* **10**, 1609–21.
Iwaki, A., Nagano, T., Nakagawa, M., Iwaki, T. and Fukumaki, Y. (1997). Identification and characterization of the gene encoding a new member of the alpha-crystallin/small hsp family, closely linked to the alphaB-crystallin gene in a head-to-head manner. *Genomics* **45**, 386–94.
Iwaki, T., Kume-Iwaki, A., Liem, R. K. and Goldman, J. E. (1989). Alpha B-crystallin is expressed in non-lenticular tissues and accumulates in Alexander's disease brain. *Cell* **57**, 71–8.
Iwaki, T., Wisniewski, T., Iwaki, A., Corbin, E., Tomokane, N., Tateishi, J. and Goldman, J. E. (1992). Accumulation of alpha B-crystallin in central nervous system glia and neurons in pathologic conditions. *Am. J. Pathol.* **140**, 345–56.
Jacks, T., Fazeli, A., Schmitt, E. M., Bronson, R. T., Goodell, M. A. and Weinberg, R. A. (1992). Effects of an Rb mutation in the mouse. *Nature* **359**, 295–300.
Jacobs, D. B., Ireland, M., Pickett, T., Maisel, H. and Grunberger, G. (1992). Functional characterization of insulin and IGF-1 receptors in chicken lens epithelial and fiber cells. *Curr. Eye Res.* **11**, 1137–45.
Jacobson, A. (1958). The roles of neural and non-neural tissues in lens induction. *J. Exp. Zool.* **139**, 525–57.
Jacobson, A. (1966). Inductive processes in embryonic development. *Science* **152**, 25–52.
Jacobson, A. and Sater, A. (1988). Features of embryonic induction. *Development* **104**, 341–58.
Jacobson, A. G. (1963a). The determination and positioning of the nose, lens and ear. I. Interactions within the ectoderm, and between the ectoderm and underlying tissues. *J. Exp. Zool.* **154**, 273–84.
Jacobson, A. G. (1963b). The determination and positioning of the nose, lens and ear. II. The role of the endoderm. *J. Exp. Zool.* **154**, 285–92.
Jacobson, A. G. (1963c). The determination and positioning of the nose, lens and ear. III. Effects of reversing the antero-posterior axis of epidermis, neural plate and neural fold. *J. Exp. Zool.* **154**, 293–303.
Jacobson, M. and Hirose, G. (1978). Origin of the retina from both sides of the embryonic brain: a contribution to the problem of crossing at the optic chiasma. *Science* **202**, 637–39.
Jakob, U., Gaestel, M., Engel, K. and Buchner, J. (1993). Small heat shock proteins are molecular chaperones. *J. Biol. Chem.* **268**, 1517–20.
Jakobiec, F. A., ed. (1982). *Ocular Anatomy, Embryology, and Teratology*. Philadelphia: Harper & Row.
Jakobs, P. M., Hess, J. F., FitzGerald, P. G., Kramer, P., Weleber, R. G. and Litt, M. (2000). Autosomal-dominant congenital cataract associated with a deletion mutation in the human beaded filament protein gene BFSP2. *Am. J. Hum. Genet.* **66**, 1432–6.
Janicke, R. U., Ng, P., Sprengart, M. L. and Porter, A. G. (1998). Caspase-3 is required for alpha-fodrin cleavage but dispensable for cleavage of other death substrates in apoptosis. *J. Biol. Chem.* **273**, 15540–5.
Jansen, G., Groenen, P., Bachner, D., Jap, P. H. K., Coerwinkel, M., Oerlemans, F., Vandenbroek, W., Gohlsch, B., Pette, D., Plomp, J. J., et al. (1996). Abnormal myotonic dystrophy protein kinase levels produce only mild myopathy in mice. *Nat. Genet.* **13**, 316–24.

Janssens, H. and Gehring, W. J. (1999). Isolation and characterization of *drosocrystallin*, a lens crystallin gene of *Drosophila melanogaster*. *Dev. Biol.* **207**, 204–14.

Jasoni, C., Hendrickson, A. and Roelink, H. (1999). Analysis of chicken Wnt-13 expression demonstrates coincidence with cell division in the developing eye and is consistent with a role in induction. *Dev. Dyn.* **215**, 215–24.

Jean, D., Ewan, K. and Gruss, P. (1998). Molecular regulators involved in vertebrate eye development. *Mech. Dev.* **76**, 3–18.

Ji, X., Rosenvinge, E. C. v., Johnson, W. W., Tomarev, S. I., Piatigorsky, J., Armstrong, R. N. and Gilliland, G. L. (1995). Three-dimensional structure, catalytic properties, and evolution of a sigma class glutathione transferase from squid, a progenitor of the lens S-crystallins of cephalopods. *Biochemistry* **34**, 5317–28.

Jiang, Z., Chung, S. K., Zhou, C., Cammarata, P. R. and Chung, S. S. M. (2000). Overexpression of Na^+-dependent myo-inositol transporter gene in mouse lens led to congenital cataract. *Invest. Ophthalmol. Vis. Sci.* **41**, 1467–72.

Jiang, J. X. and Goodenough, D. A. (1996). Heteromeric connexons in lens gap junction channels. *Proc. Natl. Acad. Sci. USA.* **93**, 1287–91.

Jiang, J. X. and Goodenough, D. A. (1998). Phosphorylation of lens fiber connexins in lens organ cultures. *Eur. J. Biochem.* **255**, 37–44.

Jiang, J. X., White, T. W., Goodenough, D. A. and Paul, D. L. (1994). Molecular cloning and functional characterization of chick lens fiber connexin45.6. *Mol. Biol. Cell* **5**, 363–73.

Jiang, R. and Grabel, L. B. (1995). Function and differential regulation of the alpha 6 integrin isoforms during parietal endoderm differentiation. *Exp. Cell Res.* **217**, 195–204.

Jimenez-Asensio, J. and Garland, D. (2000). A lens glutathione S-transferase, class mu, with thiol-specific antioxidant activity. *Exp. Eye Res.* **71**, 255–65.

Jin, S., Shimizu, M., Balasubramanyam, A. and Epstein, H. F. (2000). Myotonic dystrophy protein kinase (DMPK) induces actin cytoskeletal reorganization and apoptotic-like blebbing in lens cells. *Cell Motil. Cytoskeleton* **45**, 133–48.

Johnson, M. C. and Beebe, D. C. (1984). Growth, synthesis and regional specialization of the embryonic chicken lens capsule. *Exp. Eye Res.* **38**, 579–92.

Jones, S. E., Jomary, C., Grist, J., Makwana, J. and Neal, M. J. (1999). Retinal expression of γ-crystallins in the mouse. *Invest. Opthalmol. Vis. Sci.* **40**, 3017–20.

Jordan, T., Hanson, I., Zaletayev, D., Hodgson, S., Prosser, J., Seawright, A., Hastie, N. and van Heyningen, V. (1992). The human PAX6 gene is mutated in two patients with aniridia. *Nat. Genet.* **1**, 328–32.

Joshi, H. C. (1994). Microtubule organising centres and γ-tubulin. *Curr. Opin. Cell Biol.* **6**, 55–62.

Jurand, A. and Yamada, T. (1967). Elimination of mitochondria during Wolffian lens degeneration. *Exp. Cell Res.* **46**, 636–8.

Kaestner, K. H., Knochel, W. and Martinez, D. E. (2000). Unified nomenclature for the winged helix/forkhead transcription factors. *Genes Dev.* **14**, 142–6.

Kaiser, H. W., O'Keefe, E. and Bennett, V. (1989). Adducin: Ca^{++}-dependent association with sites of cell-cell contact. *J. Cell Biol.* **109**, 557–69.

Kaiser-Kupfer, M. I., Freidlin, V., Datiles, M. B., Edwards, P. A., Sherman, J. L., Parry, D., McCain, L. M. and Eldridge, R. (1989). The association of posterior capsular lens opacities with bilateral acoustic neuromas in patients with neurofibromatosis type 2. *Arch. Ophthalmol.* **107**, 541–4.

Kaltner, H. and Stiersdorfer, B. (1998). Animal lectins as cell adhesion molecules. *Acta Anat.* **161**, 162–79.

Kamachi, Y. (1996). Involvement of SOX proteins in activation of crystallin genes and lens development [in Japanese]. *Tanpakushitsu Kakusan Koso* **41**, 1113–23.

Kamachi, Y., Cheah, K. S. E. and Kondoh, H. (1999). Mechanism of regulatory target selection by the SOX high-mobility group domain proteins as revealed by comparison of SOX1/2/3 and SOX9. *Mol. Cell. Biol.* **19**, 107–20.

Kamachi, Y. and Kondoh, H. (1993). Overlapping positive and negative regulatory elements determine lens-specific activity of the δ1-crystallin enhancer. *Mol. Cell. Biol.* **13**, 5206–15.

Kamachi, Y., Sockanathan, S., Liu, Q., Breitman, M., Lovell-Badge, R. and Kondoh, H. (1995). Involvement of SOX proteins in lens-specific activation of crystallin genes. *EMBO J.* **14**, 3510–19.

Kamachi, Y., Uchikawa, M., Collignon, J., Lovell-Badge, R. and Kondoh, H. (1998). Involvement of

Sox1, *2* and *3* in the early and subsequent molecular events of lens induction. *Development* **125**, 2521–32.

Kamachi, Y., Uchikawa, M. and Kondoh, H. (2000). Pairing SOX off with partners in the regulation of embryonic development. *Trends Genet.* **16**, 182–7.

Kamachi, Y., Uchikawa, M., Tanouchi, A., Sekido, R. and Kondoh, H. (2001). Pax6 and SOX2 form a co-DNA–binding partner complex that regulates initiation of lens development. *Genes Dev.* **15**, 1272–86.

Kamijo, T., Zindy, F., Roussel, M. F., Quelle, D. E., Downing, J. R., Ashmun, R. A., Grosveld, G. and Sherr, C. J. (1997). Tumor suppression at the mouse INK4a locus mediated by the alternative reading frame product p19ARF. *Cell* **91**, 649–59.

Kammandel, B., Chowdhury, K., Stoykova, A., Aparicio, S., Brenner, S. and Gruss, P. (1999). Distinct *cis*-essential modules direct the time-space pattern of the *Pax6* gene activity. *Dev. Biol.* **205**, 79–97.

Kannabiran, C., Rogan, P. K., Olmos, L., Basti, S., Rao, G. N., Kaiser-Kupfer, M. and Hejtmancik, J. F. (1998). Autosomal dominant zonular cataract with sutural opacities is associated with a splice mutation in the betaA3/A1-crystallin gene. *Mol. Vis.* **4**, 21.

Kannan, R., Yi, J. R., Tang, D., Zlokovic, B. V. and Kaplowitz, N. (1996). Identification of a novel, sodium-dependent, reduced glutathione transporter in the rat lens epithelium. *Invest. Ophthalmol. Vis. Sci.* **37**, 2269–75.

Kantorow, M., Becker, K., Sax, C. M., Ozato, K. and Piatigorsky, J. (1993a). Binding of tissue-specific forms of alpha A-CRYBP1 to their regulatory sequence in the mouse alpha A-crystallin–encoding gene: double-label immunoblotting of UV-crosslinked complexes. *Gene* **131**, 159–65.

Kantorow, M., Cvekl, A., Sax, C. M. and Piatigorsky, J. (1993b). Protein-DNA interactions of the mouse alpha A-crystallin control regions: differences between expressing and non-expressing cells. *J. Mol. Biol.* **230**, 425–35.

Kantorow, M., Horwitz, J. and Carper, D. (1998). Up-regulation of osteonectin/SPARC in age-related cataractous human lens epithelia. *Mol. Vis.* **4**, 17.

Kantorow, M., Horwitz, J., Sergeev, Y., Hejtmancik, J. F. and Piatigorsky, J. (1997). Extralenticular expression, cAMP-dependent kinase phosphorylation and autophosphorylation of βB2-crystallin. *Invest. Opthalmol. Vis. Sci.* **38**, S205.

Kantorow, M., Horwitz, J., van Boekel, M. A., de Jong, W. W. and Piatigorsky, J. (1995). Conversion from oligomers to tetramers enhances autophosphorylation by lens alpha A-crystallin: specificity between alpha A- and alpha B-crystallin subunits. *J. Biol. Chem.* **270**, 17215–20.

Kantorow, M., Huang, Q., Yang, K. J., Sage, E. H., Magabo, K. S., Miller, K. M. and Horwitz, J. (2000). Increased expression of osteonectin/SPARC mRNA and protein in age-related human cataracts and spatial expression in the normal human lens. *Mol. Vis.* **6**, 24–9.

Kantorow, M. and Piatigorsky, J. (1994). Alpha-crystallin/small heat shock protein has autokinase activity. *Proc. Natl. Acad. Sci. USA* **91**, 3112–6.

Kappelhof, J. P. and Vrensen, G. F. (1992). The pathology of after- cataract: a minireview. *Acta Ophthalmol. Suppl.* **205**, 13–24.

Kappelhof, J. P., Vrensen, G. F., de Jong, P. T., Pameyer, J. and Willekens, B. (1986). An ultrastructural study of Elschnig's pearls in the pseudophakic eye. *Am. J. Ophthalmol.* **101**, 58–69.

Karasaki, S. (1964). An electron microscopic study of Wolffian lens regeneration in the adult newt. *J. Ultrastruc. Res.* **11**, 246–73.

Karim, A. K. A., Jacob, T. J. C. and Thompson, G. M. (1987). The human anterior lens capsule: cell density, morphology and mitotic index in normal and cataractous lenses. *Exp. Eye Res.* **45**, 865–74.

Karkinen-Jääskeläinen, M. (1978a). Permissive and directive interactions in lens induction. *J. Embryol. Exp. Morph.* **44**, 167–79.

Karkinen-Jääskeläinen, M. (1978b). Transfilter lens induction in avian embryos. *Differentiation* **12**, 31–7.

Karpinski, B. A., Morle, G. D., Huggenvik, J., Uhler, M. D. and Leiden, J. M. (1992). Molecular cloning of human CREB-2: an ATF/CREB transcription factor that can negatively regulate transcription from the cAMP response element. *Proc. Natl. Acad. Sci. USA* **89**, 4820–4.

Kasper, M., Moll, R., Stosiek, P. and Karsten, U. (1988). Patterns of cytokeratin and vimentin expression in the human eye. *Histochemistry* **89**, 369–77.

Kasper, M. and Viebahn, C. (1992). Cytokeratin expression and early lens development. *Anat. Embryol (Berlin)* **186**, 285–90.

Kastner, P., Grondona, J. M., Mark, M., Gansmuller, A., LeMeur, M., Decimo, D., Vonesch, J. L., Dolle, P. and Chambon, P. (1994). Genetic analysis of RXR alpha developmental function: convergence of RXR and RAR signaling pathways in heart and eye morphogenesis. *Cell* **78**, 987–1003.

Kataoka, K., Fujiwara, K. T., Noda, M. and Nishizawa, M. (1994a). MafB, a new Maf family transcription activator that can associate with Maf and Fos but not with Jun. *Mol. Cell. Biol.* **14**, 7581–91.

Kataoka, K., Han, S. I., Shioda, S., Hirai, M., Nishizawa, M. and Handa, H. (2002). MafA is a glucose-regulated and pancreatic beta-cell-specific transcriptional activator for the insulin gene. *J. Biol. Chem.* **277**, 49903–10.

Kataoka, K., Nishizawa, M. and Kawai, S. (1993). Structure-function analysis of the maf oncogene product, a member of the b-Zip protein family. *J. Virol.* **67**, 2133–41.

Kataoka, K., Noda, M. and Nishizawa, M. (1994b). Maf nuclear oncoprotein recognizes sequences related to an AP-1 site and forms heterodimers with both Fos and Jun. *Mol. Cell. Biol.* **14**, 700–12.

Katar, M., Alcala, J. and Maisel, H. (1993). NCAM of the mammalian lens. *Curr. Eye Res.* **12**, 191–6.

Kato, J., Matsushime, H., Hiebert, S. W., Ewen, M. E. and Sherr, C. J. (1993). Direct binding of cyclin D to the retinoblastoma gene product (pRb) and pRb phosphorylation by the cyclin D-dependent kinase CDK4. *Genes Dev.* **7**, 331–42.

Kato, K., Shinohara, H., Kurobe, N., Goto, S., Inaguma, Y. and Ohshima, K. (1991). Immunoreactive alpha A crystallin in rat non-lenticular tissues detected with a sensitive immunoassay method. *Biochim. Biophys. Acta* **1080**, 173–80.

Katoh, A. and Yashida, K. (1973). Delta crystallin synthesis during chick lens differentiation. *Exp. Eye Res.* **15**, 353–60.

Kaufmann, E. and Knochel, W. (1996). Five years on the wings of fork head. *Mech. Dev.* **57**, 3–20.

Kaulen, P., Kahie, G., Keller, K. and Wlooensak, J. (1991). Autoradiographic mapping of the glucose transporter with cytochalasin B in the mammalian eye. *Invest. Ophthalmol. Vis. Sci.* **32**, 1903–11.

Kaverina, I., Rottner, K. and Small, J. V. (1998). Targeting, capture, and stabilization of microtubules at early focal adhesions. *J. Cell Biol.* **142**, 181–90.

Kawakami, K., Ohto, H., Takizawa, T. and Saito, T. (1996). Identification and expression of *Six* family genes in mouse retina. *FEBS Lett.* **393**, 259–63.

Kawakami, K., Sato, S., Ozaki, H. and Ikeda, K. (2000). *Six* family genes-structure and function as transcription factors and their roles in development. *BioEssays* **22**, 616–26.

Kawauchi, S., Takahashi, S., Nakajima, O., Ogino, H., Morita, M., Nishizawa, M., Yasuda, K. and Yamamoto, M. (1999). Regulation of lens fiber cell differentiation by transcription factor c-Maf. *J. Biol. Chem.* **274**, 19254–60.

Kelley, P., Sado, Y. and Duncan, M. (2002). Collagen IV in the developing lens capsule. *Matrix Biol.* **21**, 415–23.

Kenworthy, A. K., Magid, A. D., Oliver, T. N. and McIntosh, T. J. (1994). Colloid osmotic pressure of steer alpha- and beta-crystallins: possible functional roles for lens crystallin distribution and structural diversity. *Exp. Eye Res.* **59**, 11–30.

Kenyon, K. L., Moody, S. A. and Jamrich, M. (1999). A novel *forkhead* gene mediates early steps during *Xenopus* lens formation. *Development* **126**, 5107–16.

Kern, H. L. and Ho, C. K. (1973). Localization and specificity of the transport system for sugars in the calf lens. *Exp. Eye Res.* **15**, 751–65.

Kerppola, T. K. and Curran, T. (1994a). A conserved region adjacent to the basic domain is required for recognition of an extended DNA binding site by Maf/Nrl family proteins. *Oncogene* **9**, 3149–58.

Kerppola, T. K. and Curran, T. (1994b). Maf and Nrl can bind to AP-1 sites and form heterodimers with Fos and Jun. *Oncogene* **9**, 675–84.

Kibbelaar, M. A., Ramaekers, F. C., Ringens, P. J., Selten-Versteegen, A. M., Poels, L. G., Jap, P. H., van Rossum, A. L., Feltkamp, T. E. and Bloemendal, H. (1980). Is actin in eye lens a possible factor in visual accommodation? *Nature* **285**, 506–8.

Kibbelaar, M. A., Selten-Versteegen, A. M., Dunia, I., Benedetti, E. L. and Bloemendal, H. (1979). Actin in mammalian lens. *Eur. J. Biochem.* **95**, 543–9.

Kim, A. S., Anderson, S. A., Rubenstein, J. L., Lowenstein, D. H. and Pleasure, S. J. (2001). *Pax-6* regulates expression of *SFRP-2* and *Wnt-7b* in the developing CNS. *J. Neurosc.* **21**, RC132.

Kim, J. I., Li, T. S., Ho, I. C., Grusby, M. J. and Glimcher, L. H. (1999). Requirement for the c-Maf transcription factor in crystallin gene regulation and lens development. *Proc. Natl. Acad. Sci. USA* **96**, 3781–5.

Kim, R. Y., Lietman, T., Piatigorsky, J. and Wistow, G. J. (1991). Structure and expression of the duck α-enolase/τ-crystallin–encoding gene. *Gene* **103**, 193–200.

King, H. (1905). Experimental studies on the eye of the frog embryo. *Arch. Entw.-Mech.* **19**, 85–107.

King, R. W., Jackson, P. K. and Kirschner, M. W. (1994). Mitosis in transition. *Cell* **79**, 563–71.

Kioussi, C., O'Connell, S., St-Onge, L., Treier, M., Gleiberman, A. S., Gruss, P. and Rosenfeld, M. G. (1999). *Pax6* is essential for establishing ventral-dorsal cell boundaries in pituitary gland development. *Proc. Natl. Acad. Sci. USA* **96**, 14378–82.

Kirsch, T., Koyama, E., Liu, M., Golub, E. E. and Pacifici, M. (2002). Syndecan-3 is a selective regulator of chondrocyte proliferation. *J. Biol. Chem.* **277**, 42171–7.

Kishi, M., Mizuseki, K., Sasai, N., Yamazaki, H., Shiota, K., Nakanishi, S. and Sasai, Y. (2000). Requirement of *Sox2*-mediated signaling for differentiation of early *Xenopus* neuroectoderm. *Development* **127**, 791–800.

Kistler, J. and Bullivant, S. (1987). Protein processing in lens intercellular junctions: cleavage of MP70 to MP38. *Invest. Ophthalmol. Vis. Sci.* **28**, 1687–92.

Kistler, J. and Bullivant, S. (1988). Dissociation of lens fibre gap junctions releases MP70. *J. Cell Sci.* **91**, 415–21.

Kistler, J., Christie, D. and Bullivant, S. (1988). Homologies between gap junction proteins in lens, heart, and liver. *Nature* **331**, 721–3.

Kistler, J., Goldie, K., Donaldson, P. and Engel, A. (1994). Reconstitution of native-type noncrystalline lens fiber gap junctions from isolated hemichannels. *J. Cell Biol.* **126**, 1047–58.

Kistler, J., Kirkland, B. and Bullivant, S. (1985). Identification of a 70,000-D protein in lens membrane junctional domains. *J. Cell Biol.* **101**, 28–35.

Kistler, J., Schaller, J. and Sigrist, H. (1990). MP38 contains the membrane-embedded domain of the lens fiber gap junction protein MP70. *J. Biol. Chem.* **265**, 13357–61.

Kleiman, N. J., Chiesa, R., Gawinowicz-Kolks, M. A. and Spector, A. (1988). Phosphorylation of β-crystallin B2 (βBp) in the bovine lens. *J. Biol. Chem.* **263**, 14978–83.

Klein, M., Moore, B., Rothstein, H., Hayden, J., Gordon, S., Holsclaw, D. and Sobrin, J. (1989). A comparison of growth regulation of mammalian with amphibian lens epithelium. *Lens Eye Toxic Res.* **6**, 675–86.

Kleinjan, D. A., Seawright, A., Schedl, A., Quinlan, R. A., Danes, S. and van Heyningen, V. (2001). Aniridia-associated translocations, DNase hypersensitivity, sequence comparison and transgenic analysis redefine the functional domain of PAX6. *Hum. Mol. Genet.* **10**, 2049–59.

Klement, J. F., Cvekl, A. and Piatigorsky, J. (1993). Functional elements DE2A, DE2B, and DE1A and the TATA box are required for activity of the chicken alpha A-crystallin gene in transfected lens epithelial cells. *J. Biol. Chem.* **268**, 6777–84.

Klement, J. F., Wawrousek, E. F. and Piatigorsky, J. (1989). Tissue-specific expression of the chicken αA-crystallin gene in cultured lens epithelia and transgenic mice. *J. Biol. Chem.* **264**, 19837–44.

Klemenz, R., Frohli, E., Steiger, R. H., Schafer, R. and Aoyama, A. (1991). Alpha B-crystallin is a small heat shock protein. *Proc. Natl. Acad. Sci. USA* **88**, 3652–6.

Klesert, T. R., Cho, D. H., Clark, J. I., Maylie, J., Adelman, J., Snider, L., Yuen, E. C., Soriano, P. and Tapscott, S. J. (2000). Mice deficient in Six5 develop cataracts: implications for myotonic dystrophy. *Nat. Genet.* **25**, 105–9.

Klok, E., Lubsen, N. H., Chamberlain, C. G. and McAvoy, J. W. (1998). Induction and maintenance of differentiation of rat lens epithelium by FGF-2, insulin and IGF-1. *Exp. Eye Res.* **67**, 425–31.

Klok, E. J., van Genesen, S. T., Civil, A., Schoenmakers, J. G. and Lubsen, N. H. (1998). Regulation of expression within a gene family: the case of the rat gammaB- and gammaD-crystallin promoters. *J. Biol. Chem.* **273**, 17206–15.

Klopp, N., Favor, J., Loster, J., Lutz, R. B., Neuhauser-Klaus, A., Prescott, A., Pretsch, W., Quinlan, R. A., Sandilands, A., Vrensen, G. F., et al. (1998). Three murine cataract mutants (Cat2) are defective in different gamma-crystallin genes. *Genomics* **52**, 152–8.

Klopp, N., Loster, J. and Graw, J. (2001). Characterization of a 1-bp deletion in the gammaE-crystallin gene leading to a nuclear and zonular cataract in the mouse [in process citation]. *Invest. Ophthalmol. Vis. Sci.* **42**, 183–7.

Kluwe, L., Pulst, S. M., Koppen, J. and Mautner, V. F. (1995). A 163-bp deletion in the neurofibromatosis 2 (NF2) gene associated with variant phenotypes [corrected; published erratum appears in *Hum. Genet.* **96**, 254]. *Hum. Genet.* **95**, 443–6.

Kobayashi, M., Toyama, R., Takeda, H., Dawid, I. B. and Kawakami, K. (1998). Overexpression of the forebrain-specific homeobox gene *six3* induces rostral forebrain enlargement in zebrafish. *Development* **125**, 2973–82.

Koch, C. and Nasmyth, K. (1994). Cell cycle regulated transcription in yeast. *Curr. Opin. Cell Biol.* **6**, 451–9.

Koch, P. J., Goldschmidt, M. D., Zimbelmann, R., Troyanovsky, R. and Franke, W. W. (1992). Complexity and expression patterns of the desmosomal cadherins. *Proc. Natl. Acad. Sci. USA* **89**, 353–7.

Kodama, R. and Eguchi, G. (1994). The loss of gap junctional cell-to-cell communication is coupled with dedifferentiation of retinal pigment epithelial cells in the course of transdifferentiation into lens. *Int. J. Dev. Biol.* **38**, 357–64.

Koff, A., Giordano, A., Desai, D., Yamashita, K., Harper, J. W., Elledge, S., Nishimoto, T., Morgan, D. O., Franza, B. R. and Roberts, J. M. (1992). Formation and activation of cyclin E–cdk2 complex during the G1 phase of the human cell cycle. *Science* **257**, 1689–94.

Kohno, T., Sorgente, N., Ishibashi, T., Goodnight, R. and Ryan, S. J. (1987). Immunofluorescent studies of fibronectin and laminin in the human eye. *Invest. Ophthalmol. Vis. Sci.* **28**, 506–14.

Kok, A., Lovicu, F. J., Chamberlain, C. G. and McAvoy, J. W. (2002). Influence of platelet-derived growth factor on lens epithelial cell proliferation and differentiation. *Growth Factors* **20**, 27–34.

Kolkova, K., Novitskaya, V., Pedersen, N., Berezin, V. and Bock, E. (2000). Neural cell adhesion molecule-stimulated neurite outgrowth depends on activation of protein kinase C and the Ras-mitogen-activated protein kinase pathway. *J. Neurosci.* **20**, 2238–46.

Komori, N., Usukura, J. and Matsumoto, H. (1992). Drosocrystallin, a major 52 kDa glycoprotein of the *Drosophila melanogaster* corneal lens: purification, biochemical characterization, and subcellular localization. *J. Cell Sci.* **102**, 191–201.

Kondoh, H. (1999). Transcription factors for lens development assessed *in vivo*. *Curr. Opin. Genet. Dev.* **9**, 301–8.

Kondoh, H., Araki, I., Yasuda, K., Matsubasa, T. and Mori, M. (1991). Expression of the chicken "delta 2-crystallin" gene in mouse cells: evidence for encoding of argininosuccinate lyase. *Gene* **99**, 267–71.

Kondoh, H., Katoh, K., Takahashi, Y., Fujisawa, H., Yokoyama, M., Kimura, S., Katsuki, M., Saito, M., Nomura, T., Hiramoto, Y., et al. (1987). Specific expression of the chicken δ-crystallin gene in the lens and the pyramidal neurons of the piriform cortex in transgenic mice. *Dev. Biol.* **120**, 177–85.

Kondoh, H., Uchikawa, M., Yoda, H., Takeda, H., Furutani-Seiki, M. and Karlstrom, R. O. (2000). Zebrafish mutations in gli-mediated hedgehog signaling lead to lens transdifferentiation from the adenohypophysis anlage. *Mech. Dev.* **96**, 165–74.

Kondoh, H., Yasuda, K. and Okada, T. S. (1983). Tissue-specific expression of a cloned chick delta-crystallin gene in mouse cells. *Nature* **301**, 440–2.

Konig, N., Zampighi, G. A. and Butler, P. J. (1997). Characterisation of the major intrinsic protein (MIP) from bovine lens fibre membranes by electron microscopy and hydrodynamics. *J. Mol. Biol.* **265**, 590–602.

Korchynskyi, O., Landstrom, M., Stoika, R., Funa, K., Heldin, C. H., ten Dijke, P. and Souchelnytskyi, S. (1999). Expression of Smad proteins in human colorectal cancer. *Int. J. Cancer* **82**, 197–202.

Koretz, J. F., Doss, E. W. and LaButti, J. N. (1998). Environmental factors influencing the chaperone-like activity of alpha-crystallin. *Int. J. Biol. Macromol.* **22**, 283–94.

Koroma, B. M., Yang, J. M. and Sundin, O. H. (1997). The Pax-6 homeobox gene is expressed throughout the corneal and conjunctival epithelia. *Invest. Ophthalmol. Vis. Sci.* **38**, 108–20.

Kosaka, M., Kodama, R. and Eguchi, G. (1998). *In vitro* culture system for iris-pigmented epithelial cells for molecular analysis of transdifferentiation. *Exp. Cell Res.* **245**, 245–51.

Koster, R. W., Kuhnlein, R. P. and Wittbrodt, J. (2000). Ectopic *Sox3* activity elicits sensory placode formation. *Mech. Dev.* **95**, 175–87.

Kostrouch, Z., Kostrouchova, M., Love, W., Jannini, E., Piatigorsky, J. and Rall, J. E. (1998). Retinoic acid X receptor in the diploblast, *Tripedalia cystophora*. *Proc. Natl. Acad. Sci. USA* **95**, 13442–7.

Kowalik, T. F., DeGregori, J., Schwarz, J. K. and Nevins, J. R. (1995). E2F1 overexpression in quiescent fibroblasts leads to induction of cellular DNA synthesis and apoptosis. *J. Virol.* **69**, 2491–500.

Kozmik, Z., Czerny, T. and Busslinger, M. (1997). Alternatively spliced insertions in the paired domain restrict the DNA sequence specificity of Pax6 and Pax8. *EMBO J.* **16**, 6793–803.

Kozmik, Z., Daube, M., Frei, E., Norman, B., Kos, L., Dishaw, L. J., Noll, M. and Piatigorsky, J. (2003). Role of Pax genes in eye evolution: a cnidarian PaxB gene uniting Pax2 and Pax6 functions. *Dev. Cell.* **5**, 773–85.

Kralova, J., Czerny, T., Spanielova, H., Ratajova, V., and Kozmik, Z. (2002). Complex regulatory element within the gammaE- and gammaF-crystallin enhancers mediates Pax6 regulation and is required for induction by retinoic acid. *Gene* **286**, 271–82

Kramer, P. L., LaMorticella, D., Schilling, K., Billingslea, A. M., Weleber, R. G. and Litt, M. (2000). A new locus for autosomal dominant congenital cataracts maps to chromosome 3. *Invest. Ophthalmol. Vis. Sci.* **41**, 36–9.

Krek, W. G., Ewan, M., Shirodkar, S., Arany, Z. Z., Kaelin, W. G. and Livingston, D. (1994). Negative regulation of the growth-promoting transcription factor E2F-1 by stably bound cyclin A–dependent protein kinase. *Cell* **78**, 161–72.

Krek, W. G., Xu, G. and Livingston, D. M. (1995). Cyclin A kinase regulation of E2F-1 DNA binding function underlies suppression of an S phase checkpoint. *Cell* **83**, 1149–58.

Kreutziger, G. O. (1968). Freeze etching of intercellular junctions of mouse liver. In *Proceedings of the 26th Annual Meeting of the Electron Microscopy Society of America*, ed. C. J. Arcenaux. New Orleans and Baton Rouge: Clator's Publishing Division, pp. 234–5.

Kubota, Y., Kleinman, H. K., Martin, G. R. and Lavoley, T. J. (1988). Role of laminin and basement membrane in the morphological differentiation of human endothelial cells into capillary-1 structures. *J. Cell Biol.* **107**, 1589–98.

Kulyk, W. M., Zalik, S. E. and Dimitrov, E. (1987). Hyaluronic acid production and hyaluronidase activity in the newt iris during lens regeneration. *Exp. Cell Res.* **172**, 180–91.

Kumagai, A. and Dunphy, W. G. (1992). Regulation of the cdc25 protein during the cell cycle in *Xenopus* extracts. *Cell* **70**, 139–51.

Kumagai, A. K., Glasgow, B. J. and Pardridge, W. M. (1994). GLUT1 glucose transporter expression in the diabetic and nondiabetic human eye. *Invest. Ophthalmol. Vis. Sci.* **35**, 2887–94.

Kumar, L. V., Ramakrishna, T. and Rao, C. M. (1999). Structural and functional consequences of the mutation of a conserved arginine residue in alphaA and alphaB crystallins. *J. Biol. Chem.* **274**, 24137–41.

Kumar, N. M., Jarvis, L. J., TenBroek, E. and Louis C. F. (1993). Cloning and expression of a major rat lens membrane protein, MP20. *Exp. Eye Res.* **56**, 35–43.

Kumar, J. P. and Moses, K. (2001). EGF receptor and notch signaling act upstream of eyeless/Pax6 to control eye specification. *Cell* **104**, 687–97.

Kurosaka, D., Kato, K., Nagamoto, T. and Negishi, K. (1995). Growth factors influence contractility and alpha-smooth muscle actin expression in bovine lens epithelial cells. *Invest. Ophthalmol. Vis. Sci.* **36**, 1701–8.

Kurosaka, D. and Nagamoto, T. (1994). Inhibitory effect of TGF-beta 2 in human aqueous humor on bovine lens epithelial cell proliferation. *Invest. Ophthalmol. Vis. Sci.* **35**, 3408–12.

Kushmerick, C., Rice, S. J., Baldo, G. J., Haspel, H. C. and Mathias, R. T. (1995). Ion, water and neutral solute transport in *Xenopus* oocytes expressing frog lens MIP. *Exp. Eye Res.* **61**, 351–62.

Kuszak, J. R. (1995a). Development of lens sutures. *Prog. Ret. Eye Res.* **14**, 567–91.

Kuszak, J. R. (1995b). Ultrastructure of epithelial and fiber cells in the crystalline lens. In *International Review of Cytology: A Survey of Cell Biology*, ed. K. W. Jeon. San Diego: Academic Press, pp. 305–50.

Kuszak, J. R. (1997). A re-examination of primate lens epithelial cell size, density and structure as a function of development, growth and age. *Nova Acta Leopoldina* **NF 75**, 45–66.

Kuszak, J., Alcala, J. and Maisel, H. (1980). The surface morphology of embryonic adult chick lens-fiber cells. *Am. J. Anat.* **159**, 395–410.

Kuszak, J. R., Al-Ghoul, K. J. and Costello, M. J. (1998). Pathology of age-related human cataracts. In *Duane's Clinical Ophthalmology*, ed. W. Tasman and E. A. Jaeger. Philadelphia: Lippincott, Williams & Wilkins, pp. 1–14.

Kuszak, J. R., Bertram, B. A. and Rae, J. L. (1986). The ordered structure of the crystalline lens. In *Cell and Developmental Biology of the Eye: Development of Order in the Visual System*, ed. S. R. Hilfer and J. B. Sheffield. New York: Springer-Verlag, pp. 35–60.

Kuszak, J. R. and Brown, H. G. (1994). Embryology and anatomy of the lens. In *Principles and practice of ophthalmology: Basic Sciences*, ed. D. M. Albert and F. A. Jakobiec. Philadelphia: Saunders, pp. 82–96.

Kuszak, J. R. and Costello, M. J. (2003). Embryology and anatomy of human lenses. In *Duane's Clinical Ophthalmology* (on CD-ROM), ed. W. Tasman and E. A. Jaeger. Philadelphia: Lippincott, Williams & Wilkins.

Kuszak, J. R., Ennesser, C. A., Bertram, B. A., Imherr-McMannis, S., Jones-Rufer, L. S. and Weinstein, R. S. (1989). The contribution of cell-to-cell fusion to the ordered structure of the crystalline lens. *Lens Eye Toxic Res.* **6**, 639–73.

Kuszak, J. R., Ennesser, C. A., Umlas, J., Macsai-Kaplan, M. S. and Weinstein, R. S. (1988). The ultrastructure of fiber cells in primate lenses: a model for studying membrane senescence. *J. Ultrastruct. Mol. Struct. Res.* **100**, 60–74.

Kuszak, J. R., Macsai-Kaplan, M., Bloom, K. J., Rae, J. L. and Weinstein, R. S. (1985). Cell-to-cell fusion of lens fiber cells in situ: correlative light, scanning electron microscopic and freeze-fracture evidence. *J. Ultrastruct. Res.* **93**, 144–60.

Kuszak, J. R., Maisel, H. and Harding, C. V. (1978). Gap junctions of chick lens fiber cells. *Exp. Eye Res.* **27**, 495–8.

Kuszak, J. R., Novak, L. A. and Brown, H. G. (1995). An ultrastructural analysis of the epithelial-fiber interface (EFI) in primate lenses. *Exp. Eye Res.* **61**, 579–97.

Kuszak, J. R., Peterson, K. L. and Brown, H. G. (1996). Electron microscopic observations of the crystalline lens. *Microsc. Res. Tech.* **33**, 441–79.

Kuszak, J. R. and Rae, J. L. (1982). Scanning electron microscopy of the frog lens. *Exp. Eye Res.* **35**, 499–519.

Kuszak, J. R., Rae, J. L., Pauli, B. U. and Weinstein, R. S. (1982). Rotary replication of lens gap junctions. *J. Ultrastruct. Res.* **81**, 249–56.

Kuszak, J. R., Shek, Y. H., Carney, K. C. and Rae, J. L. (1985). A correlative freeze-etch and electrophysiological study of communicating junctions in crystalline lenses. *Curr. Eye Res.* **4**, 1145–53.

Kuszak, J. R., Sivak, J. G. and Weerheim, J. A. (1991). Lens optical quality is a direct function of lens sutural architecture. *Invest. Ophthalmol. Visual Sci.* **32**, 2119–29.

Kuwabara, T. (1968). Microtubules in the lens. *Arch. Ophthalmol.* **79**, 189–95.

Kuwabara, T. (1975). The maturation of the lens cell: a morphological study. *Exp. Eye Res.* **20**, 427–43.

Kuwabara, T. and Imaizumi, M. (1974). Denucleation process of the lens. *Invest. Ophthalmol. Vis. Sci.* **13**, 973–81.

Kuwada, S. K. and Li, X. (2000). Integrin alpha5/beta1 mediates fibronectin-dependent epithelial cell proliferation through epidermal growth factor receptor activation [in process citation]. *Mol. Biol. Cell* **11**, 2485–96.

Lacy, K. R., Jackson, P. K. and Stearns, T. (1999). Cyclin-dependent kinase control of centrosome duplication. *Proc. Natl. Acad. Sci. USA* **96**, 2817–22.

Lagunowich, L. A. and Grunwald, G. B. (1989). Expression of calcium-dependent cell adhesion during ocular development: a biochemical, histochemical and functional analysis. *Dev. Biol.* **135**, 158–71.

Lagunowich, L. A. and Grunwald, G. B. (1991). Tissue and age-specificity of post-translational modifications of N-cadherin during chick embryo development. *Differentiation* **47**, 19–27.

Lagutin, O., Zhu, C. Q. C., Furuta, Y., Rowitch, D. H., McMahon, A. P. and Oliver, G. (2001). *Six3* promotes the formation of ectopic optic vesicle–like structures in mouse embryos. *Dev. Dyn.* **221**, 342–9.

Lahoz, E. G., Liegeois, N. J., Zhang, P., Engelman, J. A., Horner, J., Silverman, A., Burde, R., Roussel, M. F., Sherr, C. J., Elledge, S. J., et al. (1999). Cyclin D- and E-dependent kinases and the p57KIP2 inhibitor: cooperative interactions in vivo. *Mol. Cell Biol.* **19**, 353–63.

Lampi, K. J., Ma, Z., Hanson, S. R., Azuma, M., Shih, M., Shearer, T. R., Smith, D. L., Smith, J. B. and David, L. L. (1998). Age-related changes in human lens crystallins identified by two-dimensional electrophoresis and mass spectrometry. *Exp. Eye Res.* **67**, 31–43.

Lampi, K. J., Ma, Z., Shih, M., Shearer, T. R., Smith, J. B., Smith, D. L. and David, L. L. (1997). Sequence analysis of βA3, βB3 and βA4 crystallins completes the identification of the major proteins in young human lens. *J. Biol. Chem.* **272**, 2268–75.

Lampi, K. J., Shih, M., Ueda, Y., Shearer, T. R. and David, L. L. (2002). Lens proteomics: analysis of rat crystallin sequences and two-dimensional electrophoresis map. *Invest. Opthalmol. Vis. Sci.* **43**, 216–24.

Land, M. F. (1988). The optics of animal eyes. *Contemp. Phys.* **29**, 435–55.

Landel, C. P., Zhao, J., Bok, D. and Evans, G. A. (1988). Lens-specific expression of recombinant ricin induces developmental defects in the eyes of transgenic mice. *Genes Dev.* **2**, 1168–78.

Lang, R. A. (1999). Which factors stimulate lens fiber cell differentiation in vivo? *Invest. Ophthalmol. Vis. Sci.* **40**, 3075–8.

Lawson, K. A., Meneses, J. J. and Pedersen, R. A. (1991). Clonal analysis of epiblast fate during germ layer formation in the mouse embryo. *Development* **113**, 891–911.

Lazarides, E. (1980). Intermediate filaments as mechanical integrators of cellular space. *Nature* **283**, 249–56.

Le, A. C. and Musil, L. S. (2001). FGF signaling in chick lens development. *Dev. Biol.* **233**, 394–411.

Lecoin, L., Sii-Felice, K., Pouponnot, C., Eychene, A. and Felder-Schmittbuhl, M. P. (2004). Comparison of maf gene expression patterns during chick embryo development. *Gene Expr. Patterns* **4**, 35–46.

LeCron, W. (1907). Experiments on the origin and differentiation of the lens in *Amblystoma. Am. J. Anat.* **6**, 245–57.

Le Douarin, N. (1969). Particularités du noyau interphasique chez la caille japonique (*Coturnix coturnix japonica*): utilasation de ces particularités comme "marquage biologique" dans les recherches sur les interactions tissulaires et les migrations cellulaires au cours de l'ontogenèse. *Bull. Biol. Fr. Belg.* **103**, 435–52.

Lee, A., Fischer, R. S. and Fowler, V. M. (2000). Stabilization and remodeling of the membrane skeleton during lens fiber cell differentiation and maturation. *Dev. Dyn.* **217**, 257–70.

Lee, A. Y. W., Chung, S. K. and Chung, S. S. M. (1995). Demonstration that polyol accumulation is responsible for diabetic cataract by the use of transgenic mice expressing the aldose reductase gene in the lens. *Proc. Natl. Acad. Sci. USA* **92**, 2780–4.

Lee, A., Morrow, J. and Fowler, V. (2001). Caspase remodeling of the spectrin membrane skeleton during lens development and aging. *J. Biol. Chem.* **276**, 20735–42.

Lee, D. C., Gonzalez, P. and Wistow, G. (1994). Zeta-crystallin: a lens-specific promoter and the gene recruitment of an enzyme as a crystallin. *J. Mol. Biol.* **236**, 669–78.

Lee, E. H. and Joo, C. K. (1999). Role of transforming growth factor-beta in transdifferentiation and fibrosis of lens epithelial cells. *Invest. Ophthalmol. Vis. Sci.* **40**, 2025–32.

Lee, E. H., Seomun, Y., Hwang, K., Kim, J., Kim, J. and Joo, C. (2000). Overexpression of the transforming growth factor-beta–inducible gene betais-h3 in anterior polar cataracts. *Invest. Ophthalmol. Vis. Sci.* **41**, 1840–5.

Lee, E. Y., Chang, C. Y., Hu, N., Wang, Y. C., Lai, C. C., Herrup, K., Lee, W. H. and Bradley, A. (1992). Mice deficient for Rb are nonviable and show defects in neurogenesis and haematopoiesis. *Nature* **359**, 288–94.

Lee, G. J., Roseman, A. M., Saibil, H. R. and Vierling, E. (1997). A small heat shock protein stably binds heat-denatured model substrates and can maintain a substrate in a folding-competent state. *EMBO. J.* **16**, 659–71.

Lee, M. D., King, L. S. and Agre, P. (1998). Aquaporin water channels in eye and other tissues. *Curr. Topics Membr.* **45**, 105–34.

Lee, M.-H., Williams, B. O., Mulligan, G., Mukai, S., Bronson, R. T., Dyson, N., Harlow, E. and Jacks, T. (1996). Targeted disruption of p107: functional overlap between p107 and Rb. *Genes Dev.* **10**, 1621–32.

Lee, M. M., Fink, B. D. and Grunwald, G. B. (1997). Evidence that tyrosine phosphorylation regulates N-cadherin turnover during retinal development. *Dev. Genet.* **20**, 224–34.

Lee, W.-H., Bookstein, R., Hong, F., Young, L.-J., Shew, J.-Y. and Lee, E. Y.-H. P. (1987). Human retinoblastoma susceptibility gene: cloning, identification and sequence. *Science* **235**, 1394–9.

Leenders, W. P., van Genesen, S. T., Schoenmakers, J. G., van Zoelen, E. J. and Lubsen, N. H. (1997). Synergism between temporally distinct growth factors: bFGF, insulin and lens cell differentiation. *Mech. Dev.* **67**, 193–201.

Leimeister, C., Bach, A. and Gessler, M. (1998). Developmental expression patterns of mouse sFRP genes encoding members of the secreted frizzled related protein family. *Mech. Dev.* **75**, 29–42.

Lengler, J., Krausz, E., Tomarev, S., Prescott, A., Quinlan, R. A. and Graw, J. (2001). Antagonistic action of Six3 and Prox1 at the γ-crystallin promoter. *Nucleic Acids Res.* **29**, 515–26.

Leone, G., DeGregori, J., Sears, R., Jakoi, L. and Nevins, J. R. (1997). Myc and Ras collaborate in inducing accumulation of active cyclin E/Cdk2 and E2F. *Nature* **387**, 422–6.

Leong, L., Menko, A. S. and Grunwald G. B. (2000). Differential expression of N- and B-cadherin during lens development. *Invest. Ophthalmol. Vis. Sci.* **41**, 3503–10.

Lewis, G. P., Erickson, P. A., Kaska, D. D. and Fisher, S. K. (1988). An immunocytochemical comparison of Müller cells and astrocytes in the cat retina. *Exp. Eye Res.* **47**, 839–53.

Lewis, K. E., Drossopoulou, O., Paton, I. R., Morice, D. R., Robertson, K. E., Burt, D. W., Ingham, P. W. and Tickle, C. (1999). Expression of ptc and gli genes in talpid3 suggests bifurcation in Shh pathway. *Development* **125**, 2397–407.

Lewis, S. A. and Donaldson, P. J. (1990). Ion channels and cell volume regulation: chaos in an organized system. *NIPS* **5**, 112–19.

Lewis, W. (1904). Experimental studies on the development of the eye in amphibia. I. On the origin of the lens. *Rana palustris. Am. J. Anat.* **3**, 505–36.

Lewis, W. H. (1909). The experimental production of cyclopia in the fish embryo (*Fundulus heteroclitus*). *Anat. Rec.* **3**, 175–81.

Li, H. S., Yang, J. M., Jacobson, R. D., Pasko, D. and Sundin, O. (1994). Pax-6 is first expressed in a region of ectoderm anterior to the early neural plate: implications for stepwise determination of the lens. *Dev. Biol.* **162**, 181–94.

Li, W. C., Kuszak, J. R., Dunn, K., Wang, R.-R., Ma, W., Wang, G.-M., Spector, A., Leib, M., Cotliar, A. M., Weiss, M., et al. (1995). Lens epithelial cell apoptosis appears to be a common cellular basis for non-congenital cataract development in humans and animals. *J. Cell Biol.* **130**, 169–81.

Li, X., Cvekl, A., Bassnett, S. and Piatigorsky, J. (1997). Lens-preferred activity of chicken δ1- and δ2-crystallin enhancers in transgenic mice and evidence for retinoic acid–responsive regulation of the δ1-crystallin gene. *Dev. Genet.* **20**, 258–66.

Li, X., Zelenka, P. S. and Piatigorsky, J. (1993). Differential expression of the two delta-crystallin genes in lens and non-lens tissues: shift favoring delta 2 expression from embryonic to adult chickens. *Dev. Dyn.* **196**, 114–23.

Liao, G. and Gundersen, G. G. (1998). Kinesin is a candidate for cross-bridging microtubules and intermediate filaments: selective binding of kinesin to detyrosinated tubulin and vimentin. *J. Biol. Chem.* **273**, 9797–803.

Liegeois, N. J., Horner, J. W. and DePinho, R. A. (1996). Lens complementation system for the genetic analysis of growth, differentiation, and apoptosis in vivo. *Proc. Natl. Acad. Sci. USA* **93**, 1303–7.

Lieska, N., Shao, D., Kriho, V. and Yang, H. Y. (1991). Expression and distribution of cytoskeletal IFAP-300kD as an index of lens cell differentiation. *Curr. Eye Res.* **10**, 1165–74.

Lillien, L. and Wancio, D. (1998). Changes in epidermal growth factor receptor expression and competence to generate glia regulate timing and choice of differentiation in the retina. *Mol. Cell. Neurosci.* **10**, 296–308.

Lim, D. J., Rubenstein, A. E., Evans, D. G., Jacks, T., Seizinger, B. G., Baser, M. E., Beebe, D., Brackmann, D. E., Chiocca, E. A., Fehon, R. G., et al. (2000). Advances in neurofibromatosis 2 (NF2): a workshop report. *J. Neurogenet.* **14**, 63–106.

Lin, J. S., Eckert, R., Kistler, J. and Donaldson, P. J. (1998). Spatial differences in gap junction gating in the lens are a consequence of connexin cleavage. *Eur. J. Cell Biol.* **76**, 246–50.
Lin, J. S., Fitzgerald, S., Dong, Y., Knight, C., Donaldson, P. J. and Kistler, J. (1997). Processing of the gap junction protein connexin50 in the ocular lens is accomplished by calpain. *Eur. J. Cell Biol.* **73**, 141–9.
Liou, W. (1990). Whole-mount preparations of mouse lens epithelium for the fluorescent cytological study of actin. *J. Microsc.* **157**, 239–45.
Litt, M., Carrero-Valenzuela, R., LaMorticella, D. M., Schultz, D. W., Mitchell, T. N., Kramer, P. and Maumenee, I. H. (1997). Autosomal dominant cerulean cataract is associated with a chain termination mutation in the human β-crystallin gene CRYBB2. *Hum. Mol. Genet.* **6**, 665–8.
Litt, M., Kramer, P., LaMorticella, D. M., Murphey, W., Lovrien, E. W. and Weleber, R. G. (1998). Autosomal dominant congenital cataract associated with a missense mutation in the human alpha crystallin gene CRYAA. *Hum. Mol. Genet.* **7**, 471–4.
Liu, D., Matzuk, M. M., Sung, W. K., Guo, Q., Wang, P. and Wolgemuth, D. J. (1998). Cyclin A1 is required for meiosis in the male mouse. *Nat. Genet.* **20**, 377–80.
Liu, J., Chamberlain, C. G. and McAvoy, J. W. (1996). IGF enhancement of FGF-induced fiber differentiation and DNA synthesis in lens explants. *Exp. Eye Res.* **63**, 621–9.
Liu, J., Hales, A. M., Chamberlain, C. G. and McAvoy, J. W. (1994). Induction of cataract-like changes in rat lens epithelial explants by transforming growth factor beta. *Invest. Ophthalmol. Vis. Sci.* **35**, 388–401.
Liu, Q., Ji, X., Breitman, M. L., Hitchcock, P. F. and Swaroop, A. (1996). Expression of the bZIP transcription factor gene Nrl in the developing nervous system. *Oncogene* **12**, 207–11.
Liu, X., Zou, H., Slaughter, C. and Wang, X. (1997). DFF, a heterodimeric protein that functions downstream of caspase-3 to trigger DNA fragmentation during apoptosis. *Cell* **89**, 175–84.
Liu, Y. and Zacksenhaus, E. (2000). E2F1 mediates ectopic proliferation and stage-specific p53-dependent apoptosis but not aberrant differentiation in the ocular lens of Rb deficient fetuses. *Oncogene* **19**, 6065–73.
Lo, W. (1988). Adherens junctions in the ocular lens of various species: ultrastructural analysis with an improved fixation. *Cell Tissue Res.* **254**, 31–40.
Lo, W. K. and Harding, C. V. (1983). Tight junctions in the lens epithelia of human and frog: freeze-fracture and protein tracer studies. *Invest. Ophthalmol. Vis. Sci.* **24**, 396–402.
Lo, W. K. and Harding, C. V. (1984). Square arrays and their role in ridge formation in human lens fibers. *J. Ultrastruct. Res.* **86**, 228–45.
Lo, W. K. and Harding, C. V. (1986). Structure and distribution of gap junctions in lens epithelium and fiber cells. *Cell Tissue Res.* **244**, 253–63.
Lo, W. K., Shaw, A. P., Paulsen, D. F. and Mills, A. (2000). Spatiotemporal distribution of zonulae adherens and associated actin bundles in both epithelium and fiber cells during chicken lens development. *Exp. Eye Res.* **71**, 45–55.
Lo, W. K., Shaw, A. P. and Wen, X. J. (1997). Actin filament bundles in cortical fiber cells of the rat lens. *Exp. Eye Res.* **65**, 691–701.
Lo, W. K. and Wen, X. J. (1999). Microtubule polarity and molecular motors associated with organelle/proteins transport during lens differentiation. *Invest. Ophthalmol. Vis. Sci.* **40**, S881.
Lobsiger, C. S., Magyar, J. P., Taylor, V., Wulf, P., Welcher, A. A., Program, A. E. and Suter, U. (1996). Identification and characterization of a cDNA and the structural gene encoding the mouse epithelial membrane protein-1. *Genomics* **36**, 379–87.
Lohka, M. J., Hayes, M. K. and Maller, J. L. (1988). Purification of maturation-promoting factor, an intracellular regulator of early mitotic events. *Proc. Natl. Acad. Sci. USA* **85**, 3009–13.
Lohnes, D., Kastner, P., Dierich, A., Mark, M., LeMeur, M. and Chambon, P. (1993). Function of retinoic acid receptor gamma in the mouse. *Cell* **73**, 643–58.
Lohnes, D., Mark, M., Mendelsohn, C., Dolle, P., Dierich, A., Gorry, P., Gansmuller, A. and Chambon, P. (1994). Function of the retinoic acid receptors (RARs) during development. I. Craniofacial and skeletal abnormalities in RAR double mutants. *Development* **120**, 2723–48.
Lois, N., Dawson, R., McKinnon, A. D. and Forrester, J. V. (2003). A new model of posterior capsule opacification in rodents. *Invest. Ophthalmol. Vis. Sci.* **44(8)**, 3450–7.
Lok, S., Stevens, W., Breitman, M. L. and Tsui, L. C. (1989). Multiple regulatory elements of the murine gamma 2-crystallin promoter. *Nucleic Acids Res.* **17**, 3563–82.

Lonchampt, M. O., Laurent, M., Courtois, Y., Trenchev, P. and Hughes, R. C. (1976). Microtubules and microfilaments of bovine lens epithelial cells: electron microscopy and immunofluorescence staining with specific antibodies. *Exp. Eye Res.* **23**, 505–18.

Loosli, F., Winkler, S. and Wittbrodt, J. (1999). *Six3* overexpression initiates the formation of ectopic retina. *Genes Dev.* **13**, 649–54.

Louis, C. F., Hur, K. C., Galvan, A. C., TenBroek, E. M., Jarvis, L. J., Eccleston, E. D. and Howard, J. B. (1989). Identification of an 18,000-dalton protein in mammalian lens fiber cell membranes. *J. Biol. Chem.* **264**, 19967–73.

Lovicu, F. J., Chamberlain, C. G. and McAvoy, J. W. (1995). Differential effects of aqueous and vitreous on fiber differentiation and extracellular matrix accumulation in lens epithelial explants. *Invest. Ophthalmol. Vis. Sci.* **36**, 1459–69.

Lovicu, F. J., de Iongh, R. U. and McAvoy, J. W. (1997). Expression of FGF-1 and FGF-2 mRNA during lens morphogenesis, differentiation and growth. *Curr. Eye Res.* **16**, 222–30.

Lovicu, F. J. and McAvoy, J. W. (1989). Structural analysis of lens epithelial explants induced to differentiate into fibers by fibroblast growth factor (FGF). *Exp. Eye Res.* **49**, 479–94.

Lovicu, F. J. and McAvoy, J. W. (1993). Localization of acidic fibroblast growth factor, basic fibroblast growth factor, and heparan sulphate proteoglycan in rat lens: implications for lens polarity and growth patterns. *Invest. Ophthalmol. Vis. Sci.* **34**, 3355–65.

Lovicu, F. J. and McAvoy, J. W. (1999). Spatial and temporal expression of p57(KIP2) during murine lens development. *Mech. Dev.* **86**, 165–9.

Lovicu, F. J. and McAvoy, J. W. (2001). FGF-induced lens cell proliferation and differentiation is dependent on MAPK (ERK1/2) signaling. *Development* **128**, 5075–84.

Lovicu, F. J. and Overbeek, P. A. (1998). Overlapping effects of different members of the FGF family on lens fiber differentiation in transgenic mice. *Development* **125**, 3365–77.

Lovicu, F. J., Schulz, M. W., Hales, A. M., Vincent, L. N., Overbeek, P. A., Chamberlain, C. G. and McAvoy, J. W. (2002). TGFbeta induces morphological and molecular changes similar to human anterior subcapsular cataract. *Br. J. Ophthalmol.* **86**, 220–6.

Lu, S. F., Pan, F. M. and Chiou, S. H. (1996a). Characterization of gamma-crystallin from the eye lens of bullfrog: complexity of gamma-crystallin multigene family as revealed by sequence comparison among different amphibian species. *J. Protein Chem.* **15**, 103–13.

Lu, S. F., Pan, F. M. and Chiou, S. H. (1996b). Sequence analysis of four acidic beta-crystallin subunits of amphibian lenses: phylogenetic comparison between beta- and gamma-crystallins. *Biochem. Biophys. Res. Commun.* **221**, 219–28.

Lubsen, N. H., Aarts, H. J. M. and Schoenmakers, J. G. G. (1988). The evolution of lenticular proteins: the β- and γ-crystallin super gene family. *Prog. Biophys. Mol. Biol.* **51**, 47–76.

Lucas, V. A. and Zigler, J. S. (1987). Transmembrane glucose carriers in the monkey lens: quantitation and regional distribution as determined by cytochalasin B binding to lens membranes. *Invest. Ophthalmol. Vis. Sci.* **28**, 1404–12.

Lucas, V. A. and Zigler, J. S. (1988). Identification of the monkey lens glucose transporter by photoaffinity labelling with cytochalasin B. *Invest. Ophthalmol. Vis. Sci.* **29**, 630–5.

Lukas, J., Parry, D., Aagaard, L., Mann, D. J., Bartkova, J., Strauss, M., Peters, G. and Bartek, J. (1995). Retinoblastoma-protein–dependent cell-cycle inhibition by the tumour suppressor p16. *Nature* **375**, 503–6.

Lukas, J. T., Herzinger, K., Hansen, M. C., Moroni, D., Resnitzky, K., Helin, K., Reed, S. I. and Bartek, J. (1997). Cyclin E–induced S phase without activation of the pRb/E2F pathway. *Genes Dev.* **11**, 1479–92.

Lundberg, A. S. and Weinberg, R. A. (1999). Control of the cell cycle and apoptosis. *Eur. J. Cancer* **35**, 1886–94.

Luo, R. X., Postigo, A. A. and Dean, D. C. (1998). Rb interacts with histone deacetylase to repress transcription. *Cell* **92**, 463–73.

Lux, S. E., Tse, W. T., Menninger, J. C., John, K. M., Harris, P., Shalev, O., Chilcote, R. R., Marchesi, S. L., Watkins, P. C., Bennett, V., et al. (1990). Hereditary spherocytosis associated with deletion of human erythrocyte ankyrin gene on chromosome 8. *Nature* **345**, 736–9.

Ma, T., Van Tine, B. A., Wei, Y., Garrett, M. D., Nelson, D., Adams, P. D., Wang, J., Qin, J., Chow, L. T. and Harper, J. W. (2000). Cell cycle-regulated phosphorylation of p220(NPAT) by cyclin E/Cdk2 in Cajal bodies promotes histone gene transcription. *Genes Dev.* **14**, 2298–313.

Maandag, E. C., van der Valk, M., Vlaar, M., Feltkamp, C., O'Brien, J., van Roon, M., van der Lugt, N., Berns, A. and te Riele, H. (1994). Developmental rescue of an embryonic-lethal mutation in the retinoblastoma gene in chimeric mice. *EMBO J.* **13**, 4260–8.

Mackay, D., Ionides, A., Berry, V., Moore, A., Bhattacharya, S. and Shiels, A. (1997). A new locus for dominant "zonular pulverent" cataract, on chromosome 13. *Am. J. Hum. Genet.* **60**, 1474–8.

Mackay, D., Ionides, A., Kibar, Z., Rouleau, G., Berry, V., Moore, A., Shiels, A. and Bhattacharya, S. (1999). Connexin46 mutations in autosomal dominant congenital cataract. *Am. J. Hum. Genet.* **64**, 1357–64.

Macioce, P., Gandolfi, N., Leung, C. L., Chin, S. S., Malchiodi-Albedi, F., Ceccarini, M., Petrucci, T. C. and Liem, R. K. (1999). Characterization of NF-L and betaIISigma1-spectrin interaction in live cells. *Exp. Cell Res.* **250**, 142–54.

Mackic, J. B., Jinagouda, S., McComb, J. G., Weiss, M. H., Kannan, R., Kaplowitz, N. and Zlokovic, B. V. (1996). Transport of circulating glutathione at the basolateral side of the anterior lens epithelium: physiological importance and manipulations. *Exp. Eye Res.* **62**, 29–37.

Macleod, K. F., Hu, Y. and Jacks, T. (1996). Loss of RB activates both p53-dependent and independent cell death pathways in the developing mouse nervous system. *EMBO J.* **15**, 6178–88.

Maden, M. (1997). Retinoic acid and its receptors in limb regeneration. *Cell. Dev. Biol.* **8**, 445–52.

Magabo, K. S., Horwitz, J., Piatigorsky, J. and Kantorow, M. (2000). Expression of betaB(2)-crystallin mRNA and protein in retina, brain, and testis. *Invest. Ophthalmol. Vis. Sci.* **41**, 3056–60.

Magnaghi-Jaulin, L., Groisman, R., Naguibneva, I., Robin, P., Lorain, S., Le Villain, J. P., Troalen, F., Trouche, D. and Harel-Bellan, A. (1998). Retinoblastoma protein represses transcription by recruiting a histone deactylase. *Nature* **391**, 601–5.

Magnus, H. (1998). *Ophthalmology of the ancients.* In J. Hirschberg (Ed.), The History of Ophthalmology: The Monographs, Vol. 4, Part 1 (F. C. Blodi, Trans.). Bonn: Wayenborgh.

Maillard, C., Malaval, L. and Delmas, P. D. (1992). Immunological screening of SPARC/osteonectin in nonmineralized tissues. *Bone* **13**, 257–64.

Maisel, H. and Atreya, P. (1990). N-cadherin detected in the membrane fraction of lens fiber cells. *Experientia* **46**, 222–3.

Maness, P. F., Beggs, H. E., Klinz, S. G. and Morse, W. R. (1996). Selective neural cell adhesion molecule signaling by Src family tyrosine kinases and tyrosine phosphatases. *Perspect. Dev. Neurobiol.* **4**, 169–81.

Mann, I. C., ed. (1928). *The Development of the Human Eye.* Cambridge: Cambridge University Press.

Mann, I. (1950). *The Development of the Human Eye.* 2nd edition. Grune and Stratton, Inc. New York.

Mann, I. C., ed. (1964). *The Development of the Human Eye*, 3rd ed. London: British Medical Association.

Manns, M. and Fritzsch, B. (1991). The eye in the brain: retinoic acid effects morphogenesis of the eye and pathway selection of axons but not differentiation of the retina in *Xenopus laevis. Neurosci. Lett.* **127**, 150–4.

Mansouri, A., Goudreau, G. and Gruss, P. (1999). Pax genes and their role in organogenesis. *Cancer Res.* **59** (Suppl), 1707–9.

Mantych, G. J., Hageman, G. S. and Devaskar, S. U. (1993). Characterization of glucose transporter isoforms in the adult and developing human eye. *Endocrinology* **133**, 600–7.

Marcantonio, J. M. and Vrensen, G. F. (1999). Cell biology of posterior capsular opacification. *Eye* **13**, 484–8.

Mardon, G., Solomon, N. M. and Rubin, G. M. (1994). *Dachshund* encodes a nuclear protein required for normal eye and leg formation in *Drosophila. Development* **120**, 3473–86.

Marquardt, T., Ashery-Padan, R., Andrejewski, N., Scardigli, R., Guillemot, F. and Gruss, P. (2001). Pax6 is required for the multipotent state of retinal progenitor cells. *Cell* **105**, 43–55.

Martinez-Morales, J. R., Signore, M., Acampora, D., Simeone, A. and Bovolenta, P. (2001). Otx genes are required for tissue specification in the developing eye. *Development* **128**, 2019–30.

Maruno, K. A., Lovicu, F. J., Chamberlain, C. G. and McAvoy, J. W. (2002). Apoptosis is a feature of TGF beta–induced cataract. *Clin. Exp. Optom.* **85**, 76–82.

Massague, J. (1998). TGF-beta signal transduction. *Annu. Rev. Biochem.* **67**, 753–91.

Masuda, A. and Eguchi, G. (1982). Effects of thioureas, inhibitors of melanogenesis on lens transdifferentiation of cultured chick embryonic retinal pigment cells. *Dev. Growth Differ.* **24**, 589–99.

Masui, Y. and Markert, C. L. (1971). Cytoplasmic control of nuclear behavior during meiotic maturation of frog oocytes. *J. Exp. Zool.* **177**, 129–45.

Mathers, P. H., Grinberg, A., Mahon, K. A. and Jamrich, M. (1997). The *Rx* homeobox gene is essential for vertebrate eye development. *Nature* **387**, 603–7.

Mathers, P. H. and Jamrich, M. (2000). Regulation of eye formation by the *Rx* and *Pax6* homeobox genes. *Cell. Mol. Life Sci.* **57**, 186–94.

Mathias, R. T. and Rae, J. L. (1985). Steady-state voltages in the frog lens. *Curr. Eye Res.* **4**, 421–30.

Mathias, R. T., Rae, J. L. and Baldo, G. J. (1997). Physiological properties of the normal lens. *Physiol. Rev.* **77**, 21–50.

Mathias, R. T., Rae, J. L. and Ebihara, L. and McCarthy, R. T. (1985). The localization of transport properties in the frog lens. *Biophys. J.* **48**, 423–34.

Mathias, R. T., Rae, J. L. and Eisenberg, R. S. (1979). Electrical properties of structural components of the crystalline lens. *Biophys. J.* **25**, 181–201.

Mathias, R. T., Rae, J. L. and Eisenberg, R. S. (1981). The lens as a nonuniform spherical syncytium. *Biophys. J.* **34**, 61–83.

Mathias, R. T., Riquelme, G. and Rae, J. L. (1991). Cell to cell communication and pH in the frog lens. *J. Gen. Physiol.* **98**, 1085–103.

Mathias, R. T., Sun, X., Gao, J., Baldo, G. J. and Kushmerick, C. (1999). Intercellular pH of the normal lens. *Invest. Ophthalmol. Vis. Sci.* **40**, S885.

Matsumoto, Y., Hayashi, K. and Nishida, E. (1999). Cyclin-dependent kinase 2 (Cdk2) is required for centrosome duplication in mammalian cells. *Curr. Biol.* **9**, 429–32.

Matsuo, I., Kitamura, M., Okazaki, K. and Yasuda, K. (1991). Binding of a factor to an enhancer element responsible for the tissue-specific expression of the chicken alpha A-crystallin gene. *Development* **113**, 539–50.

Matsuo, I., Kuratani, S., Kimura, C., Takeda, N. and Aizawa, S. (1995). Mouse *Otx2* functions in the formation and patterning of rostral head. *Genes Dev.* **9**, 2646–58.

Matsuo, I., Takeuchi, M. and Yasuda, K. (1992). Identification of the contact sites of a factor that interacts with motif I (alpha CE1) of the chicken alpha A-crystallin lens-specific enhancer. *Biochem. Biophys. Res. Commun.* **184**, 24–30.

Matsuo, I. and Yasuda, K. (1992). The cooperative interaction between two motifs of an enhancer element of the chicken alpha A-crystallin gene, alpha CE1 and alpha CE2, confers lens-specific expression. *Nucleic Acids Res.* **20**, 3701–12.

Matsuoka, S., Edwards, M. C., Bai, C., Parker, S., Zhang, P., Baldini, A., Harper, J. W. and Elledge, S. J. (1995). p57KIP2, a structurally distinct member of the p21CIP1 Cdk inhibitor family, is a candidate tumor suppressor gene. *Genes Dev.* **9**, 650–62.

Matsushima-Hibiya, Y., Nishi, S. and Sakai, M. (1998). Rat Maf-related factors: the specificities of DNA binding and heterodimer formation.

Matsushime, H., Roussel, M. F., Ashmun, R. A. and Sherr, C. J. (1991). Colony-stimulating factor 1 regulates novel cyclins during the G1 phase of the cell cycle. *Cell* **65**, 701–13.

Mattern, R. M., Zuk, A. and Hay, E. D. (1993). Retinoic acid inhibits formation of mesenchyme from lens epithelium in collagen gels. *Invest. Ophthalmol. Vis. Sci.* **34**, 2526–37.

Mazaki, Y., Mochii, M., Kodama, R. and Eguchi, G. (1996). Role of integrins in differentiation of chick retinal pigmented epithelial cells *in vitro*. *Dev. Growth Differ.* **38**, 429–37.

McAvoy, J. W. (1978). Cell division, cell elongation and the co-ordination of crystallin gene expression during lens morphogenesis in the rat. *J. Embryol. Exp. Morphol.* **45**, 271–81.

McAvoy, J. (1978a). Cell division, cell elongation and distribution of α, β, and γ-crystallins in the rat lens. *J. Embryol. Exp. Morph.* **44**, 149–65.

McAvoy, J. W. (1978b). Cell division, cell elongation and the co-ordination of crystallin gene expression during lens morphogenesis in the rat. *J. Embryol. Exp. Morphol.* **45**, 271–81.

McAvoy, J. W. (1980). Cytoplasmic processes interconnect lens placode and optic vesicle during eye morphogenesis. *Exp. Eye Res.* **31**, 527–34.

McAvoy, J. (1988). Cell lineage analysis of lens epithelial cells induced to differentiate into fibres. *Exp. Eye Res.* **47**, 869–83.

McAvoy, J. W. and Chamberlain, C. G. (1989). Fibroblast growth factor (FGF) induces different responses in lens epithelial cells depending on its concentration. *Development* **107**, 221–8.

McAvoy, J. W. and Chamberlain, C. G. (1990). Growth factors in the eye. *Prog. Growth Factor Res.* **2**, 29–43.

McAvoy, J. W., Chamberlain, C. G., de Iongh, R. U., Hales, A. M., Lovicu, F. J. (2000). Peter Bishop Lecture: growth factors in lens development and cataract: key roles for fibroblast growth factor and TGF-beta. *Clin. Experiment. Ophthalmol.* **28** 133–9

McAvoy, J. W., Chamberlain, C. G., de Iongh, R. U., Hales, A. M. and Lovicu, F. J. (1999). Lens development. *Eye* **13**, 425–37.

McAvoy, J. W., Chamberlain, C. G., de Iongh, R. U., Hales, A. M. and Lovicu, F. J. (2000). Growth factors in lens development and cataract: key roles for fibroblast growth factor and TGF-β. *Clin. Experiment. Ophthalmol.* **28**, 133–9.

McAvoy, J. W., Chamberlain, C. G., de Iongh, R. U., Richardson, N. A. and Lovicu, F. J. (1991). The role of fibroblast growth factor in eye lens development. *Ann. NY Acad. Sci.* **638**, 256–74.

McAvoy, J. W. and McDonald, J. (1984). Proliferation of lens epithelial explants in culture increases with age of donor rat. *Curr. Eye Res.* **3**, 1151–3.

McAvoy, J. and Richardson, N. (1986). Nuclear pyknosis during lens fibre differentiation in epithelial explants. *Curr. Eye Res.* **5**, 711–5.

McAvoy, J. W., Schulz, M. W., Maruno, K. A., Chamberlain, C. G. and Lovicu, F. J. (1998). TGF-β–induced cataract is characterised by epithelial-mesenchymal transition and apoptosis. *IVOS* **39**, S7.

McCaffery, P., Lee, M., Wagner, M. A., Sladek, N. E. and Drager, U. C. (1992). Asymmetrical retinoic acid synthesis in the dorsoventral axis of the retina. *Development* **115**, 371–82.

McCaffery, P., Posch, K. C., Napoli, J. L., Gudas, L. and Drager, U. C. (1993). Changing patterns of the retinoic acid system in the developing retina. *Dev. Biol.* **158**, 390–9.

McCaffrey, J., Yamasaki, L., Dyson, N. J., Harolw, E. and Griep, A. E. (1999). Disruption of retinoblastoma protein family function by human papillomavirus type 16 E7 oncoprotein inhibits lens development in part through *E2F-1*. *Mol. Cell Biol.* **19**, 6458–68.

McCartney, B. M., Kulikauskas, R. M., LaJeunesse, D. R. and Fehon, R. G. (2000). The neurofibromatosis-2 homologue, Merlin, and the tumor suppressor expanded function together in *Drosophila* to regulate cell proliferation and differentiation. *Development* **127**, 1315–24.

McDermott, J. B., Cvekl, A. and Piatigorsky, J. (1996). Lens-specific expression of a chicken beta A3/A1-crystallin promoter fragment in transgenic mice. *Biochem. Biophys. Res. Commun.* **221**, 559–64.

McDermott, J. B., Cvekl, A. and Piatigorsky, J. (1997). A complex enhancer of the chicken beta A3/A1-crystallin gene depends on an AP-1-CRE element for activity. *Invest. Ophthalmol. Vis. Sci.* **38**, 951–9.

McDermott, J. B., Peterson, C. A. and Piatigorsky, J. (1992). Structure and lens expression of the gene encoding chicken beta A3/A1-crystallin. *Gene* **117**, 193–200.

McDevitt, D. S. and Brahma, S. K. (1982). Alpha-, beta- and gamma-crystallins in the regenerating lens of *Notophthalmus viridescens*. *Exp. Eye Res.* **34**, 587–94.

McDevitt, D. S. and Brahma, S. K. (1990). Ontogeny and localization of alpha A- and alpha B-crystallins during regeneration of the eye lens. *Exp. Eye Res.* **51**, 625–30.

McDevitt, D. S., Brahma, S. K., Courtois, Y. and Jeanny, J. C. (1997). Fibroblast growth factor receptors and regeneration of the eye lens. *Dev. Dyn.* **208**, 220–6.

McKeehan, M. S. (1951). Cytological aspects of embryonic lens induction in the chick. *J. Exp. Zool.* **117**, 31–64.

McKeehan, M. S. (1958). Induction of portions of the chick lens without contact with the optic cup. *Anat. Rec.* **132**, 297–305.

McKeehan, M. S. (1961). The capacity for lens regeneration in the chick embryo. *Anat. Rec.* 141–3.

McNutt, N. S. and Weinstein, R. S. (1970). The ultrastructure of the nexus: a correlated thin-section and freeze-cleave study. *J. Cell Biol.* **47**, 666–88.

Meads, T. and Schroer, T. A. (1995). Polarity and nucleation of microtubules in polarized epithelial cells. *Cell Motil. Cytoskeleton* **32**, 273–88.

Meakin, S. O., Du, R. P., Tsui, L.-C. and Breitman, M. L. (1987). γ-Crystallins of the human eye lens: expression analysis of five members of the gene family. *Mol. Cell. Biol.* **7**, 2671–9.

Medema, R. H., Herrera, R. E., Lam, F. and Weinberg, R. A. (1995). Growth suppression by p16ink4 requires functional retinoblastoma protein. *Proc. Natl. Acad. Sci. USA* **92**, 6289–93.

Meech, R., Kallunki, P., Edelman, G. M. and Jones, F. S. (1999). A binding site for homeodomain and Pax proteins is necessary for L1 cell adhesion molecule gene expression by Pax-6 and bone morphogenetic proteins. *Proc. Natl. Acad. Sci. USA* **96**, 2420–5.

Mehta, P. D. and Lerman, S. (1972). Alpha crystallin subunits in the bovine lens. *Can. J. Ophthalmol.* **7**, 218–22.

Mencl, E. (1903). Ein Fall von beiderseitiger Augenlinsenausbildung wahrend der Abwesenheit von Augenblasen. *Arch. Entw.-Mech.* **16**, 328–39.

Mendelsohn, C., Lohnes, D., Decimo, D., Lufkin, T., LeMeur, M., Chambon, P. and Mark, M. (1994). Function of the retinoic acid receptors (RARs) during development. II. Multiple abnormalities at various stages of organogenesis in RAR double mutants. *Development* **120**, 2749–71.

Menko, A. S. and Boettiger, D. (1988). Inhibition of chicken embryo lens differentiation and lens junction formation in culture by pp60v-src. *Mol. Cell Biol.* **8**, 1414–20.

Menko, A. S., Klukas, K. A. and Johnson, R. G. (1984). Chicken embryo lens culture mimic differentiation in the lens. *Dev. Biol.* **103**, 129–41.

Menko, A. S., Kreidberg, J. A., Ryan, T. T., Van Bockstaele, E. and Kukuruzinska, M. A. (2001). Loss of 31 integrin function results in an altered differentiation program in the mouse submandibular gland. *Dev. Dyn.* **220**, 337–49.

Menko, A. S. and Philp, N. J. (1995). Beta 1 integrins in epithelial tissues: a unique distribution in the lens. *Exp. Cell Res.* **218**, 516–21.

Menko, S., Philp, N., Veneziale, B. and Walker, J. (1998). Integrins and development: how might these receptors regulate differentiation of the lens. *Ann. NY Acad. Sci.* **842**, 36–41.

Mercader, N., Leonardo, E., Azpiazu, N., Serrano, A., Morata, G., Martinez, C. and Torres, M. (1999). Conserved regulation of proximodistal limb axis development by Meis1/Hth. *Nature* **402**, 425–9.

Mercader, N., Leonardo, E., Piedra, M. E., Martinez, A. C., Ros, M. A. and Torres, M. (2000). Opposing RA and FGF signals control proximodistal vertebrate limb development through regulation of Meis genes. *Development* **127**, 3961–70.

Merdes, A., Brunkener, M., Horstmann, H. and Georgatos, S. D. (1991). Filensin: a new vimentin-binding, polymerization-competent, and membrane-associated protein of the lens fiber cell. *J. Cell Biol.* **115**, 397–410.

Merdes, A., Gounari, F. and Georgatos, S. D. (1993). The 47-kD lens-specific protein phakinin is a tailless intermediate filament protein and an assembly partner of filensin. *J. Cell Biol.* **123**, 1507–16.

Merriman-Smith, R., Donaldson, P. J. and Kistler, J. (1999). Differential expression of facilitative glucose transporters GLUT1 and GLUT3 in the lens. *Invest. Ophthalmol. Vis. Sci.* **40**, 3224–30.

Merriman-Smith, R., Tunstall, M., Kistler, J., Donaldson, P., Housley, G. and Eckert, R. (1998). Expression profiles of P2-receptor isoforms P2Y(1) and P2Y(2) in the rat lens. *Invest. Ophthalmol. Vis. Sci.* **39**, 2791–6.

Mertens, C., Kuhn, C., Moll, R., Schwetlick, I. and Franke, W. W. (1999). Desmosomal plakophilin 2 as a differentiation marker in normal and malignant tissues. *Differentiation* **64**, 277–90.

Meyerson, M., Enders, G. H., Wu, C. L., Su, L. K., Gorka, C., Nelson, C., Harlow, E. and Tsai, L. H. (1992). A family of human cdc2-related protein kinases. *EMBO J.* **11**, 2909–17.

Mikami, Y. (1941). Experimental analysis of the Wolffian lens-regeneration in the adult newt, *Triturus pyrrhogaster*. *Jpn. J. Zool.* **9**, 269–302.

Miki, H., Setou, M., Kaneshiro, K. and Hirokawa, N. (2001). All kinesin superfamily protein, KIF, genes in mouse and human. *Proc. Natl. Acad. Sci. USA* **98**, 7004–11.

Mikulicich, A. G. and Young, R. W. (1963). Cell proliferation and displacement in the lens epithelium of young rats injected with tritiated thymidine. *Invest. Ophthalmol. Vis. Sci.* **2**, 344–54.

Millar, A., Hooper, A., Copeland, L., Cummings, F. and Prescott, A. (1997). Reorganisation of the microtubule cytoskeleton and centrosomal loss during lens fibre cell differentiation. *Nova Acta Leopoldiana* **299**, 169–83.

Miller, C. and Sassoon, D. A. (1998). Wnt-7a maintains appropriate uterine patterning during the development of the mouse female reproductive tract. *Development* **125**, 3201–11.

Miller, D. L., Ortega, S., Bashayan, O., Basch, R. and Basilico, C. (2000). Compensation by fibroblast growth factor 1 (FGF1) does not account for the mild phenotypic defects observed in FGF2 null mice [published erratum appears in *Mol. Cell. Biol.* 2000;20:3752]. *Mol. Cell. Biol.* **20**, 2260–8.

Miller, T. and Goodenough, D. A. (1986). Evidence for two physiologically distinct gap junctions expressed by the chick lens epithelial cell. *J. Cell Biol.* **102**, 194–9.

Milstone, L. M. and Piatigorsky, J. (1975). Rates of protein synthesis in explanted embryonic chick lens epithelia: differential stimulation of crystallin synthesis. *Dev. Biol.* **43**, 91–100.

Milstone, L. M., Zelenka, P. and Piatigorsky, J. (1976). Delta-crystallin mRNA in chick lens cells: mRNA accumulates during differential stimulation of delta-crystallin synthesis in cultured cells. *Dev. Biol.* **48**, 197–204.

Mitashov, V. I. (1966). Comparative study of lens regeneration in Cobitid fishes [in Russian]. *Dokl. Akad. Nauk. SSSR.* **170**, 1439–42.

Mitashov, V. I., Kazanskaya, O. V., Luk'yanov, S. V., Dolgilevich, S. M., Zarayskii, A. G., Znoiko, S. L. and Gause, G. G. (1992). Activation of genes coding for gamma-crystallins during lens regeneration in the newt. *Monogr. Dev. Biol.* **23**, 139–45.

Miyashita, T. and Reed, J. C. (1995). Tumor suppressor p53 is a direct transcriptional activator of the human bax gene. *Cell* **80**, 293–9.

Mizuno, N., Agata, K., Sawada, K., Mochii, M. and Eguchi, G. (2002). Expression of crystallin genes in embryonic and regenerating newt lenses. *Dev. Growth Differ.* **44(3)**, 251–6.

Mizuno, N., Mochii, M., Yamamoto, T. S., Takahashi, T. C., Eguchi, G. and Okada, T. S. (1999). Pax-6 and Prox 1 expression during lens regeneration from *Cynops* iris and *Xenopus* cornea: evidence for a genetic program common to embryonic lens development. *Differentiation* **65**, 141–9.

Mizuno, S., Ikeda, K. and Hirabayashi, Y. (1995). Histochemical studies of the separation of the lens vesicle in the mouse. *Jpn. J. Ophthalmol.* **39**, 340–6.

Mizuno, T. (1972). Lens differentiation in vitro in the absence of optic vesicle in the epiblast of chick blastoderm under the influence of skin dermis. *J. Embryol. Exp. Morph.* **28**, 117–32.

Mizuno, T. (1973). Induction de cristallin in vitro dans l'ectoblaste de tronc presomptif ou dans l'aire opaque chez le blastoderme de poulet en l'absence de la vésicule optique. *CR. Acad. Sc. Paris, Ser. D.* **277**, 1229–32.

Mizuno, T. and Katoh, Y. (1972a). Présence de protéines de cristallin dans des cristallins induits in vitro en l'absence de la vésicule optique chez l'embryon de poulet. *CR Acad. Sci. Paris, Ser.D.* **274**, 1086–8.

Mizuno, T. and Katoh, Y. (1972b). Immunohistological studies on lens differentiation experimentally induced in vitro in the epiblast of chick blastoderm. *Proc. Jpn. Acad.* **48**, 522–7.

Mizuseki, K., Kishi, M., Matsui, M., Nakanishi, S. and Sasai, Y. (1998). Xenopus Zic-related-1 and Sox-2, two factors induced by chordin, have distinct activities in the initiation of neural induction. *Development* **125**, 579–87.

Mochii, M., Agata, K. and Eguchi, G. (1991). Complete sequence and expression of a cDNA encoding a chick 115-kDa melanosomal matrix protein. *Pigment Cell Res.* **4**, 41–7.

Mochii, M., Agata, K., Kobayashi, H., Yamamoto, T. S. and Eguchi, G. (1988b). Expression of gene coding for a melanosomal matrix protein transcriptionally regulated in the transdifferentiation of chick embryo pigmented epithelial cells. *Cell Differ.* **24**, 67–74.

Mochii, M., Takeuchi, T., Kodama, R., Agata, K. and Eguchi, G. (1988a). The expression of melanosomal matrix protein in the transdifferentiation of pigmented epithelial cells into lens cells. *Cell Differ.* **23**, 133–42.

Modak, S. P. and Bollum, F. J. (1972). Detection and measurement of single-strand breaks in nuclear DNA in fixed lens sections. *Exp. Cell Res.* **75**, 544–61.

Modak, S. P., Morris, G. and Yamada, T. (1968). DNA synthesis and mitotic activity during early development of the chick lens. *Dev. Biol.* **17**, 544–61.

Modesto, E., Lampe, P. D., Ribeiro, M. C., Spray, D. C. and Campos de Carvalho, A. C. (1996). Properties of chicken lens MIP channels reconstituted into planar lipid bilayers. *J. Membr. Biol.* **154**, 239–49.

Mogensen, M. M., Malik, A., Piel, M., Bouckson-Castaing, V. and Bornens, M. (2000). Microtubule minus-end anchorage at centrosomal and non-centrosomal sites: the role of ninein [in process citation]. *J. Cell Sci.* **113**, 3013–23.

Montagnoli, A., Fiore, F., Eytan, E., Carrano, A. C., Draetta, G. F., Hershko, A. and Pagano, M. (1999). Ubiquitination of p27 is regulated by CDK-dependent phosphorylation and trimeric complex formation. *Genes Dev.* **13**, 1181–9.

Montgomery, M. K. and McFall-Ngai, M. J. (1992). The muscle-derived lens of a squid bioluminescent organ is biochemically convergent with the ocular lens: evidence for recruitment of aldehyde dehydrogenase as a predominant structural protein. *J. Biol. Chem.* **267**, 20999–1003.

Morgenbesser, S. D., Williams, B. O., Jacks, T. and DePinho, R. A. (1994). p53-Dependent apoptosis produced by Rb-deficiency in the developing mouse lens. *Nature* **371**, 72–4.

Morimura, H., Shimada, S., Otori, Y., Saiahin, Y., Yamauchi, A., Minami, Y., Inoue, K., Ishimoto, I., Tano, Y. and Tohyama, M. (1997). The differential osmoregulation and localization of taurine transporter mRNA and Na^{+}/myo-inositol cotransporter mRNA in rat eyes. *Mol. Brain Res.* **44**, 245–52.

Morrow, E. M., Furukawa, T. and Cepko, C. L. (1998). Vertebrate photoreceptor cell development and disease. *Trends Cell. Biol.* **8**, 353–8.

Moscona, A. A., Fox, L., Smith, J. and Degenstein, L. (1985). Antiserum to lens antigens immunostains Müller glia cells in the neural retina. *Proc. Natl. Acad. Sci. USA* **82**, 5570–3.

Moseley, A. E., Dean, W. L. and Delamere, N. A. (1996). Isoforms of Na,K-ATPase in rat lens epithelium and fiber cells. *Invest. Ophthalmol. Vis. Sci.* **37**, 1502–8.

Moser, M., Imhof, A., Pscherer, A., Bauer, R., Amselgruber, W., Sinowatz, F., Hofstädter, F., Schüle, R. and Buettner, R. (1995). Cloning and characterization of a second AP-2 transcription factor: AP-2β. *Development* **121**, 2779–88.

Mousa, G. Y. and Trevithick, J. R. (1977). Differentiation of rat lens epithelial cells in tissue culture. II. Effects of cytochalasin B and D on actin organisation and differentiation. *Dev. Biol.* **60**, 14–25.

Mukhopadhyay, P., Bhattacherjee, P., Andom, T., Geoghegan, T. E., Andley, U. P. and Paterson, C. A. (1999). Expression of prostaglandin receptors EP4 and FP in human lens epithelial cells. *Invest. Ophthalmol. Vis. Sci.* **40**, 105–12.

Mulders, S. M., Preston, G. M., Deen, P. M. T., Guggino, W. B., van Os, C. H. and Agre, P. (1995). Water channel properties of major intrinsic protein of lens. *J. Biol. Chem.* **270**, 9010–16.

Mulders, J. W., Voorter, C. E., Lamers, C., de Haard-Hoekman, W. A., Montecucco, C., van de Ven, W. J., Bloemendal, H. and de Jong, W. W. (1988). MP17, a fiber-specific intrinsic membrane protein from mammalian eye lens. *Curr. Eye Res.* **7**, 207–19.

Mullen, L. M., Bryant, S. V., Torok, M. A., Blumberg, B. and Gardiner, D. M. (1996). Nerve dependency of regeneration: the role of distal-less and FGF signaling in amphibia limb regeneration. *Development* **122**, 3487–97.

Mulligan, R. and Jacks, T. (1998). The retinoblastoma gene family: cousins with overlapping interests. *Trends Genet.* **14**, 223–9.

Munger, K., Werness, B. A., Dyson, N., Phelps, W. C., Harlow, E. and Howley, P. M. (1989). Complex formation of human papillomavirus E7 proteins with the retinoblastoma tumor suppressor gene product. *EMBO J.* **8**, 4099–105.

Murphy, M., Stinnakre, M. G., Senamaud-Beaufort, C., Winston, N. J., Sweeney, C., Kubelka, M., Carrington, M., Brechot, C. and Sobczak-Thepot, J. (1997). Delayed early embryonic lethality following disruption of the murine cyclin A2 gene. *Nat. Genet.* **15**, 83–6.

Murphy-Erdosh, C., Napolitano, E. W. and Reichardt, L. F. (1994). The expression of B-cadherin during embryonic chick development. *Dev. Biol.* **161**, 107–25.

Murray, A. W. and Marks, D. (2001). Can sequencing shed light on cell cycling? *Nature* **409**, 844–6.

Muthukkaruppan, V. (1965). Inductive tissue interaction in the development of the mouse lens in vitro. *J. Exp. Zool.* **159**, 269–88.

Nagamoto, T., Eguchi, G. and Beebe, D. C. (2000). Alpha-smooth muscle actin expression in cultured lens epithelial cells. *Invest. Ophthalmol. Vis. Sci.* **41**, 1122–9.

Nakamura, T., Donovan, D. M., Hamada, K., Sax, C. M., Norman, B., Flanagan, J. R., Ozato, K., Westphal, H. and Piatigorsky, J. (1990). Regulation of the mouse alpha A-crystallin gene:

isolation of a cDNA encoding a protein that binds to a cis sequence motif shared with the major histocompatibility complex class I gene and other genes. *Mol. Cell. Biol.* **10**, 3700–8.
Nakamura, T., Mahon, K. A., Miskin, R., Dey, A., Kuwabara, T. and Westphal, H. (1989). Differentiation and oncogenesis: phenotypically distinct lens tumors in transgenic mice. *New Biol.* **1**, 193–204.
Nakayama, K., Nagahama, H., Minamishima, Y. A., Matsumoto, M., Nakamichi, I., Kitagawa, K., Shirane, M., Tsunematsu, R., Tsukiyama, T., Ishida, N., et al. (2000). Targeted disruption of Skp2 results in accumulation of cyclin E and p27(Kip1), polyploidy and centrosome overduplication. *EMBO J.* **19**, 2069–81.
Nantel, F., Monaco, L., Foulkes, N. S., Masquilier, D., LeMeur, M., Henriksen, K., Dierich, A., Parvinen, M. and Sassone-Corsi, P. (1996). Spermiogenesis deficiency and germ-cell apoptosis in CREM-mutant mice. *Nature* **380**, 159–62.
Nead, M. A., Baglia, L. A., Antinore, M. J., Ludlow, J. W. and McCance, D. J. (1998). Rb binds c-Jun and activates transcription. *EMBO J.* **17**, 2342–52.
Needham, J. (1959). *A History of Embryology*. New York: Abelard-Schuman.
Neff, N. F., Thomas, J. H., Grisafi, P. and Botstein, D. (1983). Isolation of the beta-tubulin gene from yeast and demonstration of its essential function in vivo. *Cell* **33**, 211–9.
Nemeth-Cahalan, K. L. and Hall, J. E. (2000). pH and calcium regulate the water permeability of aquaporin 0. *J. Biol. Chem.* **275**, 6777–82.
Neumann, C. J. and Nuesslein-Volhard, C. (2000). Patterning of the zebrafish retina by a wave of sonic hedgehog activity. *Science* **289**, 2137–9.
Nevins, J. R. (1998). Towards an understanding of the functional complexity of the E2F and retinoblastoma families. *Cell Growth Differ.* **9**, 585–93.
Newport, J. W. and Kirschner, M. W. (1984). Regulation of the cell cycle during early *Xenopus* development. *Cell* **37**, 731–42.
Nguyen, M. M., Nguyen, M. L., Caruana, G., Bernstein, A., Lambert, P. F. and Griep, A. E. (2003). Requirement of PDZ-containing proteins for cell cycle regulation and differentiation in the mouse lens epithelium. *Mol. Cell Biol.* **23**, 8970–81.
Nguyen, M. M., Potter, S. J. and Griep, A. E. (2002). Deregulated cell cycle control in lens epithelial cells by expression of inhibitors of tumor suppressor function. *Mech. Dev.* **112**, 101–13.
Niazi, I. A. (1967). A contribution to the study of lens regeneration capacity in chick embryos. *Experientia* **23**, 970–2.
Nickerson, J. M., Wawrousek, E. F., Borras, T., Hawkins, J. W., Norman, B. L., Filpula, D. R., Nagle, J. W., Ally, A. H. and Piatigorsky, J. (1986). Sequence of the chicken delta 2 crystallin gene and its intergenic spacer: extreme homology with the delta 1 crystallin gene. *J. Biol. Chem.* **261**, 552–7.
Nickerson, J. M., Wawrousek, E. F., Hawkins, J. W., Wakil, A. S., Wistow, G. J., Thomas, G., Norman, B. L. and Piatigorsky, J. (1985). The complete sequence of the chicken delta 1 crystallin gene and its 5′ flanking region. *J. Biol. Chem.* **260**, 9100–5.
Nieuwkoop, P. (1952). Activation and organization of the central nervous system in amphibians. *J. Exp. Zool.* **120**, 1–108.
Nieuwkoop, P. (1963). Pattern formation in artificially activated ectoderm. *Dev. Biol.* **7**, 255–79.
Nieuwkoop, P. D. and Faber, J. (1956). *Normal Table of Xenopus Laevis*. Amsterdam: North Holland.
Niiya, A., Ohto, H., Kawakami, K. and Araki, M. (1998). Localization of *Six4*/AREC3 in the developing mouse retina: implications in mammalian retinal development. *Exp. Eye Res.* **67**, 699–707.
Nilsson, D. E. (1990). From cornea to retinal image in invertebrate eyes. *Trends Neurosci.* **13**, 55–64.
Nishiguchi, S., Wood, H., Kondoh, H., Lovell-Badge, R. and Episkopou, V. (1998). Sox1 directly regulates the γ-crystallin genes and is essential for lens development in mice. *Genes Dev.* **12**, 776–81.
Nishikawa, S., Ishiguro, S., Kato, K. and Tamai, M. (1994). A transient expression of alpha B-crystallin in the developing rat retinal pigment epithelium. *Invest. Ophthalmol. Vis. Sci.* **35**, 4159–64.

Nishizawa, M., Kataoka, K., Goto, N., Fujiwara, K. T. and Kawai, S. (1989). *v-Maf*, a viral oncogene that encodes a "leucine zipper" motif. *Proc. Natl. Acad. Sci. USA* **86**, 7711–15.

Noll, M. (1993). Evolution and role of *Pax* genes. *Curr. Opin. Genet. Dev.* **3**, 595–605.

Nornes, S., Clarkson, M., Mikkola, I., Pedersen, M., Bardsley, A., Martinez, J. P., Krauss, S. and Johansen, T. (1998). Zebrafish contains 2 *Pax6* genes involved in eye development. *Mech. Dev.* **77**, 185–96.

Norose, K., Clark, J. I., Syed, N. A., Basu, A., Heber-Katz, E., Sage, E. H. and Howe, C. C. (1998). SPARC deficiency leads to early-onset cataractogenesis. *Invest. Ophthalmol. Vis. Sci.* **39**, 2674–80.

Norose, K., Lo, W. K., Clark, J. I., Sage, E. H. and Howe, C. C. (2000). Lenses of SPARC-null mice exhibit an abnormal cell surface–basement membrane interface. *Exp. Eye Res.* **71**, 295–307.

Nottoli, T., Hagopian-Donaldson, S., Zhang, J., Perkins, A. and Williams, T. (1998). AP-2-null cells disrupt morphogenesis of the eye, face, and limbs in chimeric mice. *Proc. Natl. Acad. Sci. USA* **95**, 13714–19.

Novotny, G. E. and Pau, H. (1984). Myofibroblast-like cells in human anterior capsular cataract. *Virchows. Arch. A. Pathol. Anat. Histopathol.* **404**, 393–401.

Nurse, P. (1994). Ordering S phase and M phase in the cell cycle. *Cell* **79**, 547–50.

Nutt, S. L., Eberhard, D., Horcher, M., Rolink, A. G. and Busslinger, M. (2001). Pax5 determines the identity of B cells from the beginning to the end of B-lymphopoiesis. *Int. Rev. Immunol.* **20**, 65–82.

Obata, H., Kaburaki, T., Kato, M. and Yamashita, H. (1996). Expression of TGF-beta type I and II receptors in rat eyes. *Curr. Eye Res.* **15**, 335–40.

Oda, S.-I., Watanabe, K., Fujisawa, H. and Kameyama, Y. (1980). Impaired development of lens fibers in genetic micropthalmia eye lens obsolescence, Elo, of the mouse. *Exp. Eye Res.* **31**, 673–81.

Offield, M. F., Hirsch, N. and Grainger, R. M. (2000). The development of *Xenopus tropicalis* transgenic lines and their use in studying lens developmental timing in living embryos. *Development* **127**, 1789–97.

Ogawa, M., Takabatake, T., Takahashi, T. C. and Takeshima, K. (1997). Metamorphic change in EP37 expression: members of the $\beta\gamma$-crystallin superfamily in newt. *Dev. Genes Evol.* **206**, 417–24.

Ogden, A. T., Nunes, I., Ko, K., Wu, S., Hines, C. S., Wang, A. F., Hedge, R. S. and Lang, R. A. (1998). GRIFIN, a novel lens-specific protein related to the galectin family. *J. Biol. Chem.* **273**, 28889–96.

Ogino, H. and Yasuda, K. (1996). Involvement of maf gene family in crystallin gene regulation [in Japanese]. *Tanpakushitsu Kakusan Koso* **41**, 1050–7.

Ogino, H. and Yasuda, K. (1998). Induction of lens differentiation by activation of a bZIP transcription factor, L-Maf. *Science* **280**, 115–8.

Ogino, H. and Yasuda, K. (2000). Sequential activation of transcription factors in lens induction. *Develop. Growth Differ.* **42**, 437–48.

Oharazawa, H., Ibaraki, N., Lin, L. R. and Reddy, V. N. (1999). The effects of extracellular matrix on cell attachment, proliferation and migration in a human lens epithelial cell line. *Exp. Eye Res.* **69**, 603–10.

Ohtaka-Maruyama, C., Hanaoka, F. and Chepelinsky, A. B. (1998a). A novel alternative spliced variant of the transcription factor AP2-α is expressed in the murine ocular lens. *Dev. Biol.* **202**, 125–35.

Ohtaka-Maruyama, C., Wang, X., Ge, H. and Chepelinsky, A. B. (1998b). Overlapping Sp1 and AP2 binding sites in a promoter element of the lens-specific *MIP* gene. *Nucl. Acids Res.* **26**, 407–14.

Ohto, H., Kamada, S., Tago, K., Tominaga, S., Ozaki, H., Sato, S. and Kawakami, K. (1999). Cooperation of Six and Eya in activation of their target genes through nuclear translocation of Eya. *Mol. Cell. Biol.* **19**, 6815–24.

Ohto, H., Takizawa, T., Saito, T., Kobayashi, M., Ikeda, K. and Kawakami, K. (1998). Tissue and developmental distribution of *Six* family gene products. *Int. J. Dev. Biol.* **42**, 141–8.

Ohtsubo, M. and Roberts, J. M. (1993). Cyclin-dependent regulation of G1 in mammalian fibroblasts. *Science* **259**, 1908–12.

Ohtsubo, M., Theodoras, A. M., Schumacher, J., Roberts, J. M. and Pagano, M. (1995). Human cyclin E, a nuclear protein essential for the G1-to-S phase transition. *Mol. Cell Biol.* **15**, 2612–24.

Okamoto, M. (1987). Induction of ocular tumor by nickel subsulfide in the Japanese common newt, *Cynops pyrrhogaster*. *Cancer Res.* **47**, 5213–7.

Okamoto, M. (1997). Simultaneous demonstration of lens regeneration from dorsal iris and tumour production from ventral iris in the same newt eye after carcinogen administration. *Differentiation* **61**, 285–92.

Okladnova, O., Syagailo, Y. V., Mössner, R., Riederer, P. and Lesch, K.-P. (1998). Regulation of *Pax6* gene transcription: alternate promoter usage in human brain. *Mol. Brain Res.* **60**, 177–92.

Olitsky, S. E., Waz, W. R. and Wilson, M. E. (1999). Rupture of the anterior lens capsule in Alport syndrome. *J. AAPOS* **3**, 381–2.

Oliver, G., Loosli, F., Koster, R., Wittbrodt, J. and Gruss, P. (1996). Ectopic lens induction in fish in response to the murine homeobox gene *Six3*. *Mech. Dev.* **60**, 233–9.

Oliver, G., Mailhos, A., Wehr, R., Copeland, N. G., Jenkins, N. A. and Gruss, P. (1995a). *Six3*, a murine homologue of the *sine oculis* gene, demarcates the most anterior border of the developing neural plate and is expressed during eye development. *Development* **121**, 4045–55.

Oliver, G., Sosa-Pineda, B., Geisendorf, S., Spana, E. P., Doe, C. Q. and Gruss, P. (1993). Prox 1, a *prospero*-related homeobox gene expressed during mouse development. *Mech. Dev.* **44**, 3–16.

Oliver, G., Wehr, R., Jenkins, N. A., Copeland, N. G., Cheyette, B. N., Hartenstein, V., Zipursky, S. L. and Gruss, P. (1995b). Homeobox genes and connective tissue patterning. *Development* **121**, 693–705.

Olivero, D. K. and Furcht, L. T. (1993). Type IV collagen, laminin, and fibronectin promote the adhesion and migration of rabbit lens epithelial cells in vitro. *Invest. Ophthalmol. Vis. Sci.* **34**, 2825–34.

Ookata, K., Hisanaga, S., Okano, T., Tachibana, K. and Kishimoto, T. (1992). Relocation and distinct subcellular localization of p34cdc2–cyclin B complex at meiosis reinitiation in starfish oocytes. *EMBO J.* **11**, 1763–72.

Ornitz, D. M. and Itoh, N. (2001). Fibroblast growth factors. *Genome Biol.* **2**(3): reviews 3005.1–reviews3005.12.

Ortiz, J. R., Vigny, M., Courtois, Y. and Jeanny, J. C. (1992). Immunocytochemical study of extracellular matrix components during lens and neural retina regeneration in the adult newt. *Exp. Eye Res.* **54**, 861–70.

Ortwerth, B. J. and Byrnes, R. J. (1971). Properties of a ribonuclease inhibitor from bovine lens. *Exp. Eye Res.* **12**, 120–7.

Ortwerth, B. J. and Byrnes, R. J. (1972). Further studies on the purification and properties of a ribonuclease inhibitor from lens cortex. *Exp. Eye Res.* **14**, 114–22.

Ostrer, H. and Piatigorsky, J. (1980). Beta-crystallins of the adult chicken lens: relatedness of the polypeptides and their aggregates. *Exp. Eye Res.* **30**, 679–89.

Overbeek, P. A., Chepelinsky, A. B., Khillan, J. S., Piatigorsky, J. and H. Westphal. (1985). Lens-specific expression and developmental regulation of the bacterial chloramphenicol acetyltransferase gene driven by the murine alpha A-crystallin promoter in transgenic mice. *Proc. Natl. Acad. Sci. USA* **82**, 7815–9.

Ozaki, L., Jap, P. and Bloemendal, H. (1985). Electron microscopic study of water-insoluble fractions in normal and cataractous human lens fibers. *Ophthalmic Res.* **17**, 257–61.

Ozaki, H., Watanabe, Y., Takahashi, K., Kitamura, K., Tanaka, A., Urase, K., Momoi, T., Sudo, K., Sakagami, J., Asano, M., et al. (2001). *Six4*, a putative *myogenin* gene regulator, is not essential for mouse embryonal development. *Mol. Cell. Biol.* **21**, 3343–50.

Paffenholz, R., Kuhn, C., Grund, C., Stehr, S. and Franke, W. W. (1999). The arm-repeat protein NPRAP (neurojungin) is a constituent of the plaques of the outer limiting zone in the retina, defining a novel type of adhering junction. *Exp. Cell Res.* **250**, 452–64.

Pagano, M., Pepperkok, R., Lukas, J., Baldin, V., Ansorge, W., Bartek, J. and Draetta, G. (1993). Regulation of the cell cycle by the cdk2 protein kinase in cultured human fibroblasts. *J. Cell Biol.* **121**, 101–11.

Pagano, M., Pepperkok, R., Verde, F., Ansorge, W. and Draetta, G. (1992). Cyclin A is required at two points in the human cell cycle. *EMBO J.* **11**, 961–71.

Pal, J. D., Berthoud, V. M., Beyer, E. C., Mackay, D., Shiels, A. and Ebihara, L. (1999). Molecular mechanism underlying a Cx50-linked congenital cataract. *Am. J. Physiol.* **276**, C1443–6.

Palmade, F., Sechoy-Chambon, O., Coquelet, C. and Bonne, C. (1994). Insulin-like growth factor-1 (IGF-1) specifically binds to bovine lens epithelial cells and increases the number of fibronectin binding sites. *Curr. Eye Res.* **13**, 531–7.

Palmiter, R. D. and Carey, N. H. (1974). Rapid inactivation of ovalbumin messenger ribonucleic acid after acute withdrawal of estrogen. *Proc. Natl. Acad. Sci. USA* **71**, 2357–61.

Pan, F. M., Chang, W. C., Chao, Y. K. and Chiou, S. H. (1994). Characterization of gamma-crystallins from a hybrid teleostean fish: multiplicity of isoforms as revealed by cDNA sequence analysis. *Biochem. Biophys. Res. Commun.* **202**, 527–34.

Pan, H. (1995). Regulation of cell proliferation and cell death in the developing mouse lens by tumor suppressors Rb and p53. Unpublished doctoral dissertation. Madison: University of Wisconsin.

Pan, H. and Griep, A. E. (1994). Altered cell cycle regulation in the lens of HPV-16 E6 or E7 transgenic mice: implications for tumor suppressor gene function in development. *Genes Dev.* **8**, 1285–99.

Pan, H. and Griep, A. E. (1995). Temporally distinct patterns of p53-dependent and p53-independent apoptosis during mouse lens development. *Genes Dev.* **9**, 2157–69.

Papaconstantinou, J. (1967). Molecular aspects of lens cell differentiation. *Science* **156**, 338–46.

Pardee, A. B. (1989). G1 events and regulation of cell proliferation. *Science* **246**, 603–8.

Park, J. H. and Saier, M. H. (1996). Phylogenetic characterization of the MIP family of transmembrane channel proteins. *J. Membr. Biol.* **153**, 171–80.

Parmelee, J. T. and Beebe, D. C. (1988). Decreased membrane permeability to potassium is responsible for the cell volume increase that drives lens fiber cell elongation. *J. Cell. Physiol.* **134**, 491–6.

Parmigiani, C. and McAvoy, J. (1984). Localisation of laminin and fibronectin during rat lens morphogenesis. *Differentiation* **28**, 53–61.

Parmigiani, C. M. and McAvoy, J. W. (1991). The roles of laminin and fibronectin in the development of the lens capsule. *Curr. Eye Res.* **10**, 501–11.

Patek, C. E. and Clayton, R. M. (1990). Age-related changes in the response of chick lens cells during long-term culture to insulin, cyclic AMP, retinoic acid and a bovine retinal extract. *Exp. Eye Res.* **50**, 345–54.

Patil, R. V., Saito, I., Yang, X. and Wax, M. B. (1997). Expression of aquaporins in the rat ocular tissue. *Exp. Eye Res.* **64**, 203–9.

Pau, H., Novotny, G. E. and Arnold, G. (1985). Ultrastructural investigation of extracellular structures in subcapsular white corrugated cataract (anterior capsular cataract). *Graefes. Arch. Clin. Exp. Ophthalmol.* **223**, 96–100.

Paul, D. L., Ebihara, L., Takemoto, L. J., Swenson, K. I. and Goodenough, D. A. (1991). Connexin46, a novel lens gap junction protein, induces voltage-gated currents in non-junctional plasma membrane of *Xenopus* oocytes. *J. Cell Biol.* **115**, 1077–89.

Paul, D. L. and Goodenough, D. A. (1983). Preparation, characterization, and localization of antisera against bovine MIP26, an integral protein from lens fiber plasma membrane. *J. Cell Biol.* **96**, 625–32.

Pearce, T. L. and Zwaan, J. (1970). A light and electron microscopic study of cell behavior and microtubules in the embryonic chicken lens using Colcemid. *J. Embryol. Exp. Morphol.* **23**, 491–507.

Peek, R., McAvoy, J. W., Lubsen, N. H. and Schoenmakers, J. G. G. (1992). Rise and fall of crystallin gene messenger levels during fibroblast growth factor induced terminal differentiation of lens cells. *Dev. Biol.* **152**, 152–60.

Peek, R., van der Logt, P., Lubsen, N. H. and Schoenmakers, J. G. (1990). Tissue- and species-specific promoter elements of rat gamma-crystallin genes. *Nucleic Acids Res.* **18**, 1189–97.

Pelton, R. W., Saxena, B., Jones, M., Moses, H. L. and Gold, L. I. (1991). Immunohistochemical localization of TGF beta 1, TGF beta 2, and TGF beta 3 in the mouse embryo: expression patterns suggest multiple roles during embryonic development. *J. Cell Biol.* **115**, 1091–105.

Pera, E. and Kessel, M. (1999). Expression of DLX3 in chick embryos. *Mech. Devel.* **89**, 189–93.

Peracchia, C., Girsch, S. J., Bernardine, G. and Peracchia, L. L. (1985). Lens junctions are communicating junctions. *Curr. Eye Res.* **4**, 1155–69.

Perez-Castro, A. V., Tran, V. T. and Nguyen-Huu, M. C. (1993). Defective lens fiber differentiation and pancreatic tumorigenesis caused by ectopic expression of the cellular retinoic acid-binding protein I. *Development* **119**, 363–75.

Perillo, N. L., Marcus, M. E. and Baum, L. G. (1998). Galectins: versatile modulators of cell adhesion, cell proliferation, and cell death. *J. Mol. Med.* **76**, 402–12.

Perng, M. D., Muchowski, P. J., van Den, I. P., Wu, G. J., Hutcheson, A. M., Clark, J. I. and Quinlan, R. A. (1999). The cardiomyopathy and lens cataract mutation in alphaB-crystallin alters its protein structure, chaperone activity, and interaction with intermediate filaments in vitro. *J. Biol. Chem.* **274**, 33235–43.

Perry, M. M., Tassin, J. and Courtois, Y. (1981). Fine structure of bovine lens epithelial cells in vitro in relation to modifications induced by a retinal extract (EDGF). *Exp. Cell Res.* **136**, 379–90.

Peterson, C. A. and Piatigorsky, J. (1986). Preferential conservation of the globular domains of the beta A3/A1-crystallin polypeptide of the chicken eye lens. *Gene* **45**, 139–47.

Petkovich, M., Brand, N. J., Krust, A. and Chambon, P. (1987). A human retinoic acid receptor which belongs to the family of nuclear receptors. *Nature* **330**, 444–50.

Pevny, L. H. and Lovell-Badge, R. (1997). *Sox* genes find their feet. *Curr. Opin. Genet. Dev.* **7**, 338–44.

Pevny, L. H., Sockanathan, S., Placzek, M. and Lovell-Badge, R. (1998). A role for SOX1 in neural determination. *Development* **125**, 1967–78.

Philpott, G. W. and Coulombre, A. J. (1965). Lens development: the differentiation of embryonic chick lens epithelial cells in vitro and in vivo. *Exp. Cell Res.* **38**, 635–44.

Piatigorsky, J. (1973). Insulin initiation of lens fiber differentiation in culture: elongation of embryonic lens epithelial cells. *Dev. Biol.* **30**, 214–16.

Piatigorsky, J. (1975). Lens cell elongation in vitro and microtubules. *Ann. NY Acad. Sci.* **253**, 333–47.

Piatigorsky, J. (1984). Lens crystallins and their gene families. *Cell* **38**, 620–1.

Piatigorsky, J. (1993). Puzzle of crystallin diversity in eye lenses. *Dev. Dyn.* **196**, 267–72.

Piatigorsky, J. (1998). Gene sharing in the lens and cornea: facts and implications. *Prog. Ret. Eye Res.* **17**, 145–74.

Piatigorsky, J. and Horwitz, J. (1996). Characterization and enzyme activity of argininosuccinate lyase/delta-crystallin of the embryonic duck lens. *Biochim. Biophys. Acta* **1295**, 158–64.

Piatigorsky, J., Horwitz, J., Kuwabara, T. and Cutress, C. E. (1989). The cellular eye lens and crystallins of cubomedusan jellyfish. *J. Comp. Physiol. A* **164**, 577–87.

Piatigorsky, J., Horwitz, J. and Norman, B. L. (1993). J1-crystallins of the cubomedusan jellyfish lens constitute a novel family encoded in at least three intronless genes. *J. Biol. Chem.* **268**, 11894–901.

Piatigorsky, J., Kozmik, Z., Horwitz, J., Ding, L., Carosa, E., Robison, W. G., Steinbach, P. J. and Tamm, E. R. (2000). Ω-Crystallin of the scallop lens: a dimeric aldehyde dehydrogenase class 1/2 enzyme-crystallin. *J. Biol. Chem.* **275**, 41064–73.

Piatigorsky, J., Norman, B. and Jones, R. E. (1987). Conservation of delta-crystallin gene structure between ducks and chickens. *J. Mol. Evol.* **25**, 308–17.

Piatigorsky, J., Norman, B., Dishaw, L. J., Kos, L., Horwitz, J., Steinbach, P. J. and Kozmik, Z. (2001). J3-crystallin of the jellyfish lens: similarity to saposins. *Proc. Natl. Acad. Sci. USA* **98**, 12362–7.

Piatigorsky, J., O'Brien, W. E., Norman, B. L., Kalumuck, K., Wistow, G. J., Borras, T., Nickerson, J. M. and Wawrousek, E. F. (1988). Gene sharing by δ-crystallin and argininosuccinate lyase. *Proc. Natl. Acad. USA* **85**, 3479–83.

Piatigorsky, J., Rothschild, S. S. and Milstone, L. M. (1973a). Differentiation of lens fibers in explanted embryonic chick lens epithelia. *Dev. Biol.* **34**, 334–45.

Piatigorsky, J., Rothschild, S. S. and Wollberg, M. (1973b). Stimulation by insulin of cell elongation and microtubule assembly in embryonic chick-lens epithelia. *Proc. Natl. Acad. Sci. USA* **70**, 1195–8.

Piatigorsky, J., Webster, H. and Wollberg, M. (1972). Cell elongation in the cultured embryonic chick lens epithelium with and without protein synthesis: involvement of microtubules. *J. Cell Biol.* **55**, 82–92.

Piatigorsky, J. and Wistow, G. J. (1989). Enzyme/crystallins: gene sharing as an evolutionary strategy. *Cell* **57**, 197–9.

Piatigorsky, J. and Wistow, G. (1991). The recruitment of crystallins: new functions precede gene duplication. *Science* **252**, 1078–9.

Piatigorsky, J. and Zelenka, P. S. (1992). Transcriptional regulation of crystallin genes: cis elements, trans-factors and signal transduction systems in the lens. In *Advances in Developmental Biochemistry*, vol. 1, ed. P. M. Wasserman. JAI Press, pp. 211–56.

Pichaud, F., Treisman, J. and Desplan, C. (2001). Reinventing a common strategy for patterning the eye. *Cell* **105**, 9–12.

Pignoni, F., Hu, B., Zavitz, K. H., Xiao, J., Garrity, P. A. and Zipursky, S. L. (1997). The eye-specification proteins So and Eya form a complex and regulate multiple steps in *Drosophila* eye development. *Cell* **91**, 881–91.

Pignoni, F. and Zipursky, S. L. (1997). Induction of *Drosophila* eye development by decapentaplegic. *Development* **124**, 271–8.

Pineda, D., Gonzalez, J., Callarets, P., Ikeo, K., Gehring, W. J. and Salo, E. (2000). Searching for the prototypic eye genetic network: *sine oculis* is essential for eye regeneration in planarians. *Proc. Natl. Acad. Sci. USA* **97**, 4525–9.

Pines, J. (1995). Cyclins and cyclin-dependent kinases: a biochemical view. *Biochem. J.* **308**, 697–711.

Pines, J. and Hunter, T. (1991). Human cyclins A and B1 are differentially located in the cell and undergo cell cycle-dependent nuclear transport. *J. Cell Biol.* **115**, 1–17.

Pinson, K. I., Brennan, J., Monkley, S., Avery, B. J. and Skarnes, W. C. (2000). An LDL-receptor–related protein mediates Wnt signaling in mice. *Nature* **407**, 535–8.

Pizette, S., Coulier, F., Birnbaum, D. and DeLapeyriere, O. (1996). FGF6 modulates the expression of fibroblast growth factor receptors and myogenic genes in muscle cells. *Exp. Cell Res.* **224**, 143–51.

Planque, N., Leconte, L., Coquelle, F. M., Benkhelifa, S., Martin, P., Felder-Schmittbuhl, M. P. and Saule, S. (2001). Interaction of Maf transcription factors with Pax-6 results in synergistic activation of the glucagon promoter. *J. Biol. Chem.* **276**, 35751–60.

Plaza, S., Dozier, C. and Saule, S. (1993). Quail Pax-6 (Pax-QNR) encodes a transcription factor able to bind and trans-activate its own promoter. *Cell Growth Differ.* **4**, 1041–50.

Pomerantz, J., Schreiber-Agus, N., Liegois, N. J., Silverman, M. A., Alland, L., Chin, L., Potes, J., Chen, K., Orlow, I., Lee, H. W., et al. (1998). The Ink4a tumor suppressor gene product, p19ARF, interacts with MDM2 and neutralizes MDM2's inhibition of p53. *Cell* **92**, 713–23.

Porter, F. D., Drago, J., Xu, Y., Cheema, S. S., Wassif, C., Huang, S.-P., Lee, E., Grinberg, A., Massalas, J. S., Bodine, D., et al. (1997). *Lhx2*, a LIM homeobox gene, is required for eye, forebrain, and definitive erythrocyte development. *Development* **124**, 2935–44.

Potts, J. D., Bassnett, S., Kornacker, S. and Beebe, D. C. (1994). Expression of platelet-derived growth factor receptors in the developing chicken lens. *Invest. Ophthalmol. Vis. Sci.* **35**, 3413–21.

Potts, J. D., Kornacker, S. and Beebe, D. C. (1998). Activation of the Jak-STAT-signaling pathway in embryonic lens cells. *Dev. Biol.* **204**, 277–92.

Pozzi, A., Wary, K. K., Giancotti, F. G. and Gardner, H. A. (1998). Integrin alpha1beta1 mediates a unique collagen-dependent proliferation pathway in vivo. *J. Cell Biol.* **142**, 587–94.

Prahlad, V., Yoon, M., Moir, R. D., Vale, R. D. and Goldman, R. D. (1998). Rapid movements of vimentin on microtubule tracks: kinesin-dependent assembly of intermediate filament networks. *J. Cell Biol.* **143**, 159–70.

Pras, E., Frydman, M., Levy-Nissenbaum, E., Bakhan, T., Raz, J., Assia, E. I. and Goldman, B. (2000). A nonsense mutation (W9X) in CRYAA causes autosomal recessive cataract in an inbred Jewish Persian family. *Invest. Ophthalmol. Vis. Sci.* **41**, 3511–5.

Prescott, A. R., Duncan, G., Rawlins, D. and Shaw, P. J. (1991). Dye communication properties in three regions of the intact frog lens. In *Eye Lens Membrane and Aging*, ed. G. F. J. M. Vrensen and J. Clauwert. Leiden: Eurage, pp. 59–71.

Prescott, A., Duncan, G., Van Marle, J. and Vrensen, G. (1994). A correlated study of metabolic cell communication and gap junction distribution in the adult frog lens. *Exp. Eye Res.* **58**, 737–46.

Prescott, A. R., Sandilands, A., Hutcheson, A. M., Carter, J. M. and Quinlan, R. A. (1996). The intermediate filament cytoskeleton of the lens: an ever changing network through development and differentiation [minireview]. *Ophthalmic Res.* **28**, 58–61.

Prescott, A. R., Stewart, S., Duncan, G., Gowing, R. and Warn, R. M. (1991). Diamide induces reversible changes in morphology, cytoskeleton and cell-cell coupling in lens epithelial cells. *Exp. Eye Res.* **52**, 83–92.

Price, L. S., Leng, J., Schwartz, M. A. and Bokoch, G. M. (1998). Activation of Rac and Cdc42 by integrins mediates cell spreading. *Mol. Biol, Cell* **9**, 1863–71.

Priolo, S., Sivak, J. G. and Kuszak, J. R. (1999a). Effect of age on the morphology and optical quality of the avian crystalline lens. *Exp. Eye Res.* **69**, 629–40.

Priolo, S., Sivak, J. G. and Kuszak, J. R. (1999b). Effect of experimentally induced ametropia on the morphology and optical quality of the avian crystalline lens. *Invest. Ophthalmol. Vis. Sci.* **41**, 3516–22.

Prosser, J. and van Heyningen, V. (1998). PAX6 mutations reviewed. *Hum. Mutat.* **11**, 93–108.

Pullan, S., Wilson, J., Metcalfe, A., Edwards, G. M., Goberdhan, N., Tilly, J., Hickman, J. A., Dive, C. and Streuli, C. H. (1996). Requirement of basement membrane for the suppression of programmed cell death in mammary epithelium. *J. Cell Sci.* **109**, 631–42.

Qi, Y., Jia, H., Huang, S., Lin, H., Gu, J., Su, H., Zhang, T., Gao, Y., Qu, L., Li, D. et al. (2004). A deletion mutation in the betaA1/A3 crystallin gene (CRYBA1/A3) is associated with autosomal dominant congenital nuclear cataract in a Chinese family. *Hum. Genet.* **114**, 192–7.

Qin, X. Q., Livingston, D. M., Kaelin, W. G. J. and Adams, P. D. (1994). Deregulated transcription factor E2F-1 expression leads to S-phase entry and p53-mediated apoptosis. *Proc. Natl. Acad. Sci. USA* **91**, 10918–22.

Quax-Jeuken, Y., Janssen, C., Quax, W., van den Heuvel, R. and Bloemendal, H. (1984). Bovine beta-crystallin complementary DNA clones: alternating proline/alanine sequence of beta B1 subunit originates from a repetitive DNA sequence. *J. Mol. Biol.* **180**, 457–72.

Quelle, D. E., Ashmun, R. A., Shurtleff, S. A., Kato, J. Y., Bar-Sagi, D., Roussel, M. F. and Sherr, C. J. (1993). Overexpression of mouse D-type cyclins accelerates G1 phase in rodent fibroblasts. *Genes Dev.* **7**, 1559–71.

Quelle, D. E., Zindy, F., Ashmun, R. A. and Sherr, C. J. (1995). Alternative reading frames of the INK4a tumor suppressor gene encode two unrelated proteins capable of inducing cell cycle arrest. *Cell* **83**, 993–1000.

Quinlan, G. A., Williams, E. A., Tan, S.-S. and Tam, P. P. L. (1995). Neuroectodermal fate of epiblast cells in the distal region of the mouse egg cylinder: implication for body plan organization during early embryogenesis. *Development* **121**, 87–98.

Quinlan, R., Hutchison, C. and Lane, B. (1995). Intermediate filament proteins. *Protein Profile* **2**, 801–952.

Quinlan, R. A. (1991). The soluble plasma membrane–cytoskeleton complexes and aging in the lens. In *Eye Lens Membranes and Aging*, vol. 15, ed. G. F. J. M. Vrensen and J. Clauwaert. Leiden: Eurage, pp. 171–84.

Quinlan, R. A., Carter, J. M., Sandilands, A. and Prescott, A. R. (1996). The beaded filament of the eye lens: an unexpected key to intermediate filament structure and function. *Trends Cell Biol.* **6**, 123–6.

Quinlan, R. A., Schiller, D. L., Hatzfeld, M., Achstatter, T., Moll, R., Jorcano, J. L., Magin, T. M. and Franke, W. W. (1985). Patterns of expression and organization of cytokeratin intermediate filaments. *Ann. NY Acad. Sci.* **455**, 282–306.

Quinn, J. C., West, J. D. and Hill, R. E. (1996). Multiple functions for Pax6 in mouse eye and nasal development. *Genes Dev.* **10**, 435–46.

Quiring, R., Walidorf, U., Kloter, U. and Gehring, W. J. (1994). Homology of the *eyeless* gene of *Drosophila* to the *Small eye* gene in mice and *Aniridia* in humans. *Science* **265**, 785–9.

Rabaey, M. (1962). Electrophoretic and immunoelectrophoretic studies on the soluble proteins in the developing lens of birds. *Exp. Eye Res.* **1**, 310–6.

Rabl, C. (1899). Über den Bau und die Entwicklung der Linse. III. Die Linse der Säugetiere: Ruckblick und Schluss. *Z. Wiss. Zool.* **67**, 1–138.

Rae, J. L. (1994). Outwardly rectifying potassium currents in lens epithelial cell membranes *Curr. Eye. Res.* **13**, 679–86.

Rae, J. L., Bartling, C., Rae, J. and Mathias, R. T. (1996). Dye transfer between cells of the lens. *J. Membr. Biol.* **150**, 89–103.

Rae, J. L. and Cooper, K. (1990). New techniques for the study of lens electrophysiology. *Exp. Eye Res.* **50**, 603–14.

Rae, J. L., Dewey, J., Rae, J. S. and Cooper, K. (1990). A maxi calcium-activated potassium channel from chick lens epithelium. *Curr. Eye Res.* **9**, 847–61.

Rae, J. L. and Kuszak, J. R. (1983). The electrical coupling of epithelium and fibers in the frog lens. *Exp. Eye Res.* **36**, 317–26.

Rae, J. L., Mathias, R. T., Cooper, K. and Baldo, G. (1992). Divalent cation effects on lens conductance and stretch-activated cation channels. *Exp. Eye Res.* **55**, 135–44.

Rae, J. L. and Rae, J. S. (1992). Whole-cell currents from noncultured human lens epithelium. *Invest. Ophthalmol. Vis. Sci.* **33**, 2262–8.

Rae, J. L. and Shepard, A. R. (1998a). Identification of potassium channels in human lens epithelial cells. *Curr. Top. Membr. Trans.* **45**, 69–104.

Rae, J. L. and Shepard, A. R. (1998b). Inwardly rectifying potassium channels in lens epithelium are from the IRK1 (Kir 2.1) family. *Exp. Eye Res.* **66**, 347–59.

Rae, J. L. and Shepard, A. R. (1998c). Molecular biology electrophysiology of calcium-activated potassium channels from lens epithelium. *Curr. Eye Res.* **17**, 264–75.

Rae, J. L. and Shepard, A. R. (2000). Kv3.3 potassium channels in lens epithelium and corneal endothelium. *Exp. Eye Res.* **70**, 339–48.

Rae, J. L. and Stacey, T. R. (1979). Lanthanum and procion yellow as extracellular markers. *Exp. Eye Res.* **28**, 1–21.

Rae, J. L., Truitt, K. D. and Kuszak, J. R. (1982). A simple fluorescence technique for light microscopy of the crystalline lens. *Curr. Eye Res.* **2**, 1–5.

Rafferty, N. S. and Goossens, W. (1978). Cytoplasmic filaments in the crystalline lens of various species: functional correlations. *Exp. Eye Res.* **26**, 177–90.

Rafferty, N. S. and Rafferty, K. A., Jr. (1981). Cell population kinetics of the mouse lens epithelium. *J. Cell. Physiol.* **107**, 309–15.

Rafferty, N. S. and Scholz, D. L. (1984). Polygonal arrays of microfilaments in epithelial cells of the intact lens. *Curr. Eye Res.* **3**, 1141–9.

Rafferty, N. S. and Scholz, D. L. (1985). Actin in polygonal arrays of microfilaments and sequestered actin bundles (SABs) in lens epithelial cells of rabbits and mice. *Curr. Eye Res.* **4**, 713–18.

Rafferty, N. S. and Scholz, D. L. (1989). Comparative study of actin filament patterns in lens epithelial cells: are these determined by the mechanism of lens accommodation? *Curr. Eye Res.* **8**, 569–79.

Rafferty, N. S., Zigman, S., McDaniel, T. and Scholz, D. L. (1993). Near-UV radiation disrupts filamentous actin in lens epithelial cells. *Cell Motil. Cytoskeleton* **26**, 40–8.

Ramaekers, F. C. S., Boomkens, T. R. and Bloemendal, H. (1981). Cytoskeletal and contractile structures in bovine lens cell differentiation. *Exp. Cell Res.* **135**, 454–61.

Ramaekers, F. C. S., Dunia, I., Dodemot, H. J., Bendetti, E. L. and Bloemendal, H. (1982). Lenticular intermediate-sized filaments: biosynthesis and interaction with plasma membrane. *Proc. Natl. Acad. Sci. USA* **79**, 3208–12.

Ramaekers, F. C. S., Osborn, M., Schmid, E., Weber, K., Bloemendal, H. and Franke, W. W. (1980). Identification of the cytoskeletal proteins in lens-forming cells, a special epitheloid cell type. *Exp. Cell Res.* **127**, 309–27.

Ramaekers, F. C., Selten-Versteegen, A. M. and Bloemendal, H. (1980). Interaction of newly synthesized alpha-crystallin with isolated lens plasma membranes. *Biochim. Biophys. Acta* **596**, 57–63.

Rampalli, A. M., Gao, C. Y., Chauthaiwale, V. M. and Zelenka, P. S. (1998). pRb and p107 regulate E2F activity during lens fiber cell differentiation. *Oncogene* **16**, 399–408.

Rao, P. V., Robison, W. G., Jr., Bettelheim, F., Lin, L. R., Reddy, V. N. and Zigler, J. S., Jr. (1997). Role of small GTP-binding proteins in lovastatin-induced cataracts. *Invest. Ophthalmol. Vis. Sci.* **38**, 2313–21.

Rasmussen, J. T., Deardorff, M. A., Tan, C., Rao, M. S., Klein, P. S. and Vetter, M. L. (2001). Regulation of eye development by frizzled signaling in *Xenopus*. *Proc. Natl. Acad. Sci. USA* **98**, 3861–6.

Rath, P. C. and Aggarwal, B. B. (1999). TNF-induced signaling in apoptosis. *J. Clin. Immunol.* **19**, 350–64.

Ray, M. E., Wistow, G., Su, Y. A., Meltzer, P. S. and Trent, J. M. (1997). AIM1, a novel non-lens member of the betagamma-crystallin superfamily, is associated with the control of tumorigenicity in human malignant melanoma. *Proc. Natl. Acad. Sci. USA* **94**, 3229–34.

Redden, J. R. and Dziedzic, D. C. (1982). Insulin-like growth factors, IGF-1, IGF-2 and somatomedin c trigger cell proliferation in mammalian epithelial cells cultured in a serum-free medium. *Exp. Cell Res.* **142**, 293–300.

Redden, J. R. and Wilson-Dziedzic, D. C. (1983). Insulin growth factor and epidermal growth factor trigger mitoses in lenses cultured in a serum-free medium. *Invest. Ophthalmol. Vis. Sci.* **24**, 409–16.

Reddy, S., Smith, D. B. J., Rich, M. M., Leferovich, J. M., Reilly, P., Davis, B. M., Tran, K., Rayburn, H., Bronson, R., Cros, D., et al. (1996). Mice lacking the myotonic dystrophy protein kinase develop a late onset progressive myopathy. *Nat. Genet.* **13**, 325–35.

Reed, N. and Gutmann, D. H. (2001). Tumorigenesis in neurofibromatosis: new insights and potential therapies. *Trends Mol. Med.* **7**, 157–62.

Reed, U. C., Tsanaclis, A. M., Vainzof, M., Marie, S. K., Carvalho, M. S., Roizenblatt, J., Pedreira, C. C., Diament, A. and Levy, J. A. (1999). Merosin-positive congenital muscular dystrophy in two siblings with cataract and slight mental retardation. *Brain Dev.* **21**, 274–8.

Reese, D. H., Puccia, E. and Yamada, T. (1969). Activation of ribosomal RNA synthesis in initiation of Wolffian lens regeneration. *J. Exp. Zool.* **170**, 259–68.

Relaix, F. and Buckingham, M. (1999). From insect eye to vertebrate muscle: redeployment of a regulatory network. *Genes Dev.* **13**, 3171–8.

Ren, Z., Li, A., Shastry, B. S., Padma, T., Ayyagari, R., Scott, M. H., Parks, M. M., Kaiser-Kupfer, M. I. and Hejtmancik, J. F. (2000). A 5-base insertion in the gammaC-crystallin gene is associated with autosomal dominant variable zonular pulverulent cataract. *Hum. Genet.* **106**, 531–7.

Reneker, L. W. and Overbeek, P. A. (1996). Lens-specific expression of PDGF-A alters lens growth and development. *Dev. Biol.* **180**, 554–65.

Renkawek, K., de Jong, W. W., Merck, K. B., Frenken, C. W., van Workum, F. P. and Bosman, G. J. (1992). Alpha B-crystallin is present in reactive glia in Creutzfeldt-Jakob disease. *Acta Neuropathol.* **83**, 324–7.

Resnitsky, D., Hengst, L. and Reed, S. I. (1995). Cyclin A–associated kinase activity is rate limiting for entrance into S phase and is negatively regulated in G1 by p27Kip1. *Mol. Cell Biol.* **15**, 4347–52.

Resnitsky, D. and Reed, S. I. (1995). Different roles for cyclins D1 and E in regulation of the G1-to-S transition. *Mol. Cell Biol.* **15**, 3463–9.

Revel, J. P., Yee, A. G. and Hudspeth, A. J. (1971). Gap junctions between electrotonically coupled cells in tissue culture and in brown fat. *Proc. Natl. Acad. Sci., USA* **68**, 2924–7.

Reyer, R. W. (1962). Differentiation and growth of the embryonic nose, lens and corneal anlagen implanted into the larval eye or dorsal fin in *Amblystoma punctatum*. *J. Exp. Zool.* **151**, 123–49.

Reyer, R. W. (1948). An experimental study of lens regeneration in *Triturus viridescens*. I. Regeneration of a lens after lens extirpation in embryos and larvae of different ages. *J. Exp. Zool.* **107**, 217–68.

Reyer, R. W. (1953). Lens regeneration from heteroplastic iris grafts between *Triturus viridescens* and *Amblystoma punctatum*. *Anat. Rec.* **115**, 362–3.

Reyer, R. W. (1954). Regeneration of the lens in the amphibian eye. *Q. Rev. Biol.* **29**, 1–46.

Reyer, R. W. (1956). Lens regeneration from homoplastic and heteroplastic implants of dorsal iris into the eye chamber of *Triturus viridescens* and *Amblystoma punctatum*. *J. Exp. Zool.* **133**, 145–90.

Reyer, R. W. (1961). Lens regeneration from intra-ocular, iris implants in the presence of the host lens. *Anat. Rec.* **139**, 267.

Reyer, R. W. (1962). Regeneration in the amphibian eye. In *Regeneration* (20th Growth Symposium), ed. D. Rudnick. New York: Ronald Press, pp. 211–65.

Reyer, R. W. (1966a). DNA synthesis and cell movement during lens regeneration in adult *Triturus viridescens*. *Am. Zool.* **6**, 329.

Reyer, R. W. (1966b). The influence of neural retina and lens regeneration from dorsal iris implants in *Triturus viridescens larvae*. *Dev. Biol.* **14**, 214–45.

Reyer, R. W. (1971). DNA synthesis and the incorporation of labeled iris cells into the lens during lens regeneration in adult newts. *Dev. Biol.* **124**, 533–58.
Reyer, R. W. (1977). The amphibian eye: development and regeneration. In *Handbook of Sensory Physiology*. Vol. 2. *The Visual System in Vetrebrates*, ed. F. Crescitelli. Berlin: Springer-Verlag, pp. 309–90.
Reyer, R. W. (1982). Dedifferentiation of iris epithelium during lens regeneration in newt larvae. *Am. J. Anat.* **163**, 1–23.
Reyer R. W. (1990a). Macrophage invasion and phagocytic activity during lens regeneration from the iris epithelium in newts. *Am. J. Anat.* **188**, 329–44.
Reyer, R. W. (1990b). Macrophage mobilization and morphology during lens regeneration from the iris epithelium in newts: studies with correlated scanning and transmission electron microscopy. *Am. J. Anat.* **188**, 345–65.
Reyer, R. W., Woolfitt, R. A. and Withersty, L. T. (1973). Stimulation of lens regeneration from the newt dorsal iris when implanted into the blastema of the regenerating limb. *Dev. Biol.* **32**, 258–81.
Reynhout, J. K., Lampe, P. D. and, Johnson, R. G. (1992). An activator of protein kinase C inhibits gap junction communication between cultured bovine lens cells. *Exp. Eye Res.* **198**, 337–42.
Riach, R. A., Duncan, G., Williams, M. R. and Webb, S. F. (1995). Histamine and ATP mobilize calcium by activation of H1 and P2U receptors in human epithelial cells. *J. Physiol.* **486**, 273–82.
Richardson, J., Cvekl, A. and Wistow, G. (1995). *Pax-6* is essential for lens-specific expression of ζ-crystallin. *Proc. Natl. Acad. Sci. USA* **92**, 4676–80.
Richardson, N. A. and McAvoy, J. W. (1990). Age-related changes in fibre differentiation of rat lens epithelial explants exposed to fibroblast growth factor. *Exp. Eye Res.* **50**, 203–11.
Richardson, N. A., McAvoy, J. W. and Chamberlain, C. G. (1992). Age of rats affects response of lens epithelial explants to fibroblast growth factor. *Exp. Eye Res.* **55**, 649–56.
Richiert, D. M. and Ireland, M. E. (1999). TGF-beta elicits fibronectin secretion and proliferation in cultured chick lens epithelial cells. *Curr. Eye Res.* **18**, 62–71.
Rieger, D. K., Reichenberger, E., McLean, W., Sidow, A. and Olsen, B. R. (2001). A double-deletion mutation in the *Pitx3* gene causes arrested lens development in aphakia mice. *Genomics* **72**, 61–72.
Ring, B. Z., Cordes, S. P., Overbeek, P. A. and Barsh, G. S. (2000). Regulation of mouse lens fiber cell development and differentiation by the *Maf* gene. *Development* **127**, 307–17.
Ritz-Laser, B., Estreicher, A., Klages, N., Saule, S. and Philippe, J. (1999). Pax-6 and Cdx-2/3 interact to activate glucagon gene expression on the G1 control element. *J. Biol. Chem.* **274**, 4124–32.
Robanus-Maandag, E., Dekker, M., van der Valk, M., Carrozza, M. L., Jeanny, J. C., Dannenberg, J. H., Berns, A. and te Riele, H. (1998). p107 is a suppressor of retinoblastoma development in pRb-deficient mice. *Genes Dev.* **12**, 1599–609.
Robinson, G. W. and Mahon, K. A. (1994). Differential and overlapping expression domains of *Dlx-2* and *Dlx-3* suggest distinct roles for *Distal-less* homeobox genes in craniofacial development. *Mech. Dev.* **48**, 199–215.
Robinson, K. M., Taube, J. R., Reed, N. A. and Duncan, M. K. (2003). Role of betaB2-crystallin in fertility **44**, [ARVO E-abstract 2136].
Robinson, K. R. and Patterson, J. W. (1983). Localization of steady-state currents in the lens. *Curr. Eye Res.* **2**, 843–7.
Robinson, M. L., MacMillan-Crow, L. A., Thompson, J. A. and Overbeek, P. A. (1995b). Expression of a truncated FGF receptor results in defective lens development in transgenic mice. *Development* **121**, 3959–67.
Robinson, M. L., Ohtaka-Maruyama, C., Chan, C.-C., Jamieson, S., Dickson, C., Overveek, P. A. and Chepelinsky, A. B. (1998). Disregulation of ocular morphogenesis by lens-specific expression of FGF-3/Int-2 in transgenic mice. *Dev. Biol.* **198**, 13–31.
Robinson, M. L. and Overbeek, P. A. (1996). Differential expression of αA- and αB-crystallin during murine ocular development. *Invest. Ophthalmol. Vis. Sci.* **37**, 2276–84.
Robinson, M. L., Overbeek, P. A., Verran, D. J., Grizzle, W. E., Stockard, C. R., Friesel, R., Maciag, T. and Thompson, J. A. (1995a). Extracellular FGF-1 acts as a lens differentiation factor in transgenic mice. *Development* **121**, 505–14.

Rodokanaki, A., Holmes, R. K. and Borras, T. (1989). Zeta-crystallin, a novel protein from the guinea pig lens is related to alcohol dehydrogenases. *Gene* **78**, 215–24.

Rossant, J., Zirngibl, R., Cado, D., Shago, M. and Giguere, V. (1991). Expression of a retinoic acid response element-hsplacZ transgene defines specific domains of transcriptional activity during mouse embryogenesis. *Genes Dev.* **5**, 1333–44.

Roth, H. J., Das, G. C. and Piatigorsky, J. (1991). Chicken βB1-crystallin gene expression: presence of conserved functional polyomavirus enhancer–like and octamer binding–like promoter elements found in non-lens genes. *Mol. Cell. Biol.* **11**, 1488–99.

Rothstein, H., Van Wyk, J. J., Hayden, J. H., Gordon, S. R. and Weinsieder, A. (1980). Somatomedin C: restoration in vivo of cycle traverse in G0/G1 blocked cells of hypophysectomized animals. *Science* **208**, 410–2.

Rup, D. M., Veenstra, R. D., Wang, H.-Z., Brink, P. R. and Beyer, E. C. (1993). Chick connexin-56, a novel lens gap junction protein: molecular cloning and functional expression. *J. Biol. Chem.* **268**, 706–12.

Russell, P., Qin, C., Garland, D., Tabor, Y. and Zigler, J. S. (1996). RNA and protein synthesis in the primate lens. *Exp. Eye Res.* **63**, 121–4.

Ryerse, J. S. and Nagel, B. A. (1991). Gap-junction quantification in biological tissues: freeze-fracture replicas versus thin sections. *J. Microsc.* **163**, 65–78.

Saavedra, H. I., Wu, L., De Bruin, A., Timmers, C., Rosol, T. J., Weinstein, M., Robinson, M. L. and Leone, G. (2002). Specificity of E2F1, E2F2, and E2F3 in mediating phenotypes induced by loss of Rb. *Cell Growth Diff.* **13**, 215–25.

Sage, J., Mulligan, G. J., Attardi, L. D., Miller, A., Chen, S., Williams, B., Theodorou, E. and Jacks, T. (2000). Targeted disruption of the three Rb-related genes leads to loss of G1 control and immortalization. *Genes Dev.* **14**, 3037–50.

Saha, M. S., Spann, C. L. and Grainger, R. M. (1989). Embryonic lens induction: more than meets the optic vesicle. *Cell Diff. Dev.* **28**, 153–72.

Saika, S., Kawashima, Y., Miyamoto, T., Tanaka, S., Okada, Y., Yamanaka, O., Katoh, T., Ohnishi, Y., Ohini, S., Ooshima, A., and Yamanaka, A. (1998). Immunolocalization of prolyl 4-hydrolase subunits, alpha-smooth muscle actin, and extracellular matrix components in human lens capsules with lens implants. *Exp. Eye Res.* **66**, 283–94.

Saika, S., Okada, Y., Miyamoto, T., Ohnishi, Y., Ooshima, A. and McAvoy, J. W. (2001). Smad translocation and growth suppression in lens epithelial cells by endogenous TGFbeta2 during wound repair. *Exp. Eye Res.* **72**, 679–86.

Sakai, M., Imaki, J., Yoshida, K., Ogata, A., Matsushima-Hibiya, Y., Kuboki, Y., Nishizawa, M. and Nishi, S. (1997). Rat maf related gene: specific expression in chondrocytes, lens and spinal cord. *Oncogene* **14**, 745–50.

Samejima, K. and Earnshaw, W. C. (2000). Differential localization of ICAD-L and ICAD-S in cells due to removal of a C-terminal NLS from ICAD-L by alternative splicing. *Exp. Cell Res.* **255**, 314–20.

Sandilands, A., Hutcheson, A. M., Long, H. A., Prescott, A. R., Vrensen, G., Loster, J., Klopp, N., Lutz, R. B., Graw, J., Masaki, S., et al. (2002). Altered agregation properties of mutant gamma-crystallins cause inherited cataract. *Embo. J.* **21**, 6005–14.

Sandilands, A., Prescott, A. R., Carter, J. M., Hutcheson, A. M., Quinlan, R. A., Richards, J. and FitzGerald, P. G. (1995a). Vimentin and CP49/ filensin form distinct networks in the lens which are independently modulated during lens fibre cell differentiation. *J. Cell Sci.* **108**, 1397–406.

Sandilands, A., Prescott, A. R., Hutcheson, A. M., Quinlan, R. A., Casselman, J. T. and FitzGerald, P. G. (1995b). Filensin is proteolytically processed during lens fiber cell differentiation by multiple independent pathways. *Eur. J. Cell Biol.* **67**, 238–53.

Sandilands, A., Prescott, A. R., Wegener, A., Zoltoski, R. K., Hutcheson, A. M., Masaki, S., Kuszak, J. R. and Quinlan, R. A. (2003). Knockout of the intermediate filament protein CP49 destabilises the lens fibre cell cytoskeleton and decreases lens optical quality, but does not induce cataract. *Exp. Eye Res.* **76**, 385–91.

Sandilands, A., Wang, X., Hutcheson, A. M., James, J., Prescott, A. R., Wegener, A., Pekny, M., Gong, X. and Quinlan, R. A. (2004). Bfsp2 mutation found in mouse 129 strains causes the loss of CP49 and induces vimentin-dependent changes in the lens fibre cell cytoskeleton. *Exp. Eye Res.* **78**, 109–23.

Sanyal, S. and Hawkins, R. K. (1979). *Dysgenetic lens (dyl)*: a new gene in the mouse. *Invest. Ophthalmol. Vis. Sci.* **18**, 642–5.

Sardet, C., Vidal, M., Cobrinik, D., Geng, Y., Onufryk, C., Chen, A. and Weinberg, R. A. (1995). E2F-4 and E2F-5, two novel members of the E2F family, are expressed in the early phases of the cell cycle. *Proc. Natl. Acad. Sci. USA* **92**, 2403–7.

Sarkar, P. S., Appukuttan, B., Han, J., Ito, Y., Ai, C. W., Tsai, W. L., Chai, Y., Stout, J. T. and Reddy, S. (2000). Heterozygous loss of *Six5* in mice is sufficient to cause ocular cataracts. *Nat. Genet.* **25**, 110–14.

Sas, D. F., Sas, J., Johnson, K. R., Menko, A. S. and Johnson, R. G. (1985). Junctions between lens fiber cells are labeled with a monoclonal antibody shown to be specific for MP26. *J. Cell Biol.* **100**, 216–25.

Sasaki, K., Kojima, M., Nakaizumi, H., Kitagawa, K., Yamada, Y. and Ishizaki, H. (1998). Early lens changes seen in patients with atopic dermatitis applying image analysis processing of Scheimpflug and specular microscopic images. *Ophthalmologica* **212**, 88–94.

Sato, T. (1930). Beitrage zur Analyse der Wolffschen Linssenregeneration. I. *Wilhem Roux Arch. Entwickl.-Mech. Org.* **122**, 451–493.

Sato, T. (1935). Beitrage zur Analyse der Wolffschen Linssenregeneration. Pt. 3. *Wilhem Roux Arch. Entwickl.-Mech. Org.* **133**, 323–48.

Sato, T. (1940). Vergleichende Studien uber die Geschwindigkeit der Wolffschen Linsenregeneration bei *Triton taniatus* und bei *Diemyctylus pyrrhogaster. Wilhem Roux Arch. Entwickl.-Mech. Org.* **140**, 573–613.

Sato, T. (1961). Uber die Linsen-Regeneration bei den Cobitiden Fischen *Misgurnus Anguillicaudatus*. *Embryologia* **6**, 251–91.

Saunders J. B. de C. M (John Bertrand de Cusance Morant), and O'Malley, C. D. (1950). The Illustrations from the Works of Andreas Vesalius of Brussels. New York: World Publishing.

Sawhney, R. S. (1995). Identification of SPARC in the anterior lens capsule and its expression by lens epithelial cells. *Exp. Eye Res.* **61**, 645–8.

Sawhney, R. S., Wood, L. S. and Vogeli, G. (1997). Molecular cloning of the bovine 1 (IV) procollagen gene (COL4A1) and its use in investigating the regulation of expression of type IV procollagen by retinoic acid in bovine lens epithelial cells. *Cell. Biol. Int.* **21**, 501–10.

Sax, C. M., Cvekl, A., Kantorow, M., Gopal-Srivastava, R., Ilagan, J. G. Ambulos, N. P., Jr. and Piatigorsky, J. (1995). Lens-specific activity of the mouse alpha A-crystallin promoter in the absence of a TATA box: functional and protein binding analysis of the mouse alpha A-crystallin PE1 region. *Nucleic Acids Res.* **23**, 442–51.

Sax, C. M., Cvekl, A., Kantorow, M., Sommer, B., Chepelinsky, A. B. and Piatigorsky, J. (1994). Identification of negative-acting and protein-binding elements in the mouse alpha A-crystallin −1556/−1165 region. *Gene* **144**, 163–9.

Sax, C. M., Cvekl, A. and Piatigorsky, J. (1997). Transcriptional regulation of the mouse alpha A-crystallin gene: binding of USF to the −7/+5 region. *Gene* **185**, 209–16.

Sax, C. M., Ilagan, J. G. and Haynes, J. I., II. (1996). Lens-preferred activity of the −1809/+46 mouse αA-crystallin promoter in stably integrated chromatin. *Biochim. Biophys. Acta* **1305**, 49–53.

Sax, C. M., Ilagan, J. G. and Piatigorsky, J. (1993). Functional redundancy of the DE-1 and alpha A-CRYBP1 regulatory sites of the mouse alpha A-crystallin promoter. *Nucleic Acids Res.* **21**, 2633–40.

Sax, C. M. and Piatigorsky, J. (1994). Expression of the alpha-crystallin/small heat-shock protein/molecular chaperone genes in the lens and other tissues. *Adv. Enzymol. Relat. Areas Mol. Biol.* **69**, 155–201.

Scheffner, M., Werness, B. A., Huibregtse, J. M., Levine, A. J. and Howley, P. M. (1990). The E6 oncoprotein encoded by human papillomavirus types 16 and 18 promotes the degradation of p53. *Cell* **63**, 1129–36.

Scheiner, C. (1619). Oculus hoc est fundamentum opticum Innsbruck, Austria: Agricola.

Scherr, M., Morgan, M. A. and Eder, M. (2003). Gene silencing mediated by small interfering RNAs in mammalian cells. *Curr. Med. Chem.* **10**, 245–56.

Schlotzer-Schrehardt, U. and Dorfler, S. (1993). Immunolocalization of growth factors in the human ciliary body epithelium. *Curr. Eye Res.* **12**, 893–905.

Schmidt, A., Heid, H. W., Schafer, S., Nuber, U. A., Zimbelmann, R. and Franke, W. W. (1994). Desmosomes and cytoskeletal architecture in epithelial differentiation: cell type–specific plaque components and intermediate filament anchorage. *Eur. J. Cell Biol.* **65**, 229–45.

Schmitt, G. A., Pau, H., Spahr, R., Piper, H. M., Skalli, O. and Gabbiani, G. (1990). Appearance of alpha-smooth muscle actin in human eye lens cells of anterior capsular cataract and in cultured bovine lens-forming cells. *Differentiation* **43**, 115–22.

Schorle, H., Meier, P., Buchert, M., Jaenisch, R. and Mitchell, P. J. (1996). Transcription factor AP-2 essential for cranial closure and craniofacial development. *Nature* **381**, 235–8.

Schuetze, S. M. and Goodenough, D. A. (1982). Dye transfer between cells of the embryonic chick lens becomes less sensitive to CO_2 treatment with development. *J. Cell Biol.* **92**, 694–705.

Schulz, M. W., Chamberlain, C. G., de Iongh, R. U. and McAvoy, J. W. (1993). Acidic and basic FGF in ocular media and lens: implications for lens polarity and growth patterns. *Development* **118**, 117–26.

Schulz, M. W., Chamberlain, C. G. and McAvoy, J. W. (1996). Inhibition of TGFB-induced cataractous changes in lens explants by ocular media and a2-macroglobulin. *Invest. Ophthalmol. Vis. Sci.* **37**, 1509–19.

Schulz, M. W., Chamberlain, C. G. and McAvoy, J. W. (1997). Binding of FGF-1 and FGF-2 to heparan sulphate proteoglycans of the mammalian lens capsule. *Growth Factors* **14**, 1–13.

Schulze, E. and Kirschner, M. (1987). Dynamic and stable populations of microtubules in cells. *J. Cell Biol.* **104**, 277–88.

Schwartz, J. S., Lee, D. A. and Isenberg, S. J. (1989). Ocular size and shape. In *The Eye in Infancy*, ed. S. J. Isenberg. Chicago: Year Book Medical Publishers, pp. 164–84.

Seimiya, M. and Gehring, W. J. (2000). The *Drosophila* homeobox gene *optix* is capable of inducing ectopic eyes by an *eyeless*-independent mechanism. *Development* **127**, 1879–86.

Sekido, R., Murai, K., Funahashi, J., Kamachi, Y., Fujisawa-Sehara, A., Nabeshima, Y. and Kondoh, H. (1994). The delta-crystallin enhancer-binding protein delta EF1 is a repressor of E2-box–mediated gene activation. *Mol. Cell. Biol.* **14**, 5692–700.

Sekido, R., Murai, K., Kamachi, Y. and Kondoh, H. (1997). Two mechanisms in the action of repressor deltaEF1: binding site competition with an activator and active repression. *Genes Cells* **2**, 771–83.

Sellers, W. R., Notvich, B. G., Miyake, S., Heith, A., Otterson, G. A., Kaye, F. J., Lassar, A. B. and Kaelin, W. G., Jr. (1998). Stable binding to E2F is not required for the retinoblastoma protein to activate transcription, promote differentiation, and suppress tumor cell growth. *Genes Dev.* **12**, 95–106.

Semina, E. V., Brownell, I., Mintz-Hittner, H. A., Murray, J. C. and Jamrich, M. (2001). Mutations in the human forkhead transcription factor *FOXE3* associated with anterior segment ocular dysgenesis and cataracts. *Hum. Mol. Genet.* **10**, 231–6.

Semina, E. V., Ferrell, R. E., Mintzhittner, H. A., Bitoun, P., Alward, W. L. M., Reiter, R. S., Funkhauser, C., Daackhirsch, S. and Murray, J. C. (1998). A novel homeobox gene *PITX3* is mutated in families with autosomal-dominant cataracts and ASMD. *Nat. Genet.* **19**, 167–70.

Semina, E. V., Murray, J. C., Reiter, R., Hrstka, R. F. and Graw, J. (2000). Deletion in the promoter region and altered expression of *Pitx3* homeobox gene in *aphakia* mice. *Hum. Mol. Genet.* **9**, 1575–85.

Semina, E. V., Reiter, R. S. and Murray, J. C. (1997). Isolation of a new homeobox gene belonging to the *Pitx/Rieg* family: expression during lens development and mapping to the *aphakia* region on mouse chromosome 19. *Hum. Mol. Genet.* **6**, 2109–16.

Seo, H. C., Curtiss, J., Mlodzik, M. and Fjose, A. (1999). *Six* class homeobox genes in *Drosophila* belong to three distinct families and are involved in head development. *Mech. Dev.* **83**, 127–39.

Seto, Y., Nagawa, H., Mori, M., Tsuruo, T. and Muto, T. (1999). Effect of 5-fluorouracil on gastrointestinal carcinogenesis induced by N-methyl-N′-nitro-N-nitrosoguanidine in rats. *Dig. Dis. Sci.* **44**, 75–8.

Servetnick, M., Cook, T. L., Jr. and Grainger, R. M. (1996). Lens induction in axolotls: comparison with inductive signaling mechanisms in *Xenopus laevis*. *Int. J. Dev. Biol.* **40**, 755–61.

Servetnick, M. and Grainger, R. M. (1991). Changes in neural and lens competence in *Xenopus* ectoderm: evidence for an autonomous developmental timer. *Development* **112**, 177–88.

Shan, B. and Lee, W. H. (1994). Deregulated expression of E2F-1 induces S-phase entry and leads to apoptosis. *Mol. Cell Biol.* **14**, 8166–73.
Sharma, Y., Rao, C. M., Narasu, M. L., Rao, S. C., Somasundaram, T., Gopalakrishna, A. and Balasubramanian, D. (1989). Calcium ion binding to δ- and to β-crystallins: the presence of the EF-hand motif in δ-crystallin that aids in calcium ion binding. *J. Biol. Chem.* **264**, 12794–9.
Sharon-Friling, R., Richardson, J., Sperbeck, S., Lee, D., Rauchman, M., Maas, R., Swaroop, A. and Wistow, G. (1998). Lens-specific gene recruitment of zeta-crystallin through Pax6, Nrl-Maf, and brain suppressor sites. *Mol. Cell. Biol.* **18**, 2067–76.
Shaw, L. M., Lotz, M. M. and Mercurio, A. M. (1993). Inside-out integrin signaling in macrophages: analysis of the role of the alpha 6A beta 1 and alpha 6B beta 1 integrin variants in laminin adhesion by cDNA expression in an alpha 6 integrin–deficient macrophage cell line. *J. Biol. Chem.* **268**, 11401–8.
Shaw, L. M., Messier, J. M. and Mercurio, A. M. (1990). The activation dependent adhesion of macrophages to laminin involves cytoskeletal anchoring and phosphorylation of the alpha 6 beta 1 integrin. *J. Cell Biol.* **110**, 2167–74.
Shen, L., Shrager, P., Girsch, S. J. and Peracchia, C. (1991). Channel reconstitution in liposomes and planar lipid bilayers with HPLC-purified MIP26 of bovine lens. *J. Membr. Biol.* **124**, 21–32.
Shen, W. and Mardon, G. (1997). Ectopic eye development in *Drosophila* induced by directed dachshund expression. *Development* **124**, 45–52.
Sheng, G., Harris, E., Bertuccioli, C. and Desplan, C. (1997a). Modular organization of Pax/homeodomain proteins in transcriptional regulation. *Biol. Chem.* **378**, 863–72.
Sheng, G., Thouvenot, E., Schmucker, D., Wilson, D. S. and Desplan, C. (1997b). Direct regulation of *rhodopsin 1* by *Pax-6/eyeless* in *Drosophila*: evidence for a conserved function in photoreceptors. *Genes Dev.* **11**, 1122–31.
Shepard, A. R. and Rae, J. L. (1998). Ion transporters and receptors in cDNA libraries from lens and cornea epithelia. *Curr. Eye Res.* **17**, 708–19.
Shepard, A. R. and Rae, J. L. (1999). Electrically silent potassium channel subunits from human lens epithelium. *Am. J. Physiol.* **46**, C412–24.
Sherr, C. J. (1993). Mammalian G1 cyclins. *Cell* **73**, 1059–65.
Sherr, C. J. (1994). G1 phase progression: cycling on cue. *Cell* **79**, 551–5.
Sherr, C. J. (1996). Cancer cell cycles. *Science* **274**, 1672–7.
Sherr, C. J. and Roberts, J. M. (1995). Inhibitors of mammalian G1 cyclin–dependent kinases. *Genes Dev.* **9**, 1149–63.
Shestopalov, V. I. and Bassnett, S. (1999). Exogenous gene expression and protein targeting in lens fiber cells. *Invest. Ophthalmol. Vis. Sci.* **40**, 1435–43.
Shestopalov, V. I. and Bassnett, S. (2000a). Expression of autofluorescent proteins reveals a novel protein permeable pathway between cells in the lens core. *J. Cell Sci.* **113**, 1913–21.
Shestopalov, V. I. and Bassnett, S. (2000b). Three-dimensional organization of primary lens fiber cells. *Invest. Ophthalmol. Vis. Sci.* **41**, 859–63.
Sheterline, P., Clayton, J. and Sparrow, J. (1995). Actin. *Protein Profile* **2**, 1–103.
Shiels, A. and Bassnett, S. (1996). Mutations in the founder of the MIP gene family underlie cataract development in the mouse. *Nat. Genet.* **12**, 212–15.
Shiels, A., Mackay, D., Ionides, A., Berry, V., Moore, A. and Bhattacharya, S. (1998). A missense mutation in the human connexin50 gene (GJA8) underlies autosomal dominant "zonular pulverent" cataract, on chromosome 1q. *Am. J. Hum. Genet.* **62**, 526–32.
Shiloh, Y., Donlon, T., Bruns, G., Breitman, M. L. and Tsui, L. C. (1986). Assignment of the human gamma-crystallin gene cluster (CRYG) to the long arm of chromosome 2, region q33–36. *Hum. Genet.* **73**, 17–9.
Shimada, N., Aya-Murata, T., Reza, H. M. and Yasuda, K. (2003). Coorperative action between L-Maf and Sox2 on delta-crystallin gene expression during chick lens development. *Mech. Dev.* **120**, 455–65.
Shinohara, T. and Piatigorsky, J. (1976). Quantitation of delta-crystallin messenger RNA during lens induction in chick embryos. *Proc. Natl. Acad. Sci. USA* **73**, 2808–12.
Shinohara, T., Singh, D. P. and Fatma, N. (2002). LEDGF, a survival factor, activates stress-related genes. *Prog. Retin. Eye Res.* **21**, 341–58.
Shirke, S., Faber, S. C., Hallem, E., Makarenkova, H. P., Robinson, M. L., Overbeek, P. A. and

Lang, R. A. (2001). Misexpression of IGF-I in the mouse lens expands the transitional zone and perturbs lens polarization. *Mech. Dev.* **101**, 167–74.

Shirodkar, S., Ewen, M., DeCaprio, J. A., Morgan, J. and Livingston, D. M. (1992). The transcription factor E2F interacts with the retinoblastoma product and a p107–cyclin A complex in a cell cycle-regulated manner. *Cell* **68**, 157–66.

Shroff, N. P., Cherian-Shaw, M., Bera, S. and Abraham, E. C. (2000). Mutation of R116C results in highly oligomerized alpha A-crystallin with modified structure and defective chaperone-like function. *Biochemistry* **39**, 1420–6.

Siegner, A., May, C. A., Welge-Lussen, U. W., Bloemendal, H. and Lutjen-Drecoll, E. (1996). Alpha B-crystallin in the primate ciliary muscle and trabecular meshwork. *Eur. J. Cell Biol.* **71**, 165–9.

Siezen, R. J., Bindels, J. G. and Hoenders, H. J. (1978). The quaternary structure of bovine alpha-crystallin: size and charge microheterogeneity: more than 1000 different hybrids? *Eur. J. Biochem.* **91**, 387–96.

Siezen, R. J. and Shaw, D. C. (1982). Physicochemical characterization of lens proteins of the squid *Nototodarus gouldi* and comparison with vertebrate crystallins. *Biochim. Biophys. Acta* **704**, 304–20.

Siezen, R. J., Wu, E., Kaplan, E. D., Thomson, J. A. and Benedek, G. B. (1988). Rat lens γ-crystallin: characterization of the six gene products and their spatial and temporal distribution resulting from differential synthesis. *J. Mol. Bio.* **199**, 475–90.

Simeone, A., Acampora, D., Gulisano, M., Stornaiuolo, A. and Boncinelli, E. (1992). Nested expression domains of four homeobox genes in developing rostral brain. *Nature* **358**, 687–90.

Simeone, A., Acampora, D., Mallamaci, A., Stornaiuolo, A., D'Apice, M. R., Nigro, V. and Boncinelli, E. (1993). A vertebrate gene related to *orthodenticle* contains a homeodomain of the *bicoid* class and demarcates anterior neuroectoderm in the gastrulating mouse embryo. *EMBO J.* **12**, 2735–47.

Singer, C. (1921). Steps leading to the invention of first optical apparatus. In C. Singer, ed., *Studies in the History and Method of Science*. Oxford: Clarendon, pp. 385–413.

Singer, B. and Grunberger, D. (1983). *Molecular Biology of Mutagens and Carcinogens.* New York: Plenum, pp. 55–78.

Sinha, D., Esumi, N., Jaworski, C., Kozak, C. A., Pierce, E. and Wistow, G. (1998). Cloning and mapping the mouse Crygs gene and non-lens expression of [gamma]S-crystallin. *Mol. Vis.* **4**, 8.

Sirotkin, A. M., Edelmann, W., Cheng, G., Klein-Szanto, A., Kucherlapati, R. and Skoultchi, A. I. (1995). Mice develop normally without the H1(0) linker histone. *Proc. Natl. Acad. Sci. USA* **92**, 6434–8.

Sivak, J. G., Gershon, D., Dovrat, A. and Weerheim, J. (1986). Computer assisted scanning laser monitor of optical quality of the excised crystalline lens. *Vision Res.* **26**, 1873–9.

Sivak, J. G., Herbert, K. L., Peterson, K. L. and Kuszak, J. R. (1994). The inter-relationship of lens anatomy and optical quality. I. Non-primate lenses. *Exp. Eye Res.* **59**, 505–20.

Skow, L. C., Donner, M. E., Huang, S. M., Gardner, J. M., Taylor, B. A., Beamer, W. G. and Lalley, P. A. (1988). Mapping of mouse gamma crystallin genes on chromosome 1. *Biochem. Genet.* **26**, 557–70.

Slack, J. M., Isaacs, H. V., Song, J., Durbin, L. and Pownall, M. E. (1996). The role of fibroblast growth factors in early *Xenopus* development. *Biochem. Soc. Symp.* **62**, 1–12.

Slee, E. A., Adrain, C. and Martin, S. J. (2000). Executioner caspase-3, -6 and -7 perform distinct, non-redundant roles during the demolition phase of apoptosis. *J. Biol. Chem.* **276**, 7320–6.

Slingsby, C. and Bateman, O. A. (1990). Rapid separation of bovine beta-crystallin subunits βA3 and βA4. *Exp. Eye Res.* **51**, 21–6.

Smith, B. S. (1989). Histochemical analysis of extracellular matrix material in embryonic trisomy 1 mouse eye. *Dev. Genet.* **10**, 287–91.

Smith, F. J., Eady, R. A., Leigh, I. M., McMillan, J. R., Rugg, E. L., Kelsell, D. P., Bryant, S. P., Spurr, N. K., Geddes, J. F., Kirtschig, G., et al. (1996). Plectin deficiency results in muscular dystrophy with epidermolysis bullosa. *Nat. Genet.* **13**, 450–7.

Smith, R. S., Hawes, N. L., Chang, B., Roderick, T. H., Akeson, E. C., Heckenlively, J. R., Gong, X., Wang, X. and Davisson, M. T. (2000). Lop12, a mutation in mouse Crygd causing lens opacity similar to human Coppock cataract. *Genomics* **63**, 314–20.

Smolich, B. D., Tarkington, S. K., Saha, M. S. and Grainger, R. M. (1994). *Xenopus* g-crystallin gene expression: evidence that the g-crystallin gene family is transcribed in lens and non-lens tissues. *Mol. Cell. Biol.* **14**, 1355–63.

Smulders, R. H., van Boekel, M. A. and de Jong, W. W. (1998). Mutations and modifications support a "pitted-flexiball" model for alpha-crystallin. *Int. J. Biol. Macromol.* **22**, 187–96.

Somasundaram, T. and Bhat, S. P. (2000). Canonical heat shock element in the alpha B-crystallin gene shows tissue-specific and developmentally controlled interactions with heat shock factor. *J. Biol. Chem.* **275**, 17154–9.

Song, S., Pitot, H. C. and Lambert, P. F. (1999). The human papillomavirus type 16 E6 gene alone is sufficient to induce carcinomas in transgenic animals. *J. Virol.* **73**, 5887–93.

Soriano, P. (1997). The PDGF alpha receptor is required for neural crest cell development and for normal patterning of the somites. *Development* **124**, 2691–700.

Spector, A., Chiesa, R., Sredy, J. and Garner, W. (1985). cAMP-dependent phosphorylation of bovine lens alpha-crystallin. *Proc. Natl. Acad. Sci. USA* **82**, 4712–6.

Spemann, H. (1901). Ueber Corelationen in der Entwicklung des Auges. *Verh. Anat. Ges.* **15**, 61–79.

Spemann, H. (1907). Neue Tatsachen zum Linsenproblem. *Zool. Anz.* **31**, 379–86.

Spemann, H. (1912). Zur Entwicklung des Wirbeltierauges. *Zool. Jahrb.* **32**, 1–98.

Spemann, H. (1938). *Embryonic Development and Induction.* New York: Hafner.

Spiewak Rinaudo, J. A. and Zelenka, P. S. (1992). Expression of c-fos and c-jun mRNA in the developing chicken lens: relationship to cell proliferation, quiescence, and differentiation. *Exp. Cell Res.* **199**, 147–53.

Spivak-Kroizman, T., Lemmon, M. A., Dikic, I., Ladbury, J. E., Pinchasi, D., Huang, J., Jaye, M., Crumley, G., Schlessinger, J. and Lax I. (1994). Heparin-induced oligomerization of FGF molecules is responsible for FGF receptor dimerization, activation, and cell proliferation. *Cell* **79**, 1015–24.

Srinivasan, Y., Lovicu, F. J. and Overbeek, P. A. (1998). Lens-specific expression of transforming growth factor beta1 in transgenic mice causes anterior subcapsular cataracts. *J. Clin. Invest.* **101**, 625–34.

Stamer, W. D., Snyder, R. W., Smith, B. L., Agre, P. and Regan, J. W. (1994). Localization of aquaporin CHIP in the human eye: implications in the pathogenesis of glaucoma and other disorders of ocular fluid balance. *Invest. Ophthalmol. Vis. Sci.* **35**, 3867–72.

Steele, E. C., Kerscher, S., Lyon, M., Glenister, P. H., Favor, J., Wang, J. H. and Church, R. L. (1997). Identification of a mutation in the MP19 gene, Lim2, in the cataractous mouse mutant To3. *Mol. Vis.* **3**, 5. Available at http://www.molvis.org/molvis/v3/steele.

Steele, E. C., Lyon, M. F., Favor, J., Guillot, P. V., Boyd, Y. and Church, R. L. (1998). A mutation in the connexin50 (Cx50) gene is a candidate for the No2 mouse cataract. *Curr. Eye Res.* **17**, 883–9.

Steinert, P. M., Chou, Y. H., Prahlad, V., Parry, D. A., Marekov, L. N., Wu, K. C., Jang, S. I. and Goldman, R. D. (1999). A high molecular weight intermediate filament–associated protein in BHK-21 cells is nestin, a type VI intermediate filament protein: limited co-assembly in vitro to form heteropolymers with type III vimentin and type IV alpha-internexin. *J. Biol. Chem.* **274**, 9881–90.

Steno, N. (1910). *Opera Philosophica.* Copenhagen: Christian Christensen.

Stephan, D. A., Gillanders, E., Vanderveen, D., Freas-Lutz, D., Wistow, G., Baxevanis, A. D., Robbins, C. M., VanAuken, A., Quesenberry, M. I., Bailey-Wilson, J., et al. (1999). Progressive juvenile-onset punctate cataracts caused by mutation of the gammaD-crystallin gene. *Proc. Natl. Acad. Sci. USA* **96**, 1008–12.

Stillman, B. (1996). Cell cycle control of DNA replication. *Science* **274**, 1659–64.

Stolen, C. M. and Griep, A. E. (2000). Disruption of lens fiber cell differentiation and survival at multiple stages by region-specific expression of truncated FGF receptors. *Dev. Biol.* **217**, 205–20.

Stolen, C. M., Jackson, M. W. and Griep, A. E. (1997). Overexpression of FGF-2 modulates fiber cell differentiation and survival in the mouse lens. *Development* **124**, 4009–17.

Stone, L. S. (1943). Factors controlling lens regeneration from the dorsal iris in de adult *Triturus viridescens* eye. *Proc. Soc. Exp. Biol. Med.* **54**, 102–03.

Stone, L. S. (1952). An experimental study of the inhibition and release of lens regeneration in adults eyes of *Triturus viridescens*. *J. Exp. Zool.* **121**, 181–23.

Stone, L. S. (1953). An experimental analysis of lens regeneration. *Am. J. Ophthalmol.* **36**, 31–9.
Stone, L. S. (1954a). Further experiments on lens regeneration in eyes of the adult newt *Triturus v. viridescens*. *Anat. Rec.* **120**, 599–624.
Stone, L. S. (1954b). Lens regeneration in secondary pupils experimentally produced in eyes of the adult newt *Triturus v. viridescens*. *J. Exp. Zool.* **127**, 463–92.
Stone, L. S. (1958a). Lens regeneration in adult newt eyes related to retina pigment cells and the neural retina factor. *J. Exp. Zool.* **139**, 69–84.
Stone, L. S. (1958b). Inhibition of lens regeneration in newt eyes by isolating the dorsal iris from the neural retina. *Anat. Rec.* **131**, 151–72.
Stone, L. S. (1966). Experiments dealing with the inhibition and release of lens regeneration in eyes of adult newts. *J. Exp. Zool.* **161**, 83–93.
Stone, L. S. (1967). An investigation recording all salamanders which can and cannot regenerate a lens from the dorsal iris. *J. Exp. Zool.* **164**, 87–103.
Stone, L. S. and Steinitz, H. (1953). The regeneration of lenses in eyes with intact and regenerating retina in adult *Triturus v. viridescens*. *J. Exp. Zool.* **124**, 435–67.
Stone, L. S. and Vultee, J. H. (1949). Inhibition and release of lens regenaration in the dorsal iris of *Triturus v. viridescens*. *Anat. Rec.* **103**, 560.
St-Onge, L., Sosa-Pineda, B., Chowdhury, K., Mansouri, A. and Gruss, P. (1997). *Pax6* is required for differentiation of glucagon-producing α-cells in mouse pancreas. *Nature* **387**, 406–9.
Straub, B. K., Boda, J., Kuhn, C., Schnoelzer, M., Korf, U., Kempf, T., Spring, H., Hatzfeld, M. and Franke, W. W. (2003). A novel cell-cell junction system: the cortex adhaerens mosaic of lens fiber cells. *J. Cell Sci.* **116**, 4985–95.
Streeten, B. W. and Eshaghian, J. (1978). Human posterior subcapsular cataract: a gross and flat preparation study. *Arch. Ophthalmol.* **96**, 1653–8.
Streuli, C. H., Bailey, N. and Bissell, M. J. (1991). Control of mammary epithelial differentiation: basement membrane induces tissue-specific gene expression in the absence of cell-cell interaction and morphological polarity. *J. Cell Biol.* **115**, 1383–95.
Stump, R. J., Ang, S., Chen, Y., van Bahr, T., Lovicu, F. J., Pinson, K., de Iongh, R. U., Yamaguchi, T. P., Sassoon, D. A., and McAvoy, J. W. (2003). A role for Wnt/beta-catenin signaling in lens epithelial differentiation. *Dev. Biol.* **259**, 48–61.
Sugimura, T., Fujimura, S. and Baba, T. (1970). Tumor production in the glandular stomach and alimentary tract of the rat by N-methyl-N′-nitro-N-nitrosoguanidine. *Cancer Res.* **30**, 455–65.
Sullivan, C. H., Marker, P. C., Thorn, J. M. and Brown, J. D. (1998). Reliability of delta-crystallin as a marker for studies of chick lens induction. *Differentiation* **64**, 1–9.
Sullivan, C. H., Norman, J. T., Borras, T. and Grainger, R. M. (1989). Developmental regulation of hypomethylation of delta-crystallin genes in chicken embryo lens cells. *Mol. Cell. Biol.* **9**, 3132–5.
Sullivan, C. H., O'Farrell, S. and Grainger, R. M. (1991). Delta-crystallin gene expression and patterns of hypomethylation demonstrate two levels of regulation for the delta-crystallin genes in embryonic chick tissues. *Dev. Biol.* **145**, 40–50.
Sveinsson, O. (1993). The ultrastructure of Elschnig's pearls in a pseudophakic eye. *Acta Ophthalmol. (Copenhagen)* **71**, 95–8.
Svennevik, E. and Linser, P. J. (1993). The inhibitory effects of integrin antibodies and the RGD tripeptide on early eye development. *Invest. Ophthalmol. Vis. Sci.* **34**, 1774–84.
Svitkina, T. M., Verkhovsky, A. B. and Borisy, G. G. (1996). Plectin sidearms mediate interaction of intermediate filaments with microtubules and other components of the cytoskeleton. *J. Cell Biol.* **135**, 991–1007.
Swamynathan, S. K., and Piatigorsky, J. (2002). Orientation-dependent influence of an intergenic enhancer on the promoter activity of the divergently transcribed mouse Shsp/alpha B-crystallin and Mkbp/HspB2 genes. *J. Biol. Chem.* **277**, 49700–6.
Swenson, K. I., Jordan, J. R., Beyer, E. C. and Paul, D. L. (1989). Formation of gap junctions by expression of connexins in *Xenopus* oocyte pairs. *Cell* **57**, 145–55.
Takagi, T., Moribe, H., Kondoh, H. and Higashi, Y. (1998). DeltaEF1, a zinc finger and homeodomain transcription factor, is required for skeleton patterning in multiple lineages. *Development* **125**, 21–31.
Takahashi, Y., Hanaoka, K., Goto, K. and Kondoh, H. (1994). Lens-specific activity of the chicken delta 1-crystallin enhancer in the mouse. *Int. J. Dev. Biol.* **38**, 365–8.

Takano, K., Yamanaka, G. and Mikami, Y. (1958). Wolffian lens-regeneration in the eye containing a full brown lens in *Triturus pyrrhogaster*. *Mie Med. J.* **8**, 177–82.

Takata, C., Albright, J. F. and Yamada, T. (1964). Lens antigens in the lens-regenerating system studied by immunofluorescent technique. *Dev. Biol.* **9**, 385–97.

Takata, C., Albright, J. F. and Yamada, T. (1966). Gamma-crystallins in Wolffian lens regeneration demonstrated by immunofluorescence. *Dev. Biol.* **14**, 382–400.

Takeda, H., Nagafuchi, A., Yonemura, S., Tsukita, S., Behrens, J. and Birchmeier, W. (1995). V-src kinase shifts the cadherin-based cell adhesion from the strong to the weak state and beta catenin is not required for the shift. *J. Cell Biol.* **131**, 1839–47.

Takei, K., Furuya, A., Hommura, S. and Yamaguchi, N. (2001). Ultrastructural fragility and type iv collagen abnormality of the anterior lens capsules in a patient with Alport syndrome. *Jpn. J. Ophthalmol.* **45**, 103–4.

Takeichi, M. (1991). Cadherin cell adhesion receptors as a morphogenetic regulator [review]. *Science* **251**, 1451–5.

Takeichi, M. (1995). Morphogenetic roles of classical cadherins. *Curr. Opin. Cell Biol.* **5**, 806–11.

Tamm, E. R., Russell, P., Johnson, D. H. and Piatigorsky, J. (1996). Human and monkey trabecular meshwork accumulate alpha B-crystallin in response to heat shock and oxidative stress. *Invest. Ophthalmol. Vis. Sci.* **37**, 2402–13.

Tamura, R. N., Cooper, H. M., Collo, G. and Quaranta, V. (1991). Cell type specific integrin variants with alternative alpha chain cytoplasmic domains. *Proc Nat. Acad. Sci. USA* **88**, 10183–7.

Tanaka, T., Tsujimura, T., Takeda, K., Sugihara, A., Maekawa, A., Terada, N., Yoshida, N. and Akira, S. (1998). Targeted disruption of ATF4 discloses its essential role in the formation of eye lens fibres. *Genes Cells* **3**, 801–10.

Tang, S.-S., Lin, C.-C. and Chang, G.-G. (1994). Isolation and characterization of octopus hepatopancreatic glutathione S-transferase: comparison of digestive gland enzyme with lens S-crystallin. *J. Prot. Chem.* **13**, 609–18.

Tao, Q. F., Hollenberg, N. K. and Graves, S. W. (1999). Sodium pump inhibition and regional expression of sodium pump α-isoforms in the lens. *Hypertension* **34**, 1168–74.

Tardieu, A. and Delaye, M. (1988). Eye lens proteins and transparency: from light transmission theory to solution X-ray structural analysis. *Ann. Rev. Biophys. Biochem.* **17**, 47–70.

Tardieu, A., Veretout, F., Krop, B. and Slingsby, C. (1992). Protein interactions in the calf eye lens: interactions between beta-crystallins are repulsive whereas in gamma-crystallins they are attractive. *Eur. Biophys. J.* **21**, 1–12.

Taube, J. R., Gao, C. Y., Ueda, Y., Zelenka, P. S., David, L. L. and Duncan, M. K. (2002). General utility of the chicken betaB1-crystallin promoter to drive protein expression in lens fiber cells of transgenic mice. *Transgenic Res.* **11**, 397–410.

Taylor, V., Welcher, A. A., Program, A. E. and Suter, U. (1995). Epithelial membrane protein-1, peripheral myelin protein 22, and lens membrane protein 20 define a novel gene family. *J. Biol. Chem.* **270**, 28824–33.

Taylor, V. L., Al-Ghoul, K. J., Lane, C. W., Davis, V. A., Kuszak, J. R. and Costello, M. J. (1996). Morphology of the normal human lens. *Invest. Ophthalmol. Visual Sci.* **37**, 1396–410.

Taylor, V. L. and Costello, M. J. (1999). Fourier analysis of textural variations in human normal and cataractous lens nuclear fiber cell cytoplasm. *Exp. Eye Res.* **69**, 163–74.

Tebar, M., Destree, O., de Vree, W. J. and Ten Have-Opbroek, A. A. (2001). Expression of Tcf/Lef and sFrp and localization of beta-catenin in the developing mouse lung. *Mech. Dev.* **109**, 437–40.

TenBroek, E. M., Louis, C. F. and Johnson, R. (1997). The differential effects of 12-O-tetradecanoylphorbol-13-acetate on the gap junctions and connexins of the developing mammalian lens. *Dev. Biol.* **191**, 88–102.

Thiery, J. P., Delouvee, A., Gallin, W. J., Cunningham, B. A. and Edelman, G. M. (1984). Ontogenetic expression of cell adhesion molecules: L-CAM is found in epithelia derived from the three primary germ layers. *Dev. Biol.* **102**, 61–78.

Thomas, G., Zelenka, P. S., Cuthbertson, R. A., Norman, B. L. and Piatigorsky, J. (1990). Differential expression of the two delta-crystallin/argininosuccinate lyase genes in lens, heart, and brain of chicken embryos. *New Biol.* **2**, 903–14.

Thomas, G. R., Duncan, G. and Sanderson, J. (1998). Acetylcholine-induced membrane potential oscillations in the intact lens. *Invest. Ophthalmol. Vis. Sci.* **39**, 111–19.

Thomson, I., Wilkinson, C. E., Burns, A. T. H., Truman, D. E. S. and Clayton, R. M. (1978). Characterization of chick lens soluble proteins and the control of their synthesis. *Exp. Eye Res.* **26**, 351–62.

Thyagarajan, T. and Kulkarni, A. B. (2002). Transforming growth factor-betal negatively regulates crystallin expression in teeth. *J. Bone. Miner. Res.* **17**, 1710–7.

Tini, M., Fraser, R. A. and Giguere, V. (1995). Functional interactions between retinoic acid receptor–related orphan nuclear receptor (ROR-α) and the retinoic acid receptors in the regulation of the γF-crystallin promoter. *J. Biol. Chem.* **270**, 20156–61.

Tini, M., Otulakowski, G., Breitman, M. L., Tsui, L.-C. and Giguere, V. (1993). An everted repeat mediates retinoic acid induction of the γF-crystallin gene: evidence of a direct role for retinoids in lens development. *Genes Devel.* **7**, 295–307.

Tini, M., Tsui, L.-C. and Giguere, V. (1994). Heterodimeric interaction of the retinoic acid and thyroid hormone receptors in transcriptional regulation on the γF-crystallin everted retinoic acid response element. *Mol. Endocrinol.* **8**, 1494–506.

Tomarev, S. I. (1997). Pax-6, eyes absent, and Prox 1 in eye development. *Int. J. Dev. Biol.* **41**, 835–42.

Tomarev, S. I., Callaerts, P., Kos, L., Zinovieva, R., Halder, G., Gehring, W. and Piatigorsky, J. (1997). Squid Pax-6 and eye development. *Proc. Natl. Acad. Sci. USA* **94**, 2421–6.

Tomarev, S. I., Chung, S. and Piatigorsky, J. (1995). Glutathione S-transferase and S-crystallins of cephalopods: evolution from active enzyme to lens-refractive proteins. *J. Mol. Evol.* **41**, 1048–56.

Tomarev, S. I., Duncan, M. K., Roth, H. J., Cvekl, A. and Piatigorsky, J. (1994). Convergent evolution of crystallin gene regulation in squid and chicken: the AP-1/ARE connection. *J. Mol. Evol.* **39**, 134–43.

Tomarev, S. I. and Piatigorsky, J. (1996). Lens crystallins of invertebrates: diversity and recruitment from detoxification enzymes and novel proteins. *Eur. J. Biochem.* **235**, 449–65.

Tomarev, S. I., Sundin, O., Banerjeebasu, S., Duncan, M. K., Yang, J. M. and Piatigorsky, J. (1996). Chicken homeobox gene *Prox1* related to *Drosophila prospero* is expressed in the developing lens and retina. *Dev. Dyn.* **206**, 354–67.

Tomarev, S. I. and Zinovieva, R. D. (1988). Squid major lens polypeptides are homologous to glutathione S-transferase subunits. *Nature* **336**, 86–8.

Tomarev, S. I., Zinovieva, R. D., Chang, B. and Hawes, N. L. (1998). Characterization of the mouse *Prox1* gene. *Biochem. Biophys. Res. Comm.* **248**, 684–9.

Tomarev, S. I., Zinovieva, R. D., Guo, K. and Piatigorsky, J. (1993). Squid glutathione S-transferase: relationships with other glutathione S-transferases and S-crystallins of cephalopods. *J. Biol. Chem.* **268**, 4534–42.

Tomarev, S. I., Zinovieva, R. D. and Piatigorsky, J. (1991). Crystallins of the octopus lens: recruitment from detoxification enzymes. *J. Biol. Chem.* **266**, 24226–31.

Tomarev, S. I., Zinovieva, R. D. and Piatigorsky, J. (1992). Characterization of squid crystallin genes: comparison with mammalian glutathione S-transferase genes. *J. Biol. Chem.* **267**, 8604–12.

Tomei, L. D., Cope, F. O. and Barr, P. J. (1994). Apoptosis: aging and phenotypic fidelity. In *Apoptosis II: The Molecular Basis of Apoptosis in Disease*, ed. L. D. Tomei and F. O. Cope. Plainview, New York: Cold Springs Harbor Laboratory Press, pp. 377–96.

Ton, C. C. T., Hirvonen, H., Miwa, H., Weil, M. M., Monaghan, P., Jordan, T., van Heyningen, V., Hastie, N. D., Meijers-Heijboer, H., Drechsler, M., et al. (1991). Positional cloning and characterization of a paired box- and homeobox-containing gene from aniridia region. *Cell* **67**, 1059–74.

Topaloglu, H., Yetuk, M., Talim, B., Akcoren, Z. and Caglar, M. (1997). Merosin-positive congenital muscular dystrophy with mental retardation and cataracts: a new entity in two families. *Eur. J. Paediatr. Neurol.* **1**, 127–31.

Torriglia, A., Chaudun, E., Chany-Fournier, F., Jeanny, J. C., Courtois, Y. and Counis, M. F. (1995). Involvement of DNase II in nuclear degeneration during lens cell differentiation. *J. Biol. Chem.* **270**, 28579–85.

Toy, J. and Sundin, O. H. (1999). Expression of the *Optx2* homeobox gene during mouse development. *Mech. Dev.* **83**, 183–6.

Toy, J., Yang, J. M., Leppert, G. S. and Sundin, O. H. (1998). The *Optx2* homeobox gene is expressed in early precursors of the eye and activates retina-specific genes. *Proc. Natl. Acad. Sci. USA* **95**, 10643–8.

Trautman, M. S., Kimelman, J. and Bernfield, M. (1991). Developmental expression of syndecan, an integral membrane proteoglycan, correlates with cell differentiation. *Development* **111**, 213–20.

Treisman, J. (2001). *Drosophila* homologues of the transcriptional coactivation complex subunits TRAP240 and TRAP230 are required for identical processes in eye–antennal disc development. *Development* **128**, 603–15.

Treisman, J. and Lang, R. (2002). Development and evolution of the eye: Fondation des Treilles, September 2001. *Mech. Dev.* **112**, 3–8.

Treisman, J. E. (1999). A conserved blueprint for the eye? *Bioessays* **21**, 843–50.

Treton, J. A., Jacquemin, E. and Courtois, Y. (1988). Variation in the relative abundance of gamma-crystallin gene transcripts during development and ageing. *Exp. Eye Res.* **46**, 405–13.

Treton, J. A., Jacquemin, E., Courtois, Y. and Jeanny, J. C. (1991). Differential localization by in situ hybridization of specific crystallin transcripts during mouse lens development. *Differentiation* **47**, 143–7.

Treton, J. A., Jones, R. E., King, C. R. and Piatigorsky, J. (1984). Evidence against gamma-crystallin DNA or RNA sequences in the chicken. *Exp. Eye Res.* **39**, 513–22.

Treton, J. A., Shinohara, T. and Piatigorsky, J. (1982). Degradation of delta-crystallin mRNA in the lens fiber cells of the chicken. *Dev. Biol.* **92**, 60–5.

Trokel, S. (1962). The physical basis for transparency of the crystalline lens. *Invest. Ophthalmol.* **1**, 493–501.

Trusolino, L. and Comoglio, P. M. (2002). Scatter-factor and semaphorin receptors: cell signaling for invasive growth. *Nat. Rev. Cancer* **2**, 289–300.

Tsai, K. Y., Hu, Y., Macleod, K. F., Crowley, D., Yamasaki, L. and Jacks, T. (1998). Mutation of E2F-1 suppresses apoptosis and inappropriate S phase entry and extends survival of Rb-deficient mouse embryos. *Mol. Cell* **2**, 293–304.

Tsai, K. Y., MacPherson, D., Robinson, D. A., Crowley, D. and Jacks, T. (2002). *ARF* is not required for apoptosis in Rb mutant mouse embryos. *Curr. Biol.* **12**, 159–63.

Tsonis, P. A. (2000). Regeneration in vertebrates. *Dev. Biol.* **221**, 273–84.

Tsonis, P. A. (2001). Regeneration of the vertebrate lens and other eye structures. In *Embryonic Encyclopedia of Life Sciences*. London: Nature Publishing Group, pp. 1–6

Tsonis, P. A., Jang, W., Del Rio-Tsonis, K. and Eguchi, G. (2001). A human pigment epithelium cell line as a model for lens transdifferentiation. *Int. J. Dev. Biol.* **45**, 753–8.

Tsonis, P. A., Tromblay, M. T., Rowland T., Chanrdaratna, R. A. S. and Del Rio-Tsonis, K. (2000). Role of retinoic acid in lens regeneration. *Dev. Dyn.* **21**, 588–93.

Tsonis, P. A., Tsavaris, M., Call, M. K., Chandraratna, R. A. S. and Del Rio-Tsonis, K. (2002). Expression and role of retinoic acid receptor alpha in lens regeneration. *Dev. Growth Differ.* **44**, 391–4.

Tsujimoto, A., Nyunoya, H., Morita, T., Sato, T. and Shimotohno, K. (1991). Isolation of cDNAs for DNA-binding proteins which specifically bind to a tax-responsive enhancer element in the long terminal repeat of human T-cell leukemia virus type I. *J. Virol.* **65**, 1420–6.

Tucker, R. P. (1991). The distribution of J1/tenascin and its transcript during the development of the avian cornea. *Differentiation* **48**, 59–66.

Tunstall, M. J., Eckert, R., Donaldson, P. J. and Kistler, J. (1999). Localised fibre cell swelling characteristic of diabetic cataract can be induced in normal rat lens using the chloride channel blocker 5-nitro-2-(3-phenylpropylamino) benzoic acid. *Ophthalmic Res.* **31**, 317–20.

Turner, C. E. (1998). Paxillin. *Int. J. Biochem. Cell. Biol.* **30**, 955–9.

Uchikawa, M., Kamachi, Y. and Kondoh, H. (1999). Two distinct subgroups of Group B *Sox* genes for transcriptional activators and repressors: their expression during embryonic organogenesis of the chicken. *Mech. Dev.* **84**, 103–20.

Ueda, Y., Chamberlain, C. G., Satoh, K. and McAvoy, J. W. (2000). Inhibition of FGF-induced alphaA-crystallin promoter activity in lens epithelial explants by TGFbeta. *Invest. Ophthalmol. Vis. Sci.* **41**, 1833–9.

Ueda, Y., Duncan, M. K. and David, L. L. (2002). Lens proteomics: the accumulation of crystallin modifications in the mouse lens with age. *Invest. Ophthalmol. Vis. Sci.* **43**, 205–15.

Uno, M. (1943). Zur Frage des Mechanismus der Wolffschen Linsenregeneration. *Jpn. J. Med. Sci. I. Anat.* **11**, 75–100.
Vallari, R. C. and Pietruszko, R. (1982). Human aldehyde dehydrogenase: mechanism of inhibition of disulfiram. *Science* **216**: 637–9.
Vallejo, M., Ron, D., Miller, C. P. and Habener, J. F. (1993). C/ATF, a member of the activating transcription factor family of DNA-binding proteins, dimerizes with CAAT/enhancer-binding proteins and directs their binding to cAMP response elements. *Proc. Natl. Acad. Sci. USA* **90**, 4679–83.
van den Heuvel, S. and Harlow, E. (1993). Distinct roles for cyclin-dependent kinases in cell cycle control. *Science* **262**, 2050–4.
van den Oetelaar, P. J., van Someren, P. F., Thomson, J. A., Siezen, R. J. and Hoenders, H. J. (1990). A dynamic quaternary structure of bovine alpha-crystallin as indicated from intermolecular exchange of subunits. *Biochemistry* **29**, 3488–93.
van der Starre, H. (1977). Biochemical investigation of lens induction in vitro. I. Induction properties of the eye cup and ectodermal response. *Acta Morphol. Neerl. Scand.* **15**, 275–86.
van der Starre, H. (1978). Biochemical investigation of lens induction in vitro. II. Demonstration of the induction substance. *Acta Morphol. Neerl. Scand.* **16**, 109–20.
van Deth, J. H. M. G. (1940). Induction et régénération du cristallin chez l'embryon de la poule. *Acta Neerl. Morph.* **3**, 151–69.
van Heyningen, R. (1976). Experimental studies on cataract. *Invest. Ophthalmol. Vis. Sci.* **15**, 685–97.
Vanita, Sarhadi, V., Reis, A., Jung, M., Singh, D., Sperling, K., Singh, J. R. and Burger, J. (2001). A unique form of autosomal dominant cataract explained by gene conversion between beta-crystallin B2 and its pseudogene. *J. Med. Genet.* **38**, 392–6.
van Kamp, G. J., Boudier, H. A. and Hoenders, H. J. (1974). Specific polypeptides in prenatal bovine alpha-crystallin. *Int. J. Pept. Protein Res.* **6**, 75–8.
vanLeen, R. W., vanRoozendal, K. E. P., Lubsen, N. H. and Schoenmakers, J. G. G. (1987). Differential expression of crystallin genes during development of the rat eye lens. *Dev. Biol.* **120**, 457–64.
Van Montfort, R. L., Bateman, O. A., Lubsen, N. H. and Slingsby, C. (2003). Crystal structure of truncated human betaB1-crystallin. *Protein Sci.* **12**, 2606–12.
Van Noort, J. M., van Sechel, A. C., van Stipdonk, M. J. and Bajramovic, J. J. (1998). The small heat shock protein alpha B-crystallin as key autoantigen in multiple sclerosis. *Prog. Brain Res.* **117**, 435–52.
van Raamsdonk, C. D. and Tilghman, S. M. (2000). Dosage requirement and allelic expression of *Pax6* during lens placode formation. *Development* **127**, 5439–48.
vanRens, G. L., deJong, W. W. and Bloemendal, H. (1991). One member of the gamma-crystallins, gamma S_i is expressed in birds. *Exp. Eye. Res.* **53**, 135–8.
vanRens, G. L. M., Driessen, H. P. C., Nalini, V., Slingsby, C., deJong, W. W. and Bloemendal, H. (1991). Isolation and characterization of cDNAs encoding βA2- and βA4-crystallins: heterologous interactions in the predicted βA4-βB2 heterodimer. *Gene* **102**, 179–88.
Varadaraj, K., Kushmerick, C., Baldo, G. J., Bassnett, S., Shiels, A. and Mathias, R. T. (1999). The role of MIP in lens fiber cell membrane transport. *J. Membr. Biol.* **170**, 191–203.
Venter, J. C., Adams, M. D., Myers, E. W., Li, P. W., Mural, R. J., Sutton, G. G., Smith, H. O., Yandell, M., Evans, C. A., Holt, R. A., et al. (2001). The sequence of the human genome. *Science* **291**, 1304–51.
Veretout, F., Delaye, M. and Tardieu, A. (1989). Molecular basis of eye lens transparency: osmotic pressure and X-ray analysis of alpha-crystallin solutions. *J. Mol. Biol.* **205**, 713–28.
Vicart, P., Caron, A., Guicheney, P., Li, Z., Prevost, M. C., Faure, A., Chateau, D., Chapon, F., Tome, F., Dupret, J. M., et al. (1998). A missense mutation in the alphaB-crystallin chaperone gene causes a desmin-related myopathy. *Nat. Genet.* **20**, 92–5.
Vogel-Höpker, A., Momose, T., Rohrer, H., Yasuda, K., Ishihara, L. and Rapaport, D. H. (2000). Multiple functions of fibrobalst growth factor-8 (FGF-8) in chick eye development. *Mech. Dev.* **94**, 25–36.
Voges, D., Zwicki, P. and Baumeister, W. (1999). The 26S proteasome: a molecular machine designed for controlled proteolysis. *Ann. Rev. Biochem.* **68**, 1015–68.

Volberg, T., Geiger, B., Dror, R. and Zick, Y. (1991). Modulation of intercellular adherens-type junctions and tyrosine phosphorylation of their components in RSV-transformed cultured chick lens cells. *Cell Regul.* **2**, 105–20.

Volk, T. and Geiger, B. (1986a). A-CAM: a 135-kD receptor of intercellular adherens junctions. I. Immunoelectron microscopic localization and biochemical studies. *J. Cell Biol.* **103**, 1441–50.

Volk, T. and Geiger, B. (1986b). A-CAM: a 135-kD receptor of intercellular adherens junctions. II. Antibody-mediated modulation of junction formation. *J. Cell Biol.* **103**, 1451–64.

Von, D. M. K. and Ocalan, M. (1989). Antagonistic effects of laminin and fibronectin on the expression of the myogenic phenotype. *Differentiation* **40**, 150–7.

von Sallman, L. (1957). The lens epithelium in the pathogenesis of cataract. *Amer. J. Ophthalmol.* **44**, 159–70.

von Woellwarth, C. (1961). Die rolle des neuralleistenmaterials und der temperatur bei der determination der augenlinse. *Embryologia* **6**, 219–42.

Voorter, C. E., Kistler, J., Gruijters, W. T. M., Mulders, J. W., Christie, D. and de Jong, W. W. (1989). *Curr. Eye Res.* **8**, 697–706.

Voorter, C. E., Mulders, J. W., Bloemendal, H. and de Jong, W. W. (1986). Some aspects of the phosphorylation of alpha-crystallin A. *Eur. J. Biochem.* **160**, 203–10.

Voorter, C. E. M., Bleomendal, H. and deJong, W. W. (1989). In vitro and in vivo phosphorylation of chicken βB3-crystallin. *Curr. Eye Res.* **8**, 459–65.

Wachs, H. (1914). Neue Versuche sur Wolffschen Linsenregeneration. *Wilhelm Roux Arch. Entwickl.-Mech. Org.* **39**, 384–451.

Waddington, C. (1936). The origin of competence for lens formation in the amphibia. *J. Exp. Biol.* **13**, 86–91.

Waddington, C. H. (1932). Experiments on the development of chick and duck embryos, cultivated in vitro. *Proc. Roy. Soc. Lond. B.* **221**, 179–230.

Waddington, C. H. and Cohen, A. (1936). Experiments on the development of the head of the chick embryo. *J. Exp. Biol.* **13**, 219–36.

Wade, N. J. (1998a). Early studies of eye dominances. *Laterality* **3**, 97–109.

Wade, N. J. (1998b). *A Natural History of Vision.* Cambridge, Massachusetts: The MIT press.

Wagner, E., McCaffery, P. and Drager, U. C. (2000). Retinoic acid in the formation of the dorsoventral retina and its central projections. *Dev. Biol.* **222**, 460–70.

Wakabayashi, Y., Kawahara, J., Iwasaki, T. and Usui, M. (1994). Retinoic acid transport to lens epithelium in human aqueous humor. *Jpn. J. Ophthalmol.* **38**, 400–6.

Walker, J. L. and Menko, A. S. (1999). Alpha6 integrin is regulated with lens cell differentiation by linkage to the cytoskeleton and isoform switching. *Dev. Biol.* **210**, 497–511.

Walker, J. L., Zhang, L. and Menko, A. S. (2002a). A signaling role for the uncleaved form of $\alpha 6$ integrin in differentiating lens fiber cells. *Dev. Biol.* **251**, 195–205.

Walker, J. L., Zhang, L. and Menko, A. S. (2002b). Transition between proliferation and differentiation for lens epithelial cells is regulated by Src family kinases. *Dev. Dyn.* **224**, 361–72.

Walker, J. L., Zhang, L., Zhou, J., Woolkalis, M. J. and Menko, A. S. (2002c). Role for alpha6 integrin during lens development: evidence for signaling through IGF-IR and ERK. *Dev. Dyn.* **223**, 273–84.

Wallace, P., Signer, E., Paton, I. R., Burt, D. and Quinlan, R. (1998). The chicken CP49 gene contains an extra exon compared to the human CP49 gene which identifies an important step in the evolution of the eye lens intermediate filament proteins. *Gene* **211**, 19–27.

Wallis, D. E., Roessler, E., Hehr, U., Nanni, L., Wiltshire, T., Richieri-Costa, A., Gillessen-Kaesbach, G., Zackai, E. H., Rommens, J. and Muenke, M. (1999). Mutations in the homeodomain of the human *SIX3* gene cause holoprosencephaly. *Nat. Genet.* **22**, 196–8.

Walls, G. L., ed. (1963). *The Vertebrate Eye and Its Adaptive Radiation.* New York and London: Hafner.

Walsh, F. S. and Doherty, P. (1997). Neural cell adhesion molecules of the immunoglobulin superfamily: role in axon growth and guidance. *Annu. Rev. Cell. Dev. Biol.* **13**, 425–56.

Walther, C. and Gruss, P. (1991). *Pax-6*, a murine paired box gene, is expressed in the developing CNS. *Development* **113**, 1435–49.

Walther, C., Guenet, J. L., Simon, D., Deutsch, U., Jostes, B., Goulding, M. D., Plachov, D., Balling, R. and Gruss, P. (1991). Pax: a murine multigene family of paired box–containing genes. *Genomics* **11**, 424–34.

Walton, J. and McAvoy, J. (1984). Sequential structural response of lens epithelium to retina-conditioned medium. *Exp. Eye Res.* **39**, 217–29.

Wang, J. Y. (1997). Retinoblastoma protein in growth suppression and death protection. *Curr. Opin. Genet. Dev.* **7**, 39–45.

Wang, K., Gawinowicz, M. A. and Spector, A. (2000). The effect of stress on the pattern of phosphorylation of alphaA and alphaB crystallin in the rat lens [in process citation]. *Exp. Eye Res.* **71**, 385–93.

Wang, K. and Spector, A. (2000). Alpha-crystallin prevents irreversible protein denaturation and acts cooperatively with other heat-shock proteins to renature the stabilized partially denatured protein in an ATP-dependent manner. *Eur. J. Biochem.* **267**, 4705–12.

Wang, L. F., Dhir, P., Bhatnagar, A. and Srivastava, S. K. (1997). Contribution of osmotic changes to disintegrative globulization of single cortical fibers isolated from rat lens. *Exp. Eye Res.* **65**, 267–75.

Wang, S. Z. and Adler, R. (1994). A developmentally regulated basic-leucine zipper-like gene and its expression in embryonic retina and lens. *Proc. Natl. Acad. Sci. USA* **91**, 1351–5.

Wang, Y., He, H., Zigler, J. S., Iwata, T., Ibaraki, N., Reddy, V. N. and Carper, D. (1999). bFGF suppresses serum-deprivation–induced apoptosis in a human lens epithelial cell line. *Exp. Cell Res.* **249**, 123–30.

Wanko, T. and Gavin, M. A. (1959). EM of the lens. *J. Biophys. Biochem. Cytol.* **6**, 97–102.

Wannemacher, C. F. and Spector, A. (1968). Protein synthesis in the core of calf lens. *Exp. Eye Res.* **7**, 623–5.

Wary, K. K., Mainiero, F., Isakoff, S. J., Marcantonio, E. E. and Giancotti, F. G. (1996). The adaptor protein Shc couples a class of integrins to the control of cell cycle progression. *Cell* **87**, 733–43.

Wary, K. K., Mariotti, A., Zurzolo, C. and Giancotti, F. G. (1998). A requirement for caveolin-1 and associated kinase Fyn in integrin signaling and anchorage-dependent cell growth. *Cell* **94**, 625–34.

Watanabe, M., Kobayashi, H., Rutishauser, U., Katar, M., Alcala, J. and Maisel, H. (1989). NCAM in the differentiation of embryonic lens tissue. *Dev. Biol.* **135**, 414–23.

Watanabe, M., Kobayashi, H., Yao, R. and Maisel, H. (1992). Adhesion and junction molecules in embryonic and adult lens cell differentiation. *Acta Ophthalmol. Suppl.* **107**, 46–52.

Waterman-Storer, C. M., Salmon, W. C. and Salmon, E. D. (2000). Feedback interactions between cell-cell adherens junctions and cytoskeletal dynamics in newt lung epithelial cells. *Mol. Biol. Cell* **11**, 2471–83.

Watsky, M. A., Cooper, K. and Rae, J. L. (1991). Sodium channels in ocular epithelia. *Pflügers Arch.* **419**, 454–9.

Wawersik, S. and Maas, R. L. (2000). Vertebrate eye development as modeled in *Drosophila. Hum. Mol. Genet.* **9**, 917–25.

Wawersik, S., Purcell, P. and Maas, R. L. (2000). *Pax6* and the genetic control of early eye development. *Results Probl. Cell. Differ.* **31**, 15–36.

Wawersik, S., Purcell, P., Rauchman, M., Dudley, A. T., Robertson, E. J. and Maas, R. (1999). BMP7 acts in murine lens placode development. *Dev. Biol.* **207**, 176–88.

Wawrousek, E. F., Chepelinsky, A. B., McDermott, J. B. and Piatigorsky, J. (1990). Regulation of the murine alpha A-crystallin promoter in transgenic mice. *Dev. Biol.* **137**, 68–76.

Weaver, M. and Hogan, B. (2001). Powerful ideas driven by simple tools: lessons from experimental embryology. *Nat. Cell. Biol.* **3**, E165–7.

Webster, E. H., Silver, A. F. and Gonsalves, N. I. (1983). Histochemical analysis of extracellular matrix material in embryonic mouse lens morphogenesis. *Dev. Biol.* **100**, 147–57.

Webster, E. H., Jr. and Zwaan, J. (1984). The appearance of alpha, beta and gamma crystallins in an anophthalmic strain of mice. *Differentiation* **27**, 53–8.

Wedlock, D. E. and McCallion, D. J. (1968). The question of lens regeneration from parts of the optic vesicle in the chick embryo. *Experientia* **24**, 620–1.

Wegner, M. (1999). From head to toes: the multiple facets of Sox proteins. *Nucl. Acids Res.* **27**, 1409–20.

Wei, J., Shaw, L. M. and Mercurio, A. M. (1998). Regulation of mitogen-activated protein kinase activation by the cytoplasmic domain of the alpha6 integrin subunit. *J. Biol. Chem.* **273**, 5903–7.

Weil, D., Blanchard, S., Kaplan, J., Guilford, P., Gibson, F., Walsh, J., Mburu, P., Varela, A., Levilliers, J., Weston, M. D., et al. (1995). Defective myosin VIIA gene responsible for Usher syndrome type 1B. *Nature* **374**, 60–1.

Weinberg, R. A. (1995). The retinoblastoma protein and cell cycle control. *Cell* **81**, 323–30.

Weis, V. M., Montgomery, M. K., and McFall-Ngai, M. J. (1993). Enhanced production of ALDH-like protein in the bacterial light organ of the Sepiolid squid *Euprymna scolopes*. *Biol. Bull.* **184**, 309–21.

Weisblat, D., Sawyer, K. and Stent, G. (1978). Cell lineage analysis by intracellular injection of a tracer. *Science* **202**, 1295–8.

Weisstein, E. W. (2003). "Alcmaeon of Croton (ca. 535-unknown BC)." Retrieved February 3, 2004 from Eric Weisstein's World of Biography Web site: http://scienceworld.wolfram.com/biography/Alcmaeon.html

Weitzer, G. and Wiche, G. (1987). Plectin from bovine lenses: chemical properties, structural analysis and initial identification of interaction partners. *Eur. J. Biochem.* **169**, 41–52.

Welsh, M. J. and Gaestel, M. (1998). Small heat-shock protein family: function in health and disease. *Ann. NY Acad. Sci.* **851**, 28–35.

Weng, J., Liang, Q., Mohan, R. R., Li, Q. and Wilson, S. E. (1997). Hepatocyte growth factor, keratinocyte growth factor, and other growth factor-receptor systems in the lens. *Invest. Ophthalmol. Vis. Sci.* **38**, 1543–54.

Werner, M. and Burley, S. K. (1997). Architectural transcription factors: proteins that remodel DNA. *Cell* **88**, 733–6.

West, J. A., Sivak, J. G. and Doughty, M. J. (1995). Microscopic evaluation of the crystalline lens of the squid (*Loligo opalescens*) during embryonic development. *Exp. Eye Res.* **60**, 19–35.

West, J. A., Sivak, J. G., Pasternak, J. and Piatigorsky, J. (1994). Immunolocalization of S-crystallin in the developing squid (*Loligo opalescens*) lens. *Dev. Dyn.* **199**, 85–92.

West-Mays, J. A., Zhang, J., Nottoli, T., Hagopian-Donaldson, S., Libby, D., Strissel, K. J. and Williams, T. (1999). AP-2α transcription factor is required for early morphogenesis of the lens vesicle. *Dev. Biol.* **206**, 46–62.

White, T. W., Bruzzone, R., Goodenough, D. A. and Paul, D. L. (1992). Mouse Cx50, a functional member of the connexin family of gap junction proteins, is the lens fiber protein MP70. *Mol. Biol. Cell* **3**, 711–20.

White, T. W., Bruzzone, R., Wolfram, S., Paul, D. L. and Goodenough, D. A. (1994). Selective interactions among multiple connexin proteins expressed in the vertebrate lens: the second extracellular domain is a determinant of compatibility between connexins. *J. Cell Biol.* **125**, 879–92.

White, T. W., Goodenough, D. A. and Paul, D. L. (1998). Targeted ablation of connexin50 in mice results in microphthalmia and zonular pulverulent cataracts. *J. Cell Biol.* **143**, 815–25.

Widlak, P., Li, P., Wang, X. and Garrard, W. T. (2000). Cleavage preferences of the apoptotic endonuclease DFF40 (caspase-activated DNase or nuclease) on naked DNA and chromatin substrates. *J. Biol. Chem.* **275**, 8226–32.

Wigle, J. T., Chowdhury, K., Gruss, P. and Oliver, G. (1999). *Prox1* function is crucial for mouse lens-fibre elongation. *Nat. Genet.* **21**, 318–22.

Willekens, B. and Vrensen, G. (1982). The three-dimensional organization of lens fibers in the rhesus monkey. *Graefe's Arch. Clin. Exp. Ophthalmol.* **219**, 112–20.

Williams, B. O., Schmitt, E. M., Remington, L., Bronson, R. T., Albert, D. M., Weinberg, R. A. and Jacks, T. (1994). Extensive contribution of Rb-deficient cells to adult chimeric mice with limited histopathological consequences. *EMBO J.* **13**, 4251–9.

Williams, L. A. and Higginbotham, L. T. (1975). The role of a normal lens in Wolffian lens regeneration. *J. Exp. Zool.* **191**, 233–52.

Williams, M. R., Duncan, G., Riach, R. A. and Webb, S. F. (1993). Acetylcholine receptors are coupled to mobilization of intracellular calcium in cultured human lens cells. *Exp. Eye Res.* **57**, 381–4.

Williams, S. C., Altmann, C. R., Chow, R. L., Hemmati-Brivanlou, A. and Lang, R. A. (1998). A highly conserved lens transcriptional control element from the Pax-6 gene. *Mech. Dev.* **73**, 225–9.

Williams, T., Admon, A., Luscher, B. and Tjian, R. (1988). Cloning and expression of AP-2, a cell-type–specific transcription factor that activates inducible enhancer elements. *Genes Dev.* **2**, 1557–69.

Williams, T. and Tjian, R. (1991). Characterization of a dimerization motif in AP-2 and its function in heterologous DNA-binding proteins. *Science* **251**, 1067–71.

Wilson, D., Sheng, G., Lecuit, T., Dostatni, N. and Desplan, C. (1993). Cooperative dimerization of paired class homeo domains on DNA. *Genes Dev.* **7**, 2120–34.

Wilson, D. S., Guenther, B., Desplan, C. and Kuriyan, J. (1995). High resolution crystal structure of a paired (Pax) class cooperative homeodomain dimer on DNA. *Cell* **82**, 709–19.

Winchester, C. L., Ferrier, R. K., Sermoni, A., Clark, B. J. and Johnson, K. J. (1999). Characterization of the expression of *DMPK* and *SIX5* in the human eye and implications for pathogenesis in myotonic dystrophy. *Hum. Mol. Genet.* **8**, 481–92.

Wisniewski, T. and Goldman, J. E. (1998). Alpha B-crystallin is associated with intermediate filaments in astrocytoma cells. *Neurochem. Res.* **23**, 385–92.

Wistow, G. (1990). Evolution of a protein superfamily: relationships between vertebrate lens crystallins and microoganism dormancy proteins. *J. Mol. Evol.* **30**, 140–5.

Wistow, G. (1995). Peptide sequences for beta-crystallins of a teleost fish. *Mol. Vis.* **1**, 1.

Wistow, G. and Kim, H. (1991). Lens protein expression in mammals: taxon-specificity and the recruitment of crystallins. *J. Mol. Evol.* **32**, 262–9.

Wistow, G., Mulders, J. W. M. and Jong, W. W. d. (1987). The enzyme lactate dehydrogenase as a structural protein in avian and crocodilian lenses. *Nature* **326**, 622–4.

Wistow, G. and Piatigorsky, J. (1987). Recruitment of enzymes as lens structural proteins. *Science* **236**, 1554–6.

Wistow, G. and Piatigorsky, J. (1988). Lens crystallins: the evolution and expression of proteins for a highly specialized tissue. *Ann. Rev. Biochem.* **57**, 479–504.

Wistow, G., Roquemore, E. and Kim, H. S. (1991). Anomalous behavior of βB1-crystallin subunits from avian lenses. *Curr. Eye Res.* **10**, 313–19.

Wistow, G., Sardarian, L., Gan, W. and Wyatt, M. K. (2000a). The human gene for gammaS-crystallin: alternative transcripts and expressed sequences from the first intron. *Mol. Vis.* **6**, 79–84.

Wistow, G., Wyatt, M. K., Sinha, D., Sardarian, L. and Lyon, M. (2000b). γS-crystallin: a key component of the adult lens. *Exp. Eye Res.* **71**, (Suppl 1), S148.

Wolff, G. (1895). Entwicklungsphysiologische Studien. I. Die Regeneration der Urodelenlinse. *Wilhelm Roux Arch. Entwickl.-Mech. Org.* **1**, 380–90.

Woo, M. K. and Fowler, V. M. (1994). Identification and characterization of tropomodulin and tropomyosin in the adult rat lens. *J. Cell Sci.* **107**, 1359–67.

Woo, M. K., Lee, A., Fischer, R. S., Moyer, J. and Fowler, V. M. (2000). The lens membrane skeleton contains structures preferentially enriched in spectrin-actin or tropomodulin-actin complexes. *Cell Motil. Cytoskeleton* **46**, 257–68.

Worgul, B. V. (1982). Lens. In *Ocular Anatomy, Embryology, and Teratology*, ed. F. A. Jakobiec. Philadelphia: Harper & Row.

Wormstone, I. M. (2002). Posterior capsule opacification: a cell biological perspective. *Exp. Eye Res.* **74**, 337–47.

Wormstone, I. M., Tamiya, S., Marcantonio, J. M. and Reddan, J. R. (2000). Hepatocyte growth factor function and c-Met expression in human lens epithelial cells. *Invest. Ophthalmol. Vis. Sci.* **41**, 4216–22.

Wride, M. A., Parker, E. and Sanders, E. J. (1999). Members of the bcl-2 and caspase families regulate nuclear degeneration during chick lens fibre differentiation. *Dev. Biol.* **213**, 142–56.

Wride, M. A. and Sanders, E. J. (1998). Nuclear degeneration in the developing lens and its regulation by TNFα. *Exp. Eye Res.* **66**, 371–83.

Wright, N. A. and Appleton, D. R. (1980). The metaphase arrest technique: a critical review. *Cell Tiss. Kinet.* **13**, 643–63.

Wu, J. E. and Santoro, S. A. (1994). Complex patterns of expression suggest extensive roles for the alpha 2 beta 1 integrin in murine development. *Dev. Dyn.* **199**, 292–341.

Wunderlich, K. and Knorr, M. (1994). Effect of platelet-derived growth factor PDGF on replication of cultivated bovine lens epithelial cells [in German]. *Ophthalmologe* **91**, 98–102.

Wunderlich, K., Pech, M., Eberle, A. N., Mihatsch, M., Flammer, J. and Meyer, P. (2000). Expression of connective tissue growth factor (CTGF) mRNA in plaques of human anterior subcapsular cataracts and membranes of posterior capsule opacification. *Curr. Eye Res.* **21**, 627–36.

Xie, M. H., Holcomb, I., Deuel, B., Dowd, P., Huang, A., Vagts, A., Foster, J., Liang, J., Brush, J., Gu, Q., et al. (1999). FGF-19, a novel fibroblast growth factor with unique specificity for FGFR4. *Cytokine* **11**, 729–35.

Xiong, Y., Hannon, G. J., Zhang, H., Casso, D., Kobayashi, R. and Beach, D. (1993). p21 is a universal inhibitor of cyclin kinases. *Nature* **366**, 701–4.

Xu, H. E., Rould, M. A., Xu, W. Q., Epstein, J. A., Maas, R. L. and Pabo, C. O. (1999a). Crystal structure of the human Pax6 paired domain–DNA complex reveals specific roles for the linker region and carboxy terminal subdomain in DNA binding. *Genes Dev.* **13**, 1263–75.

Xu, L., Overbeek, P. A. and Reneker, L. W. (2002). Systematic analysis of E-, N- and P-cadherin expression in mouse eye development. *Exp. Eye Res.* **74**, 753–60.

Xu, P. X., Adams, J., Peters, H., Brown, M. C., Heaney, S. and Maas, R. (1999b). *Eya1*-deficient mice lack ears and kidneys and show abnormal apoptosis of organ primordia. *Nat. Genet.* **23**, 113–17.

Xu, P. X., Cheng, J., Epstein, J. A. and Maas, R. L. (1997a). Mouse *Eya* genes are expressed during limb tendon development and encode a transcriptional activation function. *Proc. Natl. Acad. Sci. USA* **94**, 11974–9.

Xu, P. X., Woo, I., Her, H., Beier, D. R. and Maas, R. L. (1997b). Mouse *Eya* homologues of the *Drosophila eyes absent* gene require *Pax6* for expression in lens and nasal placode. *Development* **124**, 219–31.

Xu, P. X., Zhang, X., Heaney, S., Yoon, A., Michelson, A. M. and Maas, R. L. (1999c). Regulation of *Pax6* expression is conserved between mice and flies. *Development* **126**, 383–95.

Xu, Y., Baldassare, M., Fisher, P., Rathbun, G., Oltz, E. M., Yancopoulos, G. D., Jessell, T. M. and Alt, F. W. (1993). LH-2: a LIM/homeodomain gene expressed in developing lymphocytes and neural cells. *Proc. Natl. Acad. Sci. USA* **90**, 227–31.

Yamada, T. (1966). Control of tissue specificity: the pattern of cellular synthetic activities in tissue transformation. *Am. Zool.* **6**, 21–31.

Yamada, T. (1967). Cellular and subcellular events in Wolffian lens regeneration. *Curr. Top. Dev. Biol.* **2**, 247–83.

Yamada, T. (1977). *Control Mechanisms in Cell-Type Conversion in Newt Lens Regeneration*. Basel: Karger. Monographs in Developmental Biology No. 13.

Yamada, T. (1982). Transdifferentaitaion of lens cells and its regulation. In *Cell Biology of the Eye*, ed. D. S. McDevitt. New York: Academic Press, pp. 193–242.

Yamada, T. and Dumont, J. N. (1972). Macrophage activity in Wolffian lens regeneration. *J. Morph.* **136**, 367–84.

Yamada, T. and Karasaki, S. (1963). Nuclear RNA synthesis in newt iris cells engaged in regenerative transformation into lens cells. *Dev. Biol.* **7**, 595–604.

Yamada, T. and McDevitt, D. S. (1984). Conversion of iris epithelial cells as a model of differentiation control. *Differentiation.* **27**, 1–12.

Yamada, T., Reese, D. H. and McDevitt, D. S. (1973). Transformation of iris into lens *in vitro* and its deependency on neural retina. *Differentiation* **1**, 65–82.

Yamada, T. and Roesel, M. E. (1969). Activation of DNA replication in the iris epithelium by lens removal. *J. Exp. Zool.* **171**, 425–31.

Yamada, T. and Roesel, M. E. (1971). Control of mitotic activity in Wolffian lens regeneration. *J. Exp. Zool.* **117**, 119–28.

Yamamoto, Y. (1976). Growth of lens and ocular environment: role of neural retina in the growth of mouse lens as revealed by an implantation experiment. *Dev. Growth Differ.* **18**, 273–8.

Yan, Q., Clark, J. I. and Sage, E. H. (2000). Expression and characterization of SPARC in human lens and in the aqueous and vitreous humors. *Exp. Eye Res.* **71**, 81–90.

Yan, Q. and Sage, E. H. (1999). SPARC, a matricellular glycoprotein with important biological functions. *J. Histochem. Cytochem.* **47**, 1495–506.

Yang, D. I. and Louis, C. F. (1996). Molecular cloning of sheep connexin49 and its identity with MP70. *Curr. Eye Res.* **15**, 307–14.

Yang, H. Y., Lieska, N., Goldman, A. E. and Goldman, R. D. (1985). A 300,000-mol-wt intermediate filament–associated protein in baby hamster kidney (BHK-21) cells. *J. Cell Biol.* **100**, 620–31.

Yang, J., Bian, W., Gao, X., Chen, L. and Jing, N. (2000). Nestin expression during mouse eye and lens development. *Mech. Dev.* **94**, 287–91.

Yasuda, K. (1979). Transdifferentiation of "lentoid" structures in cultures derived from pigmented epithelium was inhibited by collagen. *Dev. Biol.* **68**, 618–23.

Ye, X., Zhu, C. and Harper, J. W. (2001). A premature termination mutation in the Mus musculus cyclin-dependent kinase 3 gene. *Proc. Natl. Acad. Sci. USA* **98**, 1682–6.

Yeaman, C., Grindstaff, K. K., Hansen, M. D. and Nelson, W. J. (1999). Cell polarity: versatile scaffolds keep things in place. *Curr. Biol.* **9**, R515–7.

Yin, X., Jedrzejewski, P. T. and Jiang, J. X. (2000). Casein kinase II phosphorylates lens connexin 45.6 and is involved in its degradation. *J. Biol. Chem.* **275**, 6850–6.

Yokohama, S. (2000). Molecular evolution of vertebrate visual pigments. *Prog. Ret. Eye Res.* **19**, 385–419.

Yoshida, K., Imaki, J., Koyama, Y., Harada, T., Shinmei, Y., Oishi, C., Matsushima-Hibiya, Y., Matsuda, A., Nishi, S., Matsuda, H., et al. (1997). Differential expression of *maf*-1 and *maf*-2 genes in the developing rat lens. *Invest. Ophthalmol. Vis. Sci.* **38**, 2679–83.

Yoshida, T. and Yasuda, K. (2002). Characterization of the chicken L-Maf, MafB and c-Maf in crystallin gene regulation and lens differentiation. *Genes Cells* **7**, 693–706.

Younan, C., Mitchell, P., Cumming, R. G., Panchapakesan, J., Rochtchina, E. and Hales, A. M. (2002). Hormone replacement therapy, reproductive factors and the incidence of cataract and cataract surgery: the Blue Mountain Eye Study. *Am. J. Epidemiol.* **155**, 997–1006.

Young, M. A., Tunstall, M. J., Kistler, J. and Donaldson, P. J. (2000). Blocking chloride channels in the rat lens: localized changes in tissue hydration support the existence of a circulating chloride flux. *Invest. Ophthalmol. Vis. Sci.* **41**, 3049–55.

Young, R. W. and Ocumpaugh, D. E. (1966). Autoradiographic studies on the growth and development of the lens capsule in the rat. *Invest. Ophthal.* **5**, 583–93.

Yu, C. C.-K., Tsui, L.-C. and Breitman, M. L. (1990). Homologous and heterologous enhancers modulate spatial expression but not cell-type specificity of the murine γF-crystallin promoter. *Development* **110**, 131–9.

Zalik, S. E. and Meza, I. (1968). *In vitro* culture of the regenerating lens. *Nature* **217**, 179–80.

Zalik, S. E. and Scott, V. (1971). Development of ^{3}H-thymidine-labelled iris in the optic chamber of the lentectomized newts. *Exp. Cell Res.* **66**, 446–8.

Zalik, S. E. and Scott, V. (1973). Sequential disappearance of cell surface components during lens dedifferentiation in lens regeneration. *Nat. New Biol.* **244**, 212–14.

Zalokar, M. (1944). Contribution a l'etude de la regeneration du cristallin chez le *Triton*. *Rev. Suisse Zool.* **51**, 443–521.

Zampighi, G. A., Eskandari, S. and Kreman, M. (2000). Epithelial organization of the mammalian lens. *Exper. Eye Res.* **71**, 415–35

Zampighi, G. A., Hall, J. E., Ehring, G. R. and Simon, S. A. (1989). The structural organization and protein composition of the lens fiber junctions. *J. Cell Biol.* **108**, 2255–75.

Zarina, S., Abbasi, A. and Zaidi, Z. H. (1992). Primary structure of bS-crystallin from human lens. *Biochem. J.* **287**, 375–81.

Zelenka, P. (1978). Phospholipid composition and metabolism in the embryonic chick lens. *Exp. Eye Res.* **26**, 267–74.

Zelenka, P. S. (1983). Phosphatidylcholine and phosphatidylethanolamine metabolism during lens fiber cell formation. *Biochim. Biophys. Acta* **752**, 145–52.

Zelenka, P. S., Beebe, D. C. and Feagans, D. E. (1982). Transmethylation of phosphatidylethanolamine: an initial event in embryonic chicken lens fiber cell differentiation. *Science* **217**, 1265–7.

Zenjari, C., Boilly, B., Hondermarck, H. and Boilly-Marer, Y. (1997). Nerve-blastema interactions induce fibroblast growth factor-1 release during limb regeneration in *Pleurodeles waltl. Dev. Growth Differ.* **39**, 15–22.

Zhang, J., Hagopian-Donaldson, S., Serbedzija, G., Elsemore, J., Plehn-Dujowich, D., McMahon, A. P., Flavell, R. A. and Williams, T. (1996). Neural tube, skeletal and body wall defects in mice lacking transcription factor AP-2. *Nature* **381**, 238–41.

Zhang, J. J. and Jacob, T. J. C. (1994). ATP-activated chloride channel inhibited by an antibody to p-glycoprotein. *Am. J. Physiol.–Cell Physiol.* **36**, C1095–102.

Zhang, J. J. and Jacob, T. J. C. (1996). Volume regulation in the bovine lens and cataract: the involvement of chloride channels. *J. Clin. Invest.* **97**, 971–8.

Zhang, J. J., Jacob, T. J. C., Valverde, M. A., Hardy, S. P., Mintenig, G. M., Sepulveda, F. V., Gill, D. R., Hyde, S. C., Trezise, A. E. O. and Higgins, C. F. (1994). Tamoxifen blocks chloride channels: a possible mechanism for cataract formation. *J. Clin. Invest.* **94**, 1690–7.

Zhang, P., Liegeois, N. J., Wong, C., Finegold, M., Hou, H., Thompson, J. C., Silverman, A., Harper, J. W., DePinho, R. A. and Elledge, S. J. (1997). Altered cell differentiation and proliferation in mice lacking $p57^{KIP2}$ indicates a role in Beckwith-Wiedemann syndrome. *Nature* **387**, 151–8.

Zhang, P., Wong, C., DePinho, R. A., Harper, J. W. and Elledge, S. J. (1998). Cooperation between the Cdk inhibitors p27(KIP1) and p57(KIP2) in the control of tissue growth and development. *Genes Dev.* **12**, 3162–7.

Zhang, W. Z. and Augusteyn, R. C. (1995). Glucose supply and enzyme activities in the lens. *Exp. Eye Res.* **61**, 633–5.

Zhang, X., Friedman, A., Heaney, S., Purcell, P. and Maas, R. L. (2002). Meis homeoproteins directly regulate Pax6 during vertebrate lens morphogenesis. *Genes Dev.* **16**, 2097–107.

Zhang, Y., Xiong, Y. and Yarbrough, W. G. (1998b). ARF promotes MDM2 degradation and stabilizes p53: ARF-INK4a locus deletion impairs both Rb and p53 tumor suppression pathways. *Cell* **92**, 725–34.

Zhao, C., Takita, J., Tanaka, Y., Setou, M., Nakagawa, T., Takeda, S., Yang, H. W., Terada, S., Nakata, T., Takei, Y., et al. (2001). Charcot-Marie-Tooth disease type 2A caused by mutation in a microtubule motor KIF1Bbeta. *Cell* **105**, 587–97.

Zhao, J., Kennedy, B. K., Lawrence, B. D., Barbie, D. A., Matera, A. G., Fletcher, J. A. and Harlow, E. (2000). NPAT links cyclin E-Cdk2 to the regulation of replication-dependent histone gene transcription. *Genes Dev.* **14**, 2283–97.

Zhao, S., Hung, F. C., Colvin, J. S., White, A., Dai, W., Lovicu, F. J., Ornitz, D. M. and Overbeek, P. A. (2001). Patterning the optic neuroepithelium by FGF signaling and Ras activation. *Development* **128**, 5051–60.

Zhou, C., Agarwal, N. and Cammarata, P. R. (1994). Osmoregulatory alterations in myo-inositol uptake by bovine lens epithelial cells: cloning of a 626 bp cDNA portion of a Na^+/myo-inositol cotransporter, an osmotic shock protein. *Invest. Ophthalmol. Vis. Sci.* **35**, 1236–46.

Zhou, Y., Ching, Y. P., Kok, K. H., Kung, H. F. and Jin, D. Y. (2002). Post-transcriptional suppression of gene expression in *Xenopus* embryos by small interfering RNA. *Nucleic Acids Res.* **30**, 1664–9.

Ziebold, U., Reza, T., Caron, A. and Lees, J. A. (2001). E2F3 contributes both to the inappropriate proliferation and to apoptosis arising in Rb mutant embryos. *Genes Dev.* **15**, 386–91.

Zimmerman, L. and Font, R. (1966). Congenital malformations of the eye: some recent advances in knowledge of the pathogenesis and histopathological characterstics. *J. Am. Med. Assoc.* **196**, 684–92.

Zimmerman, L. B., De Jesus-Escobar, J. M. and Harland, R. M. (1996). The Spemann organizer signal noggin binds and inactivates bone morphogenetic protein 4. *Cell* **86**, 599–606.

Zindy, F. E. L., Chemivesse, X., Sobczak, J., Wang, J., Fesquet, D., Henglein, B. and Brechot, C. (1992). Cyclin A is required in S phase in normal epithelial cells. *Biochem. Biophys. Res. Comm.* **182**, 1144–54.

Zinovieva, R. D., Tomarev, S. I. and Piatigorsky, J. (1993). Aldehyde dehydrogenase-derived O-crystallins of squid and octopus: specialization for lens expression. *J. Biol. Chem.* **268**, 11449–55.

Zuber, M. E., Perron, M., Philpott, A., Bang, A. and Harris, W. A. (1999). Giant eyes in *Xenopus laevis* by overexpression of *XOptx2*. *Cell* **98**, 341–52.

Zuk, A. and Hay, E. D. (1994). Expression of beta 1 integrins changes during transformation of avian lens epithelium to mesenchyme in collagen gels. *Dev. Dyn.* **201**, 378–93.

Zumbrunn, J., Kinoshita, K., Hyman, A. A. and Nathke, I. S. (2001). Binding of the adenomatous polyposis coli protein to microtubules increases microtubule stability and is regulated by GSK3 beta phosphorylation. *Curr. Biol.* **11**, 44–9.

Zutter, M. M., Santoro, S. A., Wu, J. E., Wakatsuki, T., Dickeson, S. K. and Elson, E. L. (1999). Collagen receptor control of epithelial morphogenesis and cell cycle progression. *Am. J. Pathol.* **155**, 927–40.

Zwaan, J. (1968). Lens-specific antigens and cytodifferentiation in the developing lens. *J. Cell. Physiol.* **72** (Suppl 1), 47–71.

Zwaan, J. (1983). The appearance of alpha-crystallin in relation to cell cycle phase in the embryonic mouse lens. *Dev. Biol.* **96**, 173–81.

Zwaan, J. and Kenyon, R. E., Jr. (1984). Cell replication and terminal differentiation in the embryonic chicken lens: normal and forced initiation of lens fibre formation. *J. Embryol. Exp. Morphol.* **84**, 331–49.

Zwaan, J. and Silver, J. (1983). Crystallin synthesis in the lens rudiment of a strain of mice with congenital anophthalmia. *Exp. Eye Res.* **36**, 551–7.

Zwaan, J. and Webster, E. H. (1984). Histochemical analysis of extracellular matrix material during embryonic mouse lens morphogenesis in an aphakic strain of mice. *Dev. Biol.* **104**, 380–9.

Zygar, C. A., Cook, T. L., Jr. and Grainger, R. M. (1998). Gene activation during early stages of lens induction in *Xenopus*. *Development* **125**, 3509–19.

Index

2NI-36, 305, 310

α2-macroglobulin, 288
α-actinin, 256, 257
αCE1, 136
αCE2, 136, 139
αCE3, 136
acid phosphatase, 94
actin, 184, 226, 233, 257, 258, 310
  cytoskeleton, 185
  isoforms, 185
  α–smooth muscle, 185, 243, 255, 286, 287
Ad Vitellionem Paralipomena, 9
adenomatous polposis coli (APC), 183, 184
adherens junction, 183, 185, 186, 215, 228, 230, 257, 259
α-enolase, 124, 144
α-helix, 120
AIM1, 123
ak. See aphakia
Alcmaeon, 4
aldehyde dehydrogenase, 127, 145
ALDH1/2. *See* aldehyde dehydrogenase
Alexander's disease, 122
Alexander the Great, 4
Alk3, 43
Alk6, 46, 281
alkaline phosphatase, 292
Alports syndrome, 247
alternate translational initiation, 122, 140
Ammar, 6
aniridia, 41, 265
annular pad, 21, 22, 217, 282
anophthalmia, 266
anterior chamber, 25, 217, 297
anterior pad, 21
anterior vascular capsule *see* pupillary membrane
antioxidant responsive element. *See* ARE
AP1, 133, 134, 140, 141, 144–145, 146, 148
AP2, 67, 269–270, 285
APC. *See* adenomatous polposis coli
APC (anaphase promoting complex). *See* ubiquitin, enzymes
aphakia, 61, 245
apoptosis, 93–94, 121, 123, 195, 203, 205, 206, 207, 209, 236, 237, 238, 239, 240, 255, 286
Appendix Repetitas Auctasque De ovo Incubato Observationes Continens, 11
apterous, 62
AQP. *See* aquaporin
AQP0. *See* MIP
AQP1, 154
aquaporin, 163
aquaporin0. *See* MIP
aqueous humour, 25, 85, 199, 217, 247, 272–273, 276–279, 284
ARE (antioxidant responsive element), 144
arginosuccinate lyase, 124, 142
Aristotle, 4
ATF, 134
axons, 181

β-sheet, 120, 122
Baer, Karl Ernst von, 12–13
BAP37, 234
Bartisch, Georg, 7–8
basal lamina, 293
basal membrane complex (BMC), 186, 226, 227, 228, 229, 257, 259
basal promoter, 129
basal transcriptional machinery, 129
basement membrane, 245, 253, 285
basic helix-loop-helix, 135, 143
basic region/leucine zipper genes, 65–66, 134, 140
BAX, 206, 207
beaded filament, 105, 173, 273
beaded filament structural protein 2. *See* intermediate filament, CP49
Beiträge zur Gewebelehre des Kristallkörpers, 13
BM40. *See* SPARC
BMC. *See* basal membrane complex
BMP. *See* bone morphogenetic proteins
Boc-D-FMK, 238
bone morphogenetic proteins, 270, 281, 282
  BMP 4, 45, 46, 57, 267–269
  BMP 7, 45, 46, 57, 266, 267, 270
  receptor, 281
Bonnet, C, 290
bow region, 26, 55, 60
Brachio-Oto-Renal syndrome, 67
BrdU, 209, 216
bZIP domain, 65, 66
bZIP genes. *See* basic region/leucine zipper genes

c-fos, 66, 67
c-met, 284
CAAT, 145
Cadherin, 184, 186, 187, 229, 257
  B-cadherin, 170, 186, 257, 259
  E-cadherin, 60, 187, 257
  N-cadherin, 170, 179, 186, 227, 228, 229, 257, 258, 259
caldesmon, 227
calmodulin, 171
calpain, 123, 158
cAMP, 282
casein kinase, 158
caspase, 236, 237, 238, 240
cataract
  abnormal suture development, 84
  aculeiform, 124
  aftercataract. *See* cataract, posterior subcapsular opacification
  age-related, 118
  anterior subcapsular, 169, 175, 185
  apoptosis, 94
  associated with muscular dystrophy, 248
  associated with mutations in β/γ-crystallins, 124
  Cat^{elo}, 223
  $Cat2^{nop}$, 123
  $Cat2^{ns}$, 123
  Cat Fraser mutation, 247
  caused by capsule thinning/rupture, 247, 249, 253
  caused by disruption of calcium homeostasis, 123
  caused by *Sox1* mutations, 64
  cell globulization, 156
  cerulean blue dot, 124
  congenital, 160, 232
  Coppock-like, 124
  Cortical, 242
  couching, 4, 6, 12
  expression of collagen type I, 250
  hypochyma, 6
  inactivation of pRb, 211
  in ancient Egypt, 3–4
  in *Elo* mice, 123
  in mice lacking αA-crystallin, 121
  in mice lacking connexins, 159, 160, 167
  in mice lacking SPARC, 249
  in mice overexpressing PDGF-A, 281
  in patients with αA-crystallin mutations, 122
  in patients with *EYA1* mutations, 67
  in patients with *FOXE3* mutations, 65
  in patients with *PITX3* mutations, 61
  in *Opj* mice, 124
  in *Philly* mice, 124
  in *Sey* mice, 57
  in *Six5* mutant mice, 60
  integrins, 187
  lens depolarization, 156
  loss of transparency, 211
  misexpression of TGFß, 185
  models, 250
  nuclear, 159, 160, 242
  osmotic, 167, 168
  posterior subcapsular, 212, 214, 242
  posterior subcapsular opacification (PCO), 212, 214, 243, 255, 287, 289
  role of actin, 184
  role of Pax6, 187
  secondary. *See* posterior subcapsular opacification
  subcapsular, 287, 289
  variable zonular pulverulent, 124
catenin
  $\alpha-$, 186, 257, 259
  $\beta-$, 186, 257, 259, 285
  $\gamma-$(plakoglobin), 170, 186, 257
CBP, 147
Cdc42, 249
Cdx-2, 135
C/EBPs, 208
cell adhesion, 67
cell adhesion molecules (CAM). *See also* neural cell adhesion molecule
  L-CAM, 186
cell cycle, 26, 128, 192, 193, 293
  G0 phase, 195, 199
  G1 phase, 192, 193, 195, 196
  G1/S transition, 193, 195, 196, 205
  S phase, 192, 193, 195, 205, 209
  G2 phase, 192, 193, 195, 196
  G2/M transition, 193
  M phase (mitosis), 192, 193, 195, 196, 197, 216, 217
    metaphase, 196
    prophase, 196
    restriction point, 192, 193, 205
cell theory, 13
Celsus, Aulus Cornelius, 4–5
central sulcus, 84
central zone. *See* lens epithelium, central zone
centrosome. *See* microtubules
chaperone, 121, 122
Choice of Eye Diseases, 6
choroid fissure
  discovery of, 11
  formation of, 13
chromatin, 129–130, 149, 291
ciliary body, 12, 199, 275, 283
  αB-crystallin expression in, 136
  invertebrate, 17
  muscle, 19, 22, 121
  of birds, 22
  of reptiles, 21, 22
  Pax6 expression in, 55
ciliary zonule. *See* zonular fibers
cis-regulatory element, 129, 130, 145
cis-regulatory site, 141
clathrin coated vesicles, 97
ClC. *See* Cl^- channel
Cl^- channel, 156
  blockers, 157
  inhibition, 165, 167
cochlea, 181
coilin, 240
colchicine, 180, 220
collagen, 246, 248, 251, 310
  type I, 243, 250, 251, 255, 285, 286, 287, 310
  type II, 310

type III, 286, 287
type IV, 245, 246, 247, 250, 251, 285
increased synthesis, 247
mutations, 247
subunits, 247
Collucci, V. L., 290, 294
Conductance
K^+, 154, 155
Na^+, 154, 155
Cl^-, 154, 157
connective tissue growth factor (CTGF), 287
connexins, 236
connexin (43), 94, 157, 158, 159, 287
connexin (44), 157
connexin (45.6), 157, 158
connexin (46), 108, 156, 157, 158, 160
knockout, 159
connexin (49), 157, 158
connexin [50 (α8)], 108, 157, 158, 160, 236
connexin (56), 157, 158, 259
consensus DNA binding sequence, 55, 65
core promoter. *See* basal promoter
core promoter element, 129
cornea, 13, 19, 245, 272, 275
αB-crystallin expression in, 136
clarity, 71
δ-crystallin expression in, 142
discovery of layers, 12
endothelium, 121
epithelium, 23, 24, 216, 265
in lens regeneration, 304, 305
of amphibians, 20–21
of cephalopods, 126
of cyclostomes, 20
of fish, 20
Pax6 expression in, 55
S-crystallin expression in, 126
Coulombre, Jane and Chris, 241
CP49. *See* intermediate filament, CP49
CRABP, 269, 285
CRE, 133, 134, 135, 146
CREB, 134, 135–136, 140, 145, 148
CREB-2, 66
CREM, 134, 135, 148
Cruetzfeldt-Jacob disease, 122
Crx, 61
crystallins, 119–150, 199, 208, 218, 269, 299, 303, 308
as markers of lens differentiation, 120, 128, 138, 271
as markers of lens induction, 30, 32, 33, 34, 35, 42, 43
α-crystallin, 119, 120–122, 123, 142, 147, 148, 199, 219, 222, 223, 271, 287, 292
αA-crystallin, 56, 61, 65, 120, 121–122, 130–136, 137, 139, 145, 147, 148, 149–150, 203, 273, 281, 284, 292, 308
αA-CRYBP1, 133, 134, 135
αB-crystallin, 56, 120, 121–122, 136–138, 147, 150, 269, 281, 285, 303
β-crystallin, 43, 57, 120, 122–124, 130, 138–140, 142, 147, 148, 199, 203, 208, 223, 236, 271, 275, 279, 292
βA1-crystallin, 122, 123, 124, 140, 144, 308
βA2-crystallin, 122, 123
βA3-crystallin, 122, 123, 124, 140, 144, 308
βA4-crystallin, 119, 122
βB1-crystallin, 122, 138–139, 145, 149, 292, 308
βB2-crystallin, 122, 123, 124, 140, 148, 308
βB3-crystallin, 122, 123
βHigh, 122, 123
βLow, 122, 123
proteolysis, 123
ϵ-crystallin, 124, 127
enzyme-crystallins, 124–125, 126, 127, 142
δ-crystallin, 30, 42, 45, 56, 63, 65, 120, 124, 130, 142–144, 147, 148, 150, 221, 241, 268, 273, 285, 308
drosocrystallin, 127–128, 146
γ-crystallin, 66, 119, 122–124, 125, 140–142, 147, 148, 199, 203, 208, 223, 271, 292, 308
γA-crystallin, 122, 124, 140, 141
γB-crystallin, 60, 122, 123, 124, 140, 141–142, 148
γC-crystallin, 122, 124, 140
γD-crystallin, 60, 122, 124, 140, 141–142
γE-crystallin, 122, 123, 124, 140, 303
γF-crystallin, 44, 63, 66, 122, 140, 141, 147, 285, 303
γG-crystallin, 140
γS-crystallin, 122, 124, 142, 148
J-crystallins, 127, 145
L-crystallin, 127
invertebrate crystallins, 125–128
Ω-crystallin, 126–127, 145
refractive index, 104, 120, 128
regulation by *Maf* genes, 65–66
regulation by *Pax6*, 56–57
regulation by retinoic acid receptors, 68
regulation by *Sox* genes, 44, 63, 268
S-crystallins, 125–126, 144–145
τ-crystallin, 124, 144
taxon-specific, 119, 124, 128, 142, 144
transcriptional regulation, 120, 128, 129–150
ζ-crystallin, 56, 144, 150
crystallin cone. *See* lens, crystallin cone
CTGF. *See* connective tissue growth factor
cup eye. *See* eye, cupulate
cyclins, 192, 202, 205, 211, 218
A-type, 193, 195, 200
B-type, 193, 195, 196, 200, 201
D-type, 192, 193, 196, 197, 200, 201, 208, 209
E-type, 192, 193, 195, 196, 197, 200, 205, 206, 209
Cyclin-dependent kinases (Cdks), 192, 205, 208, 218
Cdc2, 193, 195, 196, 200, 201
Cdk1. *See* Cdc2
Cdk2, 193, 195, 200, 209
Cdk3, 193
Cdk4, 193, 196, 200, 201, 209
Cdk5, 201
Cdk6, 209
Cdk7, 201
Cdk8, 201

Cyclin-dependent kinase inhibitors (CKIs), 196, 208
  p21CIP1 family, 196
    p21, 196, 201
    p27 (Kip1), 60, 196, 197, 201, 202, 209, 218, 259
    p57 (Kip2), 60, 129, 196, 201, 202, 208, 209, 218, 259
  p16INK4 family, 196, 206
    p15, 196
    p16, 196, 201
    p18, 196
    p19, 196
cyclopic monster, 27
cytochalasin, 185
cytochrome C, 240
cytoplasmic shedding, 292
cytoskeleton
  lens epithelium, 94
cz. *See* lens epithelium, central zone

δEF1, 143
δEF2, 143
δEF3, 63, 143
dac. See dachshund
dachshund, 49–51
dark cells, 93
Daviel, Jacques, 6
da Vinci, Leonardo, 6–7
DC5, 143
DE1, 133, 134, 135
decapentaplegic, 185
De Corporis Humani Structura et Usu, 8
De Humani Corporis Fabrica, 7
De Medicina, 4
De Formatione Pulli in Ovo, 11
dedifferentiation, 290, 292, 293, 294, 299, 300, 308–309, 310
depigmentation, 291, 292, 295
dermatitis, atopic, 287
dermis, 34
Descartes, René, 119
Descripto, 12
desmin, 286
desmin-related myopathy, 122
desmoglein/desmocollin complex, 179
desmoplakin, 179
desmosome, 94, 179
dextran, 230
DFF, 238
  DFF40, 238
  DFF45, 237, 238
Dickkopf, 285
dihydrofolate reductase (dhfr), 193
$DiOC_6$, 234
Dioptrice, 9
Disheveled, 285
disulfiram, 303
DLG, 211
Dlx3/dll2, 41
DNAse. *See* nuclease
DNAseI hypersensitive region, 129
donnan forces, 152
Duke-Elder, Stewart, 15
dynein, 180, 228
  glued, 181
dysgenetic lens, 45, 65, 265

E6, 207, 211
E7, 203, 205, 206, 207, 208, 211
E-box, 133
EBS. *See* Epidermolysis, Bullosa Simplex
ECM. *See* extracellular matrix
ectoderm, 306
  animal cap, 37, 39
  head, 27, 31, 32, 33, 34, 35, 36, 39, 40, 41, 43, 45, 46, 246, 299
  nasal, 40
  neural, 23, 35, 55, 59, 299
  placodal, 34–35
  presumptive lens, 27, 147, 261, 262–263, 265, 267, 270
  surface, 17, 23, 24, 25
  trunk, 32, 33–34
ectodermal enhancer, 265, 266, 267
ectopic eye, 263
ectopic lens, 58, 59, 64, 263, 300, 304
E2F, 193, 195, 197, 202, 205, 206, 207, 208, 209, 210
  DP partner, 193, 195
EE. *See* ectodermal enhancer
efi. *See* epithelial-fiber interface
EGF. *See* epidermal growth factor
Eguchi, G., 292, 295, 304, 305
elb-B, 308
Elo, 123
Elschnig's Pearls, 243
embryonic induction, 15, 27
EMT, 255, 285, 287
endoplasmic reticulum, 214, 234, 241
end point definition, 29–30
Engrailed, 54
enhancer, 55, 61, 63, 65, 129, 141, 142, 143–144, 146, 150, 264
Entwickelungsgeschichte der Thiere, 13
EP37, 123
epidermal growth factor (EGF), 200, 282, 284, 289, 309
  receptor, 170, 284
Epidermolysis Bullosa Simplex (EBS), 174, 178
epithelial-fiber interface, 97–103, 183, 185, 251
  gap junctions, 99
  role in transport, 97
epithelial-mesenchymal transition. *See* EMT
equatorial zone, 249, 251, 292
ERK, 174, 255, 275–276
  inhibitor, 175
EST. *See* expressed sequence tag
estrogen, 288
exon shuffling, 126
expressed sequence tag, 123
extracellular matrix, 245, 261, 285, 286, 293, 310
extracellular signal-regulated kinase. *See* ERK
ey. See eyeless
eya. See eyes absent
Eya genes, 67
Eya1, 44, 67

eye
  camera-like, 17, 19
  compound, 16, 18–19
  cupulate, 17
  eye-spot, 15, 16
  faceted. *See*, eye compound
  flat, 16
  pinhole camera, 17, 19
  simple, 15, 16
  structural diversity, 15
  vesicular, 16, 17–18
eyeless, 41, 49–51, 55, 263
eyes absent, 49–51, 67

FAC. *See* focal adhesion complex
FAK. *See* focal adhesion kinase
fasciae adherens, 185
F-box, 196, 197
  SKP2, 197
fetal calf serum, 219, 220, 221
FGF. *See* fibroblast growth factor
FGFR. *See* fibroblast growth factor receptor
Fabricius ab Aquapendente, Falloppio Hieronymus, 8
fibrillarin, 240
fibronectin, 286, 287
fibroblast growth factor, 129, 149, 200, 217, 222, 267, 270, 273, 289, 309
  as a lens competence factor, 39
  association with HSPG, 248
  FGF-1, 274–275, 277, 279, 280, 300, 305, 309
  FGF-2 (bFGF), 149, 274–275, 277, 279, 280, 309
  FGF-3, 275, 277, 279
  FGF-4, 279, 305
  FGF-5, 275, 277, 279
  FGF-7, 279
  FGF-8, 45, 275, 277, 279
  FGF-9, 275, 277, 279, 280
  FGF-11, 275
  FGF-12, 275, 277
  FGF-13, 275
  FGF-15, 275, 277
  FGF-19, 275
  gradient, 217, 276–279, 282, 299
  induced proliferation, 217, 283–284
  induced migration, 217, 219
  induced fiber differentiation, 219, 273–280, 282
  in regeneration, 294, 295, 299–300
  modulation by syndecan, 246
  overexpression, 217
  transgenic mice, 45
fibroblast growth factor receptor, 169, 267, 276, 279, 294, 295, 299–300, 305
  dominant negative, 267, 279
  secreted, 279
fibronectin, 169, 171, 245, 246, 247, 248, 249, 250, 252, 255, 293
filensin. *See* intermediate filaments, filensin
fixatives, first use of, 12
Fleshes, 4
focal adhesion complex (FAC), 250, 251, 256, 257
focal adhesion kinase (FAK), 256, 257, 310
follistatin, 45
forgut endoderm, 34–35, 45
forkhead, 64, 265
forkhead genes, 64–65
FoxE3, 45, 57, 64–65, 265–266, 267, 270
frameshift mutation, 123
frizzled. *See* Wnt, receptor

galectin-3, 171, 259
Galen, Claudius, 5–6, 7, 12
gap junction, 154, 157, 158, 228, 230, 259, 292
gene duplication, 120, 124–125, 126
gene sharing, 124–125, 128
geodome, 94
germinative zone. *See* lens epithelium, germinative zone
GFP. *See* green fluorescent protein
Gli2, 42
glucagon, 135
glucose
  lens homeostasis, 163
  levels in aqueous, 163, 164
  transport, 163
  transporters, 163, 165
glucose facilitated transporter. *See* GLUT
GLUT, 164
  absence of, 164
  expression, 164
glutathione S-transferase, 125–126, 144
glutathione transporter, 165
glycolysis, 180, 239
glycoprotein, 305
glycosaminoglycans (GAGs), 245
  chondroitin sulfate, 246
  heparan sulfate, 246, 279
G-protein receptors
  beta-andrenergic, 168
  muscarinic acetylcholine, 168
  purinergic, 168
Grb2, 255
Greek key motif, 122, 123, 124
green fluorescent protein, 230, 241
GRIFIN, 172, 259
growth shell, 73
GST. *See* glutathione S-transferase
gz. *See* lens epithelium, germinative zone

haemolytic anemia, 183
Haller, Albrecht von, 12
Harvey, William, 8
HAT, 147
Hayashi, T., 294
HDAC. *See* histone deacetylase
head mesenchyme, 42, 45
heart mesoderm, 34
heat shock element, 137
heat shock factors, 137, 148
hedgehog signaling, 42, 43
hemi-channel, 155, 156
hemidesmosome, 178, 248
hepatocyte growth factor, 284, 289
  receptor. *See c-met*
HGF. *See* hepatocyte growth factor
high mobility group. *See* HMG box
Hippocrates, 4

histone acetyltransferase. *See* HAT
histone deacetylase (HDAC), 193
HNK-1, 42
HMG box, 62–63, 268
homeobox gene, 45, 51, 270, 299
  paired-type, 41
homeodomain, 51, 55, 59, 60, 134, 142, 143, 263, 265
  paired-type, 54, 55, 62
  bicoid-like, 61
homothorax, 265
hornblatt, 14
host/donor marking, 29, 30
HPV16. *See* human papillomavirus type 16
HS. *See* glycosmainoglycans, heparan sulfate
HSB2, 121, 136, 138
HSF. *See* heat shock factor
hsp27, 150
hsp29, 150
HSPG. *See* proteoglycans, heparan sulfate proteoglycan
human dissection, 4, 6–12
human genome project, 123
human papillomavirus type 16, 203, 207
Huntington's disease, 122
Huschke, Emil, 13
hyaloid artery, 24
hyaluronidase, 293
hypoblast, 34

IGF. *See* Insulin-like growth factor
Immunoglobulin (IgG), 241, 249
Imokawa, Y., 305
In Ovo et Pullo Observationes, 11
initiator element, 129, 133, 135, 137
Innocent XII, pope, 11
Inr. *See* initiator element
insulin, 135, 174, 219, 220, 273, 284, 289
  receptor, 219
insulin-like growth factor (IGF), 174, 200, 211, 273, 282, 284, 289
  IGF-1, 149, 217, 220
  receptor, 170, 219, 255
integrin, 149, 170, 178, 226, 229, 245, 249, 250, 251, 255, 310
  isoforms, 170
  $\alpha_1\beta_1$, 251
  $\alpha_2\beta_1$, 251
  α_3, 310
  $\alpha_3\beta_1$, 253, 256
  α_5, 255, 257
  $\alpha_5\beta_1$, 252, 253, 256
  α_6, 253, 255, 310
  α_6A, 253
  α_6B, 253
  $\alpha_6\beta_1$, 226, 253, 255
  $\alpha_6\beta_4$, 178, 186, 253, 255
  α_8, 310
  α_V, 255
  $\alpha_V\beta_3$, 249, 255
  β_1, 251, 253, 310
  β_3, 255
  β_4, 253
intermediate filaments, 94, 105, 174, 180, 183
  associated proteins (IFAPs)
    adducin, 179
    ankyrin, 179, 184
    band 3, 179
    band 4.1, 179, 185, 186
    band 4.9, 179
    caldesman, 179
    α-crystallin, 174
    IFAP300. *See* plectin
    plakoglobin, 179, 186
    plectin, 178
      mutations, 178
    spectrin, 179, 184, 186, 236, 237, 238
    tropomodulin, 179, 186
    tropomyosin, 179
  CP115. *See* filensin
  CP49, 174, 176, 233, 236
    expression, 174, 175, 176, 177, 273
    mutations, 174, 176, 178
    phosphorylation, 177
  CP49ins, 175
  CP49/filensin network, 176, 177
  desmin, 175
  expression in different species, 175
  filensin, 174, 176
    expression, 174, 175, 177, 282
    proteolytic processing, 177, 232
  GFAP, 175
  nestin, 175, 176
  overexpression, 178
  phakinin. *See* CP49
iris, 275, 283
  dorsal, 290, 292, 294–295, 297, 299, 303, 305, 306–311
  epithelium, 291, 293
  invertebrate, 17
  papillary margin, 19, 291
  ventral, 292, 294, 295, 297, 299, 300, 304–305, 306–311

Jun, 145
  c-jun, 66, 208

K^+-channel. *See potassium channel*
K_B, 91–93
K_L, 91–93
Kepler, Johannes, 9, 119
keratin, 174, 176
  keratin (14), 174, 211
Kessler, Paul Leonhard, 14
kinesin, 180, 182, 183, 228
Kreisler. See MafB

L'Occhio della Mossca, 10
lacrimal gland, 265
lactate dehydrogenase, 124
lamellipodia, 226
Lamin B, 237, 238
Laminin, 171, 245, 246, 247, 250, 251, 253, 255, 293
  Laminin-1, 248, 253
  Laminin-5, 248, 253
  Laminin-10/11, 253
  Merosin, 248

Lectin, 259
LDL-related proteins. *See* LRP
Leeuwenhoek, Antonie van, 12
LEDGF. *See* lens epithelium-derived growth factor
lens
  accommodation, 10, 17, 19, 21, 22, 105, 183, 217, 226, 230
  adult nucleus, 90
  as center of eye, 4, 7
  as center of visual function, 4, 5, 9
  as liquid, 4, 13
  as stratified epithelium, 72, 73
  calcite, 19
  capsule, 12, 13, 14, 20, 24, 71, 94, 128, 149, 226, 229, 246, 271, 274, 279, 292
    abnormalities, 247, 249, 253
    components, 246, 247, 248, 249
    detachment from, 257
  cell membranes, 151
  cell migration, 217, 226, 243, 248
  circulating current, 151, 153, 154, 255, 259
  comparison to skin, 23, 73
  cortex, 26, 87, 90, 115, 120, 148, 161
  crystallin cone, 18–19
  cuticular, 17
  diabetic, 167
  embryonic nucleus, 26, 77, 87, 103, 115
  fetal nucleus, 87, 89, 90, 115
  fibers. *See* lens fibers
  free, 31, 34, 35, 40, 42, 44
  gap junction, 86, 94, 97, 105–108
  growth rate, 212
  internal circulatory system, 86
  invagination of surface ectoderm, 13
  inversion, 272, 283
  juvenile nucleus, 90, 115
  membrane potential, 154, 155
  nucleus, 87, 89, 90, 115, 120, 148, 161
  opacification, 157
  of amphibians, 20–21
  of avians, 76
  of bovines, 90
  of cephalopods, 17, 126
  of cyclostomes, 20
  of fish, 20, 86
  of mice, 90
  of *Peripatus*, 17
  of reptiles, 21, 76
  of scallops, 127
  polarity, 261, 272, 275, 279, 281, 295, 299, 300, 305
  regeneration, 60, 290–311
  structural diversity, 15, *see also eye, structural diversity*
  transparency, 151, 152, 225
  thickness, 86
  volume, 103
  width, 86
lens determination, 27
  bias, 35, 36, 39–42, 262
  competence, 35, 36, 37–39, 262
  differentiation, 36, 37, 44–45
  inhibition, 36, 37, 42–43
  specification, 35, 37, 43
lens epithelium, 20, 25, 27, 261, 271–272, 306
  α-crystallin expression in, 120, 136
  cell density, 93
  cell line
    αTN4-1, 130, 133, 135
    N/N1003a, 130, 133, 140
  central zone, 72–73, 92–93
  dependence on CREB-2, 66
  dependence on *FoxE3*, 65
  explantation, 32, 44
  explants, 32, 130, 149, 199, 218, 219, 220, 222, 250, 273, 274, 275, 277, 282, 283, 286
  gene expression restricted to, 45, 64, 66, 67, 283
  germinative zone, 26, 72, 73, 92–93, 199, 215, 216, 217, 247, 251, 252, 261, 273, 283–284
  homeostasis, 92–93, 283–289
  of amphibians, 21
  of fish, 21
  of reptiles, 21
  patch clamp, 154, 155, 157
  pre-germinative zone, 72, 73
  presumptive, 263
  ultrastructure, 94
lens epithelium-derived growth factor, 285–286
lens fiber, 271–272
  α-crystallin expression in, 120, 136
  arrangement of, 20
  ball and socket junctions, 108, 230, 233
  compaction, 115
  cortical fibers, 85, 87, 90, 124, 136, 140, 144, 148
  cytoplasm, 103–105
  cytoskeleton, 105, 121
  differentiation, 201, 217, 242
  discovery of, 12
  early descriptions of, 13, 14
  elongation, 73, 128, 215, 220, 221, 222, 223, 226, 273, 277, 279, 281, 292, 299
  block to, 220, 222, 223, 274
  fusion, 110–115
  globulization, 154, 155, 156
  homeostasis, 91
  loss of organelles, 26, 73, 87
  mature, 216, 231, 232
  maturing, 216
  membranes, 105–115, 121
    apical, 227, 228
    basal, 226, 228
    lateral, 228, 230
  membrane potential, 155
  nuclear degradation, 158, 234, 236, 239, 257, 274
  nuclear fibers, 85, 87, 90, 103, 108
    juvenile, 89, 90
  nucleus, 214
  organelle degradation, 234, 236, 238, 239, 257, 274
  patch clamp, 154
  Pax6 expression in, 55, 139
  polarization, 225, 228
  precursors, 215
  primary, 25, 72, 87, 123, 197, 232, 261, 271, 291, 292
  secondary, 26, 72–73, 87, 89, 93, 130, 261, 291, 292–293
  size and shape, 214, 225, 231

lens fiber (*cont.*)
S-shaped, 79
straight, 77
structure of primary fibers, 87, 103
structure of secondary fibers, 103
thickness, 87
ultrastructure, 103–115
uncoupling differentiation, 175
width, 87
lens pit, 23, 26, 59, 63, 130, 143, 147, 148, 197, 199, 246, 266, 267, 269, 271
lens placode, 14, 23, 27, 30, 32, 36, 41, 42, 45, 46, 55, 57, 58, 59, 60, 63, 65, 66, 67, 71, 120, 136, 137, 139, 142, 143, 147, 186, 246, 261, 263, 264, 265, 266, 267, 269, 270, 299
lens stalk, 23, 263, 265, 267
lens sutures, 20, 128, 223, 226, 227, 228, 229, 246, 248, 249
formed by human juvenile nuclear fibers, 89
in fetal nucleus, 89
line, 229
optical qualities of, 84–85
structure of, 77–79
of fish, 20
of reptiles, 21
star, 229
optical qualities of, 85
structure of, 79–84
suture patterns, 79, 90
umbilical, 228
optical qualities of, 84–85
structure of, 76–77
Y, 229
optical qualities of, 84–85
structure of, 77–79
lens vesicle, 187, 197, 232, 245, 249, 267, 271, 281, 291
α-crystallin expression in, 120, 130
abnormal separation from surface ectoderm, 45
βB1-crystallin expression in, 138
disruption, 251
early descriptions of, 13, 14
ectopic induction of, 34
embryology of, 23, 25, 27, 71–72, 148, 261
during regeneration, 292, 299, 300
FoxE3 expression in, 55
γ-crystallin expression in, 141
Maf gene expression in, 65–66
mitotic activity in, 26, 32
Pax6 expression in, 55, 139
Pitx3 expression in, 60
Proxl expression in, 60
retinoid signaling in, 147
Six3 expression in, 59
Sox gene expression in, 63
lentigenic cells, 126, 145
lentoid, 33, 46, 136, 158, 218, 243, 282, 306, 308
lentropin, 273
leucine zipper. *See* basic region/leucine zipper genes
Lewis, Warren, 3
Lhx2, 42, 62, 263
light organ, squid, 127
LIM domain, 62
locus control region, 129, 150
LRP, 285
Lrp6, 270
LRS1, 138, 150
LRS2, 138, 150

macrophage, 292
Maf, 63, 65–66, 136, 139, 140, 141, 142
c-Maf, 45, 65–66, 134, 135, 136, 138, 139, 140, 141–142, 143, 146, 148, 202, 208, 223
L-Maf, 45, 65, 135, 136, 139, 143, 148
MafB, 45, 66, 134, 143
NRL, 66, 140, 144, 148
v-maf, 65
Maître-Jan, Antoine, 12
major intrinsic polypeptide. *See* MIP
Malpighi, Marcello, 10–11
MAPK, 39, 211, 275
MARE, 65, 133, 139, 144, 146
MDRI see multidrug resistance gene
maturation promoting factor (MPF), 195
Meis, 265, 269, 270
Melencephalic bleb (*my*), 245
Mencl, Ernst, 3
meridional rows, 26, 223, 271
merlin, 185
tumor suppressor gene, 242, 243
mesenchymal cells, 250
myofibroblasts, 243, 287
methylation, 142
MHL, 137
Microtubules, 179, 180, 184, 220, 228, 233
epithelial cells, 180
fiber cells, 180
function, 180, 181
organizing center (MTOC), 179, 180
centrosomes, 180, 181, 195
polarity, 179
microphthalmia, 57, 61, 64, 160, 171, 263, 266, 267
micropinocytosis, 97
microscope
invention of, 10
Mikroskopische Untersuchungen über die Uebereinstimmung in der Struktur und dem Wachsthum der Thiere und Pflanzen, 13
minimal lens-specific promoter, 129, 135, 136, 137
minimal promoter, 129, 130, 138, 140, 141, 149
MIP, 67, 90, 105, 108, 110, 161, 231, 236
activity, 162
function, 163
mutations, 163, 167
structure, 162, 163
mitochondria, 180, 214, 234, 240
mitogen-activated protein kinase. *See* MAPK
mitosis. *See* cell cycle, M phase
MMP115, 308
MNNG. *See* N-methyl-N′-nitro-N-nitrosoguanidine
modulatory element, 129
morpholino, 294, 311
MP20, 171, 172, 259
MP70. *See* Connexin 50
MPF. *See* maturation promoting factor
MTOC. *See* Mictrotubule organizing center

Müller, Johannes, 13–14
multiple sclerosis, 122
multidrug resistance gene, 157
muscular dystrophy, 178, 248
B-myb, 193
Myc, 196
  c-myc, 193, 308, 309
  N-myc, 193
MyoD, 208
myo-inositol
  transporter, 165, 167
  expression, 167
myofibroblasts. *See* mesenchymal cells
myosin, 184, 227, 241
myotonic dystrophy, 59–60

N-methyl-N′-nitro-N-nitosoguanidine, 304–305, 306
NADPH: quinone oxidoreductase, 144
Na^+ pump, 160, 161
NaK-ATPase, 94, 160, 220
Na^+/K^+ pump, 152, 153, 154, 160
nasal pit, 35, 40
nasal placode, 41–42
N-CAM. *See* neural cell adhesion molecule
NCD-2, 259
neural cell adhesion molecule, 170
  glycosylation, 259
  isoforms, 170, 259
neural crest, 35, 37, 42
neural plate, 27, 31–34, 35, 36–37, 40
NF2 (neurofibromatosis type 2), 242
nickel subsulfide, 305
nidogen-entactin, 293
nocodazole, 220
noggin, 281
NPAT, 192
NPPB. *See* Cl^- channel blockers
nuclear lamina, 237
nuclease, 238, 239
  DNAseI, 239
  DNAseII, 239
nucleolar activation, 291
nucleosome, 129, 146

ocelli, 16, 127, 128, 146
octapeptide, 54
Ocula hoc est, 9
ocular media, 23, 273, 274, 276–279, 284, 285, 288
Odierna, Giovanni Batttista, 10
OFZ (organelle free zone), 234, 236, 237, 239, 241
Opj, 124
ommatidia, 18–19, 127
Ophthalmodouleia: das ist Augendienst, 7
optic cup, 71, 246, 261
  as lens inducer, 31, 32, 33, 261
  as a lens trophic factor, 43, 44
  development of, 27
  discovery of, 13
  disruption, 251
  failure in *Lhx2*mutants, 62
  gene expression in, 55
  induction of *Sox* gene expression, 63
  of nautilus, 17
optic nerve, 7, 10
optic pit, 23, 269, 270
optic vesicle, 71, 245, 246, 263
  BMP expression in, 55, 57
  discovery of, 11, 12–13
  failure in *Rx1* mutants, 62
  formation of, 12, 23, 27, 261
  induction of the lens, 15, 29–34, 57, 261–263, 267, 270, 293, 299, 306
  Lhx2 expression in, 62
  Otx gene expression in, 61
  Pax6 expression in, 55, 57–58
  requirement for *Sox* expression, 63–64
orthodenticle, 61
osteonectin. *See* SPARC
otic placode, 41
otic vesicle, 40
opposite-end curvature, 79
optix, 49–51, 59
Optx2. *See Six6*
organelle free zone. *See* OFZ
Otx1, 61–62
Otx2
  dependence on Pax6, 41
  expression pattern, 40, 61
  function, 61–62
  mutations, 61
  restriction during inhibition period, 42
  upregulation during bias period, 40
OV. *See* optic vesicle
Ovalbumin, 241
oxygen, 240

p19ARF, 206
$p27^{KIP1}$ see Cyclin-dependent kinase inhibitors
p53, 66, 206, 207, 209, 211
$p57^{KIP2}$. *See* Cyclin-dependent kinase inhibitors
p73, 207
p130Cas, 310
p300, 135, 147
paired, 54, 134
paired domain, 54, 55, 263
Pander, Christian, 12, 13
PARP, 237, 238
Pax6
  as vertebrate ortholog of *eyeless*, 51
  binding to pRB, 208
  determines N-cadherin expression, 186
  dynamic expression pattern during bias period, 40
  effect of mutations, 41, 57–58
  expression during lens formation, 54–55, 262
  expression in *Lhx2* null mice, 62
  in lens induction, 263–265
  in lens regeneration, 297, 299, 305
  interaction with Sox proteins to activate gene expression, 63, 64, 268–269
  lack of requirement for BMP4, 46, 268
  requirement to maintain lens gene expression, 44, 46
  regulation by BMP7, 45, 266
  regulation by FGF, 267
  regulation by Meis transcription factors, 265
  regulation by TGFβ, 287

Pax6 (*cont.*)
regulation of target gene expression, 55–57, 133, 134–136, 137, 139, 140, 141, 143, 144, 145, 146–148, 269, 270
restriction during inhibition period, 42
splice variant (5a), 54, 141, 255, 257
structure and evolutionary conservation, 54, 125
upregulation during bias period, 40
Pax6$^{Sey\text{-}1Neu}$. *See small eye*
Pax6$^{Sey/Sey}$. *See small eye*
paxillin, 186, 228, 230, 256, 257, 310
PDGF. *See* Platelet-derived growth factor
PDZ proteins, 211
PE1, 133
PE2, 133, 135
PECs. *See* pigment epithelial cells
Peters' anomaly, 147, 263, 265, 269
Petit, circle of, 12
p-glycoprotein
function, 157
overexpression, 168
inhibitor, 168
Philly, 124
phosphatidylethanolamine, 219
phosphorylation, 121, 123, 149, 310
photoreceptors, 181
pigment epithelial cells, 290, 291, 292, 293–295, 299, 304–305, 306–311
pituitary gland, 217
Pitx1, 60
Pitx3, 60–61, 67
PL1, 138–139
PL2, 138–139
placodal zone, 40, 41
plakoglobin. *See* catenin
Platelet-derived growth factor (PDGF), 200, 211, 281, 282, 283–284, 289
receptor, 169, 282, 283
plakophilin, 179
Platter, Felix, 8–9
Platter, Felix II, 9
PLE. *See* ectoderm, presumptive lens
Pliny the Elder, 4
posterior vascular capsule, 25
potassium channel, 152, 154, 155
species differences, 154
Maxi-K, 154, 155
pP344, 308
Praxeos Medicae, 9
preinitiation complex, 129
presbyopia, 118
prohibitin, 234
proline-serine-threonine domain, 54
prostaglandin
receptor, 170
prospero, 45, 60, 146, 270
26S proteasome, 196
protein disulphide isomerase, 234
protein kinase C, 158, 220
protein S, 123
proteoglycans, 245, 248, 249
heparan sulphate proteoglycan (HSPG), 246, 250, 279, 293, 300
syndecan, 246
Prox1, 45, 57, 60, 141, 187, 202, 223, 270, 297, 299, 305
PST domain. *See* proline-serine-threonine domain
pseudogene, 119, 140
pupillary membrane, 23–25

RA. *See* retinoic acid
Rabl, Carl, 14
RAR. *See* retinoic acid receptors
RARE. *See* retinoic acid response element
Ras, 196, 308
H-ras, 211
Rathke's pouch, 42, 43, 143
RB see retinoblastoma, pRb
Remak, Ernst Julius, 14
Remak, Robert, 14
Remak, Robert II, 14
retina, 275, 300
αA-crystallin expression in, 121
ablation of, 27
as center of visual perception, 9
conditioned-medium, 222
δ-crystallin expression in, 142
degeneration, 184, 242
ectopic expression of *L-Maf*, 45
effect of overexpression of *Rx1* in, 62
in lens regeneration, 295, 307
Lhx2 expression in, 62
Muller cells, 121
NRL expression in, 66
Otx2 expression in, 61
Six3 expression in, 59
Sw3-3 expression in, 66
Pax6 expression in, 55, 58
Prox1 expression in, 60
retinitis pigmentosa, 242, 287
retinoblastoma transcription factor family, 193
pRb, 193, 195, 196, 201, 202, 203, 205, 206, 207, 208, 209, 218
deficiency, 205, 206, 207, 209, 210, 211
minigene, 207, 210
p107, 193, 195, 201, 202, 207, 208, 209, 210
p130, 193, 195, 201, 202, 207, 209, 210
retinoic acid, 289
as lens induction factor, 45–46, 67, 269–270
dependence on *Pax6*, 44, 57, 67
induction of αB-crystallin, 137
in lens epithelial maintenance, 283, 285, 289
in lens regeneration, 300–304, 305
regulation of δ-crystallin, 143–144
regulation of γF-crystallin, 141, 142
retinoic acid receptors, 67–68, 137, 141, 144, 145, 146–148, 269, 285, 300–304
retinoic acid response element, 141, 145, 147
retinoid. *See* retinoic acid
retinule, 19
Reyer, R. W., 290
rhabdomere, 19
rhodopsin, 125
ribosomal RNA, 291
Rieger syndrome, 60
RNA polymerase, 129, 130, 146–147
RNAi, 294, 311
RNAse activity, 241

rods, discovery of, 6–12
Roesel, M. E., 292
Rufus of Ephesus, 5
Rx1, 62, 263

Sato, T., 290
Sato stages, 290
Scheiner, Christoph, 9
Schleiden, Matthias Jakob, 13
Scott, V., 295
Schwann, Theodor, 13
Schwannomin. *See* merlin
Scribble, 211
Sey. See Small eye
Shc, 255
Shingai, R., 292
silencers, 130, 141–142
SIMO element, 265, 266, 267
sine oculis, 49, 57, 58–59, 270
Six1, 59
Six2, 58
Six3, 67
  expression during bias period, 41
  gain or loss of function, 59
  homology to other transcription factors, 59
  lack of requirement for BMP4, 46
  ocular expression pattern, 59
  regulation by *Pax6*, 41, 44, 57, 58, 59, 270
  regulation of *Pax6*, 57, 59
  repression of γF-crystallin promoter, 141
Six4, 59
Six5, 59–60
Six6, 59
Six domain, 59
SL11/Lops4, 125–126
Smad, 281
Small eye, 41, 42, 44, 45–46, 57–58, 186, 263, 264, 265, 266, 269–270
small heat shock protein, 119, 120, 121
so. See sine oculis
Soemmering's Ring, 243
somatatropin, 217
somatomedin-C, 217
somatostatin, 135
sorbitol, 167
Sox1, 45, 63, 64, 141, 143, 146–148, 223, 268
Sox2, 39
  activation of δ-crystallin expression, 64
  dependence on BMPs, 45, 46, 64, 268
  dependence on the optic vesicle, 64
  dependence on *Pax6*, 41, 44, 45, 57
  downregulation upon *Sox1* expression, 45
  expression during bias period, 40, 45
  expression in chick head ectoderm, 63
  regulation of crystallin expression, 141, 143, 146–148
  restriction during inhibition period, 42
  targeted mutation, 64
Sox3, 268
  as lens competence factor, 39, 45
  binding to γF-crystallin promoter, 141
  dependence on the optic vesicle, 64
  downregulation during bias period, 40
  expression in the lens placode, 63
  induction of ectopic lens, 64
  interaction with *Pax6* to activate gene expression, 64
  re-expression during specification, 44
  restriction during inhibition period, 42
Sox genes, 62–64, 142, 143, 148
SPARC, 172, 246, 249
Spemann, Hans, 3, 12, 14–15, 27, 29, 31
spermatogenesis, 195
spherical aberration, 120
spheroid, 86
spherulin 3a, 123
square array junction, 108
square array membrane, 87, 90, 97, 108
Src, 258, 310
Sry, 62
Steinitz, H., 290
stem cells, 72
Steno, Nicolaus, 4
Stone, L. S., 290, 305
succinic dehydrogenase, 234
suspensory ligaments. *See* zonnular fibers
SV40, 203
  large T-antigen, 203
  trunc T-antigen, 208
Sw3-3, 45, 66
Synemin, 175
Synctium, 214, 231

T-MARE, 141–142
TALE-class transcription factor, 265
Talin, 256
talpid, 43
tamoxifen see Cl^- channel blockers, p-glycoprotein inhibitor
TATA-box, 129, 133, 134, 135, 137, 144, 145
TATA-rich region, 137
TBP, 135
tenascin, 246, 249, 255, 286, 287
tensin, 229, 310
TFIIA, 146
TFIIB, 129, 146
TFIID, 135, 146–147
TGF-α. *See* transforming growth factor-alpha
TGF-β. *See* transforming growth factor–beta
TGFβ superfamily, 281, 282
Theoria Cataracta, 9
thin symmetric junctions, 105, 108
thrombospondin, 255
tight junction, 94
TNFα. *See* tumor necrosis factor-alpha
toy. See twin of eyeless
trabecular meshwork, 121, 136
Tractatus de Oculo Visuque Organo, 8
Traité des Maladies des Yeux, 12
transcriptional activators, 130, 133, 139, 141, 143, 149
transcriptional repressors, 130, 133, 139, 140, 143, 146–148
transcriptional start site, 129
transcytosis, 97
transdifferentiation, 290, 293, 299, 305, 306–311
transforming growth factor-alpha, 282, 284, 289
transforming growth factor–beta, 175, 185, 243, 281, 282, 284, 286–288, 289
  receptor, 169, 281, 284

transgenic mice, 29, 45, 55, 57, 129, 130, 133, 135, 137, 138–139, 140, 141, 142, 144, 146, 148, 167, 168, 169, 202, 203, 205, 206, 207, 210, 217, 267, 269, 273, 276, 279, 281, 283, 286, 300
transitional zone, 26, 72, 73, 131, 138, 199, 216, 217, 223, 261, 273, 282, 284, 285
transplantation, 29, 30, 61, 294, 295, 297
tritiated thymidine, 197, 216
Triton X-100, 253
tropomodulin, 227
tubulin, 180, 182
γ-tubulin/pericentrin, 181
tumor necrosis factor
  -alpha (TNFα), 240
  receptor (TNFR), 170, 240
TUNEL, 236
tunica vasculosa lentis, 24–25
twin of eyeless, 49

Uber den Bau und die Entwicklung der Linse, 14
Ubiquitin, 196, 201, 207
  enzymes
    E1, 196
    E2, 196
    E3, 196
      HECT-domain proteins, 196
      APC, 196
      SCF, 196
Ubiquitination, 196
Ueber die erste Entwinkenlung des Auges und die damit zusammenhängende Cyclopie, 13
Untersuchungen über die Entwickelung der Wirbethiere, 14
USF, 134, 135, 148
Ushers 1B syndrome, 184

vasa hyaloidea propria, 25
Vesalius, Andreas, 7, 8, 9
vesicular lens rudiment, 291
vimentin, 174, 175, 176, 178, 232
vinculin, 186, 228, 229, 256
vitreous, 4, 5, 12, 85, 199, 217, 218, 219, 220, 221, 250, 261, 272–273, 276–279, 288, 297
  formation of, 12
  hyaloid fossa, 19
  invertebrate, 17
vitronectin, 255

Watanabe, K., 304
water channel. *See* aquaporin
water permeability, 161
Werneck, Wilhelm, 13
winged helix genes. *See* forkhead genes
Wigle, J. T., 299
Wnt, 57, 270, 285, 289
  receptor, 270, 285
Wolff, G., 290, 294, 295
Wolffian regeneration, 290

xenobiotic responsive element, 144
Xlens1, 40, 64
β-D Xyloside, 248

Yamada, T., 290, 292, 295
yot, 42

z-DEVD-FMK, 238
Zalik, S. E., 295
zinc finger, 135, 143
Zinn, Johann Gottfried, 12
zones of discontinuity, 84
zonular fibers, 12, 19, 22, 297
ZPE, 144, 150
Zur Entwickelung des Auges der Wirbelthiere, 14
zVAD-FMK, 237, 238